Advances in Intelligent a
Soft Computing

Editor-in-Chief: J. Kacprzyk

Advances in Intelligent and Soft Computing

Editor-in-Chief

Prof. Janusz Kacprzyk
Systems Research Institute
Polish Academy of Sciences
ul. Newelska 6
01-447 Warsaw
Poland
E-mail: kacprzyk@ibspan.waw.pl

Further volumes of this series can be found on our homepage: springer.com

Vol. 94. J.M. Molina, J.R. Casar Corredera,
M.F. Cátedra Pérez, J. Ortega-García, and
A.M. Bernardos Barbolla (Eds.)
*User-Centric Technologies and
Applications, 2011*
ISBN 978-3-642-19907-3

Vol. 95. R. Burduk, M. Kurzyński,
M. Woźniak, and A. Żołnierek (Eds.)
Computer Recognition Systems 4, 2011
ISBN 978-3-642-20319-0

Vol. 96. A. Gaspar-Cunha, R. Takahashi,
G. Schaefer, and L. Costa (Eds.)
*Soft Computing in Industrial
Applications, 2011*
ISBN 978-3-642-20504-0

Vol. 97. W. Zamojski, J. Kacprzyk,
J. Mazurkiewicz, J. Sugier,
and T. Walkowiak (Eds.)
Dependable Computer Systems, 2011
ISBN 978-3-642-21392-2

Vol. 98. Z.S. Hippe, J.L. Kulikowski, and
T. Mroczek (Eds.)
*Human – Computer Systems Interaction:
Backgrounds and Applications 2, 2011*
ISBN 978-3-642-23186-5

Vol. 99. Z.S. Hippe, J.L. Kulikowski, and
T. Mroczek (Eds.)
*Human – Computer Systems Interaction:
Backgrounds and Applications 2, 2011*
ISBN 978-3-642-23171-1

Vol. 100. S. Li, X. Wang, Y. Okazaki, J. Kawabe,
T. Murofushi, and Li Guan (Eds.)
*Nonlinear Mathematics for Uncertainty and
its Applications, 2011*
ISBN 978-3-642-22832-2

Vol. 101. D. Dicheva, Z. Markov,
and E. Stefanova (Eds.)
*Third International Conference on Software,
Services and Semantic Technologies
S3T 2011, 2011*
ISBN 978-3-642-23162-9

Vol. 102. R.S. Choraś (Ed.)
*Image Processing and Communications
Challenges 3, 2011*
ISBN 978-3-642-23153-7

Vol. 103. T. Czachórski, S. Kozielski, and
U. Stańczyk (Eds.)
Man-Machine Interactions 2, 2011
ISBN 978-3-642-23168-1

Vol. 104. D. Jin and S. Lin (Eds.)
*Advances in Computer Science, Intelligent System
and Environment, 2011*
ISBN 978-3-642-23776-8

Vol. 105. D. Jin and S. Lin (Eds.)
*Advances in Computer Science, Intelligent System
and Environment, 2011*
ISBN 978-3-642-23755-3

Vol. 106. D. Jin and S. Lin (Eds.)
*Advances in Computer Science, Intelligent System
and Environment, 2011*
ISBN 978-3-642-23752-2

Vol. 107. P. Melo-Pinto, P. Couto, C. Serôdio,
J. Fodor, and B. De Baets (Eds.)
Eurofuse 2011, 2011
ISBN 978-3-642-24000-3

Vol. 108. Y. Wang (Ed.)
Education and Educational Technology, 2011
ISBN 978-3-642-24774-3

Vol. 109. Y. Wang (Ed.)
*Education Management, Education Theory
and Education Application, 2011*
ISBN 978-3-642-24771-2

Yuanzhi Wang (Ed.)

Education Management, Education Theory and Education Application

 Springer

Editor

Prof. Yuanzhi Wang
Anqing Teachers College
128#, Linghu S Road
Anhui Province
Anqing
China
E-mail: wyz1970_cn@yahoo.cn

ISBN 978-3-642-24771-2   e-ISBN 978-3-642-24772-9

DOI 10.1007/978-3-642-24772-9

Advances in Intelligent and Soft Computing   ISSN 1867-5662

Library of Congress Control Number: 2011938972

Typeset & Cover Design: Scientific Publishing Services Pvt. Ltd., Chennai, India

Printed on acid-free paper

5 4 3 2 1 0

springer.com

Preface

We are delighted to invite you to participate 2011 2nd International Conference on Education and Educational Technology (EET 2011) in Chengdu, China, October 1–2.

Chengdu is the capital of Sichuan Province, which is known as the "Heavenly State" (Tian Fu Zhi Guo). Being the natural habitat of cute giant pandas, Chengdu is located in the west of Sichuan Basin and in the center of Chengdu Plain. It covers a total area of 12.3 thousand square kilometres (4,749 square miles) with a population of over 11 million.

Benefiting from Dujiangyan Irrigation Project which was constructed in 256 B.C., Sichuan Province is reputed as "Tian Fu Zhi Guo", literally a place richly endowed with natural resources. Chengdu, as the capital, is extremely productive. The Min and Tuo Rivers, two branches of the Yangtze River, connected to forty other rivers, supply an irrigation area of more than 700 square kilometres (270.27 square miles) with 150-180 million kilowatts of water. Consisting of abundant mineral resources, the land is extremely fertile. The history of Chengdu can be traced back 2,400 when the first emperor built his capital here and named the city. Through thousands of years its original name has been kept and its position as the capital and as the significant center of politics, commerce and military of the Sichuan area (once called Shu) has remained unchanged. Since the Han (206B.C.-220) and Tang (618-907) Dynasties when its handicraft industry flourished, Chengdu has been famous for its brocades and embroideries. Shu embroideries still enjoy a high reputation for their bright colors and delicate designs, ranking among the four main embroideries in China. Chengdu was the place where the bronze culture, an indispensable part of ancient Chinese culture, originated, the place where the Southern Silk Road started, and the place where the earliest paper currency, Jiaozi (not the dumpling!), was first printed. It is listed among the first 24 state-approved historical and cultural cities and owns 23 state and provincial cultural relic units.

The objective of EET 2011 is to provide a forum for researchers, educators, engineers, and government officials involved in the general areas of Education Management, Educational Theory to disseminate their latest research results and exchange views on the future research directions of these fields.

2011 2nd International Conference on Education and Educational Technology (EET 2011) is sponsored by Beijing Normal University. The mission is to bring together innovative academics and industrial experts in the field of Education Management, Educational Theory. The EET 2011 will also include presentations of contributed papers and state-of-the-art lectures by invited keynote speakers. The conference will bring together leading researchers, engineers and scientists in the domain of interest.

We would like to thank the program chairs, organization staff, and the members of the program committees for their hard work. Special thanks go to Springer Publisher.

We hope that EET 2011 will be successful and enjoyable to all participants. We look forward to seeing all of you next year at the EET 2012.

Yuanzhi Wang, Intelligent Information Technology Application Research Association, Hong Kong.

EET 2011 Organizing Committee

General Chair

Prof. Honghua Tan   Wuhan Instititue of Technology, China

General Co-chair

Prof. Qihai Zhou   Southwestern University of Finance and Economics, China
Prof. Junwu Zhu   Yangzhou, University, China
Prof. David Zhang   Hong Kong University, China

TPC Chair

Prof. Ming Zhang   Beijing Normal University, China

TPC Co-chair

Prof. Yuanzhi Wang   Intelligent Information Technology Application Research Association, Hong Kong

TPC

Dr. Yi-chuan Zhang   Henan Institute of Science and Technology, China
Prof. Jun Wang   The Chinese University of Hong Kong, Hong Kong
Prof. Chin-Chen Chang   Feng Chia University, Taiwan
Dr. Tianshu Zhou   George Mason University, USA
Dr. Kurt Squire   University of Wisconsin-Madison, USA
Prof. Toshio Okamoto   University of Electro-Communications, Japan

Contents

The Research on Supplement of Glutamine in Sports 1
Dunhai Wang, Yubo Han, Ancun Jiao

Innovative Practices of IC Design and Application Courses 7
Jianping Hu, Yinshui Xia, Haiyan Ni, Lunyao Wang

**Relationship between the Academic Achievement and Social
Responsibility for Teachers in University** 15
Junyi Chen, Li Xu, Haifeng Wu, Zhiguo Yang

Investigation for Occupational Interests of University Students 21
Bing An, Hui Cao, Yun Zhai, YanGao

Chinese Olympic Education and Olympic Publicity 29
Hua Pan, Xiaowei Liu

**Research on Evaluation of Laboratories in Independent Institutes Based
on the Elman Neural Network** 37
Yangyong Jiang, Chaoyang He, Jianbo Ding

**Risk and Prevention Mechanism of University Student Loans from the
Public Administration Perspective** 45
Bing An, Yun Zhai, YanGao

**A Study of the Characteristics of Civil Engineering and Architecture
Talents and the Way of Talents Training** 51
Peng Zhang, Changming Liu

Study on the Moral Internalization Mechanism for Young Teachers 59
Changming Liu

Mediating Effects of Computer Self-efficacy between Learning
Motivation and Learning Achievement   67
I-Hui Hwang, Chi-Cheng Chang, Hsin-Ling Wang, Shang-Jiun Tsai,
Tuo-Yu Chen

Research on Shipping Route Optimization Allocation   75
Degang Fu

The Effectiveness of Social Networking Applications in E-Learning   79
Rossafri Mohamad

The Application of Online Machine Translation System in Translation
Teaching   87
Xiaowei Guan

Research on Sports Industry System in Germany   93
Hua Pan

Review on Chinese Women Athletes' Involvement in the Olympic Games . . .   101
Hua Pan, Xiaowei Liu

Study on Game Decision of BOT Financing Model of Public-Rent
Housing   107
Liming Wang, Fangqiang Liu, Jiansheng Dai

The Design and Development of Web-Based Examination System   117
Jun Ma, Qinghua Liu

Institute-Industry Co-operation: A Comparison of Two Educational
Modes of School of Software and Higher Vocational Education in China . . .   123
Ziru Li, Yanqing Wang, Yangyang Han, Ligeng Guo, Yundi Zhang

The Practice of Talents Cultivation and Scientific Research Cooperation
Scheme Based on Interdiscipline   129
Jing-chang Pan, Xuemei Lin, Peng Wei, A-li Luo

How Can One Reflect upon the People's Republic of China's Totalitarian
Political Economy from a Root Perspective of Moral Philosophy?   137
Yu Mao, Lijuan Hou

A Study on Chinese Culture Shortage in CELT and Countermeasure   141
Quanzheng Zhang

To Discuss the Guiding Role of Examination Combining
"Software Engineering" Course   147
Yu Gao, Zhen-bo Bi

Preliminary Study of Dialectical Relationship between the Credit System
and Tutorial System in Our Higher Education   155
Wang Jian, Sun Lin

On the Phenomenon of China English in Chinese-English Interpretation ... 163
Xiaoya Qin

The Application of Formative Evaluation in Integrated English Teaching ... 171
Tianshu Xu, Lei Zhu

Research on the Embedded Professional Talent Training Mode of
Chinese-Foreign Cooperation Education 177
Binbin Shi

Study on Cultivation Mode of Independent Innovative Talents 185
Guihua Ma

Characteristics and Enlightenment of Faculty Building for Vocational
Education in Developed Countries 189
Zeying Zhang

Study on the Motivation for Saving Behavior of Chinese Urban
Residents 195
Honghui Deng, Jirong Su

To Arouse the Passion of Teachers to Become Educators 209
Yong Ao

Studies on Graduate Innovation Training Mode and Interactive
Education Platform Based on Internet Environment 215
Ruzhi Xu, Heli Li, Peiguang Lin

Exploration and Practice of Training "Applied Talents" in Universities 225
Yude Liu, Fang Wang, Qifeng Zhao, Jianhui Li, Lizi Liu, Yuanyuan Liu

The Exploration on Hierarchical and Progressive Bilingual Education
Based on College Students' English Skills 235
Zhonghao Cheng, Yuchai Sun

Teaching Exploration on Software Outsourcing Talents Training 241
Zhiyu Zhou, Yubo Jia

Effectiveness of an English Course with Motive Regulation Strategies on
In-Service Adult Learners 247
Yu-Ping Chang

Application of RMI Thinking Method Based on Mathematical
Optimization 253
Wei Wang, Xiaohui Liu, Dawei Zhao

The Chance and Challenge of Economic Recovery in Tourism of Mianzhu
after Earthquake Disaster 259
Shu Jianping, Zhang Linling, Wen Xiaoyuan

Information Systems: New Demands, New Proposals 265
Rosângela Lopes Lima, Isabel Cafezeiro, Luiz Valter Brand Gomes

Electronic Technology Course Teaching Mode Based on Practice 273
Dongli Jiao, Jinsheng Li, Yongqiang Zhao

Research and Implementation of "Project-Driven" Teaching Method in
<Web Systems Development and Practice> 277
Ruxin Ma, Yu Liu, Xiao Wang

Auto-assessment System of Ship Craft Electrical Engineering
Technology 283
Guangzheng Li, Huimin Lu

Construction of Practice Teaching System for Training Innovative Talents
Based on the Idea of Large-Scale Engineering 289
Aijun Tang, Hailong Ma, Deming Sun, Yuchao Niu, Mingdi Li

Study to the Development of Management Ideology of Physical Education
Since Reform and Opening Up in the China 297
Liu Fusheng, Lu Chun, Ma Meiyu, Li Xiang

Exploration and Practice of Physics Teaching for Liberal Arts
Undergraduate 303
Lizhen Ma, Jun Ma, Shukuan Wang, Yurong Shi, Yongjian Gu

Talent-Cultivation Mode Research on Direction of Services and
Outsourcing-Oriented Software Engineering 307
Yubo Jia, Qi Sun, Na Zhang, Hongdan Fan

Case Teaching Based on the Network 313
Yang Huansong, Liang Lu

On How to Efficiently Carry out Multi-media Teaching 319
Xiaojing Wang

The Mode Analysis and Enlightenment of Cooperation of Production,
Teaching and Research in Japanese Universities 325
Liping Han

Teaching Innovation and Course Practice in "Electrical Control
Technology" 333
Xiaoqian Hu, Lian Zhang, Shan Li

The Function of the University Libraries in Constructing Lifelong
Education System 341
Tian Lizhong, Zhang Aichen, Sun Yongjie

Innovation and Talent Training Mechanism for the Law in the Mode of Production and Research Strategy 347
Hui Ma, WeiWei Yang

Thinking of High-Quality Courses Construction 355
Dongping Yang, Juanjuan Li, Zhengyan Li, Guoqiang Sun

Transformation of Education on the Course of Control Theory 361
HaiSen Ke, Dong Wei, Min Xie, EnHui Zheng, XiuYing Zhou

The Study of Specialized Courses Using the PDCA Cycle 367
Chi BaoQuan, Huang ZhenHai, Zheng EnHui, Wang GuiRong

Discussion on Experiment Course of Signal and System Reform 371
Huang Xiaohui, Jia Zhenhong, Li Xingang

Survey on the Concept of Shaping Behavior and Cognition of Female Flight Attendants 375
Song Faming, Shu Jianping, Wen Xiaoyuan

The Observable Characteristics of Effective Teaching in Professional Experimental Courses 383
Hongmei Ai, Baomin Wang, Junying Bai, Puguang Lu

Optimal Design of High-Radix Router's Switching Fabrics Based on Tile ... 389
Xian-Wen Wu, An-Hua Mo, Li-Quan Xiao

Forecast of Employment Situation and Countermeasure Research on Twelfth "Five-Year Plan" Period in Hebei Province of China 397
Xiaomei Shang, Zheng Li, Shituan Zhou, Guilan Song

Analysis of Employment Situation of Shandong Province Based on Factor Analysis 405
Ma Jun, Han Xin

Research on the Discipline Construction of Medical Information Retrieval Based on the Evidence-Based Decision-Making 419
Wei-Li Chen

The Development of Scientific Activities among University Students Majoring in Logistics 427
QiuHong Zhang, Lianying Cao, Yujian Lv, Qi Qian

The Review and Prospect of Physical Education under China's Accession to the WTO within Last 10 Years 433
Jiasi Luo

Financial Incentive Mechanism Effects of Low-Renting Public Housing with BOT Mode 439
Fangqiang Liu

Research on Psychological Health of Poor College Students 445
Qiuhong Zhang, Xiaoli Ni, Jian Liu, Junni Zhao

**Reform and Practice of Talents Training Model of Electronic Information
Engineering Majors** 451
Xia Zhelei, Xiao Binggang, Wang Xiumin

**The Exploration and Practice in the Training Model for Interdisciplinary
Science Professionals** 457
Baohua Tan, Fei Yang, Chuyun Huang, Guowang Xu

Enabling Nursing Students' Critical Thinking with Mindtools 465
Chin-Yuan Lai, Sheng-Mei Chen, Cheng-Chih Wu

**Canonical Chinese Syntax Awareness Facilitated by an e-Learning
Program** 473
C.C. Lu, C.H. Lu, M.M. Lu, C.H. Hue, W.L. Hsu

**Multilayer Fundamental Physics Curriculum-Group-Construction with
Opening and Mutual Learning in Selective Instruction and Discussion** 481
Shi-jun Xu, Xiao-ling Ren, Jia-qing Cui

Research on Automation Specialty Application Talents Training 489
Rongmin Cao, Denghua Li, Zhong Su, Yingnian Wu

**On the Developmental Stages and Cultivation of Academic Awareness for
Graduate Students in China** 497
Kaige Ren, Guonian Wang

**Exploration on Construction of Scientific Research Network Platform in
Colleges and Universities** 503
Fanmei Liu, Yunxiang Liu

**The Entrepreneurial Motivations and Barriers for Technical University
Students in Taiwan** 511
*Su-Chang Chen, Hsi-Chi Hsiao, Chin-Pin Chen, Chun-Mei Chou,
Jen-Chia Chang, Chien-Hua Shen*

**The Development of Architectural Design Management System Based on
Petri Nets** 519
Wen Ding

Inquiry-Based Education and Its Implication to Education in China 527
Tu Huiwen, Xie Feng

**Study on Construction and Management of Innovational GIS
Laboratory** 535
Lifeng Yuan, Xingfei Liu

Analysis of Professional Skill and the Teaching of Effective Interface
Ordnance N.C.O 543
Zhu Tian-yu, Jiang Zhong-bao, Liu Yun, Duan Shao-li

Research on the Construction of the Clothing Video Database 547
Jianping Liu, Lu Chang, Huilan Chen

Strengthening Cooperation with IT Enterprise, Promote the Practical
Teaching of Information Specialty 555
Wang Liejun, Jia Zhenhong

A Laboratory Measurement Method of Antenna Radiation Pattern 561
Hui Xie, Yujun Liang, Qin Wang

On the Model of Postgraduate Student's Self-management from the
Perspective of Self-Organization Theory 567
Wang Jijun, Zhao Long

An Fast Max-Min Ant Colony Optimization Algorithm for Solving the
Static Combinational Optimization Problems 575
Zeng Lingguo

Practice Study on Integrating Teaching and Research in a Graduate
Course 583
Xiao-Qun Dai

The Development and Application of Virtual Instrument Technology in
the Experimental Teaching 589
Jianqiang Liu, Xingqi Fu, Xingcheng Zhang, Jianye Song

Changes and Characteristics of Vocational Curriculum in Taiwan:
1964-2010 595
Chuan-Yuan Shin, Kung-Huang Lin, Hung-Min Lin

Development Analysis of Featured Industrial Base in Shijiazhuang Based
on SWOT Analysis 607
Wei-li Shi

Coordination Development Prospects of Rural Education and
Community in China 615
Jing Tian, Ling Wang, Zongling Zheng

The Three Basic Working Attitudes That College Teachers Should Highly
Emphasize 623
Shouhui Chen, Zheng Guo, Yanjie Zhang

Application of Project Teaching Method in Higher Project Business
Website Development 627
Songjie Gong

Exploration and Practice on Project Curriculum of Business Website
Development 631
Songjie Gong

On the Training Mode of Professional Printing Talents Based on the
Social Demand and Employment-Orientation 635
Hong He, Haiyan Wang, Yulong Yu, Jieyue Yu

Think in Higher Education Administration Based on Quality Project 641
Hou Xianjun, Peng Wuliang

Comparison of Statistical Clustering Techniques for Correction Analysis
of Achievements of the College Entrance Examination 649
Hu Xifeng

Literature Review on Research of Real Estate Taxation in China 655
Xie Feng

The Time Effect of DNA Damage and Oxidative Stress on Mice Liver
Cells Induced by Exercise Fatigue 661
Su Meihua

The Application of "Functional Equivalence" in Trade Mark
Translation 669
Zhuo Wang

A Probe into Image Shift in Translation 675
Zhuo Wang

The Construction of Management System for Combination of Sports and
Education 683
Shu Gang Li, Peng Feng Huo, Hai Jun Wang

Reflection on Golf Education Development in China under Leisure Sports
Perspective 689
Xueyun Shao, Zhenming Mao, Xiaorong Chen

On the Intellectual Property Right Protection Issue of Digitized Resources
in a University Library 697
Tingrong Liu

Research and Exploration of Light Chemical Engineering Specialty
Excellent Engineers School-Enterprise Cooperative Education 705
Lizheng Sha, Huifang Zhao

The Application of Fast Fourier Transform Algorithm in WiMAX
Communications System 711
Zhiling Tang

The Study on Voltage Controlled Oscillator in Electronic Applications 721
Wu Wen

Optimization Design Method of Mountain Tunnel Lining Based on Stress
Mapping Return Arithmetic 729
Zelin Niu, Zhanping Song

Pareto Analysis of Learning Needs about Adult Courses 737
Shuang Li, Ling Zhang

Virtual Experimental Platform in the Network Database Application
Development 741
Xiaoyu Wu, Wei Dong, Guowei Tang, Huyong Yan, Liquan Yang

An Empirical Analysis on the Interdependence Relation between Higher
Education Tuition Expenditure and Urban Residents' Income—Based on
ECM Model 747
Renjing Xu

Outdoor Sports and Teenagers' Moral Education...................... 757
Tao Yuping

Language, Culture and Thought from a Perspective of English Teaching ... 765
Yingbo Liu

Cultural Differences at the Discourse Level in TEFL in Chinese Class 771
Yingbo Liu

The SWOT Analysis of New Practical English 777
Jin-jing Zheng, Xue-shen Liu

Research on New Mode of University Study Style Construction Based on
Party Construction Blog 785
Chunlin Li, Shuhong Ge

The Application of CVAVR, AVRstudio, Proteus in MCU Teaching 791
Lee Xingang, Jia Zhenhong, Wang Liejun, Huang Xiaohui

The Balanced ScoreCard and Educational Technology
Management —- Take Research on the Hangzhou College Student
Probation Quality Assessment as the Example 799
Xiao-jun Chen, Xiao-yun Yan

Fully Understanding Vocabulary in Five Steps 807
Zhao-jun Liu

Study and Practice in Major Diversity of Undergraduates Program 813
Ge Baojun, Wang Junming, Li Shanqiang

Analysis of Comparing Mulan-Boxing with Other Aerobic Exercises to
Impact Physically on the Old and Middle-Aged Women   821
Xu Cai-yan

A Study of the Application of English Listening Strategies by College
Students   831
Hong Dang

Practice and Reflection of Computer-Aided Chemical Analysis
Experiment   839
Li Liu, Jianzhong Guo, Bing Li, Yanlong Feng

Research on the Design of Function Module of Petroleum Engineering
Practice Base   843
Fengxia Li, Wenhua Li, Wang Li, Zhengku Wang

Research on Constructing the Practical Teaching Base of Petroleum
Engineering   851
Fengxia Li, Wenhua Li, Wang Li, Zhengku Wang

Study on Constructing the Practice Teaching Base of Petroleum
Engineering   857
Zhengku Wang, Wenhua Li, Wang Li, Fengxia Li

A Development Method of Resources for the E-Learning Based
on VRML   865
Xin Huang, Yuxing Peng

Comparative Study on Training Patterns of Entrepreneurial Talents both
at Home and Abroad   873
Shufeng Sun, Xiaoman Chen, Pingping Wang

Analysis of Current Strategic Modes of Chinese Higher Education
Internationalization   879
Wenzhong Zhu, Dan Liu, Yi Wang, Ming Zhang

The Reform and Practice of Automation Excellent Engineers Training
Program   889
Haizhu Yang, Jie Liu

Development of the DSP Experiment System Based on the Emulator of
XDS510   897
Jie Liu, Haizhu Yang

Research on Golf Education in China: Its Significance, Characteristics
and Future   905
Xueyun Shao, Xiaochun Zhang, Xiaorong Chen

The Prediction of Publishing Scale of Literature Books in China—Based on GM (1, 1) Model 913
Renjing Xu

The Preference of Computers over Books and Anxiety among Iranian College Students: The Moderating Role of Demographic Factors 923
Sima Sharifirad, Mohammad Sadegh Sharifirad

Aligning COBIT and ITIL with an IT Academic Courses 933
Vanco Cabukovski, Vase Tusevski

Investigation and Analysis of Current Situations of Participation in Traditional Sports Activities of Ethnic Minorities by Urban and Rural Kazak, Kyergyz, Mongol and Tajik Residents in Xinjiang 939
EnLi Han

Author Index 945

The Research on Supplement of Glutamine in Sports

Dunhai Wang, Yubo Han, and Ancun Jiao

Rizhao Polytechnic
Rizhao, China

Abstract. By the method of documents and datum, This article describe the characteristics of metabolism of Glutamine and its physiological function. Its discuss the impact of exercise performance on Glutamine and the impact of exogenous Glutamine on improving exercise performance. The supplementary of Glutamine is very important in some area,such as antioxidant capacity for the body, immune function, measure the over-training, enhance hormone levels and other aspects.This article offers references for researching the application of Glutamine in sports.

Index Terms: Glutamine, Exercise, Metabolize, Supplement.

1 Introduction

Glutamine is a very important condition for the body of essential amino acids, the body is the most abundant amino acids. Games of the metabolism of glutamine, and glutamine metabolism in the body's movement in turn will affect the ability of exogenous glutamine in the sports supplement and the premise of material after prolonged exercise can prevent the content of plasma glutamine decline, it is now increasingly recognized that the relationship between glutamine and exercise and glutamine supplementation in the importance of the movement.

2 Its Physiological Role in the Metabolism of Glutamine

Glutamine accounts for about 60% of free amino acids in the body, containing five carbon amino acids and two amino acids, the plasma concentration of normal human body 500 ~ 700mmol / L, the concentration in human skeletal muscle which is 20m mol / Kg wet weight [1], we can see the changes it can directly affect the level of amino acids in the body. Synthesis and release of human skeletal muscle is the main organ of glutamine is the main source of plasma glutamine. Small intestine is the major organ of glutamine utilization. Intestinal mucosal cells by ceramide Valley helium high blood circulation through the digestive tract of each one, there is 20% to 30% of the metabolism of endogenous glutamine is absorbed and used, of which 90% of cells in the intestinal mucosa Metabolism of [2]. When the blood will be transported to the liver, glutamine, glutaminase within the liver cells will be broken down into glutamate and ammonia, glutamine, followed by the liver synthesis of urea excreted by the kidneys.

Y. Wang (Ed.): Education Management, Education Theory & Education Application, AISC 109, pp. 1–6.
springerlink.com   © Springer-Verlag Berlin Heidelberg 2011

Glutamine has a very wide range of physiological role in maintaining immune function plays an important role, because the glutamine is a lot of immune cells and intestinal epithelial cells of the important energy source. Second, glutamine is an important regulator of protein synthesis, can be adjusted in the movement of protein synthesis and reduce muscle protein breakdown, in order to maintain the body's physiological functions. Third, glutamine can be used as the carrier of renal ammonia generated directly involved in the metabolism of ammonia, which played Victoria.

Support the important role of acid-base balance. Fourth, the body glutamine is an important means of delivery of nitrogen and carbon. Fifth, the glutamine can be generated by gluconeogenesis glucose to maintain blood glucose balance, regulate glucose metabolism.

3   The Impact of Exercise on Glutamine Metabolism

3.1   Short-Term High Intensity Exercise on Glutamine

Many studies abroad, found the body during the short high-intensity exercise, plasma glutamine concentration increased. Bergstrom J et al [13] to four healthy volunteers to 70% VO2max for cycling, the sport's first 10min and 20min for testing blood and found that plasma glutamine concentration. KatzA [14] found that the 8 healthy male subjects, 50% and 97% VO2max for cycling took 10min and 5.2min, respectively, to fatigue, resulting significant increase in arterial plasma glutamine concentration. Pring-Billings et al [15] found that 10 × 6s sprint Games, the plasma glutamine concentration from 556μmol / L up to 616μmol / L. Therefore, the short-term high intensity exercise produced the greatest impact on the glutamine concentration is to make it up, it produces short-term strenuous exercise may be the mechanism of muscle and blood ammonia levels increase, and accompanied by lactic acid concentration, lactic acid increased to 14mmol / L, the increase in plasma ammonia, then will promote the adenine nucleotide decomposition of ammonia, so that muscle glutamate glutamine help increase plasma glutamine. Similarly, exercise training to adapt to make blood glutamine levels.

3.2   Prolonged Endurance Exercise on Glutamine

Many studies showed that prolonged exercise (including low-intensity and high intensity) in plasma glutamine concentrations significantly decreased. Castell et al [16] study found that marathon 1h, plasma glutamine level declined by approximately 20%. Smith et al [17] The study also found that high-intensity training cycle, the plasma glutamine level from before training (585 ± 54) μmol / L decreased significantly from the trained (522 ± 53) μmol / L, which identified the plasma glutamine content can reflect the body's tolerance for the amount of training. After prolonged exercise decreased plasma glutamine concentrations may be generated by the mechanism of glutamine during prolonged exercise to reduce the generation of the release, while increased absorption of various tissues and organs, it is caused by decreased blood glutamine in the second, reduced movement maximal activity of glutamine synthetase, it could be one of the reasons. Third, the ultra-long exercise in the body glucocorticoid concentration, is also a reason for the decline of glutamine.

Our study can be drawn from the above, glutamine is closely linked with the exercise, glutamine supplementation has a very important role.

4   Effect of Glutamine Supplementation and Exercise Performance

Study found that exercise duration, intensity, the size of the impact of plasma glutamine concentration is different, indicating that the movement of glutamine and human, has a very important relationship, then the increase will become supplementary glutamine human exercise capacity in an important part.

4.1   Glutamine Supplementation and Exercise Antioxidant Capacity

During exercise, the body produces a large number of free radicals, damage of membrane structure and affect athletic ability. Glutathione is an important protective factor in vivo, could protect the activity of thiol enzymes, glutathione peroxidase in the (GSH-Px) catalyzed protection of the biofilm and biological macromolecules from oxidative damage [3] 274. Also in the spring Jian et al [4] found that glutamine can also increase glutathione synthesis of organizations to protect the liver from free radical damage. Glutamine increased the stability not only to increase the cell membrane, more importantly, can increase the glutathione content of liver and reduce the oxygen free radicals in liver damage. Glutamine helps maintain the immune cells in plasma and glutathione levels, antioxidant capacity enhancement to improve exercise capacity. SPORTS MEDICINE other studies have shown that [5], glutamine supplement can significantly reduce the high-intensity exercise the level of lipid peroxidation. Although the direction of these two studies, conclusions are different, but the same can be seen that the relationship between the two, glutamine to protect liver, improve the ability of human motion has very important significance.

Renal production of ammonia or glutamine precursors, the body plays an important role in environmental regulation in vivo. Exercise-induced changes in renal function forms, varying degrees of renal dysfunction caused by the Games, glutamine supplementation can improve renal metabolism.

4.2   Glutamine Supplementation and Exercise Influence Immune Function

Prolonged high intensity exercise can lead to immune suppression, which may lead to a decline in glutamine sports one of the mechanisms of immunosuppression. Study found that exercise immunology in recent years, exercise and secretory immunoglobulin SlgA close [3] 277. Studies have shown that SlgA can effectively prevent bacterial adhesion in the intestinal mucosa, and its ability to other immune globulin 7 to 10 times [6]. However, in the extraordinary exercise or heavy exertion SlgA saliva after exercise was significantly reduced. Glutamine supplementation can prevent SlgA reduce and prevent intestinal mucosal plasma cells and lymphocytes decreased SlgA enhance the organ - gut associated lymphoid tissue function; is conducive to glutathione storage, and enhance antioxidant capacity and the host defense capacity to maintain intestinal mucosal immune function is normal, increase the body's ability to fight infection.

In addition, severe stress can cause the immune system significantly increased utilization of glutamine, which is conducive to maintaining a variety of immune cells after trauma and effective response of immune injury. Glutamine supplementation can increase the production of 1L-2 to improve the membrane inhibited the expression of 1L-2R, reducing trauma to the inhibition of T lymphocyte proliferation. Thus, exogenous glutamine can improve immune function and intestinal immune function, as during exercise or after exercise and nutritional intake of the immune regulation provides a certain degree of protection is conducive to the improvement of exercise capacity.

In animal studies also found that exogenous amide nitrogen supplement Valley, lymphocyte proliferation and stimulation of value-added natural white are enhanced, B lymphocytes increased ability to secrete antibodies, phagocytic activity increased, the immune function of red blood cells significantly increased, indicating that glutamine on immune function has significantly increased its role in [7].

4.3 Glutamine Supplementation and Over-training

Glutamine is the body condition of a multi-purpose essential amino acids, which not only early diagnosis as an indicator of overtraining, but also a measure of intermittent hypoxic training and continuous hypoxic training advantages and disadvantages of a ruler. Bu in Jun and other studies have shown that [8], before and after training the ratio of plasma glutamine and glutamine (GM / GA) is reflected in the size of the amount of training experiment proves that when GM / GA <3.58, it will appear over the so-called training. Also, by observing the changes in plasma glutamine concentration found that intermittent hypoxia training group than in plasma glutamine concentration and VO2max sustained hypoxia training group significantly improved. Glutamine and shows the close relationship between exercise training, in addition, white blood cell count and infection with no significant differences before training, indicating that intermittent hypoxic training exercise is more conducive to maintaining immunity.

4.4 Body Movement for Restoration of Glutamine Supplementation

High-intensity exercise, increasing the body's free radicals, which lead to exercise-induced fatigue and the causes of sports injuries. Qiaoyu Cheng [9] by way of exhaustive swimming in rats, glutamine supplementation compared through experiments 18h after exhaustive exercise observed in rat liver tissue malondialdehyde (MDA), GSH, superoxide dismutase (SOD) content changes. The results showed that: simple exercise in liver tissue MDA content was significantly higher than the quiet control group (P <0.01), glutamine supplementation did not change significantly the exercise group, but significantly lower than the exercise group (P <0.01); simple exercise in liver tissue GSH content was significantly lower than the sedentary control group (P <0.01)), glutamine supplementation exercise group significantly higher than the exercise group (P <0.05); liver SOD activity: simple exercise group was significantly higher than that of sedentary control group (P < 0.05)), glutamine supplementation did not change significantly the exercise group, but significantly lower than the exercise group (P <0.01). These results suggested that exogenous glutamine after exercise helps to maintain liver GSH level, and increase its antioxidant ability, reduce lipid peroxidation in the liver, reducing liver lipid peroxidation. Therefore, glutamine supplementation can promote the movement through the free radical produced to help the body recover after exercise.

4.5 Glutamine Supplementation and Increased Hormone Levels

Maintain a certain hormone levels, can help athletes maintain a high state of activity level, glutamine supplement to some extent, improve the athlete's hormone levels. Yu-Chen chain, Ming-Fang et al [10] in order to participate in summer training camp in a college track athletes 12 men and women of all six human subjects, the glutamine supplement the test experiment. The results show that glutamine supplementation can inhibit the prolonged exercise and training the large decline in blood testosterone, so that T / C value is maintained at a relatively high level of protein synthesis and metabolism helps the body to effectively delay the occurrence of fatigue.

4.6 Glutamine Supplementation and Sports Injury Recovery

Xu Kai et al [11] found that a serious surgical trauma, the body and the tissue concentration of glutamine decreased. On the one hand surgery, trauma and other serious stress cases, increased secretion of glucocorticoids lead to increased release of glutamine, but on the other hand, the absorption of the digestive tract of glutamine increased nearly 2-fold, so the overall performance of blood the decline of free glutamine. Similarly, for patients with surgical trauma glutamine supplementation, for mucosal cells, lymphocytes, fibroblasts provide a rich energy supply, promote wound healing. Glutamine and because the improvement of nutritional status and immune capabilities, which further enhances the body's ability to adapt to stress, promote the body recover faster.

4.7 Glutamine Supplementation and Exercise Skill Formation

Skill formation is a prerequisite and basis for learning and memory, glutamate as the main central excitatory amino acid neurotransmitter, mediated a series of higher nervous activity. Glutamine as a precursor of glutamate, which can freely enter and leave the brain through the blood-brain barrier, it is more than the glutamate regulating the function of the nervous system, Jinghong Jiang et al [12] studies have shown that glutamine supplement can improve the brain nitric oxide synthase (NOS) activity and hippocampal N-methyl-D-aspartate receptor number is important for learning and memory, is conducive to the formation of motor skills. Currently, the research in this area is also less.

5 Summary

Glutamine is the body condition of a multi-purpose essential amino acids, their synthesis can be either exogenous supplement. It is in the antioxidant capacity, immune, measuring over-training to improve hormone levels, body movement for restoration of motor skills in areas such as the formation of a very important role.

Exercise can influence the concentration of glutamine, depending on the duration and intensity of exercise. In sports training, supplemented with glutamine for enhancing athletic ability is very important, it should be further study the relationship between exercise capacity and body.

References

1. King, M., Zhu, B., Tang, S.: Optimal path planning. Mobile Robots 8(2), 520–531 (2001)
2. Scripts before. Glutamine and immune function after exercise. Fuyang Teachers College, 21 (4):28-32 (2004)
3. Ma, F.A.: Morals Sports and glutamine metabolism. Physical 18(5), 56–57 (1997)
4. Zeyi, Y.: Exercise and glutamine metabolism of the penetration of gold. Chinese Journal of Sports Medicine 22(3), 273–278 (2003)
5. Chun, Y.J., Zhu, J., Demin, L.: Parenteral nutrition with glutamine dipeptide when the regulation of glutathione in the metabolism, mechanism of protection of the liver. Chinese Academy of Medical Sciences 20(2), 104–108 (1998)
6. SPORTS MEDICINE banners. Glutamine and Exercise Research. Xi'an Institute of Physical Education18 (4), 30–32 (2001)
7. Huang, Y., Xie, Y.: Glutamine on Immune Regulation. FJ Medical Journal 27(2), 273–278 (2005)
8. Huang, Y.: Glutamine on immune function. Nutrition 23(4), 363–364 (2001)
9. Bu, B. J., WangHui, Y.: Hui glutamine and Exercise Research. PHYSICAL 24(3), 56–59 (2003)
10. Qiao, Y., et al.: Exhausted swimming glutamine on liver MDA, GSH, SOD content. Chinese Journal of Sports Medicine 12(1), 91–92 (2002)
11. Lian, Y.-C., Shen, M.-F., Wei, Y., Ming, Y., Xu, J.-r.: Glutamine supplementation on college students summer training on serum T / C value of Urea in Blood. Nanjing Institute of Physical Education (Natural Science) 5(1), 13–16 (2006)
12. Xu, K., kun: Glutamine supplementation and exercise capacity in Research. Nanjing Institute of Physical Education (Natural Science) 3(1), 16–21 (2004)
13. Jiang, J., Yiyong, C., Shutian, L., et al.: Glutamate and glutamine on learning and memory. Health Research 29(1), 40–42 (2000)
14. Bergstrom, J.: Free amino acids in muscle tissue and plasma during exercise in man. Clin. Physiol. 5(2), 155–160 (1985)
15. Katz, A.: Muscle ammonia and amino acid metabolish during dynamic exercise in man. Clin. Physiol. 5, 365–379 (1986)
16. Parry-Billings, M.: Plasma amino acid concertration in over-traning syndrom: Possible effects on the immune system. Med. Sci. Sports Exerc. 24, 1353–1358 (1992)
17. Castell, L.M.: The effects of oral glutamine supplementaion on athletes after prolonged, exhaustive exercise. Nutrition 13(7-8), 738–742 (1997)
18. Rowbottom, D.G.: Training adaptation and biological changes among well-trained made trained made triathletes. Med. Sci. Sports Exerc. 8, 1233–1239 (1997)

Innovative Practices of IC Design and Application Courses

Jianping Hu, Yinshui Xia, Haiyan Ni, and Lunyao Wang

Faculty of Information Science and Technology, Ningbo University
315211 Ningbo City, China
nbhjp@yahoo.com.cn

Abstract. The IC design and application undergraduate program is an inter-discipline emerging one with high technology integration. This paper presents innovative practices of the IC design and application undergraduate program at Ningbo University to meet strong demand for designers of ASIC chips and electronic engineers of home electronic products. The course system, teaching content, teaching methods, and practice teaching modes are explored for the undergraduate specialty, which are developed by referencing developing experiences of IC postgraduate students at Ningbo University and advanced teaching experiences of microelectronics. The study results have some value for IC design and application undergraduate professional construction of other similar universities.

Keywords: Teaching practices, innovative practices, IC design, professional construction.

1 Introduction

Integrated circuits are the foundation and core of the information technologies. In the next few years, there exist huge demands for integrated circuit design talents. However, in the traditional education systems in China, there isn't an IC design undergraduate specialty, although many colleges and universities have undergraduate programs on semiconductor technology and microelectronics. In order to meet the urgent needs for IC design and application talents, Chinese Education Department approved "integrated circuit design and integrated systems" undergraduate program from 2001.

The integrated circuit design represents the key competitiveness of a country or region. Therefore, undergraduate specialty construction of IC design and application can help develop the national and regional information science, and has extremely vital significance for enhancing original innovation ability in a country or region. However, the integrated circuit design undergraduate program was developed in only a few of universities in China in recent years. From 2006, Ningbo University has being offered the IC design and application programs for the undergraduate students of the Faculty of Information Science and Technology, which is one of the earliest ones in China.

Y. Wang (Ed.): Education Management, Education Theory & Education Application, AISC 109, pp. 7–14.
springerlink.com   © Springer-Verlag Berlin Heidelberg 2011

The IC design and application undergraduate specialty is a recent new undergraduate specialty all over the world, although some courses have begun for microelectronic undergraduate and postgraduate program for a long time [1-5]. Therefore, course system, teaching contents, teaching methods and experimental modes of its core courses are not mature experience for reference. One of the keys of the specialty construction is how to set teaching contents, teaching methods and experimental modes of the core curriculums reasonably and effectively.

A series of IC core courses for IC design and application undergraduate program have been developed at Ningbo University. In this paper, the innovative practices of the IC design and application undergraduate specialty are presented. The course system, teaching content, teaching methods, and practice teaching modes are explored for the undergraduate program, which are developed by referencing developing experiences of IC postgraduate students at Ningbo University and advanced teaching experiences of microelectronics. Teaching innovation and practices for IC design and application undergraduate specialty have been carried out, which includes course syllabus, lecture notes, experiment guide books, etc. The study results have some value for IC design and application undergraduate professional construction of other similar universities.

2   Curriculum System of Undergraduate Program

The training goal of IC design and application undergraduate specialty is ASIC design and application ability. Through four years of learning and training, the students should have a solid mathematical and physical basis. They should master theory, principle and design methods related on circuit principle, analog circuits, digital circuits, IC design, and IC manufactures with the related knowledge and ability. They should also be skillful at IC design EDA tools such as function modeling, logic simulations, logic synthesis, layout design and layout verification (DRC, LVS, and LPE), layout place and route, post-simulation, and have design capabilities of ASIC chips by using these IC design tools. Moreover, students should have basic skills at literature search, strong ability of self-study, practice ability, and certain scientific research ability and innovative consciousness.

The setting principles of the course system include systematicness, completeness, advancement, and practicability related on IC design and applications. To achieve the training target, the course system of the IC design and application undergraduate program should cover all main IC design knowledge. In order to make students adapt to future work, the courses should reflect the advanced knowledge related on IC designs. The course system should also emphasize the engineering practice ability and team cooperation spirit of students, and trains students the capability of solving problems with the IC design knowledge that they can easily transfer to a future workplace.

After careful thorough investigation and demonstration, finally we determine the curriculum system of the IC design and application undergraduate program, as shown in Fig. 1, which is composed of four parts, named as comprehensive education courses, compulsory courses in big academic subjects, professional education platform courses, and professional module courses of IC designs and applications.

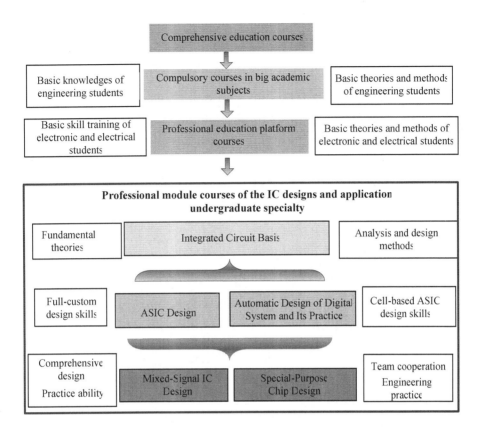

Fig. 1. The curriculum system of the IC design and application undergraduate program at Ningbo University

Comprehensive courses such as college Chinese, foreign language, economics, etc are covered by comprehensive education course platform at Ningbo University. Basic knowledge and abilities of engineering students such as mathematics and physics, etc are covered by compulsory courses in big academic subjects for engineering students at Ningbo University. Basic theories and methods of electronic and electrical courses such as circuit theory, analog & digital circuits, computer, etc are covered by professional education platform courses for basic skill training of electronic and electrical students.

The core courses in professional module of the IC design and application undergraduate program are shown in Table 1. The core professional courses in IC design and application undergraduate specialty include mainly: Integrated Circuit Basis, ASIC Design, Automatic Design of Digital System and Its Practice, Mixed-Signal IC Design, Analysis and Design of Digital System, and Special-Purpose Chip Design.

Table 1. The core professional courses in IC design and application undergraduate program at Ningbo University

Course name	Teaching hours per week
Integrated Circuit Basis	4.0
ASIC Design	4.0
Automatic Design of Digital System and Its Practice	5.0
Mixed-Signal IC Design	5.0
Analysis and Design of Digital System	4.0
Special-Purpose Chip Design	Three weeks (Short term)

3 Innovation Practice of the Core Professional Courses in IC Design and Application Undergraduate Program

The organization structure of the core professional courses in IC design and application undergraduate program is shown in Fig. 2.

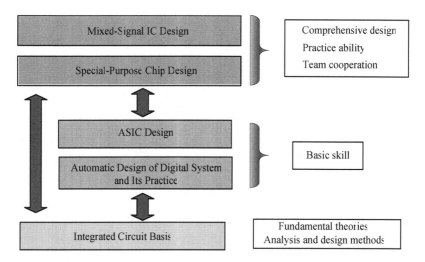

Fig. 2. The organization structure of the core professional courses in IC design and application undergraduate specialty

The Integrated Circuit Basis course is one of the professional core curriculums in the integrated circuit design and application undergraduate program. The goal of this course is an understanding of IC fabrication process, semiconductor devices, layouts, transistors, and CMOS logic design. The course trains students to develop skills at analysis and design of CMOS logic gates, combination logic blocks, and sequential circuits, and use IC design tools for carrying out schematic input, simulation and optimization of IC digital circuits. The course also introduces design methods of basic digital modules including flip-flops, adders, shifters, etc.

The ASIC Design course introduces simulation, design, verification, and testing of IC chips, which are a professional core course in IC design and application undergraduate program. The goal of this course is an understanding of full-custom layout design and verification. The course trains students to develop skills at full-custom layout design and layout verification by using IC EDA tools such as Virtuoso Layout Editor, Diva verification, etc. The course also introduces design methods of basic digital modules including multipliers, memory, etc.

The Automatic Design of Digital System and Its Practice course is also one of the professional core courses in IC design and application undergraduate program. The goal of the course is an understanding of the cell-based IC design. The course introduces Verilog HDL hardware description language, logic synthesis, layout place and route, etc. The course trains students to develop skills at cell-based IC design and layout verification by using IC EDA tools such as DC, Encounter, etc.

The Mixed-Signal IC Design course introduces basic knowledge and skill of mixed-signal IC devices, CMOS operational amplifier, CMOS OTA, filter, ADC, DAC, mathematical model, mixed-signal IC design method, etc. The goal of this course is an understanding of fundamental theories, basic knowledge, and analysis and design methods of the mixed-signal integrated circuits. The course trains students to develop skills at layout design and layout verification of mixed-signal integrated circuits.

The Special-Purpose Chip Design course is also one of the professional core curriculums in the integrated circuit design and application undergraduate program, and it also is the only undergraduate course design in the undergraduate program. This course is a comprehensive application training course, and it emphasizes the engineering practice ability and team cooperation spirit of students, and trains students to resolve the actual engineering problems with the comprehensive adoption of fundamental theories, basic knowledge and basic technical ability through several projects, and develop student's team cooperation sprits. Through this course, the student should grasp full-custom ASIC design and cell-based ASIC design methods by using IC EDA tools.

4 Experimental Platform of the Integrated Circuit Design

In the course teaching, course design, and undergraduate thesis, the integrated circuit design tool platforms is needed for experiment teaching related on IC designs. As mentioned above, in the traditional education system in China, there isn't an IC design undergraduate specialty, thus few experiment conditions can be provided for IC design and application undergraduate students.

After years of technology preparation from 2001, basic integrated circuit design tools have set up for IC scientific researches of the main members of our IC design and research group. Because the existing IC design experimental platform is mainly used for scientific researches, is not suitable for the experiment teaching of undergraduate courses. Therefore, we have constructed the experimental platform of the integrated circuit design including full-custom design tools and cell-based IC design tools for the experiment teaching of IC design and application undergraduate students.

The constructed experimental platform of the integrated circuit design provides the experiment teaching of the six core professional course: Integrated Circuit Basis, ASIC Design, Automatic Design of Digital System and Its Practice, Mixed-Signal IC Design, Analysis and Design of Digital System, and Special-Purpose Chip Design. The IC design tool experimental platform includes the following contents:

The first part of the Integrated circuit design tools experimental platform is the full-custom IC design tools that are used for the experiment teaching of the Integrated Circuit Basis ASIC Design, and Special-Purpose Chip Design courses. The design tools include IC5141, HSPICE, Caliber, etc.

The second part of the integrated circuit design tool experimental platform is the cell-based IC design tools. It is used for the experiment teaching of Automatic Design of Digital System and Its Practice, Analysis and Design of Digital System, and Special-Purpose Chip Design courses. The IC design tools include Cell-based design tools and validation tool such as DC, Encounter, etc.

The third part of the integrated circuit design tool experimental platform is the mixed-signal IC design tools. It is used for the experiment teaching of Mixed-Signal IC Design, and Special-Purpose Chip Design courses.

In order to make students master the operations of IC design tools, lecture notes, experiment guide books, etc of IC design tools have been developed, which is helpful to grasp IC design tools well.

5   Project-Driven Teaching Method of the Core Courses with Powerful Practice Ability

To achieve teaching goad of the IC design undergraduate specialty, the graduate students should have a series of knowledge and abilities from the system's ideas, design methods to the realization of the required chips. They should master theory, principle and design methods related on IC designs, and also have the related knowledge and ability of IC designs. Moreover, students should also be skillful at IC design EDA tools such as function modeling, logic simulations, logic synthesis, layout design and verification, layout place and route, post-simulation, and have design capabilities of ASIC chips by using these IC design tools, so that they have the skill and practice ability from the system's ideas to the realization of the required chips. Therefore, in the course teaching contents and modes, the ability practices of integrated circuit designs are strengthened, and project-driven teaching method is constructed for the core courses of the IC designs.

5.1 Strengthening Practice Abilities of IC designs

Experiment training is an important teaching step, which teaches actual operation ability and scientific research methods of IC designs. It provides students with powerful learning experiences that they can easily transfer to a future workplace. In several courses, the experiment teaching hours are all not less than theory ones, as shown in Table 2.

Through a lot of experimental, the students are trained to complete IC designs by using mainstream design tools. For example, in the ASIC Design course, the experiment contents are strengthened for design and practice ability of ASIC, and the experiment teaching trains students to complete layout design and verification skills by using IC EDA tools to master the design ability of the ASIC chip that they can easily transfer to a future workplace. The Special-Purpose Chip Design course emphasizes the engineering practice ability and team cooperation spirit of students, and trains students the capability of solving problems with the IC design knowledge. Through the experiment training of these courses, the student should grasp full-custom ASIC design and cell-based ASIC design methods by using mainstream IC design EDA tools.

Table 2. Lecture and experiment teaching hours of the core professional courses in IC design and application undergraduate specialty

Course name	Lecture hours per week	Experiment hours per week
Integrated Circuit Basis	2.0	2.0
ASIC Design	2.0	2.0
Automatic Design of Digital System and Its Practice	1.0	4.0
Mixed-Signal IC Design	2.0	3.0
Analysis and Design of Digital System	2.0	2.0
Special-Purpose Chip Design	Three weeks (Short term)	Three weeks (Short term)

5.2 Project-Driven Teaching Methods

For the core courses of IC designs, we increased comprehensive experiment projects on the basis of basic experiment skills to improve students' design and practice abilities that they can easily transfer to a future workplace. For example, in the ASIC Design course, experiment teaching contents are composed of three basic practices and five projects. The experiment contents strengthen design and practice ability of ASIC. In the five projects, the students are demanded to complete schematic input and pre-simulation for the CMOS modules by using Cadence composer, then layout design and layout verification (DRC, LVS, and LPE), and post-simulation by using Cadence Virtuoso Layout Editor with HSpice simulation tool, and finally add I/O

PAD. Through a lot of experiments and projects, the experiment teaching of the ASIC Design course train students to complete layout design and verification skills by using IC EDA tools to master the design ability of the ASIC chip that they can easily transfer to a future workplace.

In the Mixed-Signal IC Design and Special-Purpose Chip Design courses, students are trained to resolve the actual engineering problems with the comprehensive adoption of fundamental theories, basic knowledge and basic technical ability through several projects, and develops student's team cooperation sprits. Through the two courses, the students have greatly promoted knowledge application ability and project practice ability, and these can provide good foundation for their further study and develop.

6 Conclusion

IC design core courses for the IC design and application undergraduate program have been developed at Ningbo University to meet strong demand for IC designers. After several years of construction and practice, the teaching practices have obtained many innovation achievements in course system, teaching content, teaching method, experiment mode, courseware construction, and construction of IC design experimental platform. For experiment teaching, we constructed the progressive hierarchy of curriculum experiment system, strengthened experiment skill training, and developed their abilities to analyze and solve problems. For content organization of the experiment teaching, we increased comprehensive experiment projects on the basis of basic experiment skills to improve students' design and practice abilities that they can easily transfer to a future workplace. The teaching practice results have some value for other similar universities.

Acknowledgments. Project is supported by Zhejiang Undergraduate Key Professional Construction Project, Ningbo Undergraduate Key Professional Construction Project (Szdzy200711), Zhejiang New Century Higher Education Teaching Reform Project (zc2010015), Ningbo key construction service-oriented professionals (Sfwxzdzy200903).

References

1. http://www.cs.eee.ntu.edu.sg/students/Pages/undergrad.aspx
2. Rabaey, J., Chandrakasan, A., Nikolic, B.: Digital Integrated Circuits: A Design Perspective, 2nd edn. Prentice-Hall (2003)
3. Rabaey, J.: Low-Power Design Essentials. Springer, Heidelberg (2009)
4. Smith, M.: Application-Specific Integrated Circuits. Addison-Wesley Pub. Co. (1997)
5. Behzad, R.: Design of Analog CMOS Integrated Circuits. McGraw-Hill Inc., US (2000)

Relationship between the Academic Achievement and Social Responsibility for Teachers in University

Junyi Chen, Li Xu, Haifeng Wu[*], and Zhiguo Yang[*]

Institute of Medical Biotechnology in Chongqing/Institute of Chinese Medicine in Chongqing, Chongqing Medical and Pharmaceutical College, Chongqing 401331, P.R. of China
chenjunyi3@126.com, lionspaper@foxmail.com

Abstract. The correlation and path analysis on relationship between the academic achievement and social responsibility for teachers were studied in university in Chongqing. The results showed that the social responsibility for teachers had significant positive correlations with the title of outstanding teachers, to upgrading diploma , paper, and revenue in last 3 years. And revenue had the most positive influence on the social responsibility for teachers with the direct path coefficient 0.7891, the average number of teaching hours had the most negative influence on the social responsibility for teachers with the direct path coefficient-0.0184. The influence order of other characters was the title of outstanding teachers (0.2438), paper (0.0535), to upgrading diploma (0.0329), and as academic leaders or key teachers (0.0301).

Keywords: University, Teachers, Academic achievement, Social Responsibility.

1 Introduction

After rapid development in nearly 20 years, college had reached the extent of the initial mass popularity. To construction of a high level of university teachers, that is needed for proper management-oriented, establishing good management practices of university teachers, Teachers first need to find out the sources of social responsibility. Relationship was studied between the academic achievement and social responsibility for teachers in universit by correlation and path analysis (granted of outstanding teachers, the average number of teaching hours, to upgrading diploma, as academic leaders or key teachers, paper, and revenue in last 3 years). To clarify the correlation between each factor and the social responsibility of university teachers gain size, for the management of university teachers and the development of management measures to provide a theoretical basis, see references [1-5].

2 Research Objects and Methods

The study objects was teachers of universities in Chongqing of China, To protect the privacy of teachers, the survey taken anonymously, issued 331 questionnaires, 248

---

[*] Corresponding authors.

Y. Wang (Ed.): Education Management, Education Theory & Education Application, AISC 109, pp. 15–19.
springerlink.com © Springer-Verlag Berlin Heidelberg 2012

copies of questionnaires, of which 211 valid questionnaires, a random sample of 41 was analysized, Correlation and path analysis were conducted on the social responsibility of university teachers. Table description of the project: (1) Evaluation methods for 3 years received the title of outstanding teachers: 1 point for school level , 3 points for provincial level, and 10 points for national level; (2) to upgrading diploma in the last 3 years 1 point for for bachelor, Master 2 points, 3 for Dr, postdoctoral 4, if the foreign diploma, the bachelor 2 points, 3 MA, PhD 4, postdoctoral 5; (3) as academic leaders or key teachers in the last 3 years : 2 point for school level , 4 points for provincial level, and 10 points for national level; (4) paper in the last 3 years, the Chinese core papers 1 point, authoritative core 2 points, SCI or EI 3 points; (5) revenue means the average annual income of nearly 3 years; (6) Social responsibility is the teachers themselves rate, the 100 system.

3 Results

3.1 The Correlation Analysis

Table 1. The correlation between the academic achievement and social responsibility for teachers in university

Item	X_1	X_2	X_3	X_4	X_5	X_6
X_2	-0.2253					
X_3	0.4690**	0.1667				
X_4	0.4725**	-0.1182	0.1969			
X_5	0.8591**	-0.3028	0.4259**	0.3342		
X_6	0.1938	0.3152	0.6364**	-0.0606	0.2448	
Y	0.4764**	0.1611	0.6750**	0.1240	0.4857**	0.8627**

Note: the title of outstanding teachers（X1）、 the average number of teaching hours （X2）、 to upgrading diploma（X3）、 as academic leaders or key teachers（X4）、 paper （X5); revenue（X6);

The results showed that responsibility for university teachers was significantly related to the title of outstanding teachers, to upgrading diploma, paper, and revenue. It suggested that responsibility was affected by the title of outstanding teachers, to upgrading diploma, paper, and revenue.

3.2 The Path Analysis

The direct path of different factors coefficient may reflect the relative importance to social responsibility of university teachers.

Table 2. The path analysis between the academic achievement and social responsibility for university teachers

path coefficience	Direct	Indirect					
		1→Y	2→Y	3→Y	4→Y	5→Y	6→Y
X_1	0.2438		0.0042	0.0154	0.0142	0.0459	0.1530
X_2	-0.0184	-0.0549		0.0055	-0.0036	-0.0162	0.2490
X_3	0.0329	0.1143	-0.0031		0.0059	0.0228	0.5020
X_4	0.0301	0.1152	0.0022	0.0065		0.0179	-0.050
X_5	0.0535	0.2095	0.0056	0.0140	0.0101		0.1930
X_6	0.7891	0.0472	-0.0058	0.0209	-0.0018	0.0131	

Note: the title of outstanding teachers (X1)、the average number of teaching hours (X2)、to upgrading diploma (X3)、as academic leaders or key teachers (X4)、paper (X5) ; revenue (X6) ; responsibility (Y)

3.2.1 The Direct Effect of Factor of Revenue to Social Responsibility

The direct path coefficient of revenue of teachers to social responsibility was 0.7891. it was the directly positive head factor among all relative factors, it also had indirectly nega tive effects by nearly 3-year average number of teaching hours and nearly 3 years as the backbone of academic leaders or teachers, which obscures annual income of college teachers of social responsibility of the direct contribution, through the last 3 years received the title of outstanding teachers, nearly 3 years to improve levels of academic degrees and nearly three years of papers published on the university teachers, which have a positive indirect effects of social responsibility. So annual income can be used as a college teacher's instructions with high sense of social responsibility factors. Teachers in the higher income, a higher sense of social responsibility.

3.2.2 The Direct Effect of Factor of the Title of Outstanding Teachers to Social Responsibility

The direct path coefficient of the title of outstanding teachers of teachers to social responsibility was 0.2438. it was the directly positive second factor, it also had indirectly effects by the average number of teaching hours, to upgrading diploma, as academic leaders or key teachers, paper and revenue for social responsibility. Thus, the teachers who had received the title of outstanding in past 3 years had capacity, popularity, and good relations. The school leaders and teachers both considered that it was important index as evaluation of teachers

3.2.3 The Direct Effect of Factor of Paper to Social Responsibility

The direct path coefficient of the title of outstanding teachers of teachers to social responsibility was 0.0535. it was the directly positive factor after revenue and the title of outstanding teachers, it also had indirectly effects by the title of outstanding teachers, the average number of teaching hours, to upgrading diploma, as academic leaders or key teachers, and revenue for social responsibility. It showed that the teachers who had more paper in past 3 years had relative high responsibility, but its role inferior to revenue and the title of outstanding teachers.

3.2.4 The Direct Effect of Factor of Upgrading Diploma to Social Responsibility

The direct path coefficient of upgrading diploma of teachers to social responsibility was 0.0329. it had indirectly negative effects by nearly 3-year average number of teaching hours, which obscures role of upgrading diploma to social responsibility of the direct contribution, it had positive indirect effects of social responsibilit through the title of outstanding teachers, as academic leaders or key teachers , paper, and revenue on the university teachers y. So factor of upgrading diploma had role for social responsibility.

3.2.5 The Direct Effect of Factor of Academic Leaders or Key Teachers to Social Responsibility

The direct path coefficient of factor of academic leaders or key teachers of teachers to social responsibility was 0.0301. it had indirectly negative effects by revenue, which obscures role of factor of academic leaders or key teachers to social responsibility of the direct contribution, it had positive indirect effects of social responsibilit through the average number of teaching hours, upgrading diploma , and paper.

3.2.6 The Direct Effect of Factor of the Average Number of Teaching Hours to Social Responsibility

The direct path coefficient of factor of the average number of teaching hours to social responsibility was -0.0184. it was the only directly negetive factor to social responsibility among all factors, it also had indirectly positive effects by upgrading diploma and revenue, it had indirect negative effects of social responsibilit through the title of outstanding teachers, academic leaders or key teachers, and paper. Thus, In comparison, teachers in the higher the average number of teaching hours, a weaker sense of social responsibility, reflecting occupational fatigue

4 Discussion

Through path analysis, it reveals the social responsibility of university teachers in the interaction between the various relevant factors and their role of teachers in the size of social responsibility.It suggested that teacher of the factors associated with intensity of social responsibility is not the same. Annual income of college teachers of social responsibility is the most direct effect of the strong factors, it showed that teaching faculty is the main means of livelihood for teachers, it had relationship to atmosphere of the market economy. In addition, the results suggestted that an average of teaching hours had direct negative effect of responsibility. It generally believed that teaching students was the basic responsibility of university teachers, but relation between the number of teaching hours and college teachers of social responsibility was not made clear. Some university teachers had occupational fatigue.

Acknowledgements. The study was funded by the humanities and social sciences project of Chongqing municipal education commission (No. 2011SKU10) and education reform project of Chongqing medical and pharmaceutical college, Chongqing 401331, P. R. of China (No.201012).

References

1. Tong, F.Y.: Current Problems in College Teachers and Countermeasures. Journal of WenZhou University 03, 48–49 (2009)
2. Zuo, T.X.: Course and Teacher, pp. 210–211. Education Science Press, Beijing (2010)
3. Xu, S.: Teacher Training Information theory and the experiment. Higher Education Research 11, 31–33 (2010)
4. Huang, L.H., Liu, Q.H.: Analysis of College Teachers and Reflection. Journal of Beijing Normal University 25, 56–59 (2010)
5. Chen, J.Y.: The study of Accomplishment of college teachers of teachers. Journal of Beijing Normal University 22, 59–65 (2008)

Investigation for Occupational Interests of University Students

Bing An[1], Hui Cao[2], Yun Zhai[3], and YanGao[1]

[1] School of Environment and planning, Liaocheng University,
252000 Liaocheng, China
[2] Student affairs Department, Liaocheng University,
252000 Liaocheng, China
[3] School of Computer, Liaocheng University,
252000 Liaocheng, China

Abstract. According to Holland's Statistics for occupational interest, we compiled the Preliminary Questionnaire of the Undergraduates' Occupational Interests. Then questionnaire of the Undergraduates' Occupational Interests was formed by the confirmatory factor analysis and correct of the questionnaire. Finally, it has been used to conduct a poll among some students of Liao Cheng University, and the result has been analyzed by SPSS statistics software. The result shows that the situation of career interest development from high to low include: art type, no differentiation, enterprise type, research-based type, conventional type, reality type and social type. More importantly, it is good to improve the guidance for the orientation of undergraduates' occupational interest.

Keywords: Self-compiled Scale, university students, occupational interest, employment guidance.

1 Introduction

While the international financial crisis affect China's economic development, it made a serious career challenge to the employment of college students. The professional two-way choice - people choose or chosen, we must understand ourselves, and then the right career choices and ultimately achieving the best "person level match "and a good grasp of his career. To choose a suitable career needs full understanding of their interests, personality, ability, physical condition and the circumstances and requirements of various occupations. And a person's interested in a profession is one of the factors to be considered first. Interests can enhance the future adaptation of a career, studies have shown that if a person is interested in his career, he can play his full 80% -90% potential, and stay productive longer without feeling tired, and vice versa you can only play a full 20% -30% talent, and can be easy to feel fatigued and tired of [1]. Interest can also affect a person's sense of job satisfaction and stability. China is currently in a period of rapid economic development, social competition, pressure on employment, a wide variety of new careers, which requires us to give students some career guidance to help them understand their own interests, and to understand some characteristics of professional, providing the necessary psychological preparation for future studies and employment . This study combines the theory of vocational interest research and the research of assessment tools at home and abroad, to make the vocational interest survey of college students as an effective tool for measuring the basic status of the students of vocational interests in Liaocheng University, . So as to we can provide evidence of employment data and theoretical basis to help students recognize their own career interests and orientation, making them more confident in the choice of career planning and design. The type of individual career interests and the corresponding This study divides the types of interests according to the degree of interests of in a variety of professions simplifying the definition of professional interest.

Six occupational types are:

(1) Real type: it mainly refers to the various types of engineering work, agricultural work. Usually they require a certain physical, need to use tools or operate machinery. Major occupations are: engineers, technicians; mechanical operation, maintenance, installation workers, miners, carpenters, electricians, shoemakers, etc.; drivers, surveyors, tracing officers; farmers, pastoralists, fishermen and so on.

(2) Social type: it mainly refers to a variety of direct services to others, such as medical services, educational services and life services.

Main Occupation: teachers, nurses, administrative staff; health carer; managers of basic needs of the service sector, managers and service personnel; welfare personnel.

(3) Research type: it mainly refers to scientific research and experimental work. Main Occupation: natural science and social science researchers, experts; engineers and technical personnel of chemistry, metallurgy, electronics, radio, television, aircraft and other aspects ; pilots, computer operators and so on.

(4) Business Type: it mainly refers to the job impacting and organization others to work together to complete the organization's objectives. Main Occupation: Manager, entrepreneurs, government officials, businessmen, leaders of industry sector and unit, managers and so on.

(5) Art type: it mainly refers to the various types of art work. Main Occupation: actors of music, dance, drama and other aspects, artists, director, teacher; literature, arts critic; radio show host, editor, author; painting, calligraphy, photographer; art, furniture, jewelry, houses decoration designers and other industries.

(6) Conventional type: it mainly refers to the wide range of paper files, books, statistical reports and others related to the work of various departments. Main Occupation: accounting, treasury, statisticians; typists; office staff; secretarial and clerical; librarians; tourism, foreign trade staff, custodians, mail carriers, auditors, personnel and other staff.

2   The Research Methods

2.1 Objectives

This study used the method of stratified random sampling, chose 260 students in 4 grades from various disciplines Liaocheng University to answer the anonymous questionnaires, and a total of 215 as 82.69% valid questionnaires were returned., with the boys 112 (52.09%), girls 103 (47.91%).

2.2 Tools

Holland vocational interest questionnaire under the revised "students vocational interest questionnaire."

2.3 Experimental Procedure

The experimental procedures include questionnaire preparation and test statistics.

1) Students Vocational Interest Questionnaire for first test. According to the interviews and all possible situation⬜ sample of 60 occupational titles from the Holland vocational interest questionnaire, we chose the reorder, grading with a 5 points and then prepare this study tools - Students from the vocational interests first test questionnaire with 60 occupational titles constituting a representative survey of the project.

2) The determination of students formal research of career interests. According to the analysis of the questionnaire first test result, we first select and modify the individual questionnaire items, finally forming 50 items ,including a formal questionnaire. Analysis of the questionnaire obtained by testing the various dimensions and sub-sub-questionnaire survey of internal consistency coefficient (0.840) shows that it can achieve an acceptable value, indicating that the questionnaire has good reliability and is stable credible to regard it as professional measuring tools in college students interests .

Questionnaire items are mainly from Holland Career Orientation Scale, and after many revisions and review we guarantee good content validity of the questionnaire.

In this study, the correlation matrix between factors and the confirmatory factor analysis of the side shows questionnaire has good construct validity.

3) Surveying the questionnaire.

(1) questionnaire survey are mainly in two forms: random sample of quarters to quarters ,regarding it as the investigation unit; use of the time of student learning in the classroom after a meal to survey the collective with unified guidance language.

(2) pretreatment of the questionnaire: to answer the questionnaire on the integrity and authenticity of the check and number it in order to facilitate statistical analysis, if the following occurs, we need to standardize the processing of the questionnaires, the answer to a law of the questionnaire appears, we regard it as a marker of waste volume and remove it; leakage answering the questionnaire items also, as the volume of waste tag, and remove it.

4) Data processing. Test data can be obtained by processing the test results for each student, after finishing it, all the data used SPSS13. 0 statistical software for data processing and analysis of results.

3   Results and Analysis

3.1   The Students' Basic Situation of Vocational Interests

According to measurement results, each student in their own professional interest has in the 6 highest scores on the subscales as their career interest type, if the score difference between the two types is less than 2 points ,it was considered as non-interest type differentiation.

   We brained by statistical analysis that 27.1% of students do not differentiate types of vocational interest; 29.3% of the students for the artistic types of vocational interest, the highest proportion, followed by business type (14.1%) and research type (11.4%) , while the conventional type, reality-based and community-based account for small proportion of three types of vocational interest, a total of 18.1% (see Figure 1).

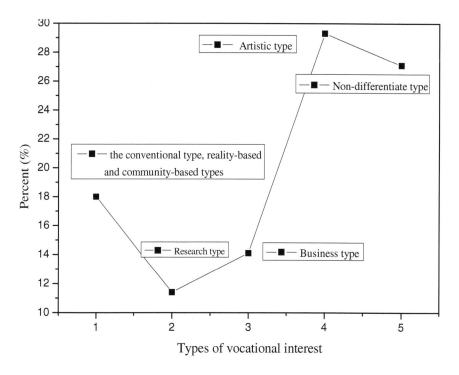

Fig. 1. The distribution of types of students' interests

3.2   Test of Differences in Vocational Interest of Students

1) Test of students interests in gender differences in occupational .The inspection found that men and women in the scores were overall significantly different, women being slightly higher than men (see Table 1).

Table 1. Test of students interests in gender differences in occupational

Sex	N	Mean	SD	t	Sig.
Male	112	4.52	1.93		
Female	103	5.25	1.47	-2.831	.005

Further test of scores for each gender on the difference in the significance. The results showed that women in the real-type dimension scores much lower than men (t = 3.484, p <.01), while women in the community-based dimension scored higher than men (t =- 3.217, p <.01) , in the artistic dimension , women scored higher than men (t =- 4.098, p <.01), women in the conventional type dimension also scored higher than males (t =- 2.103, p <.05).

2) Grade test of vocational interest . The average number of multiple comparisons showed that scores on the questionnaire did not show significant differences between grades (see Table 2). Further scores for each factor in the year of significant differences on the test show the difference between grades not significant.

Table 2. Grade test of vocational interest

Grade	N	Mean	SD	F	Sig.
One	25	4.50	1.99		
Two	43	5.04	1.79		
Three	63	4.64	1.80		
Four	84	4.79	1.76	.465	.803

3) Test of professional differences of students career interests The average number of multiple comparisons on the questionnaire showed that students in vocational interest in different types of subjects were significant differences. (p <.05) (Table 3).

Table 3. The scores of professional factors in the questionnaire

Profession	N	Mean	SD	F	Sig
Liberal	59	4.83	1.80		
Science	78	4.62	1.94		
Arts	78	5.16	1.46	1.431	.042

The scores of professional factors in Table 3 in the questionnaire

Further significance test of each factor in the professional differences showed that type in the real dimension of literature and history and art students scored significantly lower than that science and engineering students (p <.05), in the research-based maintenance history degree students and art class scored significantly lower than the science and engineering students .(p <.01).

The regional differences of students test career. Analysis of variance showed that vocational interest questionnaire scores was not significant at the district level. (Table 4).

Table 4. Scores of in the regional factors in the questionnaire

Region	N	Mean	SD	t	Sig.
City	44	4.61	1.67		
countryside	171	4.90	1.79	-.835	.405

Further significant test of each factor in the region differences found that in the arts constitute a significant difference (p <.05), college students living in urban than students in rural areas showed great interest in the artistic career.

4   Career Interest of Characteristics of College Students

According to preliminary research data, we explore the characteristics of the students' vocational interest. The following results will be the meaning of the above interpretation, evaluation, and evaluating its significance, in order to better provide recommendations for the vocational guidance.

4.1   The General Characteristics of Vocational Interest Type

From a variety of Interest types we can see that 27.1% of the students do not differentiate in vocational interest, there are several possible reasons: Some students are basically not interested in various vocational (drawn from the original status to answer the questionnaire); Some students are interested in many career and score high, and interest differentiation was not obvious; there are still some students blind to answer because of not understanding the nature of the various occupational . Only 6.5% of students interested in social interaction, tend to deal with people engaged in the occupation, which is not consistent with the findings carried by Li Jianying [3], he derived that most of the students were interested in the Professional Society.The two studies may be slightly different, but this study shows that we should take some effective measures to improve the ability of the communication to facilitate future employment. Research-based, arts business type reached 54.8% ,showing that our school give students enough free space and atmosphere of freedom and strict co-existence influence student' self-confidence, independence and freedom . Only by matching up the type of individual career interests with the type of environment can we reflect its value [4].

4.2   Gender Differences of Students Interest in Occupational

The official results of the questionnaire analysis showed significant differences in gender psychology: in the real dimensions ,female-type scores far lower than men's, which consists with the results of other studies carried out by scholars abroad. For example, Peter'[5] studies have shown that male students are not interested in community-based vocational and female students are not interested in the reality-based career. Yong Hua Su [6] according to the choice of career self-revised scale, analysize adult subjects with the test results of students studies have shown that adult male in the real type, research and business type, whether in ability or interest are higher than women. Shi Li [7] after comparing the results showed that gender differences in the conventional type, girls are more dominant than boys in outdoor type, social type, artistic areas.

Boys are good at thinking and reasoning, pay attention to reality and experimental work needing the skills, girls tend to take the initiative with people, express themselves in the community-based area, it also illustrates the problem. On the other hand, girls account up to 8.2% on the conventional type, also showing that some females influenced by the traditional female occupations and are interested in the systematic, methodical work of interest, such as office staff, secretaries, librarians and so on. The reason may be related to the differences of men and women in temperament, character and abilities. In general, women have an advantage in the memory, computation, intuitive thinking, speech ability ,delicate feelings, and patience is good, but the power a bit weak, so women interest in the activities related with the artistic and professional aspect and are good at abstract thinking, taking risks, more rational, powerful, but somewhat less patience, less delicate, and therefore men are reality-and research-related activities and professional interested in, which makes the male and female students produced a significant difference in some types.

4.3 Students Professional Differences in Career Interests

In reality-and research-based dimensions, literature and history and art students scored significantly lower than the science and engineering students. Wu Junhua' study shows that in reality-based aspect, literature and history science scored far lower than engineering students in arts-based social type and conventional type, Literature and History students scored higher than students in science and engineering, which is consisted with part of the results of this study. This is because the reality-based vocational and related activities and the nature of the discipline of science and engineering more closely linked, science and engineering students tend to select the type. This indicates that professional interest in college is different from the obvious professional division, students of different professions have different interests, and the "professional counterparts," Vocational Interests of College Students is still an important factor of influencing and literature and history students to the breadth in student career interests much broader than science and engineering students. At present, the social division of labor is getting smaller, emerging cross-disciplinary, there is a a large number of social needs of innovative potential, multi-disciplinary background and ability of the compound talents. Therefore, universities should be actively involved in academic reform, dilute departmental system, in-depth follow-up and improve the two-degree system, place the students in the center of education, professional education ,change from a narrow into general education and professional education. This will help students develop a wide range of career interests and broaden the scope of their careers, and then ease the employment pressure.

4.4 Regional Differences of Students' Occupational Interest

The results show that the location of different family types have basically no difference in other than the arts students in the five types of vocational interest. The reason, in addition to showing that regional differences may be shrinking, but also probably because this survey students from rural areas accounts for the proportion of large (79.53%), makes the results a certain bias. But regional factors constitute a significant difference in the type of art .Students in urban areas clearly demonstrated a great interest than students living in rural areas, showing that the effects of the family environment on the career interests can not be ignored.

References

1. Chen, G.P.: Psychological tests and scales used. Shanghai Popular Science Press, Shanghai (2005)
2. Kou, J.L.: Talent Assessment. China Development Press, Beijing (2006)
3. Li, J.Y.: An investigation on college students' vocational interest and vocational instruction. Journal of Fujian Agriculture and Forestry University (Philosophy and Social Sciences) 9(5), 66–69 (2006)
4. Ling, W.Q., Bai, L.G., Fang, L.L.: Initial Exploration on the Departments Map of Chinese Universities According to Vocational Interest. Acta Psychologica Sinica 30(1), 78–84 (1998)
5. Peter, Y.J., Richard, T.: Vocational Interests and Career Efficacy Expectations in Relation to Occupational Sex-Typing Beliefs for Eighth Grade Students. Journal of Career Development 31(2), 143–154 (2004)
6. Su, Y.H.: The Comparison Between Adults and Middle School Students on Holland's Self-Directed Search(SDS). Ergonomics 6(3), 10–14 (2000)
7. Shi, L.: The Occupational Interest Research on Contemporary College Student. Su Zhou University (2004)
8. Wu, J.H.: A Survey on Status and Designing Questionnaire of the Undergraduates Occupational Interests. Southwestern University (2006)

Chinese Olympic Education and Olympic Publicity[*]

Hua Pan[1] and Xiaowei Liu[2]

[1] Institute for Sports History, Chengdu Sport University, Chengdu Sichuan China
huapan2002@yahoo.com.cn
[2] Foreign Language Department, Chengdu Sport University, Chengdu Sichuan China
xiaoweiliu2004@yahoo.com.cn

Abstract. By means of literature, mathematical analysis and other methods, the paper analyzes the undertaking of Olympic education and Olympic publicity in China in detail. In Olympic education, the relevant work carried out for primary and middle school students during the organizing and the hosting periods of the Beijing Olympic Games and Paralympics Games is expounded. In the aspects of Olympic publicity, what is mainly discussed here is the relevant key priorities of the propaganda organs of government sports departments and mass media. The conclusion is that China's Olympic education and the Olympic publicity have achieved initial results. Such education and publicity have enabled the Olympic spirit to spread in China, the Olympic movement and the Olympic culture to become popular and develop in China.

Keywords: 1. Olympic Education, 2. Olympic Publicity, 3. China.

1 Introduction

The essence of Olympic is education. The Olympic Charter states clearly at the beginning: " Olympism is a philosophy of life, exalting and combining in a balanced whole the qualities of body, will and mind. Blending sport with culture and education, Olympism seeks to create a way of life based on the joy of effort, the educational value of good example and respect for universal fundamental ethical principles."[1] These remarks explain that Olympism is based on education and publicity as its core content. So, how do China's Olympic education and the Olympic publicity go? The author is to provide a solution in this article according to the historical facts.

2 China's Olympic Education

2.1 Olympic Education

To discuss Olympic education in China, we must first understand what the Olympic education is. Olympic education is an important part of the Olympic movement.

---

[*] This work is partially supported by Chengdu Sports University for doctoral station construction (Grant #:BSZX1038).

Coubertin always based on education as his main theme from the day he restored the modern Olympic movement. He thought of education as the starting point and the end-result of Olympism. In modern life, the most important thing is education [2] .Based on this idea, he believed: "Olympic movement can build a high-soiled and pure-hearted school" [3]. The Olympic movement has to blend sports with education and culture in order to make it exert its function of promoting the all-round development of humans and to achieve its goal of reforming society. Olympic education consists of extremely rich contents, chiefly ranging from Olympic knowledge education, education of the goal of Olympics, the Olympic ideal education and the Olympic spirit education, etc.

2.2 China's Olympic Education

The Olympic movement has not only brought about wonderful sports competition, but also abundant education resources and good education opportunities. It is one of the important tasks of the Chinese Olympic Committee to carry out the Olympic education in adolescents, advocate the Olympic spirit and ideal, spread Olympic knowledge and promote school sports [4]. This section, due to space limitation, will mainly discuss the relevant work carried out in primary and middle schools during Beijing Olympic Games and the organizing period of Paralympics Games.

2.2.1 "Beijing 2008" Olympic Education Plan for Primary and Middle School Students

In order to better implement the Olympic education, BOCOG and the Ministry of Education of the People's Republic of China jointly formulated the ″ Beijing 2008 Olympic education program for primary and middle school students ″ [5].

The Beijing 2008 Olympic education program for primary and middle school students revolved closely around the " unique and high-level " Olympics working target, the comprehensive implementation of the three principles of " green Olympics, hi-tech Olympics, humanist Olympics " and the propaganda theme of " one world, one dream ". It blended the Olympic education with cultivating teenagers' comprehensive quality, promoting primary and secondary school sports, carrying forward the Olympic spirit, the international spirit and patriotism, and strengthening the juvenile moral education in an effort to create a good human atmosphere for the successful hosting of Beijing Olympics, thus forming the Olympic education legacy with Chinese characteristics.

The "Beijing 2008 Olympic education program for primary and middle school students ″ enables students to understand the Olympic movement and to actively participate in the preparations for the games by means of popularizing knowledge about the Olympics, carrying forward the Olympic spirit and publicizing the Beijing Olympic concept, targets and organizing work among primary and middle school students.

Primary and middle school students' involvement in the Olympics international exchange activities has widened their international vision and enabled them to pursue world peace and progress, establish the consciousness of Olympics hosts, which shows that Chinese students are peace-loving, friendly and enthusiastic.

By including Olympic events and the competition knowledge into school education and teaching practice and publicizing the competition events and the speciation tips of Beijing Olympic and Paralympics games, the distance between Olympics and the students is shortened, the Olympic movement and Olympic spirit are promoted in school physical education, thus promoting school sports and developing student's good habit of involving in physical exercise.

By holding colorful theme activities of Olympic education, we can improve the primary and middle school students' understanding of the Olympic movement, so that the students can benefit and obtain great fun through their own personal experience and involvement in Olympic Games and Olympic education activities, which will eventually constitute the vigorous education heritage of the Beijing Olympics.

2.2.2 Education Heritage Activities of Beijing Olympics

In specific Olympic education program, the Olympic education heritage activities of Beijing Olympics are also advocated as follows:

First, the Olympic education is put into normal school education and teaching with three main "inclusions"--- the inclusion of Olympic education into moral education work, the inclusion of Olympic education into the course system and the inclusion of Olympic education into physical education.

Second, education activities of Olympic theme are launched. In August and September during 2006 and 2008, education activities with the theme of "green Olympics, hi-tech Olympics, humanistic Olympics" and "one world, one dream" were launched nationwide in primary and secondary schools, such as photography, painting, poetry, calligraphy, seal cutting, composition, foreign language speech contests, organization of Olympic summer camp and other activities.

Third, the international exchange activities of "knot of one heart" were conducted. Depending on the experience of each organizing committee of Olympic Games, qualified primary and middle schools in Beijing were organized to be paired off with teams and the schools of the countries (regions) for the 2008 Olympic Games and the Paralympics games. The students were encouraged in various ways to understand each other's language, culture, history, geography, art, sports, and the history of the countries in the Olympics. In the meanwhile, various international exchange activities were carried out to enhance the mutual understanding and friendship and convey Chinese youth's peace will. During the Olympics, student representatives were organized to take part in the flag-raising ceremonies in the Olympic village with the athletes of the paired-off countries or regions. Students were also organized to cheer for the paired-off athletes at the competition arena.

Fourth, "the Model School of 2008 Beijing Olympic Education" was built to launch and name the activities of "Beijing 2008 Olympic Education Model Schools", which promoted Olympic education in primary and middle schools nationwide.

Fifth, Olympic education curriculum resources were organized and developed. To cater for the implementation of the Olympic education in schools, BOCOG complied and published the Outline of Beijing 2008 Olympic Education, Olympic Knowledge Reader for Middle School Students, Olympic Knowledge Reader for Primary School Pupils, Flip Charts of Olympic Knowledge, Introduction Handbook of the Beijing

Olympic Games and Beijing Paralympics Events as well as co-ops and other curriculum resources. In addition, BOCOG, through cooperation with all kinds of media, launched the Olympic education column for students and teachers, providing more media courses of Olympic education.

Sixth, the research of Olympic education was strengthened. Experts and scholars from colleges, universities and scientific research departments were organized to conduct research on Olympic education. After the successful bidding of Beijing Olympics, some related and special research institutions were successively established to actively conduct humanistic Olympics research in universities, with the China People's University, Beijing Normal University, Beijing Union University, Beijing Sport University as representatives. Beijing Olympic Art Research Center was established in the Central Academy of Fine Arts. Beijing Olympic Economy Research Center launched by Beijing Socialist Institute, Beijing Economic and Social Development Center and Beijing Beiao Group was established in September 2003 and became the only social institution dedicated to the research of Olympic economy.

Jacques Rogge, president of International Olympic Committee, concluded in Beijing during the Olympic Games that the Olympic education program would spread the Olympic value to 400 million students in China [6].

3 China's Olympic Publicity

With the extensive development of Olympic movement, it is increasingly important to carry out systematic and complete Olympic publicity. It is provided in The Olympic Charter that the first task of NOC (the national Olympic Committee) is to "publicize the basic principle of olympism within the scope of national sports activities and to promote the spreading of olympism in the sports teaching plan in schools and universities." [7]. The Chinese Olympic Committee responded positively to the IOC's call by combining the actual situation in our country and regarded it as important work and part of the Chinese Olympic cause to promote the Olympic spirit and ideal and to popularize the Olympic movement. The Chinese Olympic Committee mainly relied on government sports propaganda organs and mass media to implement Olympic publicity and spreading work.

3.1 The Propaganda Organs of National Sports Departments

In China, the Propaganda Department of the State General Administration of Sports is the national professional department in charge of the implementation and management of sports and the Olympic publicity. It also bears the responsibility of propaganda work of the NOC and Chinese Sports. There are propaganda organs in the sports departments across the country to undertake the corresponding sports and Olympic publicity work.

The Propaganda Department of the State General Administration of Sports is the main functional department with its main responsibilities of Olympic publicity as follows: to be responsible for the foreign propaganda work of State General Administration of Sports, Chinese Olympic Committee and the All-China Sports

Federation and to be responsible for the work of Chinese Sports Journalists Association and China IOC press center, etc.

3.2 Mass Media's Olympic Publicity

China's mass media, including carriers of medium of communications such as radio, television, newspapers, books and network play an inestimably active role in the publicity, spreading and popularization of Olympics. Every development of the Chinese Olympics cause and Chinese masses' changes from knowing of and understanding Olympics to loving and supporting it are all largely indebted to Chinese mass media's enthusiastic publicity and spreading efforts. It is just because of the active participation and promotion of mass midis that has enabled the Olympic spirit to spread in China and Olympic movement and culture to develop and popularize in China.

3.2.1 Chinese TV Media and Olympic Publicity

In China, CCTV is the level-1national TV station, the major Olympic TV broadcasting institution in Chinese mainland. It was China's first time to take part in the 13th Winter Olympic Games in 1980 after it resumed its lawful seat in the international Olympic Committee. CCTV assigned four members to Lake Placid to film the Games live with 16mm cameras and broadcasted it on TV as special programs after editing it at home. It was the first time for Chinese mainland audiences to watch and understand Olympic Games through TV. Since then, CCTV has broadcasted every summer and winter Olympic Games.

3.2.2 Xinhua News Agency and Olympic Publicity

Since the 23rd Los Angeles Olympic Games in 1984, Xinhua News Agency has been the other major media to report the Olympic Games decides CCTV. The 1984 Olympics in Los Angeles is the new China's first participation in summer Olympic Games after its resumption of its lawful seat in the International Olympic Committee. The Xinhua News Agency obtained 30 quota for its first large-scale live coverage of the Olympic Games. During the Games, what was most impressive of the Chinese media was that the Xinhua News Agency was the first of all media in the world to report Chinese athlete Xu Haifeng's grabbing of the first gold medal, which was 20 minutes faster than the Associated Press and 15 minutes faster than Reuters, thus obtaining "the first gold medal" for the fastest news release of the Games.

3.2.3 Newspapers and Olympic publicity

Newspapers are another important medium of Chinese sports communication and Olympic publicity. There are sports of Chinese newspapers. During the Olympic Games and other major tournament periods, all daily newspapers, from the national ones such as the People's Daily to local ones, cover the concerning events and the related news. In addition, in China there are numerous sports journals. These publications, including integrated sports journals, numerous special sports journals and those featuring in promoting the Olympic spirit of "faster, higher, stronger", such as

Five Rings, Olympic Review, etc., are equally important media to spread the Olympic culture

3.2.4 New Media and Olympic Publicity

Compared with traditional media of newspapers, radio, and television, new media refer to the high-tech based emerging media such as cell phones and networked. On December 18, 2007, China Central Television signed contracts with the International Olympic Committee formally in Beijing and announced the new media platform of CCTV. Com as the official Internet/mobile broadcast institution of the Beijing Olympic Games. This is CCTV's another important broadcasting right of the Olympic Games after obtaining the broadcasting right of the 2008 Beijing Olympic Games, which means that CCTV. Com has become the only institution of new media with the broadcasting right of the Olympic Games in mainland China and Macao.

3.2.5 Books and Olympic Publicity

Books play a unique role in China's Olympic publicity and spreading efforts. Due to the characteristics of book publishing, books attach more importance to the mental shaping and cultural transmission in Olympic publicity. Early in 1980 when China resumed its legitimate seat in the International Olympic Committee, the State Sports History Material Committee compiled and published the " Album of China and Olympics " and so on. According to incomplete statistics, from 1979 to 2010, only in mainland China, thousands of Olympic books were published. These books, to a certain extent, have deepened and broadened the Olympic culture research and publicity in China and promoted Chinese people's understandings of the Olympic culture and knowledge, which has resulted in great achievements in inheriting the fine cultures of the human beings.

4 Conclusion

To sum up, we can see that China's Olympic education and the Olympic publicity have achieved initial results. Such education and publicity have enabled the Olympic spirit to spread and the Olympic movement and spirit to develop and popularize in China. In order for China's Olympic movement to develop sustainably and soundly, we will have to strive to expand the Olympic education and strengthen the Olympic propaganda and publicity.

References

1. I O C.: The Olympic Charter. Olympic Publishing House of China, Beijing (2001)
2. Hao, Q.: 200 Questions on the Olympic Movement. Shu Rong Qiyi Press, Chengdu (2001)
3. Wang, Y.M.: Strengthening Education and Reflections on the Management of Violation Handling. China Sports Daily (6) (January 4, 2011)

4. Zhang, C.Y.: Significance and Implementation of Olympic Education in College students' Quality Education (unpublished)
5. Luo, S.M., Tan, H.: The Olympic Science. High Education Press, Beijing (2007)
6. Rogge, J.: Preface. Olympic Comments (69), 7 (2008)
7. Ren, H.: The Olympic Movement. Olympic Publishing House of China, Beijing (2005)

Research on Evaluation of Laboratories in Independent Institutes Based on the Elman Neural Network

Yangyong Jiang, Chaoyang He, and Jianbo Ding

Zhijiang College, Zhejiang University of Technology, Hangzhou. Zhejiang, 310024 China
328468030@qq.com

Abstract. Through the research on the status of laboratory evaluation system in college, this paper drafts an evaluation system which is applicable to the characteristics of labs development. It also introduces the architecture of the Elman neural network and study process. Taking a certain Independent Institutes for example, it uses the Elman neural network to the evaluation of the labs' assessment. The conclusion shows that the system is applicability and easy to operate. So we believe it should be used and popularizes in college.

Keywords: Independent Institutes, laboratory assessment, Elman neural network.

1 Introduction

Independent Colleges is the implementation of the bachelor degree or above education in colleges and universities or national organizations or individuals outside the community, that takes advantage of funding non-state financial holding the implementation of the undergraduate academic education in colleges and universities, mainly develops a "professional, applied, compound" talent for the educational goals. The key to achieve this goal is the training on student practical ability and creative ability, laboratories and practical teaching system play an important role.

To strengthen the teaching of Independent Colleges and laboratory experimental management, in addition to the conceptual, policy-driven changes on the "emphasis on the theory and practice of light" phenomenon, further explore and gradually improve the scientific laboratory evaluation system should be done. Take objective evaluation on the laboratory, turn the passive activity to an active routine laboratory management tools is necessary, and then a long-term mechanism of the laboratory evaluation in Independent Colleges will be established.

In view of the laboratory evaluation system of nonlinear, complexity and uncertainty of the basic features, BP network can be used for laboratory assessment, but there are slow convergence and local minimum of the shortcomings. Elman neural network is a dynamic feedback system, first used by Elman in 1990's. Compared to the BP network, it also has a very strong nonlinear mapping ability, but the convergence rate is much faster than BP neural network. With more computing power at the same time, prior to the neural network to overcome the possible shortcomings of local minimum, it pays more attention to global stability.

According to the laboratory evaluation index system for the assessment of deficiencies caused by migration and other issues-oriented in Independent Colleges, drawing on the Ministry of Education, "Professional Laboratory Evaluation Standards of Colleges and Universities", this paper defines an evaluation index system for the development characteristics of Independent Colleges. Taking a certain Independent Colleges in Zhejiang Province for example, this paper uses the Elman neural network to the labs' assessment, promotes a comprehensive laboratory evaluation system, a reasonable construction, with a view to strengthen the experiment teaching management, and effectively improve the experiment teaching quality.

2   Construction of Laboratory Evaluation System

2.1   Research Status on Evaluation System

Many scholars have put forward their views on evaluation criteria in the laboratory. For example, Yao Liu, Zhejiang Normal University, pointed out that higher education should be assessed with western trends, the quantitative assessment and peer assessment of qualitative should be combined to improve the quality of education assessment[1]. Ying Lv, Guangcheng Cui, Qiqihar Medical University, also suggested that the key projects and the general allocation of items to be appropriate to make some adjustments in the elements of the current evaluation, such as experimental research should be adjusted to focus on the project, as a guide, encourage experimental study, thereby increasing the level of experimental teaching[2].

In the laboratory evaluation method, AHP, fuzzy comprehensive evaluation method, data envelopment analysis and other modern statistical methods achieved a certain effect, but there are still many imperfections, such as difficulties in determining the weight of second-grade indexes, usually with expertise in assessment, leading to arbitrary and subjective assessment, evaluation results and the actual value exist a certain degree of error and so on.

Qian Cai, Kunming University of Science and Technology, proposed the use of gray clustering method to evaluate the laboratory, he thought the universities which have more laboratories should use the gray clustering idea first for "clustering" before evaluation, which can greatly improve the assessment efficiency and simplify the assessment process[3]. In addition, according to the uncertainty on weight calculation of the pairwise matrix in the traditional AHP, many scholars and foreign experts proposed the introduction of fuzzy set theory called the fuzzy analytic hierarchy process (Fuzzy-AHP) to solve this problem. In addition to the specific methodology of the study, there are some scholars have suggested that we should gradually establish an open database to provide a scientific basis for the operation of assessment mechanism.

As can be seen from the above research, the assessment of university laboratory still lacks a simple and effective evaluation system. Laboratory assessment is a scientific and systematic engineering, the existing systems always express as a simple laboratory compliance assessment. There are more or less problems on part of the evaluation indicators and rating standards, and the evaluate method lacks a certain science and rationality.

2.2 Evaluation Index System of Laboratory

In the multi-index comprehensive evaluation, the evaluation index system is the core content, which relates to comprehensive assessment whether accurately reflect the essence of the problem. This article sets up a assessment and forecasting group of six members that engaged in the construction and management of laboratory research, full of education and management experience, have a strong prediction, evaluation and analysis capabilities to this study.

Through the analysis of construction characteristics of laboratory in Independent Colleges, following the basic principles to construct index system, the project team builds evaluation index system of six compositions[4], as shown in Table 1, which includes 6 first-grade indexes and 32 second-grade indexes.

Table 1. Evaluation index system of laboratory in Independent Colleges

first-grade index	second-grade index
Experiment system and management	Establishment of laboratory X_1, Administration and management tools X_2, Laboratory system X_3, Construction plan X_4、 Opening experimental teaching X_5
Experimental teaching	Experimental teaching task X_6、 Research tasks (social services) X_7, Research on experimental teaching X_8, Experimental project management X_9, Experimental materials and test reports X_{10}, Experimental assess X_{11}、 Comprehensive design of experiment X_{12}
Equipment management	Common device management X_{13}, Management of large-scale instruments X_{14}, Equipment maintenance X_{15}、 Equipment update X_{16}, Conventional experiment teaching instrument configuration X_{17}
Experimental teachers	Laboratory director X_{18}, Staff appointment and appraisal X_{19}, Personnel training X_{20}, Experimental instructor X_{21}
Environment and security	Experimental space X_{22}, Facilities and the environment X_{23}, Safety measures X_{24}, Environmental protection X_{25}、 Cleanliness and hygiene X_{26}
Management rules and regulations	Equipment management system X_{27}, Safety and environmental protection system X_{28}, Experimental rules X_{29}, File management system X_{30}, Personnel management system X_{31}, Basic information gathering system X_{32}

3 Elman Neural Network Model

3.1 Structure of the Elman Neural Network

Elman neural network is generally divided into four layers: input layer, hidden layer, following layer and output layer, as shown in Fig. 1. The connection of input layer, hidden layer and output layer is similar to feed-forward network. The input layer unit only plays the role of signal transmission while the output layer unit for the linear weighted, the transfer function of hidden layer units can be linear or non-linear function, furthermore, the following layer takes the memory of previous time output value of the hidden layer units, is a step delay operator, which allows the network have dynamic memory function[5].

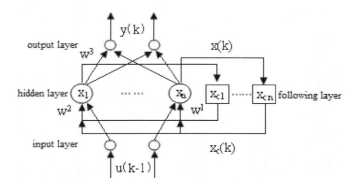

Fig. 1. Elman neural network structure

3.2 Learning Process of the Elman Neural Network

Elman neural network sets input vector u(k), output vector y(k), hidden layer vector x(k), following layer vector x_c(k), so the input-output function of Elman neural network was:

$$x(k)=f(w^1x_c(k)+w^2u(k-1)) \qquad (1)$$

$$x_c(k)=x(k-1) \qquad (2)$$

$$y(k)=g(w^3x(k)) \qquad (3)$$

Which, w^1, w^2, w^3, respectively expresses the connection weight of following layer to hidden layer, input layer to hidden layer, hidden layer to output layer, while f (•) and g (•) were the transfer function of neurons about the hidden layer and output layer. By (1), (2), we obtain:

$$x_c(k)=x(k-1)= f(w^1_{k-1}x_c(k-1)+w^2_{k-1}u(k-2)) \qquad (4)$$

Since x_c(k)=x(k-1)=x(k-2), so equation (4) to continue. This shows that x_c(k) depends on the connection weight of the w^1_{k-1}, w^2_{k-1}, ... ,w^n_{k-1} at different times, so x_c(k) is a dynamic recursive process.

4   Empirical Analysis of the Evaluation Model

In order to promote the construction of the laboratory better, taking a certain Independent Colleges in Zhejiang Province for example, 15 laboratories were reviewed by the experts with long-term related work experience who come from the first-line of the teaching and management positions.

4.1   Data Preprocessing

Because in the current evaluation index system, there are not only quantitative indicators but also qualitative indicators, in order to make each index comparable in the entire system, the indexes have to be processed by standardization.

For quantitative indicators, when the greater the target value, the higher the evaluation, are handled as follows: $F_i=(x_i-x_{imin})/(x_{imax}-x_{imin})$; when the smaller the target value, the higher the evaluation, $F_i=1-(x_i-x_{imin})/(x_{imax}-x_{imin})$. Where, F_i is the standardized value for target value x_i, x_{imin} and x_{imax} are the minimum and maximum for the n-th predetermined indicator, i is the number of evaluation. For qualitative indicators, Delphi method is always used to its quantification, and then be normalized, the same way as quantitative indicators.

4.2   Matlab Simulation of the Model Assessment

This article sets up 32 second-grade indexes as input, an output layer neurons, the output interval is [0,1], five levels of evaluation are as follows: excellent, good, medium, poor, bad, corresponding to the interval evaluation of [0.8,1], [0.6,0.8), [0.4,0.6), [0.2, 0.4), [0,0.2).

Considering the network performance and speed, we determined the hidden layer nodes is 26 by repeated testing. In this paper, author uses the function programming test in the neural network toolbox Matlab7.0. Among all the normalized data, the first 11 sets of data are used as training samples, the other 4 sets of data are used as test samples. Before entering to the Elman neural network, all the datas are handled through standardization. Network input layer to hidden layer uses Tansig() function, while hidden layer to output layer uses Logsig() function, give 0.001 as the learning precision. When the training times reach 139, the standard of error is achieved, and now error is 0.00099. The curves of the decreasing error of training and test error are shown below in Fig. 2 and Fig. 3.

At the same time, during the empirical investigation, we invited 9 experts to give marks of the 15 laboratories, and then compared the Elman neural network test results and expert evaluation, the results of the comparison are shown below in Table 2. In the four test samples, the network assessment is excellent 2, good 2, which is consistent with the results of special inspection on the laboratory. Through several follow-up evaluation, discussions with the experts in school, the result agrees with the experimental operation, indicating that this model has a good fault-tolerance and generalization ability. Experimental data shows that the training and prediction accuracy of the experiment assessment model based on Elman neural network are in full within the acceptable range, so it is a reasonable, feasible predictive model.

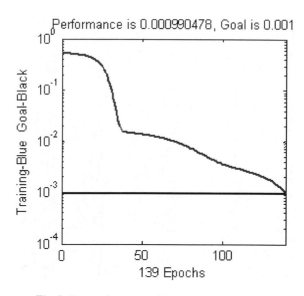

Fig. 2. Decreasing error of the training experiment

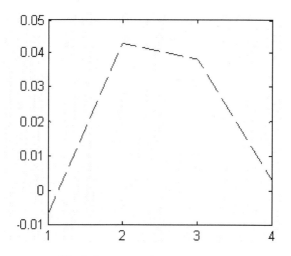

Fig. 3. Test error of the experiment

Table 2. Comparison and classification of the test results

Sample	1	2	3	4
Network test results	0.827	0.789	0.858	0.698
Evaluation results	excellent	good	excellent	good
Specialist grading	0.833	0.746	0.821	0.695
Evaluation results	excellent	good	excellent	good

5 Discussion and Conclusion

Get an evaluation value is not the ultimate goal of our evaluation, according to the evaluation results, combined with the indicators of the evaluation index system, think about how to strengthen experiment teaching management, and constantly improve the level and efficiency to ensure the quality of experiment teaching is the fundamental goal. According to the first-grade index for the assessment, taking six aspects into account, the author proposes several policy recommendations to improve the construction and management of laboratories in Independent College.

(1) Only when we grasp the tendency of the development of experimental technology, we can ensure the advance of the lab construction. We should straighten management structure and rationalize management system. Combined with the actual condition of opening experimental teaching in the college, it saves an effective mode of experimental education in an open environment to cultivate innovative talents in new era. Because experimental teaching administration requires complexity and accuracy, to establish the experimental teaching administration information system based on the net to realize the modernization of the school experimental teaching administration is an effective way to improve the experimental teaching administration.

(2) Along with the new trend of the experimental teaching system, the combination of the base and innovation, personality and sociality, the experimental teaching system should be reformed and innovated. The use of modern educational technique could improve informational attainment of students, teachers and administrative staff, optimize the procedure and improve the environment of educating and learning, expand learning resource and could form students independence and innovation. Flexible teaching methods should be used to the experiments, such as curricular and extra-curricular integration, laboratory fully open, choice or self-proposition, etc., so self-learning ability of students will be trained to promote the improvement of experimental interest.

(3) With the enlargement of the laboratory scale, the personnel and the equipment increase correspondingly, so the difficulty of the management on the instrument and equipment become greater and greater. The more administrators are needed, but the instrument and equipment control system is not perfect. In the process of building the school's laboratory, equipment investment should be increased, good equipment maintenance and record in logbook is necessary. The efficiency of the equipments and labs will be greatly improved when running under a strict constitution and scientific managing.

(4) In order to strengthen the experimental team building, improve the quality of laboratory personnel, we must take a combination of introduction and in-service training to continuously optimize the structure of experimental teaching staff. The establishment of experimental teachers that have high level of teaching and technology, reasonable structure, a combination of full-time and part-time is necessary. At the same time, certain incentives in education, such as awards, appraised, promoted as well as specific treatment, should also be established.

(5) In the experimental project, we should put the experimental environment in an important position to ensure the safe operation of experiments, so that the reasonable laboratory layout and safe facilities in place is very important to prevent security incidents. In the daily management process, self-examination by faculty and school inspection should be strengthened for the laboratory environment improvement.

(6) Rules and regulations are the basic requirements of the laboratory building and scientific management, while colleges and universities attach importance to laboratory hardware construction, the software development should also be taken seriously. Timely modification of rules and regulations involving personnel management, security equipment management is very important.

Because of the current guidance shift, emphasis on the results but not on the process in the laboratory evaluation, this paper has established evaluation index system, while Elman neural network was used for empirical assessment. The conclusion shows that the system is applicability and easy to operate. So we believe it should be used and popularizes in college.

References

1. Liu, Y.: Western Evaluation Modal of Education Quality and Its Development Trend. Journal of Yangzhou University (Higher Education Study Edition) 10(3), 19–21 (2006)
2. Lv, Y., Cui, G.: Practice and Thinking about Evaluation on Laboratories of Fundamental Course at Institution of Higher Education. China Higher Medical Education (2), 43–44 (2000) (in Chinese)
3. Cai, Q., Chen, R.: Application of Grey Classification Method in the Evaluation of Laboratory Work. Journal of Kunming University of Science and Technology 21(4), 104–110 (1996) (in Chinese)
4. Pu, D.: A New Method of Fuzzy Comprehensive Evaluation and Its Application in Evaluating Laboratory. Journal of Southwest Jiaotong University 34(3), 354–359 (1999) (in Chinese)
5. Hansen, L.K., Salamon, P.: Neural Network Ensembles. IEEE Trans. Pattern Analysis and Machine Intelligence 12(10), 993–1001 (1990)

Risk and Prevention Mechanism of University Student Loans from the Public Administration Perspective

Bing An[1], Yun Zhai[2], and YanGao[1]

[1] School of Environment and planning, Liaocheng University,
252000 Liaocheng, China
[2] School of Computer, Liaocheng University,
252000 Liaocheng, China

Abstract. As an important component of social order, the students' credit has become an important issue. The credit of college students is one of the most important credits, also, an important part of social trust. For the risk of loans for students, the paper proposed a novel loan risk prevention mechanism named "Six Components into One", which not only needs the public administrators to standardize funding legislation, establish and improve the personal credit system, but more importantly, to strengthen the moral education such that it improves students credit level.

Keywords: Public administration, Loan risk, Belief, Prevention mechanism.

1 Introduction

Since China launched the policy of national student loan in 1999,it has examined and approved the sum total up to 6.52 billion and funded to 7.91 billion students[1], playing a positive role in the development of higher education. However, a considerable number of students did not fulfill the loan commitments, leading to the high rate of arrears push the government and bank into a considerable embarrassment and the national student loans into a corner [2]. Comprehensive analysis of risk in the current student loans, which can so as to effectively avoid the phenomenon of student loan defaults, not only the credibility of university students but also whether or not it can enter the scientific, standardized and sustainable development track. The author proposes the plan on how to reduce the risk and build a corresponding repayment system, based on public management combined with the actual work on the current national student loan and proposed repayment plan.

2 Risk and Prevention Mechanism of University Student Loans from the Public Administration Perspective

To build a sound national student loan repayment system, we must take full account of the principal parts of public management (the state government), banks (People's Bank of China, Industrial and Commercial Bank of China), universities, parents, employers and individual students and other comprehensive factors together, positively building a comprehensive, gradual penetrative so as six integrated feedback system (see Figure 1) to form strong leadership, well-coordination and supervision in place to ensure harmonious and scientific development of the management.

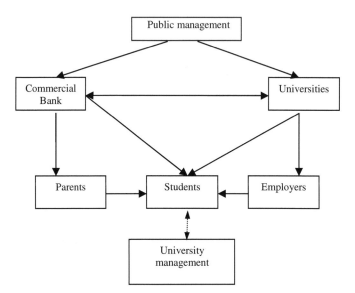

Fig. 1. The six integrated feedback system.

2.1 Government-Led and Funded Higher Education Legislation Is the Key Theory

According to public management theory, government is the main body of public management [3]. At the national student loan system, the government should give full play to its main function, through that legislation clearly stipulates the rights and obligations that government, banks, students have, legally ensure the safety of banks credit fund and raise the enthusiasm of bank lending. Legislation not only protects the use and safe operation the student loan fund, but also provides a legal guarantee for Student Loan Risk and Prevention. In the long run, establishing student loan risk prevention based on the law than to regard ethical guidelines as a repayment guarantee is more effective in preventing student loan default risk.

The current national student loans issued by the state policy banks, the National Development Bank to grant the students credit student loan, take loans for a policy model. Loan interest subsidy funds and risk compensation is borne by the central finance department, the State should ensure full and discount funds, appropriated in a timely manner. For provinces lack of revenue, the central government should allocate special funds for subsidized student loans and a variety of compensation. In the practice of student loan system, whether it is loans recovery of the loans or the disbursement of funds and the loan losses, all aspects of the initiative require government commitment. Moreover, other stakeholders also need to perform duties in the national macro-control policies and supervision.

2.2 The Multi-channel Regulation of Strengthening the Banks, Universities and Employment Units "Three-Pronged Approach" of Is an Effective Way

From the current situation analysis, student loans defaulted student default lies in not only the students for the subjective reasons, but also the banks, universities and employers' poor management and lax supervision for the objective factors. The multi-channel regulation of "three-pronged" the student Loan Bank to strengthen countries, the universities and the graduate student loans employment units is imminent.

First, establishing regular contact and communication system between banks, universities and employment units. Among them, the colleges and universities should perform their main function and play a role as a bridge for passing the examination by the qualified tuition and state college graduates of student loans to establish compensatory complete and accurate files and loans to students studying in the school section and the state will receive a fee compensation to notify the student loans graduate personnel employment department and the National Student Loan Bank [4].

Second, banks should remain connection with the college student loan management departments to timely feedback repayment condition of graduates to inform changes of the home address and contact information, and changes in circumstances promptly notify the bank in order to grasp handling bank student dynamic situation and do the work of post-loan of the student loan business well. I believe that the bank can take the following specific measures:

First, take flexible repayment methods. According to analysis, employment difficulties is the main reason for students failing to repay on schedule, the bank should consider the reality condition of the graduates, such as employment rate, unit properties, and the wages and treat the actual situation discriminatingly, flexibility. The current monthly payments can be change d into quarterly, annual payments, or one-time paying off, or we can adjust the amount and repayment period according to actual conditions in real time also.

Second, the implementation of certain policies. Bank should return student loans ahead of the time or pay off student loans in one-time implementation and carry out certain preferential financial costs sustained by the Government. For some specific industries, namely hard industries in remote underdeveloped areas, countries have also introduced a policy of, compensatory funding, with preferential policies of national repaying the loan instead, but for the current employment situation in the country, compensation funding policy for student loans shows the lack of touch with reality of the situation. The credit for good behavior in school and after graduation supporting voluntary work in remote areas, the students may enjoy a reduction of preferential policies. The current deployment of institutions have introduced policies and the provincial Institutions are also required to provide preferential policies to support student services in the western region. This can reduce the proportion of students in the repayment pressure to obtain better social benefits; it can also play a role in macro control meeting the demand for talent shortage in some sectors, thus creating social and personal the fortunate "win-win" situation.

Again, employment units should give full play to supervisory functions. As part of broader public sector, employment units should try to maintain social and economic order, and ensure the role of socio-economic benefits. Pursuing recovery of defaulted student loans is mainly the responsibility of bank managers, but the responsibility does not only lies in banks, universities and employers also bear some responsibility. Therefore, the employment unit should formulate relevant rules and regulations, and sign with the staff have not yet paying off the loan repayment to regulate the deadline of payment or work for the loan from the salary directly. Meanwhile, the employment unit relevant departments should not grasp the information of the staff of the paying off student loans repayment timely communicate with the bank.

2.3 Strengthen the Sincerity to Improve Their Credit Is the Fundamental Method

Higher education should be fully aware of the importance of integrity, carry out extensive "good education" series of activities to promote students to develop moral values of the "good faith, the word honor, promises disgraceful". The implementation of the national student loan is not just to help students to go through the temporary difficulties, more importantly to help students through the activities more fully feel the warmth of the socialist family. Therefore, university education is not only to impart cultural knowledge, but also should pay attention to thanksgiving education, attention to the cultivation of moral education for students. College students are actively encouraged to the rational use of credit for the successful completion of their studies, and to be educated to maintain personal credit through the implementation of education student loans. It is the only way to effectively curb the phenomenon of the current lack of credit loans to the spread of the students.

First, uphold the education of integrity in the classroom. Create a comprehensive, three-dimensional integrity of the education system. Integrity of education should not only be included in moral character curriculum, targeted education for college students in credit and responsibility of education, in classroom teaching as an important part of the ideological content of education into the school to help students to establish a deep sense of integrity and a strong sense of social responsibility; but also by the full use of the campus network, we should establish special education sites and pages for the purpose of good faith , target to release relevant information and data directly for the effective extension of credit education to the network and make the network as the second class education of the students integrity .

Second, uphold the integrity of educational practice. Integrity of education is not just a slogan, the key is to implement in practice. Honesty educational content, essentially including daily life, study examination, interpersonal relationships, social activities and other aspects. Thus, in practice the integrity of college students should be high goal of education, low starting point, start from the basic behavior of individuals, so that the students can in their daily interaction, learning and examination process achieve his word, honesty and credibility; in the same time, develop appropriate incentives for good performance, good faith, timely recognitize awards, such as establishing a typical propaganda.

Third, uphold the integrity of data files. Credibility is a valuable resource for students and is an intangible property of students. The smooth operation of student loan system depends largely on the student's awareness of good faith. The establishing of the integrity of the students file is the first honest student records, and credit records reflect the level of a student of trustworthiness is symbol of the students having a credit of wealth. College student credit file by the teachers or students in charge of part of the responsibility, mainly during the school recording of students of the complying with school rules, student loan situation, payment of tuition and fees, the responsibility to fulfill the relevant obligations and commitments and so on. Colleges should record to student file trustworthy situation of students. and truthfully provide the students credit rating to the bank, the employer, so that banks and employers to can keep abreast of the situation and the trustworthy degree of student loan repayment, fully realizing information sharing across systems and is conducive to banks to reduce lending risks and costs of information gathering..

2.4 Establishment of a Sound System of Personal Credit Is Important

Lack of personal credit system is an important factor that restrict our personal credit loans and especially the lack of good personal credit system is bottlenecks which slow and restrain the development of personal credit, so establishing and improving personal credit system can not wait. Thus, I believe that at this stage, to solve this problem, we can take advantage of college graduates academic inquiry system and the relevant government departments such as the Home Department, household management, and banking and schools work together to build a personal credit system, collecting and storing lenders personal information. The bank can at any time to conduct real-time access of individual students, thus increasing the transparency of the breach and the breach in any place queried via the Internet. Meanwhile, we should issued violations in various ways, in order to attract public attention, forming an objective opinion on a certain binding.

3 Conclusion

In short, in the circumstances of China's education resources limited and the relative shortage of financial strength, how to effectively solve the problem of student loans becomes important issues in college work. In the new era, we must adapt to the new changing situation, and continuously improve management of the new mode of student loans, so that the management of college student loans took a healthy track of the scientific, standardized and sustainable development.

References

1. Cumulative Loan for 206.8 million Students in six-year Implementation of the policy of National Student Loan. News of Chinese Youth (March 29, 2006)
2. Guo, H.X.: Implement Status and Efficient Compensation Institution of China's Student Loan System. Journal of Technology College Education 23(5), 15–17 (2004)

3. Cao, X.Q., Wang, D.L.: General Introduction to Public Management. China Renmin University Press, Beijing (2005)
4. Ministry of Finance Ministry of Education. the notice about Interim Measures for Students in higher Education Tuition Fees and the National Student Loan Compensatory (April 9, 2009), http://www.work.gov.cn/Article/Details.aspx?id=289

A Study of the Characteristics of Civil Engineering and Architecture Talents and the Way of Talents Training

Peng Zhang and Changming Liu

School of urban Construction, Yangtze University, Jingzhou, Hubei, China, 434023
zhangpeng535207@126.com

Abstract. The development of civil engineering and architecture talents plays a important role in the career of infrastructure construction. With the rapid development of construction trades of our country, there is a larger and larger demand of civil engineering and architecture talents in the society and the requirements are also increasing. Therefore, we should actively cultivate new ways to bring up civil engineering and architecture talents in order to foster more such talents who can meet the demands of construction trades. This article puts forwards the working thoughts of educating civil engineering and architecture talents by a deep research in the characteristics of civil engineering and architecture talents and comes to a main measure to cultivate civil engineering and architecture talents. It provides guidance and reference to to the education of civil Engineering and Architecture talents.

Keywords: Civil engineering and architecture, talents with applied competence, training, characteristics, methods.

1 Introduction

The higher education law clearly stipulated: " higher education's mission is fostering the advanced experts with ability of creativity and practice, developing science technology, culture and socialist modernization. " From the Higher Education law and Social Development perspective, the development of technology-based education is an inevitable result of the higher education in social and economic development to a certain stage. Under the guide of this general approach, the colleges and universities are facing two key issues in fostering current college civil engineering talents, that is how to further clarify the application of civil engineering talents of the training objectives and gradually explore the scientific methods of applied civil engineering talent training.

2 The Characteristics

The training goal is the core concept with the nature of prerequisite in the process of educational practice, the starting point and basis as well as the ultimate destination of the whole activities of school education . Thus, the key to establish the goal of raising civil engineering applied talents is to figure out what characteristics applied civil engineering talents should have in the new period.

2.1 Innovative

Engineering innovation is the main battlefield of innovation activities, and its progress will be one of the most important symbols to test and to decide the progress of building an innovative country of significant strategy . The application of civil engineering construction talents as the main force of China is to achieve strong guarantees of engineering innovation. Meanwhile, there are many levels of innovation, civil engineering applications of innovative talent does not mean the main theoretical knowledge of innovation, but for "the secondary innovation" of turning theoretical knowledge and technology of civil engineering construction into productive forces. Such an innovation is the process of implementation and verification of science and technology, which are the primary productive forces.

2.2 Knowledge

The application of civil engineering talents not only have a certain breadth of knowledge, but also have a certain depth of knowledge. As to the breadth of knowledge, both the solid expertise in civil engineering and strong applied knowledge are necessary. In further, they should have some scientific human knowledge and related financial, management and interpersonal aspects of knowledge; In the depth of knowledge, they should turn the requirements of the "good enough" and "practical" which is limited gradually into the "solid foundation, potentially very strong" and change the knowledge of mastering professional and technical job skills requirements of gradually into grasping the complete, systematic and scientific civil specialty knowledge system in the area.

2.3 Ability

The application of civil engineering talents requires the ability to focus on practical ability, that is to turn the knowledge of scientific principles and the discipline they have learned into a design or design drawings., relying on the professional theoretical knowledge ,basic skills and system The practical abilities especially performance in the following areas: First, the design capacity. Be familiar with Office software, AutoCAD, Tengen construction, PMPK and other professional applications, with some engineering capability of drawings and paintings; Second, writing ability, it mainly means the ability to finish the project contract, project bidding and other documents prepared by engineering review; Third, bargaining power, mainly refer to possess excellent communication skills, to participate in the negotiation of engineering activities; Fourth, organization and management ability. To independently lead the team to complete the design or construction projects; Fifth, the ability to quickly adapt to the rapid completion of civil engineering industry, changing their status from students to a worker and quickly adapting to the industrial environment of civil engineering industry.

2.4 Composite

"Professional qualities decide the starting point, and non-professional qualities decide the height ".Application of civil engineering talents not only have the high professional qualities, but also have certain non-professional qualities. In fact, p during the course of construction, professional knowledge, skills and personal responsibility of application of civil engineering talents, are often closely related to the quality of professional, moral, psychological and other non-physical conditions. These non-professional qualities will directly affect the effectiveness and quality of the project is completed.

3 The Basic Idea

The basic idea of training of the civil engineering applied talents is to foster "the future of civil engineering applications outstanding talents" as a goal, "for the purpose of ideological and political education and comprehensive development as the core theory of teaching-based, practical training for the focus, innovation and design for the soul, to schools and businesses rely on two kinds of teaching resources, "the cultivation of ideas and efforts to build civil engineering application of the quality of education and training platform for talent, nurturing the spirit of innovation with a certain, profound professional knowledge of skills, excellent as the practice of high quality and rapid adaptation of applied civil engineering personnel.

Adhering to the ideological and political education is to put the improvement of the students of civil engineering education in the first place, and actively carry out the "Eight Honors and Eight Shames" as the main content of the socialist concept of honor education and develop the Chinese revolution, construction and reform history education, carry out situation and policies of education, the civil engineering firm ideals and beliefs of students, establish a correct outlook on life and values, a high political consciousness, responsibility, and moral quality of a good application of civil engineering talent, to make a consruction to the development of industrustry of China.

Adhering to the comprehensive development of the core is to thoroughly implement the people-oriented philosophy, the civil engineering overall development of students as the training of civil engineering applications talents starting point and develop training plans to plan long-term, action should be out of human nature, focusing on civil engineering .Students of all-round development, to create a good educational environment, the civil engineering student's personal development and overall development of university organic integration of units, the formation and development of a good cycle.

Adhering to the theory-based teaching is to pay close attention to theoretical knowledge of civil engineering students to learn the education sector, specifically the classroom application of civil engineering education is still the main front Talents.

Theory of teaching through systematic professional knowledge content, Yan Jin, classroom organization, so vivid and wonderful explanation of civil engineering students with profound professional theoretical knowledge.

Adhering to the practice of training as the key is to fully implement the theory with practical ideas, adhere to sterile artificial combination of socialist modernization, the principles of service, pay attention to cultivate and foster the development of the construction industry needs to adapt to the application of civil engineering talents. Through a series of practical activities, so that civil engineering from the classroom to learn the large number of theoretical knowledge combined with social practice, it really will learn to understand and master the theoretical knowledge, really to turn knowledge into capability.

Adhering to the school and teaching resources as the basis of two enterprises, that is, it is necessary to attach importance to use the school's own teaching resources for students of civil engineering training, but also focus on strengthening the cooperation of school types of construction enterprises, the formation of the extra provide students a practice base, we must also give full play to the students' autonomy and initiative and encourage them to contact their own internships.

4   The Major Methods

The application of talent in today's society is a comprehensive high-quality personnel who has solid theoretical knowledge, excellent practical abilities and strong innovation abilities, but personnel training is inseparable from two major environments ▯ the school and society(mainly enterprises). Therefore, the application of civil engineering personnel training model should be based on the students' ability to learn professional knowledge, engineering practice, innovation quality and ability of non-professional capacity-building and rely on schools and enterprises, in this way the application of civil engineering can be better to meet the modern construction industry talent requirements.

4.1   Theoretical Knowledge Teaching System

Application of civil engineering talents in the knowledge structure has a "wide knowledge, based on solid foundations of applied strong" characteristics. So in theory teaching system ,it is in the need to strengthen the integration of theoretical courses, to highlight the applicability of theoretical teaching, to build the target clear, logical theory of teaching system. In general design, students need to master the knowledge and capacity development requirements of points later, the principle according to some things on a bold cut down and merge a variety of courses, integration, and content updates, design, organization and the creation of the whole new courses, build a more complete and systematic theoretical knowledge of teaching system.

Meanwhile, we should take into full account the requirements of the full range of talents, like expanding the students knowledge, improving their comprehensive quality, widening open cultural quality courses, and having vocational courses for students freedom of choice, allowing multi-disciplinary and other specialized courses as a public elective courses. Therefore, students have more room for character development, and lay the foundation for the sustainable development of students. Teaching courses in the theoretical structure gradually form a "base & professional" in the platform architecture, "main courses& optional multi-disciplinary courses" portrait of the theory of modular teaching and training system , to ensure that the basic specifications and diverse talent, personality development, and to enhance the adaptability of students in the community.

4.2 Teaching System Constructed Experimental Operation

Around the construction industry for the innovative spirit of civil engineering talent and practical ability of the requirements of the system need to build the classroom, comprehensive practical skills training and extracurricular self-opening experiment of combining culture system. To integrate the experiment (practice), which established the basis of the corresponding experimental (practical) technical series, Experiment (practice) technology series and module test (practice) technology series and other content, reducing the proportion of confirmatory experiments, increased design and comprehensive experimental content ratio. Validation experiments and related theory courses to maintain close contact. Independent experimental course offered comprehensive training to reflect the skills, reflect the design and comprehensive abilities, but also reflects the unity of the stage of experimental course requirements. To highlight the various stages of capacity-building focus, form validation experiments students the basic operation of the experimental capacity, integrated comprehensive experimental ability students, students design experiments innovative awareness and ability of the experimental teaching system.

Meanwhile, for the implementation of personalized training, the nature of the setting in the experimental course, the civil engineering profession can be set to master the skills required test points (practice), for the extension of the skills test (practice) or sub-skill point experiments (practice), is set to select the experiment (practice), free choice for students, students with personal development space. To basic and advanced, integrated the principle of combining and screen out some of the traditional curriculum of classic experiments in the validation, change, or the thread into a comprehensive design experiments so that students fully understand the experimental principles, methods, answers, handle problems that may arise experiment to improve the analysis of problems and problem-solving skills, strengthen the students ability to innovate, technological development ability. In the experimental teaching methods, we should focus on cause and effect style guide, results-based training, help to stimulate the students a sense of achievement, professional learning in order to stimulate students interest and curiosity of study, which independent study habits and develop a spirit of innovation active role. This comprehensive, three-dimensional systematic, standardized training model of innovation and practice of civil engineering personnel can meet the training needs of the construction industry.

4.3   Off-Campus Internship Appointment System

The purpose of applied personnel training should focus on teaching the theory and experiment in teaching, especially enhancing students practical skills and vocational skills training. And the best way of improving their practical skills and vocational skills is to build off-campus internship appointment system. By allowing students to use summer and winter to train themselves at the Design Institute and the construction site, students can linked a large number of theory learned from classroom to social practice, to improve their professional skills.

As to the methods of constructing induction training system on the campus of specific initiatives, I believe that Yangtze University, School of Urban Construction practice should be learned by its principal, including four comprehensive practice 〖 summer vacations when they were freshman, sophomore, junior and graduation work term when they were senior . Freshman Summer (the first semester of work): to enable students to construction workers on site when the trainee to understand the types of work, be familiar with the building materials,to exercise social interpersonal communication skills; Sophomore summer (the second work term): to make students be the trainee foreman in the construction site and engaging in site construction, quality inspection, project management, graphic design, or project planning and real estate development; Junior summer (third working semester): let students as trainee project manager, involved in the construction organization and management,through the preparation of unit engineering construction design, in quality testing, inspection, attending production meetings and training management and coordination of organizational skills; Senior (fourth work term) : a comprehensive design. All the knowledge will be used to complete the graduate design for students to improve their ability of appling the knowledge integrated to analyze and solve practical problems.

4.4   Construct the Second Class Quality Development System

Applied talents are not "narrowly on technology," the craftsman, but also should have a high overall quality. Thus, in theory teaching system, teaching system experimental operation, off-campus internship appointment system, we have to build the second class quality development system. The second class includes not only the quality of the content development to students in professional skills, technological innovation, the expansion of professional knowledge, should also include comprehensive ability to expand the training of social, spiritual and mental and physical qualities of the mold to enhance the quality. Quality development requirements should be standardized as a means of comprehensive development of students, paying attention to a sound education to the students the role of Personality, pay attention to the scientific spirit of the enlightenment of adults, talent, innovation and entrepreneurial spirit of Han Yu. Expand the system to be fully integrated into the quality of personnel training programs, build the project management, credit certification, standardized operation of the education system, thus contributing to the "first class" and "second class" of the organic integration of science and technology to promote cultural activities and extra-curricular student The combination of curricular and teaching, standardizing and strengthening the management of extra-curricular education.

Specific initiatives include: ① the students were able to carry out a variety of professional competition race, such as model design contest, design contest quick question for students to enhance professionalism at the same time, in the race to get exercise, improve the quality of non-professional; ② Regular organize students to listen to the topic of Eminent Persons report, the report will not only include professional reports, but also including the report of the humanities to improve the overall quality of students; ③ organized students to expand the base of professional development training quality so that students non-professional training to enhance the quality of the system, such as teamwork, hard work ahead, and so consciousness; ④ encourage interest in, the ability of students to participate in academic research, students in the process of doing academic research and creative spirit at the same time, further enhance professionalism.

5 Conclusions

Training of civil engineering talents is an applied far-reaching project of educating people and the cause of national construction development. At the same time, it is a pioneering and exploratory work. We need to work together to explore in practice, continuous review and improve. Universities and society should pay full attention to the civil engineering work of applied training of personnel, clearly applying the civil engineering training objectives, clarifying the application of civil engineering talents ideas, researching appropriate to the current civil engineering personnel training methods to construct a scientific and systematic application of civil engineering talents training system, in order to cultivate a large number of applied talents in the construction industry, for greater contributions to the construction of our country.

References

1. Order of the president of the people's republic of china: The higher education law (1998)
2. Engineering innaovation the battleground of innaovation activities, Sina news center (2005), http://news.sina.com.cn/c/2005-10-13/06247154130s.shtml
3. Song, S.: The construction of Applied undergraduate talents training. XuZhou Engineering College Journal (2005)
4. Qian, G., Wang, G., Xu, L.: The characteristics of Undergraduate application talents and the construction of its training system. China University Teaching (2005)
5. Liu, C., Zeng, L.: Local colleges and universities in the Civil Engineering College engineering practice research and practice. Hubei Province Education Association in the Civil Engineering Majors Engineering Practice Seminar Papers (July 26, 2009)

Study on the Moral Internalization Mechanism for Young Teachers

Changming Liu

School of urban Construction, Yangtze University, Jingzhou, Hubei, China, 434023
jyylcm@163.com

Abstract. This paper uses integrated research methods, the integration of pedagogy, psychology knowledge to discuss the internalization mechanism for young teachers' professional ethics and puts forward that young teachers professional moral internalization of psychological mechanism mainly is the motive mechanism of point of view. Think psychological motive mechanism is divided into internal motivation mechanism and external dynamic mechanism, from dynamic mechanism in progressive mental mental psychological constraints, cultivates, obligations of catalysis, proud conscience baptism psychological motivation, psychological happiness, external dynamic mechanism mainly constraint mechanisms, evaluation and incentive mechanism.Conclusion : to promote the internalization of young teacher's professional ethics must be comprehensive use of training strategies, it is necessary to instil professional ethics knowledge, establish a model of professional ethics, strengthen habits of professional ethics and incentives in a timely manner, activates the inherent power, a solid professional ethics emotion, of constructing a harmony of internal and external, psychological environment.

Keywords: Young teacher professional ethics, Internalization mechanism, dynamic mechanism.

1 Introduction

Youth psychological mechanism of teacher professional ethics within the main mechanism is the driving force, including those outside the inner motive force mechanism and dynamic mechanism.

2 Youth Teacher's Professional Moral Internalization Inside Dynamic Mechanism

To youth teacher's professional moral internalization plays an important role of psychological factors, mainly is the cognitive elements, emotional factors and the willingness to work hard.. Cognitive elements refers to the young teachers contact objective thing but in the heart produces perception, memory, imagination and thinking activity. Affective factors of young teachers in certain refers to hold things attitude of the experience. Will factor is young teachers for certain purposes dominated and regulating the behavior. If no one with surrounding environment of contact, there would be no human psychology, so the person's psychology to objective reflection is positive, but also actively move the doer. Youth teacher's professional moral internalization, is in the whole process of young teachers psychological activities under the control of realization. Young teachers to teacher's professional moral internalization of power, mainly from their psychological factors effect. The psychological factors mainly for the following aspects.

2.1 Enterprising Psychological Catalytic

An outstanding young teachers, affirmation is a high-level pursuit and strong entrepreneurial spirit. High-level pursuit and enterprising spirit of performance: firm political direction and beliefs, broad ideal and ambition, strong dedication to work and the sense of responsibility, innovation of courage, indomitable perseverance, practical and realistic attitude and dedication character, etc. High-level pursuit and enterprising spirit is young teachers of value-oriented, are they progress and growth of internal force source. This kind of psychological power comes from the country to train talents, revitalize the sincere affection, that kind of feeling drive, whip and encouraging young teachers in ideological quality, work quality and the moral quality unceasingly enterprising, leading the they are not content with general to deal with the job, but in knowledge, academic, personality have higher pursuit, the pursuit is they constantly design oneself, motivate yourself, develop their own consciousness and upward momentum.

2.2 Proud Psychological Influence

Proud psychological from young teachers to cultivate successor passion for work, to inherit and spread and innovation of human culture great understanding of the significance. Students of adult stone-sculpture, is through teacher's teaching labor and their own effort, students then enter society, with their intelligence and ability for social progress and development makes the contribution, and use their contribution to Alma mater and teachers. Students' success is the best consolation to teacher education workers, is the embodiment of the social value, is a teacher work produces social benefits should certificate. Teacher's great, is by its hard work produced society needs the builders and successors, with their own candle spirit lit up young students forward life mileage . Therefore, they are worthy of "human souls engineer" title. In this proud psychological domination, young teachers will produce firmly, conscious teaching consciousness and behavior, eliminate all sorts of interference of the motherland, for the development of education enterprise struggle for life.

2.3 Obligations Psychological Constraint

Obligation is the responsibility of citizens or legal persons in accordance with the legal provisions should be made to individual under certain moral principles and norms, moral responsibility for others or society. Which concentrated on society and others to personal proposed various moral requirement, also reflect his own to society and others shall bear the ethical responsibilities. Due to the moral obligation to show how personal consciously processing of personal interests and the interests of the collective relationship problems, so it shows that individuals for the interests of the collective true attitudes. In fact, compulsory content is from a certain social moral principles and norms decision. In education work, compulsory content is moral principles and norms by teachers of the decision. This obligation is by teachers social morality on individual body internalize and produce a moral responsibility. Obligation as high ethical responsibility, is their will obey with certain moral value of inner source task. A moral obligation, although it has the nature of moral orders, but to perform the obligations of the subject that it isn't a kind of external compulsory, but profound understanding moral requirement, consciously undertake morality duties formed on the basis of an inner beliefs and moral responsibility. Therefore, the young teachers in moral internalization process, focusing on cultivation of the sense of moral obligation, not only to make clear their young teachers of moral responsibility, and to make the young teachers formed the noble sense of obligation and strong responsibility, so as to maintain a mature attitude towards obligations, improve compliance with the consciousness of teachers moral principles and norms.

2.4 Conscience Psychological Stimulation

Conscience as a moral category, refers to the personal and social performance of an obligation of moral responsibility to others and self evaluation capacity, is also self-control role of personal moral consciousness structure of moral psychology one. If it violates the guidelines and norms of professional ethics of teachers, will be blamed by others or peers, the individual will feel uneasy, the so-called conscience condemned. The person's mood changes in addition to the objective factors, but also by subjective factor. To a large extent, the man's conscience by understanding and attitudes towards the left and right sides, conscience can either by external objective things, but also because of self-consciousness aroused by the balance, the manipulation of consciousness and regulation. A young teacher if the job irresponsible, perfunctory, drift along, on the one hand, will be affected by students' improper reflect, on the other hand, he himself by moral conscience of blame, feel psychological uneasy, nervousness in the psychological state of fear. His behavior transgressed if the minimum of human morality and conscience, will from the charge and cast off, its personality is insulting, his conscience will consciously feel lost balance, resulting in a mental stress, and even cause all sorts of disease. Therefore, conscience is the result of the moral internalization of young teachers, but also in the young teachers moral internalization plays an important role. Only a sound conscience mechanism, which can effectively achieve moral self control.

2.5 Happiness Psychological Baptism

Happiness is the realm of yearning for all, is everyone pursuit of the ideal. Different concept of happiness, tends to determine the attitudes and values, but also determine the patterns of behavior. From a rational point of view, happiness based on happy happy and above. Happiness is the attachment inside joy, is flowing and lasting good pursuit. In the work of young teachers in the teaching, through their own efforts to get the students ' recognition and peer praise, it will have a sense of satisfaction and happiness, emotion so as to further strengthen its professional ethics, in the promotion of professional ethics. Repeated stimulation and enhanced it is this sense of happiness, forged the teachers of the people "soul agent" image, has guided generations of young teachers to join the cause of education, adhering to imparting knowledge and educating people, contribute to prosperity and education intelligence. Thus, mental happiness is an important source of professional ethics of young teachers.

3 External Dynamic Mechanism of Young Teacher's Professional Ethics

Moral internalization of external impetus mechanism for young teachers mainly constraint mechanisms, evaluation and incentive mechanism.

3.1 Restriction Mechanism

According to the constraint different nature, can be divided into rigid constraints (constraint institutional constraints) and flexible constraints (moral restraint, public opinion constraint). Rigid constraint, a professional moral internalization of important external mechanism. Young teachers into teachers on teachers professional jobs, demand is not very clear, need to have the pre-job training, needs to have the standardization of the institutional constraints. In the realistic society in power super role, under the situation of moral educational association received some effects. However, for a long time, people misunderstood "moral education teacher" and "moral" relations, the teachers to high standards of personal morals request misplace for universal ethics, thereby, make the teacher moral research and moral construction astray, also make teachers had long suffered the burden of moral. Actually, moral restraint and public opinion constraint is essential, institutional constraints is very important also. For teaching requirements, through system terms, urging teachers, not only helps regulate implement practicing teaching behavior, is more advantageous to the young teachers of learning, follow and internalizes the individual consciousness. Therefore, institutional constraints in young teachers' professional ethics play an important role.

However, no matter be rigid constraint, or flexible constraint, only external stimulation and requirements. External effect, still need to have internal mechanism response, not an inner world of young teachers docking, have no body inner response, the external requirement also can become a mere formality, can't realize the positive role. Say, rigid institutional constraints must with the flexible public opinion constraints and self-discipline combined. Public opinion constraint, mainly in accordance with school requirement, build a kind of good public opinion atmosphere, using the power of the press, urging young teachers practice teachers' professional ethics, accomplish practices good self-discipline, imparts knowledge and educates people, be a qualified people's teacher, for cultivating builders and successors for the socialist cause make useful contribution. Nevertheless, no matter be institutional constraints, or public opinion constraint, the need for young teachers of subjective identity and acceptance. Therefore, subject aware moral constraint is the key vocational moral internalization. Improve individual moral cognition capability and level, enhance the moral responsibility, young teachers is to promote the healthy growth of young teachers of primary path.

3.2 Evaluation Mechanism

Evaluation mechanism is important the outside dynamic mechanism, is the effect of youth teachers' professional ethics of environmental constraints. According to the evaluation of the main body division, the evaluation mechanism can be divided into he review mechanism and self-evaluation mechanism of two kinds. He review mechanism can divide again for expert evaluation, leading students to evaluate and teacher evaluation and proved. Expert evaluation is by experts attend classes and practice.the check and other forms of understanding of young teachers teaching situation after giving comments and judgment. Such an evaluation, for promoting the role of the young teachers are large, young teachers to help improve the sense of professional identity and pride, improve the ability of self-recognition, learning disadvantages, take the superior and eliminating the inferior, strengthening teachers ' moral cultivation, conscious teaching, an excellent teacher of the people. Leadership assessment and evaluation expert evaluation is not only different, the leadership in class and after inspection lesson plan, and they tend to peacetime work performance are combined. Therefore, the leadership evaluation is not pure teaching comment on sex, but with a comprehensive evaluation colour. Therefore, the leadership evaluation of young teachers tend to produce larger stimulation, this stimulation of young teachers's mind produces an effect, in their psychological lay important imprints, be they judge oneself in the leading position the main basis of mind. This kind of appraisal to youth the role of teachers' professional ethics, to some extent, is more than experts brit young teachers' professional ethics of main external power. So, unit the leadership, learn to master evaluation yardstick, exquisite evaluation art, pay attention to evaluation method, it is very important, not only can reflect the level of leadership, and can achieve morale, mobilize the enthusiasm of the goal. Evaluation is the main part of the teaching evaluation of students, many schools teaching assessment scores of students to teachers, as a measure of teacher's teaching quality main basis. Therefore, students, teaching evaluation in the realistic society, to teacher plays an important role in promoting. In order to obtain the positive opinion of students, the teacher must rigorous reserch, improve teaching ability and the level, improve teaching methods, ensure the quality of teaching, at the same time, the teaching and the nurture organically, students doing bosom friends, wins student life. Otherwise, it is difficult to obtain a student praise. Of course, there are some teachers, to cater to students, dare not strict requirement, afraid of tube strict the causes students dislike and be student evaluation of low grade. This is a mistake. In fact, students more love that practices good self-discipline, rigorous doing scholarly research, strict teacher. Teachers proved, it is a kind of viewing sex of evaluation, this evaluation, which has important reference value and promoting effect, especially experienced experienced teachers of young teachers lecture appraisal of youth teachers improve teaching ability is crucial to the ethics of young teachers also plays an active role. Therefore, effective utilization evaluation mechanism, strengthen the young teachers professional ability and the ethics of teachers training, with regular, durability and regulatory role, he shall carefully studied in a scientific and reasonable use.

3.3 Incentive Mechanism

Motivation is a common method. Incentive can be divided into positive incentive and negative motivations two kinds. There are incentive awards motivation,promotion stimulation, further incentive etc, negative motivations are warning, criticism, disposition, dismissal, etc. Awards incentive is the most commonly used the means of incentive, also be pair of young teachers has function method, even if we do not evaluate the advanced, verbally shall praise promptly, also of young teachers has a good stimulation. But, praise and decide, be sure to do fair, avoid by all means is injustice, slant listen both.in believe and eccentric partial. Promotion stimulation is an effective means of incentive, promotion titles, promotion position, for young teachers stimulation, compared with praise, appraisal is more apparent, it is a kind of lasting target motivation and intrinsic motivation. Use good promotion stimulation, for youth teacher's individual growth bilingually, for the whole of the construction of the teaching staff, has a positive role in promoting this. Study motivation is a new means of incentive, using good, has the very big positive role. Young teachers,self-motivated, we hope to go further study, improve education background, enhance the ability and level. Therefore, clever use of study motivation, can yet be regarded as scientific and effective way and method. To form good atmosphere, let those who diligently study, work hard, the performance of outstanding young teachers early education study, not only can build a good work atmosphere, but also for poor performance of young teachers is also a whip, have with one action a few effect. However, for those who don't work hard, and the teaching effect is bad, or illegal discipline of teachers, also need to undertake warning, criticism, serious and even give disciplinary until dismissal. Only in this way, to create a positive atmosphere and a centralizer cure evil environment. With such atmosphere and environment, is conducive to young teachers pay attention to professional morals, toward integrity direction diligently, growth for conduct and optimal backbone teachers.

4 Conclusions

To promote youth of teachers' professional morality, must pay attention to internalize method, comprehensive training strategies, both must strengthen training, instilling the professional ethics knowledge, again want model demonstration, establish professional moral models, both must standard management, enhancing professional ethical behavior habits, and timely excitation, build the occupation morals atmosphere, construct a union of inside and outside, psychological harmony the internalization of environment.

References

1. KangZhi, Z.: For moral education the effectiveness of doubt. Academia (5), 125–135 (2003)
2. The Chinese academy of social sciences language institute dictionary templates plait, modern Chinese dictionary, 2nd edn. (January 1983); Beijing the 69th times printing first 1370 page (November 1985)
3. Gan, J.: Education should be DaoDeGu? – about teacher moral philosophy of education research and experiment reflection, vol. (3) (2003)

Mediating Effects of Computer Self-efficacy between Learning Motivation and Learning Achievement

I-Hui Hwang[1], Chi-Cheng Chang[2], Hsin-Ling Wang[1],
Shang-Jiun Tsai[2], and Tuo-Yu Chen[2]

[1] Department of Management Information System, Takming University of Science and
Technology, Taipei, Taiwan
{mary,sandia}@takming.edu.tw
[2] Department of Technology Application and Human Resource Development, National Taiwan
Normal University, Taipei, Taiwan
samchang@ntnu.edu.tw, cklight228@gmail.com, aq1723@hotmail.com

Abstract. From the perspective of pedagogy, this study aims to recognize the influence of students' learning motivations and computer self-efficacy when learning through online platforms, with learning achievement as the criterion for future instruction and related studies. By literature review, this study probes into related learning theories and actual operations as the base of research development, and elaborates on the students' differences of learning achievements according to questionnaire and learning performance. The results show that: 1. Learning motivation positively influences computer self-efficacy. 2. Learning motivation positively influences learning effectiveness. 3. Computer self-efficacy positively influences learning effectiveness. 4. Computer self-efficacy reveals mediating effects between learning motivation and learning effectiveness.

Keywords: Learning Motivation, Computer self-efficacy, Learning Achievement.

1 Introduction

With the progress of time, the Internet has gradually become closely related to daily life. Blogging is popular on the Internet, a trend that changes public habits of information acquisition and media use. In addition to learning in class, students have further opportunity to study at home through the Internet, a phenomenon that indirectly influences on-site instruction. Students' learning not only relies on teachers' lectures or textbook reading; they gradually adopt technology-based learning. By flexible and interactive multi-media teaching tools, such as rich pictures, audio, and images on blogs, instructors allow learners to actively receive knowledge and information.

Blogs and traditional websites are not entirely the same, as they have the characteristics of interaction, real dialogue, connection, and low threshold[1]. Blogs reveal more advantages than traditional media, such as telephone interviews, e-mails, forums, etc. Through blogs teachers can enter directly into discussions with students,

Y. Wang (Ed.): Education Management, Education Theory & Education Application, AISC 109, pp. 67–73.
springerlink.com   © Springer-Verlag Berlin Heidelberg 2011

without limitations of location or time, in order to recognize students' learning and demands. Students can learn other classmates' thoughts and ideas through the Internet. In recent years, the authority of education actively promotes information technology-based instruction to develop e-courses. However, digital learning is not simply developing traditional teaching materials on a computer, but should enhance learning effectiveness and combine digital teaching materials with creative instructional strategies and activities. Thus, it can reveal the advantages of technology-based instruction[2].

This study attempts to determine if students' learning motivations will influence computer self-efficacy when learning through online platforms, and if the enhancement of computer self-efficacy will influence students' learning achievements. This research also aims to determine if the influence of students' learning motivation on learning effectiveness in online platform learning is related to previous learning through non-Internet platforms. In addition, the role of online learning self-efficacy between learning motivation and learning effectiveness is also determined.

2 Literature Review

2.1 Learning Motivation

The ARCS model proposed by Keller (1983) is a tool to analyze learning motivation demands in order to indicate instructional strategies for students. The model combines cognition, humanity, behavior, and the social learning theory, and is constructed through reorganization. It aims to enhance systematic instructional designs that trigger learners' participation and interaction through the design of teaching materials for teachers' practical implementation and application [3]. The content includes attention, relevance, confidence, and satisfaction. Attention refers to the teaching process leading to learners' curiosity, and interest; relevance refers to connecting teaching design with learners' demands, goals, and motivations; confidence means that prior to assigning learning tasks, teachers must recognize learners' levels and help learners construct learning confidence; satisfaction refers to the sense of achievement obtained by learners. With the above, learners will be more willing to follow the teachers and have high learning motivation.

Through research on learning motivations and self-efficacy, many scholars have demonstrated their relationship. Schunk (1996) suggested that self-efficacy would influence learners' task selection, academic efforts, persistence, and learning achievements. In comparison to learners who question their own learning competence, students with high self-efficacy are more persistent when encountering difficulties and challenges, and work harder in academic studies [4]. Pajares (1996) indicated views on motivation and self-efficacy, and suggested that motivation would be influenced by individuals' expectations regarding results. When expected outcomes and goals can be successfully fulfilled, motivation will be increased. Expected outcome is related to individuals' self-efficacy, as learners with high self-efficacy are more confident in their abilities to fulfill goals and are more persistent in learning. Therefore, self-efficacy is the key factor of expected outcomes and motivations [5]. Thus, H1 is proposed: learning motivation positively influences computer self-efficacy.

Han(2008) conducted experiments on 343 grade-2 students from nine different departments of vocational schools. The results showed that there is a positive and significant relationship between learning motivations and learning achievements [6]. Chou (2008) argued that learning motivation can significantly predict learning achievement [7]. Based on related researches, learning motivation is an important factor of learning achievement. As to instruction, teachers can create warm and comfortable learning atmospheres through different tools or activities to enhance learners' motivations. Thus, learning achievement can be effectively enhanced. Therefore, H2 is proposed: learning motivation positively influences learning effectiveness.

As to the scale of learning motivation, upon literature reviews, this study adopts the Instructional Materials Motivation Survey (IMMS) of Keller[3]. The scale includes 36 items, including four sub-scale items, namely, "attention" (12 items), "relevance" (9 items), "confidence" (9 items), and "satisfaction" (6 items).

2.2 Learning Motivation

Bandura (1989) indicated that self-efficacy refers to a person's belief in the successful execution of tasks. Belief in success is based on four kinds of experiences, namely, (1) positive experience: means a person's cognition toward performance of previous similar behaviors; (2) observation learning: evaluates and influences self-efficacy by learning models; (3) verbal affirmation: peer praise can enhance a person's confidence in their competence and belief in successful actions; and (4) a person can avoid obstacles of action and increase self-efficacy by reinforcing emotional weakness [8].

As to studies on self-efficacy and learning achievements, Bandura (1986) suggested that, performance is the most reliable and direct factor of self-efficacy, and it would influence learners' confidence and determination in accomplishing tasks [9]. Chou (2006) conducted a questionnaire survey on 1002 grade-2 students in 12 senior high schools, and found that there is a positive correlation between self-efficacy and English proficiency [10]. In other words, if students have higher perceived English self-efficacy, their English proficiency is higher. Therefore, H3 is proposed: computer self-efficacy positively influences learning achievements. In addition, according to previous studies, learning motivation influences computer self-efficacy, which affects learning achievements. Therefore, H4 is proposed: computer self-efficacy has a mediating effect between learning motivations and learning achievements.

As to scale of computer self-efficacy, this study adopts the scale designed by Compeau and Higgins (1995), which includes five items [11]. The operational definition is shown as follows. Computer self-efficacy means learners' belief in capability in situations, such as participating in, or fulfilling, online learning behaviors and goals.

3 Research Method

3.1 Research framework

Research framework is shown below:

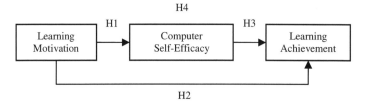

Fig. 1. Research Framework

3.2 Research Method and Subjects

This study treated one class of freshmen enrolled in the Department of Information Management for the academic year of 2010 in a university of technology in Taipei City as subjects. In addition to learning in class, the students learned with a double enhancement interactive e-portfolio learning system. The platform includes two parts: the first is an interactive learning process e-portfolio, where students record weekly learning processes and obstacles of the platform. The second part allows it to save supplementary teaching materials by FTP. Then, 8 weeks later, a questionnaire survey is conducted on students' use and cognition.

Based on literature review, this study developed variables for measurement. The questionnaire includes scales of computer self-efficacy and learning motivation. It is based on a Likert 5-point scale, with scoring as follows: 1 is totally disagree; 2 is disagree; 3 is no comment; 4 is agree; and 5 is totally agree.

4 Data Analysis

4.1 Sample Description, Reliability and Validity

Among the valid samples, there are 32 males (60.4%) and 21 females (39.6%), and the average age is 18.08 years old. Students spend an average of 9.4 hours at part-time jobs every week and 7.5 hours in clubs.

Reliability of scales is analyzed by Cronbach's α. According to Table 3, reliabilities of internal consistency of scales are above 0.9. According to Nunnally (1978), when Cronbach's α is above 0.7, it means there is high reliability [12]. Therefore, items in the questionnaire of this study reveal a high degree of internal consistency. As to the measurement of validity, since all questionnaires in this study are based on English scales developed in previous studies, which are translated into Chinese and modified by experts and professors in related fields, it should involve a certain degree of content validity. The table below shows the means, standard deviation, reliability of internal consistency, and correlation coefficients of variables.

In order to avoid common method variance (CMV), as caused by the self-reported scale, Harman's one-factor test is conducted. After factor analysis, and without rotation, 9 factors, instead of one factor, are extracted. Moreover, the accumulated explained variance of the first factor is 39.754% (<50%), meaning there is no CMV in the questionnaires.

Table 1. Mean, SD and Correlation Coefficient

Research variables	Mean	Standard deviation	Learning motivation	Computer self-efficacy	Learning effectiveness
Learning motivation	3.33	0.43	(0.923)		
Computer self-efficacy	3.14	1.07	.572**	(0.967)	
Learning effectiveness	84.55	11.76	.360**	.381**	

Notes: ** p<0.01; N=53; ( )= Cronbach's α.

4.2 Regression Analysis

Single regression analysis is conducted to validate H1~H3. First, according to model 1 in the table below, learning motivation significantly influences computer self-efficacy ($\beta=0.572$, p<.001), and the explanatory power is 32.8%. Therefore, H1 (learning motivation positively influences computer self-efficacy) is supported.

According to model 2 in the table below, learning motivation significantly and positively influences learning effectiveness ($\beta=0.36$, p<.001), and the explanatory power is 12.9%. Thus, H2 (learning motivation positively influences learning effectiveness) is supported.

According to model 3 in the table below, computer self-efficacy significantly influences learning effectiveness ($\beta=0.381$, p<.001), and the explanatory power is 14.5%. Therefore, H3 (self-efficacy positively influences learning effectiveness) is supported.

Table 2. Regression Table

Model	Dependent variables	Independent variables	β	T	F	R^2
1	Computer self-efficacy	Learning motivation	0.572***	4.9862***	24.859***	0.328
2	Learning effectiveness	Learning motivation	0.36***	2.753***	7.579***	0.129
3	Learning effectiveness	Computer self-efficacy	0.381***	2.941***	8.649***	0.145
4	Learning effectiveness	Learning motivation Computer self-efficacy	0.211 0.323*	1.345	4.659*	0.157

Notes: ** p<0.01; N=53.

4.3 Mediating Effect of Computer Self-efficacy

In order to determine if there is mediating effect of knowledge sharing, validation is conducted according to the test proposed by Baron and Kenny (1986)[13]. Moreover, variance inflationary factor (VIF) is conducted to determine if the variables reveal a high degree of multiple co-linearity.

According to model 1 in the table above, learning motivation ($\beta = 0.572$, p<0.001) significantly and positively influences computer self-efficacy; in addition, based on models 2 and 3 in the table above, learning motivation ($\beta = 0.36$, p<0.001) and computer self-efficacy ($\beta = 0.381$, p<0.001) significantly and positively influence learning effectiveness. Therefore, analysis meets the two conditions suggested by Baron and Kenny (1986). According to regression model 3 and 4: Learning motivation ($\beta = 0.36$, p<0.001) significantly and positively influences learning effectiveness. With the mediating variable of computer self-efficacy ($\beta = 0.323$, p<0.05), learning motivation becomes insignificant ($\beta = 0.211$, p>0.05). Thus, it meets the third condition. In other words, computer self-efficacy has a mediating effect between learning motivation and learning effectiveness. As to the VIF of the regression models, the test finds that they are below 10. Therefore, there is no covariance among variables, and H4 is supported. In other words, learning motivation influences learning effectiveness through computer self-efficacy.

5 Conclusions and Suggestions

5.1 Research Results and Implicaions

After data analysis, the four hypotheses are completely supported. First, learning motivation positively influences computer self-efficacy, which is consistent with the results of Pajares(1996)[5]. It shows that students' learning motivation influences their attitude toward online platform learning. The higher the learning motivation, the more confident they are to accomplish online learning. Therefore, when course design is based on an online platform, teachers must first enhance students' learning motivation in order to allow students to fulfill learning goals through an online platform.

In addition, learning motivation and computer self-efficacy positively influence learning effectiveness, and the result is consistent with that of Han (2008) and Chou (2006)[6][7]. In other words, high learning motivations or individuals' confidence in online platform use will result in prominent learning performance. Therefore, in order to enhance students' learning effectiveness, teachers can enhance their learning motivations and computer self-efficacy, in order to help students succeeding in learning and improving the learning effectiveness.

This study finds that learning motivation affects students' learning effectiveness through computer self-efficacy. Since this aspect is not discussed in previous research, it is the most valuable finding of this study. Regarding the application on-site, learning motivation can influence learning effectiveness by computer self-efficacy. Thus, teachers can increase students' learning effectiveness through three paths, namely, the direct effect of learning motivation or computer self-efficacy and the indirect effect through computer self-efficacy to enhance learning motivation.

5.2 Research Limitations and Suggestions

Although this study makes efforts on scientific principle and preciseness, there are limitations regarding research time, funding, and the researcher's knowledge. The subjects of this study are limited to freshmen in a university of technology in Taipei City. Future studies can include students of different levels, or multiple cases, in order to completely probe into different schools and department for more complete analysis and results.

Acknowledgment. We thank the National Science Council of the Republic of China for financial support (NSC 99-2511-S-147 -001).

References

1. Wu, Y.C.: Virtual Community Development and Value Co-Creation with Customers of Corporate Blogs. Unpublished doctoral dissertation. National Chengchi University, Taipei (2008)
2. Wen, M.C.: The Effect of E-generation On Campus. Information and Education 79, 20–30 (2000)
3. Keller, J.M.: Motivational design of instruction. In: Reigeluth, C.M, ed. (1983)
4. Schunk, D.H.: Self-efficacy for learning and performance. Paper Presented at the Annual Meeting of the American Educational Research Association, New York (1996)
5. Pajares, F.: Self-efficacy beliefs in academic settings. Review of Educational Research 66, 543–578 (1996)
6. Han, W.H.: A Study of Vocational High School Students'English Learning Motivation, Learning Style, Learning Strategy and English Learning Achievement. Unpublished doctoral dissertation. National Chung Cheng University, Chiayi (2008)
7. Chou, S.P.: The Study of the correlation Among Learning Motivation, Learning Strategy, Emotional Intelligence and Learning Achievement for Vocational High School students– case of central area in Taiwan. Unpublished doctoral dissertation. National Changhua University of Education, Changhua (2008)
8. Bandura, A.: Human agency in social cognitive theory. American Psychologist 44, 1175–1184 (1989)
9. Bandura, A.: Social foundations of thought and action:A social cognitive theory. Prentice Hall, Englewood Cliffs (1986)
10. Chou, C.T.: A Study on the Relationships among English Self-efficacy, English Learning Anxiety, English Learning Strategies and English Learning Achievement of Senior High School Students. Unpublished doctoral dissertation. National Taiwan Normal University, Taipei (2006)
11. Compeau, D.R., Higgins, C.A.: Computer self-efficacy: development of a measure and initial test. MIS Quarterly 19(2), 189–211 (1995)
12. Nunnally, J.C.: Psychometric Theory, 2nd edn. McGraw-Hill, New York (1978)
13. Baron, R.M., Kenny, D.A.: The moderator-mediator variable distinction in social psychological research: conceptual, strategic, and statistical considerations. Journal of Personality and Social Psychology 51(6), 1173–1182 (1986)

Research on Shipping Route Optimization Allocation

Degang Fu

Navigation Department of Qingdao Ocean Shipping Mariners College,
Qingdao, China
xiafei1204@126.com

Abstract. Dynamic response to emergencies requires real time information from transportation agencies, public safety agencies and hospitals as well as the many essential operational components. In emergency response operations, good vehicle dispatching strategies can result in more efficient service by reducing vehicles' travel times and system preparation time and the coordination between these components directly influences the effectiveness of activities involved in emergency response. In this chapter, an integrated emergency response fleet deployment system is proposed which embeds an optimization approach to assiwww.lw20.comst the dispatch center operators in assigning emergency vehicles to emergency calls, while having the capability to look ahead for future demands. The mathematical model deals with the real time vehicle dispatching problem while accounting for the service requirements and coverage concerns for future demand by relocating and diverting the on-route vehicles and remaining vehicles among stations. A rolling-horizon approach is adopted in the model to reduce the relocation sites in order to save computation time. A simulation program is developed to validate the model and to compare various dispatching strategies.

Keywords: Dynamic Fleet Management, Optimization, Simulation.

1 Introduction

In other transportation field, about route choice, vehicle configuration optimization problem has been widely research, however, relevant shipping route planning problem, have few research. The main reason is shipping with more uncertain factors, often made shipping route planning issues than other transportation model, lack of structural analysis model is hard to build up.

Although the world liner shipping by accounted for only the total turnover of the world shipping ten percent, its freight cost one-fifth of the world shipping freight but about half of total consumption. Liner shipping involves the capital cost and operation cost, therefore, shipping line selection, container ship configuration directly related to the liner companies operating performance. In addition, shipping line once was determined, in a particular operating period, so difficult to adjust, must pass system the research to find optimal route the fitting decisions. Meanwhile, in order to adapt to the change of environment, such as shipping cargo demand changes, freight rates change, change of shipping regulations, and regular through the system analysis method to optimize the whole route network system is very necessary.

Y. Wang (Ed.): Education Management, Education Theory & Education Application, AISC 109, pp. 75–78.

In the past, shipping company's decision makers usually according to practical experience to cope with external environment factors change, however, with the company fleet expansion and increased, decision-making course should consider more complex factors and feasible scheme also exceeded people out ability. In this case, through the system optimization model can help policymakers make better decisions.

In recent years, the optimization analysis model was gradually used to solve liner vessel fleet routes. The fitting If the United States and a university of Michigan professor A.N.Perakis professor A.I.Jaramillo proposed routes the fitting linear programming model[1], they assumed that there are several prior already certain routes, then puts forward a model will be assigned to each ship have been confirmed in these routes, is planning to go within the total operating costs and idle fleet minimum cost. Professors Krishna Rana and professor R.G.Vichson proposed a linear programming model, they tried to find the optimal each ship by port and the order of the biggest, total benefits in solution, they used the Lagrangian relaxation decomposition method[2].

Compared with the above model, this paper puts forward the model simple easy to handle, suitable for practical applications, can obtain solutions through computer simulation.

2 Ease of Use

Assume that the plan period all can use ships are known and fixed, the shipping company has decided the new before withdrawal, namely, what kind of ship or the shipping company has proposed a rented boat, what about what ships, buy and rent what ship plan to build. Given this capital investment plan and demand forecasting conditions, this decision problem is from consider alternative routes and find a group of routes ship with boat sets and plan period the expected profit maximization[3]. Now put each ship assigned to the company considering the limitation of alternative routes concentration, and make sure that every time liner time interval..

The conditions of the Model as following

1. According to the experience and the judgment for future major goods, shipping company's policymakers can put forward a series of alternative routes, it can be developed or previously[4].
2. In a plan period, the cargo from port i to port j demand is known.
3. A linear can be sail in many routes if need.

Establish the objective function

$$\max\left(\sum_{r=1}^{R} \sum_{k \in K_r} \pi_{rk} x_{rk} - \sum_{k=1}^{K} h_k y_k \right) \tag{1}$$

Where x_{rk} represents r routes every time on k ship operating income expected, π_{rk} is for k ship's expected profit, h_k represents k ship unit idle cost, y_k represents plan period of idle time of k ship.

Assuming w_{ij} represents that according to historical data of every voyage average volume from port I to port j, $a_{ij,rk}$ represents all flights of k ship, d_{ij} and represents the goods flow of port I to port j, So get below function:

$$w_{ij}\left(\sum_{r=1}^{R}\sum_{k\in K_r}a_{ij,rk}x_{rk}\right)\geq d_{ij} \tag{2}$$

Meanwhile, the object function should be meet the constraints as following:

$$\sum_{r\in k}t_{rk}x_{rk}+yk=t_k \quad k=1,2\ldots\ldots K \tag{3}$$

The model can use standard linear programming MATLAB package solution. The optimal scheme $x^* = (x_{rk})$, Represent the optimization sheet after the ship choice. For each ship k, x_{rk} Represents ship k's voyages in one routes.

Because of course the fitting question belongs to the long-term planning, model parameter will inevitably is uncertain, through sensitive du analysis, we can easily be dealing with such uncertainty. Tiny of operating costs, such as fuel prices fluctuate or future freight rate uncertainty, freight demand uncertainty can be processed through the analysis of the sensitivity and adjust.

3 Simulation Results

With software MATLAB to simulate the function. Comparative statistics and forecast data, we can see this model is beneficial to solve the problem.

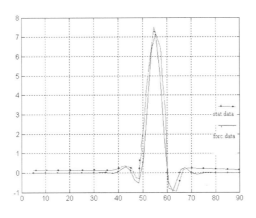

Fig. 1. Statistical data Forecast data

4 Conclusion

Aiming at the shipping line the fitting and proposes a practical linear programming model, determine the optimal route choice and routes the fitting scheme is total profit maximization. Through the model aided decision, which helps to improve the company's operation efficiency and service level. Through the simulation verified the feasibility of the model.

References

1. Council Regulation (EEC). No 4057/86 on unfair pricing practices in maritime transport (2003) (December 22, 1986)
2. MeiWeiming Industry Concentration in Japan. Japan Economy Press, Tokyo (1983)
3. Boffey, T.B., Edmond, E.D.: Two APProaches To Scheduling Containerships With An Application To North Atlantic Route. Operational Research Quarterly 42(12), 35–45 (1979)
4. Ronen, D.: Cargo ships Routing And Scheduling: Survey of Models And problems. European Journal of Operational Research 78(5), 46–49 (1983)

The Effectiveness of Social Networking Applications in E-Learning

Rossafri Mohamad

Centre for Instructional Technology and Multimedia,
Universiti Sains Malaysia,
11800 Penang, Malaysia
rossafri@usm.my

Abstract. This study aimed to see and explain the effectiveness of ICT social networking application in e-learning in the subject of History among Form 4 students. The objective was to identify methods of collaborative learning by applying the use of Facebook application in order to enhance student motivation and understanding on the topic of Greece and Roman Civilization. Samples of 60 Form 4 students are randomly selected from a rural school in Penang, Malaysia. The questionnaire has been used as an instrument and data were analyzed by using descriptive statistics and the statistics of intervention. This study was carried out because most students are not interested in the subjects of history and consider it very boring to studying history. In addition, approaches and techniques less attractive version cause students to lose interest and not motivated to learn when it is a compulsory subject. Therefore, researchers conducted this study to in order this subject will became a subject of interest and thus provide a good impact on learning. Overall, the findings show that e-learning method by applying the use of social networking application to increase student motivation and understanding on gender and class stream. The study also found that there was a significant difference in the improvement of gender and motivation on the students' understanding based on the class stream. The overall mean was 3.81. However, the understanding of students in a e-learning method by applying the use of social networking does not depend on students' computer skills. The result showed that there were no significant differences in improvement based on student understanding of computer skills. Therefore, some suggestions and implications are discussed for effective learning is produced mainly in the subjects of history.

Keywords: E-Learning, Use sage of social networking application, History Subject, Motivation, Computer Skills.

1 Introduction

The teaching and learning process using the approach discussed in the group are one of the methods used by teachers during teaching sessions. This is because it can encourage students to formulate ideas and opinions freely. However, the discussion often fails to function, as a unit for effective interaction among students does not happen. Therefore, this opportunity is used by researchers to promote e-learning by

Y. Wang (Ed.): Education Management, Education Theory & Education Application, AISC 109, pp. 79–85.
springerlink.com
© Springer-Verlag Berlin Heidelberg 2011

applying the use of social networking to improve the method of discussion groups. In this context, the research done to examine the effectiveness of collaborative teaching and learning in the subjects of Form 4 History students by applying the use of Facebook in order to enhance understanding and motivate students and to overcome problems that exist.

2 Problem Statement

Teaching and learning (T&L) process history subject in school often confronted with various problems caused learning and performance history is less encouraging. According to the *Sijil Pelajaran Malaysia* (SPM) examination results for the year 2008 shows that the performance of students in history is not consistent and its very poor, namely of 64.0% (2004), 52.4% (2005), 59.62% (2006), 58.0 % (2007) and 59.8% (2008). It shows that the performance of subjects was low both compared to other subjects. This situation is one of the most significant challenges to teachers as it is a core subject in the SPM examination.

Trigwell, Prosser & Waterhouse[4], considered that the school should change the way teaching and learning process that focused on student compared to teachers by collaborative learning as more favorable to students. Therefore, collaborative learning should be implemented to activate the learning environment and increase students' interest in the History subject. This approach is expected to be an alternative in the study of history by transferring a number of titles and sub titles in the form of a computer display that can be accessed at any time. In this method, the use of facebook manipulated in line with the interest of a generation that likes to explore new surroundings. The new environment will cause students to focus on the use and exploration (Wlodkowski, RJ) [6] thus creating a learning environment that is fun and challenging.

3 Significant Case Study

This study was carried out since there are no research on the use of collaborative methods of facebook application in learning history, especially on the topic Greece and Rome Civilization that will make the teaching and learning process more meaningful and enjoyable. Furthermore, in accordance with the development of ICT it will enable students to access information at their own time. This situation will indirectly promote the improvement and proliferation of knowledge. Among the objectives to be achieved through this research is to determine whether the teaching and learning process by using the method of collaborative on facebook can: a) increase students' motivation towards the subject of History; b) to improve technique and style of teaching and learning of history; c) increase students' understanding on the topic of study and; d) to overcome problems or constraints of time in learning history.

4 Research Methodology

This study was conducted using a survey method involving the use of Likert scale survey instrument fifth stage consists of four constructs. Two variables are used in this study that the dependent variable and moderator variable. Dependent variable is the motivation and understanding, and moderator variables were gender, class stream, and skills. A sample of four students from a secondary school, which is common in rural areas in Penang. Sample age is between 16 to 17 years. 60 respondents were randomly selected. All respondents were selected from the same school because they have the same characteristics in terms of culture and learning environment. All samples had a prior knowledge of the Early Human Civilization in the previous chapter in Upper Secondary History syllabus.

5 Data Analysia

a. Statistical Analysis of Intervention in Motivational Enhancement Based on Gender

Ho1 There is no difference significant on the collaborative method with the use of facebook applications in increasing motivation based on gender. Table 1 shows the mean motivation for male respondents was 3.71 and the standard deviation is 0.12. While the mean motivation for female respondents is 3.63 and the standard deviation is 0.16. This shows that the level of mean high level.

Table 1. Motivation Mean and Standard Deviation by gender

	Gender	N	Mean	Standard Dev.
Motivation	Male	31	3.71	0.12
	Female	29	3.63	0.16

Table 2. T-Test result comparison by gender

	Item	Levene's Test for Equality of Variances		t-test for Equality of Means		
		F	Sig.	t-Value	df	Sig. (2-tailed)
Motivation	Equal variances assumed	.945	.335	2.086	58	0.041
	Equal variances Not assumed			2.069	52.80	0.043

* $p < 0.05$.

Levene's Test for Equality of Variances in Table 2 shows the probability p is .335 ie p> 0.05. This means that the variance is not homogeneous between the two groups. In the independent sample t tests show significant differences between the means of increasing students' motivation based on gender. Mean motivation for the boys was 3.71 and the standard deviation is 0.12. While the mean motivation for girls are 3.63 and the standard deviation of 0:16 in which t = 2086, df = 58, p = 0.041 ie p < 0.05. Therefore, there is a statistically significant difference. Thus, the null hypothesis is rejected.

b. Statistical Analysis of Increase Students Understanding Based on Class Stream

Ho2 There is no difference significant the collaborative method with the use of facebook applications in enhancing students' understanding based on the class stream.

Table 3. Mean and Standard Deviation by Class Stream

Content	Stream	N	Mean	Standard Dev.
Understanding	Science	25	4.85	0.16
	Arts	35	4.76	0.18

Table 3 shows the mean on respondents' understanding the content of topic based on class stream (science and arts). Mean in the understanding of content show that science stream class is 4.85 and the standard deviation is 0.16. While mean for arts stream class is 4.76 and the standard deviation is 0.18. This shows that the level of mean in high level, it is more than 3.66.

Table 4. Results of t Test for Understanding the Topic Content Base on Class Stream

Item		Levene's Test for Equality of Variances		t-test for Equality of Means		
		F	Sig.	t-Value	df	Sig. (2-tailed)
Content Under standing	Equal variances assumed	2.211	.142	2.047	58	0.045
	Equal variances Not assumed			2.099	55.924	0.040

* p < 0.05.

Levene's Test for Equality of Variances in Table 4 shows the probability p is .142 ie p> 0.05. This shows that the variance is not homogeneous between the two groups. In the independent sample, t test show there is a difference in improving student understanding based on the class stream. Mean motivation for science stream is 4.85 and standard deviation is 0.16. While the motivation for the mean of arts stream is 4.76 and the standard deviation is 0:18 where t-Value is 2.047, df = 58, p = 0.045 ie p < 0.05. Therefore, there is a statistically significant difference. Thus, the null hypothesis is rejected.

c. **Statistical Analysis of Intervention in the Skills Enhancement Based on Students' Understanding Computers**

Ho3 There is no difference significant on the collaborative method with the use of facebook applications in improving the understanding of students on computer skills.

Table 5. One Way ANOVA test based on the increased understanding of students in computer skills

	Sum of Squares	Df	Mean Square	F	Sig
Between Groups	1.71	2	.085	2.872	.065
Within Groups	1.696	57	.030		
Total	1.867	59			

$* p < 0.05.$

4 Based on Table 5 One-way ANOVA results showed no significant differences between the understanding of students with computer skills (F = 2.872, p = 0.065). Therefore, the analysis failed to show the hypothesis was rejected because there was no significant difference in improving student understanding based on computer skills.

6 Discussion

The overall findings of this study showed a mean total of 74 items is at 15.24 with the overall average mean of 3.81. This finding indicates that respondents agree to the method of collaborative teaching and learning in the subjects of history by applying the use of Facebook is able to enhance motivation, learning styles, understanding and overcoming the problems / constraints faced by students as the level of the mean level at the high level of more than 3.66 (Table 6). Overall, the finding of this data has to answer all research questions and hypotheses.

Table 6. Mean total and average

Mean Total	Mean Average	Learmimg Method
15.24	3.81	Good

Overall findings indicate that the research question has been answered. The first research question is whether the collaborative method with the use of facebook application to increase students' motivation towards learning History? The results showed that the collaborative method with the use of facebook application to increase students' motivation towards the subject History. This is because of the 36 items tested with overall mean was 3.67.

The second research question is whether the collaborative method with the use of facebook application techniques and styles can improve the teaching and learning process in history?. Yes, the results shows that collaborative teaching and learning by using facebook to improve the technique and style of teaching and learning history is successful. These learning styles include how a group (collaborative) can learn, two-way interaction and learning that eventually acquire effective learning. The mean overall technique and style of learning is 3.86 out of 10 items tested. Study Martin [2] states that motivation and achievement will be higher if the teaching methods tailored to the students' preferred learning mode.

To answer the third research question is, whether the collaborative method with the use of facebook application can get the better understanding of these topics?. Of the 12 items tested found the overall mean for the understanding of the topic was 4.80. This suggests that a collaborative method with the use of facebook application is better on understanding related to the topic of Greece and Roman Civilization. Two-way interaction, collaboration, and independent learning will enhance students' understanding of the topic. This view was supported by Windle,R. and Waren,S. [5], the theory of collaborative learning can raise interest and motivation. Teachers guide individuals in a collaborative process in order to build a relationship of interaction between students to develop thinking skills of a student and achieve an understanding of the concept.

The last question is whether the collaborative method with the use of facebook application can overcome problems or constraints in the teaching and learning of history? The mean total problem / constraint are 3.86 out of nine items that were tested showed that the problem is moderate. This means that students have different views on the problems that occurred during the process of teaching and learning. Feelings of fear exist in students when using facebook as waste of time.

7 Conclusion

Overall, the study of effectiveness of collaborative teaching and learning in the subject of Form 4 History by applying the use of facebook to increase student motivation and understanding. This approach is a process of teaching and student-centered learning by providing opportunities for students to engage in activities with an active and enabling social learning occurs through collaborative discussion when a problem or misunderstanding occurs. Hence, their motivations for teaching and learning materials are formed indirectly through the use of facebook. Therefore, effective learning will occur, followed by the increase in student achievement in the subjects of history.

References

1. Keller, J.M.: Development and Use of the ARCS Model of Motivational Design. Journal of Instructional Development 10(3), 2–10 (1987)
2. Martin, K.: Issues of Teaching and Learning - Focussing on the student. In: Proceedings of the 11th Annual Teaching Learning Forum. Edith Cowan University, Perth (2002)
3. Kurikulum, P.P.: Huraian Sukatan Pelajaran Kurikulum Bersepadu Sekolah Menengah Sejarah Tingkatan, vol. 4. Kementerian Pendidikan Malaysia, Kuala Lumpur (2002)
4. Trigwell, K., Prosser, M., Waterhouse, F.: Relations between teachers' approaches to teaching and students' approaches to learning. Higher Education 37, 57–70 (1999)
5. Windle, R., Warren, S.: Collaborative Problem Solving and Dispite Resolution in Special Education: Training Manual (1999),
   http://www.directionservices.org/cadre/contents.cfm
   (Retrived November 27, 2009)
6. Wlodkowski, R.J.: Motivation With A Mission: Understanding Motivation And Culture In Workshop Design. In: New Directions for Adult and Continuing Education, vol. 76, pp. 19–31 (1997)

The Application of
Online Machine Translation System
in Translation Teaching[*]

Xiaowei Guan[**]

School of Foreign Languages, Dalian University of Technology,
116023 Dalian, China

Abstract. Machine translation (MT) technology has developed for many decades and can provide high quality reference translation for certain fields. The main MT methods, on-line MT systems and their defects are introduced by analyzing the necessity of combing modern information technology with translation teaching. The application of the online MT systems in translation teaching is discussed in the aspect of the objectives and patterns. To apply computer-aided techniques into translation teaching will promote the reform of translation teaching and meet the social requirements for the training of translators.

Keywords: Online machine translation system, translation teaching, the application of machine translation.

1   Introduction

Machine Translation (MT), the earliest branch in Natural Language Processing (NLP), is the translation by computer from one natural language (the source language) to the other natural language (the target language). The software to fulfill this process is called Machine Translation System [1]. The history of machine translation is more than a half century long, but its quality is still poor and the machine translation systems still haven't reached the ideal phase of high-quality, which is mainly due to the complexity of languages, and the limited knowledge of humans to the linguistics, that is, the knowledge of language cannot be expressed in precision.

With the rising of the status of translation as a subject, more and more people begin to focus on the study and practice of translation, because of the large demand for technical translators by the present society and the market. However, the higher educational institutions in China have so far failed to lay enough emphasis on teaching new technologies of translation. The cultivation mode for the students who major or minor in translation basically follows the translation theory and practical

---

[*] This work is supported by the Fundamental Research Funds for the Central Universities (DUT10RC(3)46).

[**] Xiaowei Guan: Lecturer, Doctor. Main research areas are machine translation and comparison of C-E and E-C translation.

methods of traditional literature and linguistics. The teaching materials can not offer up-to-date texts and the course rarely involves translation technology, other new technologies and tools [2]. The students, after a period of learning, still can not meet the market demand for translation. They can not deal with the specific translation tasks properly, but often give poor quality translation using a long time.

The 21st century is highly information-oriented society, so the integration of modern information technology with translation teaching has become an important task of the reform of translation teaching [3]. Translation teaching should be improved and innovated with the development of IT industry in order to meet the social requirements for the training of translators. Therefore to apply computer-aided technologies into translation teaching can not only ease the embarrassed situation but also improve the quality of the translators [2].

This paper offers an account of the application of online MT system to translation teaching. In section 2, some online MT systems and their defects are introduced. Section 3 discusses the objectives and teaching patterns of translation teaching by using online MT system.

2   Online MT System

2.1   Introduction of Online MT System

The existing machine translation systems can be divided into two categories: rule-based MT and corpus-based MT. The former can be divided into the method based on transformation and on inter-language, while the latter can be divided into statistical and example-based method [1]. Here we mainly introduce the rule-based machine translation (RBMT) and the statistical machine translation (SMT). The two are distinctly different, in which the main difference lies in whether the language knowledge base is used or not. RBMT mainly depends on various types of language knowledge bases of artificially extracted rules, including lexical, syntactic, semantic and other linguistic knowledge. However, SMT does not need language knowledge base, but relies on a large number of bilingual corpora, using an algorithm to automatically learn translation knowledge and build statistical models to develop algorithms to fulfill the translation.

The representative of RBMT is SYSTRAN, which has been applied for more than 30 years and is also most widely used in Europe and US. SYSTRAN offers online translation service of articles and web pages, which can realize the mutual translation between English and other European languages [4].

The representative of SMT is Google online translation system. In 2005, U.S. National Institute of Standards and Technology (NIST) tested 20 kinds of machine translation systems, and the SMT developed by Google got the first.

These online translation engines provide basic terminology and vocabulary queries by the function of quick information retrieval, which to some extent play the role of dictionaries and encyclopedias. They also provide reference translation, and in some specific fields can provide high quality translation. Moreover, they are able to provide help in many aspects of translation teaching and learning, therefore they deserve enough attention by teachers and researchers.

2.2 Defects of Online MT System

So far, most practical MT systems are based on a single sentence, and its analysis of the source language is only confined to an isolated sentence. Not full consideration is given to the effect of the discourse and context on the semantics of words and sentences. Therefore the accuracy of the source language analysis can not be ensured. The system can not generate all the information of a correct translation, so the quality of translation is significantly constrained. The ignorance of the language element in a discourse or context will results in difficulties in semantic understanding and ambiguity.

There are a variety of means to realize the coherence and cohesion among elements within the discourse, which include two levels, grammar and vocabulary, such as reference, substitution, ellipsis, conjunction and lexical cohesion, etc. Ambiguity and polysemy, the common phenomena in natural languages, also bring a great obstacle for machine translation. Natural languages are filled with ambiguity in vocabulary, syntax, semantics and pragmatics. The polysemy and obscurity in sentence caused by pragmatic factors is one of the reasons for polysemy.

A machine translation system which can only provide one translation is not enough. Although it can also provide a reference translation, which quality is not stable enough and can only meets the requirement of some fields, occasions and styles, it is difficult to be used as a direct reference for translation teaching [3]. However, we can use several different MT systems at the same time. Each MT system use different training corpus, MT theory or method, so they will give different reference translations.

3 The Application of Online MT System in Translation Teaching

It is evident that MT system has brought impact to the translation industry. The MT quality is being improved. It will be an innovation to the translation practice, research and teaching. It not only helps users to quickly read and understand the general content of a document or a web page, and can even play a supporting role in translation for some fields. More importantly, it brings challenge and thinking to translation teaching.

With the development and progress of the society, the integration of modern information technology with translation teaching to cultivate technical translators to meet the market demand will be the future task of translation teaching [4]. In the construction of translation discipline, the teachers should use modern educational technology to enrich their translation teaching methods.

3.1 The Objectives of the Application

The emphasis of translation course should not be laid on the disciplinary system, but on students' activities in and out of class. In other words, students should be the main part of the course, whose learning initiative should be emphasized. Its concern is not about presenting the students what kind of content they should learn, but encouraging students to actively engage in various activities [3]. In the translation class, the students will not only do the exercise on the textbook, but also on the internet. The knowledge of computer-assisted translation should be increased. The teaching content should be regarded as a process of learning experience; the emphasis is students' interest and need.

The translation course in the university aims at enabling students to master the knowledge and translation skills to develop translating abilities in practice. Therefore, in addition to the traditional teaching method focusing on faithfulness, more attention should be paid to training the students how to communicate [3]. In traditional translation teaching, the teaching of translation skills constitutes an important part of classroom teaching. The teachers comment on the typical translation examples so that the students will learn by analogy, combined with classroom or after-class practice to deepen and consolidate their knowledge of translation skills. In the online teaching environment, teachers are no longer the center of classroom teaching, and how to make students receive timely and effective guidance on translation skills is a challenge for online translation teaching.

The translation course is required not only to help students to understand and master a variety of translation skills, but more importantly to develop students' abilities to analyze and solve problems and to apply the knowledge into practice. When we organize the specific content of translation course, we should let students distinguish the applicability of MT system in different texts, and develop their operation ability. We should encourage students to discover and identify the semantic features which are hard to find in traditional translation teaching method, and to actively study on the style, codes of language (lexical co-occurrence, syntax pattern, coherence, redundancy, and, punctuation features, etc). The translation norms and universality are discovered and proved quickly and reliably in order to help students to choose the appropriate translation strategies [3].

3.2   The Application Patterns

In assisting translation, the network technology can provide a reference translation for learners to learn and use. Learners can search relevant text through the search engine, analyze and learn the special format and expression by the critical analysis and study on relevant reference translation, which can help learners to learn the standard and authentic expressions to improve the quality of translation. Learners can also compare their own translation with that of MT [5]. For example, sometimes the translation given by the students seems correct, but in fact in a particular context, it may be totally inconsistent with the translation requirement of this kind of style. Now, learners only need to enter this sentence in the translation engine to quickly get a very authentic translation.

In the self-learning environment, students can consult the translation engine. The original text, the translations of MT engines and the students are compared with each other. Therefore, the students can gain useful information and inspiration, make translation skills learned in practice, and really understand what they have learned, which will provide an effective instruction for translation learners. In addition, learners can easily realize the function of "back translation", which is an effective means of testing the quality of translation. By the "back translation" and comparison of their own translations in MT engine, learners can assess the quality of translation and techniques more effectively. As a result, MT engine has partially played a role of a teacher and a good supporting role for students' self-learning [5].

At present, there are a variety of network resources to provide vocabulary, terms, sentences, cultural background and other information retrieval functions, which greatly facilitate learning and research of translation and provide a new way of teaching translation. Human-computer interaction techniques essentially change the traditional method of looking up words in dictionary, and the online MT system can provide a large amount of vocabulary substantial to give instant and effective support, which will reduce the labor intensity of the translator, improve the translation quality, continue to add new vocabulary and accumulate their resources.

We must let students fully realize the importance of the cooperation of human brain and computer, so that students are able to think, participate in the process of knowledge acquisition by themselves. In teaching and learning process, students play an active role of researchers; while teachers play the role of promoters who provide lead and instruction. In short, the classroom teaching should be student-centered, and students should actively discover, analyze and solve problems and apply knowledge under the guidance of teachers. Students must learn to proofread and improve the translation with their own judgment, selection and organization to produce real high quality translation. Then teachers can compare, evaluate and proofread students' translation.

4 Conclusion

MT can provide a wide range of assistance to translation teaching, although its functions are limited. After all, the MT engine is just a tool, and there still exist many problems so far. Therefore MT systems can not completely replace human translation. But we can make use of the MT engine to provide new tools and ideas for translation teaching, to improve the self-learning ability of students, to promote the reform of translation teaching, and to cultivate high-quality translators.

References

1. Feng, Z.W.: Study on Machine Translation. China Translation & Publishing Corporation, Beijing (2004)
2. Cai, J.: A Bibliometric Analysis of the Studies on Translation Teaching with Information Technology in China from 2000 to 2009. Foreign Language World 2, 8–18 (2010)
3. Wu, Y.: The Application of Computer-Aided Translation Systems in Translation Teaching. Media in Foreign Language Instruction 6, 55–69 (2006)
4. Lv, L.S., Mu, L.: CAT Technology and Translation Teaching. Foreign Language World 3, 35–43 (2007)
5. Wang, Z., Sun, D.Y.: The Application of Statistical Machine Translation in Web-based Translation Teaching. Shanghai Journal of Translators 2, 73–77 (2009)

Research on Sports Industry System in Germany[*]

Hua Pan

Sports History Institute, Chengdu Sport University, Chengdu, Sichuan, China
huapan2002@yahoo.com.cn

Abstract. German sports industry, a new economic growth point, ranks among the best in the world and plays a proper role in its national economy. Then how is German sports industry system composed? How did Germany manage to build itself into such a grand sports power in the world? Depending on the research approaches of literature, mathematical analysis, etc., the author analyzes the vitals of Germany sports industry with the purpose of providing some beneficial references for the sound development of China's sports industry.

Keywords: 1. sports industry, 2. sports system, 3. Germany.

1 Introduction

In the sports industry, Germany is one of world powers. The 2008 German sports industry accounts for1.99percentage of German GDP [1]. In 2010, there were more than 90,000 sports only in Olympic league of Germany with more than 2763 members, accounting for 34% of its total population [2]. According to the latest statistics, the main branches of German sports industry system is analyzed and discussed herein to provide references for the sound and sustainable development of sports industry in China.

German sports industry includes two categories of professional sports industry and leisure sports industry involving the subsections of sporting goods industry, fitness entertainment, sports sponsorship industry, sports lottery sales industry, sports expo industry and sports television industry and other areas, which are to be separately discussed below.

2 Sporting Goods Industry

Sporting goods industry is the pillar industry in German sports industry. Germany has a population of about 82 million and it is Europe's largest sporting goods market, with the sports market share amounting to $8.25 billion in Germany in 2005. In recent years, sakes in sports markets has been growing steadily and the import gross amounts have reached 3.65 billion dollars. Adidas products and market share represent the overall level of German sports industry, with the enterprise's brand ranking 67 in World Brand Top 500 complied by the World Brand Lab in 2006.

[*] This work is partially supported by Chengdu Sports University for doctoral station construction (Grant #:BSZX1013).

German professional and leisure sports equipment includes the corresponding equipment concerning golf, fishing, tennis, fitness, gymnastics, archery, bowling, winter and summer sports, amusement sports, beach sports, wall tennis, outdoor and indoor sports and the group sports, excluding here the sports apparel and footwear, hunting equipment and the traffic tools for leisure sports, such as ships, bicycles, motorcycles and sleds, etc.

Experts believe that the equipment concerning skating, outdoor sports, football, fitness, golf, bicycle, tennis, snowboarding, amusement sports, skateboarding, skiing, jogging, basketball, beach sports, badminton, handball, and other related sports will become the fastest growing sports equipment in the near future.

Tennis, as an unseasonal sport has more and more popular in Germany. There are more than 32 million people participating in tennis in Germany, so the sales in tennis facilities reaches 260 million dollars. Every German tennis enthusiast will spend an average amount of 800 dollars on tennis shows and rackets and 300 dollars on sportswear per year.

Sports clothing is an important part of sports industry in Germany. A survey of 20,000 German consumers shows that, in 2005, German consumers consumed more than 2.9 billion US dollars in sports clothing, among which German sportswear retailers account for about 80 percent of the total market share.

Germany is not only Europe's leading sports goods producers, but also an important importer of sports equipment. 25% of its sports products are exported, and at the same time, large scale of foreign sports equipment and products are imported. In 2005, Germany imported the total worth of 3.65 billion of sports product equipment, ninety percent from the United States, China, Taiwan, Austria, Italy, France and other countries. Even though Germany is now in economic recession, the 2008 sports equipment imports also increased one ~ two-percentage points.

Family sports consumption expenditure amounted to $14.9 billion in Germany only in 1990, averaging about $19 person[3]. In 1996, the amount reached as high as $34 billion, averaging nearly $50 per capita [4]. In the 21st century, sports consumption is unabated and 2005 witnessed 790 dollars of consumption in per capita in terms of only sports goods in Germany [5].

3   Fitness Entertainment Industry

Fitness entertainment industry is also very developed in Germany. At present there are more than 90.000 sports clubs in Germany, most of which are related to national fitness entertainment. Since the 1990s, profitable private clubs has been developing rapidly in Germany and the Germans called them as "commercial sports enterprises". These private clubs, mainly including fitness centers, fitness training rooms and sports schools, charge 5 times higher than non-profit public clubs [6]. The Germans attach great importance to bodybuilding. Among its total population of 82 million, about 2/3 take part in certain sports, which creates a high demand for fitness equipment and a very promising market prospect. In the 21st century, body-building entertainment consumption is unabated, with 2005 witnessing 790 dollars per capita in sports goods only consumption. Again, at the 2010 European Expo of Outdoor Equipment, German marketing research organization NPD data shows that German outdoor market scale has reached 1.66 billion Euros. In 2009, the Germans spent 92 Euros per person on outdoor products [7]. It can be seen form the statistics of German Olympic Sports League that since 1979, the number of the members in various kinds of sports clubs has been rising and in 2010 the number has reached more than 27.63 million, among which the highest number is taken up by football 6.75 million), then respectively by gymnastics (4.97 million), tennis (1.55 million), shooting (1.43 million), athletics (8.8 million), handball (8.4 million), mountaineering (8.3 million), and equestrian and swimming (7.3 million people), etc[8]. Germany's outdoor sports also develops is rapidly. According to statistics by German sports product experts, there are about 10 million people participating in outdoor sports, with the expenditure in this regard amounting to several billion dollars. German Sports Gym Association (DSSV, 2007 data) determined that in 2007 there were 5950 fitness venues in which about 5.25 million members received training and that there were about 7.1 fitness venues for every 100,000 residents. The fitness industry sales averaged about 3 billion Euros, an increase of nearly 1 billion Euros compared with 2000 [9]. These data show that the fitness industry is a market that can not be underestimated and the market is of important economic significance.

4 Sports Sponsorship Industry

Sports sponsorship in Germany is also unprecedentedly active. In 1996, the total amount of German sports sponsorship reached $1.648 billion, secondary only to America and Japan and ranking world third, while in European countries, it ranked the first , accounting for 30% of the total $5.5 billion of European sports sponsorship. In 2000, the trademark-related sponsorship to major sports games and sports clubs in Germany reached as high as $1.1 billion, with the amount secondary to the United States ($5.2 billion) and Japan (2.2 billion) and ranking the third in the world. Dpa statistics says that the shirt sponsorship volume of the 17 clubs in Bundesliga in 2009 reached 129.1 million Euros, with the finally- announced sponsor being a cheap supermarket of Nett that was signed in Bochum. Although the contract amount was not been released, they previously required 2.5 million Euros of shirt sponsorship, which means that shirt sponsorship in 2009/10 season would certainly be more than 130 million in total. The Bundesliga shirt sponsorship fees created record high in successive years, 122.7 million Euros for 2007/08 season, 129.65 million Euros for 2008/09 season. The growing trend is obviously seen [10].

5 Sports Lottery Industry

Sports lottery industry (SportWetten) belongs to oncologic industry in German sports industry. The roar of sales of sports lotteries also can promote the prosperity of some relevant industries, such as printing, banking. It can also stimulate consumption and boaster demand. Germany's quiz type of sports lottery is primarily the football lottery ticket (Toto) just as football lottery ticket companies and SZR company started issuing football tournament quizzes from 1949(Ergebniswette). According to Germany's relevant laws, 25% of the gambling income of various gaming activities should be returned to the participants; another 25% of the income must be provided to all kinds of social organizations, including various welfare institutions, sports organizations and sports organizations obtain 50%. In addition, there is a kind of special lottery "good luck" (GlucksSpirale) in which all the income is dominated by sports organizations. The federal government and the state governments except through sports lottery provide sports organizations with certain financial aid by means of sports lotteries. In other words, the sport lottery is one of the main sources of German sports funds and German sports gets big money form all kinds of sports contests, gaming industry, etc. It was reported that from 1970 to 2005, German Sports Association obtained a cumulative amount of 750 million Euros from sports lottery [11], most of which was used in various sports activities.

The 2006 Germany World Cup of soccer is not only a world cup of beer, but also a World Cup of gambling. In Hessen, the location of the German World Cup Organizing Committee, there are as many as 170 lottery shops [12]. Again, lotto lottery is one of the important financial sources for every German state. Every year, the investment of the gamblers of lotto lotteries in Germany reaches 40 billion Marks, among which half of the money is used as bonuses, the other half is paid to the state governments as tax revenues and the state governments set aside a certain amount to sponsor sports or cultural activities. It is estimated that among Germany's total population of 82 million, 20 million are lotto customers. In order to serve all these customers, the 25,000 lottery shops in Germany employ about 50,000 staff. The Germans have such a strong interest in buying lottery tickets that they also bet their luck in Austria or Britain. Lotteries on the result of sports competition, such as the Tour DE France or football matches, can always attract many gamblers to pay [13].

Lottery fund is the major source of income for German federal, state, and local sports organizations of all levels. German lottery market will affect the existence of sports clubs and other non-profit organizations, and these sports clubs play an important role of communication in public welfare field. In 2003, the German national franchised sports lottery injected 250 million Euros directly into non-league sports mainly to develop different sports activities, sports infrastructure construction, and to provide various sports jobs, organized youth sports activities and provide free volunteer service, etc.

6   Sports Television Industry

German sports television industry is not only very flourishing but also full of fierce competition. RTL, the leading radio and Television Company in Germany, had its budget of 220 million German Mark for sports programs in 1996. RTL Radio and Television Station only broadcast four sports programs: football (including Bundesliga), car racing, tennis and boxing. However, it really costs tremendously to seize the market! In 1996, every minute of sports program would cost RTL 19,100 Mark, 450% higher than five years before. The rising situation of the cost of sports programs per minute for the previous 5 years was as follows: 3,500 marks in 1991, 5000 Mark in 1992, 9300 marks in 1993, 14,500 marks in 1994 and 17,400 marks in

1995. At present, among various types of programs, sports programs cost the most (film programs only 3,200 marks per minute). German sports television (DSF) has brought changed its bowling transmission from 12:00 late night to 10:15. The all-embracing sports TV programs of the station include lots of sports dance broadcast. In 1995, 151 hours of dance programs were broadcasted at the station. The largest broadcasting volume in German sports TV programs is football: 391 hours in 1993, 420 hours in 1994 and 521 hours in 1995 [14]. Combining the news reports on 11 April 2000 of German Info seek, Yahoo and Le Monde: there was an intense race concerning the transmission rights of 2001--2003 Series A and B in German football league among Kirch Gruppe that owned German SAT1, German Sports TV, Premiere World, Sport 1. De and other Internet media, and Michael Kohlberg (Cinema World Co., LTD) and Haierbold Kloyper(Munich TV Telecommunications Germany). Finally, Michael Kohlberg, Grams Roy and Primacom teamed up to offer German Football Association 1,5 billion Mark for three years of transmission, while Kirch Gruppe offered 1.65 billion Mark.[15] According to the reports of German "Sponsorship News" reports, Kirch Gruppe again bought four years of transmission rights of German Bundesliga for 3 billion Mark, in which the Internet transmission rights were also bound. Later, the 5 broadcasts of German Premiere on Friday afternoon were conducted by means of PTV or VOD while "Ran" of German SAT 1 broadcasted game reviews as usual. Gone were the days when people could watch live Bubdesliga free [15]!

7 Sports Expo Industry

There are numerous sports exposed in Germany in all seasons. Here we will focus on the German Munich International Sports Utilities and Sports Fashion Trade Fair (German Sportartikel fur Internationale Fachmesse name und Sportmode, abbreviation ISPO). ISPO began in 1970, and is currently the biggest one in the world in terms of its scale and influence. ISPO is composed of several different types of "galleries" which are exhibited together with the theme emphasized. ISPO is only open to professional visitors, which shows that the ISPO is professional, modern, informative, encouraging and pioneering. Organizers of ISPO maintain a very good contact and communication at any time with decision makers of the industry, and keep a watchful eye on the changes in sports industry and outdoor industry. ISPO is the meeting ground of the new trend and new products of the industry every year. In outdoor sports and snow sports exhibition areas, ISPO has always kept its authoritative and leading position. Appearance of new brands, demonstration of new products, release of new trends, innovations of technical functions, etc. Have all impressed visitors of the latest trends.2010 ISPO, with a total exhibition area of 180,000 square meters, witnessed 2,111 companies from 45 countries including professional 64,000 visitors from 177 countries. Exhibited brands and visitors outside Germany account for 83% and 66% respectively. During the exhibition, ISPO also hold all kinds of new product launches, trend forums, summit seminars and new brand competitions to attract more professionals and policymakers. The 2011 ISPO includesthe following core galleries: skiing (ski_ ISPO), board (board_ISPO), outdoor sports (outdoor_ISPO), fashion sports and life style (ISPO_vision), leisure sports clothing (sportstyle_ ISPO), sports function galleries: running, fitness, team sports, pat class, ball and triathlon, etc. (performance_ISPO), trend and innovation galleries: the new brand, new products, new technology (trends & innovation) and export processing galleries (SSE). As the world's No.1 sporting goods and fashion fair, the ISPO gathered purchasing managers and decision-makers. What is more important is that it provides the overall development trend for sports industry and serves as the top platform that reflects the current development trend of global sports fashion industry. The highlights of the 2011 ISPO include: industry event (annual winter sports, outdoors, snow sports industry event, attracting decision-makers from various fields of the industry); information platform (largest scale of press releases and innovation displays, industry trends, field crowds and information exchanges; new product launches (product launches, fashion shows and seminars of all the famous brands of sports field); encouragement of innovation (more attractive new brand promotion plans and awarding activities, encouragement of growth and promotion of the new technology, new brands, new products); the theme focus (closely related with society hoots issues and market demand, a carter theme every year. The theme of 2011 remains focused on environmental protection and "sustainable development", and "environmental responsibility award" is established) and the trend leading (ISPO vision galleries will have more trendy and designing product displays to provide visitors a brand-new experience of the lifestyle of sports and fashion.) After more than 40 years, ISPO has been successfully held for 74 times and has grown into one of the ten largest trade fairs in the world. The exhibition covers all the important parts sports market. ISPO is located in Munich in the center of Europe, which has long been the trade center of sports goods in East Europe and it influences more than 400 million consumers. Now, in the field of international sporting goods, including sporting goods manufacturers, brands, retailers and distributors, designers, media and athletes, ISPO has become the most important trade exhibition and communication platform. It is also the world is largest and most comprehensive trade fair of sporting goods and sports fashion. In the whole world, ISPO represents the highest level of sports goods exhibition, which is characterized by a high degree of internationalization, large numbers of international enterprises, high level of specialized exhibition management and strong trade function [16]

8   Professional Sports Industry

Germany's professional sports industry is also highly developed. German racing cars, soccer and tennis are the most commercialized sports. In football, German Bundesliga is a little inferior to Series A and EPL overall, but it is properly managed and there are fewer clubs in charge of the management than those in Series A and EPL. Germany's commercialized professional sports is no longer simple competitive sport, but is becoming more and more closely related with social life and national economy. The successful commercialized operations of numerous enterprises and companies, plus the further efforts of TV, radios, magazines, newspapers, etc., have brought about unusually brilliant and extremely exciting competitions which have attracted countless enthusiasts and created an infinitely vast sports market in Germany. A professional sport industry is Germany is the most active, most positive factor in sports industry. The existence of professional sports stimulates the continuous development of sports industry, thus leading to a chain economic effect.

9 Conclusion

To sum up, we can conclude that German sports industry is worth our learning and reference in many ways. Today, China is making its efforts toward a world sports power from a major sports country. However, China's sports industry development is still at its early stage, so it is faced with such problems as irrational industrial structure, imperfect policies and systems, nonstandard market management, limited market scale, inefficient market mechanism to allocate sports resources, etc. In order to realize the goals of the "12th five-year plan" [17], we also have to guide and expand sports consumption demand, further optimize the structure of sports industry so as to enable sports industry to develop rapidly, promote the development of sports industry with regional characteristics, strengthen the sports market to regulate management, expand and revitalize the sports facility resources, guide and regulate professional sports development, strengthen sports intangible assets development and protection, and accelerate the implementation of brand strategy of sports products and complete sports lottery tasks before we can build a sports industry system with Chinese characteristics in the near future.

References

1. Yang, Y.: Demand for Sports of China's Economic and Social Development for the Next 10 Years. People's Sports Publishing House, Beijing (2010)
2. Pan, H.: Mass Sports in Germany. People's Sports Publishing House, Beijing (2011)
3. Tian, H., Sun, S.G., Song, Y.M., et al.: The Current Situation and Trend of the Main World Sports Powers. People's Sports Publishing House, Beijing (2010)
4. Tang, H.G.: Study on Chinese Citizens' Sports Consumption Needs. Legislative Affairs Office of the State General Administration of Sports Information,
   http://www.sport.gov.cn/n16/n1152/n2523/n377568/n377613/n377748/390529.html
5. Tian, S.H.: Role of sports industry in national economy. Securities Daily (8) (July 10, 2001)
6. Liu, B.L., Li, W.L.: An Introduction of Sports industry. People's Sports Publishing House, Beijing (2005)
7. Mu, Q.: Half Germans Love to Seek Excitement Outdoors. Global Times (20) (August 20, 2010)
8. The Fitness Industry under Financial Crisis in German Website of IST Sports Leisure Tourism Institute in Germany,
   http://www.istchina.com/zhuantidetail.aspid=39
9. DOSB.: Bestanderhebung. 2010 Aktualisierte Fassung vom, DOSB, Fannkfurt am Main (December 15, 2010)
10. Ai, W.: The Bundesliga 130 Million Shirt Sponsorship Taking the Lead in Europe Bremen Assumes the Task of Brand Cognition (July 27, 2009),
    http://sports.sina.com.cn/g/2009-07-27/14194503141.shtml
11. Hou, H.B.: German Sports Association and Commercial Banks Reward Volunteer Service. Mass Sports Information (1), 13–18 (2006)
12. Wang, Q.: The World Cup Stimulates Gambling Industry 2006 will become a Gaming World Cup (December 25, 2004),
    http://www.sports.cn.yahoo.com/041225/346/2637z.html

13. Baidu Encyclopedia, Sports Lottery,
    http://baike.baidu.com/view/119062.htm
14. Zheng, B.: World TV Sports Programs Broadcast. Journal of Foreign Sports (31), 23–25 (1998)
15. Yang, Y.W.: German Football TV Broadcast Right War. Journal of Foreign Sports (4), 12–15 (2000)
16. Pan, H., Song, L.: Comparison between Chinese and German Sports expositions. Journal of Physical Education 15(1), 51–55 (2008)
17. State General Administration of Sports, The 12th five-year Plan for Sports Development. China Sports Daily (2-3) (April 1, 2011)

Review on Chinese Women Athletes' Involvement in the Olympic Games[*]

Hua Pan[1] and Xiaowei Liu[2]

[1] Institute for Sports History, Chengdu Sport University, Chengdu Sichuan China
huapan2002@yahoo.com.cn
[2] Foreign Language Department, Chengdu Sport University, Chengdu Sichuan China
xiaoweiliu2004@yahoo.com.cn

Abstract. By means of documentary and logic analysis and other methods, the paper summaries and analyzes the four stages of Chinese female athletes' in involvement in all the previous summer and winter Olympic Games and their excellent achievements, with the purpose of providing some beneficial enlightenment for Chinese women athletes' further achievements in future Olympic Games.

Keywords: 1. The Olympic Games, 2. Women players, 3. China.

1 Introduction

Since 1896, 29 summer Olympic Games and 20 winter Olympic Games have been held. In this period, Chinese women athletes first appeared in the Olympic field from 11th Olympic Games and they experienced the process from being weak to being strong [1]. This paper will be divided into several stages to summarize Chinese women athletes' participations in the summer, winter Olympic Games (including Taiwan and Hong Kong's most outstanding female athletes), and analyze their successes and failures to provide references for the further development of Chinese women's sports.

2 Development Process

2.1 Initial Stage (1936-1949)

Chinese women began to appear in the Olympic Games at the 8th Olympic Games in 1924, but not one woman athlete participated in games for 8th, 9th or 10th Olympic Games. The 11th Olympic Games in Berlin in 1936 finally witnessed Chinese women athletes' debut. They are Li Seng for official track and field competition, Yang Xiu qiong for swimming competition, Fu Shuyun and Zhai Lianyuan and LiuYuhua for martial arts performances.[2] Although Li Seng and Yang Xiuqiong did not perform well enough to enter the semi, which indicated a distinct distance between Chinese

---

[*] This work is partially supported by Chengdu Sports University for doctoral station construction (Grant #: BSZX1013).

women athletes and world top ones, they were still the first and the unforgettable Chinese women athletes to appear in the Olympics. In addition, Zhai Lianyuan, Fu Shuyun and LiuYuhua's martial arts performances were very popular with the people in Germany and Europe, which could be described as an encouraging step for Chinese martial arts to go out of the country. This was the first time for Chinese women athletes to go out of the country to compete in the Olympics, the significance of which was far more than the game itself, because they played a major role then in encouraging women to go out of their homes to participate in physical activities in modern China. They became not only the images of women's liberation, but also the outstanding representatives of women's sports in modern China.

At the 14th London Olympic Games in 1948, China sent only 33 male athletes to participate, not a single Chinese woman athlete was seen [3].

2.2 The Early Stage (1950-1979)

After the founding of the People's Republic in 1949, China sent 26 athletes (no females) to the 15th Olympic Games. After that, a minority of people in some international sports organizations attempted to create "two Chinas", which led to twists and turns in the relationships between the new China and the international Olympic Committee (IOC), and Chinese athletes were blocked from taking part in the Olympic Games. But Chinese Taipei, representing China, participated in the Olympic Games many times during this time. For example, at the 1968 Mexico Olympic Games, Chinese Taipei's Ji zheng won a bronze medal in the women's 80m hurdles as Chinese women's first Olympic medal [4].

2.3 The Striving Stage (1980-1999)

The 1979 Nagoya resolution of restoring the legitimate seat of the People's Republic of China in the international Olympic committee (IOC), plus the historic decision of the third plenary session to reform and open to the outside world and the good environment both at home and abroad provided the favorable safeguard for Chinese female athletes to take part in the Olympic Games.

In February 1980, the 13th Olympic winter games was held in Lake Placid in the United States. China sent 28 athletes (16cmales and 12 females) to attend speed skating, figure skating, alpine skiing, cross-country skiing and biathlon, totaling 5 major and 18 minor events. This was China's first winter Olympics. Although also-rans again, we were able to see the gap between China's winter sports and that of the world developed countries

In 1984, China sent 225 athletes to attend the 23rd summer Olympic Games in Los Angeles and won 15 gold medals, among which female athletes pocketed 5 (Wu Xiaoxuan in standard rifle, Zhou Jihong in platform diving, Ma Yanhong in uneven bars, Luan Jujie in foil and the women's volleyball team, especially Wu Xiaoxuan who realized the "zero" breakthrough of gold medals for Chinese female athletes in the Olympic Games. From then on, Chinese women athletes showed their sports powers of elegant appearance in the Olympic Games. In the same year, the 14th winter Olympic Games was held in Yugoslavia's Sarajevo, China and Taiwan both sent female athletes.

They still did not win any medals, but progress could be witnessed in China's competitive level in winter sports compared to the 1980 winter Olympics.

In 1988, China selected 301 candidates to participate in the24th Seoul Olympic Games. Chinese athletes won five gold medals, three of which were won by women (XuYanmei in platform diving, Gao Min in springboard diving and Chen Jing in table tennis singles) [5]. In the same year, the 15th winter Olympic Games was held in Calgary, Canada. In the contest, big distances in the sports level could be seen clearly between Chinese delegation, Chinese Taipei and the world tops. But in the exhibition events, Li Yan won a gold medal in women's 1000m short track speed skating by 1'39'', and two bronzes in 1500m and 50m. China's five-star red flag rose for the first time in the Olympic winter games and Chinese women speed skaters exhibited their debut in the world

At the 25th Olympic Games in Barcelona, Spain,1992, Chinese athletes won a total of 16 gold medals, including 12 ones won by women (Fu Mingxia in springboard diving, Gao Min in platform diving, Zhuang Yong in 100 meters freestyle, Qian Hong in 100-meter butterfly, Lin Li in 200m individual medley, Yang Wenyi in 50m freestyle, Deng Yaping/QiaoGong in table tennis doubles, Deng Yaping in table tennis singles, LU Li in uneven bars, Zhang Shan in skeet, Chen Yueling in 10km race walking and Zhuang Xiaoyan in judo of 72kg plus classs). In the same year, the 16th winter Olympic Games was held in Albertville, France, and the Chinese delegation realized the "zero" breakthrough of medals in the winter Olympics, with a total of three silver medals, all of which were contributed by women (Ye Qiaobo, 2 and Li Yan, 1).

In 1994, the 17th Olympic winter games in Lillehammer, Norway. China sent out 27 athletes (19 women) to compete in speed skating, short track speed skating, figure skating, biathlon and freestyle skiing. They won one silver medal (Zhang Yanmei in 500m shorttrack speed skating), two bronze medals (Ye Qiaobo in 1000m short track speed skating, Chen Lu in figure skating singles). Since then China has been listed in the world advanced levels in the three ice events[6].

At the 26th Atlanta Olympic Games in 1996, the Chinese delegation was composed of 495 people, 310 of whom were athletes (200female athletes) to take part in the competition of 22 major events and 153 minor events. The athletes obtained 16 gold medals, 9 of which were won by women (Sun Fuming in judo of 72kg plus class, Fu Mingxia in 10m and 3m springboard diving, Deng Yaping in table tennis singles, Deng Yaping/QiaoGong in table tennis doubles, Ge Fei/Gu Jun in badminton doubles, Le Jingyi in 100m freestyle, Wang Junxia in 5000m running and Li Duihong in sport pistol). Chinese Hong Kong (Li Lishan in sailing boat, also first Olympic gold medal won by Hong Kong Olympic Committee) and Chinese Taipei athletes (Chen Jing in table tennis singles) won a gold and bronze medal at the games respectively.

In 1998, the 18th Winter Olympic Games was held in Nagano, Japan. China sent 60 players(including 44 female athletes)to participate in the competition of speed skating, shorttrack speed skating, figure skating, women's hockey, cross-country skiing and freestyle skiing and biathlon (7 major and 29 minor events) and won six silver medals (4 by female athletes: speed skating relay, Yang Yang in 500m and 1000m speed skating and Xu Nannan in freestyle mogul), 2 bronze medals (1 by Chen Lu in figure skating). The number of silver medals and bronze medals exceeded the previous total, which marked the best performance of China after its first participation of the games in 1980.

2.4 The Prosperity Stage (2000-Present)

From September 15 to October 1, 2000, the 27th summer Olympic Games was held in Sydney Australia. China sent a delegation of 311 athletes (108male athletes, 203female athletes) to compete in 24 major and 166 minor events, with the result of 28 gold medals, among which 16.5 were won by women(Tao Luna in shooting, Liu Xuan in balance beam, Fu Mingxia in 3m springboard, Li Na/Sang Xue in 10m platform doubles, Wang Nan/Li Ju in table tennis doubles, Wang Nan in table tennis singles, Ge Fei/GuJun in badminton doubles, Zhang Jun/Gao Ling in badminton mixed doubles, Gong Zhichao in badminton singles, Yang Xia in 53kg class weightlifting, Chen Xiaomin in 63kg class weightlifting, Lin Weining in 69kg class weightlifting, Ding Meiyuan in 75kg plus class weightlifting, Wang Liping in 20km race walking, Tang Lin in 78kg judo, Yuan Hua in 78kg plus class judo, Chen Zhong in tae kwon do), 16 silver and 15 bronze medals, the total medals reaching 59 [7]. Not only did the Chinese delegation make a record high in its gold medals and medal tally in a single games after taking part in the Olympic Games, but also it jumped into the world top three in gold medals and medal tally, second only to the United States and Russia. China leaped into the first group and made a historic breakthrough in that year.

From February 8 to 24, 2002, the 19th Olympic winter games was held in Salt Lake City of the United States. A Chinese delegation composed of 133 members, among whom 72 were athletes (49 women) was sent to compete in short track speed skating, speed skating, figure skating, freestyle skiing, biathlon, cross-country skiing, women's hockey, etc. (7 major and 38 minor events). Chinese Hong Kong also sent its athletes for the first time. Yang Yang, who won 2 gold medals in 500m and 1000m short track speed skating, became the first woman athlete in Olympic winter games to win two personal gold medals in the same games. After 22 years of waiting, the Chinese delegation finally realized the "zero" breakthrough in gold medals in the winter Olympic Games in its seventh participation in the games and realized the dream of several generations of Chinese athletes in winter sports.

In 2004, 28th Olympic Games was in Athens, Greece. This was the first time for the Olympic family to come back to the Olympic hometown a hundred years later after the first Olympic Games in 1896. China sent a delegation of 407 athletes (269 female athletes) to compete in 26 major events, pocketing 32 gold medals (with 19.5 gold medals won by women: Du Li in shooting, Guo Jingjing/Wu Mingxia in 3m board doubles, Guo Jingjing in 3m board singles, Li Ting/Lao Lishi in 10m platform doubles, Chen Yanqing in 58kg class weightlifting, Liu Chunhong in 69kg weightlifting, Tang Gonghong in 75 kg plus class weightlifting, Wang Nan/Zhang Yining in table tennis doubles, Zhang Yining in table tennis singles, Zhang Ning in badminton singles, Zhang Jun/Gao Ling in badminton mixed doubles, Yang Wei/Zhang Jiewen in badminton doubles, Xing Huina in 10000 metres, Luo Xuejuan in 100m breaststroke, Li Ting/Sun Tiantian in tennis doubles, Xian Dongmei in 52kg class judo, Wang Xu in 72kg class wrestling, Luo Wei in 67kg tae kwon do, Chen Zhong in 67kg plus class of tae kwon do and the Chinese women's volleyball team in volleyball) [8], 17 silver and 14 bronze medals. The outstanding achievements with a total of 63 medals pushed China to the second in medal tally. The two indexes of gold medals and medal tally both created the highest record since China's first participation in the Olympic Games.

Athletes from Chinese Taipei also gained their success in the Olympic Games. Chen Shixin won the first gold medal for Chinese Taipei in 49kg minus class kickboxing, which was a historic breakthrough [9].

From February 10 to 26, 2006, the 20th winter Olympic Games was held in Turin, Italy. China sent a delegation of 151 members, including 76 athletes (36 men and 40 women)to compete in short track speed skating, figure skating, speed skating, freestyle skiing, cross-country skiing, alpine skiing, snowboarding, ski jumping and biathlon(9 major and 47 minor events). After 17 days of competition, the Chinese sports delegation achieved a good result of winning 5 gold medals (including a gold medal won by Wang Meng, a female, in 500m speed skating), 4 silver and 5 bronze medals.

From August 8 to 24, 2008, the 29th summer Olympic Games was in China's seven cooperative cities of Beijing, Shanghai, Tianjin, Shenyang, Qinhuangdao, Harbin, Qingdao and Hong Kong city.The Chinese delegation, composed of 639 athletes (331male athletes and 308 female athletes), won a total of 51 gold medals(27 were won by women: Chen Xiexia in 48kg class weightlifting, Chen Yanqing in 58kg weightlifting, Liu Chunhong in 69kg weightlifting, Cao Lei in 75kg weightlifting, Guo WenJun in pistol shooting, Du Li in 50m rifle shooting, Guo Jingjing/ Wu Minxia in 3m board doubles, Wang Xin/Chen Ruolin in 10m platform diving doubles, Guo Jingjing in 3m board singles, Chen Ruolin in 10m platform diving, gymnastics team, He Kexin in uneven bars, table-tennis group, Zhang Yilin in table-tennis singles, Gang Yang/Du Jing in badminton doubles, Zhang Ning in badminton singles, Liu Zige in 200m butterfly, Xian Dongmei in 52kg judo, Yang Xiuli in 78kg judo, Tong Wen in 78kg plus judo, Zhang Juanjuan in individual archery, He Wenna in trampoline, Yin Jian in RS: X windsurfing, Tang Bin/Jin ZiWei / Xi Aihua/Zhang Yangyang in women's quadruple sculls, Wang Jiao in wrestling and Wu Jingyu in 49kg taekwondo), 21 silver medals and 28 bronze medals, with the medal tally amounting to 100 and four world records created. The total number of gold medals surpassed that of the United States and the medal tally exceeded that of Russia. China ranked first in gold medals and second in medal tally, which was the best achievement of China since its participation in the Olympic Games and this created a new success in the history of China's competitive sports. In addition, China is the first among all Asian and developing countries to rank first in gold medals in Olympic history. China has made outstanding contributions to the Olympic movement, thus writing a new page.

From February 12 to 28, 2010, 2632 athletes from 82 countries and areas gathered in Vancouver, Canada for the 21st Winter Olympic Games with the largest number of athletes compared with all the previous winter games. At this winter games, the Chinese delegation won five gold medals (Shen Xue/ Zhao Hongbo in pair figure skating, Wang Meng in 500m, 1000m short track speed skating, Zhou Yang in 1500m short track speed skating and 300m relay), four silver and two bronze medals, with the number of gold medals reaching a historic high and the medal tally equaling the highest record of the 2006 Turin winter Olympics. [10]

At Vancouver Olympic winter games, the Chinese athletes achieved a series of breakthroughs: gold medals amounting to a record high compared with the previous games, ranking top ten in gold medals for the first time, first gold medal won in figure skating and admirable performances in women curling and other events. However, we should also be aware of the fact that China lacks a solid foundation in winter sports. Compared with such strong teams as Canada, Norway and the United States, China is still weak in the traditional events of the winter games. It is still not competitive in alpine skiing, cross-country skiing and ice hockey, and there have never been any Chinese athletes in platform skiing, Northern Europe biathlon, sledding or ice sledding. Even in the potential gold-winning events such as figure skating and curling, a dramatic gap still exists in terms of the number of people involved in such sports compared with that of the traditional powers in Europe and America.

3 Conclusion

The development of the world women Olympic sports has promoted the development of Chinese women's sports, so that Chinese women have entered international sports and made great contributions to the Olympic family. Thanks to the correct sports development strategy of the government and the full exploitation of Chinese women's potential, Chinese women have achieved great success in sports in just a few decades. With the rapid development of Chinese women's sports, Chinese women have being playing a more and more important role and more and more Chinese women have participated in the Olympic movement in China, which has contributed greatly to devloping China's sports industry, promoting international sports exchanges and the Olympic sports.

References

1. Zhang, Z.Z.: Review and prospect of China Women sports. The International Journal of Sport History and Culture (4), 3–6 (1994)
2. Han, Z.F.: The Relationship between the Emancipation Movement of Chinese Women and the Development of Chinese Women in Athletics. Journal of Hebei Institute of Physical Education (2), 27 (1999)
3. Wang, Y.L.: The Rising and Growing of Women Sports of Modern China. Shandong Sports Science & Technology (2), 40 (2004)
4. Luo, S.M., Wang, Y.: Analysis on the Rise of Modern China Women Sports. Journal of Chengdu Sport University 32(1), 18–20 (2006)
5. Ma, Z., Zhang, C.Y.: Review and Forecast of Chinese Female Sports Participation in Olympics Games. Journal of China Women's University 20(4), 112–116 (2008)
6. Dong, J.X.: Continuity and Change: Gender Relations in Chinese Elite Sport. Journal of Sports and Science (2), 20 (2006)
7. Wei, W.Z.: On the cause of Formation of Rapid Development of China Women Sports. Sichuan Sports Science (1), 2 (2006)
8. Song, L., Pan, H.: On Chinese Female Sports in the 20th Century. Sports Culture Guide (6), 124–125 (2008)
9. Gong, T.S.: Taiwan Sports. China Sports (1-3), 32–39 (2011)
10. Long, J.Y., Liang, J.P.: Research on Trends and Distribution of China's Gold Medals in Summer Olympic Games List. Journal of Xi'an Physical Education University 28(2), 246–252 (2011)

Study on Game Decision of BOT Financing Model of Public-Rent Housing

Liming Wang[1], Fangqiang Liu[2], and Jiansheng Dai[1]

[1] College of Economic & Industry and Business Management, Chongqing University,
Chongqing 400044, China
[2] College of Construction Management and Real Estate, Chongqing University,
Chongqing 400044, China
wangliming2006@126.com

Abstract. It is an important measure to construct public-rent housing for our housing security system. Nowadays it is explored to induce civil capital into public-rent housing construction through BOT in a number of areas. In this paper an investment decision model about Government and civil capital is set up by game theoretical method and the relations between public-rent housing and real estate market, concision period and rent as decision variables. It theoretically indicates that it is hard to induce civil capital into construction of public-rent housing owning to prosperity of real estate market and lower return of public-rent housing. We suggest that Government should design incentive mechanism to induce civil capital into public-rent housing.

Keywords: Public-rent housing, BOT, financing.

1 Introduction

The Ministry of Housing and Urban-Rural Development plans to build 10 million units of affordable housing in 2011 and 36 million units in the next 5 years in China, making the coverage rate of affordable housing more than 20%. According to the estimation of the Ministry of Housing and Urban-Rural Development that the construction funds needed for 10 million units of affordable housing this year alone will be at least 1.3 trillion yuan and after taking all sources of funds of the state and local government into account, there is still a funding gap of nearly trillions of yuan. How to raise funds? Construction funds are not in place, how to make construction? How to finish the delivery in time? The series of problems is the key to hindering the construction of affordable housing in China. To explore the construction effect of affordable housing in recent years in China we can see it is not satisfactory, there are limited financial resources and insufficient capital investments in many parts, which is the main reason [1]. This paper takes public-rent housing in affordable housing for example to explore the possibility of inducing civil capital into public-rent housing construction through BOT model to address the insufficient funds for construction.

To solve this capital problem, 7 ministries of the Ministry of Housing and Urban-Rural Development and so on jointly issued "Guidance on how to speed up the development of public-rent housing" (Construction and Security[2010]No. 87) which clearly stated: each region should increase the government investment on public-rent housing, at the same time, it is necessary to take the land, taxation, financial and other support policies to fully incentivize the various types of enterprises and other organizations to invest on and operate public-rent housing. In some areas people begin to draw on the experience of BOT models of roads, bridges and other infrastructures and actively explore the possibility of inducing civil capital into public-rent housing construction [2] [3].

Many domestic scholars have conducted a preliminary discussion on this, but in the aspect of research on financing model of public-rent housing, the majority of the literature are mainly about qualitative analysis[4] [5] [6], few about quantitative investigation, thus making there is clearly insufficient guidance. Nevertheless, a large number of literatures studied the decision problem of concession periods in BOT projects. Nombela et al [7] take maximizing social welfare as the objective to meet the conditions for constraints of private participation, making the establishment of a function between the optimal price charged by and the best concession period of BOT project. Li Qiming[8]、Yang Hongwei[9,10]、Shen et al[11] established the decision model of concession period of BOT project from a cash flow perspective. These studies have great reference significances on the BOT financing model of public-rent housing.

2 Analysis on the Factors Influencing on Project Companies' Revenue under the BOT Model

The so-called BOT is a financing model which makes the government grant the investment, construction, operation and other concession rights of an infrastructure project to the project units in a certain period (i.e. the concession period) and after that period the projects be handed over unconditionally to the government by the project units.

BOT projects usually go through several periods of construction, operation and transferring. Make T_b represent the construction starting period, T_o represent the operation starting period, and T_t represent the transferring time, then T_e represent the last time when the project is abandoned.

Make $\pi(P_t, Q, b_t, I, r, T_t, T_o)$ represent the profit function of the project units' taking the opportunity cost into account to participate in public-rent BOT projects, where: P_t is the rental price of the time of t, Q is the built area of public-rent housing(we assume it has nothing to do with the time, that means the built area of public-rent housing remains the same during the projects' operating period), b_t means maintenance cost units required by the normal operation of projects during the projects' operating period, I means the total amount of investment for the public-rent housing, and r means the revenue rate required for project units. The profit function of project units can be expressed as:

$$\pi(P_t,Q,b_t,I,r,T_t,T_o) = \sum_{t=T_0}^{T_t} (\frac{PQ}{(1+r)^t} - \frac{b_tQ}{(1+r)^t}) - I(Q) \tag{1}$$

Noted that $\pi(P_t,Q,b_t,I,r,T_t,T_o)$ is the profit obtained under the premise of having considered the opportunity cost of project units, rather than the usual accounting profits. Private capital can not only participate in BOT projects, but also have many other investment opportunities. In its other alternative projects, the best interests that can bring to the project units constitute the opportunity cost of project units' participating in the public-rent housing. The r in equation (1) represents the opportunity cost, i.e. the investment revenue rate obtained by project units participating in other projects. The $\pi > 0$ represents the construction of public-rent housing participated by the project units will receive greater benefits; and $\pi = 0$ represents the compensation just obtained by the opportunity cost through the construction of public-rent housing participated by the project units, whether the project units are participating in the project or not has no difference for the project units. In summary, to attract the project units to participate in the construction of public-rent housing we must make $\pi \geq 0$.

To simplify the discussion, we assume that the project units obtain a loan from the bank in the early construction of the projects, the loan period is T^r, and the repayment of the loan is as follows: payment of interest is only made during the period of the loan and the principal is returned in the end of the loan. Make D the total loans of project units, the $I - D$ owed funds invested by enterprises for the development of public-rent housing and the i lending rate of project unit's. Project units' profit function can be expressed as:

$$\pi(P_t,Q,b_t,I,D,i,r,T_t,T_o) = \sum_{t=T_0}^{T_t} (\frac{PQ}{(1+r)^t} - \frac{b_tQ}{(1+r)^t}) - \sum_{t=T_b}^{T^r} \frac{Di}{(1+r)^t}$$

$$- \frac{D}{(1+r)^{T^r}} - (I(Q)-D) \tag{2}$$

Proposition 1: When $r > i$ there is $\sum_{t=T_b}^{T^r} \frac{Di}{(1+r)^t} + \frac{D}{(1+r)^{T^r}} < D$

Proposition 1 has pointed out that the project units can use the loan to improve the efficiency of using its own funds. Therefore, the project units will certainly make full use of financial leverage to access to funding through other financing channels, thus improving the efficiency of using its own funds.

Proposition 2: When $P_t > b_t$ and to $\forall t$ it is tenable, the greater $T_t - T_o$ is, the greater $\pi(P_t,Q,b_t,I,D,i_0,r,T_t,T_o)$ will be. Given T_t and T_o, the greater P_t is, the greater $\pi(P_t,Q,b_t,I,D,i_0,r,T_t,T_o)$ will be. The smaller $T_o - T_b$ is, the greater the profits $\pi(P_t,Q,b_t,I,D,i_0,r,T_t,T_o)$ obtained by the project units will be.

Verification of this proposition is obvious.

Proposition 2 has pointed out that the longer the construction period is, the fewer the project units can benefit, mainly due to the use of funds with the opportunity cost. What more important is: Proposition 2 has shown two decision variables, rental price P_t and operating period $T_t - T_o$ of public-rent housing. For the given P_t, project units will receive the maximum profit which is expressed as:

$$\pi_m(P_t) = \max_{T_t - T_o} \pi$$

When $P_t > b_t$ and to $\forall t$ it is tenable, there is:

$$\pi_m(P_t) = \sum_{t=T_0}^{T_e} \left(\frac{P_t Q}{(1+r)^t} - \frac{b_t Q}{(1+r)^t} \right) - \sum_{t=T_b}^{T^r} \frac{D^* i_0}{(1+r)^t} - \frac{D^*}{(1+r)^{T^r}} - (I(Q) - D^*)$$

The D^* in the equation is the best scale of loans.

As a product of the public nature, the development of the lease price of public-rent housing must consider the ability of ordinary people to afford it, and thus being subject to strict limitations. To set $0 < P_t \leq P_t^m$ as well, in which P_t^m represents the highest price that may be afforded by public-rent housing rental households, assuming P_t is set by the upper limit, the project units will receive the maximum profit π_m which is expressed as:

$$\pi_m = \max_{P_t, T_t - T_o} \pi$$

That is: $\pi_m = \sum_{t=T_0}^{T_e} \left(\frac{P_t^m Q}{(1+r)^t} - \frac{b_t Q}{(1+r)^t} \right) - \sum_{t=T_b}^{T^r} \frac{D^* i_0}{(1+r)^t} - \frac{D^*}{(1+r)^{T^r}} - (I(Q) - D^*)$

When r is large enough, $\pi_m < 0$ is entirely possible. This means even if the concession period of project units granted by the government achieves the maximum, participating in the construction of public-rent housing by the project units is still unprofitable, and then the decision-making of the project units is not being involved in the construction of public-rent housing.

Proposition 3: The higher the market price of commercial housing is, the more profitable the construction of commercial housing can be, and the more reluctant to participate in the project of public-rent housing the project units will be.

The conclusion of this proposition has shown that: the government cannot completely cut open the public-rent market and the commercial housing market. As it can be relatively profitable to engage in the construction of commercial housing, the project units naturally consider more the development of the real estate, rather than engaging in the development of public-rent housing (the rental price is controlled by the government.) The commercial housing market not only has affected the public-rent housing but also is affected by the construction of public-rent housing to the same extant of importance. The more the outputs of public-rent housing projects are, the fewer the demands of the commercial housing market will be, thus forming downward pressures on the prices of the commercial housing, which in turn will promote the construction of public-rent housing.

3 Negotiations and Games between the Government and the Project Units

The Q of the equation(2)is the construction area of public-rent housing, in the current discussion, we assume that it is exogenously determined, that is designated by the government. And b_t is the unit maintenance cost required to maintain the normal operation of the project during the project operating period also identified by the factors out of the models. I , the total investment of public-rent housing, assuming it is the deterministic function of the output Q . Without prejudice to the general, the construction period is also considered established. As to the total amount of loans of the project units, the total assumed project units can achieve the best loans, determined not only related to the model (2), but also impacted by the actual structure of assets and liabilities of the project units and other variables, only considered as endogenous variables when needed to be handled. Without prejudice to the general, the loan interest rates i are also considered as exogenous. As to the internal rate of return r , the factors affecting it are out of the model, thus naturally being regarded as exogenous variables.

In the following discussion, in order to simplify the illustration, π will be considered only the function of Q , P_t and T_t , namely:

$$\pi(P_t,T_t,Q)=\sum_{t=T_0}^{T_t}(\frac{P_tQ}{(1+r)^t}-\frac{b_tQ}{(1+r)^t})-\sum_{t=T_b}^{T^r}\frac{Di}{(1+r)^t}-\frac{D}{(1+r)^{T^r}}-(I(Q)-D)$$

And, in order to facilitate our discussion, we won't consider other financing models of the project units, namely assuming:

$$\pi(P_t,T_t,Q)=\sum_{t=T_0}^{T_t}(\frac{P_tQ}{(1+r)^t}-\frac{b_tQ}{(1+r)^t})-I(Q)$$

In particular, there is $\pi(P_t,T_t,0)=0$.

The Government plans to deliver a public-rent housing project by way of BOT to a representative project unit for development, the proposed size of the construction of public-rent housing has been determined by the Government, under normal circumstances, the total investment of the representative project unit on the development of this project can be regarded as given. The Government and the project units are clear that to make the project units involved in the construction of the project unit must first guarantee $\pi(P_t,T_t,Q)\geq 0$. Given the Government and the project units are both clear that the company's internal rate of return r , the Government utility function is represent as u , and:

$$u(P_t,T_t,Q) = \sum_{t=T_t}^{T_e} (\frac{P_tQ}{(1+r)^t} - \frac{b_tQ}{(1+r)^t}) + v(Q)$$

To meet: $v(0) = 0$, $\partial v/\partial Q > 0$, and $v(Q) \geq N$, N is a constant.

Obviously, if $P_t, P_t > b_t$, tenable to $\forall t$ and $T_t, T_t \leq T_e$, then for any $Q > 0$, there is $u(P_t,T_t,Q) > u(P_t,T_t,0)$, this shows that: the government can get good effectiveness through providing the community a certain amount of public-rent housing by way of BOT financing model. In addition, the paper also assumes the government needs to provide public-rent housing project to the developers to develop, for organizing this event will bring the Government a negative effect $-c < 0$. The public-rent housing as a nonprofit product, its rental price is constraint by the spending of low-income residents, asked to meet $0 < P_t \leq P_t^m$, P_t^m is the maximum acceptable unit lease of the residents at time t . The range of another variable needs to meet: $T_o \leq T_t \leq T_e$ and the limiting is very natural. To the given P_t and T_t , the Government has the freedom to choose whether or not to provide the project to the representative developers to develop and the developer has the freedom to choose whether or not to participate in its construction. For the given P_t and T_t the payment matrix figure of the Government and the representative developers is shown in Figure 2:

The developers

		To participate	Not to participate
The Government	To provide	$u(P_t,T_t,Q)-c$, $\pi(P_t,T_t,Q)$	$-c$, 0
	Not to provide	0 , 0	0 , 0

Fig. 1. The game payment matrix figure of public housing between the Government and the developers

Proposition 4: When $\pi(P_t,T_t,Q) < 0$, the optimal choice of the developer is not to be involved in the construction of public-rent housing projects, and the optimal choice of the Government is to construct the public-rent housing projects without the BOT financing model.

We can express the game described above through the game tree:

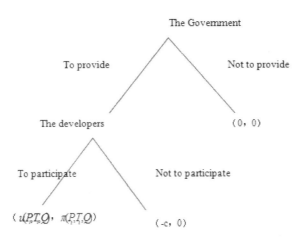

Fig. 2. The negotiating tree of the public-rent housing with the BOT model

The equilibrium of the game can be solved by backward induction. When $\pi(P_t, T_t, Q) < 0$, the project units will choose "Not to participate" between two strategies of "To participate" and "Not to participate" in the construction of public-rent housing projects; given the choice of the project unit, if the Government choose "To provide " policy, it will get paid " $-c$ ", and if you choose "Not to provide" policy, it will get "0", so the government's optimal strategy is " Not to provide". In this case, the strategy combination of the Government and the project units (not to supply, not to participate) is a strictly dominant strategy and the payment combination of the Government and the project units is (0, 0).

The foregoing discussion is under the assumption that P_t and T_t are given, it is the optimal strategy choice of the Government and the project unit. In fact, the Government and the project unit can negotiate in terms of variables P_t and T_t namely the public-rent housing lease can negotiate with the concession operating period. However, $\pi_m = \max_{P_t, T_t - T_o} \pi$ when $\pi_m < 0$, the Government and the project units can only end in failure in the negotiation we set. Since the Government is aware of the internal rate of return of the representative project units, before the Government negotiations with the project units, we can calculate the profit of project units from (P_t, T_t). Therefore, if $\pi_m < 0$, the Government would not be necessary to negotiate with the project units, because the project units will certainly reject any proposals of the Government. That being the case, the Government chose "Not to provide" strategy is optimal, or put it that way, by BOT financing model the construction of public-rent housing is not feasible.

4 Empirical Examples

From the nationwide view, the exploration and practice of Chongqing city in the construction of public-rent is at the forefront. We take the data that Chongqing municipal government announced as an example to make an empirical analysis on the preliminary model as follows:

The total investment on a public-rent housing project is 3.20037 billion yuan (including land cost of 369 million yuan), public-rent area: 954,700 square meters, the rent lease: 10 yuan / month / square meter; assuming that 80% of the total investment is from bank loans, 6% per annum. Even under the most favorable principle, set the using period of public-rent housing 70 years with no maintenance costs, assuming that the within risk free rate under the requirements of the project company was 10%. Given these data, using equation (2) to calculate the maximum possible benefits of the project company $\pi_m = -14.14$ ten thousand, far less than zero. If the project company is engaged in the construction of public-rent housing project constructing, it will face huge losses.

Therefore, in the case with the absence of the government's preferential policies to support, the optimal strategy for the project company is certainly not to be involved in the construction of public-rent housing. The empirical results are consistent with the current status of the domestic financing of public-rent housing. Although many local governments are exploring on inducing civil capital into public-rent housing construction through BOT, the actual situation is still based on financial investments.

Proposition 5: In BOT financing model of public-rent housing, if the Government proposes with the policies of the highest rent of public-rent housing and the longest operating concision period to the project units, and makes $\pi_m < 0$ be met then the BOT financing model of public-rent housing is not feasible.

5 Policy and Recommendations

The fact can be found in the theoretical analysis that public-rent housing is the commonweal project facing middle-low-income residents, because of its rental prices are constraint by residents' consuming capacity, even if the operating concision period of the public-rent housing is extended to the entire life cycle, or with certain tax incentives given, the project unit is still difficult to obtain profits through effective management, with lack of active participation.

In order to attract private capital to be introduced, the Government can draw on the back subsidies of BOT financing of sewage treatment plant. The Government authorizes private capital investment on sewage treatment plant, the main charges and payment is from the Government and to take "expenditure" manner to run. First, determine the pollutant emission standards, and then agree to running costs and security amount of water of sewage treatment of the unit, ensuring service levels and the basic income of investors. We can see the situation from the practice in recent years that the model has a strong practical value to public welfare projects of not high level of profitability.

In the construction of the public-rent housing, we should first determine building standards, the Government committed to the future rent levels and preserving occupancy rates. By way of tender we can select the main franchise of the public-rent housing, with independent investment from the private capital. After completion of the project, the property rights belong to all investors and are run and managed by investors. However, the eligibility is audited by the government-appointed management administrations for the public-rent housing, with unified collection of rent. Regardless of the actual rent payment of households and high or low rental occupancy rates, rental income of investors are paid by agreement, the lack part is paid by financial subsidies. Subsidies can take the form of cash, it also can be leased Certificates (investors can use them to offset taxes and fees), circulating in the public-housing system to avoid disrupting the normal order of the real estate market. The rent beyond the security rental occupancy rate is divided by the government and investors. It can inspire investors to improve the quality and service of the construction of public-rent housing, to attract low-income people to move in, by increasing the occupancy rate to obtain higher yields. The occupancy rate as an important indicator, it also can be used to guide the Government to determine a reasonable scale of construction of public-rent housing. If the overall occupancy rate is not high then it shows the public-rent housing is already saturated and the scale of construction should be controlled.

If some of the public-rent housing in the future construction use the BOT model, then trillions of financial payment can be invested at one-time and be resolved to tens of billions rent subsidies each year, greatly reducing the one-time investment of the Government. Moreover, tens of thousands of billions of private capital to found an investment channel with stable income. Through the introduction of private capital, the construction of public-rent housing can also be further marketization, to improve development efficiency and management level. Then low-income groups can choose their own satisfactory public-rent housing, thus better meeting the housing needs of the majority of the people.

Acknowledgment. Fund Project: Soft Science Research Program Projects of Chongqing in 2010 (No.: CSTC, 2010CE0120).

References

1. Li, D.: Analysis on the Construction of Public-rent Housing by Applying BOT Financing Model. Economic and Social Development 8, 8–11 (2010)
2. Changzhou Municipal People's Government. Management Methods for the Public-rent Housing in the Urban City of Changzhou[EB/01] (July 16, 2009),
   http://www.changzhou.gov.cn/CZPortal2008Web/Default/InfoDetail.aspx?InfoGuid=d03f8ce8-c32a-4998-9517-6669b901d0f2
3. Beijing Municipal People's Government. Management Methods for the Public-rent Housing in the Urban City of Beijing(Trial) [EB/01] (September 16, 2009),
   http://bj.9ask.cn/fangchanjiufen/fwzlsifajieshi/200909/232996.html

4. Li, Z., Tian, J.: Prediction Research on the Supply of Affordable Housing in Beijing. Harbin Institute of Technology (Social Science Edition) 11(5), 89–94 (2009)
5. Wei, D.: Argument on Establishment of Rental Housing Security System. Construction Economics 4, 81–83 (2009)
6. Jiang, J.: Research on Low-cost Housing Financing Model. Xi'an University of Architecture Technology (2004)
7. Nombela, G., Gines, D.R.: Flexible-term. Ontracts for road franchising. Transportation Research Part A 38(3), 163–179 (2004)
8. Li, Q., Shen, L.: Decision-making Model of Infrastructure BOT Projects Concession Period. Management Engineering 14(1), 43–46 (2000)
9. Yang, H., Zhou, J., He, J.: Decision-making Model of Traffic BOT Projects Concession Period Based on Game theory. Management Engineering 17(3), 93–95 (2003)
10. Yang, H., He, J., Zhou, J.: Game Decision-making Model of Road Toll Pricing and Investment under the BOT Model. China Management Science 11(4), 30–33 (2003)
11. Shen, L.Y., Wu, Y.Z.: Riskeon Cession model for buil-operation-transfer contract projects. Journal of Construetion Engineeringand Management 131(2), 211–220 (2005)

The Design and Development of Web-Based Examination System

Jun Ma and Qinghua Liu

Institute of Information Engineering of Jiaozuo University
jsjxxxy@163.com

Abstract. With the rapid development of network technology, online test has become an important feature of the modern examination. This paper presents web-based test system architecture, design concept, and on this basis, the use of new web technologies: ASP.NET, SQLserver200 and ADO.NET as a development tool to achieve the examination system.

Keywords: B/S architecture, exam system, web server.

1 Introduction

With the rapid development of Internet technology, Network examination has already become the important features of modern exam, traditional online examination systems mainly operate in the C/S mode, in which it is inconvenient to install, maintain and upgrade. But the openness and distribution of web-based online examination system make the test break through time and space limitation. The client is a standard browser and the server is a web server. The close combination of web server, database and application server expand its use field. Its main advantages are: easy expanding application, simple upgrades and maintenance and especially, its objective score assessment. So the web examination system is one of hot issues in the present examination system research. It is a typical B/S-based application, in which it is simple to install and maintain and can satisfy the needs of a variety of exams with high speed network support.

The examination system, the operation is simple, with a high efficiency test entry modification and query, management students, the teacher data and other important function, i.e. real now real paperless examination.

2 The Architecture and Development Platform of Web-Based Examination System

2.1 B/S Architecture

In the B/S three-tier architecture, the user can make a request through a browser to the server on the Internet which processes the browser's request before returning the required information to the user browser, which will simplify a lot of client processing

Y. Wang (Ed.): Education Management, Education Theory & Education Application, AISC 109, pp. 117–121.

in C/S .The client only needs to configure a small client software, or just a browser to connect while the access to database and implementation of application program will be completed on the sever. Browser makes a request, but the data request, processing, results returning and dynamic page generation etc. are all done by the sever. B/S architecture makes the transaction processing logic module of C/S two- tier structure specifically separated from the client with a layer formed by the web server to complete its task, so the pressure on client is reduced. Web examination system structure see figure 1.

B/S architecture has the following advantages over C/S architecture:

(1) no client software development, simple maintenance and upgrades
(2) cross-platform operation and only www browser needed
(3) good openness and scalability
(4) reducing network traffic, greatly easing the pressure of network band width.

The system uses B/S three-tier architecture.

2.2 System Computing Platform

Web Services are a general model for building application and a basic building block with distributed computing on the Internet. Open standards and the communication

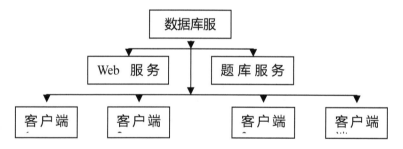

Fig. 1. System structure

collaboration between users and applications have created a new environment, in which Web Services become a platform for application integration. The current technologies to support distributed Web Services platform include: Web Services on Web Sphere of IBM, Tool Kit and Forte platform of J2EE architecture from SUN company and Visual Studio.NET development environment of Microsoft NET, formerly known as NGWS(Next Generation Windows Services) which aims to become the next generation Internet-based distributed computing application development platform to create and operate next-generation service NET infrastructure and tools: Net User Experience to start a great number of clients, millions of NET building module service of high distribution, and NET equipment software only to start new generation Internet device. NET platform includes the following components:

(1) user data access technologies, including a new XML-based, mixed information architecture with browser as its component, called universal drawing board.
(2) construction and development tools based on Windows DNA2000
(3) a series of module services, including authentication, information transmission, store, search and software delivery capabilities.
(4) client driver software.

2.3 System Development Platform

The development platforms of this system are Windows XP+IIS5 and 5+ASP.NET. Background database uses a SQL Server 2005. ASP. NET front-end development tool is Visual Studio. NET

(1) IIS 5.1 server. Windows XP is a network operating system developed by Microsoft, integrated with IIS5.1 and IIS servers, responsible for network and network information management, specifically service management (FTP, HTTP, GOPHER, SMTP protocols management), directory management (website directory structure and security management) and other functions, which own the highest rate of sever market.
(2) SQL Server 2000. SQL Server 2000 developed by Microsoft, a large–scale relative database management system, is very powerful in relative database creation, development, design and management. It connects data with the Internet, displays data operation through the Web browser with client-server structure, and has good compatibility with Microsoft's other products and third-party product to easily achieve a seamless operation.

3 The Analysis and Design of Web-Based Examination System

Test system's main users are: teachers (mainly non-computer professional teachers), students and system administrators. The main objectives of system design are: teachers can conveniently generate test questions to the teaching content. In addition, it is applied to some personal teaching sites involving online exams, exercises test and many-people discussions. Students can conveniently log on to a teacher's teaching site to study independently, participate in online examinations or many-people discussions. Besides, administrator can make management and maintenance. According to the system requirements, the system is divided into three subsystems, each of several functional modules. A brief description of each module function is as the following:

(1) test subsystem: log-on, registration, test paper generation, test starting, automatically grading score function etc.
(2) information query subsystem
(3) test management subsystem. The subsystem as an ordinary window application, is used for system maintenance, which runs on the server side and does not require browser support. The main functions of the subsystem: set configuration information such as the various document paths and database path required by system; assign different permissions for all types of personnel; check or print information of test paper, test scores or students conditions.

4   The Realization of Web-Based Examination System

4.1 Candidates Login and Information Browsing

On students register/login sub-module, Students apply for registration and log on to teachers teaching –aid site after message authentification unit audits its information Then, through access to sub-modules as course information, topic test, online test and discussion forum, they can acquire basic teaching information and CAI course wares, test their own learning content, participate in online examination and discuss problems with the teacher or other students in the forum.

4.2 Implementation of the Test

(1) Exam time setting. When the paper is formed, the test time is deposited in a variable Application while the end time in other variable End Time. The exam timer starts automatically. Once the examination time ends, the system automatically submitted the paper.

Application (Begin Time) = now

End time = hour (Begin Time) + 2(exam time is two hours)

(2) The generation of test paper.

The generation of test paper is the display of test questions, which is the major part of the examination system, involving the automatic organization of test questions. In the existing examination system, after students log on and select to test the chapter, section or the total course, test questions are generated automatically or manually. But it is difficult to achieve the goal of different users with different difficulty-level test paper and often hard to realize the desired purpose of testing. The main function of this system is to achieve a personalized Web examination system. This system can make different login users choose different difficulty-level paper. According to the paper difficulty degree submitted by users, the system can randomly extract questions and form test paper automatically to the difficulty degree chosen by users, and in the end grade the users paper.

4.3 Score Points

The examination system will automatically score. On the score page, after the submission of paper, the system compares the student's answers with those in Temporary Paper to grade the paper and display it to the user.

5   Conclusion

In the Web-based examination system, the server is designed through ASP.NET and SQL server 2005 database, students log on to system through the browser to complete the online exam and administrators make the maintenance and management by logging at distance. The system is universally strong, easy to use, efficient and has a strong application value. That is a truly paperless examination which meets the candidates' access to test without time and place limitation and greatly reduces the teacher's heavy workload in generating and grading test paper, and offers a solution to many teaching-aid issues in college.

References

[1] Hen Walther, S.: ASP. NET Unleashed, pp. 184–208. China Machine Press, Beijing (2002)
[2] Micro software company. Microsoft SQL Server. pp.152 – 162. Beijing Hope Electronic Press, Beijing (2000)
[3] Shu, Q.: Windows 2000 Server, pp. 306–344. China Machine Press, Beijing (2000)
[4] Zhang, L., Zhu, Y., Zhu, B.: Browser-based Examination System realization. Dalian Maritime University 26(3) (2000)
[5] Zheng, R.: Practical Software Engineering, 2nd edn. Tsinghua University Press, Beijing (2000)
[6] Zhang, H.: Introduction to Software Engineering, 3rd edn. Tsinghua University Press, Beijing (1999)
[7] Zhai, Y., Chen, J., Liu, L.: National Geomatics Center of data storage grid construction. Geographic Information World (4) (2007)
[8] Zhou, C.: A new era of geographic information system: GIS grid. Geographic Information World (4) (2007)
[9] Cui, T., Li, G.: Grid-based geo-spatial information services key technologies. Surveying and Mapping (05)(2007)
[10] Wang, J.-Y., Zhu, Y.: On the grid and grid GIS. Surveying and Mapping (01) (2006)

Institute-Industry Co-operation: A Comparison of Two Educational Modes of School of Software and Higher Vocational Education in China

Ziru Li, Yanqing Wang, Yangyang Han, Ligeng Guo, and Yundi Zhang

School of Management, Harbin Institute of Technology, Harbin 150001, China
yanqing@hit.edu.cn

Abstract. With the development of engineering education, industry-institute co-operation has now become the value orientation for both the enterprises and the institutes in the new circumstances, and "strategic alliances" is becoming more and more important as the rational choice for both sides. From three major aspects: cultivating objective, educational policies and co-operation strategy, this article analyses the present situation and evaluates the intensity of the implementation of industry-institute co-operation in schools of software and higher vocational education. This work can provide reference to the reform of industry-institute co-operation in various educational patterns, and contribute to the development of engineering education in China as well.

Keywords: industry-institute co-operation (I-I-C), school of software (SoS), higher vocational education (HVE).

1 Introduction

The remarkable effects of engineering education on cultivating applied talents have been widely recognized by enterprises and institutes all over the world. More and more concerns have been drawn on industry-institute co-operation (I-I-C) as the significance of engineering education is widely identified. Engineering education has been well implemented in the majority of developed countries such as the USA, Germany and the United Kingdom, and in these countries, the gap between engineering science and engineering practice is well bridged. With the feature of pragmatism and powerful absorbing capacity, engineering education has provided timely and efficient technical support for American industrialization rise project; in Germany, engineering education has provided a batch of technicians to fully meet the quantity and quality requirements of the enterprises; in UK, the gap between Engineering council professional qualifications, Engineering education academic qualifications and Industry professional qualification has been bridged successfully, which marks the maturity of engineering education.

In China, many educators and entrepreneurs have made devotions to theoretical research and production practice of industry-institute co-operation. Many key universities and enterprises are active practitioners in I-I-C on different stages. School of software (SoS) and higher vocational education (HVE), as the two typical

Y. Wang (Ed.): Education Management, Education Theory & Education Application, AISC 109, pp. 123–128.

engineering education modes, have the responsibility to practice actively in the field of I-I-C. What have they done in the field of I-I-C? From the analysis of both sides, this article discusses these issues to inspire thoughts on how to further China's engineering education.

These two kinds of education have different guiding ideology, educational ethos, educational methods, and some other differences as listed in Table 1. Thus, the goal of the comparison of I-I-C between SoS and HVE is not determining which one is better, but providing feasible suggestions for I-I-C and better enforcement in engineering education by analyzing the situation of I-I-C implementation in China's engineering education.

Table 1. Comparison of two educational modes in engineering education field

Educational Mode	Regional	Talent Specialty	Talent Level	Emergence Time in China
SoS	China	Software	Higher	2002
HVE	Worldwide	No limited	Various	1903

However, the differences in education and learning process are not rigid and constant. Through analysis, thinking, learning from each other and establishing relationships, the uttermost goal of making the two different types of education more vibrant and active can be achieved, thus beneficial illuminations for the innovation and development of China's educational training for engineering talents can be concluded.

2   Comparison of Cultivating Objectives

SoS and HVE have significant differences on training objectives:

According to the concrete situation of China's software industry, the cultivation goals of SoSs are:

- Software developers with English proficiency, who can read and write English-written projects, and discuss problems and communicate in English [1].
- Leaders with excellent project management ability and acute commercial awareness and perspective, who can lead the team to make innovations together.
- High-quality software talents who are international-applied and available to different applications.

In China, HVE is specialized to provide batches of high-quality professional technicians for different industries and vocations, especially for the industries that are in urgent needs for skilled practical workers at present. In this way, HVE makes great contributions to the economy and social development.

3 Comparison of Educational Policies

3.1 School of Software

In August 2001, Ministry of Education of China (MOE) authorizes the establishment of 35 pilot SoSs, which is a significant attempt and action in the reform and development of higher education in the new era. Pilot SoS cultivates software talents according to the needs of software industry and the market demands[2], and discovers and develops the training system, mechanism and methods for cultivating applied talents by learning from the successful experience of software talents cultivation in foreign countries. With the spirit of reform and innovation, SoS gradually enlarge the training scale of software talents and improve the training quality with the prominent features of being practical, multiple, multi-layer and international [1].

After years of development, pilot SoS have gained satisfactory achievements and valuable experience in the aspects of practical software talents cultivation and educational system reform. In the national pilot SoS comprehensive assessment in 2006, 9 out of the 35 pilot SoSs were granted qualified, and the rest were granted basic qualified. Through evaluation, the experts from enterprises, institutes and administrative departments concludes that pilot SoSs have been taking new steps in many aspects, such as cultivation mode and the reform of higher education system, and have discovered a feasible approach in education [3].

3.2 Higher Vocational Education

In the decade from 1998 and 2007, HVE had made remarkable achievements and laid a solid foundation for the development of China-featured HVE.

HVE has provided batches of high-quality professional technicians for China, and has made great contributions to economic construction and social development of China. It also plays a fundamental and important role for China's higher education popularization. In recent years, the cultivation goal of HVE is gradually becoming clear, the characteristic of work-integrated learning gradually emerges, and the college-running mode of relying on local input primarily, part on industry participation and enterprise-institute-cooperation has been basically established, which marks the formation of the system frame of HVE and enriches higher education system structure. HVE has provided more opportunities for the whole society, especially the school-age youth, to receive higher education. It also meets the needs for high-quality professional technicians in China.

In 2006, MOE and Ministry of Finance triggered the "Construction of National Pilot Higher Vocational Colleges Project". After that, the development of HVE, especially the reform in talent structure and cultivating mode, is greatly prompted. Also, the running ideology of I-I-C, work-integrated learning and society-serving is strengthened.

4 Comparison of I-I-C Strategy

4.1 I-I-C in School of Software

Based on the development strategy of "Using IT to propel Industrialization", many SoS enrich I-I-C theory and practice to a large extent in their active participation in the cooperation with domestic and international well-known software enterprises.

Many domestic and international software enterprises have provided supports for pilot SoSs. Till 2004, SoS have established cooperative relationships and internship bases with 487 enterprises. Many international cooperators, such as Microsoft, IBM, Sun and Cisco, have donated their techniques, equipments and devices, teaching materials and faculties. One example is Microsoft Great Wall Project. Till 2004, multinational corporations have donated software and hardware equipments and devices, with the value of as much as 1.6 billion CNY, to SoS[4].

Many SoS have discovered a set of feasible approaches in I-I-C practice. For example in School of Software at Harbin Institute of Technology, entrepreneur forum, students' visit to enterprise, I-I-C summit, practical training, internship, joint project and many other forms of I-I-C practices are applied to the cultivation of applied talents[2].

4.2 I-I-C in Higher Vocational Education

The development of HVE in developed countries has suggested that, I-I-C is crucial for the development of HVE. Therefore, I-I-C is indispensable and fundamental for vocational colleges, and is the determining factor for its development.

China's HVE, which began in late 1980s, is the product of the "reform and opening up" policy. In spite of its short history, HVE has not only provided batches of applied talents for relevant industries and vocations that are in urgent needs for technicians, contributing to the local economic construction, but also got in touch and formed cooperative relationships with many enterprises in different forms, on different stages and with different features [5].

5 Evaluation on I-I-C in These Two Education Modes

SoS can be improved for a promising prospect. Although pilot SoSs have made some achievements in I-I-C and talents cultivation to some extent, the construction of SoS is still at the starting stage, and the lack of universality, thoroughness and depth are problems that need to be solved urgently. From the mid-term evaluation and the graduates' employment situations, the strategy of establishing pilot SoS is unquestionable right [3]. However, because the new education mode is still at the developing stage, its immaturity and instability determines that it cannot lead I-I-C implementation. Nevertheless, the development of SoS trend is satisfactory and promising, and it is possible that SoS can be the leader in the future. There is still a long way to go and a lot of work to be done before the goal of cultivating first-class talents with international competitive ability can be achieved, which relies on the support of administrations at all levels, the public and institutes.

Higher-vocational-education's development is limited due to lack of regulation. In China, HVE suffers from relative discrimination. This education mode bears 80% of responsibility in the field of engineering education, but only 20% attention. The development of HVE is uneven due to its irregular management and private colleges' excessive seek for profits. Some colleges violate educational aim, which discredits the reputation of HVE, and restricts its further development.

The implementation of I-I-C in some vocational colleges is successful. However, the I-I-C form of vocational colleges is still at the shallow-layer primary stage or the mid-layer starting stage. In summary there are two main reasons. First, long-term cultivation goal and I-I-C mechanism haven't been formed, and relevant policies and laws haven't been regulated. The enterprises' actions are short-term, and the colleges emphasizes on profits. The government, the enterprises and the colleges are all at the stage of "professed love for what one fears"[6]. Second, higher vocational colleges' scientific research competence is insufficient, thus high levels of both I-I-C and school-run industries cannot be achieved. At present, many colleges are forming I-I-C for their survival, development and adaptation to market economy. The patrons from the enterprises are mostly on the stages of project support, internship basis supply and personnel training[7]. This kind of cooperation is still far away from the real I-I-C goal - establishing a sustainable proper cyclic mechanism.

6 Conclusions

Pilot SoS is a burgeoning software industry oriented education mode, while the HVE takes up the mission of cultivating batches of high-quality applied technicians to meet the urgent needs of the society. I-I-C is not only the inevitable trend for the development of SoS and HVE, but also the inner needs for their survival and development, and it is also economic development requirement set for education [5]. As to implementation, both of them have defects and are not qualified to be leaders. Further discovery on reform and concrete implementation are pressing matters that needs to be dealt with.

Engineering education has gone through a bumpy journey for a century. However, China is still far behind in this field. At present, China's cultivation system for applied talents is far behind from its rapid economic development mode. I-I-C is the value orientation for both the enterprises and the institutes in the new circumstances, and "Strategic Alliances" is becoming more and more important as the rational choice for both sides. On the one hand, improving enterprises' self-directed innovative capability is the need for meeting the China' strategy of constructing an innovative country, and production-study-research cooperation is essential for enterprises' independent innovation and development. The enterprises need not only innovative talents and high-quality technicians, but also import and transformation between practical techniques and scientific achievements, and the cooperation with HVE has gradually become the enterprises' strategic requirements. On the other hand, developing engineering education is China's need for constructing a powerful IT country, and close cooperation between enterprises and institutes is institutes' development strategy. Thus, "Strategic alliance" between institutes and industries is not only strategic guidance, but also the inner demand of enterprises' and institutes' strategies.

The implementation of I-I-C directly determines the development of engineering education in China. Thus, a scientific and mature I-I-C model needs to be constructed, and its construction and implementation is bound to make great contributions to the reform and development of engineering education.

References

1. Dong, X., He, J.: Thoughts on establishing pilot school of software. Computer Education (1), 38–42 (2008)
2. Xu, X.: he approach and practice of software industry-oriented education in China. Journal of Harbin Institute of Technology (new series) 12 (suppl.), 1–3 (2006)
3. Zhang, Y.: Thoughts on establishing pilot school of software. China Higher Education (4) (2004)
4. Qi, Z.: Thoughts and experience on the assessment of pilot school of software. Computer Education (September 2006)
5. Liu, X., Yang, R.: The situation, problems and mode choices of industry-institute cooperation in higher vocational education. Vocational Education Forum (14) (2003)
6. Sun, Z.: Probe and analysis on industry-institute co-operation in vocational education in China. Higher Vocational Education (March 2009)
7. Niu, Y.: Analysis on the present situation of industry-institute co-operation in higher vocational education. Educational Theory Research 21(2) (January 2009)

The Practice of Talents Cultivation and Scientific Research Cooperation Scheme Based on Interdiscipline*

Jing-chang Pan[1], Xuemei Lin[1], Peng Wei[1], and A-li Luo[1,2]

[1] Shandong University at Weihai, Weihai China (264209)
[2] National Astronomical Observations, Chinese Academy of Sciences,
Beijing China (100012)
jingchangpan@163.com

Abstract. This paper expounds the significance and role of interdisciplinary talents training and technological innovation, and analyzes its necessity and conditions. Through many years' practice of graduate cultivation and cooperative research in computer science and astronomy, we present a feasible interdisciplinary operating scheme which results in ideal effect.

Keywords: cross-disciplinary, talents training, scientific research team.

The graduate education development policy proposed by the Program for China Academic Degrees and Graduate Education Development (2006-2020) is to *be people-oriented, to optimize structure, to improve quality, to meet demand, and to lead the future*. During the period of *Eleventh Five-Year Plan*, China established a strategy of improving self-directed innovation capability, promoting the optimization and upgrading of industrial structure, building a resource-conserving and environment-friendly society, constructing a harmonious society. Thus new requirements on academic degrees and graduate education were put forward, which is to cultivate a large number of top-notch and high-quality talents with international competitiveness, to create significant research results, to provide a powerful knowledge contribution for socialist modernization[1]. The highly integrated trend of interdisciplinary subject and cross-sectional discipline objectively require academic degrees and graduate education to cultivate interdisciplinary talents with extensive knowledge. Interdisciplinary graduate joint training was gaining more and more attentions. Through the years of practice, we explore a feasible scheme of graduate cultivation and research joint based on the interdiscipline and discipline integration, which has yielded beneficial results.

1 The Necessity and Significance of Interdiscipline

1.1 Interdiscipline

The interdiscipline refers to the new discipline emerging from intersection and infiltration of original basic subjects. Their common features are to study the object of

* This work is supported by the "Eleventh Five-Year" research project (06CB127) and the National Natural Science Foundation of China (10973021).

another subject and cross-field using concepts and methods of one or more subjects, so that the objects and methods of different disciplines can be integrated organically. The emerging interdiscipline is eliminating the traditional disciplinary boundaries, and making all the science integrated into a complete scientific knowledge system.

Interdisciplines are "intersectional" or "interdisciplinary" research activities, and their knowledge systems lead to the interdisciplinary science. Zhang Yuxian and Li lei et al. analyzed and summarized the four characteristics of cross-disciplinary subjects: (1) The speed is from the slow growth to an accelerated development; (2) the form is shifted from linear intersection to three-dimensional network intersection; (3) the theory is transformed from theory generalization to the foundation of comprehensive theory; (4) the formation is converted from natural development to organized research[2]. The history of science shows that science goes through integration, differentiation, and then integration again. Modern science is both highly differentiated and highly integrated, and so is interdisciplinary science, which achieves the science integration [3].

1.2 The Impact of Interdiscipline on Talents Training and Scientific Development

The 21st century is the era characterized by creativity, when the intersection and integration between different disciplines are becoming more and more profound and extensive. Various phenomena in the nature have always been an interrelated organic entirety, and the human society is part of the nature. Therefore scientific knowledge system derived from human's understanding of the nature is also bound to have a holistic feature. Many of the world's important scientific breakthroughs can only be solved by using multidisciplinary perspectives and a variety of knowledge, theories and methods. Just as Rosenblueth said:" Appropriate survey work in the blank areas on the science map can only be served by such a group of scientists, each of whom is the expert in his own field. But each of them has a very accurate and skilled knowledge of their adjacent fields. They are accustomed to working together, familiar with each other's thinking habits, and can understand the meaning of new ideas before colleagues' expressing their new ideas in a complete form." [4].

The tasks of talent training in higher education are to deepen the basic theory, especially paying attention to the infiltration and the exchange of inherent problems in the basic theory, to overcome the over-detailed division of professional disciplines and fragmented knowledge, to grasp intrinsic relation among various disciplines [5]. The construction of innovative talents training platform has a positive significance in the further reform of graduate training management system [6]. In recent years, this research field has become a new hot spot in scientific research and higher education [7].

The impacts of interdiscipline are becoming more and more apparent in terms of scientific progress. Heisenberg, the famous physicist, once said:" In the history of human thought, the discovery of significant results often occur in the intersection of two different thinking routes". Looking back on hundred years of the Nobel Prize, looking at the nature of Nobel prize-winning achievements, the primacy of winning rate is often interdisciplinary achievement, and in the last 25 years of 20th century, the number of cross-disciplinary award is close to 50% [8]. Liu Zhonglin et al. reviewed the new interdisciplinary progress abroad and put forward the countermeasures of development of Chinese interdisciplinary science by displaying and analyzing representative examples selected with the *point-surface-combination* method in the international conferences, publications, papers and projects perspectives[9]. Research shows that the scientific outputs brought by interdisciplinary science are very apparent in the international, interdisciplinary academic conferences, interdisciplinary academic journals, interdisciplinary scientific research and typical interdisciplinary achievements, etc.

2 Interdisciplinary Talents Training and Scientific Research Practice

2.1 Project Background

Shandong University and the National Astronomical Observatories (NAOC) have worked out construction program of space science and development plan, and built high-level talents training and scientific research center which is based on Weihai, relying on Shandong University and NAOC. An observatory was built in Majia Mountain, Weihai. The observatory successively obtained the naming rights of three asteroids. We cooperate with the National Astronomical Observatories of Chinese Academy of Sciences to establish a wide range of technical cooperation, and carry out research projects and postgraduate training, with the major national scientific project LAMOST (Large Sky Area Multi-Object Fiber Spectroscopic Telescope) as research and application background.

2.2. The Operation Mechanism of Talent Cultivation and Joint Research

An important issue drawing the attention of the current international science and education and management is how to explore the general rules of the cross-disciplinary development, to grasp the principles and methods of intersection, to build flexible management mechanism which is conducive to interdicsipline, and to give full play to innovation in interdiscipline [10]. Through many years of exploration and practice, we established a talents training and project cooperation mechanism based on interdiscipline, as shown in figure 1. The structure of this mechanism is reasonable and feasible, and has achieved good results in practice.

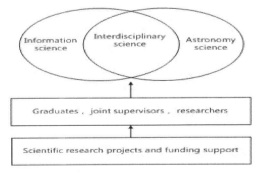

Fig. 1. Interdisciplinary talents training and research mechanism

2.2.1 The Interdisciplinary Field

The development of interdiscipline derived from practice and research [11]. And the occurrence has its inherent law of demand. The interdiscipline we are dealing with includes information science and astronomy.

Astronomy is a subject which studies cosmic objects and the formation, structure, and evolution of the universe. It plays an irreplaceable role in exploring the natural laws of the universe, promoting the development of other natural disciplines and technological progress, studying the Sun-Earth space environment, and improving the national quality-oriented education. Information science is a new emerging comprehensive discipline whose main research tools are computers and other technologies; its main target is to expand the human information capability.

The technical requirements of astronomical research, such as the measurement of precise location, brightness, fine structure of distant and dark objects, the high precision measurement of time, the large sample statistical analysis of all kinds of physical, chemical and evolutionary process and large-scale numerical calculation, require the use of information technology and methods. These research activities need both solid astronomical knowledge and advanced information analysis and processing methods; therefore there is a necessity in the integration of information science and astronomy. Based on the background and actual needs, we have established a long-term stable cooperative field with the National Astronomical Observatories, and carried out joint scientific research and graduate training.

2.2.2 The Structure of Human Resource

First, we establish a supervisor team including experts from different disciplines. Supervisor team is not only the key component to cultivating talents, but also the leading power of conducting interdisciplinary research. Supervisor team is mainly consisted of the supervisor of our information sciences and astronomy, inviting researchers of National Astronomical Observatory as our guest professors and joint supervisor, who can come to our university to hold academic seminars and give graduate useful advice on a regular basis. We cooperate in many aspects, from the training the methods to the thesis topic selection and paper defense. We explored a feasible way of joint cultivation so that graduates' computer technology can have a good application in astronomical field. At present, there are three guest adjuct professors.

Second, we formed a joint research team consisted of teachers and researchers from information and astronomy subjects. In order to carry out joint research successfully, personnel in one subject should have basic understanding of the other subject, as can provide a basic communication platform to meet the requirements of the research. The team has been in a free, democratic and harmonious atmosphere, which is helpful for generating new ideas and new methods.

At last, graduates cultivation. Interdisciplinary graduates training needs a flexible management model [12]. Interdisciplinary graduate education is a major force in promoting the interdisciplinary construction. Joint training of multidisciplinary graduates is an important way to develop interdisciplinary talents. To ensure the quality of interdisciplinary graduate training, we established the guiding mode of "one student, two supervisors". Two supervisors are from information sciences and astronomical disciplines respectively. Our approach is that in course phase, graduates study related astronomy course for the future interdisciplinary research; in the opening stage, thesis topic is determined jointly by two supervisors, and then studied by the whole team; in research stage, students can stay in the original school or go to the National Astronomical Observatory according to the research needs.

2.2.3 Project Funds Raise

Scientific and technological research in today's world has become a high investment business, and it is difficult to sustain if there is no strong economic support. So is the interdisciplinary research and talents training. To some extent, we can even say that research funding is the key element to decide whether the joint research can survive or not.

Astronomy research is the national-level strategy, and can't directly generate economic benefits. Therefore, our joint research funding mainly comes from the state fund, such as the National Natural Science Foundation of China, National Astronomical United Fund in addition to project funds entrusted from observatory. Due to the more and more solid research foundation, the project is going very well with the financial support.

3   Practice Effects of Talents Cultivation and Joint Research

To meet the requirement of graduate training and scientific research, we have conducted many years of interdisciplinary graduate training and scientific research practices, and achieved satisfactory results.

3.1   Interdisciplinary Research-Based Talents Training

In recent years, we have chosen many graduates of information sciences to participate in the astronomical information processing research. So far we have cultivated ten graduate students, all of whom have the interdisciplinary research ability, and one graduate was sent to Michigan State University (MSU) for cooperative research. In addition, there were several young teachers who obtained a doctor degree of astronomy on interdisciplinary studies. All of these laid a solid foundation for further interdisciplinary research. The results of interdisciplinary practice won the second award of graduate education and teaching achievements of Shandong University.

3.2   Establishment of Interdisciplinary Cooperation Research Team

We believe that the establishment of the research team is no longer limited to a particular unit. On the contrary, the team composition can be distributed across geographic area and disciplines. We set up a research team of astronomical data analysis and processing and carry out academic research and project development in supervisors and teachers level. In order to exert cross-disciplinary advantages, the research team is consisted of computer teachers and professionals, and astronomical experts and researchers. The research areas include processing of astronomical information, spectral analysis and automated stellar parameter measurement. By now, our team has cooperated in two research projects, both of which are the National Natural Science Foundation supported.

3.3 Development of Joint Academic Research and Project

In interdisciplinary field, the research team consisted of supervisors, teachers and graduates. We have finished two cooperative projects: *The research of automated analysis of stellar spectrum with high or low resolution* and *Automatic stellar spectral analysis*. The National Natural Science Foundation project applied by both sides, *The discovery of rare objects and variables from LAMOST spectra*, has obtained the funding support of National Natural Science Foundation (Approval Number: 10973021).We have established long-range object target tracking data exchange platform, and joined the International Virtual Observatory VOEVENT. By far, we has published more than 20 articles academic research papers, developed many practical software system for astronomical research.

The joint application of *The Research of Graduates Cultivation Mechanism Based on Interdicsipline and Discipline Integration* got approved by Institute of Graduate Education in China as "Eleventh Five-Year " project and received funding (No.: 06CB127). Through the scientific analysis and design, we built graduates cultivation mechanism based on interdicsipline and discipline integration, and eventually completed the report. This report was divided into 10 parts which had a detailed summary and analysis of all the all aspects of the project.

4 Summary

After five years of exploration and practice, we gradually formed feasible graduates' joint- training mode between computer science and astronomy. In the mean time, we improved all aspects of graduates training, such as curriculum setting, evaluation, opening, research, thesis writing, defense, etc. The practice has proved that the joint-training has many advantages. It achieved good results by stimulating the students' study interests. The interdisciplinary research team is formed through joint-training. The mode is not only limited to computer and astronomy, but also can apply to the intersection and integration of other disciplines. Therefore, this approach has a good promotional value.

References

1. Yu, J., Shi, C.: Construction of graduates training mechanism base on cross-disciplinary integration. China Education Daily 13(3) (June 2007)
2. Zhang, y., Li, l., et al.: The reviews of interdisciplinary graduates training mode of domestic famous universities. Mass Technology, 171–173 (June 2008)
3. Lu, y.: The significance of interdicsipline and interdisciplinary science. Bulletin of the Chinese Academy of Sciences 1(16), 63–65 (2005)
4. Wiener: Cybernetics Hao jiren, translator. Science Press, Beijing (1962)

5. Wu, s.: The cultivation of doctors' interdisciplinary knowledge. China Higher Education 17(8), 21–22 (2001)
6. Yu, j., et al.: Construction of graduates training mechanism base on cross-disciplinary integration. China Education Daily 13(3), 6 (2007)
7. Xiang, p.: Interdisciplinary Studies: The issues of graduates education. Journal of Guangxi University. Philosophy social science edition 02, 95–98 (2003)
8. Yuan, z.: The characteristics and enlightenment of the Nobel Prize winning. Academic Degree and Graduate Education (5) (2004)
9. Liu, z., et al.: New progress of foreign interdisciplinary science (interdisciplinary) research. Journal of Hechi University 2, 8–12 (2009)
10. Liu, z.: Interdisciplinary classification model and management meditation. Science of Sciences 6(1), 551–556 (2003)
11. Lu, j., et al.: Innovative graduate training research based on the interdiscipline. China Higher Education 1, 46–84 (2006)
12. Li, w., Li, j.: Theory of constructing interdisciplinary research center mode of innovative graduate education. Academic Degree and Graduate Education 5, 43–47 (2007)
13. Zhang, y., Li, g., Zhang, h.: The analysis of motivate factors and social conditions of interdisciplinary foundation. Academic Degree and Graduate Education 3, 48–52 (2009)

How Can One Reflect upon the People's Republic of China's Totalitarian Political Economy from a Root Perspective of Moral Philosophy?

Yu Mao and Lijuan Hou

The Australian National University, Australian
Beijing Normal University, China

Abstract. Regardless of Marshall's and Jevon's attempt to divorce economy from politics and hence morals in the 19th century, the ideals of socialism that rose at this time have similarly been steadfastly betrayed in practice and letter by the CCP. Neither is there any genuine prospect of public-choice in the sense of the Chicago School in terms of political economy in the PRC. There are merely factional and rent-collecting socialist party organ institutions that create distributional conflicts and eat up and misdirect precious resources. However, the PRC Party cadres need not concern themselves with newspapers questioning government policy as all newspapers are state-controlled, resulting in a "powerless public" oppressed by a totalitarian, socialist-come-mixed economy regime. Yet, usually, the largely rural Chinese populace remains complacent.

Keywords: political economy, economy regime, reform.

1 Introduction

When the political economy of the People's Republic of China (PRC) is reviewed, whether a perspective of political economy is taken from its original meaning of the term for analyzing production, commerce and these practices' relationships to the law, tradition and political administration (also encompassing the distribution of a county's common weal and budgetary considerations) or from the moral philosophical perspective of political economy (which developed into the 18th century analysis of the economies of states, known as *polities),* the PRC comes out sadly lacking. With both an enormous slave-labor program and army, the People Liberation Army, to administer it, the Chinese economy is firmly based upon a militaristic war economy that is transitioning to a mixed economy based on the self-same production of armaments and suppression of home and other population. From Marxist points of view and by the Chinese Communist Party's own admission, it is unlikely that a withering of the State will be seen, inaugurating any real sense of a Dictatorship of the Proletariat.

Y. Wang (Ed.): Education Management, Education Theory & Education Application, AISC 109, pp. 137–140.
springerlink.com © Springer-Verlag Berlin Heidelberg 2011

1.1 Political Economy of China Remained Stolid from a View of Production

During Maoist times, the political economy of China remained stolid from a view of production, according to most Western authors, and usually morally-bankrupt: "Mao Zedong was China's "Great Helmsman" between 1949 and 1976.Corollary to his vision for China, all major social, economic and political decisions bore his personal imprimatur until his demise. By the time of his demise in 1976, the meld had been caps. China's economy remained structurally rigid and functionally inefficient." Even though Mao's successors promised social change, they only delivered economic alterations without real political freedom: "Once Mao was out of the picture, Deng Xiaoping and his previously disgraced colleagues were promptly rehabilitated and brought back to centre stage once again. Though not as dictatorial, Deng nevertheless retained Mao's mantra of "Politics in Command." In his conversation with a delegation from the former Yugoslavia on June 12, 1987, Deng succinctly summed up his philosophy:" "I favor reform, because there is no future without reform. Reform must be comprehensive, including system reform in the economy, in the political system, and in all related domains. The overall objective of reform is to solidify our Socialist system that (concurrently) helps solidify the Party's leadership position." Therefore the CCP's aim is contrary to Marxist belief that the State should wither and be replaced by a true Dictatorship of the Proletariat, something that is not happening in the PRC.

1.2 Economic Reforms

While economic reforms took place, this was at the expense of promise social and political reforms once part of Deng's agenda for the new mixed economy: "Economic restructuring began in earnest, with reforming political system convenient ly overlooked. The economic legacies which Mao and Deng left behind painted vastly differing pictures. Where Mao failed miserably, Deng succeeded beyond expectation". Yet Deng failed to deliver on social and political reforms to his economic restructuring, buying off certain sections of an isolated but enriched entrepreneurial class in the new mixed economy of a supposedly "liberalizing" PRC. Thus certain sections of the PRC's evolving and supposedly egalitarian class system took the gold (the Chinese political and entrepreneurial classes – usually through selling unproductive but materially-sound state enterprises at rock-bottom prices to associates) while the lower classes in Beijing took the lead from People's Liberation Army guns at Tiananmen Square with the widespread shooting of students: "Deng's approach to systemic transformation in China combined tight political controls with liberalizing economic policies. Deng's three successors since 1997 have since were following closelyin his footsteps, adopting liberalizing economic policies withou t permitting democratic processes". Hence, despite a mixed economy that only certain Party cadres and their economic associates can access, little social or political reform has occurred in the post-Mao era of the PRC.

2 Other Views

Other major Occidental authors concur with Mantzopoulos, and Shen view of the sham-liberalization with the People's Republic of China, post-Mao "Although China is a historical military power with a rich tradition that goes back 5000 years, modern Chinese security perspectives are about twenty-five years old. After two and a half decades as a nation under Mao Tse-Tung's leadership, his death in 1976 brought with it initiatives for broad strategic reform. These were focused on modernization, liberalization, internal reform, opening up China to outsiders, and exposing the Chinese to the outside world. The overarching mantra has been national economic development, focused broadly on industry, agriculture, science and technology, and defense. The overall economic goal is to double the Chinese GDP by the year 2000, and to double it again by the year 2050, thus placing China squarely in the class of modern nations".

The CCP does not only use the PLA to suppress its own citizens, be they religious such as Falun Gong or otherwise. They also annexed Tibet by force during the 1950s and are trying to ethnically displace the indigenous occupants currently with Han Chinese immigrant workers as a part of their command-control come mixed economy. Meanwhile, the PRC has fought continual border skirmishes with its fellow socialist- and once-ally neighbour, the People's Republic of Vietnam, for intervening in the genocidal Khmer Rouge regime of Pol Pot of Kampuchea (Cambodia), a regime which the PRC funded and supported. This demonstrates that the CCP are morally-bankrupt in terms of political economy when one examines the sharp end of their socialist military-industrial complex. On the 29th of December, 2009, the CCP executed an obviously innocent and mentally-ill British man, Akmal Shaikh, in one of their mobile liquidation vans that tour the nation putting large numbers of people to death, similar to the supposedly-judicial mobile van killings of the disabled, deemed 'unfit for life', by the Nazis before and during World War II. Shaikh was lured to the PRC via the Internet by a Chinese criminal gang who coaxed the Briton with a ludicrous story involving a supposed and non-existent media job. The diagnosed mentally-incompetent Shaikh was induced to unknowingly carry a large amount of high-grade Chinese heroin through customs, quite openly, in a suitcase. Alert and demonstratably 'honest' CCP functionaries and custom guards (more used to fatally rifle-sniping long-range, unarmed, civilian escapee Tibetan refugees fleeing across the border) pounced (MSNBC 2006, 2009). The extensive and pervasive gulag system in the PRC runs at a profit and contributes a substantial portion through slave labour to the PRC's military-industrial complex. This is the truth of the PRC's political economy and its morals from a root perspective.

3 Overseas Ideas

Often the Chinese Communist Party (CCP) acts so capriciously and in such a dictatorial fashion, it is difficult to know when exactly they are dispensing due and honest justice. An example of this is the recent case of Stern Hu, the Australian-Chinese business official acting in lieu of the international and Australian mineral resource corporation, Rio Tinto, has been incarcerated by the Chinese Communist Party for alleged corporate espionage (Winkler. 1997, p.59) in the People's Republic of China. How much the CCP can be believed in these cases, considering their past totalitarian actions and expropriations, is left completely up to critical question and rightful, outright suspicion.

4 Conclusion

In conclusion, if the political economy of the People's Republic of China (PRC) is considered, be this a perspective of political economy taken from the original meaning of the term for investigating production, commerce and these practices' connections to the law, custom and political administration (also covering the distribution of a nation's common weal and budgetary concerns) or from the moral philosophical perspective of political economy (which grew into the 18th century investigation of the economies of states, called polities), the PRC measures up poorly. With both a huge slave-labor gulag and army, the PLA, to control it, the Chinese economy is firmly entrenched in a militaristic war economy that is morphing into a mixed economy based on the self-same production of weapons and oppression of indigenous and other populaces. From Marxist perspectives and by the CCP's own statements, it is unlikely that a withering of the State will ever be allowed to occur, inaugurating any real concept of a Dictatorship of the Proletariat.

References

1. Battilega, J.A., et al.: China Case Study from Transformation in Global Defence Markets and Industries: Implications for the Future Warfare, National Intelligence Council, Washington DC (July 16, 2001)
2. Bussière, M., Schnatz, B.: Evaluating China's Integration in World Trade with a Gravity Model Based Benchmark. Springer, Heidelberg (2009)
3. http://www.springerlink.com/content/wm65614j46126157/fulltext.html (Cited 11 times)
4. Curtain, P.A., et al.: Privileging Identity, Difference and Power: the Circuit of Culture as a Basis for Public Relations Theory. Journal of Public Relations Research 17(2) (2005)
5. Davey, G., Chen, Z., Lau, A.: Peace in a Thatched Hut–that is Happiness: Subjective Wellbeing Among Peasants in Rural China. Journal of Happiness Studies (2009), http://www.springerlink.com/content/n640860784321u21/
6. Peter, G.: The New Palgrave: A Dictionary of Economics, vol. 3, pp. 904–907, 905–906 (1987)
7. Jevons, S.: The Theory of Political Economy, 2nd edn, p. xiv (1879)
8. Mantzopoulos, V.L., Shen, R.: The Political Economy of China: Systemic Transformation: Successes and Perils. Paper Presented at the Annual Meeting of the Midwest Political Science Association 67th Annual National Conference, The Palmer House Hilton, Chicago, IL -04-02 Online <PDF>, -11-11 (November 11, 2009), http://www.allacademic.com/meta/p361997_index.html
9. Marshall, A.: Principles of Economics.MSNBC, China executes Briton said to be mentally ill, Kin: Man convicted of smuggling after 30-minute trial was lured into crime, (1890), http://www.msnbc.msn.com/id/34616375/
10. MSNBC, Chinese shooting of Tibetans sparks outcry: Survivors of attack despair for relatives; incident was filmed (September 30, 2006), http://www.msnbc.msn.com/id/15390844/

A Study on Chinese Culture Shortage in CELT and Countermeasure

Quanzheng Zhang

Henan Mechanical and Electrical Engineering College, Xinxiang, Henan,
P.R. China, 453002
zhangqz1230@126.com

Abstract. College graduates' incompetence of intercultural communication is arousing more and more interest and attention from experts and educators. Explanations for this phenomenon vary according to different people. However, the shortage of Chinese culture in college English teaching as a foreign language counts for one important reason and needs more and more studies. This essay aims to find out the truth and proposes some dependable solutions for this.

Keywords: Culture shortage, CELT, Countermeasure.

1 Introduction

In China today, more and more chances are available for college students of non-English majors to communicate with people from English-speaking countries. Moreover, chances and channels of intercultural communicating with English native speakers are available on Internet all the time without even stepping out of the door. However, it is observed that although many college graduates are good at expressing western topics, they are troubled by topics concerning Chinese ones, in particular about Chinese culture. This essay aims at finding out the problems in college English language teaching (CELT) and the countermeasure.

2 The Relationship between Language and Culture

H.G.Widdowson in his Linguistics [1] describes language as "It serves as a means of cognition and communication: it enables us to think for ourselves and to cooperate with other people in our community." Uncountable definitions about culture have been suggested by now. The definition of culture made by Bates and Plog [2] is preferred, "Culture is a system of shared beliefs, values, customs, behaviors and artifacts that the members of a society use to cope with their world and with one another, and that are transmitted from generation to generation through learning." From this, we can see the relationship between language and culture. On one hand, a language is a system in a given culture, or a symbol of a culture. On the other hand, a language is the medium of learning a culture.

Y. Wang (Ed.): Education Management, Education Theory & Education Application, AISC 109, pp. 141–145.

The relationship between culture and language is analyzed in Communication between Cultures as "language is an organized, generally agreed on, learned symbol system used to represent the experiences within a cultural community." [3]

As to the relationship between language and culture, a common belief has long been formed and agreed that they are extremely linked together and inseparable. Learning a language means learning the culture and vice versa. According to Kramsch[4], the purpose of learning a foreign language is "a way of making cultural statements" as well as "learning a new way of making communication". If we select a language in communication without being aware of its cultural literacy, we may not communicate well at least or even send the wrong message at worst. For example, if a Chinese says "where are you going?", he may just want to greet others, and it has noting to do with offence of privacy. Learning a language, in fact, is inseparable from learning its culture. [5]

3   The Present CELT in China

3.1 A Survey of CELT in China

In 2002, Zhang Weimin from Tsinghua University once made an investigation into the ability of 126 of non-English-major undergraduates to communicate in English about Chinese culture. The investigation showed that most of the undergraduates could hardly express Chinese culture in English properly. [6] Eight years later, Xiao Longfu and his fellow teachers did the same investigation in 2010, and the result was almost the same. [7] It is observed that although many college students have received formal English education for nearly ten years or even more, and most of them have even passed College English Test (CET) Band 4 or eve Band 6, they still remain deficient in the intercultural communication. As the cultivation of college students' intercultural communicative competence has been one of the most important goals of foreign language teaching in China, it is the indispensable task of the English teachers and scholars to find the main factors constraining the cultivation of college students' intercultural communicative competence.

As we all know, communications between people are the exchanges of information and culture. As the medium of communication, language and culture are interdependent on each other. Misunderstanding occurs everywhere because of the lack of necessary cultural knowledge. To some degree, the culture plays even a more important role than oral language. Hence culture teaching in college English class plays a key role.

However, in current English language teaching, most of the teachers still lay focus on the teaching of grammar or structures rather than on the cultural teaching. The first aim of English teaching for many teachers is to make their students pass various exams that students have to take. Cultural components included in language are neglected partially or totally for various reasons. Stimulating of speaking, or the cultivation of intercultural communicative competence is seldom seen in English class. When using language to communicate in practice, college students tend to have cultural shock or breakdown, or behave inappropriately that make their interlocutors embarrassed or cause misunderstanding.

3.2   A Survey of the College English Books Generally Used at Present

As we all know, the teaching syllabus and curriculum are determined before text books are compiled. And textbooks are best seen as a source in achieving aims and objectives that have already been set in terms of students' needs. Hence, selecting a suitable textbook to meet the demand of syllabus and curriculum has become crucial important. As to the role of textbooks, Cunningsworth[8] points out that "However, it has to be recognized that teaching materials can exert considerable influence over what teachers teach and how they do it. Consequently, it is of crucial importance that careful selection is made and that the materials selected closely reflect the aims, methods, and values of the teaching program."

However, according to Chen Youlin[9], one of the problems about foreign language teaching is the overstress of the target language culture and the little training of its expressing competence.

As a result, to a great degree, the neglect of native culture learning in the perception of English class leads to the college students' incompetence to express their original culture. A survey of the popular college English textbooks can show us something about this:

Table 1. Proportion of Chinese culture in college English books

Book Names	Volumes in all	Articles contained	Chinese culture topics	Partially Chinese culture topics	Proportion of Chinese culture in all
College English (revised edition)	4	40	0	0	0%
New Horizon College English	4	40	0	0	0%
Upstream	4	40	0	0	0%
New Essential College English	4	48	7	5	25%
New Century English	4	40	0	6	15%
Practical English(3rd edition)	3	32	1	3	12.5%
New Horizon English Course	4	40	0	2	5%

As the basic guidance of English teaching and the main source of the cultivation of college students' intercultural communication competence, college English textbooks should be revised to teach more original culture in the angle of foreign language or English language. Hence, college students' bicultural awareness and competence will be strengthened so as to enable them to communicate effectively and appropriately in a variety of intercultural communicative contexts. Thus the content of Chinese culture should be an indispensable part of college English teaching.

4 Causes for this Phenomenon

4.1 The Lack of Chinese Culture in College English Can Go Back to the Education Philosophy

To many people, the purpose of learning foreign languages or English is regarded as a means to learn from the west which has been taken as our model for more than one hundred years. Undoubtedly, the west is truly more advanced than China in many fields, such as in science, technology, education, etc,. So, the learning goal is determined from the very beginning: to learn things concerned the west, not China. Intercultural communication is not the learning goal at all.

The policy made by the government, the relative curriculum made by the ministry of education, and the relative s composed accordingly all follow the same education philosophy or guiding line. Western culture is preferred than Chinese culture in the selection of teaching materials. In order to learn AUTHENTIC foreign language or English, original teaching materials are chosen. All the details are made strictly according to western origin. Any offence to the western culture is regarded as incorrect.

4.2 Teachers Themselves Can Hardly Be Aware of This

Most of the foreign language teachers or English teachers themselves take teaching foreign languages or English and the relative culture as natural. Few of them have the sense of intercultural communication as most of them have few chances to communicate with foreigners. Few of them have few chances to introduce Chinese culture to foreigners and they have few chances to test their ability of expressing Chinese culture, though they must have been confident of that.

4.3 Students Have the Same Confidence of Their Ability as Their Teachers

They are so familiar with their culture and they have never doubted about their love for it, nor their ability to introduce it to their foreign friends. In fact, their confidence makes them lost.

5 Countermeasure

5.1 The Principle of Mutual Communication or Intercultural Communication Must Be Established

Intercultural needs the mutual culture input, not one-way input. Otherwise, intercultural communication can not be carried out.

5.2   The Curriculum needs Altering to Involve More Chinese Culture Elements

It should contain more Chinese teaching materials.

5.3   Teachers and Students Should Pay More Attention to Chinese Culture

They should be proud of their culture and appreciate it form their heart. Only in this way, will they be willing to study it and introduce it to foreigners.

6   Summary

College English teaching has been developed in China for tens of years. Barriers to college graduates' incompetence of intercultural communication are different to different students due to their different backgrounds, motivation, interest, talent, basis and purpose. In teaching, teachers should take different measures to suit their teaching aim and teaching contents. However, the shortage of Chinese culture elements needs to be paid more attention by teachers and experts. Only in this way, the mutual input of Chinese and western culture can be made and a successful intercultural communication can be got.

References

1. Widdowson, H.G.: Linguistics. Shanghai Foreign Language Educational Press, Shanghai (2000)
2. Bates, D.G., Plog, F.: Cultural Anthropology. Mc Graw-Hill, USA (1990)
3. Samovar, L.A., Porter, R.E.: Communication between Cultures. Foreign Language Teaching and Research Press, Beijing (2000)
4. Kramsch, C.: Context and Culture in Language Teaching. Oxford University Press, U.K (1993)
5. Deng, Y.C., Liu, R.Q.: Language and Culture– A Comparison of Language and Culture between Chinese and English. Foreign Language Teaching and Research Press, Beijing (1989)
6. Zhang, W.M., Zhu, H.M.: Chinese Culture in College English Teaching. Tsinghua University Education Research S1, 34–40 (2002)
7. Xiao, L.F., Xiao, D., Li, L., Song, Y.W.: Chinese Cultural Aphasia in College English Education. Foreign Language Theory and Practice (1), 39–46 (2010)
8. Cunningsworth, A.: Choosing Our Coursebook. Shanghai Foreign Language Education Press, Shanghai (2002)
9. Chen, Y.L.: Some Misunderstandings in College English Teaching. Sino-US English Teaching 2, 147 (2003)

To Discuss the Guiding Role of Examination Combining "Software Engineering" Course

Yu Gao and Zhen-bo Bi

Department of Computer,Zhejiang Ocean University,
Zhoushan, Zhejiang, P.R.China 316000
{Gaoyu,bzb136}@zjou.edu.cn

Abstract. For universities, the examinations have basically two roles, one is inspection role, and another is guiding role. The two kinds of role are all necessary. For a long time, inspection role of examination is focused on, and the attention of guiding role of the examinations is not enough. To consider training objective and teaching quality, people more should pay attention to guiding role of examination. To combine "software engineering" course, guiding role of examination is studied. Guiding role of examination form and examination content is discussed. The characteristics of "software engineering" were analyzed. On this basis, in order to play a guiding role of examination, several issues that teachers should pay attention were discussed.

Keywords: software engineering, examination, guiding role.

I Introduction

In the universities, examination is an important part of teaching. In each semester, we have scheduled a large number of examinations of various types. Many experts and scholars researched examination [1],[2],[3]. Through research, they proposed many proposal of examination reform [4],[5].

For universities, the examinations have basically two roles, one is inspection role, and another is guiding role.

For the inspection role, we are very familiar. Pass the examinations, teachers assess each student's grade, teachers can examine the learning situation of students. Pass the examinations, the school can examine the teaching situation of teachers, and this is another inspection role.

Guiding role is that: for students who after take an examination, now the format and content of the examination will affect their behavior. Students can refer to the current format and content of the examination, to prepare for the next examination. As the saying goes: How to test for the teachers, how to learn for the students. What contents are examined then students should study those contents.

The goal of teaching is training talents. Every teaching activity should focus on this goal. Examination is a part of teaching. The examination should serve this goal. Pass inspection role of examination, we can learn the degree of students and current teaching situation in the process of training talents. But, the examination is not only

Y. Wang (Ed.): Education Management, Education Theory & Education Application, AISC 109, pp. 147–153.

testing student's learning situation and teacher's teaching situation, and the examination should also guide students realizing the training objective.

Inspection role and guiding role of examination supplement each other. In order to achieve the goal of training talents, the two kinds of role all are necessary.

For a long time, inspection role of examination is focused on, and the attention of guiding role of examinations is not enough. Because we need to test the situation of student's learning and teacher's teaching, so we should pay attention to the inspection role of examination. However, in order to achieve training objective and improve teaching quality, we should pay attention to the guiding role of examination.

On the guiding role of examinations, people made some studies [6],[7],[8],[9],[10], but the depth of studies is not enough. Moreover, with a specific course, to study the guiding role of the examinations is very small.

With "software engineering" curricular examination, we studied the guiding role of examinations. In the following, we describe the study contents.

2   Guiding Role of Examination

2.1   Important Meaning of Emphasizing the Guiding Role of Examination

After the examination results come out, if the examination result is very good (the results meet the requirements of training objectives), we'll be happy, this result also shows that the usual teaching is very good. If examination results are not ideal, although this phenomenon can explain that there are problems in the usual teaching work, and can explain that there are a certain gap from the goal of training talents, but it was too late. At this time it was too late to do the work. Complaining is useless. The reason is that: for all students it is impossible learning and testing once again.

The guiding role is that starting point is placed before the examination. In normal times, let the students know information of about exams, these exams information reflect the requirements of training objectives, it can guide the students to learn, to take the initiative in accordance with the requirements of training objectives.

Important meaning of emphasizing the guiding role of examination is taking precautions. In question did not arise, work to be done. If before the examination, the students know the form and content of examination, and the form and content accord with the requirement of training objective, then, in the study students can be based on this information, so that the exam results of course is not bad. Such according to training objective, we actively guide the student, the requirement of training objective can achieve commendably.

2.2   Guiding Role of Examination Form

Examination form has important guiding role to student learning.

Written form is the most common form of examination. Written form is divided into two forms, they are form of open examination paper and shut examination paper. Form of open examination paper allows students to bring reference material. To form of shut examination paper, to bring reference material not allow. Main function of this exam's form is to test student's situation that they understand the theoretical knowledge.

Writing the small paper also is a common examination form. On the basis of content of the curriculum, teachers determine the scope of the paper, and put forward some demands. The students complete a small paper in accordance with the requirements of provisions. This examination primarily tests the student's situation that they understand and apply the knowledge.

Hands-on experiment is also a common examination form. According to the experimental outline of the course, teachers determine the content of the experiment, and put forward some demands. The students complete experimental project in accordance with the requirements of the experiment content. This examination form primarily tests the practical ability of students.

Questions and answers form is also a common examination form. This examination form is more flexible. First, the students explain according to examination topics. Then teachers put forward some questions around the topics, the students answered. This examination form primarily tests the student's situation that they understand and grasp the knowledge.

In the above, some common examination forms are listed. The guiding role of examination form reflected in that: "What examination form is used by teachers, students will pay attention to this examination form. So this aspect ability virtually was strengthened. For no used other examination form, students are seldom consider related content. In this way, students virtually weaken that aspect ability."

For example, for most course, teachers will adopt written form, and won't adopt other examination form. Due to written form primarily check student's situation that they understand and grasp the theoretical knowledge. Therefore, according to this information, in learning of peacetime and review of end-of-term, students will focus on backing formula, remembering definition, understanding and grasping the theoretical knowledge above. For other aspects content outside written form, the student is not efforts to study.

For a certain course, when determining the examination form, the teacher should be cautious. Teachers should carefully study training objective, and on the basis of the content of training objective, according to the requirement for student's abilities, using one or more appropriate examination form. At the beginning, teacher should let students know the examination form, to guide students achieving training objective.

2.3 Guiding Role of Examination Content

Examination content has important guiding role to student learning.

Students like asking the teacher or senior students. Students want to know that: What do examination contents include and what have type of examination questions in the past course examination? Afterwards, in the usual study and a final review, students focus on relevant content of the examination. In this way, students are guided virtually, and the study of certain aspects was strengthened.

Usually, the examination contents include that:

(1)The theory knowledge on books;

(2)The practical problems that can be solve applying theoretical knowledge (including hands-on practical issues);

(3)The problems of no standard answer, the problems need to play a creative thinking.

In preparing examination paper for most teachers, the examination content includes the above (1) and (2), the ratio of (1) part is relatively large. Some examination content does not contain (2) part. Objectively speaking, such examination content will guide the students to ignore the application of theoretical knowledge and increase of practical skills. This may cause that ability solving practical problems is low for students.

Examination content include the situation of above (3) is relatively small. The reason is that: Seeking examination subject of type (3) is compare difficult. When to grade papers, for the type (3), it is difficult to determine rating criteria, and responses from the students can not be too good. In the examination content, because the examination questions of type (3) does not appear or is very few. So, students are guided, just focus on absorption and acceptance for existing theory knowledge, and students do not want to explore new problems, lack of innovation. In this way, training innovation thinking is not conducive.

When designing examination content, the teacher must consider the guiding role of the examination content. Examination content both can guide students to learn the basic theoretical knowledge of the course, but also can guide students to improve the ability solving practical problems (including hands-on practice), but also can guide students developing innovation thinking. Examination content should be able to encourage students to think independently, and should be able to put forward their own independent views. The examination content of a course must include the above content of the (1) and (2), the proportion of content of the (2) can be different for each course, but content of the (2) must exist. Apart from some basic course, the examination content of the other course (especially those related to advanced knowledge in the professional field) should design some question of type (3), its ratio can be small, such as no more than 15%.

3   The Characteristic of "Software Engineering" Course

"Software engineering" is a major course of computer specialty. Other specialty of information type also teaches this course. The course content is to introduce theoretical knowledge of software development and maintenance, in accordance with the engineering's concepts, principles, techniques and methods, to contrapose the issues of software development and maintenance.

Training objective of "software engineering" is to let students to master theoretical knowledge of software development and maintenance, and can guide their practice activities, and to have initial ability of developing and maintaining software. In order to achieve training objectives, we require students both to learn the theoretical knowledge, but also learn to apply theoretical knowledge in practice. Because the "software engineering" is a relatively new discipline, knowledge renewal and development is relatively fast. Therefore, teachers should train students with ability of learning new knowledge, discovering new knowledge, solving new problems and innovation.

To analyse the content of "software engineering" course and the training objective, we found that "software engineering" course have the following characteristics.

(1) Practicality is quite strong. Students not only should pay more attention to understand knowledge and to master knowledge, but also should pay attention to the application of theoretical knowledge through practice.

(2) Knowledge update is quicker, and innovation is stronger. Its theoretical knowledge is in constantly updated and developing. Only learning the existing book knowledge is not enough, the students must keep up with the pace of knowledge updating, learning new knowledge. Teachers should cultivate the ability that students find the new knowledge, and the ability to solve new problems.

In the design of "software engineering" examination scheme, we must consider characteristics of "software engineering" course. In the examination form and examination content, these characteristics should be reflected, so that to guide students, to improve ability of application and practice, to improve ability of acquiring new knowledge, training innovation ability.

4   How to Play Guiding Role of Examination of "Software Engineering" Course

In the front, we studied the guiding role of examination, and studied the characteristics of "software engineering" course. On this basis, we shall study how to play guiding role of examination of "software engineering" course.

On the basis of the previous research, and teaching experience of "software engineering" course, in order to play guiding role of examination of "software engineering" course, we think should pay attention to the following questions:

First, teachers should change their philosophy of education, and change the traditional understanding to examination. Teachers should know that: teaching purpose of "software engineering" course is to require students grasping theoretical knowledge of software development and maintenance, and possessing the initial ability of development and maintenance software. Examination is only a means of teaching, and it is not teaching purposes. Examination is not just a performance evaluation to the students. Examination as a part of the teaching process, it must serve to the purpose of teaching.

Second, the examination form should be diversified. The form of written examination is necessary, so that to guide students learning and mastering basic theory knowledge of "software engineering". After the end of the course, we organized written examination. Examination form of hands-on experiment is necessary. This form reflected characteristic of "software engineering ". Our approach is to let students designing smaller software. In the beginning of the semester, the tasks are arranged down. Students are free to choose topics. Each student independently designs. To the end of the semester, students complete the design task, small software to the teacher, the teacher evaluate student's grades. Questions and answers form is also necessary. Our approach is to use the interlocution form, combining software that students design. First, students introduce case of design and program. Then, teacher asks some question, students answer questions.

Third, the examination content should be diversified. Theory knowledge of "software engineering" must be included in examination content. In the form of written examination, the main contents of examination paper belong to the basic theoretical knowledge. It must appear in the examination content to apply theory knowledge solving practical problems, so that guiding students to combine theory and practice. We are mainly to let students doing the course design, to reflect this examination content. Examination paper also contains a part of such examination questions. On the basis of characteristic of "software engineering", its knowledge innovativeness is relatively doughty. Examination subject in addition to knowledge of needing memory, examination content also contains the question of no standard answer, so that to guide students training positive thinking, training innovation awareness, to encourage student's independent views.

Fourth, when beginning the teaching activities of "software engineering" course, teachers should inform students for examination's form and content (type and range of examination). The form and content of examination is in accordance with training objectives, teachers ahead of time telling students information of the examination, so that virtually played a guiding role, to guide students in peacetime in accordance with the requirements of training objectives to study hard.

Fifth, the evaluation criteria of academic record of "software engineering" should be changed. It is not correct that an examination determines results. Academic record of student should be constituted by various factors. In general, academic record of student should include such factors: situation of the usual learning (such as assignments, lecture to answer questions), situation of answer papers, situation of doing experiment project, situation of expression in questions and answers.

5 Conclusion

In the above, we discussed the guiding role of examination. Combining "software engineering" course, we studied the problem to play guiding role of the examination. The students are the subject of learning activities. We value the guiding role of the examination, the aim is to mobilize the enthusiasm of students, so that to guide the students to achieve the learning objectives. How to play the guiding role of the examination? To teachers this is a topic worthy of study. Around the topic, clearly there are many issues that need further study.

References

1. Deng, Y.-g.: Research on Problem and Countermeasures in College Examination. Higher Agricultural Education 5, 34–37 (2008)
2. Yan, W.Q., Qing, C.: On the Function of the Examination in the Talent Fostering of Higher Education. Journal of Jixi University 7(4), 16–18 (2007)
3. Huang, Y.-h.: Overview of Reserch on Testing – Mode in University. Theory and Practice of Education 26(10), 9–10 (2006)
4. Zhou, B.: How to Develop the Function of Tests Effectively. Journal of Hunan Radio and Television University 3, 41–43 (2004)
5. Gao, Y.-y., Zhang, F.: Overview of Reserch on Reform of Examination Method. Journal of Technology College Education 22(6), 100–101 (2003)

6. Liu, X.-e.: To Discuss the Guiding Role of Examination in Quality-Oriented Education. Journal of Qingdao Unversity of Science and Technology 50(2), 51–52 (1997)
7. Pan, X.-y.: Play the Guiding role of Examinatorial to Teaching. China Higher Education Research 12, 62–63 (2004)
8. Chen, M.-w., Yuan, J.-y.: Where Examination Baton Point to — How to Play the Guiding Role of College Examination. Information of Scientific and Technical (Academic Research) 34, 445–446 (2008)
9. Wang, M.-l.: To Play the Guiding Role of Examination to Ensure the Quality of Personnel Training. Journal of Mudanjiang Normal University 4, 98–99 (2008)
10. He, H.-j., et al.: Improving Extensive Capabilities of Students through Design of Examination. Computer Education 9(20), 1–4 (2010)

Preliminary Study of Dialectical Relationship between the Credit System and Tutorial System in Our Higher Education

Wang Jian and Sun Lin

Department of Aetiology & Immunology, Medical College,
Anhui University of Science & Technology, Huainan 232001, Anhui, China

Abstract. Credit System, a new type and student-centered education system, can fully mobilize the enthusiasm and initiative of students in our higher education. Tutorial system, an also novel type education system, can play a leading full role on decisive guidance to centered students. In our country, credit management in higher education is still at trial stage and mentoring is only applied in part of colleges and universities with better conditions so that without successful and mature experience can be copied. It will be face to new multiple challenges how to carry out the tutorial system under the condition of credit system of higher education. The dialectical relationship between the credit system and tutorial system in higher education will be explored on basis of higher education reform in our country in this paper.

Keywords: High education, Credit system, Tutorial system.

Introduction

Credit system is widely praised and applied in foreign higher education, such as engineering, science, medicine and other humanities. Although higher education credit system was implemented many years ago, but it is mainly in some key universities, many colleges still have not been widely used. Because the differences in background of higher education between in Western and China, many implementation details of the credit system need to be further developed and refined.

Tutorial system is come from abroad and based on the experience of university education in recent years, many creative measures of student education and management can be explored with the development of higher education. However, more and more confusions have been gradually revealed, such as the positioning of tutorial system is not clear, very low ratio of teacher-student, assessment indicators are difficult to quantify[1].

1 Comparison of Higher Education between Domestic and Foreign

The educational system in China has been emphasized the state and national interests first for thousands years by the influence of Confucianism, and also has been

Y. Wang (Ed.): Education Management, Education Theory & Education Application, AISC 109, pp. 155–162.
springerlink.com   © Springer-Verlag Berlin Heidelberg 2011

subjected to the rules and regulations the school authority, rules and regulations. However, number of educational philosophy, such as individuality and independence, competence and creativity, can be emphasized in European, American and other Western education. In addition, a type of single education system, only taken by the Government, a variety of education plans could be controlled and interfered by planned economic system in China so that the multiple activities and efficiency could poor played. Interestingly, Western education model is gradually established, development and improvement in the market economy, and adapted to the needs of market economy. Therefore, the different ideas of personnel training, system, method are based on different educational and cultural background so that its effects would be difference[2].

1.1 Differences in Assessment Objectives

It is the major assessment of student goal in Chinese education that the students have learned the memory, understanding and skills. The examination range of most colleges and universities is the content of classroom and the focus of assessment is focused on the knowledge that has been learned to master and understand. Under normal circumstances, the type of "wishful thinking" can't be often praised and appreciated sincerely so that the enthusiasm of students must be depressed severely. Most of students one-sided have been focus on theory while ignoring the practical ability so that learning becomes boring.

Interestingly, the major assessment of student goal in European and American education is base on understanding, innovation and so on. For example, two kinds of assessment in American from high school to college have been taken via two ways, student assessment examinations and university entrance examinations, and each test will be hold every year. The test subjects typically include biology, chemistry, physics, etc., in addition to mathematics, all kinds of standard answers to question are not generally provided, so that even if more than familiar with the back, not necessarily score high. It is important that the assessment focuses in most of American universities have been on student problem-solving skills, logic, common sense and innovation, rather than memory and descriptive ability.

1.2 Differences in Training System

More education in China is particularly emphasized on the specificity of the Ideological problem, and the curriculum in our high educational system is extraordinarily careful so that few changes of curriculum and professional settings have been made for a long period of time. Fortunately, with the deepening reforms of higher education in our country, the curriculum and their professional settings have been transformed and updated in recent years so that more and more elective courses and new courses are established. However, the vast majority of our required courses and fewer elective courses have been set up in our colleges, the students can not easily select the desired professional according to their own interests, hobbies. In China, conversion professional is not an easy in universities and colleges, and conversion schools more difficult.

The high education system in European and American, curriculum and professional are closely associated with the social economy so that most of professional courses is to serve the economy. Moreover, in order to adapt to social development and progress, many American universities also offer other professional courses for student choice. Because of the required subject little, elective project more, almost students in American universities can make multiple select or a variety of choice according to their interests and hobbies, such as free choice of professional, conversion professional, conversion school and other selection. Diversity and flexibility, a type of successful model of American higher education system, is unique and effective higher education in the world so that many universities and colleges around world all want to follow and copy the education system[3].

1.3 Differences in Educational Models

For the purpose of quickly and maximize the absorption of knowledge, many teachers in Chinese universities still adopt old teaching method, such as mouthpiece or spoon-fed, a lot of textbook knowledge will quickly be taught to students in a short time, and then a lot of intensive training will be taken so as to deal with all possible examinations. Although this model that students can learn a lot and may think little, the ability of analyze and solve problems has been insufficient. Most examinations of China can directly reflect classroom content and also are regarded as the unique criteria on evaluation the ability of students. As follow the teacher, as the test very well.

Whimsical, a type of novel education model, often can not be sincerely appreciated so that their learn enthusiasm are serious blow, and their focus have been on theory and neglected practice. China's education, one of encouragement style, also a kind of high standards and strict requirements, always emphasize on a sense of crisis, and encourage every student to succeed in the first.

However, the major western education methods focus on how the brains of students are full of various questions, the courage and ability to ask questions will be consciously cultured. Teachers generally do not require students memorize a lot of formulas and theorems, but do everything possible to guide the students how to think about issues and how to find the answer to face the problem. The key of teach is to enable students to grasp the way of thinking, and to develop their creative idea. Western education style, let bygones be bygones encouragement style, is stressed from scratch, and create a free and easy learning environment and atmosphere, and encourage students to further study in their own areas of interest.

2 Application of Credit System in Higher Education

As we all know, the credit system was set up or originated in the second half of the 19th century. With the rapid development of monopoly capitalism in Europe and America, trades and technical personnel in all walks of life need to be trained as soon as possible so as to adapt to the needs of market competition. Charles Eliot, a famous principal of Harvard University, first made and established in the elective system at Harvard University during 1872, and was officially renamed as credit system after a few years later. Credit system later in our country, a few key universities, for example Nanjing University, Wuhan University, were allowed to explore a small number of credit system under the guidance of Ministry of Education in 1978. Since 1983, the application of credit system has been extended from some major universities to non-key university. In May 1985, China promulgated the "Communist Party of China Central Committee on Education Reform's decision" clearly stated that to reduce the required courses, increase elective courses, credit system and dual degree system. Nineties of the 20th century, more and more domestic colleges and universities actively have learn from the advanced teaching models of foreign universities and adopt the mature teaching experience of domestic universities on basis of the personnel training of university new demands adapted to the market economic system and the rapid development of science and technology[4].

2.1 Credit Environment in the United States, the Students Played Down the Concept of the Class System

The most of the universities in the United States are divided by grade credits, such as 24 credits or less a year, 25 to 55 credits for the second year, 56 to 89 credits for the third grade, fourth grade for more than 90 credits. Once a person complete a 120 to 180 credits, you can graduate[5]. The students may adopt different learning styles depending on the different wishes, for example, full-time mode or part of the system. Even if a student dropped out, the previous credit would be still valid after return to school. This is known as the American universities grade point system commonly used to measure the level of quality of learning.

2.2 Focus on Basic Education, Encourage Innovation and Talent

Under the guidance of advocacy of general higher education in the United States, the students usually do not have fixed the professional in first grade and second grade so as to give priority to learning the basic knowledge. After the first period of study and reflection, the third grade students in universities have a broader range of knowledge, and make right professional choices according to the future career plans.

2.3 Strong Faculty, a Strict Evaluation System

A larger proportion of the teachers are established in the most of colleges and universities in the United States, relatively strong teaching staff. The current data have been shown that the colleges and universities in American generally take the contract appointment to the teachers so that both parties can terminate the contract at any time. The professional post evaluation of American teachers are mainly based on their qualifications assessment, teaching, research and other aspects of the specific situation, and then nominated by the president approved the appointment of principals reported. Every teacher are examined and assessed by scoring combination of students and teaching leaders during each semester or each year, the terminal evaluation results can be used as the basis for one of employment in next school year.

2.4 Professional System Relatively Free Transfer

It is easier to freedom professional conversion in department of American Universities. Harvard University is the first comprehensive private universities in the United States, turn professional at Harvard University as routine, if the professional is not fit for you, you can go to another profession, as long as you fill out an application form, and are agreed and signed by two (into and out) heads of the department. Bill Gates, the famous chairman of the software company Microsoft and one of the world's wealthiest men, was a Harvard University dropout that year. The freedom of American students is the extent permitted by law, with this premise, students will receive free movement of power. For students, the choice of freedom bigger, the emotional stress smaller. However, some shortcomings are also exposed, such as the elective share is too large, the integrity and systematic of teaching programs are difficult to guarantee.

3 Credit Problems in the Application of Credit System in Higher Education

Credit system, a new type of educational management system, is a credit-based measurement, the elective system as the core, and to obtain the minimum credits to measure academic status of students. It is regarded as a student-centered, flexible, flexible study system, and based on complement of the minimum credits prevail but not be limited. It can not only mobilize the enthusiasm of teaching, promote educational reformation, improve teaching outcomes, also help to up-regulate the enthusiasm, initiative and independence of students, and also enhance student's research and innovation. In view of these advantages, credit system has been widely applied in the vast majority at home and abroad teaching universities. However, without uniform standard credit model, credit system in domestic and international higher education is different, and some problems have been found in the practical application process.

3.1 Students Do Not Attach Importance to the Elective Course and Blind Courses

As long as the school allows students to any courses selection can be reached rated scores, so there are some students blinding in course selection. Because the courses, professional, teachers do not know enough, there is a big choice of elective blindness and utilitarian. Most of students are interested in choice the easy to pass the exam courses, and get credit as main purposes. Most students usually have inadequate attention to the elective course so that some abnormal phenomena in the spread in different universities, such as elective course must be escaped, obligatory course may be selectively escaped. Even if students do not skip class to class, they are all no concentrated in class. It is contrary to the credit of the mind.

3.2 Number of Teachers and Teaching Resources Can Not Meet the Need

In order to meet the credit system characteristics, the rate of basic courses and elective courses must be relatively expanded via the pathway of reducing the required course hours. Since the increase in the total number of courses has led to inadequate number of teachers and teaching resources, most of the elective courses have to be limited the number of students and A small number of interest elective courses can not be afforded. This also led to selection of a course of students is not interested in this course, and interested students are not on the selection of a course strange phenomenon.

3.3 Lack of Flexible Educational System

After adopt a flexible educational system, the excellent students are allowed to graduate early, the less able students may be less course selection, within a specified time or extend the school system graduate. Although the flexible education system more scientific in theory, but it is also difficult to practice in China, a few particularly outstanding students who have already obtained the required credits but still can not get diploma. Some outstanding students do not attract enough attention to many universities so that the aspirations of early graduation have to finally give up. The management focus of universities is required courses and traditional classroom teaching is still the main mode of giving lessons. Almost all students are still studying in the same classroom so that the merits of credit did not play out[6].

3.4 Application of Tutorial System in High Education

Instructors, a type of excellent teachers, can guide a group students how to learn, study and scientific research. Tutorial system is established between an instructor and a group of students, also regarded as "master and apprentice" relationship. Tutorial system is early originated in the fourteenth-century Oxford University and still regarded as elite education training model, and has cultivated numerous innovative and distinguished talents, including the current Prime Minister Cameron and other more than twenty prime ministers, the winner of the Nobel Prize for Literature VS Naipaul (Vidiadhar Surajprasad Naipaul). To the seventeenth century, tutorial system in graduate education had been widely used in Oxford, Cambridge and other universities. Harvard University undergraduate students began the implementation of elective system in 1869, and then started the implementation of undergraduate credit system in 1872 on basis of the elective system being constantly improved. Until the late nineteenth century to the early twentieth century, Oxford, Cambridge and other famous universities widely began to follow the successful example of American universities in the course of the credit system.

Either credit system or tutorial system, both of them was late start in our country. In 1918, the first marked credit system, "Selection System", was presented by the professor Cai Yuanpei and was carried out in the Peking University. In 1937, the British scholar Michael Lindsay employed as a teacher of Beijing University of Economics University established the Oxford-style tutorial system, meanwhile, professor Zhu Kezhen firstly started to set up the tutorial system in Zhejiang University. After the establishment of new China, the University credit system and the tutorial system were all replaced by academic year system, and the tutorial system was only applied to graduate education, and the class teacher or counselor system were major utilized to undergraduate education. Since 1978, some domestic credit conditions in the university began to apply the tutorial system, after the 20th century nineties years, with the establish of market economy system and the rapid development of science and technology, higher education enrollment continues to expand, domestic colleges and universities have conducted a comprehensive credit system management[7].

4 Correct Handing the Relationship between the Higher Education Credit System and the Tutorial System

Credit system is also regarded as a student-centered, flexible, flexible learning system, allowing students to choose a certain extent, the need to learn and courses of interest and professional, full of personality of students, so that the enthusiasm and initiative and independence of students is mobilized and has better effect of comprehensive and healthy development for students. Credit system pay more attention to the assessment process such as the usual examination and disciplinary assessment, rather than a separate written assessment results as the sole criterion, a new type relaxed and enjoyable learning environment is gradually established. In addition, the credit system to implement a number of teachers for each course choices, students can choose the suitable teachers according their own characteristics and teaching methods of teachers. This will not only give full play to the initiative of students, but also to mobilize the enthusiasm of teachers, and ultimately target is set up a good way to improve teaching quality.

Because the majority of our universities follow the traditional teaching methods in long-term, if without the necessary guidance, most of students lack a full understanding of the credit system, they do not comprehend the structure and professional characteristics of university courses so that some abnormal phenomena may occur during the course selection, such as blind pursuit of credit, one-sided pursuit of credits, some courses with low attendance. In order to shorten the school system and early graduation, a few students unilaterally pursuit the credits via the variety of ways, emphasis on theoretical and practical light, heavy professional learning, light overall quality of training. It is concern that some students actively take care of their credits through ignoring the quality of learning so as to graduate on schedule and the training quality will be difficult to guarantee. Therefore, credit system still has not been fully mature and also has been certain blindness or random in the initial stage of the application for most of universities and especially need targeted guidance of experienced teachers.

Supervisor is the best candidate to guide students successfully complete their education. After freshness, a number of university students are not well adapted for new life and learning in university, and also can not be independent design learning programs for their own development. The core of credit system is a free elective, however, the students often do not understand the course structure integrity and their internal relations, and are easy to pursuit academic credits unilaterally. Therefore, the new students urgently need the help of experienced instructors in terms of ideology, life and academic guidance. According to student characteristics and aspirations, instructors will guide students to develop good personal learning plan, to shorten the school to university transition, to avoid the mixed credits, and to guarantee the course quality and reasonable knowledge structure, and to promote the improvement of quality of personnel training. Thus, tutorial system is an important guarantee for the implementation of credit system, and can fully mobilize the enthusiasm and initiative, and guide students to study hard and make good progress for students.

In summary, credit system is a kind of flexible and resilient learning system, and also very fit for the students to give full play to the initiative of students. Tutorial system is a student-centered self-learning model under the guidance of the teacher, and can give full play to the teachers in the key roles of the education process. Tutorial system is not only an important guarantee for the credit system, but also improves the credit system needs.

References

1. Wang, L.P., Yang, B.H., Yuan, Y.: The problems, causes and solutions in the implementation of tutorial system. Journal of Yanshan University (Philosophy and Social Science Edition) 9, 120–122 (2008)
2. Marginson, S.: Dynamics of national and global competition in higher education. Higher Education 52, 1–39 (2006)
3. Wu, J.: A Comparative Study on Chinese and American High Education. Education and Teaching Research 24, 51–55 (2010)
4. Huo, Y.: Preliminary analysis of college credit at home and abroad. Journal of Changsha University 21, 142–144 (2007)
5. Liu, L., Zhao, D.L.: Comparison higher education between in American and China. Journal of Sichuan Normal University(Philosophy and Social Sciences) 31, 49–53 (2004)
6. Li, J.: Creating conditions for optimizing the credit system. Education and Modernization 1, 35–39 (2003)
7. Yang, Q.J.: Study on the model of the tutorial system under the condition of credit system in colleges and universities. Journal of Heilongjiang College of Education 28, 40–41 (2009)

On the Phenomenon of China English in Chinese-English Interpretation

Xiaoya Qin

English Department, North China University of Technology
Beijing 100041, China

Abstract. China English is a noticeable phenomenon in Chinese-English interpretation; its two major manifestations can be seen in the use of "unnecessary/superfluous words" and in "Chinese syntax sentence structures". Beginning with a definition of the difference between China English and Chinglish, the paper analyzes the causes for this phenomenon from various angles: linguistic causes; environmental causes; habitual thinking pattern causes and situational causes. The paper proposes a number of remedies to help fix this phenomenon, and while the suggestions offered may be practical and simple, but they are demanding. It also points out that it is impractical to eliminate the existence of China English and suggests viewing this phenomenon objectively and pragmatically; people should accept China English as one of the many types of English in the family of world Englishes. Hopefully this paper may serve as a guide and reference to those who are endeavoring to do in-depth research in this field.

Keywords: China English, interpretation, remedies, suggestions.

1 Introduction

As China becomes more and more involved in the international community, Chinese interpreters' performance is drawing increasing attention in the international community of interpreters. There has been considerable discussion of the phenomenon of China English among Chinese interpreters in their work. This paper begins with the origins of China English and Chinglish, introducing their long history in China. After an analysis of the various manifestations of China English in typical day-to-day situations in which interpretation is required, the paper suggests a range of reasons why such features can be detected. In offering ways of resolving this problem, what are hoped will be workable methods and proposed suggestions. The author, advocates viewing this phenomenon objectively and practically, because China English can now be regarded as a clear variety of English existing alongside many other Englishes. Whether such a view is reasonable or scientific, heat discussion, challenge or questioning is welcomed.

2 The Definitions of Chinglish and China English

While the idea of 'Chinglish' can be defined in a number of other ways, the most authorative and useful definition is: Chinglish (also Chingrish) is a portmanteau of the

Y. Wang (Ed.): Education Management, Education Theory & Education Application, AISC 109, pp. 163–170.
springerlink.com   © Springer-Verlag Berlin Heidelberg 2011

words <u>Chinese</u> and <u>English</u> and refers to spoken or written English which is influenced by <u>Chinese</u>. [1] While for some users the word "Chinglish" may contain pejorative or derogatory connotations, [2] its history can be traced back to the 1630s when the English first arrived in China. Initially, Chinglish was known as "pidgin", or "Yangjing Bang English" in Chinese (洋涇濱, or 洋泾浜), which derives from the name of a creek near the Bund in Shanghai where local workers communicated with English-speaking foreigners in pidgin. [3] Chinese Pidgin English began to decline in the late 19th century as Standard English began to be taught in the country's education system.[4] English language teaching has been widespread throughout modern Chinese history---it was made the country's main foreign language in 1982.[5] Shortly afterwards, English became one of the compulsory examination subjects of the National College Entrance Examination. Since then, the overall English level of the Chinese has been dramatically upgraded.

China English is a new term which describes the typical Chinese way of speaking and writing English in the new era. In 1993, Li Wenzhong defined China English as follows: the vocabulary, sentences and discourses cored with Standard English, free from mother tongue's interference, with Chinese characteristics that enter communication in English by means of transliteration, interpretation and semantic regeneration. [6] In the case of interpretation, China English usually has the following features: it is grammatically correct, easily understood and pronounced reasonably, but often it does not conform to the norms and rules of authentic English. China English is, after all, not Standard English.

3   Various Manifestations of China English in Interpretation

The book The Translator's Guide to Chinglish [7] gives a full picture of Chinglish in its various aspects. Chinglish features are grouped under two broad categories: namely "unnecessary words" and "inappropriate sentence structure"; these categories apparent in Chinglish can describe accurately the features of China English as well.

3.1   First, the Manifestation of Unnecessary Words

In interpretation, adding unnecessary words is an apparent characteristic of China English. Influenced by Chinese expressions, Chinese interpreters are inclined to add an unnecessary adjective prior to a noun or a superfluous adverb before a verb when their meaning has been conveyed by the noun or the verb, for the purpose of either stressing the tone or the word-for-word interpretation. In The Translator's Guide to Chinglish, the co-authors list numerous China English phrases, such as "actual fact", "serious chaos", "new innovation", "financial expenditure", "a great historic change" and "final completion"; in all these cases the adjectives contribute little. By the same token, "blue in color" says the same thing as "blue", and "accelerate the pace of reform" carries the same meaning as "accelerate the reform". Likewise, in "totally abolish", we may take away the intensifier "totally" because its meaning is already expressed by the verb. The same applies to "firmly ban", "thoroughly eliminate", "successfully accomplish", "completely smash", "completely conquer", "extremely shameless", "fully support", "strongly advocate" and countless others of the same type. When the future tense of a verb has been used, it is pointless to specify "in the future". Similarly, when the verb is already in the past tense, reference to time like "in the past" or "previously" is generally superfluous. In Chinese expression, when speaking of a "V+N" phrase, people tend to use the omnipotent Chinese verb "做出", meaning "make" or "have". In the interpretation process, still due to the influence of the mother tongue, the interpreter prefers the form of "make an investigation" or "make an adjustment" instead of proper English "investigate" and "adjust".

3.2 Second, the Manifestation of Inappropriate Sentence Structures

In The Translator's Guide to Chinglish, the co-authors point out that the problems with sentence structure are relatively complicated. A case in the point is that influenced by the expression habit of Chinese, numerous interpreters are prone to use more "V.+N." phrases or expressions, even when a single verb can convey the meaning. This not only increases the interpreter's burden of memory but also violates the language norms of authentic English.

Take an example from Premier Wen Jiaobao's Press Conference in 2008: "Second, we need to make breakthroughs in the reform of political and economic systems. To attain this goal, we must free our minds. To free our minds, we need to have courage, resolve and a spirit of dedication. Only by freeing our minds can we succeed in carrying out reform and innovation." A more proper manner for this expression can be: "Second, we need to breakthrough in the reform of political and economic systems. To attain this goal, we must free our minds. To do so, we need courage, resolve and spirit of dedication. Only by freeing our minds can we succeed in carrying out reform and innovation."

Another manifestation of China English sentence structure is to adopt too many noun forms of the word just as the Chinese sentences do. This not only unnecessarily lengthens the sentence but make the sentence complex and hard to express as well. Take an example in the same book. "This would not only be a hindrance to the people of different nationalities in exchanging experience with and learning from each other but also a great disadvantage to the development of culture." The presence of abstract nouns ("hindrance", "exchanging", "learning" and "disadvantage") renders the conveying of meaning difficult. A more appropriate way of expressing this idea might be: "This would not only make it difficult for people of different nationalities to exchange experience and learn from each other, but would also impede the development of culture."

With a sentence structure shaped by China English, the conciseness and clarity of the sentence has been distorted by what some would regard as China English's long-windedness and obscurity.

4 Pluralistic Causes of China English in Interpretation

4.1 Linguistic Causes

Chinese and English belong to two different language families: Chinese is the representative of hieroglyphs, whereas English is a typical alphabetical language; both have distinctive word forming rules, sentence structures, grammar and ways of expression. The dramatic difference between the two languages evidently increases the workload of Chinese interpreters because as "Wilhelm von Humboldt puts it, an initiative language can even control whatever comes later and assimilate it according to its own rules. Sapir Whorf hypothesis has also pointed out that a speaker's mother tongue exerts a profound influence on any later language learning." (Guan Meng, 2007) [8].

To be more specific, Chinese is different from English in a number of significant ways. First, Chinese is parasitic and English is hypotactic. In Chinese, clauses or parts of a sentence are free-standing and equal to each other; they are commonly joined by commas, without connectives showing the logical relation between them. English is hypotactic, where one part of a sentence is subordinate to another and incomplete on its own.

Second, Chinese is "dynamic", English is "static". In English, each sentence or clause normally possesses one predicate, and verb is the only syntactical function that can serve as the predicate in the English sentence, implying each English sentence can normally possess one verb or verbal phrase. On account of the rare number of verbs in the English sentence, linguistists agree on the point that English sentences are "static". By comparison, Chinese grammar is loose, practical and flexible; it has not the limitation and regulation of the number of verbs in the sentence. Habitually, a Chinese sentence usually includes many a verb in accordance with the requirement of context. Verbs make the Chinese sentence rather more "dynamic".

Third, while Chinese is top-heavy, English is end-weight. As a widely-accepted rule, Chinese expression norm is to put the main body of the sentence at the beginning; the back end is the less important. Such structural expression makes the Chinese sentence looks like a top-heavy triangle. Whereas in English, the main part of each sentence usually goes at the end and the whole sentence looks like a pyramid or end-weight triangle.

In addition to the above, the linguistic difference between Chinese and English also includes: Chinese is simplex, English is complex; a Chinese sentence has a more personal subject, while the English sentence has a more impersonal subject.

4.2 Environmental Causes

4.2.1 Less Exposure to Authentic English

Since adopting the reform and opening up policy, China is more and more involved in the international community and Chinese people have more chances to be exposed to English. However, in general the Chinese people's exposure to authentic English is much less than people in some other Asian countries, such as South Korea, Japan and Singapore. In China, people cannot easily access English language television broadcasts like BBC, CNN, ABC, CBS and so on. Chinese remains the more dominant language. People read Chinese books, listen to Chinese radios, watch Chinese TV programs, surf on Chinese websites and talk to each other in Chinese. Social learning theory reveals human behavior in terms of continuous reciprocal interaction between cognitive, behavioral, and environmental influences. Thus, without sufficient exposure to authentic English makes it difficult for Chinese learners to model native English.

4.2.2 Chinglish Everywhere

Chinglish has its long history and a wide range of existence in China. Instead of being exposed to authentic English, Chinese people are exposed to Chinglish all the time and in many situations. Such ridiculous examples are nurturing the English learners continuously: "steek gently" for "shut the door quietly" (轻轻关门)；"To take notice of safe" for "Be careful and take care" (注意安全)⼝"Give you some colour to see see." for "Teach you a lesson." (给你点颜色瞧瞧)；"confirming distance" for "maintaining a safe distance" (保持车距). "Welcome you to Beijing" for "Welcome to Beijing"; "Welcome to ride Line 4 Bus" for "Thank you for riding Bus Line 4".

Furthermore, currently in China, there are insufficient foreign teachers. It is impractical and demanding too much for all of China's schools and colleges to employ sufficient foreign teachers. Therefore, many of China's students are learning English from their teachers who are speaking typical China English by themselves as well.

4.3 Habitual Thinking Pattern Causes

The core of Chinese philosophy is "being practical", and it is manifested in approximately all aspects of Chinese culture. Guided by such philosophy, Chinese culture attaches great importance to results, and for the sake of a good result, many other things can be sacrificed. Applying this philosophy to language, Chinese ignores the form of the sentence and more concerned with conveying meaning. In comparison, thinking patterns in English are based on strict rules. With this principle, English values highly the form of each sentence. The divergence between these two thought patterns gives rise to many of the differences between the two languages and the most distinguishable one is: in Chinese a non-subject sentence is quite normal as long as the meaning is lucid; in English, the subject can rarely be omitted from any sentence.

4.4 Situational Causes

Checkin Interpretation is a special oral practice which is characterized by its very public nature, its need for instant results and the difficulty of preparing adequately for all eventualities. The stresses and pressures on the interpreter are hard to imagine by the outsiders. Therefore, under such difficult working conditions, the interpreter has no time to meditate and to select the most suitable words, nor to organize the most appropriate sentence structure. Experience shows that the interpreter is working subconsciously and instinctively on the spot and on the run. One example in point can prove this is a well-known mistake made by a prestigious interpreter in China. On an important international conference in 2002, a European official was delivering the speech and the interpreter was working diligently and professionally. Everything went well until suddenly the whole audience burst into laughters by the mistake of the interpreter. The fact was that when the speaker said in his speech, "Are you kidding?" "No, I'm serious." Everybody at present knew the interpretation should be "你是开玩笑吗？" "不，我是认真的。" Unfortunately, the interpreter, who must have been working subconsciously, interpreted the above sentences as "你是继挺先生吗？"

"不，我是西尔瑞斯先生。"(meaning " Are you Mr. Kidding?" "No, I'm Mr. Serious.)

It is fairly evident that under such great pressure and with so little time for meditation, the interpreter may be working instinctively and the influence of the interpreter's mother tongue is indisputable and unavoidable.

Furthermore, the widely-acknowledged Chinese-English interpretation principle and technique allows that, for the sake of alleviating the work load, the interpreter should abide by the principle of "interpreting in line with the Chinese sentence structure and order"; this will inevitably result in typical China English. Let's take a sentence in Premier Wen Jiabao's Press Conference in 2008 as the example. "5 年之后，突如其来的南方冰雪灾难，人们又看到您奔走在抗击雪灾的前线"。 "Five years later, when the disaster of sleet and snowstorms hit Southern China, people once again saw you at the forefront of the fight against the disaster". The order of the English version is almost at the same order as the Chinese sentence, and China English is evident. The better version might be: "Once more you were at the forefront of the fight against disaster when the sleet and snowstroms hit southern China five years later."

5 Suggestions and Solutions to Ameliorate the Present Situation

5.1 More Active Exposure to Authentic English

Neuroscientist Patricia Kuhl from University of Washington found in her experiment that nine-month-old American infants who were exposed to Mandarin Chinese for fewer than five hours in a laboratory setting were able to distinguish phonetic elements of that language.[9] The findings indicate the importance of exposure in language learning. Nevertheless, by "exposure", I mean "being in the language", making full use of the available conditions to be in the presence of English: eavesdropping on a conversation at an adjacent table, watching English TV programs and movies, listening to English music, reading English magazines and newspapers, surfing or navigating on the English internet. Language exposure isn't just a means to refine the learner's English pronunciation and enlarge the vocabulary. It will also be easier for the learner to think subconsciously in that pattern, which is more crucial for eliminating China English.

However, it should be noted that exposure is divided into two types: active exposure and passive exposure. Active exposure entails focus for the purpose of conscious internalization and it can't be achieved by simply being around the language, which explains why living in America doesn't absolutely guarantee that one will learn English well. The significant function of active exposure is to help the learner to learn to think gradually in the language and cast off the influence of one's mother tongue.

5.2 Reading Instead of Listening

Interpreters may have thought that listening to native English is adequate for overcoming China English; however, experiments show that no amount of TV and radio listening is sufficient to keep one's second language at the level it should be. Second language acquisition researchers find that the most effective way is to read in English as much as one can (newspapers, magazines, novels, etc.) and focus on the words/phrases instead of merely listening. Only by this, can the interpreter attempt to possess authentic and natural English, reaching the standard required of a proficient interpreter.

5.3 View the Phenomenon Pragmatically and Objectively

On the one hand, apparently, China English is not properly authentic English. It should be refined and improved. David Tool, adviser to the Beijing Speaks Foreign Languages Committee, who has been closely involved in cleaning up Chinglish signs in Beijing since 2001, notes that Chinglish fades away many signs that they should convey. To a certain extent, tampering with a language at random is an abuse to it.

However, on the other hand, considering the social, cultural and psychological distance between English and Chinese, it would take years and years of hard work for a Chinese English learner to reach the native speaker's level. Evidence shows that most people reach about halfway; in other words, they stop developing while still short of target language competence.[10] As is widely acknowledged, the purpose for interpreting is for international communication, and it does not demand native-speaker competence. Therefore, from this point of view, for China English, there is no need and possibility to eliminate Chinese interference.

Furthermore, interpretation is a special language activity which requires not only high language proficiency of the interpreter but also good memory ability of the audience. The interpretation theory has approved that in Chinese-English interpretation, China English is acceptable and workable, both for the sake of clarifying the audience's comprehension and alleviating the interpreter's memory pressure. The typical China English structure such as, "because...so...", "although...but", "ever since", "no mater why", can be adopted.

Last but not least, along with the development of China economically and politically in recent years, more and more Chinese people are studying and working abroad. Some of their expressions are gradually accepted by the local native English speakers as a supplement to English. For example, "Long time no see" is often attributed as a good example of China English being used by native English speakers. The phrase is said to have originated from Chinese好久不見了. Example with the same pattern includes "lose fase" from Chinese phrase "丢脸" and so on. Some other examples like, "coolie" for "苦力", "kongfu" for "功夫", "toufu" for "豆腐", "koutou" for "叩头"and so forth.

6 Conclusion

The phenomenon of China English is a recurring problem among those who study interpretation in China to whom English is the L2, because of the considerable distinction between Chinese and English are greatly different form each other in the aspects of pronunciation, word-forming, grammar and the ways of expression. In terms of the current situation, China English is a phenomenon that cannot be avoided and the prevailing policy remains "to expose learners to more authentic English".

On the other hand, there are nearly 600 million people speak English as a first or foreign language at present. By 2010, two billion people will be studying English and half the world's population -- or three billion people -- will speak it. The popularity of English worldwide is growing [11] Many more people are speaking English as a second language than speaking it as a first. As a matter of fact, English has been combining with various languages in the world, hybridising with their language styles and cultural features. From this point of view, China English is understandable and acceptable in interpretation, for China English is an English, a member of the big family of world Englishes or, more exactly, a member of the "expanding circle", to use Kachru's term, but with Chinese characteristics.[12].

References

1. Jing, X., Zuo, N.: Chinglish in the oral work of non-English majors. CELEA Journal 29(4) (2006)
2. Vittachi, N.: From Yinglish to sado-mastication. World Englishes 19(3), 405–414 (2000), doi:10.1111/1467-971X.00189
3. Wikipedia, Chinglish #History, http://en.wikipedia.org/wiki/ (retrieved December 1, 2007)
4. McArthur, T.: Oxford Guide to World English. Oxford University Press, Oxford (2002)
5. Kam, A.: English in education in China: policy changes and learners. Experiences. World Englishes 21(2), 245–256 (2002)
6. Li, W.: China English and Chinglish, vol. (4). Foreign Language Teaching and Research Press, Beijing (1993)
7. Pinkham, J., Guihua, J.: The Translator's Guide to Chinglish.
8. Guan, M.: China-australian English, Chinglish and English Learning. US-China Foreign Language 5(5) (May 2007)
9. Schwarz, J.: Brief exposure to Mandarin can help American infants learn Chinese joels@u.washington.edu, pp. 206–543. University of Washington (2003)
10. Ellis, R.: Second language acquisition. Shanghai Foreign Language Education Press, Shanghai (2000)
11. Graddol, D.: English Next Disigned and produced by The English Company (uk) Ltd., (2006), http://www.English.co.uk
12. Kachru, B.B.: World Englishes: approaches, ussues and resources. Language Teaching 25, P1–14 (1991)

The Application of Formative Evaluation in Integrated English Teaching

Tianshu Xu[1] and Lei Zhu[2,3]

[1] School of Foreign Languages, Anhui University of Science and Technology,
Huainan, Anhui 232001, China
tsxu@yeah.net
[2] Anhui and Huaihe River Water Resources Research Institute, Bengbu Anhui 233000, China
[3] Anhui Quality Supervision and Test Station of Construction Engeering,
Hefei Anhui 230088, China
zhuleianhui@163.com

Abstract. In China, summative evaluation has long been adopted in Integrated English for English majors, which has been proved to be a failure in practice. This paper introduces the concept of formative evaluation and presents a tentative research model of teaching in Integrated English. A one-year experimental project is designed and implemented in Integrated Course. Qualitative and quantitative results of the experiment show that formative evaluation can not only effectively monitor the implementation of every stage of learning, but also successfully cultivate learners' autonomy in the control over their learning process. Besides, learners' language ability also sees improvement during the span of experiment.

Keywords: formative evaluation, self-assessment, peer-assessment, learning autonomy.

1 Introduction

As a pillar course for English majors, Integrated English aims to cultivate students to acquire necessary language knowledge and basic language skills, and through training of students' oral and written output ability to improve students' competence of communication in English. Meanwhile, this course is expected to offer guidance of learning methods and cultivation of logic thinking, laying foundation for the specialized courses opened in higher grades[1]. Therefore, effective teaching of this course is of vital importance to the following professional courses.

In view of the importance of Integrated English, for years numerous researchers have carried out various educational reforms to explore different teaching modes and means to promote learners' language output ability. Their studies have got great achievements, however, in practice, reforms on evaluation means have often been neglected. Nowadays, it is summative evaluation that is mainly adopted in Integrated English teaching. Summative evaluation is to assess the final effectiveness of learning and teaching after a period of study. It takes question& answer tests as its main form. Long-term practices have proved that this kind of evaluation has many drawbacks: it

Y. Wang (Ed.): Education Management, Education Theory & Education Application, AISC 109, pp. 171–176.
springerlink.com © Springer-Verlag Berlin Heidelberg 2011

puts too much emphasis on the final results of learning while neglecting the process of learning; the evaluating process is one-sided, in which students do not take active participation but accept the assessment passively. Therefore, summative evaluation can hardly provide an all-sided view of learners' overall ability and skills. Neither can it exert an optimal stimulation on classroom teaching and learning. Thus, this research points out that formative evaluation should be widely applied to Integrated English teaching. Through specific formative evaluation ways, learners' ability of self-assessment and peer-assessment can be developed, which further promotes their acquisition of self-reflection and self-management.

2   Formative Evaluation: Concept and Characteristics

The first one in history to put forward the concept of "formative evaluation" is Scriven, M., an American educationist. He points out in his masterpiece *The Methodology of Evaluation*: "Formative evaluation is typically conducted during the development or improvement of a program or product (or person, and so on) and it is conducted, often more than once, for the in-house staff of the program with the intent to improve."[2]

Different from traditional summative evaluation in focus and theoretical motivation, formative evaluation has its own characteristics.

1) Student-centered. Formative evaluation does not merely start from the need of evaluator, but more importantly, it takes the need of evaluatee into account and attaches importance to the learning process and students' experience in learning. In other words, formative evaluation centers around students. From the information collected in evaluating students' learning process, teachers and students can co-construct shared knowledge and develop mutual understanding, thus promoting two-sided communication and interaction in teaching-learning context.

2) Context-specific. Formative evaluation attaches importance to process and development, and puts emphasis on openness and plurality. It goes through the whole process of teaching and learning. Information collection and research need be directed to a specific class, lesson, teaching method or task, which requires flexibility and quickness. The process, from settling evaluating object, planning evaluating programs, organizing assessment, collecting information, analyzing information, to feedback, is an ongoing circle. Teachers are not confined to settled teaching programs, but can adjust and improve the previous plans and choose the best classroom activities and methods accordingly when the context changes.

3) Individual. Formative evaluation is supposed to overcome the problems in traditional evaluation system, problems such as ignorance of learners' emotional factors, and frustration of students' enthusiasm and confidence. It pays attention to students' individual differences and gap, and emphasizes students' participation, cooperation and dynamic progress. Through students' self-assessment and peer-assessment as well as teachers' close observation and feedback, the learning potential of each student can be discovered and explored. Students can also be helped to efficiently control their learning process and then get a sense of achievement and driving force.

3 Practice of Formative Evaluation

3.1 Background of Evaluation

Integrated English is a compulsory course for undergraduate English majors. It goes through four academic years (seven semesters, the eighth semester for practice). According to the Schedule for English Majors of the University (2007), this course is supposed to be given 610 periods and 36 credits.

The research takes two classes of Grade 2007 (in all 60 students) as the experimental objects, and tentatively introduces the concept of formative evaluation into Integrated English classroom to implement reflective teaching.

The textbook of this course is *Integrated Course* by Zhaoxiong He (Shanghai Foreign Language Education Press, 2007).

3.2 Ways of Evaluation

At the beginning of the course, teachers show clearly to the students that the assessment will take the form of formative evaluation, with daily records as the reference, combination of self-assessment, peer-assessment and teacher-assessment as the means, and students' portfolios as the tool.

1) Daily records. Make a loose-leaf pamphlet to record students' daily language, behavior and study. Through classroom activities, observe students' performance and make records on the pamphlet. The content of records includes students' pre-activities, in-classroom performance (initiative, pronunciation, etc.) and afterward reviews. Students are divided into several groups. The leader of each group, whose role is played by students in turn, is expected to write down the records. Teachers examine the records regularly, and in response to the specific problems of each student, communicate in time with them and help them make an improvement.

2) Self-assessment, peer-assessment, teacher-assessment. Teachers' assessment must be combined with students' self-assessment and peer-assessment, which then can construct a relatively impartial, all-round and specific evaluation system, and thus improve the efficiency of classroom teaching-learning and evaluation. In practical operation, take "daily repot" in Integrated English teaching as example. Teachers can assign topics beforehand or students choose topics by themselves and prepare before class. In class, in certain period of time, a student first delivers the speech and then makes appropriate answers when questioned by teacher and other students. Then the speaker makes a self-comment on his/her performance and points out where is satisfactory and where is not. Next the teacher chooses some student judges to assess the speaker's performance according to the settled evaluating criteria, including pronunciation, idea, appropriateness of answers, body language, etc. Finally, the teacher makes an overall comment and provides measures for improvement. Through the combination of the three assessments, students can have a relatively objective knowledge of their level, which can enlighten their studies in the near future. Meanwhile, self-assessment helps to foster students' self-reflective ability and strengthen their confidence, while peer-assessment can not only train students' autonomy but also improve their expressive and pragmatic ability, and more importantly, it can help to prevent the phenomenon that a single teacher may make unfair judgment and assessment due to various subjective factors.

3) Student portfolios. Student portfolio is used to collect students' works, namely, learning achievements, in a certain stage or a semester. Gottillieb divides it into six steps: collecting, reflecting, assessing, documenting, thinking and evaluating[3]. Through making portfolios, students can have overall knowledge and correct evaluation of their own studies in certain period of time. During this period, they can constantly review and improve the content of portfolios until what they show is the most satisfactory works.

3.3 Notes in the Process of Evaluation

In the process of implementing formative evaluation to guide teaching and learning, the following problems need to be settled.

1) Different focus in different stages. Since Integrated English for English majors goes through four academic years (seven semesters), cultivation objectives also vary for students in different stages. In accordance with this characteristic, ways of formative evaluation need to be adjusted in time, and different stages require different focus.

The first stage, usually the first academic year, is the stage for students' knowledge transformation. Students was inculcated with a lot of language rules in high school, but the knowledge has not been effectively assimilated and transformed into language ability. Therefore, the focus of this stage is to promote the transformation of students' acquired knowledge to skills (both in written and oral aspect).

The second stage, usually the second academic year, is a stage to promote the transformation of students' in-put ability to out-put ability. The main task of this stage is to provide enough attended information and notice the gap to help to promote students' output ability through a large sum of strengthened in-put means, like unit test, stage test, etc.

The third stage, namely the third and forth academic years, is to transform from controlled communication to free communication. In this stage, test should take the role to guide students from controlled communication to free communication.

2) Data collection. Evaluation of the course and students should not depend on what teachers have taught, but count on what students have gained. Teaching content and means both exert influence on students' study. Therefore, this research designs learners-centered formative evaluation tools, and collects qualitative and quantitative data in the process of the course. Data come from four sources: ① students questionnaires and following sampling talks; ②daily records; ③student portfolios; ④ stage tests. To ensure the reliability of the data, effective daily records and questionnaires should be designed.

3) Proper weight. In portfolio-based formative evaluation, the relationship between item weight and evaluation validity should be taken into consideration. According to English Syllabus (2006) and reform experience from other universities, the initial weight is set as follows.

Formative evaluation includes:

Grades of daily oral performance (including repetition, oral report, speech, discussion, etc.) 20%

Grades of daily written performance (including composition, translation, term paper, etc.) 10%

Grades of daily dictation (including dictation of words and paragraphs) 10%
Grades of daily stage tests 20%
Daily performance (based on daily records) 10%
Grades of mid-term examination 10%
Grades of final-term oral test 20%
Integrated grades: grades of formative evaluation (50%) +grades of summative test (50%)

Notes: ① Each item is graded by average grades times the percentage; ② If necessary, what are to be collected and how often collected can be settled beforehand, which can prevent inconsistent results due to different operations by different teachers.

4 Lessons and Problems

Through study of daily records and analysis of portfolios of 60 students in Grade 2007, formative evaluation is revealed to exert positive influence on Integrated English teaching in the following aspects.

1) Final-term sampling survey shows, through formative evaluation to grade students' performance in the course, 100% students believe that compared to grading through one or two tests, formative evaluation is more scientific and persuasive, although seven students (11.7%) find their integrated grades may be lower than their grades of summative test (when they happen to be good at the questions in summative test). Real practice also shows that formative evaluation can help to solve the headache problem in classroom management: some students may be absent from class, some present but absent-minded, but summative evaluation cannot truly reflect their attitudes and punish slackers. While in formative evaluation, portfolios cannot be copied since it is impossible for different students to have the same or identical learning process.

2) The application of formative evaluation in Integrated English course makes it possible for students to obtain more opportunities to behave and assess (self-assessment, peer-assessment). Students' autonomy in planning, learning and assessing also sees improvement. But in practice, some students have different ideas on self-assessment and peer-assessment. For example, some student points out that self-assessment is not so effective since he is still student and not competent enough to assess. But he still believes that he can gain a lot from peer-assessment. While some other student thinks self-assessment provides good opportunity to self-reflect and it should be implemented constantly. But he is doubtful about the effectiveness of peer-assessment since some peer students care too much about face and are not willing to point out others' weakness. These students' opinions reflect the problems to be solved in cooperative learning.

3) Under formative evaluation mechanism, students foster good learning habits and form favorable academic atmosphere. Through systematic training of Integrated course and cooperative training of other courses, students' language ability has been greatly improved. For instance, the passing rate of Tem4 and Tem8 of students in experimental classes reached 90.6% and 91.45% respectively, which were 26.15% and 49.01% respectively higher than the average passing rate of national universities

of science and technology. Moreover, in National English Contest for College Students, "The Star of Hope" National English Beauty Race, CCTV National English Speech Contest and other national contests, students in experimental classes all got various remarkable achievements.

To conclude, formative evaluation, as a teaching method and evaluation means, has obvious advantages over summative evaluation. It organically combines measurement, test and evaluation, strengthens the effective monitoring of learning process, and has positive effect on the promotion of students' language output ability. Meanwhile, the application of formative evaluation provides students with more opportunities of self-assessment and peer-assessment, and involves them into course construction. However, total replacement of large-scale summative tests with formative evaluation system would also lead to unfavorable outcome[4], such as lack of settled standard for measurement, over-subjectivity, influence of teacher-student relationship on evaluation results, different integrated grades caused by different evaluation items and weight. Moreover, differences in courses, teachers and schools would also exert influence on the evaluation result. Therefore, teachers should reasonably combine different evaluation means and set up scientific evaluation system, which then is the foundation to improve teaching quality.

References

1. English Group of National Higher Education Foreign Language Committee: English Syllabus for English Majors of Institutes of Higher Education. Foreign Language Teaching and Research Press, Beijing (2000)
2. Scriven, M.: The Methodology of Evaluation. In: Tyler, R.W., Gagné, R.M., Scriven, (eds.) Perspectives of Curriculum Evaluation. Rand McNally, Chicago (1967)
3. Gottilieb, M.: Nurturing Student Learning through Portfolio. TESOL Journal 1, 12–14 (1995)
4. MacGaw, B., et al.: Assessment in the Upper Secondary School in Western Australia. Ministry of Education, Perth (1984)

Research on the Embedded Professional Talent Training Mode of Chinese-Foreign Cooperation Education

Binbin Shi

Dept.of Electrical Engineering, Wuxi Institute of Technology, Wuxi, China
Dept.of Electrical Engineering, Suzhou Vocational University, Suzhou, China
binncn@gmail.com

Abstract. According to the current state of the domestic embedded professional status quo of chinese-foreign cooperation in running schools, this paper makes in-depth exploration and research based on the aspects of introducing foreign advanced education concept, setting school operating mode, training goal as well as course systems; besides, it analyses the existing problems of current education mode and also provides some new optimization measures; further more, this paper proposes the innovation methods to embedded professional talent training mode of the chinese-foreign cooperation in running schools.

Keywords: Foreign Cooperative Education, Teaching Method, Embedded System.

1 Introduction

By 2010, the government issued the "national medium and long term education reform and development planning outline (2010-2020)". In whole content of chapter 16, the outline proposed to further expand the opening education, and extensively carry out the international cooperation and education services[1]. This not only accords with the national economic and social needs of opening to the world, but also accords with the inevitable trend of education facing to the world. Chinese-foreign cooperation in running schools has become an important branch of the higher education internationalization. It has important practical significance for our Chinese colleges and universities, especially the higher vocational schools which aims to bring up the international applied talents.

Embedded major has started late in our country, while it was very hot in Europe and Japan. The domestic research on embedded professional education is very rare. Since the embedded system technology has become the focus of the PC era, the embedded system education has become an important subject which can not be ignored by Chinese colleges and universities. Embedded is an emerging technology for application[2]. We should both make use of chinese-foreign cooperation in running schools pattern to absorbe foreign advanced and professional technology, but also explore a mode of embedded professional training mode consistent with China's national conditions.

Y. Wang (Ed.): Education Management, Education Theory & Education Application, AISC 109, pp. 177–183.
springerlink.com © Springer-Verlag Berlin Heidelberg 2011

2   Analysis on the Current Situation of Talent Training Mode

2.1   The Status of Talent Training Mode

The process of constructing chinese-foreign cooperation in running schools' talents training mode is the talent training activities structural frame and activity programs formed under the guidance of international education thoughts and concepts. Specifically, this process contains the two aspects, one is the forming of international thoughts and concept; another is the content construction of talents training mode. While borrowing the foreign advanced vocational education mode, we should also see the differences between Chinese and overseas vocational education; Foreign vocational education more focuses on on-the-job personnel post skills training, while the Chinese vocational education mainly aims to the employment goal, taking charge of training the groups of students without work experience[3]. So we can't simply "transplant" foreign vocational education mode and teaching system into Chinese education mode. According to China's current conditions, we must carry on appropriate vocational education through international cooperation.

From the chinese-foreign cooperation mode in domestic universities, there are basically the following modes.

Project Cooperation. Both sides of cooperative education schools retain the characteristics of their own courses, they make course evaluation and credit transfer for the courses of each other school and admit the diploma and degree certificate issued by the each other school, this mode are commonly known as the "2 + 2 mode", "3 + 2 mode" and "1 + 2 + 1 mode".

Introducing Foreign Teaching Resources. Evaluation methods and management mode.

Exchanging Teachers. China school invites the foreign teachers to give lecture, Chinese teachers take further education and practice in foreign schools, and also organize students go abroad for short-term learning and practice.

2.2   The Existing Problems in the Training Mode

The Defects of the Quality Authentication System for the Chinese-Foreign Cooperation in Running Schools. With the large number of foreign education institutions entering into the Chinese education service market, the degree certificate authentication and teaching quality related problems are also highlighted. At present, because the administrative supervision system for Chinese-foreign cooperation in running schools is closed, which can not effectively promote the establishment of the social assessment system. The current situation is that, China lacks of unified reasonable evaluation and quality authentication standard for the education quality provided by foreign education providers; besides, China still lacks of international generally accepted complete index system and rigorous evaluation standard for the quality assessment of chinese-foreign cooperation in running schools. This situation not only restricts the market mechanism adjust action for the supply and demand relation of education services, but also has a bad impact on the international recognition and fair competition of education market environment formation.

The Problem of Chinese Education Sovereignty. In chinese-foreign cooperation in running schools, education sovereignty has been the most sensitive topic in education field. As for the high vocational colleges, the low levels of chinese-foreign cooperation in running schools may lead China's limited education resources lose to overseas. Meanwhile, the inferior foreign education resources will enter into Chinese existing higher vocational education environment and cause adverse effects. Just like outputing backward production line and the enterprises with serious environmental pollution, outputing education institutions has very low credibility, and its education quality is also with low level. Some foreign education institutions were running low education reputation in their own countries, so they turned to output education industry in developing countries.

Simple Assessment Method and Single Evaluation Mechanism. The assessment methods of traditional embedded teaching experiment practice were relative simple, which can be divided into two aspects: experimental operation and experiment (training) report. The former checks the operating procedures as well as its results, the latter checks the students' experimental master level through their written expression. Teachers often make unilateral evaluation on students' practical ability according to these two aspects[4]. As a result, students may not understand the operating procedures at all, but only to simply and mechanically remember the operating procedures and hand on a "beautiful" experiment report with no innovation at all in order to cope with teachers' check during the process of practical education. The traditional single evaluation mechanism, not only plays a bad role to stimulate students' learning interests, on the opposite, it effective leads to the students' learning barriers.

The Education Quality Problem of Foreign Teaching. Viewing from national level, although China has promulgated the "The regulations of chinese-foreign cooperation in running school" and the "The regulations implementing measures for chinese-foreign cooperation in running schools" in 2004, as well as established annual check system[5]. However, the chinese-foreign cooperation in running schools had started in the 1990s, it had only lasted for a short time. At present, there is not a international recognized complete supervision education quality index system and evaluation standard, even the supervision and management department also lack of necessary executive method.

3   Optimizing Measures and Countermeasures for Talents Cultivation Mode

3.1   Unify the Standards, Setting Up Quality Certification System

The supervision way of chinese-foreign cooperation in running school can be in accordance with the rules of the GATS, that the education administrative departments' approval according to law, cooperatively-run school's self-guarantee, social intermediary organizations' authentication, educatees rights protection and monitoring systems are remained to be established. At present, Shanghai has established a chinese-foreign cooperation in running schools quality certification programe, which aims to approve the existing cooperatively-run schools and their related projects, final to establish the authentication standard with Chinese connotation, and also in accordance with the international standard, so as to prevent "false diploma" phenomenon.

3.2 Elaborate Design Combining with Application so as to Reform the Experimental Contents

Abandoning the past "knowledge-oriented" teaching mode, we should not make experiment content separated from reality. Besides, the teachers should be brave to reform and innovate, meticulously designing the practical appling projects and introducing it into students courses experiment. On the one hand, it can stimulate the students' learning interest and deepen the students' understanding on embedded industry through actual cases; on the other hand, it can speed up students to master their learned knowledge and practical application skills. During the process of designing experiment teaching content, teachers can make use of "interest guide, task driven" method to improve students' enthusiasm for study and learning efficiency. In the process of teaching, a clever design is very important, teachers can divided the complex project into several specific tasks, so as to stimulate students' learning interest, achieving cooperative exploration and comprehensive application.

3.3 Set Up Fresh-New Concept of Education Sovereignty

In chinese-foreign cooperation in running schools, developed countries take advantages of their language and information technology. While outputing their high-new technology and advanced management experience, at the same time, they also bring the capitalist society inherent ideology and value concepts to China. Therefore, we must strengthen the related students' ideological education work in the Chinese-foreign cooperation in running schools project, instructing the students to initiatively against bad influence of foreign unhealthy culture, strengthening students' national self-respect and pride. To sum up. positively occuping student's ideological and political territory, standardizing the sovereignty of the higher institutions education, and actively accessing to the international community, it won't cause any hembedded to education sovereignty.

3.4 Reforming Experiment Checking Method, and Implementing Development Oriented Assessment Evaluation

In the experiment examination, giving the opportunity for the students to clearly express experiment contents and procedures while facing teacher and classmates not only effectively improved student comprehension and expression ability, but also provide a objective evaluation for each student. In the evaluation of students' practical performance, it is better to take multiple evaluation and developing evaluation. Evaluation are made up of three parts: the teacher evaluation, student self-evaluation and students mutual appraisal. The final evaluation scores are calculated by the weighted average results of the above three aspects.

3.5 Strengthening Supervision to Make Sure of the Foreign Teaching Quality

Each higher vocational college should also strength the management and monitoring for the foreign teaching quality, establishing schools and international education college two-stage of foreign teaching quality management regulations. Firstly, it is necessary to make sure clear the concept of chinese-foreign cooperation in running school, establishing a rational method for foreign teaching quality management. On the basis of fully understanding all aspects of both sides, Chinese sides should reasonably positioning quality level and make practical training program according to the characteristics of the teaching area and running school strength, so as to form distinguishing features of their own. Secondly, the Chinese colleges should select qualified students according to quality requirement, this is also premise condition to guarantee teaching quality of the project. Finally, college should strengthen self-monitoring, and find out education activity flaws through lectures, interview, questionnaire survey or other monitoring methods, so as to make timely remedy and adjustment.

4 The Reform and Innovation of Talent Training Mode

4.1 Making Use of Chinese-Foreign Cooperation in Running Schools, Optimizing Major and Courses

Professional structure and curriculum design is interfaces connecting the higher vocational education with the social economy, it is also an important link to the marketable and practical talent cultivation. As far as we can see from the current employment situation of Chinese higher educated students, on the one hand, the big problem of employment has been more and more prominent, on the other hand, because of unreasonable professional distribution, many emerging professional talent is in short supply. This phenomenon actually exposures that the higher vocational professional education are disconnected with market demand. Therefore, the higher vocational colleges and universities' embedded professional structure and course design should closely connect with the social economy development situation, and make timely adjustment according to the the actual posts of embedded industry.

4.2 Learning from Foreign Universities Advanced Teaching Management Methods

In teaching content, we must break the old textbook knowledge, increasing the proportion of vocational skills training, strengthening the practice link. The vocational education of the most developed countries have a characteristic that they all can well communicate the link between colleges and enterprises, building a close cooperative relation between enterprises and schools, advocating the ability-oriented teaching instead of Chinese knowledge-oriented teaching. They can closely combine talents training with production, especially for the embedded majors with the feature of strong application strength.Domestic institutions can make use of the opportunity of chinese-foreign cooperation in running school, take reference of their successful experience, and enhance university-enterprise cooperation and training, and the establishment of experimental base.

In teaching way, it can make up a teaching team both by foreign teachers and Chinese teachers, each of them sharing different modules, and teaching different skills, so as to make the student get maximum benefit. Teaching team members can also timely discover the problems existing in the teaching, and solve problems through negotiating, so as to ensure the quality of teaching. Chinese teachers can learn the vivid and flexible teaching mode from foreign teachers through experience exchange and listening to their classes, so as to attach more importance to cultivating students' practice ability; the foreign teachers can also better understand Chinese students' way of thinking and knowledge structure, while communicating with students, they will not forget to consolidate their language foundation.

In the teaching management, the Chinese sides should study others' advanced management experience, to guarantee the quality of running school with rigorous teaching management mode. Especially taking reference of management style of "wide into severe out, flexible credit system" from foreign cooperative colleges. The evaluation way for students can combine the continuity assessment with summative assessment.

4.3   Use Project Integrated Method to Reform Teaching Management Method

It has been proved by practice that following the traditional teaching management mode will result in hang-ups and limitations for chinese-foreign cooperation in running schools program. In order to solve the current problems, through the summary from the practical work, we make use of the program integration management way combining with operation methods of contemporary commercial projects. In project management process, it has the following several parts:

(1) The confirmation of teaching plan confirmed: time and course;
(2) The teacher confirmation and management;
(3) The single course operation process;
(4) The Chinese teaching cooperation;
(5) The problems management: establish historical questions point database.

The management and operation of these five projects push the final accomplishment of the whole project. In the teaching plan confirmation link, time and courses can be parallel. The teacher confirmation and teaching plan confirmation are not completely connected. Because the some teachers of prepared project can teach many courses. In the teaching plan, the single course may not completely restricted by more than two project in the process of the confirmation of foreign teachers, namely, as long as there is the course, it can operate according to single course operation flow, without being constrained by course content and teachers' teaching time. While the Chinese embedded teaching cooperation and teaching plan confirmation is interactive, but this kind of interactive action will not affect the teachers management and single course operation.As for the problem points management, which will exist throughout the whole operation process. The emerging problems can find out the countermeasures from the recorded history question points. Meanwhile, recording the problem points into the database can provide reference for future work. This project management approach not only avoid project ossification to a great extent. Meanwhile, because of the routinization and electronization of project management, it also avoid the personnel dependence for the director in project, thus greatly shortened the process to adapt and familiar with the project.

5 Conclusion

China's higher education is still in attempt and exploration stage to access with the international countries, in the face of the disadvantages and threats existed in chinese-foreign cooperation in running school, it should seize the opportunity of 1025 plan at the beginning, reasonably introduce high-quality education resources overseas, fully absorb and make use of foreign advanced talent, culture, management method, sharing hembeddedony and discussing together, learning and building teaching management system with Chinese characteristics according to Chinese practical education situation. As embedded major closely integrated with practical application, it is necessary to train students' technical application abilities in the process of embedded professional teaching system. Except the paper mentioned above, the higher education institutions should pay more attention to the combination of production, learning and research; focus on teachers trainning; training next on teachers, solidly carry on the enterprise practice activities; adopt advanced management method, strengthen the laboratory management and security management, strengthen the management of each link of practice teaching, so as to construct chinese-foreign cooperation practice teaching system with its own characteristics.

Acknowledgments. The research was supported by the Innovation Team of Solar Photovoltaic Grid Inverter System and IPv6 Cloud Optical Wireless Broadband Application System, and sponsored by the Project of Modern Educational Technology in Jiangsu Province(No.2011-R-19373), and Natural Science Foundation of Jiangsu Province (No.BK2009131).

References

1. C.M. Liu et al.: Analysis of the Type of China's Cooperative Education Model. Key Engineering Materials. 426, 391–394 (2010)
2. Zheng, Y.-J.: Discussion on the Application of Embedded Teaching Model in Literature Retrieval Course. Sci-Tech Information Development & Economy 1, 19–21 (2010)
3. Liu, C.M., et al.: Research on Characteristics and Models of China's Cooperative Education. Applied Mechanics and Materials 33, 598–601 (2010)
4. Ye, J., Huang, J.-H., Zhang, X.-L.: Research on Teaching Method of Embedded Internet Technology. Journal of Guilin University of Electronic Technology 4, 12–15 (2010)
5. Kell, C.: Peer review of teaching embedded practice or policy-holding complacency. Innovations in Education and Teaching International 46, 61–70 (2009)

Study on Cultivation Mode
of Independent Innovative Talents

Guihua Ma

Xi'an International University

Abstract. In the new historical era, constructing new cultivation mode of independent innovative talents in China's colleges and universities will be conducive to the healthy development of our country's economy and society. The situation of cultivating independent innovative talents in China's colleges shows that to construct a new cultivation mode of independent innovative talents we must re-mould educational concepts and objectives, reform the process of talent cultivation and perfect talent cultivation system and mechanism.

Keywords: independent innovative talents, cultivation mode, talent cultivation.

1 Introduction

China's national economic and social development for the long-term program specifies that we must improve and enhance talents' independent innovative ability, and take it as a strategic base point for the development of science and technology. The improvement of national independent innovative ability largely lies on the independent creative talents, and universities play a critical role in the cultivation of independent innovative talents. At present, our colleges and universities have made some achievements on reforming educational ideas, profession types, and talent cultivation process and mechanism, etc. However, the cultivation of independent innovative talents is mainly determined by the process of knowledge accumulation and skills training. In the new historical period, our universities must adopt a variety of methods to establish a new mode to cultivate independent innovative talents.

Cultivation of independent innovative talents is a systematic program, mainly including three subsystems: educational concepts and cultivation objectives, cultivation process, cultivation system and safeguard mechanism. Each subsystem is composed of some elements, and they have mutual connection and influence on each other. Among them, educational ideas and cultivation objectives determine every step of the cultivation process. The implementation of cultivation process verifies the accuracy of educational ideas and cultivation objectives. Finally, cultivation system and safeguard mechanism provides a powerful protection for the realization of educational concepts, cultivation objectives and cultivation process.

Y. Wang (Ed.): Education Management, Education Theory & Education Application, AISC 109, pp. 185–188.
springerlink.com   © Springer-Verlag Berlin Heidelberg 2011

2 Reshaping Educational Concepts and Cultivation Objectives

The social demand for high-quality independent innovative talents is continuously increasing, and traditional educational ideas have already failed to meet the needs of modern social development. Therefore, our universities should get rid of the outdated talents training mode and construct a new cultivation mode, that is, adhering to the market and society oriented educational concepts and cultivation objectives, focusing on the concept of people-oriented, fully respecting students' central position and individual development, vigorously promoting quality education, establishing multi-leveled, multi-dimensional and multi-sized talent cultivation mode, comprehensively stimulating students' motivation and passion to be independent and innovative, thus achieving the cultivation goals of training independent innovative talents with thick foundation, strong ability and great potential.

3 Reforming Talent Cultivation Process

Under the guidance of the above mentioned educational concepts and cultivation objectives, China's universities should carry out more reforms on cultivation mode in such aspects as curriculum system, teaching process, practice and students' evaluation system, etc.

3.1 Curriculum System

In the course of curriculum system design, it should not only emphasize the completeness of subjects, the integrity of course contents and the rigor of theoretical systems, but also stress the courses' practicability and cultivating students' independent innovative ability. At present, the general mode of course system at universities in our country is public basic courses, specialized fundamental courses plus professional class, or compulsory public basic courses, selective public basic courses, compulsory specialized fundamental courses plus selective specialized fundamental courses. Such curriculum system has certain advantages in imparting knowledge, but in order to better adapt to the training requirements for independent innovative talents, it should broaden students' knowledge foundation, dilute the awareness of professionalism, enhance quality education and cultivate independent innovative ability, and the curriculum system will eventually become a "selective public foundation courses plus selective specialized modules" model. All courses have become selective courses regardless of basic courses or professional courses. It can fully respect students' freedom of course selection, and allow students to learn their interested courses.

3.2 Teaching Process

Teaching process mainly includes the design of teaching content and teaching method. For the arrangement of teaching content, it'd better combine the thick foundation and broad profession types, and reflect the novelty of teaching content. For teaching methods, it'd better change the old mode of "teacher-centered, class-centered, and book-centered", that is, pure cram-feeding teaching method, because under this teaching mode, students are often in a passive position, and independent creative ability cannot be well developed.

Therefore, the teaching method should shift from pure knowledge imparting to independent innovative ability cultivation, from teacher-centered class to student-centered class, from class-centered to the combination of class and extracurricular time, from book-centered to problem-centered, from the concept of teachers leading absolute authority to the concept of teaching benefiting both teachers and pupils alike, from "spoon-feeding" to autonomous learning, and establish a new teaching mode of university-industry cooperation, construct a new teaching system to develop students' independence and innovation, put students' initiative to acquire knowledge, independent thinking, innovative ability at the center of teaching activities.

3.3 Practice

Practice is an important way to cultivate college students' autonomous innovative spirit, to enhance entrepreneurship awareness, and to promote students' all-round development. Most China's universities have cognition practice (or apprentice), production practice, graduation practice and graduation design practice. College students' practice activities should not become a mere formality. It should reflect the independent innovative talent cultivation process. In the first semester, sophomores should attend cognition practice, which let students understand the application of this professional direction in reality, get perceptual knowledge of this profession and lay foundation for specialized courses. Juniors at the second semester should attend production internship, and actual post training consolidates students' theoretical knowledge they have already learnt and cultivate their tentative application of this professional knowledge to practical work, and give support for subsequent specialized course study. Seniors at the second semester should attend graduation fieldwork to cultivate their ability to use comprehensive knowledge, to improve the ability of analyzing and solving practical problems, to cultivate their innovative spirit and practice ability. More importantly, it should guarantee the combination of practice and curricular content as well as cultivation of independent innovative ability.

3.4 Evaluation System

Talent cultivation assessment is a very important part in independent innovative talent training mode. At present, examination system is the main approach to evaluate students in China's colleges, but the examination content for independent innovative talents should not be limited to the assessment of book knowledge, but should make comprehensive and systematic evaluation of students' knowledge, practical application of knowledge and independent innovative ability, etc.

Each assessment aspect is given certain weight, overcoming the old way of assessing students only by test scores. It should increase the weight of evaluating students' independent innovation ability and take many reasonable evaluation forms, such as extracurricular completion, independent innovation design, patents application and so on. These various evaluation means can stimulate students to develop independent innovative consciousness and ability.

4 Perfect Talent Cultivation System and Mechanism

First, it should vigorously promote flexible credits management mode, which allows students who have obtained required credits early to graduate early or minor in a second or third profession, select intraschool courses, etc. Secondly, it should establish and improve the mentor system. In the senior year each student can choose a mentor within all the teachers at school according to their own interests and strengths, thus ensuring interdisciplinary instruction. Finally, the teaching years can use "2 plus 2" model, and implement the curriculum in accordance with the modular in different teaching years.

5 Conclusion

Cultivation of independent innovative talents is a systematic project. Our universities are facing many challenges in the new era, so we must have a deep reflection on the problems of higher education, and establish scientific educational concepts and training objectives, reform the cultivation process of independent innovative talents, strengthen the system and mechanism of independent innovative talent cultivation, and strive for developing more high-quality independent creative talents in the new century.

References

1. Yongqin, X., Li, C.: High-skilled Personnel Training Mode. Chinese Public Administration (2008)
2. Qunxiu, Y.: hinking On Innovative Talents Training Mode. China University Teaching (2007)
3. Pang, Y., Lin, Z.: Training Mode of Talents Creative Ability. Researches in Higher Education of Engineering (2008)

Characteristics and Enlightenment of Faculty Building for Vocational Education in Developed Countries

Zeying Zhang

Department of Economics and Trade, Chongqing Education College,
SiGongLi, NanAn District, Chongqing, P.R. China
zzyounger@163.com

Abstract. Faculty is undoubtedly a key factor in the success of vocational education. The experience of faculty building for vocational education in developed countries enlightens the faculty building for vocational colleges in our country. Learning from their experience, training and hiring vocational teachers in various ways, improving faculty structure, making strict management of per-hiring qualification, and strengthening post-hiring professional training will be important ways to improve the level of vocational faculty in our country.

Keywords: Vocational education, Faculty, Enlightenment.

1 Introduction

Economy and social development requires high level of innovative talents, especially for a large number of educated ordinary workers who can transfer advanced science and technology into practical productive forces. The improvement of laborer's quality has to be achieved through education and more through vocational education which is closely related with economic development. While qualified vocational teaching stuff is the key for vocational technical education quality, scale and pace of development. There is still certain gap between China's vocational education and that in the United States, Germany, UK and other developed countries, so is the teaching stuff. Therefore we can learn from the successful experiences of vocational education teaching stuff, combine it with china's practical situations to strengthen the construction of teaching stuff in vocational school especially in higher vocational schools.

2 The Characteristics of the Construction of Vocational Education Teaching Stuff in Developed Countries

2.1 Rich Sources of Vocational Teachers

The teaching stuff in developed countries is from multiple sources, apart from formal school, there are other channels.

Y. Wang (Ed.): Education Management, Education Theory & Education Application, AISC 109, pp. 189–194.

Specialized vocational education college or technical normal college. In UK, Garnett College London, Hatusfield Technology College, Wolverhampton University and Higher Education Institute at Bolton cultivate teachers for vocational schools. In French, vocational teaching stuff is trained by national apprentice normal schools and higher technology normal universities. [1]

Further education faculty, which can grant vocational technical teacher qualification certificate, should be set in comprehensive universities, technical colleges or normal schools. In Germany, vocational teaching stuff are cultivated in technical universities, some professional faculties in comprehensive universities also train certain vocational teaching stuff. Now German's comprehensive universities and universities of science and technologies have set up specialty for cultivating vocational teaching stuff.

Technical or further education college. In Australia, there are total 92 colleges of Technical and Further Educations (short for TAFE) which are sponsored by Australia governments, some of which provide vocational teaching stuff training. Those with bachelor's degree and rich working experiences can further their study while working at the same time and get vocational teacher qualification certificate after 3 years of study. [2]

Part-time teacher from enterprises. The content of vocational education is closely related with social production practice, specialized teachers and practical guidance teachers should be familiar with industry development direction and trends, and timely update professional knowledge and skills. Therefore, specialized talents from various industries are introduced into vocational teaching stuff as part-time teachers, which are especially obvious after the promotion of Work—integrated Learning teaching mode, part-time teachers from enterprises are an ignorable force in vocational teaching stuff.

Other sources. In French, there is one section in the education department of each school district responsible for recruiting folk artisan as teachers in vocational schools. All specialists who are interested in vocational teaching, if they are advanced in age, they can directly apply to regional council as teaching stuff, while those are relatively young can participate in technical teacher qualification exam to be contract teachers with French Ministry of Education. [2]

2.2 Strict Qualification Requirements for Vocational Teachers

To ensure the quality of vocational education, strict requirements have been set in developed countries, other than diploma and teaching abilities, it also emphasizes professional practice ability and practice experiences.

In Australia, highly qualified teaching stuff with theoretical and practical experiences is the key for successful vocational education. TAFE requires that full-time teachers has to own 3-5 years of professional and practice experiences, all teachers should have professional diploma, educational diploma and industrial certificate of Level 4. [3]

In Germany, there is only one professional title for higher vocational teachers, professor, others are assisted teaching stuff like in training guidance and assisted research. Personnel with professor title must own doctorate degree (If not, they have to be excel in scientific research) and more than ten years of relevant working experiences, familiar with enterprise production and management process. For culture teachers in regular vocational schools, after they take national first test, they have to take vocational educational training for half or one year practice training in enterprises. While after working in vocational school, they need to be a probationary teacher for four semesters with the guidance of formal teachers, then they can be a qualified teacher only after taking the national second test and pass it. Professional theory teachers not only need to finish the curriculum from comprehensive university, but also other three courses: one relevant to profession, one optional course and one pedagogy, after passing national examinations, they can be on duty only after 3-5 years practice in enterprise.

In Canada, the recruitment of vocational teacher depends on education and practice ability, college graduation, master's degree, doctorate degree and more than 5 years of professional practice experiences are basic conditions. [2]

2.3 Combine Full-Time and Part-Time Teachers Together, Raise the Proportion of Part-Time Teachers

In the rapidly changing information society, professional jobs are transforming in acceleration, the training for vocational teaching stuff with new professionals or new jobs is often behind the economic development, so in developed countries, in addition to full-time vocational education teaching staff, there is also a higher proportion of part-time teachers.

In United States, full-time teachers in community colleges focus on teaching basic theory and other courses with professional theories, part-time teachers are composed by entrepreneurs from the community, specialists in certain particular fields, technical stuff and management personnel, whose courses are pertinent and applicable and closely related with technical frontier. The proportion of part-time teacher in community college has exceeded full-time teachers, and took more than 60% of the total teaching stuff. Employing a large number of part-time teachers so that the curriculum become very versatile, in the mean time, the selection of proper personnel to give vocational course enables students to keep up with the latest information technology and ensure the teaching quality. [2]

There is one obvious characteristic of the higher vocational education in French that most teachers are part-time, and some of whom are graduates from previous years. These part-time teachers have grasped advanced technologies and serve as important posts in enterprises for many years, thus they are an important channel for French vocational school to connect with society. Nowadays there are more and more part-time teachers in French vocational schools, while the proportion of full-time teachers is relatively small, but it does not mean that full-time teachers are not important because basic knowledge education is negligible to cultivate prominent and priming students. French keeps adjusting the proportion of full-time and part-time teachers based on vocational education development condition in order to get an optimal teaching stuff proportion and better promote the development of French vocational education.

2.4 Value the Continuous Education of Vocational Teachers

With technology advancement and industry structure changes, enterprises put on more and more requirements on graduates' professional knowledge and skills, so knowledge update and skill improvement are especially urgent.

Germany law regulates that vocational teachers need to participate further education, besides specialized education institutions, many colleges and enterprises also involve in this training. To encourage teaching stuff's further study, Germany requires that inspectors from Bureau of Education give teachers an assessment every four years, and the test results will be related with promotion.

United States provide various advanced study chances for teachers, which are also called "flexible multiple advance study scheme". For example, six-month of night school or summer school for teachers; participate in teaching seminar or workshop; site visit or edit teaching materials, professional magazines or other publications. [2]

From above we can see that developed countries cultivate vocational teaching stuff from multiple channels and give stringent requirements on qualifications, teachers have high level of qualities in education theories, professional knowledge and practical skills, teachers of specialist courses also need a certain years of enterprise working experiences. These countries also take active measurements to establish a relatively stable part-time teaching stuff which takes a high proportion to compensate the shortage of qualified teachers and enhance the spread of production skills and experiences so as to promote the transformation from knowledge to skills. Furthermore, developed countries treat post-work education as a powerful method to improve teaching level, they establish a full set of advanced study system network for vocational teaching stuff in order to update knowledge and improve teachers' skills, which are all favorable for vocational education quality improvement.

3 Enlightenment for China's Higher Vocational Teaching Stuff Cultivation

3.1 Cultivate and Recruit Higher Vocational Teachers from Multiple Channels

Fully play the role of comprehensive universities. Nowadays, specialist teachers in higher vocational colleges are mainly from comprehensive universities, so vocational technical education major should be added in these universities to cultivate professional teaching stuff for vocational education, for example, Tianjin University set up vocational education faculty apart from science, engineering, management and agriculture colleges. When students enter in university, they can study engineering, agriculture or management, then transform into vocational technical education during postgraduate period, they are equipped with professional knowledge and education theories through systematic study and meet the basic requirements for higher vocational teachers.

Develop a well higher vocational technical normal education. China should have a unified plan and rational distribution to gradually establish higher vocational technical normal education network. Various vocational technical normal colleges should make full use of favorable factors from normal education and multiple-discipline coordinated development to speed up reform and development, further explore a characteristic running a school and cultivate highly qualified vocational teaching talents.

Attracts high-skilled personnel to teach in vocational schools. Introduce talents from enterprises to compensate the shortcoming that teaching stuff are lack of practical experiences, which is an important channel for recruiting higher vocational teaching stuff, but its proportion is still relatively low. Schools should attracts enterprise talents with bachelor's degree or above and rich practical experiences from multiple channels, for those teachers in operational post, the academic background should not be a necessity condition, so as to attract social talents with unfettered requirements.

3.2 Further Improve Part-Time Teacher Proportion and Improve the Structure of Teaching Stuff

The proportion of part-time teachers in developed countries is high, which is favorable for students to know the latest ideas, learn newest technologies, or adjust professionals to adapt to market changes and also it is good for reducing education cost and raise the overall efficiency of schools. Therefore we need to emancipate the mind and select technical or management personnel who love education cause to participate in higher vocational education system as part-time teachers. The proportion of part-time teachers can be decided according to practical situation of various faculties, it is not always the higher the better. Meanwhile, in order to form a long-term and relatively stable teaching stuff, schools should establish part-time teacher talent pool and formulate scientific incentive system and teaching quality assessment criteria to better manage part-time teachers and give full play to their initiative and specialties to impart knowledge and educate people. Furthermore, government should supports school from policy and fund perspectives to adopt flexible full-time and part-time teacher system. [4]

3.3 Strict Management on Pre-service Qualification, Strengthen Post-work Training

Developed countries are very strict about recruiting vocational teachers; they give high requirements in academic background, professional knowledge, working experience, practical skills and education theory. In order to improve China's vocational education quality, teaching stuff qualifications should be the first criteria, it needs to have strict management of teacher's pre-service qualifications. The good thing is that many higher vocational schools have put forward comprehensive requirements in academic background, specialty and working experiences in the scheme of recruiting talents into higher vocational schools. For example one national designated higher vocational school has put forward recruitment requirements for professional teaching post as full time master's degree or above and real Estate planning experiences preferred, which combines multiple aspects together like teacher's academic background, professional knowledge, working experiences and skills, thus it avoids the situation that teachers go from one school to another or from one course to another but they are lack of practical experiences and ability.

Put strict requirements on recruiting talents, pertinent further education should be conducted for various teachers during vocational education. Education and teaching courses are for graduates from comprehensive universities, full-time or part-time teachers who did not study education theory systematically, and the courses include vocational school education philosophy, moral culture, teacher etiquette and basic teaching skills. Practical ability should be strengthened for those who lack of working experiences, schools can arrange enterprise training for teachers regularly to know new knowledge and technologies, encourage teachers to take part-time job in enterprises without interfering with normal teaching activities, which would improve teacher's professional practice ability. In the mean time, professional teacher's enterprise practice and the improvement of professional practice ability are set as hard criteria during vocational assessment, then job promotion and economic treatment are related with these criteria to guild teachers into"double-qualified teacher" development.

4 Conclusion

Highly qualified vocational teaching stuff is the essential condition for good vocational education. Higher vocational education cultivates high-skilled practical talents, which put up strict requirements for the professional level, practice ability and teaching skills on teachers. To make the foreign things serve China, learn from the experiences of developed countries and improve teachers' qualification from multiple channels will build a solid foundation for professional talents cultivation.

References

1. Xu, Y.: Surpass Gap: Studies on vocational education comparison between China and foreign countries. Hunan Education Press, Changsha (2009) (in Chinese)
2. Liang, L.: Sources of Higher Vocational Education Research. Beijing Institute of Technology Press, Beijing (2010)
3. Jiang, D.: Recognition on Australian Vocational Education and Training System. Chinese Vocational and Technical Education 5(1) (2007)
4. Chen, J.: Innovate Higher Vocational Education Teachers Mechanism. Continuing Education Research 88(1) (2010)

Study on the Motivation for Saving Behavior of Chinese Urban Residents

Honghui Deng[1] and Jirong Su[2]

[1] No.403Rm, 42Bld, Yuanling Village, Futian District, Shenzhen, China
dhh1970@yahoo.com.cn
[2] No.137, Yanjiang Road, Yuexiu District, Guangzhou, China
sujirong0802@126.com

Abstract. This article uses data of Chinese urban residents in 1980-2007 to test three types of motivations of residents' saving behavior which are life-cycle motivation, bequest motivation and the precautionary motivation. It turns out these three categories of savings motivations all have an impact on the saving behavior of Chinese urban residents. In addition, with China's opening up and ensuing growing external risks, precautionary savings motivation of Chinese urban residents accordingly enhances. In this regard, this paper proposes recommendations on policies to promote consumption and reduces the savings.

Keywords: Saving, Motivation, Consumption.

1 Introduction

2008 financial crisis originated from developed countries led global major economies into recession. China experienced a sharp decline in external demand. According to the Commerce Department, in December 2008, the national import and export value fell 11.1%. Exports fell by 2.8% and imports fell by 21.3%. In January 2009, exports plummeted by 17.5%. Significant reduction in external demand forced China's macroeconomic policy to put again its emphasis on expanding domestic demand. In this round of domestic demand expanding, the increase of household consumption became the new focal point. The Chinese Government has made policies and will continue to take measures to further expand consumption. Expanding the consumption demand of residents and supporting China's economic structural transition from export-oriented economy to a consumer economy can enable people to share the fruits of economic growth[1]. In addition, expanding consumption can adjust the relationship between external demand and domestic demand in medium and long term, which is of strategic significance for the balance of international payments.

However, weak consumption and continuing savings growth are prominent issues currently facing the Chinese economy. Resident consumption rate which reflects the

---

[1] Investment and net exports are the major force for promoting China's economic growth. Consumption share of GDP is too low. In 2007, China's ultimate consumption rate was less than 50%, 28% lower than the world average.

internal consumption level of economy dropped from 53% in the 80s to 38% in 2006. And corresponding to the low consumption rate is the increasing savings of residents in China. In the 8 years from 1998-2006, balance of saving deposits of urban and rural residents grew more than twice times with an average annual increment of 14.8%. Savings growth rate is far faster than the economic growth rate over the same period. By the end of January 2009, resident savings balance in financial institutions in China had reached RMB 23.72 trillion.

Why household savings in China keeps growing so fast? Under the pressure of inflation, resident actual deposit interest rates have been converted to negative. Then what motivates Chinese residents to continue to expand savings? This article is designed to answer these questions. And unlike previous research, this article follows a basic theoretical framework, that is the standard consumer decision optimization model, uses data of Chinese urban residents during 1980-2007, and tests the three savings motivation hypothesis proposed in the text, that is, life-cycle motivation, bequest motivations and the precautionary motivation. So far, no literature has done an empirical analysis of bequest motivation to see whether it affects the saving behavior of Chinese residents.

2   Literature Review

The phenomenon of high resident savings in China has attracted the interest of scholars. Many researchers have done a lot study on China's consumption/savings issue. Corresponding to the development of the theory of consumption, domestic consumption/savings study follows two main research methods:

2.1   Study on Life Cycle Saving Motivation

Study on life cycle motivations follows LCH/PIH as the basic theoretical framework and mainly verifies with what theoretic prophecy the consumption behavior of residents in China are in line with. It seeks to find a proper theoretic pattern[2] to describe consumer behaviors. Kraay (2000) verified that future revenue growth rate has a negative impact on the rural household savings rate but has no significant impact on the population dependency ratio using interprovincial panel data of suburban and rural areas during 1978-1983. Modigliani and Cao (2004) studied the household savings rate in China using total time series data during 1953-2000,and found that there is a positive correlation between household savings rate and economic growth, and the population dependency ratio and the inflation rate also have a positive impact on household savings rate. Using updated interprovincial panel data (1995-2004), Horioka and Wan (2007) found that revenue growth rate has a significant, positive impact on the rate of savings. But population age structures have not been proved to have the expected effects. Therefore, the theory of LCH was partially supported.

One common feature of the literature above is that they use summarized data to study saving behavior. What is worth mentioning is the study on saving behavior of Chinese farmers by Gao Mengtao and etc. (2008). Based the micro panel data during

---

[2] See Sun Feng (2002) for the literature review of this pattern.

1995-2002 of 1056 farmers in eight provinces in China, they estimated the persistent income and temporary income of farmers through the four filter algorithms respectively and studied the saving behavior of Chinese farmers by non-parametric modeling method of income to test the PIH. Estimate of household savings function result unanimously rejected the certainty equivalent standard of the permanent income hypothesis reasoning. The expanded permanent income hypothesis is a more appropriate description of Chinese farmers ' saving behavior.

2.2 Study on the Precautionary Saving Motivation

From the beginning of the last century, more and more researchers are using the precautionary theory to study the problems of consumption and savings in China. Precautionary savings studies are focused on the following three aspects: whether precautionary saving motivation exists; if do, whether precautionary saving are of importance; and how to construct a suitable method of uncertainty measurement. As a matter of fact, the answer to the first two questions depends in part on the specific choice of the latter.

Song Zheng (1999) selects income standard deviation of urban residents in China as a measurement of Chinese residents of the uncertainty of future revenue, uses 1985-1997 years of annual time series data to carry on regression analysis of annual increase value of savings deposits balance of urban and rural residents and concludes that the uncertainty of future revenue is our main reason for residents to save. Wan Guanghua and etc. examine the effect of liquidity constraints and uncertainty on the consumption of Chinese residents by extending the consumption function Hall and utilizing the total consumption data from 1961 to 1998. Results show that enhancement of liquidity constraints and uncertainty increases, resulting in a shortage of China's current low-consumption and domestic demand. Guo Ruotong（2006）and etc. apply the buffer stock model and select 1991-2004 empirical tests of the Panel data of Chinese residents of educational, medical, housing and other expenses of dependency between uncertainty and savings. Results show there is significant precautionary saving behavior of residents in our country, and target savings rate as by the explanatory variables of the model can better explain "precautionary savings" motives of Chinese residents. Documents above are based on national, provincial-level total base and recently some micro-family data in literature began to emerge, such as Wan Guanghua (2003) and Luo Chuliang (2004). The results of these documents all prove that Chinese residents have precautionary saving motivation.

Importance of precautionary saving for residents in China is tested through the calculation precautionary savings strength. Long Zhi and Zhou Haoming (2000) adopt Dynan (1993) theoretical framework, using 1991-1998 panel data on relatively cautious estimation of coefficient of urban residents in China. Results indicate that in sample spacing urban residents in China's relatively cautious coefficient is 5.2 which lead to come to the conclusion that urban residents in China during this period have a strong precautionary saving motivation. Yi Xingjian (2008) applies provincial panel data of China's rural residents 1992-2006 and inspects precautionary saving strength of rural residents and its temporal variation of and regional difference characteristics.

Results show that rural residents in China have strong precautionary saving motivation, and precautionary saving motivation of rural residents is stronger in the West Central and Eastern regions. Shi Jianhuai and etc. (2004) carries quantitative analysis of 1999-2003 years' monthly data of 35 cities in China and the result of relative prudence coefficient is 0.878, which shows that the precautionary saving motivation is not as strong as people think. They think it is because of China's savings has a structural imbalance. Precautionary motive of high income groups should be weaker than that of the low-income, but their savings in the share of the total savings are greater than low-income people. Therefore when adding data to calculate preventive motivational intensity of urban residents, we will get lower preventive motivational intensity.

Although precautionary savings motivation of the high-income is less, their bequest savings propensity is relatively high (Zhu Guolin, 2002). Therefore, mix together different income groups and carry out total data inspection. When high-income group takes a larger proportion in total saving, the bequest savings propensity of the whole society will be fairly strong. At present, researches on bequest savings propensity of Chinese residents are few and there is no empirical study. The research of Zhu Guolin and etc. (2002) on relationship between the weak consumption and income distribution in China studies the relationship between income level and bequest propensity. The bequest propensity among high-income groups is higher, therefore greater gap in income distribution would lead to most wealth concentrated among a few high-income people and the bequest savings propensity of the whole society is stronger. Along this thread of thinking, this article extends the model of savings in the past and employs Gini coefficient to measure overall bequest savings motivation of urban residents.

3   Saving Motive Theory

The studies of inhabitant's consumption may be set out from the perspective of saving, because people's consumption motive is one and only, namely obtaining utility, while the motives of saving are many and varied. Except some motives that obviously resort to psychological and sociological explanation, economists divide saving motives generally into three types: life cycle motive, bequest motive and precautionary motive.

In the analysis of inhabitant's consumption based on the neoclassic economic theory, inhabitant's consumer behavior is to plan the whole life's consumption through assigning the life's wealth, under the condition of intertemporal budget constraint, in order to realize utility maximization in the anticipated life cycle. Life cycle hypothesis (LCH), permanent income hypothesis (PIH) as well as the subsequent precautionary saving hypothesis (Carroll, 1992, 1997) share a common starting point, which is the supposition that rational consumers plan the whole life's consumption through assigning the life's wealth to realize utility maximization in one's life.

3.1   Saving Motive under Definite Condition

LCH/PIH hypothesis consists of two types: the standard and the generalized. The difference between them lies in that, based on standard LCH/PIH hypothesis, the consumer does not consider the influence of posterity's welfare on his/her own utility level, while the generalized hypothesis thinks there exists dynasty utility function[3], that the consumer cares about descendants' consumption, and that besides life cycle saving, there is bequest saving. Considers one CRRC utility function including the bequest motive:

$$U = \int_0^T \frac{C_t^{1-\delta}}{1-\delta} e^{-\rho t} dt + \frac{bK_T^{1-\beta}}{1-\beta}, \delta, \beta > 0, b \geq 0 \tag{1}$$

Among them, t is this consumer's age; T is his/her life cycle; C_t is the consumption at the time of t; ρ is the time discount rate; δ is the consumption marginal utility elasticity; β is the bequest marginal utility elasticity. b is the constant, reflecting the importance level of bequest on the consumer's utility. K_T is the consumer's ending property storage quantity. The consumer's optimal consumption trajectory is (Yuan Zhigang, etc., 2002):

$$C_t = C_0 e^{gt}, g \equiv (r-\rho)/\delta$$
$$C_0 = \phi(r, \rho, \delta, T)(W - K_T e^{-rT}) \tag{2}$$
$$K_T = (be^{rT})^{1/\beta} C_0^{\delta/\beta}$$

Among them, r is the market interest rate; W is this consumer's total disposable income in his/her lifetime. $\phi(\bullet)$ is a solvable function about r, ρ, δ and T. According to equation (2), if b≠0 and δ=β, the bequest motive exists, and one's consumption trajectory in his/her life is:

$$C_t = \left[\frac{\phi e^{gt}}{1 + \phi b^{1/\delta} e^{rT(1-\delta)/\delta}} \right] W = Y_t^p \tag{3}$$
$$S_t = Y_t - C_t = Y_t - Y_t^p$$

Y_t^p is lasting income at t time; S_t indicates t time saving. Suppose the bequest motive exists, present consumption is decided by the total disposable income of a lifetime. Because the marginal utility elasticity of consumption and that of bequest are the same, one's own consumption and his/her posterity's welfare have the identical influence on the utility. Therefore, consumer's saving motive is to smooth the consumption and bequest of a lifetime.

If δ≠β, the explicit solution of C(t) can be achieved according to equation (2). Likewise, consumption is decided by the wealth of a lifetime. It is noteworthy that, the

---

[3] People's utility is not only decided by one's own total consumption, but also decided partially by one's posterity's welfare. Therefore, people all have the bequest motive to leave partial properties to their posterity, and this is the dynasty utility function.

boundary consumption tendency now is no longer a constant which has nothing to do with the wealth (Yuan Zhigang, etc., 2002). The relation of boundary consumption tendency and wealth is in connection with the value of δ and β. Some scholars have estimated the value of δ and β. It is generally viewed that δ>β, and that the more the wealth is, the lower the boundary consumption tendency is, while the higher the saving tendency is. Therefore, the more imbalanced the wealth allocation is, the higher the entire society's marginal saving tendency and the savings ratio are[4]. Based on this, two to-be-examined hypotheses are attained.

Hypothesis I, inhabitant's consumption/saving behavior is mainly decided by the lasting income, the changes of which lead to consumption/saving fluctuation.

Hypothesis II, the bequest saving motive can affect inhabitant's consumption/saving decision-making. The more imbalanced the income allocation is and the more centralized wealth is to the small group of people, the higher the entire society's bequest saving tendency is, and also the higher the savings ratio is.

3.2 Saving Motive under Indefinite Condition

Obviously, the above equations are all inferred from definite premise, therefore they are also called the definite equivalent theory. However, as a result of the uncertainty existence, inhabitant's consumption is not smooth. The precautionary saving theory, on the basis of absorbing ideas of rational anticipation, has introduced the analysis of the influence of uncertainty on the consumer intertemporal choice, holding that consumer saving is not only distributing income equally in the entire life cycle to smooth consumption, but also preventing the occurrence of indefinite event.

Suppose the risk that consumer faces comes from income, that t time labor income Yt is a random variable, and suppose Yt obeys random walk process Yt=Yt-1+et. In the equation, et is an independent identically distributed random variable, and it obeys to the normal distribution with the average value 0 and the variance σ_Y^2. Then consumer's optimization problems are (Yuan Zhigang, etc., 2002):

$$\max_{C_t} E \sum_{t=1}^{\infty} \rho^{t-1} u(C_t)$$

$$\text{s.t. } A_{t+1} = (1+r) A_t + Y_t - C_t \tag{4}$$

Among them, At+1 represents the wealth the consumer possesses at the beginning of the t+1 time. Inferred from the Bellman equation and the envelope theorem, an Euler equation can be attained. Suppose the utility function is CRAR, and substitute it in the Euler equation, the explicit solution of optimal saving function can be obtained:

$$S_t = \underbrace{(Y_t - Y_t^p)}_{\text{life circle saving+bequest saving}} + \frac{\left[\ln\left((1+r)\rho\right)\right]}{\theta r} + \underbrace{\frac{\theta r}{2} \sigma_Y^2}_{\text{precautionary saving}} \tag{5}$$

---

[4] For this reason, in the empirical study employing total quantity data, the influence of income allocation on total consumption must be taken into consideration.

θ is constant absolute risk aversion coefficient. Inferred from equation (3), Y_t^p is the lasting income, representing the consumption level under definite condition. $(Y_t - Y_t^p)$ indicates the saving of life cycle motive and bequest saving. From equation (5), it can be inferred that uncertainty can affect consumer's saving behavior. The bigger the risk is, the more saving is.

Hypothesis III, uncertainty has influence on consumer's saving behavior. The greater uncertainty is, the more precautionary saving the inhabitant has in order to deal with risks.

4 Model Setup, Estimate and Result Analysis

4.1 Model Setup and Variable Calculation

In order to examine the hypotheses proposed in the previous section, this article, taking equation (5) as theoretical foundation, proposes an extended linear saving model:

$$s_t = \alpha_0 + \alpha_1 g_t + \alpha_2 E/M_t + \alpha_3 \text{var}_t + \alpha_4 \text{var}_t \, open_t + \alpha_5 gini_t + \beta X_t + \varepsilon_t \qquad (6)$$

Thereinto, s is the savings ratio, defined as the ratio of saving and disposable income. In order to examine the life cycle motive, the model contains g^5 and E/M, namely long-term real income growth rate and population burden coefficient (Modigliani and Cao, 2004; Horioka and Wan, 2007). var indicates the uncertainty, employed to examine the precautionary saving motive. Moreover, China's reform is carried on synchronically with the opening-up, and especially after joining the WTO, economics of China and the outer world are more closely related. In this circumstance, the cities inhabitants are faced with some extra risks, which influence cities inhabitants' consumption/saving decision-making mainly through affecting their income. Therefore, this article indicates the opening degree by foreign trade dependency degree (open); and the product of the opening degree and income uncertainty shows exterior risk affects the uncertainty of inhabitant income, thus exerting influence on inhabitant saving decision-making. gini indicates the degree of income allocation inequality, employed to examine the bequest motive. The bequest saving is luxury. The saving motives of high-income people are more likely for bequest, so their bequest saving proportion is very high (Yuan Zhigang, etc., 2002). A rather large part of the present total saving of China is possessed by a very small group of high-income people, who have a relatively high inclination of bequest saving (Zhu Guolin, etc., 2002). Therefore, the bigger gini coefficient is, the more centralized wealth is to the small group of people, and the more bequest saving is. X represents other control variables, including real interest rate (RINT) and inflation (INFL). The real interest rate can examine the influence of the financial variable on saving behavior (Horioka and Wan, 2007), while inflation can affect saving behavior through various kinds of ways.

---

[5] The life cycle model points out that quick economical growth and high inhabitant savings are causally related (Deaton, 1992).

This article selects and inspects the 1980-2007 annual data of national cities inhabitants. This period of time has covered all the times of implementing reform and opening-up policy and "only child" policy, and has witnessed a series of significant reforms in middle and late 1990s. Therefore, this period of time can best reflect the characteristics of the saving behavior of Chinese cities inhabitants. Among the data, the per capita disposable income of the urban inhabitants, consumption expense, urban consumer price index, and annual birth rate, etc. comes from "New China 55 Year Statistical Data Assembly", "China Statistical Yearbook" and "China Population Statistical Yearbook". The nominal interest rate indicator is a controversial issue. According to usual practice, this article indicates the nominal interest rate with the weighted value of official one-year saving interest rate, and the data are from the official website of People's Bank of China. The real interest rate is obtained by subtracting urban consumer price index from the nominal interest rate. The gini coefficient is calculated by the author himself according to the method of Chen Xiru (2004).

How to measure the long-term real income growth rate and the population burden coefficient is the key to examine the life cycle saving motive. Referring to the practice of Modigliani and Cao (2004), the average growth rate of the last 14 years is taken as measurement for current long-term real income growth rate. E/M equals the employed population dividing juveniles under 14 years old[6]. Moreover, how to quantify the variable uncertainty is the key to examine the precautionary motive[7]. Due to the application of time series total data, referring to the processing method of Wan Guanghua and so on (2001), this article employs the income forecasting error value square as quantification indicator of variable uncertainty. This article examines the ability of the lag phase to predict current income, and according to its coefficient significance, attains the following equation:

$$\hat{y}_t = 0.707 + 2.219 y_{t-1} - 1.61 y_{t-2} + 0.492 y_{t-3} \tag{7}$$

Based on equation (7), $(y_t - \hat{y}_t)^2$ can be structured as the substitutive variable of uncertainty.

4.2 Model Estimate

In order to confirm data stability, this article carries on the ADF unit root test of the savings ratio, the long-term real income growth rate, the population burden coefficient, the income uncertainty and the gini coefficient:

---

[6] About juvenile data, official has only provided the national general survey result of the year 1953, 1964, 1982, 1990 and 2000. In terms of the calculation of E/M after 2000, referring to the result and practice of Modigliani and Cao (2004), we take the juvenile population in 2000 as foundation, add the newly born population in 2001 which is calculated based on this year's birth rate and total population, and then subtract the newly born population 15 years ago, thus finally obtaining the estimated juvenile population in 2001. The subsequent years are analogously calculated.

[7] Zhu Chunyan and so on (2001) hold that why many precautionary saving theories exist in the present literature is due to people's views of income uncertainty.

Table 1. Variables Unit Root Test Result

Original Series	ADF Value	5% Critical Value	Conclusion	First Difference	ADF Value	5% Critical Value	Conclusion
S	0.212	-2.9798	Unstable	ΔS	-5.416	-2.985	Stable
g	-1.788	-2.9798	Unstable	Δg	-1.951	-1.9564	Stable[*]
E/M	1.442	-2.9798	Unstable	ΔE/M	-2.008	-1.9564	Stable
var	-3.457	-3.5943	Unstable	Δvar	-5.191	-2.985	Stable
gini	-0.319	-2.9798	Unstable	Δgini	-3.878	-2.985	Stable

Note: * indicates being stable under 10% significance level.

The test result shows that, these variables all contain the unit root while first order difference is stable. This article then applies two steps method which Engle and Granger (1987) proposed, in order to examine if there exists co-integration relationship among these variables. The first step carries on the OLS regression of the to-be-examined variables; the second step carries on the unit root test of the OLS return residuals. The Durbin-Watson test indicates that, some models have serial correlation. According to the method of Newey and West (1987), this article adjusts the standard deviation of parameter estimated value, result shown in Table 2. The regression result indicates that, all of the three hypotheses have passed the data test; the life cycle saving motive, the bequest saving motive and the precautionary saving motive have remarkable influence on the saving behavior of Chinese urban inhabitants.

4.3 Results Analysis

Variable g and E/M tested the motive of life cycle savings of urban residents in China. As is shown in table 2, the two variables are significant in all models, showing that life cycle savings motive is an important reason for Chinese residents' high savings rate. The simplest lifecycle model predicted that the long-term economic growth rate will result in savings, with young people saving, and the elderly spending their savings. If the actual revenue grows, young people now are better off than the old people when they were young, with more savings than those old people when they were young and their net saving will be positive. The faster the economic grows, the higher the savings rate will be. Positive Factor g shows that the savings rate rises with the rise of long-term growth rate, and residents' saving behaviors are optimal decisions based on the long-term growth rate of the economy, rather than on the current revenue. Therefore, the results of the empirical research are in line with the theoretical expectation, proving the existence of life-cycle savings motive.

Worthy of note is the E/M, which represents the influence of population structure on the savings rate. Demographic burden coefficient are related to Chinese traditional culture and the Government's "one-child" policy. In traditional Chinese cultural, children have the obligation to support the elderly. Chen Zhiwu (2006) expressed that due to the poorly developed financial markets, and a lack of suitable financial products concerning old-age insurance, the best option for residents is to produce more children for old age. Although there are no formal contracts to ensure that children would support the elderly, Confucian culture impel children from moral constraints to abide by such "invisible" contracts, thus children becoming the personified investment products, a replacement for life cycle savings. However, the implementation of "one-child" policy in 1970s rendered a strict control over the fertility rate, and residents had to accumulate life cycle savings in the form of tangible wealth to replace children, protecting their life in old age. Therefore, the greater the population burden coefficient and the fewer number of children per unit of labor support, the higher life-cycle savings rate will be.

Remarks: *** stands for 1% significance level, ** stands for 5% significance level, * stands for 10% significance level. The t testing in the bracket uses the rectified standard deviation of AR(1).

Gini coefficient measures the income distribution situation of urban residents. Highe Gini coefficient means larger gap in the income distribution of Chinese urban residents, with much more income distributed to a very small number of high income makers. According to the statistics referred to by Shi Jianhuai (2004) and others, in the total RMB savings deposit and total external savings deposit of 2002, the top 20% households who had the highest savings accounted for 64.8% and 89.1% of the total respectively, while the bottom 20% households with lowest savings deposit accounted for only 1.2% and 0.2% respectively. High-income families showed low propensity to spend, but high propensity to bequeath (Zhu Guolin, 2002). Therefore, the rise of Gini coefficient will increase the savings rates of the society as a whole. According to the results from table 2, the Gini coefficient significantly affected the savings decisions of urban residents in China—the greater the Gini coefficient, the larger the chance of bequest. In the early stage of the reform and opening up, the Gini coefficient of urban residents declined year by year, but since 1986, it showed an upward trend. In 2005, the Gini coefficient was 0.3721, close to the internationally recognized cordon 0.4. Therefore, the unequal distribution of income will cause the increase in bequest savings of the entire society, and a sluggish consumption.

Table 2. Estimated Results of the Two-step Method Engle & Granger

Variable	Model 1	Model 2	Model 3	Model 4	Model 5	Model 6	Model 7
g	1.3546*** (4.92)		0.2845** (2.09)		0.0304** (2.14)	0.1244** (2.46)	0.1116** (2.41)
E/M		8.6572*** (10.79)	7.7065*** (7.12)		3.4244* (2.03)	6.0229*** (3.17)	5.9948*** (3.16)
var					0.003** (2.06)	0.0043* (1.95)	
var×open							0.000123* (1.68)
gini				54.1909*** (12.05)	34.1817*** (3.08)	49.4445*** (3.19)	49.0007*** (3.18)
RINT						0.9387** (2.14)	0.9577** (2.18)
INFL						0.7219* (1.86)	0.7384* (1.9)
ADF testing	-3.68	-3.518	-3.757	-3.943	-4.41	-5.208	-5.571
D-W testing	0.516	1.2897	1.3805	1.5083	1.7586	2.0645	2.0868

Variable measures the uncertainties in economy. From the regressed results, it can be seen that the coefficient is positive. Therefore, the rise of the uncertainty degree increases urban residents' propensity to saving, which is called precautionary savings motive. Many existing studies have also found the existence of precautionary savings motive. Since the reform and opening up, China's economy has gone through the transition from planned economy to a market economy. In the context of furthering the reform in income distribution, employment system, education system, medical system and the social security system, residents face a greater risk of future revenue and expenditure, and the vast majority of residents in cities and towns have to cope with the uncertainties of the future through depositing money. Especially since the mid 1990, the "iron rice bowls" were broken, unemployment increased greatly, and the income gap between people were widened. In 2001 China joined the WTO, so the impact of external risk on China's economic operation increased greatly. the result of model 7 also shows that the external risk influences urban residents' savings decisions by affecting the uncertainties of their income. As urban residents in China are now increasingly facing the uncertainties of future income, the precautionary savings motive are also growing.

Unfortunately, the most important reason why the results of this paper cannot be compared with other studies is that they use different types of data and different methods to measure uncertainties. Song Zheng (1999) chose urban residents' income standard deviation as a measure of the uncertainty of future revenue targets. Guo Ruotong and others (2006) used provincial sample data to test the relevance between the uncertainty of residents' educational, medical, housing and other consumptions and savings. Wan Guanghua and others(2003) used large sample peasant households survey data, and described the uncertainty degree through 3 variables, namely the reality whether any family member has a stable joy in government or State-owned enterprises, the level of wealth and the degree of non-agricultural working of the peasant households. Luo Chuliang (2004) used the survey data of urban households and measured the uncertainty via income uncertainty, risk of unemployment, medical consumption uncertainty, education expenditure and other factors. However, this article employed the total time series data of China's urban residents, and due to data constraints, the square of the fault bit of predicted income growth has been used as the quantitative criteria of uncertain variables.

To sum up, this empirical study has verified three hypotheses mentioned before: the savings of urban residents in China contain life cycle savings motive, bequest motive and precautionary saving motive savings; and with the increasing of China's opening up level, the external risks enhance the precautionary saving motive of the residents by influencing their income.

5   Conclusion

This article has analyzed the role of the life-cycle motive, bequest motive and precautionary motive they play in the changing of Chinese urban residents' saving behaviors. The study has found that the three saving motives all affect the residents' saving decisions. Moreover, with the deepening of China's economic opening up, the external risks faced by the residents in cities and towns are constantly increasing, and these risks raise the propensity of precautionary savings of the residents through exerting an impact on the uncertainties of residents' income.

Three messages can be drawn from above in order to stimulate resident consumption and cut down savings. First, predictability of relevant policies should be enhanced, and the risk of uncertainty should be reduced. The reform of social security system, such as the construction of social pension system, health care system, unemployment solving system, and other aspects of social security, should be speeded up, and a social security system that embraces all labors and covers the whole society should be established as soon as possible to reduce residents' expecting level of uncertainty concerning future income, thus directly reducing the precautionary savings of residents. Second, consumer credit services and personal finance products should be largely developed. Improve the credit environment, so that when residents face unexpected changes in income and expenditure, uncertainty can be defused to some extent through consumer credit services. Personal financial products and commercial insurance mechanism should be developed, and the risk faced by the residents can also be reduced through the risk dispersed mechanism of financial market. Third, reduce the Gini coefficient, and improve income distribution situation. Through reforming the tax system, levy a progressive inheritance tax, lift the personal income tax threshold, increase its progressive scale, and support transfer payments for those with low-income, so as to reduce the bequest savings of residents and boost consumption.

References

[1] Xiru, C.: Gini Coefficient and Its Estimation. Statistical Research 8 (2004)
[2] Zhiwu, C.: A Financial Reflection on the Confucian Culture. China Newsweek 42 (2006)
[3] Mengtao, G., Lanlan, B., Huili, S.: Permanent Income and Farmer's Saving: Evidence from the Micro-Panel Data of Eight Provinces. The Journal of Quantitative & Technical Economics 4 (2008)
[4] Ruotong, G., Wei, L.: To Empirically Test the Chinese Saving Behavior by Using Buffer-Stock Model. The Journal of Quantitative & Technical Economics 8 (2006)
[5] Long, Z., Zhou, H.: An Empirical Study on the Precautionary Savings of China's Urban Residents. Economic Research Journal 11 (2000)
[6] Luo, C.: Uncertainty During Economic Transition and Household Consumption Behavior in Urban China. Economic Research Journal 10 (2004)
[7] Shi, J., Zhu, H.: Household Precautionary Saving and Strength of the Precautionary Motive in China. Economic Research Journal 10, 1999–2003 (2004)
[8] Zheng, S.: A Study on the Saving Behaviors of China's Urban Residents. Journal of Financial Research 6 (1999)
[9] Wan, G., Shi, Q., Tang, S.: Peasant Households' Saving Behaviors in the Economic Transition: An Empirical Study on Rural China. Economic Research Journal 5 (2003)
[10] Wan, G., Zhang, Y., Niu, J.: Liquidity Constraint, Uncertainty and Chinese Residents' Consumption. Economic Research Journal 11 (2001)
[11] Yi, X., Wang, J., Yi, J.: The Time-serial Variation of Intensity of the Precautionary Savings Motive and Regional Differences. Economic Research Journal 2 (2008)
[12] Yuan, Z., Zhu, G.: The Aggregate Consumption and Income Distribution in Consumption Theories. Social Sciences in China 2 (2002)

[13] Zhu, G., Fan, J., Yan, Y.: The Sluggish Consumption and Income Distribution in China: Theories and Data. Economic Research Journal 5 (2002)

[14] Carroll, C.D.: The Buffer-Stock Theory of Saving: Some Macroeconomic Evidence. Brookings Papers on Economic Activity 23, 61–151 (1992)

[15] Carroll, C.D.: Buffer-Stock Saving and the Life Cycle/Permanent Income Hypothesis. Quarterly Journal of Economics 112, 1–55 (1997)

[16] Dynan, K.E.: How Prudent Are Consumers. Journal of Political Economy 101, 1104–1113 (1993)

[17] Engle, R.F., Granger, W.J.: Co-integration and Error Correction: Representation, Estimation, and Testing. Econometrica 55, 251–276 (1987)

[18] Horioka, C., Wan, J.: The Determinants of Household Saving in China: A Dynamic Panel Analysis of Provincial Data. Journal of Money, Credit and Banking 39, 2077–2096 (2007)

[19] Kraay, A.: Household Saving in China. World Bank Economic Review 14, 545–570 (2000)

[20] Menchik, P.L., David, M.: Income Distribution, Lifetime Savings and Bequests. American Economic Review 73, 667–683 (1983)

[21] Modigliani, F., Cao, S.L.: The Chinese Saving Puzzle and the Life-Cycle Hypothesis. Journal of Economic Literature 42, 145–170 (2004)

[22] Newey, W.K., Kenneth, D.: Hypothesis Testing with Efficient Method of Moments Estimation. International Economic Review 28, 777–787 (1987)

To Arouse the Passion of Teachers to Become Educators

Yong Ao

School of Earth Science and Resources
Chang'an University, Xi'an, China, 710054
aoyong@chd.edu.cn

Abstract. In this paper, author carried out the objective analysis to the quality situation and the problems of undergraduate teaching, emphasizing the teaching quality awareness is one of the core of teaching quality. However, the current evaluation system for teachers severely restricted the role of stimulating teaching quality awareness. To establish scientific and reasonable teacher evaluation mechanism so that undergraduate teaching should reconsider the original objective, and to arouse the passion of teachers to become educators, it will be become the "intrinsic motivation " of teaching quality assurance.

Keywords: teaching quality, teaching quality awareness, evaluation mechanisms, educators.

1 Introduction

Expansion of higher education since 1999, China's higher education from "elite education" to "mass education" accelerated forward. However, faced with the substantial expansion in the number of students , the inadequate problem of the teaching of infrastructure, the structure of teaching staff, quality and quantity stand out. the reverse sliding trend of the quality of teaching more and more prominent, which is growing all aspects of society on a higher quality of personnel training of higher education calls for a contradiction, showing one hand, society needs a lot of talent on the other hand the university is facing a huge of pressure of employment year, according to statistics of 2008 reached 5.59 million universities graduates [1] people. In Guaranteeing the quality of higher education China has been the emphasis on monitoring mechanism, universities have also established a strict quality control system of teaching, most of them are in teaching the management level. But how does "improve quality" real will internalize to the minds and to each operation of teachers, and become a conscious act, yet to be discussed in depth, but also it is one of the most important issue of higher education, with strong practical significance.

2 The Quality Status of Undergraduate Teaching and Existing Problems

The quality of teaching in universities is not only the high and low test scores of students, but also includes outline of the understanding of teachers, grasp of the

Y. Wang (Ed.): Education Management, Education Theory & Education Application, AISC 109, pp. 209–214.

material, featured organization of teaching content, the choice of teaching methods, the use of teaching methods and channels of emotional communication with students, etc. It is not only by the impact of the quality of teachers, work attitude, emotional impact, but also depending on teaching objects, teaching environment, teaching conditions and other factors. At present, the quality issues of undergraduate teaching and the reasons caused mainly in the following areas:

2.1 Teaching Objective Fuzzy or Only Targeted at the Level of Teaching Expertise and Lack of a Guide of the Development of High-Level Thinking Ability

According to author's survey results of some different types of universities students in discussion-style and daily observation ,it shows that, at present, for various reasons, many university teachers are only working to complete the established expertise of students taught, and the ability involved in other aspects of students are insufficient or do not involve. This shows that teachers will be targeting their own teaching the students understanding of materials and professional knowledge, teaching behavior, and that the only teaching the knowledge we can complete the task of teaching. This is a great distance between the current training objectives and needs of the community. Jaspers [2], about university education in the academic training and professional training, the sermon: "The ideal of professional education in university is to rely on people to achieve, but these occupations have a scientific basis. To reach this point, we need a non-professional elementary education, professor of attitude and method of scholarship. Thus, the best professional education is not only to teach a fixed set of knowledge, instead of training the development of a scientific thought framework, so that we can continue education of mental and spiritual aspects in the process of life. " Who already have a considerable reading comprehension and memory capacity of university students, the activities of teachers teaching should be more inclined to develop this ability what they have use their professional knowledge to analyze and solve problems, and by the high professional teachers the students will be trained in effective professional thinking , and acquire this ability within the discipline of independent thinking and self-exploration on issues. In the passive acceptance of knowledge-learning, this ability of students is difficult to develop. But such problems in the university teacher's daily teaching activities did not attract enough attention and reflection.

2.2 Teachers Lack of Attention to the Results of Their Teaching, Resulting in the Shortage of Classroom Interaction

Most of universities, teaching activities are mainly in the new campus, the teacher in the limited time running the new and old campus, often to come to class hurried , after class to go quickly leaving. It is a more important matter to the teachers what of attendance status of students, as such a situation is good or bad will impact self-awareness and emotions of teachers, and this was also an important part of teaching evaluation; The results of their teaching, teachers are the performance of indifference or less concern. This has been confirmed in Michael survey data (2009) of *Chinese College Graduate's Employment Annual Report* [1]. the satisfaction of the highest the first two to University in 2008 session of "211"and not "211 " universities graduates, is school spirit, books and teaching facilities and are most dissatisfied with the classroom teacher-student interaction, its satisfaction is "211"College 62% and not "211 " 55%. This shows that the majority of teacher with the rich professional knowledge , successful completion of teaching tasks, but lacking of teaching ability or without effective classroom interaction, and thus can not effectively mobilize the enthusiasm of the students. In the class, teaching knowledge is the main way and the participation of students are low-frequency behavior. Dominated the classroom by teachers, students are passive response, the class is dull and unattractive; Between teachers and students to exchange and explore the center of problems did not become one of the most important form of education. It reduce the quality of teaching of developing autonomy and creative talent..

2.3 Expansion Led to Uneven Quality of Students, and Teachers Lack of Confidence and Positive Attitude

On the one hand, with the universities enrollment, the number of students increased rapidly, due to limited teaching resources, most universities implemented class in large classes, the differences among students increases, the difficulty of teaching and management suddenly increased, thereby affecting the quality of teaching. On the other hand the problem of teaching quality attributed to the quality of students, in fact, intentionally or unintentionally denied to their own the influence in teaching activities [3] In teaching activities , to their own influence teachers lack of understanding ,it can seriously affect their efforts in teaching, so that the emotional and effort in work of teaching put into reducing.

2.4 The Center of Teaching Is under Attack, Not Busy, It Often Happens That the Phenomenon of Teaching "Important When Talking, Less Important When Executing, and Not Important at All When Busy"

Now the job classification system for university teachers, there are some drawbacks, mainly as paying attention to research results, teaching achievements overlooked. Evaluation of teachers titles universities not only have clear published papers, patent applications, published monographs , research results and other tasks, but also the corresponding reward system; The general requirements for teachers only provides that the average annual is less than some numbers of hours of class, and less pays attention to teaching quality of teachers. Teaching activities in the reward system is very poor, compared with the research of its is far from award levels. The universities is prevalence of such errors that the research paper is the first, the good research paper is ,the good teachers are; teaching is the second, to meet the minimum requirements for hours on it. The enthusiasm of teachers engaged in scientific research induced by this evaluation mechanisms, but neglected the duty of teaching and educating, Teachers are reluctant to teaching.

Han Yu, a writer in the Tang dynasty, once wrote, "Teachers are the persons who impart truth, teach students and clear up difficult questions". Imparting truth means imparting the rules of conduct and cultivating excellent moral characters, that is, educating people; teaching students means teaching cultural knowledge, that is, teaching booklore; and clearing up difficult questions means answering puzzling questions related to the rules of conduct and the cultural knowledge. These three tasks can be summarized as imparting of knowledge and education of people. The inherent definition of teachers is the persons imparting knowledge and educating people. Modern universities originated from medieval universities in Europe, of which the fundamental mission was to cultivate talents through knowledge teaching since they were established; Berlin University, founded in Humboldt in the early 19th century, adhered to "the principle of unity of university autonomy and academic freedom and also unity of teaching and scientific research," and thus universities can also carry out scientific research; and Wisconsin University in the United States developed a third function of universities—to serve the society. These three aspects constitute the function system of modern universities and are interrelated and interpenetrating, wherein the most fundamental mission of modern universities is to cultivate talents. Any university, regardless of its type or level, should always focus on cultivation of talents. Therefore, it is a priority to establish the scientific concept of functions of universities and deal with the relationship of the three aspects with the cultivation of talents as the center..

3   The Quality Awareness of Teachers Is One Core in Guaranteeing the Teaching Quality of Universities

The teaching quality awareness of teachers refers to the teachers' recognition and evaluation of the degree to which the results of teaching activities meet teaching needs, and also comprises their faith and will to pursue certain teaching results. The teaching quality awareness of teachers includes three aspects: 1, the teachers' recognition of the quality standards for particular teaching activities and the achievement of such standards; 2, the teachers' feelings, attitudes and evaluation of results of their own teaching activities; and 3, the teachers' faith and will in pursuit of certain teaching purposes and ideals[3]. These elements constitute the teaching quality awareness of teachers from three levels, namely, knowledge, feeling and will, which can guide the teachers' teaching behaviors and are important factors affecting the teaching quality. The existing quality control system of each university only achieve the first level, "knowledge," and has no significant achievement in the other two levels, feeling and will.

Then how to enhance the teachers' teaching quality awareness? In its formation and development, the teachers' recognition on teaching quality standards is the basis, their concern about teaching results is the prerequisite, and their faith influencing their teaching is the guarantee. Firstly, academic evaluation criteria established for students in each major serve as the specific basis for teachers' teaching activities, and also the objective basis for substantive supervision of education administration institutions of all levels for the teaching quality. At present, however, universities in China lack specific academic evaluation criteria, shown in considerable arbitrariness and blindness, and thus the guarantee of the teaching quality of universities lack necessary scientific basis. Social development now needs diverse and multi-level talents, and the position adapting to talent needs of universities and the society determines the academic evaluation criteria for students. Although the establishment of the academic evaluation criteria for students is a complex and difficult job and can not be easily carried out in a sufficiently scientific way, it is necessary to help university teachers establish the awareness of quality criteria and lead them to be highly concerned about this problem, in order to achieve the overall healthy development of universities focusing on the improvement of the teaching quality. Secondly, the existing evaluation system for university teachers should be reformed to highlight the original teaching objective. In the field of higher education, the State has set up the Renowned Teacher Award and the Teaching Achievement Award. The declaration and selection of such awards are in fact associated with the assessment of the teaching quality based on students' results, but needs to be more scientific. Meanwhile, the teaching effectiveness can not play an appropriate role in the performance evaluation of teachers. The evaluation and selection can not well stimulate the teaching quality awareness of university teachers. Thirdly, universities should not only widely carry out various academic exchanges, but also should organize discussion and exchange on university education and teaching of different levels. Universities should cultivate high-level innovative talents for each sector of the society through inheriting, spreading, applying and innovating professional knowledge so that such talents can serve the country and society. This also makes the scientific research of universities different from that of special research institutions. The scientific research of universities aims at teaching and cultivation, and thus should focus on both scientific research and teaching. Teaching is the core and foundation of universities. Universities should carry out academic exchanges on teachers' teaching innovation, methods and the like in combination with different aspects of teaching activities.

4   The Scientific and Reasonable Evaluation System for Teachers Should Be Established to Arouse the Passion of Teachers to Become Educators

In July 2010, the Chinese Party Central Committee and the State Council issued the National Medium and Long-term Educational Reform and Development Project Summary (2010-2020), which provide a new opportunity and direction for great development of education. In its preface, it clearly states "comprehensively implementing the quality-oriented education based on cultivation of talents, with the reform and innovation as the driving force, the fairness promotion as the key and the quality improvement as the core, in order to promote the scientific development of education at the new historical starting point, accelerate the transformation from a large country of education to a great power of education and from a large country of human resources to a great power of human resources, and make great contribution to the great rejuvenation of Chinese nation and human civilization." It also clearly proposes a 20-word working guideline, that is, taking the development as a priority, reforming and innovating based on cultivation of talents, promoting fairness and improving quality. The Summary emphasizes "to promote the inspiration, inquiry, discussion and participation type teaching, help students learn how to learn, excite the students' curiosity, develop the students' interests, and create a good environment for independent thinking, free inquiry and brave innovation," and puts forwards that "the quality improvement is the core task for education reform and development." The

National Medium and Long-term Educational Reform and Development Project Summary also highlights that teaching backbones, "double-qualified" teachers, academic leaders and principals should be fostered through advanced study and training, academic exchanges, project financing, etc. so as to bring up a group of famous teachers and leading talents. Favorable conditions should be created to encourage teachers and principals to make bold exploration in practice, to innovate educational ideologies, educational patterns and educational techniques, and to form teaching characteristics and university running styles. Moreover, a large number of educators should be brought up and are encouraged to run universities.

Universities should reconsider the original objective. In order to ensure the passion of university teachers in teaching research, the universities should establish appropriate excitation mechanisms and evaluation systems to encourage university teachers to develop teaching research; actively carry out research on reform of courses and teaching methods and continuously innovate talent fostering modes; excite teachers to participate in international exchange and cooperation; and guide teachers to learn modern education and teaching theories. The universities also should arouse the passion of teachers to become educators, and transform the ideal of teachers to become educators into the "internal motivation"[4] guaranteeing the teaching quality.

It is a basic measure to establish a scientific and reasonable evaluation mechanism for teachers[5-6], so as to improve the teaching quality of universities. The existing evaluation system should be reformed, including the professional title assessment system for teachers, so that the teaching input and teaching process of teachers are more effective and become what can be recognized by the society. When teachers' teaching achievements can obtain the same attention and affirmation of the State and society as achievements of science and research, the teachers' sense of value obtained in teaching and also their teaching passion will be greatly improved. At the same time, teachers should be motivated to execute academic research on teaching, so as to improve their own teaching skills and teaching artistic state, and achieve the harmony and unity of imparting knowledge, educating people, cultivating people with proper management and educating people through service.

References

1. Michael Research Group about Employment of Chinese Students. Chinese College Graduate'S Employment Annual Report. vol. 6, pp. 226–228. Social Sciences Academic Press (2009)
2. Zou, J.: Jaspers. What is education, translated., vol. 153. Awakening Joint Publishing, Beijing (1991)
3. Ying, M.: Improving teaching quality awareness of Universities teachers and the protection of quality. Jiangsu Higher Education 3, 74–77 (2010)
4. Zhang, Z.-h., Chen, L.: Creating a culture of teaching quality to wake up the passion of teachers educating people. China Higher Education 1, 42–44 (2010)
5. Zhou, L.: Teaching Quality Monitoring System Exploration. Education Exploration 3, 97–98 (2010)
6. Liu, Z.: Teaching Undergraduate Teaching Quality Comparison and Reflection between in American and in China. Higher Education Exploration 1, 76–79 (2010)

Studies on Graduate Innovation Training Mode and Interactive Education Platform Based on Internet Environment

Ruzhi Xu, Heli Li, and Peiguang Lin

School of Computer & information engineering,
Shandong University of Finance,
Jinan, China
xrzpuma@gmail.com

Abstract. The internet has expended the connotation and extension of early graduate training mode, training content, training tools and management methods. This paper analyzes the content, means and methods of graduate innovation training in internet environment and gives the design framework of graduate teaching and research platform under network. This paper provides a theoretical and methodological guidance and tool for the realization of the changing pattern of graduate education in internet environment and the implementation of innovative graduate education.

Keywords: Internet environment, Graduate education, Interactive platform.

1 Introduction

As the rapid development of network information technology, there are great challenges of graduate existing training mode, training content, training tools and methods in university. Since there are many differences between graduate education and undergraduate education, the former emphasizes on the students and the role of scientific and research while the latter focuses on teachers to teach and students to learn. The research on graduate innovation training mode in internet environment is not a simple imitation of graduate education. It needs to grasp the education objectives of graduate accurately, and reform graduate training mode, training content, training tools and methods in internet environment, and integrate and make full use of a variety of graduate teaching and research resources in internet environment.

The development of computer and network technology expands the network information resources, and broadens their horizons and enriches methods of access to knowledge and information. However, the existing platform of teaching resources of graduate education generally does not prominent the innovative features of graduate education, but mainly considers to meet the demands of teaching at undergraduate level. At the same time, as teaching resources greatly rely on teaching platform, the duplication is serious and it is difficult to share teaching resources between the different

Y. Wang. (Ed.): Education Management, Education Theory & Education Application, AISC 109, pp. 215–224.
springerlink.com   © Springer-Verlag Berlin Heidelberg 2011

teaching platforms and universities. The sharing of resources generally requires a lot of work of dedicated developers and the integration of teaching resources, so sharing is inconvenience.

This paper systematically studies the content, means and methods of graduate innovation training. It provides certain theoretical innovation and extensive application value for achieving the changing pattern of graduate training in internet environment, for full use of modern network environment to improve the graduate training mode, for integrating network teaching and research resources to improve teaching and research level of graduate, for promoting academic exchanges and cooperation, and for improving research and innovation capability of graduate. At the same time, the application of the research platform of graduate network teaching will make more efficient use of the current software and hardware resources in campus and establish graduate teaching and research resources libraries rapidly and enrich them. It makes important practical significance for online research, learning and communication and enhancing the ability of innovation.

2   The Features of Graduate Education and the Opportunities and Challenges in Internet Environment

In the network environment, with the increasing ways of external learning resources and the amount of information to new knowledge, universities must improve and complete graduate teaching content, teaching patterns and methods, as well as research capacity-building priorities and objectives. Graduate education has new features and faces up with opportunities and challenges in internet environment.

2.1   The Features and Advantages of Graduate Education under the Network Support

In traditional graduate education mode, most universities have adopted the apprenticeship-style education mode of "one to one" or "one to many" apart from the irreplaceable classroom teaching. The internet environment can reduce manpower and time and enrich ways of access to sources of knowledge when compared with the traditional mode and enrich ways of access to sources of knowledge. Online library has the advantages of convenience, fast and cost savings and others when compared with the old campus library. Tutors and students can select the most appropriate way of network communication to make management, teaching, research and other graduate related daily work to maximize efficiency.

Diversification of educational resources and optimization of resource utilization. They can get useful information maximally on the network. The network resources have a huge amount of information and make anyone use or consume the information easily owing to its power of quick, convenient, inexpensive and interactive function. They can communicate with many users without restrictions of location and time in network resources. It improves the communication efficiency greatly and saves money.

In particular, it focuses points of view easily and has an important role in exploring new ideas and promoting innovation.

In network environment, the traditional education resources step over the constraints of time through the network and get maximum use. Also the school staffs learn courses that interest through the network. The universities can make full use of their strengths of subjects and educational resources. They also can transfer the best faculty, the most advanced teaching methods, and the most abundant forms of courseware to learners through the network.

Autonomy of learning behavior and flexibility of communication. The emerging of online education makes students do not have to arrange their own learning time like normal school time. Students can choose their own time and their own chapter to learn according to their own personalized options instead of learning step by step in traditional teaching. Thus it has improved the learning efficiency.

The maturity of network technology makes communication between teachers and students, students and students easy and convenient. QQ, BBS, community forums, personal blog and other communication forms have flexibility, innovation and instance features. Teachers and students exchange learning confusion, difficulty via instant messaging software and teachers can answer the questions in the first time. It reduces communication costs than the face-to-face communication as before.

Personalized training forms and diversity training methods. In online education, students select traditional teaching form, a new form of network teaching or combine the two forms according to their own characters and find the most suitable form. Whatever the form of teaching students select, it is according to the students' own characters and the effect might be better.

The training mode in the network environment can be school-enterprise cooperation or international cooperation. Universities or tutors communicate with graduate that in enterprise through the internet and guide the students to put theoretical knowledge into practice in the enterprise. Tutors give "heuristic" guidance to students. Tutors shift the emphasis to indicating the direction and path of research, and there is no need to detail the content.

2.2   The Challenges of Graduate Education in Internet Environment

The challenge of graduate education in internet environment is the impetuous nature of research. The development of network technology has made it easier for graduates to collect research data, especially second-hand information is more accessible. This behavior makes some graduates impetuous in research projects. Their research was not thorough enough and some graduates even use a pile of paper paste to complete thesis. The academic standard has the trend of declining.

Both the workers of graduate training and graduates tend to rely on the network currently. Most people think that there is anything in network and they can find all information and resources that is needed. They search the internet for a direct answer rather than thinking themselves when they have problems. This way has directly led to teachers' and students' giving up active thinking and problem-solving approach but to pursue for ready-made results. This is undoubtedly very harmful for graduate training of teaching and research ability. With the advance of network technology, the popularity of search engine and electronic resources, the phenomenon of paper plagiarism has been more severe.

Secondly, because of lack of capacity to retrieve network information resources and other related ability of graduates, the resource can not be fully utilized. The collection and collation of network information resources is the basic skill in graduates learning. But some graduates can't grasp the literature retrieval methods and skills. It has become an important aspect that constraints graduate education. In addition, there will be many errors and useless information as the opening of network. Thus training of graduates' ability of identification is also important.

In addition, graduates are so independent in study and research in network environment that they neglect the establishment and guide of their tutor's direction. The emphasis on study and research independently in the network environment may cause the contempt of tutor's direction. Graduates are trapped in the "Network quagmire" and an obscure situation and make a cart behind closed doors. Therefore, both teaching and learning should dialectically recognize the advantages and disadvantages of the training of graduate teaching and research in the network environment to avoid the defects and deficiencies.

3   Graduate Teaching and Research Innovation Mode in Internet Environment

Information resources on the network only provide graduates with research material, research ideas and the related basic conditions. How to use the network resource information to cultivate independent innovation capability is an important part of graduate quality education and the important training tasks and objectives of tutors.

3.1   Graduate Teaching Mode in Internet Environment

The teaching process of network learning mode is student-centered and emphasizes the student's control over the learning process. Students are the subject of information acquisition and processing and active constructors of the significance of knowledge. The learning process becomes self-control. Teachers are directors and facilitators to students' construction of new knowledge. Students should change the traditional teacher-centered concept. Teachers are the school's "software" and provide services for students.

According to changes of learning conditions, the most appropriate teaching mode in network environment is not simply face-to-face teaching mode or purely remote network education mode. It is the mixture of "Network + face-to-face" mode. With the development of "School Link" and the construction of campus network, "Network + face-to-face" mode has become more and more popular among universities. It is meaningful for discussing this teaching mode.

Tutors should take different learning habits of learners into account when they do course design. And they should use a variety of mixed teaching methods in order to achieve good learning results. A number of effective teaching methods that formed in traditional classroom teaching also apply to the teaching in network environment.

Teachers and students can conduct the teaching portfolio that is benefit for the development of creative thinking According to the teaching needs. Thereby it breaks the rigid traditional teaching pattern. Individualized education and fully personalized learning will no longer be the issue that education workers confused and students dissatisfied. Learning to obtain, choose and use information and using computers to process information will be essential survival skills.

In order to improve the graduates' practical experience, collaboration, there are two following teaching methods can be tried in teaching:

Teaching methods of project development: students can gain practical experience and a sense of accomplishment from project development. With teaching methods of project development in network, students can cooperate, exchange different views and upload their findings and conclusions in the project development process on the internet. And they get feedback comments from peers and experts who visit the web page to assess their own learning.

Collaboration teaching method: Collaboration is a learning method that two or more students cooperate mutually to accomplish teaching tasks. This is the fastest growing teaching method currently. Collaboration teaching method is popular in university education abroad. Collaborative teaching methods cultivate the spirit of cooperation. The students who have the spirit of cooperation are welcomed in the labor market.

Tutors and their graduates form a network group in the network environment, and communicate without barrier in "one to more" or "more to more" way. They use mail and other ways to communicate delay or appoint interview time. This benefits the graduates as well as the tutors. For tutors, it is also a process of learning and progress. Tutors and graduates enrich their knowledge structure, enhance the awareness of innovation in practice, and develop innovative thinking.

3.2 Graduate Scientific and Research Mode in Internet Environment

Scientific and research is the core of graduate study. The main difference between graduate and undergraduate education is the stress of the development of research capacity. The research results during the graduate education reflect their own personal values. Cultivating and paying attention to the ability of using the rich resource effectively is an important goal and task of graduate teaching and research ability training in network environment.

In network environment, graduates' learning and research activities will generate some new phenomena: Firstly, accesses to new knowledge and scientific frontiers increase. Tutor's professional guidance and pass of knowledge is only one way of acquiring knowledge source. Network forums, meeting information, and a variety of electronic literature databases have expanded the ways to knowledge and information. Secondly, discussion and learning remotely using network technology has become possible. Graduates discuss or learn only in tutors' institute or laboratory in traditional mode, while they communicate with experts in the field "privately" in the network environment. This is largely helpful to academic exchange and thought communication. Thirdly, inquiry learning mode based on problem has become the main mode of graduate study. The tutor's main task is how to construct a sound scientific proposition. All these indicate that the network as an information tool has significant impact on learning, understanding and building new knowledge for graduates.

Tutors construct valuable issues according to learning objectives, research level of graduates and knowledge structure they have grasped. Then they plan and design the solving process in task-driven form and provide graduates with research path heuristically so as to guide students to study and solve problems with internet resources and other information. In addition, guidance of self-learning is a necessary method for self-improvement: Self-learning includes self-custom learning progress, independent and personalized learning, and self-education. Self-learning should be encouraged not only in traditional teaching but also online teaching. Network is a good support system for self-learning. Through the network, students can visit the online library, virtual museum, and the world famous institute's website and communicate with experts and read the latest academic journals.

4   Framework of Sharing and Interaction Platforms

Sharing is one of the important guiding ideologies of this platform. To combine different educational resources on the internet and achieve easy access to educational resources of this platform by other websites, this platform uses description method based on ontology. It not only achieves the description of local resources and educational resources on the internet but also achieves "machine-readable" of educational resources by Semantic Web. It can greatly facilitate the sharing of resources.

To achieve interaction between teachers and students and among students based on platform, the platform also provides users with a variety communication ways using many technical solutions: (1) Present popular "social network" mode facilitates real-time exchange for online users. (2) To provide all users with personal blog for non-real time exchanges. (3) To provide learning / interest forums to facilitate the exchange of similar users.

The general framework of interactive system of teaching is three-tier structure. This design allows more clearly of project structure and labor division. The platform uses Zend framework development and brings MVC design pattern into the system design. The platform makes full use of zend framework to achieve a favorable environment of MVC design pattern and combines the features of Ajax technology to ensure the structural stability of the system and convenience of late upgrade and maintenance.

The system architecture is shown in Figure 1. The system makes full use of sophisticated hardware and software resources of the campus network, graduate teaching resources and open technology of web to achieve integration of interactive teaching and research platform and distributed heterogeneous educational resources. It solves the problem that graduate teaching and research resources are relatively independent and it is inconvenient for sharing resources in universities.

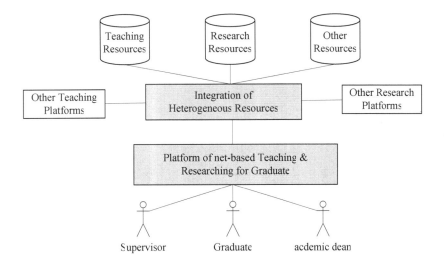

Fig. 1. The platform architecture of sharing and interaction platforms

The system is for three roles and consists of tutor management module, graduate learning modules and system administrators' management module.

Tutor management module mainly helps to do graduate teaching and manage the graduate project. Tutors can easily understand everything about their graduates through the module, including the student's learning, project information, and situation of graduates.

Graduate learning modules provide personal blog, project management, discussion and exchange and sharing of learning resources for graduate. This module can record the entire process of life and learning in school life. Graduates can get access to more and better resources through the sharing area of learning resources and exchange discussions. These resources are learning experiences, conclusion of the study and data that tutor carefully selects. Therefore, it means valuable for graduate learning. The specific function of the module is shown in Figure 2.

Administrator management module is a basic module for interactive teaching system. This module provides system administrators with the set function of blog, discussion forum, project and sharing of resources.

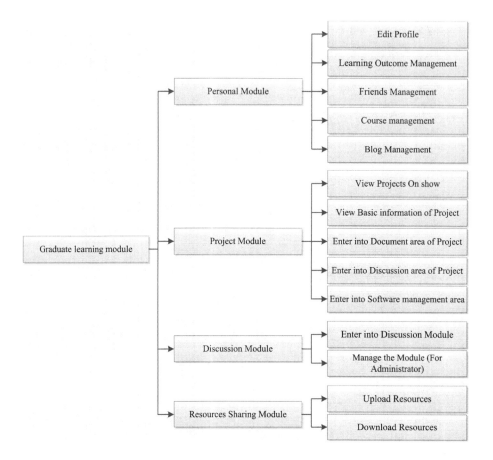

Fig. 2. The function of Graduate learning module

5   The Application of Graduate Training Mode in Internet Environment

From September 2008 to December 2010, Shandong University of Finance has opened the above platform for graduates of management science and engineering, computer application technology and other information technology majors. By tracking the use of study and discussion, experience sharing, daily guidance and thesis writing, the platform has achieved good results.

Graduate students of 2006 and 2007 increased interest in learning and research atmosphere by guidance online and project-based learning exchange. Among the four graduates who graduated in 2009, there were three pieces of excellent theses and one piece of good thesis. Among the six graduates who graduated in 2010, there were four pieces of excellent theses and two piece of good thesis. Graduates of the two grades had a high level of innovation.

Graduate students of 2008, 2009 and 2010 also increased interest in learning and scores by online courses discussions and collaborative exchanges of learning. The average test scores of students of 2009 and 2010 improved by 5.4 scores than scores of 2006 and 2007 in average. The result fully validated the application value of the platform. At the same time, the platform also increased their research interest, and grasped their own research direction early and improved the efficiency of learning and research. From the open title defense, innovation of students of 2008 and 2009 two grades has been improved significantly.

According to the application in Shandong University of Finance, graduates that were trained based on innovative platform have clear improvement in innovation. The convenient online communications between teachers and students, and communications among students not only increased the students' learning and research interest but also broadened their horizons. Students got access to more relevant knowledge and methods about frontiers so as to lay a good foundation for the innovative ability.

The above practice indicated that the platform has good practical value.

6   Conclusion

The contradictions between the need of changing learning and lack of educational resources have become more and more evident in modern society. Traditional teaching mode faces severe challenges. This is the first comprehensive and systematic study of graduate innovation training mode in internet environment. This paper provides comprehensive reference to enhance the teaching level by using the network to improve graduate training mode and integrate network teaching and research resources. It also provides comprehensive reference to improve the ability of research and innovation. In addition, the sharing interactive platform in this paper uses modern technology to combine the schools that have highly level of research and rich educational resources with technical and managerial personnel who scatter throughout the scientific research, production and management positions. It builds the "virtual university" that can meet different learning needs. Graduates and tutors as well as graduates can communicate without time constraints, which play a promoting role in improving efficiency and developing good research atmosphere.

References

1. Liang, L., Wang, F., Liu, S.: Studies on Innovative Talent Cultivation under Network. Journal of Jiangxi University of Science and Technology (12) (2008)
2. Hager, P., Sleet, R., Logan, P., Hooper, M.: Teaching Critical Thinking in Undergraduate Science Courses. Science & Education 12, 303–313 (2003)
3. Junlan, J.: Studies on the Influence of Internet Environment on Graduate Education. Journal of Guangdong University of Technology (03) (2007)
4. Li, Q.: Design and Realization of Graduate Student Interactive Teaching System Based on WEB 2.0. Northwest Universities (2008)

5. Bozeman, B., Corley, E.A.: Scientists' Collaboration Strategies: Implications for Scientific and Technical Human Capital. Research Policy 33(4), 599–616 (2004)
6. Anderson, N., De Dreu, C.K., Nijstad, B.A.: The routinization of innovation research: a constructively critical review of the state-of-the-science. Journal of Organizational Behavior 25, 147–173 (2004)
7. Yan, Y.: Study on Innovating the Model of Training Graduates. Nanchang University (2007)
8. Liu, D., Luo, Y.: Studies on Graduate teaching and learning mode based on the internet environment. Computer Education (08) (2008)
9. Jantunen, A.: Knowledge-processing capabilities and innovative performance: an empirical study. European Journal of Innovation Management 8(3), 336–349 (2005)
10. Sun, A., Mao, Y., Huang, H., Chen, W.: The discuss based on scientific and research mode under network. Journal of Shenyang Agricultural University (Social Sciences Edition) (03) (1999)
11. Zhao Z.: The Discuss Based on Creative Education Studying Pattern under Net Environment. Huazhong Normal University (2004)
12. Sun H.: Analysis of the revolution of graduate education in network environment. Science and Technology of West China (12) (2004)
13. Wang, L., Helian, Z.: Cultivating Strategy for Graduate Student's Capability of Innovation Based on Internet. Research in Teaching (06) (2007)

Exploration and Practice of Training "Applied Talents" in Universities

Yude Liu[1], Fang Wang[2], Qifeng Zhao[1], Jianhui Li[3], Lizi Liu[4], and Yuanyuan Liu[4]

[1] School of Safety and Engineering of North China Institute of Science and Technology, Beijing, 101601
[2] Foreign Languages Department of North China Institute of Science and Technology, 101601
[3] Library of North China Institute of Science and Technology, 101601
[4] Training center of North China Institute of Science and Technology, 101601
lydcumt@126.com

Abstract. Personnel training mode in universities is closely related with the development of enterprises. Based on the study of the relationship between coal mining engineering students and enterprises, which has conducted a thorough and meticulous investigation and research, collected and analyzed the training mode feedback from employers and previous graduates, this paper mainly discusses current higher education problems in training mode and proposes some countermeasures to be taken, establishing a new college's type of applied talents training mode which takes "keeping pace with times, laying solid foundation, seamless connection" as the main module, which is helpful for the reform of higher education training mode of applied talents.

Keywords: Training mode, seamless connection, keeping pace with times, laying solid foundation.

Higher education in the 21st century regards improving the quality of personnel as the starting point of personnel training. For this reason, many domestic colleges and universities have promoted engineering education reform, which focused largely on engineering practice and the practice of teaching system reform to strengthen the environment.

In order to cultivate high-quality talents, and actively explore an applied talents training model which is closely integrated with the enterprises' needs, many higher education institutions at home and abroad cooperate with enterprises, analyze the basic requirements on job competency, capacity building as a goal of college students [1~3]. Many experts believe that universities the curriculum should be in accord with the knowledge and skills required by the students and their future work and get them timely updated, ensure that they meet the requirements at present or in the future [4~7]. How to cultivate applied talents welcomed by the enterprises is an important issue for professional vitality.

This paper begins with the development need of the coal industry, takes the example of the Mining Engineering Professional Training Mode in North China Institute of Science and Technology, and based on the feedback information from the employers and the previous graduates on personnel training model, the reform on coal mining personnel training model, curriculum System updates and the adjust of teaching methods, explores the effective ways of cultivating applied talents in coal mine industry in order to establish a new "applied talents" training model, which can be applied to colleges to carry out personnel training.

1 Survey on Enterprise Needs, Work Demands and College Curriculum

The survey focuses on Luan Group, Shanxi, China Coal Pingshuo, Kailuan Group, Hebei and other employers. 200 mining graduates are surveyed from North China Institute of Science and Technology (2007 to 2010 sessions) on the demand for talents, knowledge structure, the basic quality and capacity requirements through the way of enterprise visits, interviews with graduates, network and other channels to actively carry out research questionnaires.

1.1 Investigation of the Enterprises

Enterprise leaders, experts and technical staff are mainly interviewed in the survey (Figure 1). They collect the information about enterprise demand for mining talents, the current working situation of those mining graduates in the enterprises, the expectation for the students to develop themselves on the part of universities and the enterprises accordingly.

Fig. 1. Enterprise survey talk

The survey indicates that the graduates have been acknowledged by the enterprises in terms of their dedication and stability in work and they have very good knowledge structure which meets the need of their job position.

The feedback information shows that: students' knowledge structures are out of touch with the needs of enterprises. Expertise the students are supposed to have is likely to be much-needed. Some students face an awkward situation because they can not finish their tasks no matter the job requirements are high or low. Compared with postgraduates they lack the capability to do research and development work; compared with vocational high school students they lack the capability to deal with the specific operation.

1.2 Investigation of the Students Employment

Ⅰ） *The relationship between Curriculum and employability*
92.8% of the questioned think that there is a relationship between the employability and Curriculum, of which 38% of the students think that the relationship between the two is large (Table 1).

Table 1. The relationship between Curriculum and employability

Item	sample	Proportion (%)
relevant, and big	84	38.0
relevant, but not big	121	54.8
non-relevant	16	7.2

The findings show that the curriculum plays an important role in improving the employability, which can not be ignored. A good curriculum can motivate students to improve their work capability.

Ⅱ) *Investigation on the Current Teaching Situation*

The existing curriculum satisfaction investigation shows that 27.2% of the students expressed dissatisfaction with the existing curriculum, 57.9% of the students expressed general satisfaction, as well as improvement-to-be, only 14.9% of the students say they are very satisfied (Table 2).

Table 2. The existing curriculum satisfaction

Item	Sample	Proportion (%)
very satisfied	33	14.9
generally satisfied	128	57.9
not very satisfied	51	23.1
dissatisfied	9	4.1

Ⅲ) *Investigation on the quality education*

Most mining students hope to strengthen the mining spiritual and humanistic quality education. The investigation shows that 57% of the mining graduates support the increase of humanistic quality education, 28% in favor, as shown in table 3.

Table 3. Attitudes toward mining spiritual and humanistic quality education

total approval %	try %	not voluntarily %	not necessary %	total%
28	57	12.2	2.8	100

The students' feedback concentrates on:

(a) There is no point learning the courses which are out of date.

(b) 65% of the students think that more practical knowledge and practical skills should be added into the existing curriculum, such as the pressure observation, teaching students to learn some common methods to observe pressure on the working face.

(c) Most of the mining majors think they lack mining mechanical and electrical knowledge, wanting to increase this kind of knowledge. 44.3% think that they only master some of the basic concepts.

(d) 18% find there are so many courses for them to learn that they don't have enough time to expand their knowledge vision.

(e) Its too late for them to contact with professional knowledge. And the professional direction is not specific enough, professional learning is not concentrated.

1.3 Investigation and Analysis of Training Talents

The research shows that the current main problems in Training Mode are as follows:

(a) The existing curriculum does not fully meet the employment needs of most students in such a stage where severe competition and rapid development of mechanized mining exist in today's coal industry.

(b) The talents needed badly by the coal mining enterprises are not those who are good at research and development. They need more forefront engineers and technicians, who can solve the problems on production and technology.

(c) Higher education has long emphasized on the "three attention, three contempt "training mode (pay attention to theory, knowledge, and investigation, while despise practice, operation, and application), which constitute a big contrast training model with the requirements made by the need of enterprise economic development.

2 The Cause Analysis of Current Training Mode Problems

The major causes of "Three-attention, Three-contempt" training mode lie in:

(a) Personnel training mode is out of the line of enterprise demand, the graduates' employability and job requirements do not exactly match.

There is a dislocation between the standards for employee's recruitment in enterprises and personnel training in universities. The university curriculums, educational philosophy, training methods are lagged behind enterprise need or demand. They cultivate graduates who can not meet the standards of enterprises. The reasons lie in that the school curriculum planning is meticulous, not comprehensive, and the knowledge range is narrow; in the content, it emphasizes on single discipline, without considering multi-disciplines and their penetration between courses, content, outdated content is difficult for the students to understand in order to form a good knowledge structure, to inspire students to have positive thinking and creativity.

(b) The lack of Connection between Enterprise and University

Some scholars argue that [8 ~ 10] "a sound liaison mechanism hasn't been established between universities and enterprises, thus effective communication with enterprises is not enough, even difficult to determine a clear training objectives according to market demand for college students". The combination of higher education and enterprise development is not close enough, even out of touch, while enterprises lack the initiative to provide support for the education, which comes from the consciousness of higher education.

(c) In professional personnel training, higher education adapting the narrow, relatively single training mode

The objectives which aim to foster the students' basic knowledge of the specific professional training emphasize the education in goal-oriented professional counterparts, ignoring the students' practical skills and ability to solve practical problems. Therefore, students don't have opportunities to get knowledge in economic management, human quality, the cultivation of spiritual stamina to endure hardship as well as reasonable knowledge structure and ability to build overall quality.

(d) School-enterprise cooperation is not adequate.

The benefit of the combination of university and enterprise is to reflect the enterprise need in the curriculum directly. The students (products) can be produced through the learning of courses which meet the specific needs of enterprises, which was welcomed by the enterprises [11~12]. Although the university is aware of the importance of school-enterprise cooperation, but the implementation of the measures are not in place.

(e) Lack of complex, double-qualified professional teachers

Many professional teachers mostly have doctorate degree or master degree who have done a lot of research work, but they lack mine-site experience, their professional knowledge and expertise is lagged behind, so they can not keep abreast of the developments of coal industry, Their knowledge structure and skills are out of line with current industry requirements.

3 "Applied Talents" Training Mode in University

Based on the problems discovered in the research and the cause analysis, facing the development of the enterprises, universities should change their concepts, amend their training plan, update course content, optimize course system, further school-enterprise cooperation, combine development needs with the industry, promote educational reform, speed up the training industry in order to form the new "Applied Talents" training mode in universities. There are three modules, shown in Figure 2.

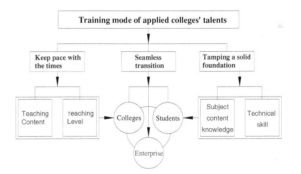

Fig. 2. "Applied Talents" training model System

3.1 Module One: Go with the Times

Including two aspects:

(a) Teaching content should meet the industry needs. In practice, the main content is to strengthen the curriculum construction and teaching reform. Mining Engineering courses should keep up with the pace of technology development of coal industry, update the professional teaching content in time, and appropriate increase the cross-professional courses between mining subject and other subjects. Meanwhile, it is necessary to appropriately increase the training of practical skills of mine for the professional courses, and lay a solid foundation for students to work in the future based on the banding together between the theory and the latest technology. In addition, it can make the professional teaching and industry development to hang closer together, and improve the industry adaptability of professional teaching that universities and coal mines construct actively the professional activities together.

(b) To improve the teacher's teaching standard and practice ability

The teaching reform of "applied talents" training modes asks the teachers for a higher request which not only have a higher teaching and researching capabilities, but have practical ability. On the one hand, colleges should provide the in-service teachers with pre-job training and real enterprise learning opportunities. The teachers may take actively themselves into the enterprise. Therefore teachers can acquaint the need of the enterprises, and set the appropriate courses for students to meet the enterprises' need. On the other hand, colleges should introduce multifunctional talents and the double-type talents into school, and hire the on-site technical management experts as part-time teachers or training teachers. Only if the teachers' abilities and vocational level rose, teachers can become the "clinical expert" in education which analysis, diagnosis and prescribing as doctors to solve the problems of students.

3.2 Module Two: Laying Solid Foundation

The aim of laying solid foundation is to enhance the students' ability to adapt to any situation they meet.

In practice, according to the development trends of industry, we shall use the ability as the backbone, take the adaptation to the demand of coal industry as the orientation, take the employing units and students' feedback information as the basis, aim at all-round development of students, construct modeling course system, and grasp the different levels of the curriculum structure between the aspects of "subject knowledge" and the aspects of "professional capacity" around the idea of "thick foundation, broad caliber, attention application, high quality". The key is to construct the rational knowledge, ability and quality structure for students. Then the talented persons who developed fully in virtue, wisdom and body may be cultivated to adapt to the socialist modernization. And then the general thought of curriculum setup i.e. "the course system of knowledge module" has been put forward.

(a) Academic knowledge module includes both public basic knowledge and professional knowledge. Courses of mechanics foundation and application, computer science, mechanical and electrical science, mining forward technology, management and monitor of mine must be strengthened.

(b) Professional knowledge module is constructed by the practical teaching links which include experiment, curriculum design, practice, training, graduation design and so on. Experimental course are designed to train students' basic skills and practical ability; Internships, training courses are set to train students' comprehensive skills in order to improve their ability of adapting to work quickly. Course design lays the foundation for the graduation design. And the graduation design may stress practicality, and aim at obtaining the basic skills of mining engineers in order to adapt quickly to work.

The modularized curriculum system is an integral designed framework which is layered implemented. This system can implement the professional teaching based on the generosity general education in order to consolidate the basic knowledge and theory, and enhance the adaptability of students.

3.3 Module Three: Seamless Connection

The purpose of seamless connection is to form a tight integration between schools, students and enterprises and to enable them to communicate with each other in time.

(a) To realize seamless connection between talent training and enterprises needs must clear the talent training goals and adapt to enterprise needs actively. The training target of mining engineering talent is taking the coal industry demand as the guidance, taking the training of quality and ability as critical, training applied senior specialized talents with safety awareness, mining spirit, and practical ability to adapt to the needs of industry.

Wide range of knowledge and profound professional foundation which provided the conditions for the mastery of knowledge and the ability of cultivation makes the mining professional students preparing for the different coal mine jobs. If only settle for the learning of professional knowledge, but do not recognize the importance of related mechanical and electrical knowledge in work, the mining major students would be difficult to stick to their own competitiveness.

(b) To realize seamless connection between practice, design and employment

In the process of teaching practice, facing the general problem of the disconnection between theory and practice, or school education and enterprise needs, we adopt a flexible strategy to actively contact enterprises, and arrange students to their employment units for graduation practice according to the phase's demand of enterprise production task. Students may advance into the working state and be familiar with the work requirements by doing graduation design of their employment units based on collecting basic geological data. Such measures may have a positive role in training good character, professionalism and professional quality, and have become a new classroom which can improve the comprehensive quality. During graduation practice and design in employment units, students are guided by the teachers both of schools and of enterprises. After graduation, graduates can adapt quickly to positions due to directly work in this enterprise. The "embedded-type" training mode based on the school-enterprise cooperation achieves the "zero distance" between school education and enterprise needs, achieves the "zero transition" from graduation to employment. The training quality of the mining engineering talents has been really improved.

(c) To realize the seamless connection between productions, learning and studying

The education of cooperation, between production and learning and studying, is an important route and effective platform to play personnel training, scientific research and social service function for applied undergraduate colleges. It can make students to understand the society, and excite the learning motivation, and play an important role for improving the students' practical ability, professional skills, and social adaptation ability. The cooperation education deepens the general sense of the production practice. For example, teachers lead part of students to participate in the enterprise's scientific research at different time, at the same time participate in the enterprise's production operation as a prospective employee identity. Teachers and students may improve the professional quality by participating in each link of the enterprise's production through replacing others work position. At the same time, teachers get the cultivation of engineering consciousness and engineering practice ability in the cooperation process. Thus the classroom teaching may be made more effective by the incorporation between the theoretical knowledge and the real work experience.

(d) To realize the se   mless connection of the communication between students, colleges and enterprises

By using the network technology and taking colleges as the platform, it may strengthen the information communication of supply and demand with enterprises based on creating the communication channel and network between students, colleges and enterprises. The seamless connection of the communication will provide the reliable basis for school education teaching reform, major setting and the direction adjustment of talents training by developing the tracking survey of graduates, coordinating the information feedback and responding to the needs of enterprises and students in time.

4 Main Conclusions

We have taken promoting enterprise development as the core, done the intensive investigation in the enterprises and studies closely around the target of training "applied talents", root-analyzed the found problems, and put forward a new college's type of applied talents training mode which takes "keeping pace with times, laying solid foundation, seamless connection" as the main module.

(a) The "three attention, three contempt" (pay attention to theory, knowledge, and investigation, while despise practice, operation, and application) training mode of higher undergraduate education exists outstanding problems, such as existed apart between the enterprise needs, training mode was single, the combination of school and enterprise did not reach the designated position, the composite and double-type professional teachers was lacking and so on.

(b) The new training mode's connotation of application talents is: to ensure the teaching content and the teaching level keeping pace with The Times, the students possessing a wide range of knowledge, and there existing a close contact between colleges, enterprises and students.

(c) Adjustment and reform the curriculum and teaching content must be adapted to the enterprise fundamental needs based on the feedback information of the employing unit, the experts and the students, then must be repeated cycle modified.

Acknowledgements. Financial support for this work, provided by NCIST Research Fund Project (Title: Preliminary Study on Mechanism and Technology of Aquifer-protective Mining in Shallow Coal Seam, 2009) and NCIST Teaching Research Project (Title: The Optimization Research of Teaching System Based on Enterprise and Student Feedback Information, 2011) is gratefully acknowledged.

References

1. Gu, D., Wu, C.: Student Training Program Combining Two Majors of Mining Engineering and Geo-Technical Engineering. Modern University Education 03, 102–104 (2004)
2. Yong, Z.: Working-learning Combination, School-enterprise Cooperation, Innovating the Construction Talents Training Mode. China Construction Education 6(11-12), 18–20 (2010)
3. Hu, B.-c., Chen, X.-c.: Appraisal of quality and promotion in higher education. Journal of Higher Education 31(02), 6–11 (2010)
4. Hao, L., Bao, L.-y.: An Enlightenment of German Dual System on Training of Applied Management Talents in Human Resources of China. Theory and Practice of Education 30(8), 24–26 (2010)
5. Wang, X.-d.: The Expansion of Social Service Functions of Local Universities. China Higher Education Research 8, 16–17 (2007)
6. Du, S., Li, Y.: Application and Practice of Production-Study-Research Cooperative Education in Application- oriented Institutions. Journal of Ningbo University of Technology 22(4), 89–92 (2010)
7. Bi, Y.-j.: On the Teaching Methods in University. Economic Research Guide 25, 256–257 (2009)
8. Zhao, T.-t., Zhang, Y.-t.: The theoretic research in assessment model in higher education. Journal of Higher Education 29(1), 38–45 (2008)
9. Jiang, L.: On Teaching Mode Reform in HEIs on the Basis of Cultivation of Innovative Talents. Modern University Education 5, 102–105 (2007)
10. Cheng, H., Ma, Y., He, K.: Study on Construction and Culture of Young Teachers Team in Pharmaceutical Engineering Specialty of Independent College. Higher Education in Chemical Engineering 4, 41–43 (2010)
11. Li, Y.: A Study of Construction of Employment Spots for Undergraduate College Students in Guangdong Province. Journal of Jiaying University (Philosophy & Social Sciences) 27(4), 82–86 (2009)
12. Kang, J.: The training of innovative talents and practice teaching reform in universities. Experimental Technology and Management 26(4), 7–9 (2009)

The Exploration on Hierarchical and Progressive Bilingual Education Based on College Students' English Skills[*]

Zhonghao Cheng[1,2] and Yuchai Sun[1,2]

[1] College of Textile & Clothing Engineering, Soochow University, Suzhou, China, 215006
[2] National Engineering Lab. of Modern Silk, Suzhou, China, 215123
{czhcheng,sunyuchai}@suda.edu.cn

Abstract. In bilingual education, foreign language is served as a tool or medium for transmitting information and knowledge. The main purpose of bilingual education is to enable students to acquire the ability of learning professional knowledge with foreign language. In this paper, several practices have been explored in our teaching process. The characteristics of hierarchical and progressing bilingual education have been analyzed. It has been found that above bilingual teaching practices in which different levels of foreign language skills of students have been taken into account will be helpful for the improvement of bilingual education.

Keywords: Bilingual education, language skills, professional knowledge learning.

1 Introduction

To face fierce challenges brought by informatization and economic globalization, international communication ability is one of the most important factors for college students to survive in current workplace. Also a prerequisite for modernization of a country is that its Citizens have higher level of English skills [1]. How to make the college student to master the bilingual learning skills is an important issue for the higher education workers and it is worthy of intensive discussing. According to the requirements proposed by China Ministry of Education in 2001, "Colleges and universities should set bilingual programs, introduce original teaching materials and improve the quality of teachers" [2], some bilingual education programs have been set up in domestic colleges and universities successively, while numerous researches were carried out. Li Mai and Zhaolin Zhou [3] classified bilingual education into three stages: the primary stage, intermediate stage and advanced stage, and proposed different measures taken in various stages of bilingual education. Xudong Wang [4], based on his teaching practice in bilingual education, after analyzing the effect of bilingual education for learning of foreign language and learning of professional

---
[*] The work was funded by the Priority Academic Program Development of Jiangsu Higher Education Institutions.

knowledge and summing the measures used in bilingual education, proposed some recommendations about selection of teaching materials, students arrangement, course selection, etc.

Over the years, authors taught bilingual lectures for the purpose of practices. At this period, the problems reflected from the practice of bilingual teaching have been rethought seriously, and in-depth discussions about its solution have been carried out.

Bilingual teaching is different from the mother-tongue teaching. Except for the ability to learn knowledge, understanding ability with foreign language has great impact on students' learning effectiveness. For engineering students, a prominent issue encountered in bilingual education is that there is a huge diversity of language proficiency among them. If the bilingual teaching was based on the average level of the students' language skill, it is obvious that only the students who have the average English proficiency will be feeling good, but for another part of students whose English proficiency is below or above the average level their needs would not be satisfied sufficiently.

In this paper, the discussion about the definition of hierarchical bilingual teaching is presented firstly. Then some measures to implement bilingual teaching based on the difference of students' English skills have been explored. Finally, the methods to further enhance the effectiveness of bilingual education have been discussed.

2 Hierarchical Design of Teaching Materials

English language has been taught as a course for Chinese students from primary school to college, i.e. so called "foreign language education". The purpose of such education is to train students to master the language skills and their background knowledge. But in the bilingual professional knowledge teaching, foreign language is just employed as "language for education" rather than "language education". Thus the language itself is not the purpose of learning.

For the purpose of reversing student's attitude towards English, the contents taught at the early stages of bilingual teaching should be easy to accept, and language used on course had better as simple as possible, i.e. the language itself should not become a difficulty so that students can't understand professional knowledge. Based on the above point of view, the content of teaching materials should have a hierarchical structure and be expanded step by step.

2.1 Own-Edited Teaching Materials

To reduce the difficulty at the beginning of bilingual teaching, less difficult own-edited teaching materials should be taken to support teaching. Those Own-edited teaching materials had better selected from original English materials and using common words and simple sentences. Meanwhile combining with characteristics of the course, its content should close to life to stimulate students' interest in learning. Its aim is to let students learning professional knowledge with English. The contents of own-edited materials should be compassed of the contents of formal textbooks and its using period should not extend too long.

2.2 Formal Textbooks

After the stage of learning with own-edited materials, students have made necessary preparations for learning professional knowledge. For in-depth, comprehensive, systematic completion of the course, teaching should return to the formal form. Because in the previous stage, part of contents has been presented to help for the understanding of professional knowledge, the contents of formal textbooks could be more easily to understand and can be accepted easily by students.

3 The Implementation of Hierarchical Teaching Based on Students' Actual English Proficiency

In the beginning of bilingual teaching, the main purpose of teaching is to build up students' awareness and abilities to learn profession knowledge with English. So it would be more appropriate to inquiry during class in order to stimulate students' interest.

3.1 The Design of Hierarchical and Progressive Teaching Methods

Bilingual learning requires not only a cognitive ability but also a positive attitude. In the bilingual teaching process, teachers should pay much attention to the needs of vast majority of students, striving to mobilize each student's enthusiasm. Therefore, the design of teaching methods should base on the needs of students' actual English proficiency, and teaching should be executed according to different levels.

3.1.1 The Hierarchical Design of Classroom Inquiries

Appropriate questioning on class would facilitate students to think actively and to focus their attention. The content and difficulty of classroom inquiries should be set differently according to the level of students' language skills.

1）For students with poor English skills

Due to lack of confidence, some students with poor English skill usually fear of bilingual learning, fear of being questioned on classroom. The primary task for teachers is to enhance learners' confidence and passion in the process of learning knowledge with English. Therefore, when to design questions, it had better to choose the question with the sample form, such as using daily vocabulary and simple sentence structures. Also the contents of the question should be easy to answer. The ultimate goal is to make them change the learning form, from teacher-forced, passive learning to positive, self-awareness study.

2) For students with good English skills

For those students with good English skills, their passion and enthusiasm to the learning is relatively high. While setting questions for those students, several measures, such as using professional vocabulary, increasing sentence difficulty, and adopting more complex questions, would be more appropriate to encourage them to improve their learning efficiency. Also that would be helpful to improve their English skills and excite their internal driving force for learning foreign language. However, the stimulus should be appropriate. If the questions were too hard to answer, this benignant stimulation would transform into an inhibition for their learning enthusiasm and shrink them back.

3.1.2 The Rhythm of Teaching, the Fatigue Cycle of Study and Attentions

Even though students usually experienced longtime English education, their language ability of English is still poorer than that of their mother tongue. So in the process of bilingual learning, students need a high concentration of their attention. Some students always translate what they have heard from English to Chinese even in their subconscious, only after that they could complete the process of knowledge learning. In the case of double pressure, their fatigue cycle is much shorter than that taught in their mother tongue. Practices have found that in the process of foreign language teaching lectures, the fatigue cycles for most students are 5~10 minutes in accordance with the difficulty of curriculum contents. If students were forced to listen and think with English continually over this time period, the effect of teaching would become worse. Therefore, the reaction of students during lesson must be carefully observed. Meanwhile, the form and content of lectures should be adjusted appropriately, for example, some of contents that could excite the students' interest should be interspersed into the lecture to refocus their attention. Otherwise, the students' learning effect would be significantly reduced.

3.1.3 The Combination of Chines and English language

"Bilingual education" refers to the process in which both of teaching and learning employ two different languages. Bilingual teaching activities should be mainly based on the use of the second language, i.e. 'foreign language'.

From our practices, it has been known that for students with poor English skills, mixed mode of teaching, i.e. teaching with mother tongue and second language simultaneously, will have a better effect to the students' understanding of the contents. If core contents and nodi of the lecture were introduced at the beginning of the class in Chinese, which will provide background knowledge to students, it would be very helpful for students to accurately understand the information. Meanwhile, a brief introduction with Chinese could give student a sense of security in learning professional knowledge. However, the amount of Chinese used in brief introduction must be strictly controlled in order not to lose the character of bilingual teaching.

3.2 The Diversification of Teaching

3.2.1 Inquiry on Class

Teachers should pay attention to foster students' consciousness to think with English, to encourage and stimulate them reducing or even eliminating the unconscious translation in the learning process. For example, if some questions that were short and easy to answer were designed, while asking student to answer them instantly, it would be helpful to let them to form such consciousness. The questions inquired in lesson should be simple and clear with the character of fast-paced rhythm between Q & As; at the same time they should have a close tie with the contents of teaching and are easy to answer. To sum up, the main purpose of inquiry is to stimulate students to answer questions quickly.

3.2.2 Group Presentation

Another issue in the bilingual teaching is that students have a poor ability in oral expression with English. Group presentation, i.e. group speech, is a higher level requirement compared with inquiry. For example, after setting a number of small topics, students initially conceive their own point of view about one problem and then express it systematically. This process will further train the habit of thinking with English for students. At a group presentation, students usually face just small number of audiences. So their psychological pressure is relatively low. They could be in the more relaxed state to complete the contents of presentation. Therefore, group speech can give students more confidence in the aspect of speaking with English.

3.2.3 Class Presentation

For students who have a better English ability, if they are chosen to give a 3 --- 5-minute speech facing their classmate, they will prepare the speech carefully and strive to express their enthusiasm. On the other side, this activity would be a great challenge for those students who have poor English ability.

The Purpose of these three forms of communication is to stimulate students to learn, to think and to effectively express their views with English directly. All this three forms should not take too much class-time. And their contents must compass the contents of textbook and should not deviate from the main line of curriculum knowledge.

3.2.4 Review and Summary

For Chinese students, the capacity of remembering professional knowledge using English will be much below than that using mother-tongue language. Therefore, giving a brief review about the last time's contents at the beginning of each class and summering the contents at the end of each class will be very helpful.

4 The Form of Assessment and the Hierarchy of Questions

The goal of bilingual education is twofold. Students are required both to complete the study of course knowledge and to master the ability of learning with a foreign language. Therefore, the form of assessment and the setting of questions should be different from the mother-tongue teaching.

1) The form of assessment: inquiry + presentation + examination

To encourage students to treat the teaching process seriously and fully mobilize them to seriously study in lessons, the performance of each link, i.e. answering questions in lecture, giving presentation, etc., should be taken into account in the assessment. The score of above link should be credited into the final results.

2) The link of examination

The main purpose of test should focus on how to foster student's ability of grasping the professional knowledge. For lots of Chinese student, answering question with English has been become a problem for a long time, especially for those questions that are descriptive. The assessment of students' ability should not focus on their English skills, but their learning ability to master the subject knowledge with English. It is more important that let those students who have poor English language skills to effectively express their understanding of the knowledge. It is permissible for students to answer question with Chinese when the question is narrative.

5   Conclusion

In bilingual education, teachers should pay particular attention on how to use of foreign language as a medium for teaching and learning. As students have been used to learning foreign language rather than to learn with foreign language, there are two challenges for students. One is how to enhance their language proficiency as soon as possible. Another is how to grasp the professional knowledge effectively. There is a huge difference between Bilingual teaching and teaching with the mother tongue language. Also there is a big diversity in the following aspect: such as learning attitude, motivation, background knowledge, understanding ability of content, to those students even they are in the same class. Therefore, Bilingual teaching has to be divided into different level according to the ability of students foreign language skills so that the maximum students can benefit from it.

In this paper, combined with the problems emerged from the process of bilingual teaching process, several aspects of hierarchical and progressive bilingual education, such as the formulation of the teaching objectives, materials selection, selection and organization of course content, curriculum implementation, and the form of test methods and test content, have been designed elaborately based on the characteristics of current time's engineering students. The measures of how to best meet the learning requirements of students according to their English level have been discussed. The purpose of this paper is to gain effective improvement for both teachers and students in bilingual education.

Acknowledgments. The authors wish to gratefully thank PAPD for its support and all anonymous reviewers who provided insightful and helpful comments.

References

1. Hu, G.: Borrowing Ideas Across Borders: Lessons from the Academic Advocacy of Chinese-English Bilingual Education. Education Across Borders, 115–136 (2009) (in China)
2. Yu, L.: English–Chinese Bilingual Education. Encyclopedia of Language and Education. Part 5., 1627–1641 (2008) (in China)
3. Li, M., Zhaolin, Z.: Discussion about methods and strategies of bilingual education in different stages. Modern Education Science 4, 173–175 (2010)
4. Wang, X.: The understanding and analyzing of bilingual teaching. China Education Innovation Herald 23, 101–102 (2010)

Teaching Exploration on Software Outsourcing Talents Training

Zhiyu Zhou and Yubo Jia

College of Information and Electronics, Zhejiang Sci-Tech University,
310018, Hangzhou, China
zhouzhiyu1993@163.com

Abstract. In construction of the state specialty computer science and technology, to train software outsourcing talents, we update teaching philosophy. Some teaching methods are implemented as follows: heuristic, discussion, inquiry, case and participation. With large professional platform and small orientation module, we optimize the combination of courses for personnel training in software outsourcing curriculum. Through cooperation between school and enterprise, from the overall, we emphasize the double qualified teacher. Establishing an open educational philosophy ensure that college teaching scientific research and enterprises production operation activities are seamless. The production, teaching and research are integrated to serve the outsourcing talents training.

Keywords: Software outsourcing, Talent training, Teaching exploration.

1 Introduction

With the vigorous development of China's software industry, software outsourcing market demand is very large, but the software personnel cannot meet it. However, the status quo is that there is a social demand for the computer professionals, but a large number of computer science graduates cannot find a suitable job, reflecting that the university's computer education is not suited to the practical needs of software outsourcing businesses. Despite government support, it emerged that a number of software parks, software export base, however, it is still unable to meet the booming software industry needs.

Software outsourcing services industry need to have compound talents with professional knowledge, communication skills, and international background of the export-oriented. Institutions of higher education is characterized by strong teachers with higher levels of theory, it has been an important front for computer software personnel training in China. But many colleges and universities in training computer professionals, lack of knowledge in terms of market demand .With a single model, the students studying in the school only have a theoretical foundation, but their grasp of basic theories and techniques of software development and the actual contact is not enough. Lack of practical experience of software development, it is difficult to meet the actual needs of enterprises.

Y. Wang (Ed.): Education Management, Education Theory & Education Application, AISC 109, pp. 241–246.
springerlink.com © Springer-Verlag Berlin Heidelberg 2011

In 2009, we successfully applied the computer science department of computer science and technology national specialty construction projects, its overall objectives are: reform the existing personnel training, practice model, and with help of enterprises we fully integrate the use of high-quality resources of companies and universities to train personnel for the goal of software outsourcing. Optimization of training program, implementation of high-level, practical, complex type of engineering science and technology education, stress the usefulness and professionalism of personnel training, and improve quality of personnel training in an effort to fight the province of professional advanced degree with a certain brand of professional domestic influence. We cultivate the compound multi-skill talents with a solid professional knowledge about compute science and technology, strong capability of software development practices, skilled in foreign language speaking and writing skills, good communication and organizational skills and strong participation in international collaboration competitiveness and innovative ability for China's information industry, especially the outsourcing business.

A key factor to enlarge and strengthen the software outsourcing industry is personnel, in order to meet today's new demand for software professionals, we have specialized computer applications for teaching reform, actively explore a sense of innovation software training model, innovative teaching philosophy, and comprehensively promote teaching and curriculum research and reform, cultivate teachers for training software outsourcing talents through school and enterprise, establish an open educational philosophy, so that students can improve the innovation consciousness and ability.

2   Update Teaching Philosophy for Software Outsourcing Talents Training

In order to develop service outsourcing talents' innovation awareness and capability, strengthen the hands-on practical ability, according to the requirements of discipline and teaching objectives ,design the content of classroom teaching instruction, imply in practice by these teaching methods, such as heuristic, discussion, inquiry, case, participatory.

Heuristic method of teaching encourages students' initiative and creativity. The emphases are students. It inspires their students to explore their own discrimination under the guidance of teachers, gain knowledge through their own life experiences. Under the heuristic teaching, students not only learn general scientific knowledge and scientific methods, but also a way of thinking, a kind of rigorous scholarship and the spirit of exploring the unknown world.

In the discussion teaching way, we achieve the multi-directional exchange of information and inspiration through discussion and dialogue between students and teachers or students themselves, clarify ideas and ambiguous things through discussion, and expand ideas, active thinking. Teachers guide students to identify problems and solve them so that they deepen their knowledge in the discussion of the understanding. In the teaching of professional courses, we allow students to search for optimum solution through discussion, the teacher evaluate these programs, analyze the advantages and disadvantages and give the idea of improvement.

Teachers can analyze and process teaching materials according to their teaching objectives, students can learn self-awareness and inquiry skills. Teachers create problem situation, guide the students in the research process, and encourage students to identify problems, guide the students to analysis and creative reasoning to solve problems, evaluate the results of the students. Through research topics and questions based on context, students understand the generation of knowledge, formation and development process, learn the knowledge to solve practical problems.

Human resources with strong practical ability, proficiency in commonly used software and software outsourcing and practical technology are the common requirements of software outsourcing business. Therefore, in practice, for project development activities, teachers set the problem, inspire students to find a solution to the problem. With the case from actual development projects, exercise the students' skills, ability to judge and solve problems. This will not only enable students to master the standard software development process, improve the students' ability to think independently, mobilize the enthusiasm of the students, but also help the students have the spirit of team communication and collaboration, comprehensive quality.

Teaching methods such as heuristic, discussion, inquiry, case, teach students to participate in the dominant position of prominence, so we combined the teaching methods to participation teaching concepts to develop students practical and innovative ability adapting to the community as software outsourcing talents.

3 Optimize Curriculum for Software Outsourcing Talents Training

In carrying out the national characteristics of the computer professional building, we are professional with large professional platform and the small direction of course modules. In the context of increasingly complex projects, we advocate for high-level engineering talent with overall quality, which requires that comprehensive curriculum also reflects the trend. On the stage of basic education, we provide students with a thick foundation, wide caliber, and curriculum system. In order to meet the employer community which hopes students specialize professional competence. In teaching we differentiate it into several different stages of little direction. Share in all directions like a professional foundation courses platform to enable students to form a complete knowledge of the system structure, and we set course modules in different directions to meet different social needs.

Increasing general education courses and broadening curriculum base develop the software outsourcing professionals who have a new vision and new thinking. We fully understand the development of cutting-edge disciplines to form a comprehensive, cross, dynamic knowledge and innovative structure, and thus lay the foundation for practical ability to meet the industrial development of China software outsourcing talents of long-term needs. We have added a computer English, computer Japanese, adapted to China's current software outsourcing services market whose main contract is from Japan, the United States, China, Hong Kong and Europe.

In order to train software outsourcing professionals better, computer professionals introduce service outsourcing, software testing and quality management in the construction of national characteristics, and project management, and case studies, trust and credibility of electronic services, software, document specification, software system design structure and system of professional courses. We also added web services, unified modeling language, theory and design of embedded systems, web programming, and database applications, and other professional elective courses, adapted to the software outsourcing needs of the small direction.

Course structure determines the students' knowledge structure, courses optimization should focus on optimizing the structure of knowledge and ability structure of the students. We build software outsourcing for personnel training curriculum which is to enable students to improve the ability of engineering practice. Our practical ability teaching is the most important which need to improve the existing curriculum further. We change the existing "emphasis on theory and basic knowledge, ignoring the content of practice and application" of the course system, establish a complete and correct view of knowledge and ability, so that the integration of knowledge and practice make a better curriculum for engineering practice to meet the goal of training services. Through the course system adjustment, theories and practice course need to be integrated to achieve the mutual penetration. Strengthen training session design and increase the proportion of practical courses and training means. The practice of engineering science and engineering courses in the whole system will be uniform, so that students can get critical thinking, analysis, design and development, innovation ability. Carrying out the construction of the national characteristics, computer professional added the database system, web projects practice, embedded system training, system development and project practice.

Curriculum integration will not only help to eliminate duplicate content among courses, so that similar or closely linked to the content of the curriculum integration, providing students with comprehensive knowledge of the background, better and easier understand the practice process, promoting the migration of practical ability, but also for teaching time compression to reduce the burden on students to theoretical study. Students have more time for project implementation training to create conditions for the formation of practical ability. In the construction of the national characteristics, computer professional introduced network protocol engineering comprehensive experiment, network security comprehensive experiment.

4   Faculty Construction for Software Outsourcing Talents Training

About the training of Software outsourcing teachers, we put emphasis on individual teacher with double qualify construction, moreover, that teaching staff need to be double qualified teachers is also important, make them help each other between different teachers. Help with enterprise, attachment training, create opportunities and conditions for teachers so that teachers can exercise and improve the practical ability to accept the influence of corporate management culture so that teachers possess the necessary practical project experience, training competent students in teaching duties. Enterprise is a good platform to develop the best double qualified teacher. The double qualified teacher enter the actual production in-depth, directly involved in engineering practice, to participate in accident investigation, analysis, processing, technical advice and training to participate in business, research, etc. It not only improve the professional skills of teachers, social work experience, but also build a closer relationship between schools and local enterprise, but also a solid base of teaching practice for the school. Schools must break a variety of rules about teachers use, and eclectic selection of all kinds of practical talents for our own use, so that students can really improve the level of practical ability and actual combat corporate officers to the podium. Employing part-time teachers, enrich the double qualified teacher. Computer professionals in carrying out the construction of national characteristics, see the enterprise software developers as a part-time teacher to practice teaching. Enterprise software developers have a wealth of experience in software development, a high level of engineering ability and a strong engineering practice, so it is easier to connect theory with practice in the teaching and to introduce new things about software updates.

5   Build the Open Education Platform for Software Outsourcing Talents Training

Establishing the correct concept of school-enterprise cooperation in running schools, we must get rid of long-established closed educational philosophy, set in line with market economy requirements for the development of modern higher education and educational philosophy, go with time, get a correct understanding of the importance of school-enterprise cooperation, comprehensively expand and play functions, establish the idea of serving the economic and social development.

Software outsourcing enterprises must realize colleges' advantages in scientific research and personnel capacity, personnel quality training and pay attention to long-term interests and change the profit first thought, do the in-depth cooperation with colleges to establish win-win situation. Healthy school-enterprise cooperation in personnel training of software service outsourcing model is based on development objectives-oriented colleges and industries of software services outsourcing industry an in training of software talent which is in the win-win driving approach. Cooperation mechanism and operation mode from the two core aspects of universities and software companies promote full cooperation, help each other. It is an effective solution to the school personnel training and software outsourcing enterprises employing.

The spirit of "mutual benefit, complementarily, sharing benefits, and risk-sharing", it fully mobilize the enthusiasm of both schools and enterprises, use their advantages, integrate the both operation resources and establish long-term, stable relations of cooperation. While ensuring that teaching and research of college software outsourcing and its production activities of enterprise can have seamless operation. The production, teaching and research are integrated in the training outsourcing process.

6 Conclusion

The existing collage personnel training mode constraints of the software outsourcing industry, it need to continuously explore a wide range of software college personnel training system. In the construction about computer science and technology of projects of the national characteristics professional, according to the characteristic which involved in software outsourcing industry, our school use advanced teaching philosophy, build personnel training curriculum for software outsourcing, and actively reform the traditional practice model, try school-enterprise integration, explore new mode of personnel training software outsourcing enterprises, not only solve the employment problem effectively, and promote outsourcing Industries, while get suitable teachers for software outsourcing talents training.

References

1. Cui, W.: Research on Software Outsourcing Expansion. In: International Conference on Management and Service Science, pp. 1–4. IEEE Press, New York (2009)
2. Ma, J., Li, J., Chen, W., Conradi, R., Ji, J., Liu, C.: An Industrial Survey of Software Outsourcing in China. In: Münch, J., Abrahamsson, P. (eds.) PROFES 2007. LNCS, vol. 4589, pp. 5–19. Springer, Heidelberg (2007)
3. Zhan, G., He, L., Zhou, Y.: Teaching Exploring of Computer Software Outsourcing Talents. Journal of ShanXi Finance and Economics University 32(2), 291 (2010)
4. Wang, B., Liu, Q., Ren, S.: The Study and Practice of the Colleges Training Mode of the Talents for Service Outsourcing. Computer Education (20), 20–23 (2010)
5. Li, Y., Feng, X., Liu, F.: Engineering Students in Teaching Methods of Self-exploration. Journal of Southwest Agricultural University (Social Science Edition) 9(2), 156–158 (2011)

Effectiveness of an English Course with Motive Regulation Strategies on In-Service Adult Learners

Yu-Ping Chang

Yu Da University, 168, Hsueh-Fu Road, Tanwen Village, Chaochiao Township,
Miaoli, Taiwan 361
ptexas@ydu.edu.tw

Abstract. This study aimed to design an English course with motive regulations strategies and to examine the effectiveness of the experimental course on adult English learners. 100 in-service adult learners participated in this study. The instruments utilized in this study were designed by the researcher included a motive regulation learning strategy scale, an English achievement test, and an experimental English course. MANCOVA was used to analyze the data. Based on the results and findings of this study, the researcher also made several suggestions for further research and implications for future EFL instruction.

Keywords: Motive regulation strategies, adult learners, English teaching.

1 Introduction

In the trend of globalization, lifelong learning is the most important measure to keep up with the world and rapidly absorb knowledge. Conditions of lifelong learning refer to "what to learn," "how to learn," and "why to learn." Learning method of "how to learn" and learning motive of "why to learn" are keys to enhance and maintain lifelong learning. Graham (1996) indicated that at present, the most common, long-term and overall educational issue is motivation.

Although "what to learn" and "how to learn" are important, without initial and persistent motives, they will be in vain; thus, "why" of "why to learn" can be the key. Regarding learning motives, self-regulated learning theory suggests that motive regulation means learners set goals, monitor, adjust, and control their cognition, motive and, behaviors to accomplish goals. It is an active construction process. Diverse and appropriate motive regulation strategies allow learners to monitor and control learning, which in turn will positively influence the learning and performance.

In order to bridge the gap between theories and practices, and respond to practice-oriented research, this study treated adult learners as subjects and English as the field to probe into the influence of experimental course on learners' English learning and try to recognize learners' learning situation. The purposes of this study were to design an English teaching experimental course with motive regulation strategies, to study the effect of the experimental course on adult learners' English learning performance, and to generate conclusions and suggestions for future studies.

Y. Wang (Ed.): Education Management, Education Theory & Education Application, AISC 109, pp. 247–252.
springerlink.com   © Springer-Verlag Berlin Heidelberg 2011

2 Review of Selected Literature

According to Pintrich's (2000b) view of motive regulation learning, learning is an actively constructive process. Therefore, motive regulation learners will actively use motive regulation learning strategy to fulfill their learning goals. Cherng (2002) suggested that self-regulated learners have the characteristics below: first of all, they had diverse strategies in learning process; second, they would select different motive regulation strategies according to the characteristics of learning situations. Newman (2002) also suggested that self-regulated learners usually have sets of strategic tools to deal with academic challenges and know how to use the proper strategies at the right time. Therefore, the use of motive regulation learning strategy will further indicate motive regulation learning process. "Motive regulation learning strategy" means learners' different strategies and approaches in the motive regulation learning process in order to monitor, regulate, and control their cognition, motives, behaviors, and situations (Lin, 2006).

There is much reserach on motive regulation learning strategies; however, each scholar suggested different classification frameworks. Based on Chang and Wu's (2010) study, they divided motive regulation learning strategies into affection/motive regulation and willpower regulation, as shown below.

2.1 Theories and Studies Related to Affection/Motive Regulation

Early educational policy and concept focused on learning cognition; thus, most research related to learning strategies referred to cognitive strategies. However, current research findings (e.g. Dornyei, 2003; Masgoret & Gardner, 2003; Peacock & Ho, 2003) indicated that negative learning affection and attitude, and the lack of learning motive would negatively influence English learning. Without students' learning motives and interests, not even a good learning strategy will lead to a positive learning performance.

According to Chang (2008), affection included individual emotion, feeling and mental states. Motive means to trigger and maintain a person's internal motive in certain activities. It can lead the individuals to certain goals. Value-expectation theory of motive indicated that learners' affection/motive of motive regulation learning includes affection, values, and expectations (Eccles, 1983; Wigfield & Eccles, 2000). According to the related literatures (e.g. Wigfield & Eccles, 2000; Wolters, 2004), self-efficacy is critical in affection/motive regulation. Thus, this study constructed a theoretical framework of English learning affection/motive regulation of students by value-expectation theory and Bandura's (1986, 1997) self-efficacy theory.

2.2 Theories and Studies Related to Willpower Regulation

The concept of willpower control regulation derived from the motive regulation learning theory; as suggested by the name it involves willpower (Pintrich, 2000a). Scholars have different views on willpower regulation strategy. For instance, Kuhl (1985) proposed action control strategy, and Corno (1989) further expanded Kuhl's action control strategy and applied it to learning situations. Pintrich (1999) proposed willpower control strategy from the perspective of willpower. Learning strategy teaching should value the cultivation of students' willpower. Learning willpower and proper strategic application enhances learning. After clarifying the classification framework of willpower control strategy, this study treated Pintrich's (1999, 2000b) classification of willpower control strategy as the basic framework and referenced Wolters' (1999) and Chang's (2008, 2010) views. The content of willpower control strategy was classified by behavior regulation, affection regulation, others regulation, and environment regulation, which further become the four measures in willpower control strategy.

3 Methodology

Participants involved in this study were 100 in-service adults from a company in Taiwan. Before this experiment, they already received at least 6 years of formal English instruction in secondary education, and they were taking an English course once a week for three months. This English course required learners and the teacher to meet three hours every week. The same instructor was teaching both classes involved in this study. The participants were randomly assigned to either a control or an experimental group. Participants in the experimental group received experimental teaching of "motive regulation learning strategies in English course;" participants in the control group received "ordinary English teaching course." The instruments employed in this study were Motive Regulation Learning Strategy Scale, English Achievement Test, and an Experimental English Course. MANCOVA was adopted to test the effect of the teaching experiment.

4 Results and Findings of the Study

This study conducted the test by MANCOVA. Since regression lines of the groups were parallel (Wilks' Lambda = .948, $p > .05$), a common regression line could indicate the relationship between the dependent variables and covariance. The researcher tested common regression line and showed that slope was not 0. Therefore, statistical analysis was based on covariance.

Table 1 showed that the experimental and control groups obtained significantly different scores on English affection/motive regulation.

Since the between-group difference was significant, a confidence interval test was conducted on the adjusted means of the four sub-scales of affection/motive regulation in two groups. The results are shown in Table 2.

According to the result of the confidence interval test, the scores of both groups in the four sub-scales of affection/motive regulation were significant. In other words, students who received the experimental course performed better than the other group in the sub-scales of affection/motive regulation.

Other results and findings of this study included the experimental group who received "motive regulation strategy in English teaching experiment course" and the control group who received "ordinary English teaching" also performed differently regarding willpower regulation and English achievement; "motive regulation strategy in English teaching experiment course" could trigger students' learning motives, and

Table 1. Summary of Affection/motive Regulation on the Experimental and Control Groups

Variance Sources	(SSCP)'					df	Λ
Covariance	20 .882					3	
	13 .934	20 .535					
	12 .296	13 .690	25 .416				
	14 .600	12 .932	9 .675	22 .535			
Between-groups	187 .585					1	.828*
	267 .674	381 .958					
	322 .217	459 .788	533 .476				
	256 .202	365 .588	440 .082	349 .920			
In-groups	2944 .295					141	
	1964 .718	2895 .451					
	1733 .721	1930 .260	3583 .614				
	2058 .570	1823 .441	1364 .123	3177 .421			

Table 2. A Confidence Interval Test of Two Groups in Four Sub-scales of Affective Reaction

	Confidence Interval Test	Difference Direction
Positive Affection	$.784 < \hat{\psi} < 3.820$	2.302*
Value	$1.779 < \hat{\psi} < 4.790$	3.285*
Expectation	$2.279 < \hat{\psi} < 5.629$	3.954*
Self-efficacy	$1.567 < \hat{\psi} < 4.721$	3.144*

maintain their learning motives, as well as enhance students' effective learning; and learning is an integrated activity and thus cannot only focus on affection/motive, willpower regulation, or cognition of the learning performance. Only the combination of these three aspects will result in the maximum learning benefit of students.

5   Suggestions

Based on the research findings, related theories and previous review were added to generate suggestions for teaching and future studies.

5.1   Integrating Motive Regulation Strategies in English Teaching Plans

In recent years, the government has been significantly promoting English teaching policy for in-service adult learners. However, English proficiencies in classes vary widely. It hampers teachers' plans and causes. Lack of confidence and failing scores on English tests are the important factors that low-achievement adult learners give up on English. In order to get them more involved in English learning, enhancement of interests and maintenance of motives are priorities in teaching. In this study, motive regulation strategy teaching effectively triggered adult learners' English learning motives, allowing them to maintain their values, expectations, and affection. It also enhanced adult learners' English learning performance. The results showed that it was more effective for state-oriented adult learners who had difficulty using motive regulation strategies. Therefore, it is suggested that when teachers are teaching English to adults should include motive regulation strategies into their course. It will enhance adult learners' learning effectiveness.

5.2 Examining Motive Regulation Strategy Use beyond the Classroom

Currently, most research on self-regulated learning are still based on the perspective of social cognition and treated self-regulation as the interaction among individuals, behaviors, and environments. Therefore, learners adopt different regulation strategies according to their learning environments. According to literature review, measurements on the use of regulation strategy were usually based on situations in the classrooms. Few of them were in informal learning environments. Motive regulation strategy teaching in this study was mainly involved in classroom teaching. For learners, learning activities and work in the classrooms were mostly arranged by teachers; they were hardly dominant. Once learners left the classrooms and could freely select their own learning activities, they were better motivated, had fewer distractions, and performed better.

5.3 Analyzing Motive Regulation Strategy Use of Learners of Different Learning Stages in Different Subjects

This study treated in-service adult learners as subjects and included motive regulation strategies in English teaching materials. In an English course, learners were instructed about motive regulation strategies. The research findings indicated that intervention in teaching would increase adult learners' use of motive regulation strategy and performance of motive belief, motive involvement, and academic performance. However, will learners use the same motive regulation strategies in different subjects? It is wroth to study because there are still few related studies in Taiwan. In the future, there should be more research study motive regulation strategies of learners in different learning stages. Future studies can continue studying the influence and enhancement on learning activity and performance after the learners absorb motive regulation strategy.

References

1. Bandura, A.: Self-efficacy: The Exercise of Control. W. H. Freeman and Company, New York (1997)
2. Bandura, A.: Social Foundations of Though and Action: A Social Cognitive Theory. Prentice-Hall, Englewood Cliffs (1986)
3. Chang, Y.P.: The Effect of an Experimental English Course on English Performance and Self-regulated Learning of EFL College Students. Taiwan ELT Publishing Co. Ltd., Taipei (2008)

4. Chang, Y.P.: A Study of EFL College Students' Self-handicapping and English Performance. Procedia-Social and Behavioral Sciences 2, 2006–2010, (2010)
5. Chang, Y.P., Wu, C.J.: Self-regulation in EFL College Students' English Learning. In: 8th Hawaii International Conference, pp. 2948–2959, Honolulu (2010)
6. Cherng, B.L.: The Interaction among Multiple Goals, Motivational Problems, and Self-regulated Learning Strategies. Journal of Normal Taiwan University 47, 36–58 (2002)
7. Corno, L.: Self-regulated learning: A volitional analysis. In: Zimmerman, B.J., Schunk, D.H. (eds.) Self-regulated Learning and Academic Achievement: Theory, Research, and Practice, pp. 83–110. Springer, New York (1989)
8. Dornyei, Z.: Attitudes, Orientations, and Motivations in Language Learning: Advances in Theory, Research, and Applications. Language Learning 53, 3–33 (2003)
9. Eccles, J.: Expectancies, Values & Academic Behaviors. In: Spence, J.T. (ed.) Achievement and Achievement Motives, pp. 75–146. Freeman, San Francisco (1983)
10. Graham, S., Weiner, B.: Theory and Principles of Motivation. In: Berliner, D.C., Calfee, R.C. (eds.) Handbook of Educational Psychology, pp. 63–84. Macmillan, New York (1996)
11. Kuhl, J.: Volitional Mediators of Cognitive-behavior Consistency: Self-regulatory Processes and Action Versus State Orientation. In: Kuhl, J., Beckman, J. (eds.) Action Control: From Cognition to Behavior, pp. 101–128. Springer, New York (1985)
12. Lin, Y.Y.: The Relation and Latent Chang Analysis among Individual Goal Orientations, Classroom Goal Structures and Self-regulated Learning Strategies. Master's thesis. National Cheng Kung University, Tainan (2006)
13. Masgoret, A.M., Gardner, R.C.: Attitudes, Motivation, and Second Language Learning: A Meta-analysis of Studies Conducted by Gardner Associates. Language Learning 53, 123–163 (2003)
14. Newman, R.S.: How Self-regulated Learning Cope with Academic Difficulty: The Role of Adaptive Help Seeking. Theory into Practice 41, 132–138 (2002)
15. Peacock, M., Ho, B.: Student Language Learning Strategies across Eight Disciplines. International Journal of Applied Linguistics 13, 179–200 (2003)
16. Pintrich, P.R.: Multiple Goals, Multiple Pathways: The Role of Goal Orientation in Learning and Achievement. Journal of Educational Psychology 92, 544–555 (2000a)
17. Pintrich, P.R.: The Role of Goal orientation in Self-regulated Learning. In: Boekaerts, M., Pintrich, P.R., Ziedner, M. (eds.) Handbook of Self-regulation, pp. 451–502. Academic Press, San Diego (2000b)
18. Pintrich, P.R.: Taking Control of Research on Volitional Control: Challenges for Future Theory and Research. Learning and Individual Difference 11, 335–355 (1999)
19. Wigfield, A., Eccles, J.S.: Expectancy-value Theory of Achievement Motivation. Contemporary Educational Psychology 25, 68–81 (2000)
20. Wolters, C.: Advancing Achievement Goal Theory: Using Goal Structures and Goal Orientation to Predict Students' Motivation, Cognition, and Achievement. Journal of Educational Psychology 96, 236–250 (2004)

Application of RMI Thinking Method Based on Mathematical Optimization[*]

Wei Wang, Xiaohui Liu, and Dawei Zhao

School of Postgraduate, Harbin University of Commerce, Harbin 150028, P.R. China
wangwei1215@yahoo.cn

Abstract. Mr. Xu Lizhi, a well-known scholar of China, proposed a method of relationship mapping inversion in late 20th century, which is a common method of dealing with the problem, a working principle of a general scope of scientific method, which solves the problem of the relationship structure taking 2 steps, mapping and inversion. Therefore we name this method as relationship, mapping and inversion, referred to as the RMI method. This paper analyzes on the core problem of optimization thinking by RMI method, mapping the answer of aim problem in mapping relation construction, give us a certain science thinking method to study the real problems by applying mathematical optimization.

Keywords: RMI method, mathematical optimization, relationship, mapping, inversion.

1 Introduction

Qiebiqifu, a Russian mathematician said, mathematics is established and developed under the influence of all the common basic issues of human activity, available data at our disposal to maximize profits. The individual, group or society as a whole, no matter how both big and small tasks operation at the time to act a certain way, always select one of the best programs from all possible options.

Any mathematical elements of two classes of mathematical object or two sets were found a correspondence, which defines a mapping. If it is one to one relationship, it is called reversible mapping, such as linear transformation in algebra, projective transformation in geometry, substitution of variables, function transformation, series transformation in analytics, which are all examples of mapping.

Let M be a map, which take elements of a set, R={a}, into another set, R*={a *} (or mapped over), where a* represents a image, a is a original image. It as follows:

$$M : R \to R^*, M(a) = a^*.$$

Particular, if M is a relational structure, M can be mapped over R*, then

$$R^* = M(R).$$

---

[*] This work is supported by Social Science Foundation under Grant 08C066 and Youth Science Technology Foundation under Grant QC07C117 and Science and Technology Research Project of Department of Education under Grant 11551112 in Heilongjiang Province.

Y. Wang (Ed.): Education Management, Education Theory & Education Application, AISC 109, pp. 253–257.
springerlink.com © Springer-Verlag Berlin Heidelberg 2011

In specific applications of solving practical problems, we need to distinguish the traits of an unknown object in the relation structure of R, which can be called the original goal image for x, and the target image for x* = M (x) which is the mapping image of x based on the map M.

RMI basic principles can be described as follows: R represents a group of the relationship structure of the original image, in which contains x, the original image, to be determined. Let M is a map, correspondence principle, by which we assume that the relation structure of original image, R, is mapped into the relation structure of mapping image, R*, which contains the mapping image x* of the unknown original image, x. If there is a way to get x*, x can also be determined by the inversion, inverse mapping. The basic principle of RMI method can be figured as follows:

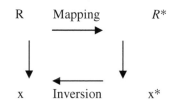

Fig. 1. The relationship structure of RMI method

The basic steps of RMI can be summarized as follows:

Relationship → Map → Mapping→ Inversion→ Solution

The map by which RMI method can be used to solve the problem must be feasible mapping. In terms of determining the original image, for the relationship structure R in which contains target original image X, if there is such a reversible mapping M, R is mapped into the image relation structure R*, in which target image, X*= M(X), can be determined by some limited mathematical procedures, can, which is regarded M as the feasible mapping.

2   Theoretical Analysis of Mathematical Optimization

In order to solve practical problems, to improve analytical scientific, simplicity and operability, it must be possible to show taking action of problem solving into mathematical relationships, obtaining the optimal solution by mathematical ways. For example, there are different methods can be used to transport some goods from A to B, along the water transport, road transport, rail transport, air transport is also available. Assume that we want to choose one of the most expensive transport routes, namely make transport values as the decision-making principle. In determining the magnitude of value may involve some mathematical problems, but the state is not for the optimal mathematical problems. Now we assume that the transport capacity on each route is limited and many different goods must be shipped from A to B. What kind of goods are along the transport route, how many do this, so that total transport cost is the minimum (in a particular route in each unit of goods, the transport cost is different.)

Some variable can be regarded as a decision-making principle, such as total costs of transportation program or the strict priority of all suggestions, which can be called the best target of selecting principle. In some problems, people try to get the maximum possible value, such as maximum value, maximum profit, etc., other problems were required to achieve their minimum value, such as minimum transportation cost, minimum production cost, etc. Maximum and minimum are often linked with a term, extreme, which is the theoretical core of optimization.

To make the final solutions into the form of mathematical problems, target value must be determined by the defined solution. In other words, target value belongs to its function, should be called objective function. People tend to choose the independent variables which are called control parameters. In the above example, the independent variables represent quantity of goods using different transport lines. The objective function represents relationship function of total transportation cost.

In addition to the objective function and the independent variables, the constraints play an important role in the problem study. In the above example, the total amount of each type of goods and the transport capacity of each transport lines are the constraints, is usually represented by mathematical relationship, equality or inequality. Summing quantity of goods transported by different lines should be equal to its total. Summing quantity of goods transported by a line should be less than or equal to its transport capacity.

3　On the Mathematical Optimization by RMI Method

Optimization problems were discussed by RMI method, that is, for a relationship structure, R, containing target problem, if there is such a reversible map, M, R will be mapped into the image relationship structure, R*, in which the corresponding image problem can be conquered effectively by some limited mathematical procedures. In terms of the determination of the original image, we can determine the original image of the required goals, x, from the target image x * practicably. To replace the object to the presentation of the problem, so target problem can be get from image problem feasibly.

In order to effectively use RMI method to solve the problem, we can introduce an appropriate map and must grasp the critical step of mapping. The so-called mapping is that the target image, x*, is determined in image relationship structure, R*. For reaching the above mapping process, the introduced map must be mapped, namely target image, x*, is found by limited mathematical procedures in image relationship structure system R*. Mathematical interpretation of optimization thinking is directly reflected from mapping process. It is typical mapping of RMI method that how to obtain the optimal solution of the actual problem based on constraints.

Assuming the objective function, f(x), is given, D is a feasible alternatives set or opportunity set, the general optimization problem is to find x to make f (x) as maximum or minimum as possible in D set. General form is as following,

$$max\,(min)\,f\,\,(x)\,\,(f \in Z)\quad constraints:x \in D.$$

The problem will be solved as that for specific f and D, f and D which conditions can ensure the existence of objective solutions? How do we get the optimal solution? Assuming f is continuous, the neighboring points in D are converted into the points in Z. If the range of f(x) is a bounded set in R, then there is a real number f (x) can not exceed. Further, if the range of f(x) is the closed set in R, then the limit of any sequence consisting of the value of f(x), f(x*), is also bound within this set, so the range of f(x) is a compact set, there must be an optimal f (x), when f is a continuous function, we can find x* while f(x*) is optimal, namely that continuous functions defined in the compact set can be achieved in the maximum or minimum.

When D is a compact set, and f is a continuous function, overall optimal value can be ensured existence. However, when D and f is more complex, it is difficult directly to find the overall maximum or minimum value in general. Thus the first feasible practice is to focus on a smaller subset D ', and to simply describe characteristics of the local optimal value used an approximate linear map, f, then compare all the local optimal value to identify the overall optimal value of f. If the function has only a local optimal value, this is the overall optimal value. Therefore, for such a function, finding the local optimal value leads to finding the overall optimal value. This result is known as the local-global theorem. For example, utility function, profit function and production function are meaningful maximization objective function, cost function is minimization.

The general steps solving practical problems based on optimization thinking is as following.

First, it is a map process that the real problem must be converted into a mathematical problem or a mathematical model. The second, to determine excellent standards or not of the various options of problem solving, and quantify these standards. Finally, it is a mapping process that finds the optimal solution.

4   An Example of Mathematical Optimization

The following is a typical stock inventory problem.

Managers need to implement a commodity purchase and warehouse inventory, if the stock is too much, we have to pay the inventory fee, but the stock is not enough, we have to re-organize human and material to purchase, then for each stock how much is the most economical?

This is an optimization problem, in order to found the objective function, some basic parameters are needed as follows:

Annual sales of the commodity is P, the amount of each purchase is Q, each purchase cost is Y, annual storage costs of each item is X. So the annual purchase expenditure is YP/Q. After purchasing goods there is Q pieces of goods in the warehouse, following the sale of goods inventory is gradually reduced to zero, in the next purchase inventory suddenly is increased to Q, the average inventory is Q/2, and assume that inventory shortages are not allowed out of stock, and the two groups commodity can not be in one warehouse, so the annual storage fee is XQ/2.

After the above analysis, we can map the original problem into a minimum mathematical formula, let C is the total cost of purchasing and inventory, then

$$C = \frac{Q}{2} \times X + \frac{P}{Q} \times Y. \tag{1}$$

Where P, X, Y are positive constant.

Now the question is that trying to find the best purchase quantity, Q*, minimizes the total cost of the objective function, C.

The local minimum value of this problem can be calculated by the method of calculus extremum, for only a local minimum value, according to local-global theorem this local optimal value is the overall optimal value.

Derivative of C for Q at both sides of the formula (1), it is as following,

$$C' = \frac{X}{2} - \frac{PY}{Q^2},$$

let $C'=0$, $\frac{X}{2} - \frac{PY}{Q^2} = 0$, $Q = \sqrt{\frac{2PY}{X}}$.

$$C'' = 2PY/Q^3 > 0,$$

when $Q* = \sqrt{\frac{2PY}{X}}$, c is minimum.

References

1. Xu, L., Zheng, Y.: The method of Relationship Mapping Inversion. Jiangsu Education Press, Nanjing (1989)
2. Shi, S.: Mathematics and Economics. Hunan Education Press, Changsha (1990)
3. Roy, W.: Economic Mathematics. Economic Science Press, Beijing (1999)
4. Li, K., Li, J.: A comprehensive mathematics education theory and practice. Mathematical Bulletin 2 (1999)
5. Lv, H.: Study on mathematical learning of asking questions and mathematical situations. Mathematics Education 4 (2002)

The Chance and Challenge of Economic Recovery in Tourism of Mianzhu after Earthquake Disaster

Shu Jianping, Zhang Linling, and Wen Xiaoyuan

Sport Department of Chengdu Sport University, Chengdu, Sichuan, China
xbjk98@163.com, zlinl@163.com, wxysfm@126.com d

Abstract. This paper make a research on the problems of economic recovery in tourism of Mianzhu after earthquake disaster through field survey, and it concludes that new Mianzhu after rebuilding offers new chances of development in its tourism with the process being more popular with world; the rebuilding of tourism attraction have made great progress in the guide of timely national instruction; the investment environment have become more attractive with the help of national policy. However it also has many problems need to be solved such as eliminating negative effects taken by earthquake, rebuilding the fame of scenic spot, improving tourism service system.

Keywords: Tourism, Recovery, Chance, Challenge.

1 Introduction

It has been nearly 3years after the 5.12 Wenchuan earthquake which caused great damages to Mianzhu's original tourism facilities. After the great supports from central and local government and Jiangsu rebuilding task have been nearly completed. However, the task of tourism recovery of Mianzhu is tough. Tourism gross output of 2010 is gradually up to 528 million yuan which far below 898 million yuan of 2007 before earthquake. On the other hand the new Mianzhu after rebuilding brings its tourism with new development chance based on which make Mianzhu set new plan for its tourism developments and highly improve tourism, followed along with challenge.

2 The Chance of Tourism after Earthquake

2.1 The Emerging and Influence of Earthquake Resource

5.12 Wenchuan earthquake caused unique earthquake ruins in the world. Dongqi factory floor of Hanwang in Mianyang, Beichuan in Mianyang, Yingxiu in Wenchuan, and Hongkou in Du jiangyan are four protected national earthquake ruins. In the destruction of earthquake, Dongqi factory floor of Hanwang is typical site whose outstanding characteristics focus on its double value as industry ruin which has the merit of history, science, economic and earthquake ruin which has the merit of social culture, sprits, earthquake science. This unique and special double-value highlights the precious earthquake resource.

2.2 The Enrichment and Improvement of Tourism Product

Hanwang earthquake ruin are carrying on the overall program of park of Hanwan earthquake ruin, the construction of earthquake relief digital museum and earthquake relief website. Jiannan old street carried with long history has been recovered with the government grant. Three Kingdom culture relic in Shuangzhong memorial temple has been rebuilt.

The village of Chinese new year picture in Miangzhu have refresh its basic and common service facilities with the aid of 20 million yuan from Suzhou.

During the time of lasting emerging of new tourism product, the existing tourism product of Mianzhu also get the unprecedented chance of improvement. Before and after Wenchuan earthquake, Chinese new year picture of Mianzhu experiences two different treatment. Before earthquake , this folk art faces the embarrassment of having no one to inherit. After earthquake , rebuilding not only miraculously bring Chinese new year picture a new birth, also make it a tourism attraction. Nowadays, a village of Chinese new year picture which is filled with poetry and full western Sichuan charm is presented to the world.

2.3 Being More Popular with the World

Mianzhu as a country-level city in the north of Sichuan has made great changes and progress in policy, economic, culture since reform and opening up. In 1999 it turned into a under construction city of eco-demonstration region and made marked achievements in the construction of eco-farming, eco-industry, eco-town and natural reserve. But on the whole, Mianzhu' popularity restrict on local area, and its influence is limited in China and abroad.

5.12 Wenchuan earthquake caused a unprecedented tragedy, also brought a special chance which make all the world know disaster-affected area. After earthquake, Mianzhu instantly became public spotlight which attracted lots of attention from China and abroad. This incident caused profound and many-sided effects, one of which improve local popularity and give unexpected and special publicity to local tourism. Because of earthquake, Mianzhu made a deep impression on people, fired up their curiosity and fostered potential tourists.

2.4 Prompt and Scientific Rebuilding Plan for Tourism

Based on the situation after earthquake, Mianzhu had made prompt and scientific rebuilding programs for tourism. All of these programs included the rebuilding plan of three ecological villages named Chinese new year picture village, clean spring village and flower village; overall plan of Jiuding mountain tourism attraction; conceptive design of Hanwang earthquake ruin park; connecting travel resource and service facilities of travel region around mountain; improving the merge of country travel and culture , forests, physical, water conservancy. In the process of country travel rebuilding, Mianzhu committee and government carried on the overall plan by order. On one hand, they put the program of county travel into the firs and second building plan in a priority especially greatly supporting important projects of country travel reception centre, village of Chinese new year picture, sightseeing agricultural garden in initiating project, planning, choosing sites, attracting money. On the other hand, they unitedly developed the villages and towns covered by travel routines. They rebuilt and recovered country travel road and tour way around mountain. The element of tourism was fully integrated into the building of farm house and market town in order to make high-end travel product. The government made a goal to build country travel around mountain into Eastern Alpes which enhanced the development of overall country travel.

2.5 The Great Help of Paring-Assistance of Scenic Spot

Mianzhu as one of the ten national worst-hit area had received great supports and attentions from nation and society in its rebuilding, especially the supports from Jiangsu gave reinforcement into fund and technique needed in rebuilding and accelerated the rebuilding process

Under the supports from Jiangsu province, the project of building water street of Jiulong travel town was finished; Suzhou water street and Mianzhu new year picture were built in order to recover reception ability of country travel; Jiansu farming sightseeing garden was constructed with the aim of showing highly scientific and environmental farming product in the way of country travel; the infrastructure and service facilities were enhanced with the operation of more than 300 joyous farmer's house and country hotel.

2.6 The Influence of National Policy and Boosting of Regional Travel Tourism

On 25th November, 2009, State Council Executive Committee passed on principle of boosting the development of tourism in which firstly upgrade tourism into strategically pillar industry from important industry, indicating tourism was taken into system of national economic. The principle provided powerful support for the development of tourism. With the continuing growth of national economy, the scale of tourism was becoming large, the potential of tourism development was huge and tourism with the character of recreation and ecology would meet greater chances of development.

The supports to tourism from policy offer a good chance for accelerating the development of Mianzhu tourism which enhance its industry status and engage into the central title of tertiary in new planning of industry by means of comparative advantages. Rebuilding after disaster will accelerate the upgrades of tourism.

2.7 The Improvement of Environment and Regional Condition

The eleventh five-year plan of Mianzhu national economic and society development and the outline for long-range objective through 2020 show the beauty blueprint of the development of Mianzhu in next 10 years. The beauty blueprint make us believe that Mianzhu' regional condition will get great improvements and its investment environment will become more attractive in next 10 years.

In the view of city development goal, the buildings on both sides of main streets need a refresh; the old house and culture relics need good protection and treatment; To speed up the construction of public facilities and improve the city public communication system; Making Mianzhu a city which is special in building style, outstanding in overall feeling, comfortable in living.

3   Challenges to the Tourism after Disaster

3.1   Removing Negative Effects of Earthquake

The earthquake had caused great damages to material foundation. The roads were broken, the buildings ware collapsed, infrastructure was paralyzed, all of these crippled the attraction of Mianzhu tourism in short time. However, with the wide transmission of earthquake information, the ecological image arose great concerns from people, the unpopular tourism could be activated in information dealt very well.

3.2   Remolding Tourism Image of Mianzhu

In order to win in the fierce market competition, Mianzhu tourism must carry on brand strategy and set up tour ism image. Only with the full and vivid tourism image and impressive brand, scenic spot can get tourists' recognition, then set as destination by tourists. All in all, Mianzhu tourism should carry on brand strategy, improve brand marketing, perfect service system and rebuild travel image.

3.3   Adjusting Strategy Goals

Travel is a high level social activity and modern life style with the melt of material civilization, sprit civilization and ecological civilization. Tourism is one kind of economic industry which pay close attention to society, civilization and ecology. Mian zhu should set on the subject of tourism after earthquake in basis of its characteristics.

3.4   Improving the System of Tourism Service

After the accomplishment of hardware infrastructure, it should improve relative system of tourism service which means to group of companies which provide relative service for tourists. Tourism is highly compound industry which not only includes transportation, accommodation, sightseeing, shopping and entertainment, but also has close connection with traffic, communication, fiancé and so on. The development of tourism expands the content and scale of these industries, and also make corresponding requests to them. The relative companies should develop corresponding tourism product on basis of tourists' needs and improve the quality and system of service.

To sum up, rebuilding after disaster will make deep effects on the development of Mianzhu tourism economic. We should meet the challenge and take the chance to improve the tourism service and develop the spirit of earthquake relief.

Acknowledgements. This paper made periodical achievements in project of Sichuan Soft Science fund that planning study on rebuilding and recover of Mianzhu tourism after disaster.

References

1. Deyang statistics website, The influences of earthquake to Mianzhu tourism and strategy research of rebuilding, 11 (2008)
2. Mianzhu travel bureau, overall planning of development of Mianzhu tourism after earthquake, 7 (2008)

3. Sichuan travel bureau, planning of rebuilding and recover of tourism after Wenchuan earthquake, 6 (2008)
4. Deyang travel bureau, planning of rebuilding and recover of tourism after Wenchuan earthquake, 7 (2008)
5. Shu, J.: Strategy research of rebuilding and recover of Mianzhu tourism, reform of economic system, 4 (2010)

Information Systems: New Demands, New Proposals

Rosângela Lopes Lima, Isabel Cafezeiro, and Luiz Valter Brand Gomes

Instituto de Computação da Universidade Federal Fluminense
{lima,isabel,lvbg}@dcc.ic.uff.br

Abstract. The current social changes that result from the advent of new forms of relationship and communication strongly based in networks begin to collapse with the rigid and inflexible academic structures which respond slowly to the new demands. Considering the experience of implementing an undergraduate course in a public university in Brazil, this article raises and discusses innovations and difficulties in achieving an academical proposal whose main focus is to perceive Organizations, Enterprises and Institutions (OEI) not as technocratic model of management but as sociotechnical entities.

Keywords: Education, Information Technology, Innovation.

1 Introduction

The rigid and hierarchical structure that still persists in most Organizations, Companies and Institutions (OIEs) in Brazil tends to lead to a technocratic, top-down management, in a model that does not give due importance to aspects inherent to individuals who work in them, such as creativity, sensitivity, autonomy and intelligence. In practicing this model is inadequate to serve a society that is characterized by exponential growth of social networks. These, created, accessed and structured through the web, as horizontal networks are being accessed at any moment, by any person or group of persons, without mechanisms of control. The construction of a new narrative that addresses the processes that operate through a network of relations becomes a necessity, in the light of the perception of a new reality that can not be simplified to be explained by models based on positivist science. Attempts to implement this new narrative confront, however, with the equally rigid academic structure of Brazilian universities, which, in theory, embraces innovative and flexible ideas, but in practice introduces a number of drawbacks that tend to derail the new proposal.

We conjecture about the need for educational efforts to enable the academy to train professionals capable of working through a new modus operandi that takes into account these changes that occur in everyday life of people. At a time when technology tends to increase exponentially the possibilities of sharing the cognitive functions of individuals through the electronic support, the management of people is one of the areas where investment is strengthened more and more in the administration of the OIE (s). The notion of the OIE (s) seen as complex systems that are people - who have a history, family life and especially dreams - demand the training of professionals with interdisciplinary, global, critical, enterprising and humanities. By understanding the OIE (s) as complex systems, we perceive the need to include in the learning of professional Information Systems thorough knowledge of the new role of human resources in contemporary OIEs.

The contemporary OIE(s) are not seen only as a set of products or their administrative divisions, but as a portfolio of resources and capabilities that can be combined in various ways [2]. And in this context it is the knowledge management of human resources that will ensure strategic and operational gains. In today's professional it becomes necessary to acquire a new worldview that incorporates the unpredictability resulting from the complexity of the world that today is more evident by the development of science and technology. Today the reality is explained by its multiple dimensions and relations since it constitutes a logical system of relationships involving the individual in being able to accept the uncertain, get a sense of unpredictability, knowing that we do not know and accept the contradiction [10]. Being prepared to be immersed in complexity gives the professional ability to connect with the unpredictability of the real.

The graduates of such training should be able to seek computational solutions to complex problems from different areas and be able to deal not only with technical issues related to information processing, but also must interact with a social context that constitutes the intelligence of the OIE. Qualified professionals to support an innovation of this sort need a training to prepare them as citizens aware of their social and professional role in a society in constant transformation and must be able to assume the role of change agents, and therefore be able to bring about change through incorporation of technology in solving everyday problems in the workplace.

In the following sections we highlight some points of the education that we consider important to be met fully these ideas, that is, the formation of professionals aware of their social and professional role in a society in constant transformation focusing the degree in Information Systems.

2   Four Essential Points to the Sociotechnical Approach in Information Systems

2.1 Interdisciplinarity

The sociotechnical nature requires interdisciplinary and must be structured on the theoretical-practical articulation to enable graduates to function effectively in the dynamics of contemporary society.

Besides focusing on the skills and abilities related to technological disciplines such as programming, software engineering, database, among other things, this proposal includes disciplines related to sociology, law, information science, management and psychology in order to assist in the formation of a professional with a more realistic and complex information technology in enterprises, organizations and institutions.

The organization of the course through the integration of these disciplines was essential, since the "real world" companies do not behave in a predictable and functional entities but as complex sociotechnical processes which are not restricted to the rational aspect. It is understood that the professional practice in Information Systems necessarily requires an understanding of the importance of treating interpersonal relationship, since it is of paramount importance for the understanding and interpretation of the social context of OIE(s).

Professionals with this understanding are able to act in such organizations as complex bodies where technical devices coexist with behavioral aspects of configuring a network interference. This conception is based only on a curriculum with a high degree of interdisciplinary, requiring the inter-relationship of content of a technological and humanistic nature, and the coexistence of practical and theoretical studies.

It is worth mentioning that interdisciplinarity is not restricted to the coexistence of various disciplines in the curriculum areas. You need a permanent work of patching the contents and teacher understanding of the importance of each content in the relevant profession. For this, the coordinating education plays a key role, and must be prepared to deal with the faculty and student resistance in relation to change their customary practices.

2.2 The Extra-Classroom Activities

In Brazil, the academic activities that have a point of contact with the community, aiming to establish a dialog and exchanging of knowledge between the academic world and its neighborhood are called extension activities. These activities tend to be more practiced in the humanities than in technical areas. In this latter they are often viewed as activity that benefit the community, without being perceived the benefits to the academic and professional training of students.

The retention in the early periods, the difficulty in basic disciplines such as programming, have become chronic problems in computer courses. Many teachers complain about the lack of preparation of students admitted, attributing this fail to the poor quality of secondary education and the existence of social programs allowing the admission of students unprepared. To deal with this situation, in order to use skills and build competences, it is necessary to implement mechanisms that create a bond of students with the University, making them want to stay in the institution and advancing the course.

The university extension, that is, those scholarly activities that promote a permanent contact with non-academic community, is an ally in this sense. The extension of scholarship programs tend to be more democratic than the scientific initiation programs: do not require a high coefficient of efficiency and can be offered to students of earlier periods. In addition, scholarship programs in Brazil usually focus on the student in the classroom and content of the classes. The extension, by contrast, aims at the performance of the student from a direct contact with the neighborhood that surrounds him, through questioning, as we emphasized in [6]: "(...) the questioning promotes the rapprochement between practice and theory, as it causes the involvement of academic (teacher and student) with their own neighborhood, their surroundings, real people, everyday situations. It is this meeting of academic and his locality that forms the professional citizen. Without it, it is formed the professional. " Thus, the extension can be used as a fixing mechanism, avoiding the search for pre-mature stage, and encouraging, awakening interest in the course. In terms of content, allows the curriculum to bring the issues of neighborhood, which in demanding computing solutions, end up tacking curricular subjects. Conversely, the extension places the student as a trainee in contact with people outside the academic space, which causes the need to deal with unforeseen situations, in colloquial speech, the scarcity of resources, that is, not idealized situations, such as those simulated in the classroom.

2.3 A Careful Look to Local Demands

There are difficulties on the part of courses in operation, in maintain the student away from the labor market only to devote himself to studies. Brazilian reality has been pushing the student pre-maturely to the labor market, what can be noted by the drop in grades, increased length of stay in University, loss of interest in theoretical content, and preference for disciplines with no schedule conflict with work. The academic community tends to see the adjustments to this reality, as for example, the option for night shift as a way to decrease the quality of courses. This ends up exacerbating the situation. Teachers tend to complain of lack of dedication on behalf of internships and jobs, but usually do not propose effective actions to justify the dedication and arouse student interest. It is important then that the educational project seeks to reverse this logic. A way to perform this is making the practice in stages and work become subject to discussion and study by academics. Thus, the use of experience in working must enrich the curriculum and establish a permanent bond between the university and the OIE(s). The approach with companies, and the constant reflection on the way of acting in Brazilian society, gives the course a realistic character, committed to the local and Brazilian situation, being thus able to form professionals able to propose changes and innovations.

2.4 Technological Mediation as a Mechanism to Approach the Contemporary Complexity

The educational project of the course must provide incentives for the use of educational methodologies that enhance technological mediation of the learning process. This approach seeks to expand the space for teaching beyond classrooms, transforming the relations of teaching and learning, and giving a new look to the meaning of didactic content. It also encourages the student's autonomy, putting him as the agent responsible for his own learning. It is important to know that educational practice should go beyond that which is based on broadcast content, and more, we must recognize that the teacher in his actions as mediator of learning must be responsible for creating ways so that the student may be able to construct his own knowledge.

For this reason, we intend to highlight as basic principle of the transformation of the structure of the classroom the thought of [12] about the pedagogy of autonomy. According to him, it is essential to the teacher learn knowledge that makes possible to the student the effective learning. A careful evaluation of the teaching action is necessary in order to seek the teacher's role as an inducer of reflective education, which transforms the relationship transmitter/receiver that is in the process of traditional education.

In the study by [7] data were collected and analyzed about the daily process of teaching and learning within the classroom. An important finding of this study is that the simplistic setting of this environment, minimizes the complexity of the process of knowledge construction and thus the effectiveness of the learning process of students. Individuals learn by relating and linking, through their senses, the information from the environment they live with the knowledge stored in their memory. In the classroom, usually characterized by the transmission of content, a teacher-pupil ratio unambiguously reduces both the possibility of creation as strengthening of interactions among students. When you do not encourage these interactions, students use partially their brains, organs that lead and connect an infinite number of complex information through their sensors.

According to [9], the individual is endowed with a brain, consisting of more than one hundred billion neurons, which allow extremely complex functions that he is constantly learning in his exposure to the world. A process of acquiring new information is not trivial because it envolves a memory storage system, which consists of neuropsychological and neurobiological processes that occur differently for each individual, since learning is dependent on the cognitive structure of each one. The understanding of the process of student learning and how this process is complex, makes it imperative a responsible and careful treatment of the learning environment. In [7] this urgency is showed by the results obtained by carrying out two surveys in the context of the classroom, through the application of a research technique called Social Network Analysis (SNA) which analyzed the behavior of flows information between students and teachers, and the other consisting of the preparation of a structured interview in which they sought the opinion of students about the process of learning in the classroom, through their verbal and behavioral manifestations. The application of ARS showed that there is need to explore information flows within a classroom, aiming to create strategies for developing links, interactions and cooperation between actors in the process of knowledge construction.

What became more evident when analyzing the results of the interview was the predominant practice of a model in which the student reproduces only the knowledge presented by the teacher. This is a practice that can survive without the technological innovations, since even when the technology is present, what is reproduced is the same model of content delivery. A model in which the teacher is the single source of knowledge, despite the information and communication technology which provides a network for sharing and collaboration for the construction of knowledge.

The constraints arising from lack of dialogue and lack of access to a variety of media, the characteristic problems of the traditional structure of the classroom, cause great damage to the construction of knowledge by the student, and influence negatively his ability to search and the information processing. The student must be guided to develop his methods of acquisition and processing of data and information from the external environment into meaningful knowledge for him. It is this process that will strengthen his capacity for dialogue, the main form of communication by humans with their peers.

Assuming the already mentioned, that teaching is not to impart knowledge, but rather a structured process based on the understanding that student learning does not happen so trivial, it is concluded that the teacher's role is not simple. In so far not uncovered much of the intricate process of human communication, the teacher's task remains extremely complex. [7] notes that a new understanding of human thought will resort to new ways of dealing with teacher/student relationship, to facilitate and guide the construction of knowledge by students so that they are capable in this process, to be autonomous and interact with others in pursuit of transforming the ocean of information that characterizes today's society in meaningful learning for life. Given the complexity of this notion the teacher needs to launch a new look at his student, other than that which sees him as one more in a class.

To find ways of living that are compatible with the new rules of the globalized world it is fundamental to accept that the vital process in the world today is shared by individuals from all parts of the planet. The Internet, which is much more than a vast compilation of data and a huge encyclopedia, is a new form of communication and network management with endless potential for cooperation. Particularly when it allows not only a particular query unlimited data and knowledge, but also placing online any creative output by anyone and without any restrictions [11]. Information technology and communication should be seen as great allies and constitute a strategic tool to support the changes that are occurring in society. It is through them that are designed hardware features and software that support the collective processes of knowledge construction, created through the network structures that can be given, independently of distance and time. These technologies besides interconnecting in network a set of individuals - also enables the meeting of ideas and resources around shared interests and projects. The collaborative work and the possibility of establishing dialogue are key factors for the enhancement of the capacity of self-organization and structuring of the teaching and learning around the construction of meaningful knowledge.

3   Conclusion

Despite the fragmentation of Brazilian established curriculum where content is distributed and isolated on black boxes, it is necessary to seek to develop in students the ability to master the cycle of problem solving in the context of IT. One way of achieving this goal is to use in the disciplines methodologies based on learning through problem solving, and link content to situations and problems of the university neighborhood, the municipality where the student resides, the workplace, etc, thus, exercising the ability to perceive the application of knowledge to real world problems. This goal is present in this educational proposal at various points: the integration with extension activities, through integration with the labor market, the imposition of extra-curricular academic activities, and for all the disciplines that are food for the development by the student of a final course project. The methodology of Action Research, and its recent developments in Brazil [1] offer an adequate support to the approach envisaged in the BSI. The methodology PBL (Problem-based Learning) also helps in the realization of intended approach in the classroom. The standard textbooks, tests applied to large number of students, classes too large and the disciplines that are offered in the same way for different courses do not motivate the involvement with the course in which the teacher operates. In this way it is not possible to effect the re-contextualization of content, and the student stays with the feeling that each course ends in a black box of curriculum content and humanities are less important than the technical content. In this respect, we follow [5:47], commenting that the loss of influence in shaping the philosophy of social science, points to the impoverishment of the humanities and the consequent difficulty of interpreting "what goes around the world." The lack of link between humanistic content and technical content introduces the risk that the course becomes a discipline "of the administration of things", which puts it at the service of the capitalist system and the globalized economy.

It is known that there always exists the possibility to pass a mismatch between what is planned and what is implemented. This argument is often a barrier to change, and ends by placing education as one of the most traditional systems and averse to new ways of acting. However, when the change is urgent due to social demand, the gap between what is planned and what is implemented can be avoided by monitoring, evaluation and adaptations. Given that the educational structure makes them shred the process of constructing knowledge, either through existing curricular structure and / or disconnected by the practice of educational proposals, it is necessary for the coordination of the course coordinator in conjunction with the college, establish mechanisms for enabling management to direct and evaluate the process of building knowledge of the student during the implementation of each semester. One of the important mechanisms used to verify the effectiveness and efficiency of the educational proposal is the evaluation of educational program in its two main parts, namely, the evaluation of product and process, which should be implemented in a complementary way by combining qualitative and quantitative methods in different areas: academic, administrative and physical infrastructure and technology.

As opposed to traditional evaluation, which seeks to verify the range of indices considered "successful", the assessment to which we refer is put toward the construction of learning personalized on the actors in the academic system: teachers, students, community. Therefore, it accompanies all stages of the course and requires participation of all actors, and not go through the usual categories of pass/fail, success/failure.

References

1. Thiolent, M.: Metodologia da Pesquisa Ação, Editora Cortez (2005)
2. Mintzberg, H., Ahlstrand, B., Lampel, J.: Safari de estratégia. Bookman, Porto Alegre (2000)
3. Currículo de referência da SBC, http://www.sbc.org.br/
4. Projeto pedagógico do curso BSI: Instituto de Computação da Universidade Federal Fluminense (2009)
5. Santos, M.: Por uma outra globalização - do pensamento único à consciência universal. Record, São Pauto (2000)
6. Cafezeiro, I.L., Lima, R., Reed, J.: Refazendo vínculos: das dicotomias locais à novas conexões acadêmicas. In: Foro Iberoamericano de Comunicación y Divulgación Científica, Campinas (2009)
7. Lima, R.L.: Uma proposta estratégica, didática e operacional para a sua unidade estrutural a sala de aula, Tese de Doutorado, UFRJ, Brasil (2008)
8. Lima, R., Cafezeiro, I.L., Gomes, L.V.B.: The technological mediation as a strategy to innovate the teacher-student interaction. In: Proceedings of the International Conference of Education, Research and Innovation, Madri (2010)

9. Lent, R.: Cem bilhões de neurônios: conceitos fundamentais de neurociências. Atheneu. São Paulo (2005)
10. Morin, E., método 3, O.: O conhecimento do conhecimento, Editora Sulina. Porto Alegre. Rio Grande do Sul (1999)
11. Moura, L.: Formigas, vagabundos e anarquia: ensaio sobre a vida artificial, arte e sociedade. Ed. AAAL – Alife Art Architecture Lab. Lisboa (2003)
12. Freire, P.: Pedagogia da Autonomia: saberes necessários à prática educativa. Editora Paz e Terra. Rio de Janeiro (1999)
13. Freire, P.: Sobre educação (diálogos), Rio de Janeiro, Paz e Terra (1984)

Electronic Technology Course Teaching Mode Based on Practice

Dongli Jiao[1], Jinsheng Li[2], and Yongqiang Zhao[2]

Taiyuan institute of industrial
(Taiyuan,China)
dongli_jiao@163.com

Abstract. Electronic technology involves many subjects (Electronics, information, communication), so it's teaching effect directly influences students' ability. The teaching mode based on practice can effectively improve teaching effect. This teaching mode is introduced in this paper. Many ways or measures are included in this mode, circuit simulation, laboratory experiments, course design,eg.

Keywords: teaching mode, effect, practice ways, example.

1 Introduction

The teaching of electronic technology, as an important basic special course, is of great help students to master basic electronic technology and train logic ability. Electronic technology includes analog electronic technology that involves analog components and analog circuits and digital electronic technology that involves digital components and digital circuits. With electronic technology developing, it's application further associated with many respects of engineer applications. This course continues reforming and refining for talent demand.

It is important that student can analysis circuits and use components after studied the design and application of electronic unit. Practice is a good way improving teaching effect. There are many paths for practice, for example: simulation, experimental, design jobs. Teaching based on practice will be a new teaching mode which adapts social and science developing.

2 The Base of Teaching Mode

The new teaching mode reforms teaching course including teaching content, practice process, measure of strengthpractice ability.

2.1 Integration about Teaching Content

Traditional electronic technology course is divided into two course: analog electronic technology and digital electronic technology. Teaching content is independent which results in contents and practice using only analog signal and component or digital signal and component, so that content wouldn't associated with practical applications.

Y. Wang (Ed.): Education Management, Education Theory & Education Application, AISC 109, pp. 273–276.

In order to improved teaching effect, analog electronic technology and digital electronic technology can be integrated to a course , which will help promoting teaching contents and practice combination and connection.

2.2 Practice Process in Teaching

There are many ways for practice based on very wide range of electronic technology application: show in class, circuit simulation, laboratory experiments, course design and design before graduation.

Show in class, as the first step which enables students percept knowledge, visuals knowledge in book and attracts students. Circuit simulation, as a key process, enables students recongniting component and circuit. Laboratory experiments leads to students from book to reality。 Students begin to study how to use electronic components and design simple electronic circuits in course design, that is knowledge integration。 Design before graduation tests ability of students apply the studied knowledge.

2.3 Two Measures for Strength the Practical Ability

Two measures are proposed in teaching course for strength the practical ability, one is laboratory assistant regulations, the other is engineer certification. Laboratory assistant that composed with students who studied electronic technology help laboratory technician preparing before experimentand daily management. The work requires assistant having a good knowledge of components and circuit. Engineer certification is a new measure of our college for improved students' abilities. A series of training are made before certification include theoretical knowledge and practical exercises.

3 Teaching Mode Based on Practice

The process of teaching mode based on practice is showed in diagram1. If a component (diode, transistor, counter, FF.eg) is studied in class, teacher can show component features through EWB or protel tool after explaing theoretical knowledge⬜ which helps students understand theoretical knowledge. On the while, leave the homework about

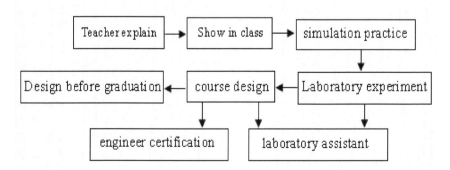

Fig. 1. Block diagram of teaching mode

simulation practice according to studied component or circuit. Students have a clear understanding about component or circuit though teacher's show and simulation practice.Third, studengs practice in laboratory, identificate and select component ,building and debug circuit, test circuit performance. Fourth, this component or circuit is further used in course design. Finally, further use is stressed in large-scall circuit design(Design before graduation).

4   Teaching Example

There are many examples that can show the teaching mode based on practice. Teaching of NE555 chip as a good example is introduced in the following.

NE555 is a common chip in circuit and is applied widely by engineer. It is also a typical chip combining   analog signal with digital signal. Teacher show NE555 ciucuit and waves in class after explained chip. The showed circuit is Smith Trigger builded with NE555 in diagram 2. The wave is displayed by virtual instruments in diagram 3.

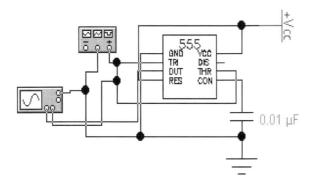

Fig. 2. Circuit of Smith Trigger

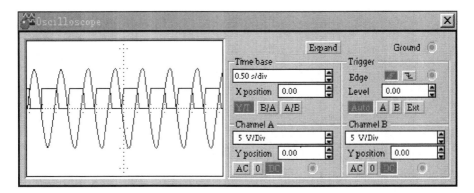

Fig. 3. Wave displayed by virtual instruments

Then students simulate this circuit with EWB or Protel, second understand this chip and thiscircuit, study other performance of this chip or this circuit.

Thus, when students experiment in laboratory, identificate chip and build crcuit themselves, they can compared r experiment result with that showed in class.

In course design, students will tend to design staircase light switch control circuit or ambulance siren circuit, or other else. NE555 is used goodly in these circuit design.]

5   Conclusion

Traching mode based on practice is always implemented on electronic technology course and leads good results. Students are encouraged to assist laboratory technician or exam engineer certification. Many ways of practice enhance teaching effect. Students not only understand books' knowledge but also can use advanced development tools, on the while, students can applicate the knowledge into practice. It have been tested that this teaching mode is a good mode in our college. We expect more teachers and students to benefit from it.

References

1. Song, E.: Reform of Electronic Technology Course Teaching. Journal of Electrical and Electronic Teaching 30(06) (October 2010)
2. Lu, E.: Virtual Electronics Laboratory. Posts & telecom press (2002)
3. Bi, M.: Electronic Experiment and curriculum design, 3rd edn. China Machine Press (June 2007)

Research and Implementation of "Project-Driven" Teaching Method in <Web Systems Development and Practice>

Ruxin Ma, Yu Liu, and Xiao Wang

School of SofteareTechnology, Dalian University of Technology.
116621 Liaoning, China
`{teacher_mrx,psodream,kara0807}@126.com`

Abstract. This paper considers the characteristics and current teaching situation of <Web systems development and practice>, comes up with a practical teaching method which based on "project-driven". By comparing the successes achieved in "project-driven" teaching methods with the traditional method, this paper proposes a relatively complete teaching proposal. Practice shows that the "project-driven" teaching model is much more propitious for improving the manipulative ability, communication skills, self-innovation, and teamwork spirit of students.

Keywords: "project-driven", Web systems development and practice; manipulative ability.

1 Introduction

A serious problem about the employment of college students is reflected by the deadlock of talent supply and demand. It becomes a matter of urgency that contemporary college students are becoming weaker and weaker in practice, manipulation and innovation.

In 2009, the Ministry of Education proposed the plan of "brilliant engineers" [1], which aimed at cultivating the type of R&D engineers. Students in new times should be able to hold global view, design the modern engineered knowledge structure and integrated planning. Therefore cultivate excellent engineers whom are rich in innovation capability and manipulative ability and adapted to Chinese remarkable economic and social development are of great importance. This requires that college students have basic skills for qualification of professional practice and qualities for engineering research and technology development.

The remainder of this paper is organized as follows. Section 2 introduces the course construction of "project driven" teaching method; Section 3 describes the aims we are going to achieve; Section 4 gives the advantages of "project-driven" method and finally, we compare the effects of two teaching methods.

Y. Wang (Ed.): Education Management, Education Theory & Education Application, AISC 109, pp. 277–281.
springerlink.com © Springer-Verlag Berlin Heidelberg 2011

2   Course Construction

Although we are limited to the current teaching system, universities like DLUT still make efforts to innovate new effective methods to inspire students' talents. For the past few years, the software school of DLUT opened up a new direction: "Financial Informationization". This major combines financial with information technology together, makes efforts to enrich students' practical operation, aims at cultivating high qualified professionals who is suitable for the actual needs of all kinds of companies. To achieve these goals, for most students, the programming ability, self-innovation, teamwork spirit and communication skills are all very important. According to the objective of undergraduate teaching goals, we open up a professional course: "Web system development and practice".

2.1   "Project-Driven" Teaching Method

Project-driven teaching method means that in the process of teaching, teachers use project as main line to expand the class training. Using this method, we integrate relevant knowledge into every aspect of the project and promote the development of project. Through the deepening of the problem and function expansion, we can expand the width and depth of knowledge, so as to get a complete project solution to achieve the goal of knowledge acquisition and ability improvement [2]. Project-driven methods meet the requirements of constructivism. The learning theory of constructivism holds the idea that the learning process is that students construct the significance of knowledge in their own mind. Students are the subjects in the whole process while teachers are instructors and subsidiary. The main task in handling the process of teaching is to pay attention to the features that students show during the construction of knowledge significance; teachers help but not replace the students to achieve the construction process. Scenarios, collaboration, conversation and significance construction are the four main points in constructing learning environment. Among them, the first three are the means of teaching and the significance construction of students in their minds is the purpose of teaching. In the project-driven teaching methods, "scenario" means that the entire procedure of study lies in the environment of the design projects proposed by teachers, "Collaboration, Conversation" implies the discussion between students and teachers, students and students and the introduction from teachers to students. The "significance construction" is that students learn through the course, complete the project and eventually hold the essentials of software development[3]..

2.2   The Characteristics of <Web Systems Development and Practice>

The wide spread of Internet has promoted the rapid development of Web technology, Web technology has become today's most influential technological mainstream at the procedure of building a modern information society. The futures of e-commerce, the all-wave Web services, the changes in lifestyle and work will rely on the rapid development of Web technology. In fact web service has shown limitless vigor and vitalities. Web technology facilitates the fast development of Internet, and provides direct services for human society; those promote the progress and development of the entire human society [4]. At the same time, the social needs in turn demands a strongly Web development technologies to further provide people with comprehensive high quality of information services, provide the developers in web technology with a more efficient and effective development platform ,therefore Web technology has become a direct access to the information society. For this reason, in recent years, various universities have opened Web-related courses, to train talents for Web development and progress to meet the demand for web talents in the domestic and oversea IT industry and even in all sectors of society. Network technology and Web technology has quietly walked out of the computer and IT field, penetrated into all fields, become a necessary part for all industries.

3 Objectives to Be Achieved

This course uses the project-driven teaching method, which is a good way to enhance students' practical skills and employability. Project Instruction emphasizes the importance of project site, knowledge of integrated and comprehensive ability of students. Projects in our course are all real projects from business, real project case throughout the entire procedure of course design. The process of project development in collaboration helps students to accumulate experience and develop collaborative project settlement capacity. Practical training in the project emphasizes that students follow the teachers to learn and explore together, asks students to join in to design the software according to the demand of the project, and makes students to discuss by groups to train their cooperation ability. At the end of courses, teachers make summary from the perspective of business, evaluate the students' design so that to enlighten students and help them to have a deeper understanding of the practical and social of the knowledge.

Goals to be achieved in project-driven course are as follows.

(1) Teachers are no longer the knowledge provider but the edifyier. They scheme the teaching content, create the study environment, allocate the study resource, instruct and monitor the development procedure and evaluate the study results without direct participation in the program.

(2) Promote self-learning, and make students from passive recipients of knowledge into learning problems of inquiry into those who make it, not depending heavily on the teacher's instruction, to have passions to learn through like libraries, networks and other partnerships.

(3) The project train system. The difficulty of projects' arrangement is from shallow to deep, from simple to complex. The academic goal is from simple knowledge verification to the design and innovation of new projects. In the entire procedure, students learn to work together efficiently and positively. In other words, project-driven method pays a lot attention to encourage student to cultivate a factual and realistic work style, cooperative team spirit and the attitude that dare to blaze new trails.

4   The Advantages of 'Project-Driven'Teaching Method

4.1   Focus on the Practical Skills

The main content of this course focus on how to choose tools that used in Web technologies, construct technical framework, design development process, coding and debugging and so on. The technology we use is supposed to be able to work to resolve the practical and actual problems.

4.2   Take Project as the Mainline

Different from traditional teaching methods, this course is demo-centered, and appropriately link cases together into a complex project. During the course, we use actual cases to organize technical content. In the theory part of this course, teachers take a small case as demo to illustrate and introduce the technical content for students; experiment class requires students to accomplish a relatively complex case which is similar to the given examples in the former class step by step depends on their own ability. For the choice of experimental case, we should not only consider the practicality of the case, but also the interests so as to strengthen the contacts with problems and phenomena encountered in daily life and to help students understand the case. By doing this, students are sure not to lose in the tremendous amount of knowledge and accumulate more industry knowledge and project experience.

4.3   Take Manipulating Ability as Breakthrough

In order to gain rich experience about project development, the ultimate learning objectives of this course are: be able to use Java or. NET; be able to participate in the entire development process based on teamwork; be able to proficiently complete coding, debugging and documentation of B / S system module in enterprise level; be an intelligent stuff who cater to the shortage for skilled technicists in business . This course takes manipulating ability as breakthrough, practices students' programming ability, and enables students to achieve the requirement fast and normatively. This requires students to do large numbers of practical practice, dare to code, be willing to code, to achieve the "hot practice" level.

4.4   Intelligent Engineers Driven

Cultivate excellent engineers is the most fundamental and main goal in this course. This course is very practical so its purpose to learn always round the goal of project practice. Course designers and business professionals tracked large numbers of enterprises and get touch with most enterprises to have further interviews; through carefully analysis and expert seminars about available data, we almost completely understood the real-time need for people in those enterprises. This course is able to train excellent software engineers who possess essential skills and have direct contacts to actual project[5].

5 Comparison of Teaching Effects

First, the enthusiasm of students to learn <Web system development and practice> becomes higher. Up to 95% of the students are very interested in the "project-driven" teaching mode. They think that this teaching method makes the development of web system closer to enterprises'; theoretical knowledge with practical assistance makes the learning process more interesting. To most students, interest is the best teacher, every design brings students the amazing sense of achievement and satisfaction, which encourages and spurs them to work harder on the next project.

Second, the concept and abilities to design, implement of students improved significantly. The "project-driven" teaching method is an improvement and complement to the traditional teaching method. Traditional teaching method focuses on the teaching of theoretical knowledge, while the "project-driven" model focuses on the practical ability of students. Students experience the process of designing, coding, and debugging, throw themselves into all aspects of the project to face the real practical problems. The unique experience quickly improves their ability to solve problems.

Third, the students see the lack of their abilities in the procedure of designing projects. This teaching method not only allows students to see the distance between themselves and the formal staff, but also to know their own shortages. Therefore it is good at stimulating students' interest in learning, and improving their learning initiative.

Fourth, unlike the traditional method that one teacher teaches, students independently study; learning by groups help students to form team spirits. Each student in group has his/hers own responsibilities and tasks such as coordinate among the team members, divide the labor in order to achieve the best results. In conclusion, students work together to enhance their ability to communicate and cooperate.

6 Conclusion

From the above comparison we can know, the project driven teaching method is much better than the traditional. The ability of college students will be greatly improved if this method can be popularized.

References

1. Wang, M., Zhou, M., Li, J.: Development of the training program for excellent advanced manufacturing technology engineers by project-teaching method. China Modern Educational Equipment 2 (2010)
2. Wu, P., Yang, Y.: Computer course in the application of the task driven method. Computer Knowledge and Technology (16) (2010)
3. Zhao, H.: Constructivism Teaching concept to the new curriculum guide and meaning. Educational Innovation (6) (2007)
4. Chen, Y.: Web2.0 and its influence on information services. Information Development and Economy (4) (2010)
5. Li, H.: Project-driven management information system course innovation research. China Education Information (5) (2010)

Auto-assessment System of Ship Craft Electrical Engineering Technology

Guangzheng Li[1] and Huimin Lu[2]

[1] Department of Marine Technology Shandong Jiaotong
University Maritime College Weihai 264200, China
ligz@sdjtu.edu.cn
[2] School of software, Changchun University of Technology,
No.2055 Yan'an Street,Changchun 130012, China
luhm.cc@gmail.com

Abstract. In order to conquer the inconvenience and complexity in modifying existing paper card of auto-assessment system of ship craft electrical engineering technology, this paper proposes one auto-assessment system of ship craft electrical engineering technology based on web technology. The auto-assessment system of ship craft electrical engineering technology consists of four parts: generation of item bank, random elaboration of test questions, print service and performance accounting. With the adjustment of assessment outline, according to the characteristics of auto-assessment system of ship craft electrical engineering technology it is of high efficiency and success rate in generating test paper when constructing a flexible item bank and adopt local random generation of test paper based on types. After that it can print test cards of auto-assessment system and finally automatically print out the statistical output of scores.

Keywords: ship craft electrical engineering technology, self-assessment system, auto generation of test paper.

1 Introduction

The master of ship craft electrical engineering technology is essential one skill for a qualified marine engineering staff. According to Convention of STCW78\95[1], the main training curriculums of ship craft electrical engineering technology consist of the following: use of commonly seen electrical instruments and measurement, use of commonly seen low-voltage electrical equipments and maintenance, repair and maintenance of marine electrical engine and safety use of ship craft power.

Concurrently assessments of ship craft electrical engineering technology is conduct under fixed paper cards of auto-assessment system. Maritime institutions have made paper cards which suit the devices in campus. Paper cards are random distributed, students conduct assessments according to contents and requirements on the selected paper card and the operation process is marked by teachers. The paper cards used in this assessment method can hardly be changed and are lack of flexibility. At the mean

Y. Wang (Ed.): Education Management, Education Theory & Education Application, AISC 109, pp. 283–288.
springerlink.com © Springer-Verlag Berlin Heidelberg 2011

time, it may lead to students mechanically memorize the contents of paper cards which means it is impossible to test the real ability of students.

This article adopts computer technology to construct the database of ship craft electrical engineering technology assessment system. It uses web technology to random generate paper cards of assessment according to different test types and print paper cards. Students random site paper cards to conduct assessment which means it is of good flexibility and convenience and can improve the effect of testing the real ability of students.

2   Architecture of Auto-assessment Ship Craft Electrical Engineering Technology

Auto-assessment system of ship craft electrical engineering technology mainly consists of four modules: item bank, random generation of paper cards, print service and performance accounting. The following figure 1 shows its structure:

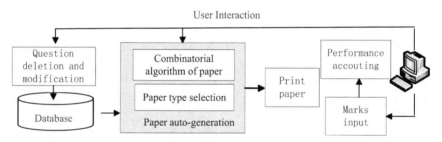

Fig. 1. Architecture of auto-assessment ship craft electrical engineering technology

Database. Item bank module consists of question database, question add, question delete and question modify. According to the requirements of convention and ship craft electrical engineering technology assessment outline, specialists elaborate item bank according to paper types. Tables in item bank are built according to training contents: use of commonly seen electrical instruments and measurement, use of commonly seen low-voltage electrical equipments and maintenance, repair and maintenance of marine electrical engine and safety use of ship craft power. For example, the database forms are as the following in Table 1.

Table 1. Database tables of ship craft electrical engineering technology

name	type	constraint	Function definition
Id	int	key,not null	Question id
Subject	nvarchar(50)	not null	Subject of a question
Student type	nvarchar(50)	not null	Postion of participant
Contentd	ntext	not null	Contents of assessment
Weight	int	not null	Degree-of-ddifficulty factor 1-10
Score	decimal(5)	not null	Score of the question

Paper auto-generation. First we should select students type, the system can auto-generate the question paper refer to students type according to certain algorithms which random choose questions complied with standards set by evaluators according to groups and weight.

Print service. The system will print assessment paper according to word template designed in advance and finally finish generating paper cards. Each score decreases gradually by 5 according to the highest score in database. For example, the highest score is set as 20; others are given due to their rank in completion.

Performance accounting. After the evaluator enters all performances of students, the computer will automatically conduct performance analysis and accounting.

3   Test Generation Algorithm

Automatic composing test paper is to use a certain set of algorithms automatically extract comprising a certain number of questions from the question library. Algorithms of composing test paper have a direct impact on the quality of the test paper, so composed of selected questions from the question library with the user requirements of the examination paper and making the composing test paper with higher efficiency and success rate are current research on hot topics[2-6].

random choice method. According to the requirements of users, randomly selected questions from the question library into the test papers. Random number is used to avoid generating the same questions and topics. Repeating this process until the end of the composing test paper, or the question library don't meet the requirements. Random algorithm is mainly suitable indicators require a single or small set of composing type, If the indicator are too much then system would repeat the treatment, and composing test paper become a long time or even fail.

Backtracking heuristics. This method is recording the records states of Randomly selected questions, and compared with the indicator requirements. If the search fails, released record status, then in accordance with certain rules transforming a new type of indicators required to continue exploration. By continuously testing until the flashback test building is completed. The regularity of the indicators required to convert, destroy the randomness of composing test paper.

Composing test paper method based on genetic algorithms. Genetic algorithm is a biomimetic algorithm, it emulates natural selection and natural process of genetic. In the iterative process this method remains some structure, while in search of a better structure. This method combines the genetic algorithm global optimization and the characteristics of fast convergence. It draws on the advantages of random choice method heuristic backtracking and can be a solution for multiple control indicator.

For the characteristics of the contents of the ship Electrical Technology Assessment, This article adopts random region program which based on the type. Specific steps are as follows:

Step 1: First selecting a database according to participate in the assessment of student positions.

Step 2: Selecting the database table according to evaluation subjects.

Step 3: Selecting questions according to the difficulty in ascending or descending order. Weights of those questions are divided into three regions (assuming only 3 weight classes). Getting the total number of each type of questions, assumption is that N1、N2、N3 as shown in Figure 2.

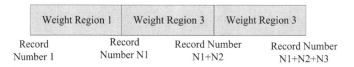

Fig. 2. The structure of weight region

Step 4: Each weight region and then divided into smaller region, according to the item number to remove : T1, T2, T3(assumed to be T1 <N1, T2 <N2, T3 <N3, otherwise they will be prompted to change the number of subjects). Record number of each smaller region is N1/T1. In this case, the number of last region may be more than the previous one, but it will not have much impact. The weight region 1, for example, shown in Figure 3.

Fig. 3. Question region of weight 1

Step 5: Using random methods to produce a value which is a random value between the upper and lower limits of each question area. This random value is record number, which is saved in an array.

Using the java.util package provided by Random class of Java. When using the class produce random values, we should firstly create Random object Random Rnd=new Random() and then produce integer random numbers by method nextInt().

int chcNum=Rnd.nextInt(lower value☐ upper value);

Step6: Searching all question regions and getting all record numbers to compose test paper.

Step7: Randomly arranging the record number in the array using "shuffle" algorithm.

```
    int len=shiti.length;
      int rnd,tmp,j;
        for(int i=0;i< len;i++)
          {
              rnd = Math.random();
```

```
j = parseInt(len * rnd);
tmp = shiti[i];
shiti[i] = shiti[j];
shiti[j] = tmp;
}
```

Step8:Getting questions from the question library according to the record number to to compose test paper.

Using this program can facilitate the realization of the random questions and no repetitive. The division of the region, On the one hand improve the speed of composing test paper, it also makes all the questions of deflecting is basically the same. Randomly arranging the record number in the array using "shuffle" algorithm enables ease of papers consisting of randomly distributed different questions.

4 Demonstration

For the marine electrical repair and maintenance of technology assessment project in ship two / three wheel electric technology ship crew jobs, we construct the evaluation questions card shown in Figure 4.

Evaluation Questions Card			
Subject	Ship motor repair process and maintenance		
Applicable Object	ship two /three wheel with 750kw and above		
Question	Judging fault properties and possible failure of links according to the symptom		
Content	Assessor specifies one of the following fault, tester points out the possible failure of links In the schematic within 10 minutes: (1) possible links of "manual operation can only be run electric" failure. (2) Power supply properly, System unresponsive and the motor does not run When you press the Start button. (3) Motor can't start correctly and has buzz sound. (4) Three-phase power supply properly, The power led isn't lit, system doesn't respond, the motor doesn't run.		
Time	10 minutes	Score	20
	Evaluation Criteria		Score
1	Operation process precisely and skilled		20
2	Operation process precisely and comparatively skilled		16
3	Operation process precisely, Proficiency is not enough, can accomplish task		12
4	Poor operation, only complete parts operation		8
5	Poor operation, can't complete task		0-4
Instruction			

Fig. 4. Evaluation questions card

5 Conclusion

Auto-assessment System of ship craft electrical engineering technology has good flexibility. This system can add, delete and modify examination questions at any time according to the new assessment training program, while it enables automatic composing test paper. Random region program which based on the type enables the non-repeating random selection of questions, and improves the efficiency of the group volume and quality, then can get the more real level of examination trainee. But the rate of assessment continues to be artificial, due to the influence of subjective factors it is difficult to ensure fair assessment completely. The future research is to use the virtual visualization technology and virtual experiment platform, which can evaluate automatically and avoid the influence of human factors. The efficiency and equity of automatic assessment of electric ship technology should be future improved.

References

1. Zhu, Y.Z., Jiang, C.F., Li, K.: The ways to ensure the competence of seafarers under the STCW 78/95 Convention. Journal of Dalian Maritime University 25(2) (1999)
2. Zhang, Y.L., Lu, L.B., Cao, X.M.: A Strategic Algorithm of Forming Test Papers. Microelectronics & Computer 20(6) (2003)
3. Yan, F., An, X.D.: Research on the Tactic of Intelligent Test-Paper Generation Based on PSO Algorithm. Journal of North University of China 29(4) (2008)
4. Xiao, Y., Wang, X., Liu, X.F.: An algorithm for setting online test papers. Journal of Beijing University of Chemical Technology(Natural Science Edition) 33(4) (2006)
5. Lin, X.M., Zhang, J.Q., Jiang, W.G.: The Foundation of Arithmetic for Organizing Examination Paper Based on Content of Exam. Computer Technology and Development 11(2) (2001)
6. Ying, J.R., Hu, L.X., Long, Y., Xu, H.T.: Creating and Realizing a Model of Selecting Items Randomly from Examination Items Library. Journal of Computer Applications 1 (2000)

Construction of Practice Teaching System for Training Innovative Talents Based on the Idea of Large-Scale Engineering

Aijun Tang[1], Hailong Ma[2], Deming Sun[2], Yuchao Niu[2], and Mingdi Li[2]

[1] College of Mechanical Engineering in Shangdong Jianzhu University,
Jinan, Shandong Province, China
[2] College of Materials Science and Engineering in Shangdong Jianzhu University,
Jinan, Shandong Province, China

Abstract. This paper constructed the "three-dimensional practice" teaching system of experimental teaching, practical teaching and technological innovation activities based on idea of large-scale engineering, which combined with the practical experience for many years in major of Materials Science and Engineering in Shangdong Jianzhu University. The teaching system runs through the training innovative practice of undergraduates in every link of the curricular teaching, extracurricular education and talents training, which improved the talent training quality and enhanced the undergraduates' technological innovation and practice abilities.

Keywords: Idea of Large-scale Engineering;Major of Materials Science and Engineering, Training of Innovative Talents, Practical Teaching System.

1 Introduction

Economic globalization is the age trend of economic development since 21th century. Globe marketization and economic globalization must bring about the educational internationalization. The international competition focuses on the competition of talents strength. As supporting the economic development, higher engineering education should train the undergraduates' innovative and practical abilities, and enhance the training of innovative quality. Shangdong Jianzhu University is one of the provincial universities and colleges, in which engineering is given priority and multi-disciplinary are coordinated development. This university trains the application talents of having innovation spirit and practical ability, which pay attention of theory with practice and push the reforms of teaching content, course system and practical teaching. Integrating the own characteristic, this university constructs the "three-dimensional practice" teaching system, which makes the platform of experimental teaching, practical teaching and technological innovation activities for the personality development of undergraduates. This system improves the talent training quality and

Y. Wang (Ed.): Education Management, Education Theory & Education Application, AISC 109, pp. 289–296.
springerlink.com   © Springer-Verlag Berlin Heidelberg 2011

enhances the application ability of technological innovation and practice of undergraduates.

2 Main Connotation of Large-Scale Engineering

Large-scale engineering is a teaching system based on the large-scale, which is a comprehensive systematical social engineering. This engineering need learn the basic theoretical knowledge and technological innovation knowledge, and adjust continually the educational goal, content, measure and method according the technology and market[1].

In order to adapt the development of modern social integration, the idea of large-scale engineering integrates the learning knowledge with training capacity, integrates the service development with comprehensive quality, integrates the scientific spirit with humane cultivation, integrates the reality adaptability with active innovation in the cycle of practice-knowledge-practice, which achieve the overall development engineering talents [2]. The traditional engineering teaching emphasizes on the engineering practice of experience stage. The science engineering teaching emphasizes on learning of scientific and technological knowledge. Taking the two teaching ideas into account, the large-scale engineering teaching emphasizes on the comprehensive subjects learning, which can avoid effectively the professional isolation. The large-scale engineering teaching emphasizes on the training of practice ability. Of course, learning knowledge and practical ability are interdependence and mutual-promotion. Learning knowledge is for the practical application. Practical training is for deepening the knowledge, in order to apply what we have learnt to train the talents[3].

In the course of training talents, idea of large-scale engineering is not limited to the knowledge of engineering theory, but emphasis on the training of innovative comprehensive ability in the engineering practice. Building the practical concept should pay attention to the flexibility, adaptability and innovation of the practice goal, which make the undergraduates contact the practical problems more early. The undergraduates can take part in the engineering practice on campus, and establish the strong engineering consciousness and innovative ability. The practical links of innovative engineering is very important in the course of training innovative talents in major of Materials Science and Engineering. The engineering practice is the basis of innovative ability. In the course of training talents, the university should take action according to the engineering practice. On the basis of engineering theory knowledge, the university should enhance the cultivation of engineering practice. According to the practical nature of large-scale engineering, the talents in major of Materials Science and Engineering can improve the innovative ability of engineering practice by practical education [4-6].

3 Construction of Practical Teaching System

The training mode is the multi-platform and specialized subjects, which strengthen the practical teaching link. The training mode integrates the extracurricular technology practice into the training system. Therefore, it drafts the "135" professional teaching system, i.e. one main subject-- Materials Science and Engineering; three-level curriculum platforms—the solid and wide public basic courses platform, the sturdy professional basic courses platform, the advanced practical professional courses and professional emphasis courses platform; five major fields—metal material, building material, liquid connection, liquid molding and plasticity molding. The material processing practical base is built. The "three-dimensional practice" teaching link is well-designed, which is divided into three parts--experimental teaching, practical teaching and technological innovation activities.

3.1 Integrating the Experimental Teaching Resources and Establishing the Experimental Teaching New System of Level-Division, Multi-module, Platform-Based and Layer-by-Layer Progressive

According to the platform of training the practical ability, experimental teaching resources is integrated. The experimental teaching process is complete experimental teaching process of material composition design—material reparation and process—material property detection and analysis—material application. In the process of experimental teaching, we establish the engineering training platform, professional experimental platform, comprehensive experimental platform, innovative experimental platform. Every platform can be divided into some experimental modules according to the demand of ability training and knowledge structure system. Every module have some experimental projects which gradually undated. The characteristic is start from the training objective, which give priority to the ability training. The experimental teaching connects closely with the engineering application. The system of level-division, multi-module, platform-based and layer-by-layer progressive trains the practical ability and shows the undergraduates' individuality.

3.2 Constructing the Comprehensive Practical Training System of Multi-Function to Increase the Undergraduates' Practical Ability

Aim at lacking of the practical operate in going out the practice, we strength the comprehensive practical training system of multi-function to increase the practical ability of the undergraduates. Such as, the undergraduates of liquid molding can accomplish the practical operate of overall process from the raw material selection, sand mulling, modeling, smelting, pouring, sand shakeout, clearing to casting defect analysis, mechanics performance testing, metallographic structure observe. This system can improve the ability of independent manufacturing engineering, organizing production, analyze and solve the problem to increase the practical ability. Hence, every undergraduate can master the whole practical teaching content of his major through the practical training link of powerful comprehension, obvious major characteristic and close production.

3.3 Constructing the Extracurricular Technological Innovation System and Carrying Out Technological Assistant Cultivating Mode to Perfect the Innovative Practical Base

(1) Extracurricular Technological Practical Activity System of Undergraduates
In order to take full advantage of the extracurricular time, we construct extracurricular technological practical activity system to form the operating mechanism which is by taking the system construction as the guarantee, taking the whole participation as the pilot, taking the practical base as carrier, taking the scientific research fund as the backing, taking the technology activities as the means. Extracurricular technological practical activity system is followed as Fig.1.

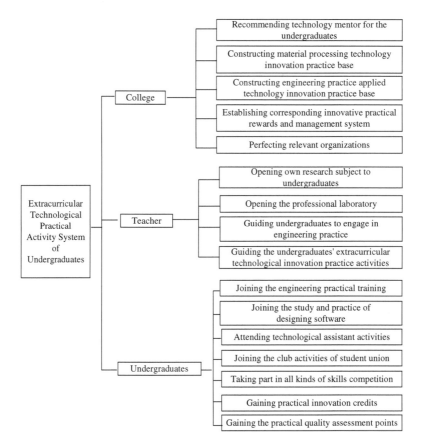

Fig. 1. Extracurricular technological practical activity system of undergraduates

(2) Technological Assistant Training Mode for Undergraduates
We recruit the excellent students to develop innovative practice, and gradually formed the features training mode of recruiting outstanding undergraduate as "technological

assistant". The activities system of extracurricular technological innovative practice is shown as Fig.2.

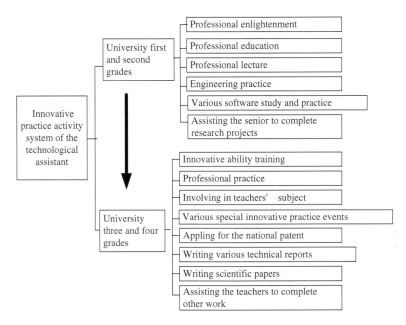

Fig. 2. Innovative practice activities system of the technological assistant

We recruit the technological assistant in autumn and spring, in which the freshmen is 5%, the sophomore is 25%, the junior is 70%. The technological activities flow of base is constituted by simulating the implementation measures of project and scientific research projects from proposing and setting up projects to the implementing and accepting projects. Through the links of public demonstration and reply to an argument, we can realize the radiation effects of base training mode. The comprehensive quality training system of technological assistant is shown as Fig.3.

3.4   Transforming the Examination Mode, Perfecting the Rules and Regulations to Promote the Innovative Ability Training for Undergraduates

In order to guarantee the characteristic of technology innovative practice and advocate the training mode of the personal cultivation target to be effectively developed, we transform the test way for the engineering practice to the network test. We establish the random papers and carry out the web-based instruction network test, which have the functions of adding, modifying or deleting the questions, setting the difficulty factor etc. In addition, our university guide the students' technology innovative activities and the teachers' guiding by "Management Method of Undergraduates' Technology Innovative Credits in Shandong Jianzhu University", which provide the institutional guarantee for the undergraduates and teachers' participation in the technology innovative activities. For some excellent design, the university can give

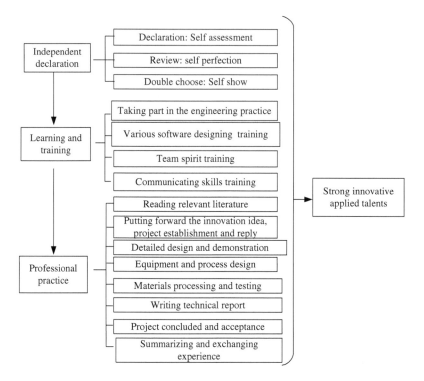

Fig. 3. Comprehensive quality training system of technological assistant

capital support. Hence, it can greatly arouse the undergraduates' enthusiasm to complete the daily practice by engineering ideas.

4   Effect of Practical Teaching System

4.1   Promoting the Undergraduates' Employment Competition

The base cultivates the undergraduates' innovation practical ability by the technological assistant mode. It gradually forms the autonomic and inquiry learning mode, which is harmony between the teachers and undergraduates, between the senior and junior in different major. This mode can ensure the undergraduates to find their characteristics of practical content. The technological assistants must guide three external-base undergraduates to learn professional knowledge, participate in technological innovation activities, which amplify the training mode of the base. This mode drives the more undergraduates to participate in professional study and technological practical activities, which promote the study style construction to steady development. For seven years, 94% undergraduates of this major participated in the technological innovation activities, and the undergraduates' employment remain above 90%, and passed the entrance exams for postgraduates is 20%, and the employment quality increased year by year.

4.2 Improving the Undergraduates' Engineering Quality and Practical Skills

Through strengthening the combination degree of experimental teaching, scientific research and engineering practice, we introduce the latest achievements and the latest technology into the experimental teaching, which raised the undergraduates' practical skills and engineering quality. Undergraduates of this major were awarded the one term national second prize, two terms national third prize in 2010, 2011 "YongGuan Cup" Casting Technique Competition in Chinese College. Undergraduates of this major were awarded honorable prize in the first National Concrete Contest (U30). Undergraduates of this major also won numerous awards in other comprehensive national competition.

4.3 Increasing the Undergraduates' Innovation Practical Ability and Winning Numerous Awards in Various Competition

Undergraduates' innovative, creative and understanding abilities have been increased. In the curriculum and graduation designs in recent years, the undergraduates' practical ability and designing thinking have a significant progress. Moreover, some undergraduates consulted the teachers some problems of engineering problems encountered. The training mode of "technological assistant" has subtly influenced students learning. Undergraduates in extracurricular technology series competitions have achieved remarkable results. In the recent three years, undergraduates of this major successively received five items national prizes and 22 items provincial awards in the "Challenge Cup" series competition. Undergraduates published 8 piece EI papers with the first author, and authorized 1 item national inventive patent and 9 items utility model patents. Undergraduates won 1 item the national special-class award, 1 item the national first award, 3 items the national third awards in the 11th "Challenge Cup" national university extracurricular technology competition.

Acknowledgments. This work was financially supported by the Higher Education Reform in Shandong Province Key Research Projects (2009045), Shandong Province Graduate Education Innovative Projects (SDYC10023, SDYY11063), Experimental Technology Research Project in Shandong Jianzhu University (2011sy004).

References

1. Liu, J., Liu, Z.: Educational Innovation of Pan-Engineering in Engineering University. Journal of JiangSu Polytechnic University 7(3), 35–38 (2006)
2. Xia, Y.: Innovative Talents Training in Major of Mechanical Engineering Based on Idea of Large-scale Engineering. The JiangSu University for the Degree of Master of Philosophy, 10–17 (2009)

3. Moskvitin, G.V.: Problems of engineering education and production of young scientists, vol. (5). Allerton Press Inc., Springer Science+Business Media LLC (2008)
4. Gu, S.: Reform of the Traditional Teaching Mode: An Exploration Practice. Educational Research (8) (2003)
5. Gig, G.: Historiography in the Twentieth Century-From Scientific Objectivity to the Postmodem Challenge. Wesleyan University Press, New England (2003)
6. Zhang, W.: A Comparative Study on the Engineering education of Five Countries-A Comparative Study Oil the Engineering of China and That of Four Western Countries and the Interaction between Engineering Education of the Five Countries and their Industrialization. Chinese Academy of Engineering (2002)

Study to the Development of Management Ideology of Physical Education Since Reform and Opening Up in the China

Liu Fusheng[1], Lu Chun[2], Ma Meiyu[3], and Li Xiang[4]

[1] Institute for Sports History, Cheng Du Sport University, Cheng Du, Sichuan, China
475638362@qq.com
[2] Institute of Liberal Arts, Yang Zhou University, China
404822141@qq.com
[3] Institute for Sports History, Cheng Du Sport University, Cheng Du, Sichuan, China
283806805@qq.com
[4] Cheng Du Sport University, Cheng Du, Sichuan, China
150671289@qq.com

Abstract. Using literature, logical analysis, expert interviews, take the development of management thinking in Physical Education in China since Reform and Opening up as object,the article summarized the progress in three stages. which is reform and opening up, the early period of the socialist market economy and the new period.

Keywords: 1. Reform and opening up, 2. Physical Education, 3. Sports Management, 4. Thought Transformation.

1 Introduction

In 1976, with the end of the decade-long Cultural Revolution, the almost paralyzed Physical education system also started to the recovery, with the "gang of four" 's crushing, the discuss in "criterion for testing truth" also carried out, in particular, the third plenary meeting's convening of the Eleventh Central Committee Bring vitality to the education sector. sports education in China is also gradually go to into the right track, with times' changing, management thought of physical education also develop rapidly.

2 Investigation for the Development of Management Ideology of Physical Education in the Early Stage of Reform and Opening Up(1978 - 1992)

With the Ideological emancipation brought by the Third Plenary Session, management thinking of physical education has also have a revival .in 1979, the Ministry of Education and the National Sports Commission jointly issued two files" Interim Regulations of Physical Education in College " (draft), and " Interim Regulations of

Physical Education Primary and secondary schools "(draft).it is the first time that new China formulated laws and regulations on school sports, the documents defined the specific aspects carried out in the process furtherly.

Subsequently, in 1985, the promulgation of "the decision on Education reform of the CPC Central Committee " and the Introduction of "Compulsory Education Law" in the following year, making the implementation of physical education in each school can find a rule to follow and provides a theoretical basis to the files " Sports Work Regulations in School "issued by national Sports Committee lately. With a clear document to follow, the development of physical education during this period showed four characteristics:

First, the construction of course is accelerating. In 1978 the promulgation of " PE syllabus for primary school " and " PE syllabus for secondary schools "contributed to the gradual formation of the core in teaching - to master the basics of sports ,to master the basic sports skills and master the basic techniques Ideas. The content contained in the physical education curriculum also seem more reasonable and rich In the promulgation of syllabus of nine-year compulsory education in 1992 lately.

Second, the formation of specialized management agencies, the formation of specialized management agencies in PE is also a sign to the development of physical education during this period .until 1990, there are 28 provinces (autonomous regions and municipalities) establish a Sports Health Department at the education sector, in which the barrier was cleaned completely for the development of PE.

Third, the establishment of testing system for physical health. A lasted two and a half years Research aimed to test the health of youngster carried out by the National Sports Commission, Ministry of Education and Ministry of Health (Nationally), found out the children's physical condition preliminarily on August 1978.August 1982, based on the results of the survey, the National Sports Council promulgated the " Training Standards for National Physical Education " and " implementation program of the health research with establishing and improve 'the physical health card' furtherly ", and determine each to conduct a small sample test for every three years, large sample test for every ten years among the Young People. It also has provided a guarantee to scientific development of physical education.

Fourth, the sports teacher's training. With the restoring of college entrance examination system in China in 1977, in the following year the National Sports Commission issued " proposal on how to manage the sports Institute well "Yangzhou Conference", was held in 1979, the Ministry of Education and issued a " teaching plan Department of Physical Education degree in High Normal School, (four-year) (draft) "and other documents in 1980. It should be said that in the first few years after the Cultural Revolution, China's PE teacher education focuses on the restoration and adjustments, and basically completed it in the early eighties. 1980--1982 the Ministry of Education issued the training program for PE teachers in college which make one-third of in-service teachers retain training by the 3 months to 1 year. In 1985, "the decision of the CPC Central Committee on Education Reform," clearly states that "the establishment of a sufficient, qualified and stable teaching staff is a fundamental policy to implement compulsory education, improve the level of basic education. " Accordingly, the State Department expand the enrollment in existing sports division , establish physical education, subjected in conditional Normal University and college, open sports classes in Secondary Schools and also proposed to open up more channels to organize different types, different levels of education, correspondence advanced studies courses, training a group of physical education talented person.

3 The Development of Management Thought in Physical Education in the Socialist Market Economy Period(1992 - 2000)

With the rapid progress of science and technology, the knowledge economy is emerging, national competition becomes increasingly fierce. Education is a foundation in the formation of comprehensive national strength. The strength of a national power depends on the quality of workers, depending on the quality and quantity of all kinds of talents, which proposed new more urgent requirements, for training and bringing up a generation of twenty-first century,

In 1992 ,from the promulgation of " the nine full-time PE syllabus (trial)," , this period of physical education is also into a new phase, The education sector begain to carry out a decade-long education experiment in China to change the exam-oriented education to Quality Education, and as a academic Branch the status of physical education has also been improved.

June 20, 1995, " National Survey of Student's Health " revealed some problems, such as students endurance, flexibility, quality, vital capacity all has a downward trend, obese children and overweight children is increased, ability to resist frustration, sense of competition and crisis, the spirit of collaboration are relatively weak. [1] This has led workers in the state and the high school sports take it seriously, have asked to strengthen and improve the work of health education in school, firmly establish the guide ideology of "health first" full implement then quality education, the country set off a wave of reform in school sports which ideas is "health, quality, environment".

1995 by the State Council promulgated the "National Fitness Program," which pointed out that students in our country at this stage is about 3 million people, accounting for a quarter of the total population, so if the school is able to smoothly carry out sports has a significant impact for the whole Fitness Program, but at this stage the question is, the students out of school at a large part did not stick to physical exercise. The introduction of "Outline" accelerated. the Speed of reform in management thought.

The promulgation and implementation of "Sports Law" and the "National Fitness Program," in1995 confirmed the students of sports rights and the basic tasks of school sports in the form of regulations creating a new era in the development of school sports, school sports in China began to move toward the systematic and scientific direction. Ideological and theoretical study of school sports, facilities of school sports construction sites, and faculty, at this stage has aslo made great achievements. [2]

In order to comply with the demand for quality education and services for the Development of socialist market economy, in1996 National Board of Education has drawn up a "full-time High School Physical Education Syllabus (for testing)", the document cancel "a purpose, three tasks" in the Physical Education Curriculum on the past point out that the purpose of physical education is three, one is a comprehensive training of students' body to ensure the health of students; Second, master the basic skills and knowledge of sports, to prepare for their life-long sport; Third, affect the idea to nurture students to cultivate good moral character. About the structure of the physical education curriculum, the document piont out that physical education courses are divided into two aspects ,one is activities and another is scholarship, This is laid a foundation. for the development of school sports in this phase. Until June 1999, the State Council issued a "decision on deepening reform to promote quality education", to promote quality education, deepen the reform of education in China has become the trend of future development.

4   The Development of Management Thought in Physical Education in the New Period(2000-Present)

With the coming of new millennium, along with the "decision" in 1999, the State paid more attention to the development of physical education, furtherly in October 2000, the state issued a new "nine-year compulsory education on Physical Education and Health in Junior High School " and "nine-year compulsory education syllabus Primary School on PE and Health Teaching. " compared the new with the previous, although they are not fully reflect the trend of curriculum reform in physical education, but from the choice of teaching content, the assessment of teaching effectiveness, or the guiding ideology of teaching, we can see the state paid more attention to establish the dominant position in students itself, to develop students' interest in sports. the consciousness and the ability of "Lifetime sports" become a major goal to value the pursuit of the school physical education. which is also the tone of management thinking. during this period [3].

Since the founding of new China, the establishment of the school curriculum has been under the direct leadership undertaken by specialized agencies in the national education department. Curriculum development in particularly seems the Government's decision, rather than academic issues, it seems that the relationship between education and government is how to understand and execute these files. The separation between academic and practice ,decision-making hindered the scientific curriculum development and nationalization to some extent "[4].

At this stage, in line with the school of Physical Education in 2005, the Ministry of Education to develop a "school sports equipment and venues", in the way of regulatory documents to ensure the implementation of compulsory physical exercise. [5]

5   Conclusion

As an important component of mass sports and the basic of development universal school sports is a key factor to the entire strategy Review the changes in management thinking in Physical Education in China since reform and opening up, you can find our physical education management ideas has independence, non-stage, progressive and other features, you also can see the thinking 's change accompany with the Changes in the values in political and economic mainstream On this basis, we need to respect the objective laws of physical education furtherly, and to strengthen the management of physical education laws and regulations, in order to provide protection for physical education management 's development.

References

1. State Board of Education and other five ministries. In: The Chinese Students and Health Survey Report. p.52. Jilin Science and Technology Press, Changchun (1995)
2. Hua, T.: Sports History, p. 426. Higher Education Press, Beijing (2005)
3. Chen, Y.Z., Xu, Q.: Research on the Value Orientation of P.E.at School and Its Future Direction during Different Historic Periods Since 1949. Journal of Beijing University of Physical Education 28(12), 1665–1668 (2005)
4. Li, B.D. (ed.): Teaching, pp. 169–170. People's Education Press, Beijing (2001)
5. Shan, X.D.: Effects of the revision of sports equipment and playground standard implemented in elementary and middle schools on the sense of success in learning of the students. Journal of Physical Education 14(4), 84–89 (2007)

Exploration and Practice of Physics Teaching for Liberal Arts Undergraduate

Lizhen Ma, Jun Ma, Shukuan Wang[*], Yurong Shi, and Yongjian Gu

Physics Department of Ocean University of China, Qingdao 266100, China
skwang@ouc.edu.cn

Abstract. According to the characteristics of liberal arts undergraduate and the reform and development needs of China higher education, we set up three physics courses as a platform, furthermore research and practice on how to infiltrate human spirit and science spirit into the teaching.

Keywords: physics teaching, human spirit, science spirit.

1 Introduction

Physics is science, and it is also culture. Physical culture is an important part of science culture between human culture and science culture. One of the basic tasks of physical education is to come down the physical culture. Science quality education based on physics among liberal arts undergraduate could create a science environment, make liberal arts undergraduate know basic science, and improve their science quality. The liberal arts undergraduates can establish science awareness, learn scientific thought, develop their exploring spirit of science and become multi-developed undergraduate. This is consistent with the current China higher education reform and development, but also conforms to the education philosophy of general education for basis and professional education for use in our university.

2 We Create a Series of Courses to Provide a Platform for Integration of Two Cultures

"The goal of school is always educating youth as a harmony people but not an expert when they leaving school." Einstein said. The harmony people mean that a man is perfect unity of science spirit and human spirit. In order to train students who have broad knowledge and reasonable knowledge structure, we have designed three general courses including "Physics for Arts Undergraduate", "Interesting Physical Experiment for Arts Undergraduate ", "Physics that Changes the World".

"Physics for Arts Undergraduate" has 34 class hours, it contains: mechanical, thermal and chaos, electromagnetic theory and relativity, optics and colorful world, the

---

[*] Corresponding author.

Y. Wang (Ed.): Education Management, Education Theory & Education Application, AISC 109, pp. 303–306.
springerlink.com © Springer-Verlag Berlin Heidelberg 2011

mysterious microscopic world of quantum information and the information era. This course takes basic physical knowledge as carrier, emphasizes human spirit of physics and takes the integration of physical thought and human spirit as mainline. This course aims to introduce arts undergraduate basic laws of physics, basic thought and method of physical investigation, and the relations of physics, technology and human civilization. This course is benefit for broadening arts student views and promoting their physical level.

"Interesting Physical Experiment for Arts Undergraduate" has 34 class hours, and it starts from simple, lively and interesting experiment, and deepens gradually. So the arts undergraduate establish an intuitive understanding of physical concept. This course aims to learn physical concept, understand the basic laws of objective world deeply and feel the application charm of science and technology. The experimental teaching emphasizes interesting, knowledge, liveliness and human background, it also makes the arts student understand theoretical knowledge deeply, improve their interesting on science and cultivate their scientific quality and physical way of thinking. Throughout operation and designing experiment themselves, the arts undergraduate will be encouraged to explore and study in the experimental process, which is a better way to cultivate their awareness of innovation and stimulate their creativity. This experimental course could also cultivate students with realistic attitude and active exploring spirit.

"Physics that Changes the World" has 34 class hours, and it introduces basic principles of physics and the world that changed by physics, and it also introduces the latest views of physics and social promotion by physics. The arts undergraduate could see why the world develops so fast in physical development views. This course provides a better way to understand the primary physical concept, latest physics in technology and physics in society.

The three courses are independent, but they are also correlative, they are an organic whole. "Physics for Arts Undergraduate" shows the basic structural system of physical knowledge, which is a preparation knowledge and direction of "Physics that Changes the World". "Physics that Changes the World" is the intensive course of "Physics for Arts Undergraduate", it adopts a scientific seminar form and takes the new physical applications in high-tech as mainline, which integrates the physical knowledge and the forefront application. The course brings the students into the world of modern science and technology and makes them feel the great promotion for social development caused by natural science. Physics is experimental science, it consists of experimental and theoretical courses. "Interesting Physical Experiment for Arts Undergraduate" is supplement and improvement to theoretical courses, and it makes the arts undergraduate understand physical knowledge by experiment.

3 Integration of Two Cultures by Course Platform

3.1 Set Up a Communication Bridge between Arts and Science Undergraduate to Guide the Collision of thought

Because of different background, arts undergraduate are good at expression and writing but weak in basis of physics and mathematics and lack of operation. The science undergraduate have better operation ability and basis of physics and mathematics. The communication between arts and science undergraduate is a better supplement for each other.

We design "Physics that Changes the World" for all arts, science and engineering undergraduate without borderline of arts and science undergraduate. The course makes all students learn in the same classroom, which is benefit for communication and development. We organize some teams with different specialties and different grades to discuss some topics, such as "chaos and the economy", "Cosmos," "social effects of internet", "global warming ", "whether mobile phone radiation is harmful or not " and so on. The arts and science undergraduate discuss from different perspectives and hand on papers at last. The teams with both arts and science undergraduate hand on much better papers than teams with all arts undergraduate or all science undergraduate. Meanwhile, this discussion is a great practice on searching papers, expression, organizing words, writing papers and cooperation.

In "Interesting Physical Experiment for Arts Undergraduate ", besides teachers, we engage several graduates and undergraduates as assistants. The assistants discuss with arts undergraduate face to face in the experiments, which help the arts undergraduate learn to observe and analyze the experiments. Many students like experiments gradually because of the mystical phenomenon of experiments, and they pay attention to the natural phenomenon in the life. The arts undergraduate are influenced by the thinking mode of science undergraduate. The arts undergraduate said in the paper, "The science undergraduate have sharp thought, objective attitude based on realistic and better ability to analyze questions, which is worthy to learn." "We appreciate that the science undergraduate have exploring spirit and high curiosity of every new and unknown thing in the experiments. I believe that is the power and resource to develop and advance all subject." Both science and arts undergraduate improve their science and human quality by learning and cooperating together in the same team.

As to "Physics for Arts Undergraduate", it has been discussed [1].

3.2   The Arts Undergraduate Go to Research Base as Second Class

We take Ministry of Education Key Laboratory of Ocean Remote Sensing and Optics and Optoelectronics Laboratory as a second class in the process of teaching. We organize arts undergraduate to visit science laboratory, such as the ocean remote sensing, laser radar, underwater robots, holographic technology, optical fiber communication, laser spectroscopy and semiconductor lighting. The professors give detailed introduction on whole process of scientific research and technology development during the visit. This could cultivate science and technology awareness and improve their science and technology quality.

3.3   Set Up a High-Quality Teacher Group That Is Good at Both Science and Arts by Improving Their Cultural Quality

Confucius said: "do best, get better; do better, get worse; do worse, get noting." Teachers must be good at both science and arts in order to cultivate multi-developed students. In order to improve human quality of teachers, we set up specific teaching group for arts undergraduate and organize them to attend many training and meeting on quality education, such as "series of master lectures", "ocean university-human-forum" and national quality education meeting. Teachers improve their human quality and have more topics that could communicate with arts undergraduate.

4 Conclusions

Physics is essential to cultivate science quality of arts undergraduate in education [2]. Physics teachers are responsible to carry out and popularize science quality education. Our series of physics courses for arts undergraduate have been elected as excellent course of Shandong Province for quality education, and we wish offer some valuable suggestion for physics teachers.

References

1. Ma, L., Ma, J., Shi, Y., et al.: Exploring and Practice of Course Teaching for Arts Undergraduate. High Science Education (3), 106–108 (2009)
2. Zhao, K.: Physics Lighten World. Peking University Press, Beijing (2000)

Talent-Cultivation Mode Research on Direction of Services and Outsourcing-Oriented Software Engineering

Yubo Jia, Qi Sun, Na Zhang, and Hongdan Fan

Institute of Information and electron
Zhejiang Sci-Tech University
Hangzhou, 310018, China
Jiayubo1964@163.com

Abstract. With the high development of software outsourcing industry, relevant majors in higher education institutions have become highly concerned about the issues regarding how to make computer teaching meet the current international market and to cultivate software outsourcing professionals for society. In this paper, we explore the talent-cultivation mode on outsourcing of services-oriented software engineering with the actual situation of our school.

Keywords: Software Outsourcing, Services Outsourcing, Talent Cultivation.

1 Introduction

Software outsourcing is an important part of services outsourcing industry[1]. It is also one of the industries that provide more jobs for graduates majored in computer science and technology. So-called software outsourcing means that software companies from developed countries ask the companies with lower labor costs in developing countries to develop the non-core software projects in form of outsourcing. The main purpose of outsourcing is to solve a temporary lack of human resources, or is a method to reduce the development costs of software developments. And human resources account for seventy percent of software development costs. Therefore, lower costs of human resources reduce the costs of software development effectively[2].

Software outsourcing is the product of continuous refinement in social division of labor. Compared with the hardware outsourcing, software outsourcing started later, but has developed rapidly. India is the current largest market of software outsourcing. In addition, Ireland, Israel and China are also the major software outsourcing markets. Building-markets of global software outsourcing mostly concentrate in North America, Western Europe and 1 3 other countries. The United States accounted for 40% and 1 3 other countries constitute 10%. Accessing-markets of outsourcing are mostly in India, Ireland and other countries. Among them, the monopoly of the U.S. market is India, 80% of the Indian software industries income depend on software outsourcing business. India has become a superpower in software outsourcing. European markets are monopolized by Irish. Nowadays, the Philippines, Brazil, Russia, Australia and other countries have joined the ranks of the world software outsourcing competition. Currently, the sales volume of global software sales is 6,000 billion dollars, of which sales of software outsourcing is about 1,000 billion dollars. Greatly increasing software outsourcing has brought China new opportunities for development.

2 The Current Situation of Software Outsourcing Talent-Cultivation

Several software outsourcing talents training modes are hard to meet the needs of enterprises in China. Specifically, firstly, the graduates can master basic software development techniques in normal colleges with diploma education training mode, but they lack work experience of software developments, project communications and teamwork practice. If enterprises choose this kind of students, they have to train students for basic knowledge and skills of projects. Secondly, talent-cultivation mode in training institutions, such as Beida Jade Bird, New Oriental, they value general skills training, however, projects have different professional requirements for different fields. So they have many weaknesses in specific areas, such as finance, logistics, and so on. Thirdly, outsourcing enterprises ask trainees to accept senior engineers' and practical outsourcing projects' training directly through internal training, and cultivate overall development and coordination skills in a team mode, but this training mode will increase the costs of outsourcing enterprises.

The trained software talents have solid basis of the theory, owe the foundation and stamina to the development because the process of cultivation emphasizes basic theories and education training research personnel in the current training modes of software outsourcing. But education teaching is limited by current macro management system, because professional knowledge and teaching materials update slowly, trained software talents are hard to satisfy ever-changing software industry for updates requirements of knowledge structure. Software knowledge and technology are obsolete for college graduates, the lack of practice education leads to the students' practical operation ability being poor, theoretical study doesn't match with practical demand. Students educated by academic education usually need to receive re-education and re-training in software companies to meet the work requirements. However, training institutions are oriented to vocational education, they aim at teaching students the latest knowledge and skills in the special industry field. The advantage of training institutions is that the software talents educated by these institutions fit the needs of enterprises, so they can start the work and create value without delay. However, the greatest weakness of these talents is they lack the basic knowledge of the relevant professional. The room of promotion as well as development potential are also poor, thus their career vitality is very limited. What's more, the qualifications and ability of the students are uneven, and there are many problems in teaching, organizing and effectiveness.

3 Construction of Talent-Cultivation Mode for Software Outsourcing

3.1 Categories of Software Outsourcing

Outsourcing Types [3] are divided into Information Technology Outsourcing (ITO) and Business Process Outsourcing (BPO). Information Technology Outsourcing ITO refers to the organization that lets all or part of the IT be work outsourced to third parties, service outsourcing providers, to complete in order to focus attention on their core business. Information Technology Outsourcing (ITO) involves three aspects, such as basic information technology outsourcing, operations outsourcing of information systems and information systems services outsourcing. BPO also has different business types, including enterprise internal management, enterprise operations, logistics management and so on.

3.2 Quality Requirements of Software Outsourcing Talents

Demands software outsourcing enterprises for talents are multifaceted, they need to grasp the software technology and have some background knowledge of industries[4]. According to the qualifications and strengths of the companies, foreign contracting companies will outsource the work at different stages. Outsourcing personnel structure should be "pyramid" type [5] to Software outsourcing companies.

The bottom is the software developers required to code and finish technology documents with team cooperation ability, also co-improve the design description with the project managers, and have strong technical capabilities and skills.

Middle level is project managers well experienced in project practice. They always have strong communication capacities and team leadership skills, accurate understanding of customers' requirements, conduct systematic and modular analysis of software projects, coordination of resources, master project schedule and conformity of quality and, also need senior technical expertise.

The top classes are proficient in management and software, and outsource marketing capabilities with high-level personnel, such personnel should be familiar with the work of international business management, to keep cutting-edge technology and trends in the world, master of international standards and norms, are familiar with each other is Culture, participation in international competition and market development software.

3.3 Reform Personnel Training Programs, Construct Curriculum for Software Outsourcing Talent-Cultivation

According to the general constructed idea of computer science and technology in our school, we adjusted the cultivation scheme of computer science and technology. The adjusted training program is mainly in two directions, one is software engineering (service-oriented outsourcing direction), and the other is a technology of network information. The formulation of the plan on the adopting of 3 +1 pattern, complete with a 3-year theory courses of study, one year for the practice of teaching. It will take two and one half years to complete the basic teaching and six months to complete the professional theory courses of study during the three years of knowledge research. One year of professional practice includes half of the professional practice training, and six months of graduation design. Combined with talents acquirement of ITO and BPO outsourcing, we have reformed the traditional curriculum, including ITO and BPO constructed two types of courses outsourcing system.

A. Determine Main *Courses in accordance with the Training Program Targeting*

To develop a reasonable course system, first we should grasp the needs of personnel outsourcing specifications and determine the main course according to training objectives. Surveys show that software companies want programmers and testers to have a basic knowledge of software engineering, master programming language proficiency and have certain reading skills in a foreign language with team spirit and strong communication skills. Currently, it is about 70% of the programmers using Java language, mainly because the Java language for building large-scale cross-platform of software and hardware for enterprise applications. At the same time, Japan's software outsourcing, requires programmers to be able to read Japanese documents, Read Japanese writing software requirements analysis and software design specification, which requires the programmer to have the Japanese reading ability. Therefore, we determine the main courses for computer software are English, Japanese, programming languages (C / C + +), Java development technology and software engineering.

B. Add Japanese Teaching *based on English with different special emphasis*

English, as a mainstream language of programming tool software nowadays, is the main programming language and the dominant Internet language. It is essential to improve the skills and raise the business level for software programmers. According to the important position of English in computer technology, lead us to keep the share of English teaching and to guarantee English teaching for at least 4 semesters. Japanese language teaching must be strengthened, especially students applying computer science. Foreign language teaching should not be interrupted during the four learning years and foster the reading capability mainly. Foreign language teaching should differ from special emphasis, such as focused on English for Europe outsourcing and Japanese for Japan outsourcing.

C. Add the Course of the *Software Development Tools on the Basic of Programming*

Students with the basic of programming, we should add the course of the software development tools. Let the students use the software development tools to finish a few small software projects to improve training students' software development ability and cooperation ability. For example, after Java Programming Language classes, we also can open development classes of Jbuilder, Eclipse, Rose, etc. These classes taught with software project examples will receive the good teaching effects.

D. Increase Quality and *Enterprise Culture Teaching*

College students' quality education is perplexing system engineering and the author's view is that professional quality education is also one of the main tasks of higher education. On the basis of extensive investigation of the need for talent in the IT industry, we know that enterprises choose the talent not only paying attention to professional skills, but also to their professional ethics, team cooperation spirit, communication ability, speech and expression ability, organization and coordination ability, innovation ability and management ability, etc. Especially, the team cooperation spirit and communication ability are more important in software engineering .In the whole course system, we should increase social etiquette, speech, expression, innovation ability training and management ability training. To make the students understand the enterprise culture better, we can add the introduction of enterprise culture of Europe and Japan courses which can help them build the foundation for communicating in the future.

3.4 Reform Practice Teaching, Establish a Practice Bases in or Outside School

In the outsourcing model talent training mode, we construct a set of scientific and perfect practice teaching system. This practice teaching system, not only conforms to the software service outsourcing personnel training but also meets to the needs for software service outsourcing enterprise unit, cultivates the student's modern engineering quality, and improves the students' practical ability and also the training of the students' innovation consciousness.

Since 2009, Microsoft (China) Co.LTD has constructed one inside campus training base and six outside campus practice bases with its partners, such as Tarena (China) established long-term cooperation relationships, Japan Totyu Soft Co. LTD and other related companies, in order to improve the standard of teaching and make students quickly adapt to the software service outsourcing unit work environment, and also to achieve the service outsourcing software development technology quickly and standard requirements in practical and scientific teaching system.

Microsoft has been committed to train more standard, practical, and the global convergence of computer technology professional software talents, adhering to the combine authentication with academic education in developing skill education of Chinese higher education institutions. According to different requirements, the training programmers finished the basic skill training in education institutions enter into the related companies with the agreements between training institutions for professional and team training, such as the financial software enterprises, logistics outsourcing software enterprises, etc. On the one hand, the college or the training institutions can provide students more practical opportunities. On the other hand, it can let enterprises, colleges and training institutions establish good talent recruitment channels. In this way, the relation between the enterprise, relevant institutions and training institutions will come more closely. It cultivates the students both professional skills and project practice experience, so that they can meet outsourcing enterprises' requirements.

4 Conclusion

At present, China's software outsourcing businesses are suffering a crisis of talents shortage. It is difficult to solve this problem by just depending on schools or training institutions, but also IT industries, governments, academic circles should be involved to build a healthy outsourcing industry chain. We should make full use of the various departments' functions and also make efforts for China to improve software outsourcing business together. In India, the information technology industry is valued as a place on the agenda of congress. Indian government provides a strong policy support for the development of information industries, so our government can learn from the experience of India.

References

1. Feng, L.: Competitiveness of China Software Outsourcing. Software World 13, 24–25 (2006)
2. Let's Talk Part. Discussion about the Shanghai Software Outsourcing. Shanghai Information 11, 14 (2006)
3. Zhang, N.: Analysis and Research of China's software outsourcing to Japan's Current Situation and Countermeasures. University of International Business and Economics (2007)
4. Wang, H.: CMM Application in Japanese Software Development. Computer and Information Technology 12, 72–74 (2006)
5. Xiao, J.: Discussion on Japan Quality of Software Outsourcing Development. Modern Business Trade Industry 9, 221–222 (2007)

Case Teaching Based on the Network

Yang Huansong[1] and Liang Lu[2]

[1] Yang Huansong, Institute of Service Engineering, HZNU, China
hzjyhs@163.com
[2] Liang Lu, School Of Primary Education, HZNU, China
lianglude@163.com

Abstract. Case Teaching is a kind of teaching methods to guide the students to discuss the special situation through describing a specific scene. It is a teaching activity that takes teaching cases as the carriers and selects certain teaching cases according to the teaching aims. Case teaching and network teaching should be combined to form a new type of teaching called Case Teaching Based on the Network which can play their roles better in teaching.

Keywords: case, teaching, network.

1 Introduction

When Harvard Graduate School of Education was created in 1920, first dean Halls put forward to use the case method which had been applied in business teaching. Case teaching aims at improving students' abilities of solving problems and judging. The teacher should distribute cases to the students to let them discuss during the teaching process. Even then, case teaching had made a great success. And now due to the popularization of computer and the rapid development of network, it is possible to combine case teaching and network teaching. And the new type of teaching called case teaching based on the Network can play an immeasurable role of raising students' abilities of exploring and cooperating.

2 The Advantages and Effects of Case Teaching Based on the Network

Case Teaching is a kind of teaching methods to guide students to discuss the special situation through describing a specific scene. It is a kind of simulate, diverse and inspiring teaching. Teaching cases are written from life by certain people according to the teaching aims. Meanwhile the students' knowledge structures and the views of the same case are different. In case teaching, the teacher should show relevant knowledge firstly, and then present the cases to inspire students to think, create and apply the cases and solve the similar problems finally.

Network teaching has the characteristics of openness and interactivity. Compared with traditional teaching, network teaching is not confined by the fixed time or fixed location. The network also provides the sharing of global and diverse educational

Y. Wang (Ed.): Education Management, Education Theory & Education Application, AISC 109, pp. 313–318.
springerlink.com   © Springer-Verlag Berlin Heidelberg 2011

resources to raise the students' interest levels and learning efficiency. Students can choose learning materials independently from a wide range of information resources according to the situation in the network environment. They can construct their knowledge systems, make the study plans according to their own actual conditions and become the real master of learning. Teachers can interact with students about the question-answering, homework, test and discussion through the network. So network teaching is an open and interactive process between humans and educational resources.

Case teaching based on the network combines the advantages of network teaching and case teaching both. It is easy to be understood and digested through group discussions to cultivate students' abilities of exploration and cooperation. It follows the students' sequence of absorbing the knowledge and improves the students' abilities of participation and flexibility in learning.

3   The Relationship between the Teacher and Students in Case Teaching Based on the Network

The relationship between the teacher and students should be explained with the theory of double subjects. They are both the subjects in teaching activities.

The teacher is the mentor and organizer in teaching. They help students understand the world, develop their emotions and improve their abilities. When students discuss the case, the teachers should build a good atmosphere for them. But there is no need to give the students any suggestions but a little inspiration during their discussions. Finally, the teachers give a summarization according to the different situations to help them master the knowledge better. At the same time, the teacher should choose good typical cases and control the teaching process depending on clear teaching aims as an organizer.

Students are participants and actors in the teaching process. They have to acquire useful knowledge, cultivate their emotions and develop their capacities with the guidance of the teacher. And in case teaching based on the network, students have to communicate with classmates and make an analysis thoroughly instead of obeying teachers' instructions blindly. In this way, students can be the real participants and actors in learning.

4   A Basic Mode and Practice of Case Teaching Based on the Network

4.1   A Basic Mode of Case Teaching Based on the Network

Show the Advance Organizer
The Advance organizer is shown earlier than cases which based on the foundation of solid theories. The advance organizer is an introductory material that presented before learning contents. It is the bridge between the old and new knowledge, including the declarative organizer (It provides students with superior ideas when students lack them.) and the comparative organizer (It provides relevant materials when students do not clear the relationship of old and new ideas.). The advance organizer can also help

students review and consolidate the knowledge that related to the case before case learning based on the network.

Assign the Subject

Introducing the advance organizer can help students know the structure of knowledge. After that, teachers can assign the subject according to the teaching aims to let the students clear the learning tasks.

Show Typical Cases

Teachers should show typical cases after the students make clear of the subject. Teachers also have to give them enough time to read and understand the cases and make a case analysis. Students can search the relevant materials and screen useful information independently with network.

Emerge Consensus From Group Discussion

The teacher should divide students into several groups after they read and understand the cases fully. And the teacher has to ensure them to have enough time to analyze and discuss the case. Then students explore independently or collaborate with a group through the network to get the solutions. Students can also communicate with the teacher or other classmates to get the perfect solutions and upload them to the teacher's computer.

Speak by Representatives and Exchange in the Whole Class

The teacher should organize the students send their representatives to speak in front of the class, when students have certain solutions. And the elected representatives present their findings, and then students communicate with each other in the whole class. In the meantime of listening, the teacher can give them visual aids with displaying the group's solutions which have been handed in.

Summarize and Get the Inductive Ascension

Different opinions appeared after the presentations and communications in class. So as an organizer, the teacher has the obligation to comment and summarize all sorts of opinions that proposed by students reasonably. The teacher should give a positive reply of all the right schemes. With teachers' affirmation, students will accept the knowledge, methods and skills happily and efficiently in case teaching based on network. They will be more engaged in a benign cycle of learning.

4.2 Practice: (An example of case teaching based on the network is given below.)

Teaching Contents: *bamboo* (science)

Teaching Aims

1. Students can master the structure of a bamboo and know the species of bamboos through case teaching based on the network.
2. Students can cultivate their abilities of exploration.
3. Students can improve their abilities of cooperation with others.

Main and Difficult Points
Students can study the species and applications of bamboos independently.

Teaching Procedures
Introduction :(The teacher shows teaching contents on students' computers.)
  (CAI displays the bamboo pictures)
T: What is this plant?
S: Bamboo.
T: Very good. Today we study the bamboo together.

Show the Advance Organizer. (5 minutes)
(CAI displays the pictures of bamboos grown in the southern mountainous area.)
Text:(1)The bamboo is a kind of common plants grown in the southern mountain area because of the fertile acidic soil there.
  (CAI displays the longitudinal section of a bamboo.) Text:(2) Bamboos' stems have obvious joints, and there are hollow sections between the joints.
  (CAI displays bamboo rhizomes.) Text: (3) The bamboo rhizomes are subterranean stems of bamboos. Bamboo shoots are the germs of the bamboo rhizomes.
T: What kinds of bamboo shoots do you know?
S: Spring bamboo shoots, rhizomes bamboo shoots and winter bamboo shoots
T: Yes, you are great. And there are so many bamboo shoots for human consumption throughout a year.

Assign the Subject. (2 minutes)
T: We have known the structure of a bamboo. Now we learn bamboo species and applications.
  The whole class is divided into several groups with four members. In each group, two of the members are in charge of finding the species of bamboos and the other two searching the applications of bamboos. They can login the teacher computer firstly to get some relevant materials and then use the internet. Four members can discuss and summarize together, and upload the final result to the teacher computer by one. The results can be present by text or pictures.
  The teacher lets students to control their own computers.

Give the Typical Case.
The teacher establishes a separate folder called 'bamboo'. There are text introduction and pictures about the species and applications of bamboos. Text: there are a lot species of bamboo. China has more than 300 species, common bamboos are as follows: moso bamboo, black bamboo, green bamboo, henon bamboo, fargesia, etc. Bamboos are the materials of construction and paper making .They can also be made into various handicrafts and supplies. And bamboo shoot is the delicious and nourishing food that fond of people. Henon bamboo can also be made into Chinese traditional medicine.
  The teacher asks students to start searching.

Emerge Consensus From Group Discussion. (15 minutes)
Students read and understand the cases fully, find the answers they want and then search more requisite information on the internet.

Students discuss and summarize the cases in a group, then upload the results to teacher's computer.

Speak by Representatives and Exchange in the Whole Class. (10 minutes)
The teacher invites the representatives to present the perfect solutions which also being shown by CAI and then communicate with classmates.

Summarize and Get the Inductive Ascension. (3 minutes)
T: Your behaviors are perfect and those you said are correctly. There are many species of Bamboos, such as moso bamboo, black bamboo, indocalamus bamboo, green bamboo, henon bamboo, Fargesia and so on.

Bamboos also have many applications. It is the material of building, papermaking, handicrafts and living goods. And the bamboo shoots can also be delicious food. Even some leaves of henon bamboos can also be Chinese traditional medicine. So in your opinion, do you think the bamboos useful?
S: They are very useful.

Practice. (5 minutes)
Let the students login teacher's computer to copy and finish the exercises.
Exercises: Let students fill out the names of bamboos with showing pictures of bamboo leaves, bamboo stems, bamboo rhizomes and bamboo shoots.

5 The Requirements of Case Teaching Based on the Network

The requirements of case teaching based on the network can be described from two respects: case teaching requirements and network teaching requirements.

5.1 Case Teaching Requirements

Whether or how often use case teaching is according to the specific content of courses, the teacher's quality and the students' actual situations. Teachers should choose the suitable, applicable and typical cases on the basis of clear teaching aims. And the cases need to be controlled flexibly by teachers and be accepted and identified easily by students. Students need to have relevant superior theories about the cases. Only for those who know the principles and basic concepts thoroughly can analyze the cases fully and get the knowledge internalization. And sufficient time should be guaranteed to let the students know the cases' backgrounds, read, analyze and discuss the cases and present and summarize what they have learned from the cases.

5.2 Network Teaching Requirements

Network teaching requires the teacher and students can operate the network. In addition, it requires schools have a certain amount of computers and internet access. So that resources of public network and general software can be applied. And the teacher could focus on the project design of inquiry learning activities and the guidance of students' learning.

6 Conclusion

Case teaching Based on the network combines the advantages of case teaching and network teaching both. But it is undeniable that it also takes the drawbacks of them. So we should pay attention to adopt their good points and avoid their shortcomings, trying to give full play of their merits and note their limitations carefully as possible as we can. Only in this way can we make the best of case teaching based on the network.

References

1. Wang, S.: The History of Case Law and Enlightenment on the Development of Case Teaching. Journal of Educational Development (10) (2000) (in Chinese)
2. Cheng, Z.: The Foundation of Network Education, vol. (5). Posts & Telecom Press (2000) (in Chinese)
3. Miao, Y.: A Research on the Typical Case Teaching Mode Based on the Network. Journal of Popular Science and Technology (6) (2007) (in Chinese)
4. Liu, Y.: A Study of Case Teaching Mode Based on the Network. Journal of Educational Informatization (17) (2006) (in Chinese)
5. Zheng, J.: Guidelines for Case Teaching, East China Normal University Press (June 2000) (in Chinese)
6. Zhang, M.: An Analysis of Cases Teaching Based on the Network with a Perspective of Problem Behavior. Journal of Educational Technology (1) (2010) (in Chinese)
7. Li, R.: An Applied Research of Network Teaching in Higher Vocational Colleges Based on Case Teaching. Journal of Modern Communication 8(2010) (in Chinese)
8. Yang, F.: A Study of Case Teaching Based on the Network of Business English. Journal of Yi Chun University (4) (2009) (in Chinese)

On How to Efficiently Carry Out Multi-media Teaching

Xiaojing Wang

Department of Mechanical Engineering,
Anyang Institute of Technology, Anyang, Henan, China
28717210@qq.com

Abstract. Multimedia replace traditional teaching has already been a trend in higher education in China. This paper analyzes the advantages and disadvantages of multimedia teaching. Teachers should adhere to people-centered concept of modern education and make excellent coursewares, meanwhile, the multimedia teaching and traditional teaching methods should complement each other. Teachers should have abilities to have timely and appropriate, the principle of appropriate for people-centered teaching.

Keywords: Multimedia teaching, College teachers, Courseware, Efficient.

1 Introduction

Multimedia teaching methods are to be recognized and gradually used by more and more teachers . The teaching methods in an intuitive, image, and other advantages of large amount of information to improve the teaching effect has played a positive role. At present, our school's multimedia teaching across all disciplines has been formed a certain scale, this paper, the author used in the actual process of teaching experience of multimedia teaching methods, analyze the pros and cons of teaching methods.

2 Existing Errors

Suhomlinski said: "Teaching should help students discover the wonder generated from, proud, happy and satisfy the desire to create a variety of emotional experience of joy, so that students will study with a soaring emotion to learn and think, teaching will become a vibrant and exciting activities. " In multimedia teaching, teachers, students and media elements are the three entities of the teaching process, in the actual teaching, I feel there may be errors in multi-media teaching:

a. Over-reliance on multimedia courseware, showing too fast

Teaching hours, all the contents are all input to the teaching of multimedia courseware the teacher became a projectionist. In teaching activities, teachers often unconsciously accelerate the pace of classroom teaching, thus ignore the rhythm in tune with the thinking of the students. They excessive depend on machinery and equipment and transfer of knowledge blindly. Teachers are busy with operating, and with no time to take into account the exchange with students, from the traditional teaching of the "people filling" model into the current "machine filling" mode. They provid a wealth of information but ignores the moral character and the character creation of students. Teachers can neither control nor take care of the students' feeling. The students can quickly browse, and do not have time to think. so that the students gradually lost the initiatives and interests in learning and motivation.

b. The cart before the horse, distracting

When courseware beautifully displayed in the classroom to the students, the feeling is pretty beautiful and fresh. But often students are attracted by beautiful courseware and distracted by the screen.so that they can't master the knowledge well..

c.Multimedia courseware have single source and production methods

Most of multimedia courseware is made by university teachers or individual teachers in collaboration with other teachers. This multimedia courseware is based on the teaching requirements and teachers "own experience and other factors. Through investigation and analysis, multimedia courseware production methods also exist some problems. All teachers have a certain ability in surveying Word, Power-point operations and other software, but the use of advanced software is still at the primary level. Teaching content of the presentation is still a lot of text-based, and even individual courseware is still the existence of the phenomenon of moving materials. If the multimedia courseware is just a pile of text, then the use of multimedia in teaching the art does not have any substantive meaning, and will become a burden for many teachers.

3 How to Efficiently Carry Out Multi-media Teaching

Excellent courseware is not put a good text, images, animation, sound combinations together but under the guidance of the concept of modern education, and follow the scientific and practical design principles. Teachers in the courseware design, according to the teaching content, teaching objectives, teaching objects, etc. choose content targeted text, graphics, animation, audio, video and other media are used to meet the teaching requirements with a properly structure. Teachers achieve standardized text, image clear and stable, composition and color correct, which will help students to observe and analyze. Teachers should pay attention to guiding inspiration, preventing frequent switching screens and general explanation, to make full use of the computers' interactive features, lost no time in learning and teaching interspersed the exchange of information. To be good at selection and arrangement, all to serve teaching as the starting point, to prevent the use centered, pursuing luxuriant and in a mere formality.

Czech educator Comenius said: "Educating people is the art of the art ." In the class, teachers are the designers of teaching situations, multimedia courseware is is the material to create teaching situation and organize learning by teachers and students and it's just a collaborative exchange of knowledge, because courseware itself does not have the idea and theory to instead the teachers. The implementation of the teaching and regulation are still leaded by teachers. Teachers should use their unique teaching methods and teaching skills to the unique charm and lively personality to explain to infected students, transfer students to actively participate in teaching to achieve good results, Although the multimedia courseware has many advantages, but we can't deny the leading role for teachers in the class. Teaching process is not only to impart knowledge of the process, but also the process of emotional communication. Secondly, teaching to the students of knowledge, and train students to analyze problems and problem-solving thinking. Moreover, teachers adjust their teaching methods and contents according to the acceptance of students, so that teachers can educate students in proper ways. Finally, teaching is a creative work, is re-processing, understanding, then play, and re-creation process.

The multimedia teaching has begun to appear in the 1980s. It refers to the process of using multimedia computer for the integrated treatment and control of symbols, language, words, sound , graphics, image, video and other media information, displaying various and organically combined elements of the multimedia according to the teaching requirements through a screen or the projector, and at the same time with the sound and the man-machine dialogue between the user and the computer, completing the teaching training and realizing the teaching activities made by multimedia teaching courseware. The multimedia teaching can not only be processed to the teaching content, display, replay, but also can display the process and the image of the phenomenon vividly with the use of simulation technology which can't achieved or observed of in normal conditions. It can greatly improve understanding and feeling of abstract things and process of the students, enhance their perceptual knowledge, and stimulate students' imagination to achieve the ultimate goal of pellucid understanding of knowledge. Take the database technology for example, the traditional teaching method can only write theory on the blackboard, paint structure of table, but can not present the operating process of "building form". And after using the multimedia teaching method, you can express different knowledge and content through text, pictures in the courseware. Emphasis the key points with the form of animation and present demo of dynamic operation process, still can show operation of the process in the actual system software according to student's master degree.

After using multimedia teaching method, the outline and key content can be broadcast through the slide packs, teachers can have more time to focus on students' feedback information and more energy to guide the students, and then make the students participate in the whole process of teaching, which reflect the concept education, must face the students. In the multimedia teaching process, all the teaching information is stored in file forms, they can be played and replayed in any need, the key points and difficulties are clear, which help the students recall teaching process out of the classroom, deepen their master of knowledge.

In multimedia teaching, teachers should pay attention to their own dominant leading and students' main role. As for teachers, to leave enough space and time for students to think, to cultivate the students' innovation ability, and to give full play to students' main body function, really through the multimedia technology make the teaching optimized, and the quality of teaching improved. College education task is not only to impart knowledge, but also to mobilize the students' subjective initiative, cultivate the sustainable learning motivation and lifelong learning desire and ability.

Modern teaching interaction theory tells us,that the study is an exchange and cooperation process, and that the interaction between teaching dynamic factors is the main way to promote students' learning.There are three main forms(of interaction): the interaction between students and content of interaction, between students and students, between teachers and students. And a good interactive classroom link is an efficiency important segment to improve classroom teaching. It should not be ignored to make full use of students' subjective initiative, let the students to participate in classroom teaching in multimedia teaching. For example,"The C language program design", in telling the cycle part, we set up two different circulation examples, and asked two groups to separately implement and discuss in groups, and finally representatives sent copleted in the computer.

4   Several Noteworthy Factors

In my opinion,there are several noteworthy factors of influence of the multimedia teaching.

Firstly, the teachers' own position in teaching influences to the results of interactive teaching.

According to LiuXiaoBao scholars of the study it is found that, there are there three ways for the teacher in the application of multimedia teaching:cabinet sitting position, cabinet standing position and standing outside the cabinet. In these three ways, 100% of the students think cabinet sitting position as the sitting position for the report, 76. 4% of the students think cabinet standing position as a report way of teaching; 96. 6% of the students think the standing by the way of teaching cabinet, be helpful for the exchange and communication between teachers and students.

Secondly, the teacher's own certain basic quality

a. Knowledge backup

Teachers must be rely to the teaching outline requirements to complete the outline of the knowledge of the provisions ability and the quality teaching task for the purpose to carry on the design.And we should extract the teaching program closely related materials, sperate the key and difficult based on widely read in search, and relevant material.It is better for the students to understand and master the knowdge.

b. The innovation ability and practice spirit

The creation and practice of interactive teaching mode reflet a kind of teaching reform consciousness and exploration the embodiment of the spirit. Teaching method should be constantly pursuing new and varied , cannot stand still, meanwhile we should continuously mobilize students' learning initiative of the subject as. Therefore, the teachers master in addition to teaching material, but also design scientificly teaching methods and teaching contents in to the classroom relying on the students and professional characteristics, and make "ask-answer" as a necessary part of the process of teaching, a flexible design interactive classroom, such as interaction between teachers and students, and between students(a pair of more one-on-one, for more debate style, etc).

c. The master degree of multimedia technology

Correct application of multi-media teaching can stimulate students' interests in study and facilitate students to master the knowledge. Because it is to present the relevant pictures, video to students, and to make difficult concepts, theoretical direct around teaching subject. So teachers must make themselves master of the authorware, and application of multimedia teaching.

Thirdly, the strengthening consciousness of participate in

It directly affects the class effect and the quality of class, whether students proactively take part in the process of class interaction. Students, as a learning agent , should advance a preview of the part to learn ,at the same time according to teaching materials , they should get varied revelant information through the labiary, internet,and magazin and ect., and think highly about the details the teacher told in class,and play an active part in interactive learning .

To sum up, no matter how education technology develops, the teacher is irreplaceable. He is not only the initiator of knowledge, but also the organizers of students' learning activities and the designer of the teaching situation. He not only needs profound knowledge and exquisite education art, but also modern education technology, more importantly he need more modern education concept. Multimedia is just one of the means to help the teachers to teach and solve teaching difficulty. In class, multimedia is only a kind of teaching tool, and like the blackboard chalk, just the level of modernization is a little bit higher. So, we can't completely dependent on multimedia computer in class. When we are in the application of multimedia assisted teaching, except working hard on courseware, making it comply with the requirements of this class, and solve the problems, we should study the basic skills in teaching, improve the teacher's lecture level and the ability to control the classroom, make ourselves on a higher level in teaching, and let the multimedia teaching play the role of gilding the lily. We want to look at problems dialectically, thing is split into two, having advantages as well as flaws, such as Modern multimedia teaching and traditional teaching. In fact, on the one hand, education needs technology, the designing education in the information age needs computers and other modern tools, but on the other hand, teachers can not be instead by any modern technology. As for multimedia teaching and traditional teaching, the writer of the article still emphasize, both should never be opposite. We should focus on how to combine the multimedia teaching and traditional teaching perfectly, making them work better for teaching service.

References

1. Yu, X.: Multimedia Technology and Applications. Science Press, Beijing (2002)
2. Ren, Y.: Correct view of multimedia. Journal of Yuncheng University 25(1), 103–104 (2007)
3. Lin, F.: Multimedia Technology. Tsinghua University Press, Beijing (2000)

The Mode Analysis and Enlightenment of Cooperation of Production, Teaching and Research in Japanese Universities

Liping Han

School of Foreign Languages, Yanshan University,
Qinhuangdao, Hebei, China, 066004
linhongju@126.com

Abstract. The tremendous successes have been achieved in the past twenty years in our country universities cooperation of production, teaching and research, at the same time some questions are exposed. The status quo and shortcomings are set forth for cooperation of production, teaching and research in domestic universities, the success mode of Japanese universities are discussed in the paper, which include in extern environment, principal cooperation path and actual effects. Experiences are drew lessons from Japanese universities and countermeasures are given for resolution shortcoming of cooperation of production, teaching and research in our country universities.

Keywords: Cooperation of Production, Teaching and Research in Japanese Universities, Mode, Shortcoming, Lessons, Countermeasure.

1 Introduction

Japan is founder of cooperation of production, teaching and research. The Japanese government set up "research council of cooperation of teaching and production" according to importance of issues in 1933. In 1956, the Japanese industry rationalization council submitted a consultancy report "education system for cooperation of production and teaching" to Japan's Ministry of Economy, Trade and Industry. Japan Cabinet pass through "Income Doubling Programme" in 1960, and it had emphasis on particular importance to cooperation of production and teaching, and strengthen cooperation of production, teaching and research. And it began to use many policies to encourage and guide to college, research institutions and industry and promote development of cooperation of production, teaching and research. In 1980s, Japan established the "national strategy of technology "which became the national policy. In 1990s, the Japanese government actively promoted integration of cooperation of production, teaching and research. The universities and research institutions and enterprises are concentrated through various means, such as cooperative research, the mandatory research, the mandatory researcher system, the scholarship donation system .And they tackled key problems of practical technology in together. The Japanese government also focused on improving technology development, and transformation period of high-tech results became shorter. Japanese universities cooperation of production, teaching and research received considerable development, and piled up certain successes experiences.

The cooperation of production, teaching and research is a key issue of common concern for the government, academic and industrial institutions in many years. Japanese college carried out it earlier, there are many cooperation paths. Our universities history of it is short; there were many problems to be further studied, such as affect factor, its development mode of reproduction. How to learn from Japan mature mode, create a Chinese-style cooperation mode. They are our discuss emphasis.

2   Connotation of Cooperation of Production, Teaching and Research

The cooperation of production, teaching and research is consist of enterprises and institutions and universities in accordance with principle of "the shared interests and shared risks, the advantages of the complementation and common development", It commonly carry out technology innovation and scientific research activities so that gradually realizing cycle of "scientific research, product and market research". It is an inevitable result of union of market economy and knowledge economy, is an important part of the national innovation system.

The cooperation of production, teaching and research has different expressions in actual, such as "connection of production, teaching and research ", "union of production, teaching and research", etc. In view of high-tech industrialization process, in fact it is integration of basic research or doctrine (teaching), application and development researches (research) and production profits (production). In view of entities organization, it is a general concept," production" means industry (enterprises or enterprises groups), "teaching" means universities, "research" means scientific institutions, further broader sense, is an organic integration of industry system, education and research system [1].

3   The Present Status and Shortcomings of Domestic Universities

Universities have functions of training intellect, scientific research and society services. Higher education is a field of high consumption and investment and lag efficiency, the shortage of funds had been main factor in its development in a long time. Universities have many talents and research results one hand, on the other hand, lack of resources allocation often constrained its developments. For the purpose of human resources and technological achievements transforming into the economic advantages, and severing for the market and social development, in 1992, the economic and trade office of State Council, the state education commission and the Chinese academy of sciences jointly organized the implementation of the project "cooperation of production, teaching and research". After nearly twenty years of exploration and practice, it is surging ahead every day, and has achieved great progress. Universities service consciousness was increased; more and more scientific and technological achievements were transformed into the market and enterprises ; universities and industry were increasingly connected closely; enterprises gain in result of universities research results; universities have also received many more resources, has greatly promoted the development of itself. But China's socialist market economy system is still in developing and perfecting stage, economic order and resource allocation of commercialization are still being standardized, so there is great disparity between the project "cooperation of production, teaching and research" and social needs ,its concrete manifestations are as follows:

Firstly, the cooperation of research and teaching was inadequate. In some universities researchers did not engaged in teaching, and teachers did not engaged in scientific research, some researches were adverse to the teaching job. Because employing mechanism was too rigidity, many teachers can only use the spare time on the part of the enterprises, full-time means separating from universities. Secondly, the cooperation of production and research was inadequate. The scientific and technological research project determination and results were most interested by researchers, interesting projects was not equal to market demand. Because the universities have not abilities for further transformation, the results of scientific and technological existed in the lab, they can't enter the market to create benefits. And enterprises' risk consciousness was not enough, did not intended to transform the higher technological achievements into productions, so it results in failure for cooperation. Furthermore, universities often emphasized vertical projects and despised Cross-cutting projects, mainly in determining technical or professional titles, payment, welfare, scientific and technology results award, tax incentives or in other aspects. In a great extent, it restricted the teachers' initiative; Finally, the cooperation of production and teaching was inadequate. At present, many universities students often study in campus by the traditional mode. Talent training base and practice teaching bases are lack that is jointed by enterprises. So it is difficult to train practical and high level talents that meet equipments of enterprises [2].

4   Mode Analysis of Cooperation of Production, Teaching and Research in Japan Universities

4.1   System Environment of Cooperation of Production, Teaching and Research in Japanese Universities

Japan congress passed through" National University Corporation Act" on July in 2003. National University Corporation is subject of setting up national university. An important feature of national university corporation is making staff not civil servants but they were civil servants ago. It means that "Law for Special Regulations Concerning Educational Public Service Personnel" is no longer applicable to the national university staff. At the same time, new national university can adopt more flexible employment mechanism, such as the appointment system etc. In addition, the national university staff part-time was limited in private enterprises ago. Now, according to the new law, they can more freely work outside of campus. Because the national university corporation replacing the government became body of the national university, main source of funds was changed. Funds raised have diversity ways. And the university has also begun to actively trying to gain more external funding by cooperation of production, teaching and research. It was one of the most important aims that the national university corporation system was designed. Thus, the national university cooperation created a more liberal and free environments for cooperation of production, teaching and research.

4.2 Main Paths of Cooperation of Production, Teaching and Research in Japanese Universities

For a long time, along with Japanese economic development, paths of cooperation of production, teaching and research in Japan universities were diversity, mainly manifested as follows:

4.2.1 Mandatory Research

Ministry of Education, Science and Culture set up Mandatory research system according to requirement of industry in 1958. That is, researchers could receive mandatory research items from private enterprises and departments and research organizations, local public entities etc by protocols that were determine by enterprises. And research fee were taken on by enterprises. The research results could be implemented by the mandatory researchers or specific personnel in the seven years after researching complete. Enterprises pay for the 30%management to the state and the remaining70%pay for the professors. In recent years, commissioned research items number has been steadily increased; it is a new trend for cooperation of production, teaching and research. In 1993, commissioned research items were 2432, amount to 69.1 trillion Yen [3].

4.2.2 Cooperative Research

Ministry of Education, Science and Culture set up "Common research system of national school and folk institutions" in 1983. It aimed at promoting universities and enterprises to common research and creating good results. Its specific provisions : university teachers and researchers of non-governmental organizations study common interesting items through university devices by contract in together. Studies were engaged in universities research institutions, they required funds, including devices maintenance and management fee that the university is responsible for providing, however private enterprises provide common study fee, and pay for rewards, traveling cost, consumable materials etc direct funds, the research results is own to both sides.

4.2.3 The Mandatory Researcher System

The technical personnel or researchers of private enterprises were accepted for mandatory researchers by universities, which were instructed on postgraduate level so as to be further improved. They were instructed by professors in the university labs together with postgraduate students and abroad students, basic research were engaged by them that would lay a foundation for their further development. Universities trained in-service engineering technical personnel and researchers on graduate level in order to improve their capacity, which worked in enterprises and were enterprise commission graduate student. The commission train time was one year in principal, some had research topic, and others had not. The commission train fee was taken on by enterprises.

4.2.4 Setting up Scholarship

The national university etc schools accepted endowing funds from private enterprise etc, setting up scholarship items on purpose of academic research and study.

4.2.5 Donation System.

Donations of enterprises were used for "donation cathedra" that mainly aimed at education in specialty or school or affiliated research institution of the university, donation cathedra or donation research institution could be given lecture by human, social and natural sciences, its name can be named by the donor name.

4.2.6 Establishing Common Research Center

In 1987, Ministry of Education, Science and Culture established" common research center" in some universities. There were 38 centers by the end of 1994. Its equipments were provided for full-time and part-time professor in together. Except for private common research and commission research being carried out in it, technical personnel could be trained and technology consults could be engaged by private development so as to boost economy for local industry. In 1995, Ministry of Education, Science and Culture established" common research center" in Osaka and Hiroshima etc universities again so as to provide more sites and facilities for cooperation of production, teaching and research.

4.2.7 Establishing Cooperation Related to Academic

There were many no-profits cooperations in Japan that aimed at revitalizing the academic research. Cooperation set up on purpose of academic research was named by academic cooperation ; Cooperation set up on purpose of endowing scientific research was named by research aid cooperation ; Cooperation set up on purpose of academic research information and knowledge information exchange was named by society cooperation. Most of these academic relevant no-profits cooperations were established based on private funds by financial institutes or juridical association. They played a vital role for cooperation of production, teaching and research. [4]

5 Lessons and Enlightenment from Cooperation of Production, Teaching and Research of Universities in Japan

Cooperation of production, teaching and research has played an important role in the economic and scientific and technological development. It is very good effective to colleges and enterprises development. We have much beneficial enlightenment from it.

1) Strengthening and improving the system and legal construction for cooperation of production, teaching and research

Japan has a complete system on cooperation of production, teaching and research, it has committed to the law amended and improved always. The complete legal environment provides the best of the premises and assurance for scientific and technological achievements transformation and science and technology progress of the enterprises. We draw lessons from Japan success experience, government should strengthen and improve system and laws and regulations for cooperation of production, teaching and research, actively play the role of public functions. Government sets up platform for universities and scientific research institutions and enterprises. Creating free external environment, and adequately paying role of government service, providing fair and impartial legal system for it. How to create law environments for cooperation between university teachers and enterprise? We draw lessons from "National University Corporation Act" in Japan. The government and universities should speed up the reform for the institutions appointment system and reform for management mechanism so as to alter teachers' single identity. Universities and research institutions should deal with relations between vertical projects and despise Cross-cutting projects. The internal management work should be strengthened, and education departments should be further distinguish from scientific research and teaching in detail.

The teacher must finish the work load (papers, works and the items) in most domestic universities; they have greater pressure, and fear for academic rank, treatment, evaluation. Therefore, university should create a good environment for cooperation of production, teaching and research in policy. It should formulate a series of affirmative action in determining of professional technology post and benefit distribution and prize awarding so that the faculties can achieve fair treatment which engaged in cooperation of production, teaching and research. It should also formulate a comprehensive allocation system for gains of cooperation of production, teaching and research. In the past, cooperation of production, teaching and research was mostly taken on part-time work, advisory services, cooperation development or single results transformation, etc. It established a relatively loosely, a short-term single goal, and the "point to point" type of partnership. We draw lessons from cooperation path "Mandatory research" in Japan, the government should guide it from loose, decentration cooperation to close, concentration cooperation pattern. And University and Technology Science Park should be established by universities domination, including the expanding size and number, scientific management and operation mechanism in order to speed up cooperation of production, teaching and research.

2) Strengthening science and technology intermediary agencies and developing and innovating risk investment mechanism

"General Liaison Committee", "the special committee of research and development "and "high-tech information center" etc intermediary agencies had played an important role in cooperation of production, teaching and research in Japan. We must focus on construction cooperation agencies of intermediary agencies in order to promote it developing rapidly. We should creatively develop and promote various intermediary agencies and improve the quality of services and improve and enhance the functions of mediation services. The intermediary agencies not only have functions of advisory services, but also have function of assurance, and can recycle research invest in time, supervise the transfer or assignment of contract in two parties. The government should take on relevant measures to promote production, teaching and research the three parties understand and demands for intermediary agencies by demonstration so that it can provide more services.

We draw lessons from setting public corporation on purpose of promoting academic research in Japan; China should speed up creation of venture capital investment mechanism. The technological innovative activities of cooperation of production, teaching and research are uncertainty and information asymmetry. They are incompatible with the banks principles of security, mobility, effictivity. So cooperation projects are difficult to obtain bank loans. Government is difficult to invest largely because great cooperation projects are long period, large investment, great risks. Therefore, it needs governments, research, enterprises and universities, private capital to organic combination, and become benign risk investment mechanism in order to speed up high technology industrialization process. The government gives high-tech enterprises and venture investment institutions financial and credit aspects preferential policies so as to encourage and stimulate venture investment development and promote high-tech industries and venture investment institutions sharing growth and boost.

3) *Clearly determining enterprise domain position of cooperation of production, teaching and research and raise their cooperation activity*

Each party spontaneously established cooperation relation at the outset, their role orientation were not clear .How to play their respective advantages was lack to long-term strategic thinking and quick success actions were more, there were many universities and research institutions from studying to the industrialization independently. As a result, scientific and technological innovation source power was vitality. We should keep the enterprises domain position and universities technique support position, special in capitals.

The talents training and personnel training were well worth learning by us on the mandatory researcher system and cooperative research in Japan. We should adequately utilize different teaching environment and resources of universities, scientific research institutions and enterprises and their respective advantages on talent training. We should integrate into knowledge in classroom and direct accessing to practical experience and practice of production. Many universities have established long period cooperation relation to enterprises on talent training. Firstly, universities directly educate technical and management personnel on the use of teaching and research. It is an important way to solute talent-lack and train comprehensive talents for enterprises. Universities trained many high level talents for enterprise that were body of technology development and technology innovation. Secondly, enterprises provided advanced production equipments and financial resources for universities with practice and experiment base, it has become an important content of talent training.

The success mode of cooperation of production, teaching and research in Japan give us much beneficial enlightenment. We should draw on Japan's experience, innovating mechanism and system; comprehensively promote organic combination on it. At the same time, we should further improve laws and regulations, determine relations and promote sustainable development on the cooperation of production, teaching and research.

References

1. Wang, F., Wang, X., Liu, W.: Retrospection and review on the cooperation of production, teaching and research in China. Industrial & Science Tribune 7(7), 148–150 (2008) (in Chinese)
2. Xie, K., Zhao, B., Zhang, L.: On the project of production, teaching and research of colleges and its new mechanism. Studies In Science of Science 20(4), 423–427 (2002) (in Chinese)

3. Li, J.: The Form and Enlightenment of the Cooperation of Production, Teaching and Research in Japanese Universities. Journal of Linyi Teachers University 28(1), 110–112 (2006) (in Chinese)
4. International cooperation and exchanges bureau the education department of the people's republic of China: Foreign higher education research report. Capital normal university press, Beijing (2001) (in Chinese)

Teaching Innovation and Course Practice in "Electrical Control Technology"*

Xiaoqian Hu, Lian Zhang, and Shan Li

College of Electronic Information & Automation
Chongqing University of Technology
Chongqing, P.R. China, 400050
hxq@cqut.edu.cn

Abstract. Aiming at the development tendency of modern industrial electrical control technology, and according to years of our teaching practice on "Electrical Control Technology" course, this paper analyzes the serious lag phenomenon existed in teaching contents of this course in detail. On the basis of the above, the idars of teaching innovation and course practice in "Electrical Control Technology" are put forward by adopting the latest knowledge of Programmable Logic Control (PLC) and Numerical Control (NC). Finally, a diversified teaching method fusing task-driven ineterst teaching, heuristic teaching, and open teaching is introduced. The practice has proved that a good teaching effect has been obtained with experience accumulated for engineering education reform.

Keywords: electrical control technology, teaching innovation, course practice, programmable control technology (PLC), Numerical Control (NC), diversified teaching method.

1 Introduction

For the majors in engineering colleges such as Electrical Engineering, Industrial Automation, Mechanical Manufacture and Automation, the "Electrical Control Technology" or its similar course.is a very imortant course[1-2]. Electrical control technology, whose control objects are transmission and drive system powered by a variety of motors, is aiming to realize the process automation. Therefore, it is the key course that can train the college undergraduates to have the ability of heavy current control. This course is a specialized course with strong practicality, and its main features are as follows, wide scope of knowledge, intense integration with real manufacturing process, and containing experiment course and course design besides theory lecture.

With the extensive application of high-tech production technology and the rapid development of modern science and technology, electrical control technology has changed fundamentally. The traditional relay control has been gradually transited to Programmable Logic Control (PLC) belonging to the field of industrial control and automation. And recently, with the high-speed development of computer technology,

---

* This paper is sponsored by "research and practice of applied talent cultivation mode for electrical information categories" (101102), the major project of Chongqing higher education teaching reform.

Numerical Control (NC) system occurred as the times require. NC technology, still in its rapid development, will become the foundation and core of modern advanced manufacturing techniques. Therefore, the course of "Electrical Control Technology" should keep innovation continuously in order to keep up with the technology development.

2 The Urgency of Course Innovation

Nowadays, the relay control is still the basic electrical control form adopted in machine tool and many other types of mechanical equipment, and is the foundation of studying the more advanced electrical control systems too. But, PLC, because of its following features, taking the microcomputer as its core technology, having the abilities of realizing different control function by software, suiting wicked industrial environment, integrating the advantages of both computer and relay control, has already become the standard and universal control equipment all over the world, and is widely applied in different industrial control fields[3].

Recently, with the development of Power Electronics, Computer, Control Theory, and Communication Network Technique, Electrical Control Technology has changed greatly. The Numerical Control machine tool, featuring with its high speed, high efficiency, high precision, low labor intensity and high automation, is a widely used automatic machine. The NC machining center based on the traditional machine is a compound NC machine with the function of automatic tool changing, so it can realize the continuous automatic processing of multi-step working procedures, so greatly improving the processing efficiency and saving the space.

Developed on the basis of rapid developments of microelectronic technique, sensor technique, automatic control technique, and artificial intelligence technique, the Electromechanical Integration technique, by fusing mechanical equipment, power equipment, control equipment and detecting sensors as a whole, can realize multi-function, high efficiency, high intelligence and high reliability, and save material and energy. In the future, the Electromechanical Integration technique must be the development trend of industrial electrical control technique.

In order to keep up with the rapid development of society and modern technology, how to cultivate high-quality talent with practice ability has become the hot spot of higher education community. That high education innovation must quicken its pace has become a consensus of education community. Teaching innovation, with teaching contents and course innovation as its most important part, is the key of education innovation. The teaching contents and course innovation must follow the teaching law and scientific development rule, and has the right guiding ideology: education should develop toward the modernism, the world and the future, by Deng Xiao ping. On the basis of the past experience, it's very important to properly balance the relationships between teaching knowledge and cultivating ability, traditional contents and modern contents, theory course and practical courses, etc.

For a long time, "Electrical Control Technology" course in our university introduced the traditional relay control system mainly, but apparently, only these contents are unsuitable for the modern industrial development. It's well known that PLC First occurred with the expectation to replace the relay control system by an universal control equipment suitable for industrial environment. It hopes to integrate computer system's advantages (multiple functions, flexibility, and universality) with relay control system's (simplicity, easy operation, and cheapness). Not long ago, in our university, besides "Electrical Control Technology" course, there is still a specialized course "PLC", which introduced PLC's hardware resource and software programming by taking PLC as a special device. In fact, from the point view of course arrangement, separating these two closely interrelated parts is apparently unreasonable.

Besides, the control of electrical equipment is of great important in machine tool numerical control system. Presently, PLC is generally applied in machine electrical control. PLC has high reliability and is easy to use, so for the complex application with many control point numbers, it can realize complex electrical control functions, by a certain amount of expanding element added to its basic unit. For example, for the sequential control functions of NC machine tool, such as spindle speed (S Function), tool base management (T Function), spindle start-stop and reversing, workpiece clamping and unclamping, moving of the cooling fluid system (M Function), etc, all the above can be realized by PLC. Currently, developing the NC system based on PLC hardware and constructing open architecture is the new tendency of developing NC system world-widely. Therefore, it's necessary to study PLC's application in NC machine tool, which can let college undergraduates' studying be tightly combined with modern industrial development.

3 Innovation Scheme of Course Teaching Contents

In order to let undergraduates not only master the analyzing and designing of the basic relay control system, but also be familiar with the latest modern electrical control technique, and by the recent years of teaching practice, we have set the following innovation scheme of teaching contents. First, by starting from the typical mechanical relay control line, give undergraduates the ideas of purposes and requirements of industrial electrical control; then, by the application of PLC in industrial electrical control, let the undergraduates master the programming of PLC control; finally, by PLC's application in NC machine tool, let the undergraduates understand NC technique furtherly.

As in Fig.1, this course content is composed by the following four parts:

- Current situation and developing trend of low-voltage apparatus, and the structure, pimple, specification and choosing of commonly used low voltage electric apparatus.
- Introduce to the basic open-loop control elements of motor, and the designing method of typical open-loop
- contactor-relay control system.
- The basic composition, operational principle and performance of PLC. The sequential control design method and closed-loop control programming.
- The PLC's control program in NC machine tool, and the basic knowledge of NC technique.

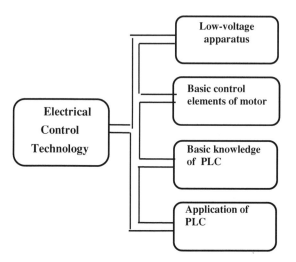

Fig 1. The plan of teaching contents

4 Study on Course Practice

Course Practice is the effective guarantee of cultivating undergraduates' practical ability, deciding the degree of undergraduates' ability to a great extent. In the Course Practice plan, we have been carefully choosing experimental contents, with especial emphasis on systematization and integrality of undergraduates' practice knowledge. Recently, after many years' construction, our laboratory (listed on Chongqing's Engineering Demonstration Center) has developed greatly, with a completed multi-functional practical teaching system, which combines scientific research with engineering application, and includes four parts of different levels: demonstration, foundation, integration and innovation. Some experimental items are available for undergraduates to choose by themselves, and can be done in the opening hours of laboratory.

As shown in Fig.2, Aiming to let undergraduates gradually master the basic machine tool control, and through the carefully choosing of experimental contents, now the course "Electrical Control Technology" has set six experiments as follows: direct starting, automatic round-trip control, Y-D reduced voltage starting of squirrel-cage induction motor, learning of PLC and programmer, programming and debugging of PLC energy consumption braking, programming and debugging of PLC speed control, integrated control. In the above-mentioned experiments, the former three experiments aim to train undergraduates to realize open-loop control by mastering the traditional relay control.

On the basis of the above, the next three experiments (learning of PLC and programmer, programming and debugging of PLC energy consumption braking, programming and debugging of PLC speed control) not only help undergraduates master the programming of PLC control, but also be familiar with the practical application of PLC in the primary components of industrial electrical control, and furthermore, master the closed-loop control by PLC. In the plan of experimental

	Direct starting	
Traditional relay control	Automatic round-trip control	
	Y-D reduced voltage starting of squirrel-cage induction motor	**Integrated control experiment**
PLC control	Learning of PLC and programmer	
	Programming and debugging of PLC energy consumption braking	
	Programming and debugging of PLC speed control	

Fig 2. The experiment arrangement

contents, the above-mentioned 6 experiments are all primary control experiments, and each one is a part of the final integrated control experiment. The final integrated control experiment can not only simulate the primary function of machine electrical control, but also train undergraduates' integrated designing and debugging ability, and consequently lay a sound foundation for the subsequent course design and graduation project.

5   Diversified Teaching Method

A good teaching method can stimulate undergraduates' study interest, help them have learning intention, and encourage them to study consciously, and consequently, obtain desirable teaching result. Therefore, in the teaching of "Electrical Control Technology", and considering the characteristics of this course, a diversified teaching method fusing task-driven ineterst teaching, heuristic teaching, and open teaching have been adopted.

5.1   Task-Driven Ineterst Teaching Method

The passive study usually makes undergraduates feel insipid and boring, reduces the teaching quality and efficiency. But if the undergraduates study with interest and question, then their study activity and enthusiasm could be increased greatly. So, before and after class we usually give undergraduates extra task highly correlated with practical engineering, in order to let them have study interest, and further, to master, synthesize and utilize the knowledge studied in this course.

For example, after the induction to "Electrical Control Technology", a task just like an overview should be given to the undergraduates, the content could be: electrical control technology in your daily life, or the influence of electrical control technology on modern lives. This task requires undergraduates summarize the application of this course in actual lives. Besides, the undergraduates are required to consult various references and understand electrical control technology's vital function and irreplaceable place in modern lives and all trades and professions. Moreover, by finishing this task, the undergraduates could contact the latest information and development trend of electrical control technology. In a word, this task aims to let undergraduates learn how to look up references and summarize knowledge, and fully know the widespread use of this course in actual lives and have deep study interest in it,

Besides, preview work will be given before starting a new chapter. The preview work requires undergraduates know the main contents of the new chapter by self-study, and summarize the difficult points. In this way, the undergraduate will study in classroom with question and deep interest.

5.2 Heuristic Teaching

In classroom teaching, the heuristic and inducing questions can concentrate undergraduates' attentions, arouse their imagination, motivate them to think actively, and build up a classroom atmosphere of concentration. When teaching the principle of a typical electrical control circuit, the teacher should give the task of this circuit at first. And in this way, help undergraduates think deeply, let them express their own ideas and thoughts, and finally apply the studied knowledge to solve this problem. Moreover, the related circuit shouldn't be given directly, but be drawn part by part according to each part's function. It's important to induce undergraduates to speak out the function of each apparatus, concentrate their attentions, and increase their interest. Then, the teacher could begin to explain the working principle, and finally, concisely and explicitly summarize the general thinking and method to solve the related problems.

5.3 Open Teaching

"Electrical Control Technology" is a basic course with strong practicality, so the experimental arrangement of it is of great significance. Aiming at the experimental arrangement, the open teaching is encouraged and advocated. Not only the time is open, but also the content. The teacher can provide some extra experiments available for talented undergraduates to choose by themselves. Through designing their own experimental system, the undergraduates could learn knowledge not only from teacher, but also from their own practice. Consequently, the undergraduates can improve their practical ability, strengthen technical training, widen professional knowledge coverage, and strengthen their adaptability. Because the young undergraduates think very actively and accept new things very rapidly, it's easy for them to have new originality, new design, new technique, and new product. The laboratory is the main battlefield to cultivate innovative talent in colleges and universities. And in history, many famous scientists were developed in laboratory[4]. Therefore, the laboratory in colleges and universities should be fully utilized to cultivate creative applied talent.

6 Conclusions

In the recent years, according to the above innovation ideas, in the teaching of "Electrical Control Technology", we have adjusted the teaching content and enhanced course practice, and obtained satisfying teaching results, with the undergraduates' satisfaction rate higher than 85%. Especially in the subsequent course design and graduation project, it is showed that the undergraduates' integrated design and debugging abilities have been apparently improved, which thoroughly demonstrates the effect of the innovation in this course. Presently, the textbook "Industrial Electrical Control Technology" written on the basis of the above research achievements has already been listed on "Eleventh Five-Year Plan" official textbooks of Ministry of Education.

References

1. Huang, X., Zeng, Q., Ling, D., et al.: The primary thought of reform on basic mechanics series course. Journal of South China University of Technology(Natural Science Edition) 24(11), 54–57 (1996)
2. Liu, X., Zhou, H.: Research and Development of Embedded PLC in Numerical Control Machine. Mechanical Science And Technology 17(6), 1008–1010 (1998)
3. Zhang, B., Liu, W., Zhu, Z.: Study on Teaching of the course Electrotechnology. Journal Of East China Shipbuilding Institute 13(6), 88–90 (1999)
4. Li, J., Xu, R.: The ability training of American higher College undergraduate and its inspiration. Journal of Hebei University (Social Science and Philosophy Edition) 4 (1998)

The Function of the University Libraries in Constructing Lifelong Education System

Tian Lizhong[1], Zhang Aichen[1], and Sun Yongjie[2,*]

[1] Library of Tianjin University of Technology, Tianjin 300191, China
[2] School of Foreign Languages, Tianjin University of Technology, Tianjin 300191, China
{tlz69,zhangaichen}eyou.com,
sunny1970-2-9@163.com

Abstract. To build a socialized lifelong learning system is the requirement of implementing the strategy that a nation thrives from science and education and sustainable development. This paper mainly explores the significance, feasibility and possible problems of the university libraries in constructing the lifelong learning system. Furthermore, it proposes the application strategies of solving these problems.

Keywords: library service, lifelong learning, application strategy.

1 Introduction

To build a socialized lifelong learning system is the requirement of implementing the strategy that a nation thrives from science and education and sustainable development. Lifelong education and learning are highly stressed by our party and government. The 21st Century Action Plan for Education Revitalization approved by State Council clearly stated: "up to 2010, a basic life-long learning system should be established to provide talents and knowledge support for national knowledge innovation system and the construction of modernization."

For university libraries, as educational functioning department, to establish and improve their educational function should be one of their developing goals. University libraries' role in life-long learning system should be realized in terms of information source, facility and space.

2 The Role University Libraries Play in Lifelong Learning System

All over the world, a very important development notion is to advocate lifelong learning and to create a society characterized by learning. Experts believe that the current undergraduates only master one-tenth of knowledge needed in one life. After graduation, it is necessary to acquire the remaining nine-tenths. The UNESCO points out in the "Research Report of Yesterday, Today, and Tomorrow of Education, we

---

* Corresponding author.

Y. Wang (Ed.): Education Management, Education Theory & Education Application, AISC 109, pp. 341–346.
springerlink.com   © Springer-Verlag Berlin Heidelberg 2011

cannot acquire knowledge once and for all, instead, we need to study during our life how to construct a ever-developing knowledge system----learn to live".

University libraries are important institutions of social education, which can store, collect, organize, transfer and develop literature information and provide service for society. University library can, as one of the exterior conditions, guarantee the realization of lifelong learning. Thus, people's interest of learning is aroused and people's quality improved.

3   Basic Conditions of University Libraries and the Feasibility of Constructing Lifelong Learning System

The ending of school education doesn't mean the ending of learning. The route to one's education should be study-practice-restudy-repractice. One cannot learn once and for all. Instead, one should continue studying and learning for all his life to meet the need of society, according to his own condition. University libraries, as the literature information center, naturally become the important sites for lifelong learning. The basic conditions of university libraries are:

3.1   The Policy Support to Research Theory and Explore the Practice

"University Libraries Requirements (revised)" issued by Education Ministry, item 21 specifies, "Some capable university libraries should try their best to be accessible to social readers and community readers". Although it does not require university libraries to be open to society, it is a trend for university libraries to serve society. For example, Beijing municipal government along with other provinces regulates its own library standards----"Beijing Library Standards", in which item 10 specifies "Beijing encourages libraries of schools, scientific institutions and other units to be open to the society". University libraries are supposed to be open to the public, while satisfying the scientific and teaching requirements of their own universities [1].

3.2   Rich Resources of Literature Information

The literature information resources of university libraries take up 40% of the social total, having complete storage system, 65% of university libraries possessing electrical journal, 30% with on-line database, greatly increasing the ability of processing, opening and distributing information resources. Meanwhile, university libraries can purchase annually a large amount of good quality text literature and electrical literature of great variety (including science, science popular, economy and popular). Aside from satisfying the need of students and stuff, the university libraries can serve the society to share resources, to guarantee literature resources of lifelong learning. This is the material foundation of lifelong learning [2].

3.3   Good Facilities

Recently, with the rapid development of information technology, the basic facilities in our country have been constructed and improved quickly. Under this condition, university libraries finish their interior technical and equipment innovations. With computer storage, transmission and other new equipment, electrical resources have been greatly developed and technical obstacle to sharing resources been partly solved. This is the good equipment foundation for learners to study via the Internet [3].

3.4 Suitable Place for Learning

Around universities, there are lots of communities. The existence of university provides strong cultural atmosphere. With the penetration of the idea of no-wall-university, university libraries will be more open to the public. With the thick air of learning, quiet surroundings, spacious room and rich resources, university libraries will provide suitable places for learners.

3.5 Librarians of All Subjects

University libraries have more than 30,000 stuff with good professional morality and business quality, 90% of whom are with college level and librarians with master's and doctor's degrees are on the rise year by year [4]. They can guide learners of different age, in all walks of life and of different subjects.

3.6 The Desire of the Learners

With the development of the social modernization, and science and technology, and the improvement of people's awareness knowledge level, the number of people who need lifelong learning is increasing rapidly. How community people need cultural education is an important problem which needs library-concerned people to focus on and study. For example, up to 2002, in Tianjin, there is no public library in many communities. People have difficulty in finding books to read.

To sum up, the university libraries can play a positive role in constructing lifelong system has good basic condition and is practicable. It has also been proved by the experiments conducted Japanese university libraries (which are open to the public). Under the precondition of not interfering the normal teaching and scientific research, university libraries can satisfy community people's need for lifelong learning by means of existing resources.

4 Problems Hindering University Library Playing Its Role in Lifelong System

Now, the obstacle which hinders university library playing its role in lifelong system does not come from technology, but from social factors. It is social notion and people's awareness that directly prevent university library playing its role. These problems are.

4.1 University Libraries belong to University and the Allocation and Use Are to Develop

University libraries mainly serve for their teaching stuff and students. How much they are open to public may influence their own function. To serve the community may be regarded as "not attending to one's proper duties". From the interior, the development of university libraries' social and cultural resources is rather left behind and some teaching resources are practically laid off and wasted.

4.2 To Serve for the Community Is Inevitably Increasing Libraries' Financial Burden

Now, there is no benefit balance mechanism to reward and compensate university libraries for their investment.

4.3 University Libraries' Function Is to Serve only for Teaching and Research and only for the Students

People's awareness of serving for readers is rather weak. There is a "wall" between the university libraries and society. Society knows a little about libraries and libraries lose their social basis[5].

5 The Application Strategies for University Libraries to Play the Role in Lifelong Learning

5.1 Reforming the Notion of University Libraries

The existing service is based on entity, stuff being the medium. With the development of internet industry, university libraries are supposed to change the traditional interactive contact between readers and stuff, turning passive service to active service, gradually forming an information mechanism which can satisfy the of society.

5.2 To Be Open to Society Should Be Based on the Principle of Mutual Benefit

The local government should grant the university libraries prizes for their quality service. For those who can do well in information service and technical consultation for communities, the government should give them allowances, encouraging university libraries to set up reading rooms in neighborhood communities.

5.3 Strengthening the Concept of Library Resources Sharing

Although it is generally acceptable that resources should be shared and constructed together, some units only want to receive rather than provide, which restricts the literature need of social readers. Modern libraries, in the system of lifelong learning system, must develop the idea of serving the society, turning the service system from storage-centered to use-centered.

5.4 Improving Basic Network Facilities, Making Information Resources Accessible

Blocked network will prevent different network readers from using libraries resources. Only the unblocked visit between different networks can realize the resources sharing. Right now, library of Tianjin University of Technology is trying to use long-distance visiting technology to meet the need of long-distance learning, fulfilling the requirement of lifelong learning.

5.5 Libraries Should Put Emphasis on the Organization Management and Development of Electrical Information Resources

Electrical information is the basis for library to serve for lifelong learning. First, finish the digitalization of stored resources. Second, finish the development of internet information resources, and build virtual library, serving better for lifelong learning. Finally, make special website to release the information demand, aiming at lifelong reader[6].

5.6 Exploring New Means of Library Information and Network Construction

With the continuous development of lifelong education, one university library cannot assume the work of providing information resources of many subjects at different levels. Therefore, different university libraries should increasingly communicate and cooperate in the development and construction of network information and stored information resources. Taking advantage of the library as a whole, a system should be established with the feature of "scattered database, centralized on-line database; scattered service, resources sharing", to guarantee literature resources for lifelong learning in the form of "mutual building and sharing" [7].

5.7 Setting Up a Service Department for the Community, Making Use of Library Equipment, Integrating Partial Literature Resources

To cooperate with community committee, build libraries in communities, in charge of borrowing, getting on-line, reading and accessing library via VPN technology and rotate text resources in multiple community libraries.

6 Conclusion

Lifelong education is a brand-new teaching mode, involving the comprehensive application of many aspects. To build a study-supporting service system of university libraries is critical to lifelong education. To develop and explore library's information resources and change the service mode of libraries and serve effectively for lifelong education are important future for university libraries. Only when they change their notions and serve actively, can they play a more important role in lifelong education.

References

1. Guo, A.: College libraries have been opening to the public. Jinghua Times 1(A8), 207–221 (2006)
2. Lu, B.: The Practice and Study of the University Library Services for the Community. Journal of Academic Libraries (1), 101–103 (2006)

3. Liu, C., Yang, Y.: Comment on service innovation of document information. Library Work and Study (1), 59–61 (2002)
4. Lu, B.: The Practice and Study of the University Library Services for the Community. Journal of Academic Libraries (1), 101–103 (2006)
5. Li, M.: Research on the University Library Facing to Social Service. Library Work and Study (5), 87–91 (2008)
6. Cao, S.: Library should give full play to the TV university course. Journal of Heibei Radio & Tv University (1), 13–14 (2001)
7. Bian, L.: Research on academic resources organizations in the net. Library Theory and Practice 1, 46–48 (2004)

Innovation and Talent Training Mechanism for the Law in the Mode of Production and Research Strategy

Hui Ma and WeiWei Yang

College of Arts and Law of Wuhan University of Technology,
Wuhan Hubei, 430070
ff6453@sohu.com

Abstract. The production and research strategy have a major significance to the global personnel training plan. In traditional legal personnel training mode, the focus is the inculcation of theoretical knowledge. Because of the limitations of the practice of docking departments, they have little chance to practice the knowledge. It is difficult to adapt to the market and meet the needs of society on graduation. The production and research strategy will be used in the cultivation of creative talents, so that legal professionals not only have a solid theoretical foundation, but also have a wealth of practical experience, which could meet the complex requirements of professionals. This is our legal personnel training objectives to be achieved in a new era.

Keywords: The production and research strategy, Legal Creative Talents, Training mechanism, Path design.

1 Overview

1.1 The Proposed and Development about Production and Research Strategy

The production and research strategy, in which we usually refer to cooperative education, is a new form of education. It is a full use of enterprises, schools, research institutes and other resources, taking their advantages in personnel training, combining production, practical experience with research practice that based on the theoretical knowledge and ability to practice [1].

The cooperative education was first proposed by the United States Institute of the University of Cincinnati Dean Herman • Schneider, who launched the first cooperative education program in 1906 at University of Cincinnati. Being emphasized, the cooperative education gradually developed into a new strategic model. At the same time, through cooperating with enterprises, the schools' contributions to the community will achieve maximum efficiency. During this period, research institutions analyze the joint utility of the organization which combine enterprises with schools .So that the objectives and effectiveness of the production and research strategy can play better. The cooperative education holds the principle of in line with the mutual benefit of all participants and market demand, and will be used in innovation and development among universities as an educational model strategic.

Y. Wang (Ed.): Education Management, Education Theory & Education Application, AISC 109, pp. 347–353.
springerlink.com   © Springer-Verlag Berlin Heidelberg 2011

1.2 The Definition of Legal Creative Talents

The people who drawn from the traditional legal education are strong in theory, but weak in the skills of practical application. [2]Legal creative talents, which called combined talents, are full of creativity, and could create a new situation. They not only have the knowledge of law, but also have other relevant knowledge and practical experience. Therefore, the cultivation of innovative law we emphasized must base on practice. The practice in training for legal personnel has two meanings: one is training of application skills, the other is innovation ability.

Emphasizing innovation is not the overthrow of all of our education system but injecting a new life into the traditional legal education, and strengthening the practice on the basis of the theory. In the trend of the production and research strategy, the legal creative talents whom we are talking about is to take advantage of this new strategic model, combine the legal education with business , use new forms of education, and train more high-quality of legal creative talents with a high speed.

2 The Status of Implementation of Legal Personnel Training in the Production and Research Strategy

2.1 The Status of Implementation in Wuhan University of Technology

As rooted in the engineering college, WHUT has its own unique approach to development. We take the legal education as our hospital-based, and focus on improving the level of practice, and combine the law with the life and production. Up to now, we have built a lots of practical platforms. In 2003, in the promotion of the first students' innovative projects, School of Arts and Law, WHUT, set up a legal services center to meet with community committees. We arrange several staff weekly for free Legal Consultation and solve the legal problems of nearby residents. In 2009, we conducted the preparatory work and the establishment for the Research Center of Legal Practice, which contain the building materials, transportation and automotive industry. We assigned special service officers to consider the current situation and problems of the three major sectors, and put forward relevant the recommendations from the scholars' point of view, meanwhile, made positive preparation for the academic conference about the three industries. At the same period, colleges and law firms get together to build a center mode, and provide employment channels for students who have a will to engage in the legal profession.

2.2 The Status of Implementation in Domestic

Close to civil law education in China, while absorbing the practice and teaching methods of two legal systems, the majority of students in colleges and universities focus on the training of vocational skills and the using of physical evidence, such as criminal investigation and technical experiments to observe the trial, moot court, and other practical teaching methods. During the late 80 era in the 20th century, we recommended cooperative education. The National Cooperative Education Association established in Shanghai in April 1991. In October 1997, Ministry of Education issued "The notice about carrying out the production and research strategy with pilot work during the "Nine • Five " " to determine the 28 pilots in Colleges and Universities carrying out the production and research strategy during the "Nine • Five". In the bias of recognition, cooperative education has long been used only as a model to be promotion of talent, but do not attract enough attention and application in the reform of undergraduate education. Since 2000, with the help of the Ford Foundation in the United States, Peking University Law School, Tsinghai University Law School and other seven key institutions created clinical legal education programs firstly. On September11, 2002, the Ministry of Education issued the "The conformity assessment of the Law education in the Undergraduate Teaching in Colleges and universities", which define the requirements about comprehensive practical capacity-building in legal aspects. From the development of legal education, some colleges and universities innovate the strategy. In 2008, China Science and Technology, Ministry of Finance, the Ministry of Education and other three departments launched "The structuring guidance regard to the construction of industrial technological innovation and strategically alliances", which clearly and comprehensively defined the technological innovation strategy. The years that followed, six departments announced a series of relevant documents, which provided a strong political and environmental protection for the construction and development of the strategy, and actively promoted the strategy into practice.

2.3 The Status of Implementation Abroad

View of legal education throughout the world, it represented by the two educational model — civil law countries and common law countries. The representative of Civil law is Germany and Japan, which is typical two-track system, and their aims are laying the legal theory, training overall quality. [3]Legal graduate education in Japan is a typical mode of production and research strategy. Graduate students are recruited by the University, and then join in the cooperative enterprises after studying the basic theory, while enterprises provide funds, places and topics, helping them complete their studies. In return, companies have the priority about employment. Meanwhile, the Japanese teacher recruit from the public, which has been a great help to the reform of the law. In common law countries, such as in United States and Britain, the main objective of legal education is training lawyers, so they underline the practical ability on the process of teaching students, and take the practical teaching model as the appropriate form of education. Law Schools have set up the appropriate legal courses, such as case teaching, legal assistance, training and some other practical aspects. The program for Undergraduate Students in Science in Massachusetts Institute of Technology where students have the task of scientific research in addition to learning the content, which will combine theory with application. In British University, which implement an institution of work-study, college students have to work in the professional sector of the enterprise a year or two during their studies, which has formed a system.

3 The Training Program of Legal Creative Talents in the Production and Research Strategy

Typically, based on the difference of applicants, the production and research strategy will be developed in the following two ways: one is the mode that school takes the initiative to contact the enterprise which has the relevant professional background, then combine theoretical teaching in classroom and the practice in business and provide technical innovation for the enterprise; the other mode is that the enterprise commissions the university to conduct scientific research, then the university trains enterprise expertise with the orders for the output of enterprises. Here are the explanation .

3.1 University as the Applicant

The university establishes the practical base in the enterprise. On the one hand, the university provides adequate theoretical support for technological innovation based on business requirements and market demand; on the other hand, the company will give a chance to take practice in the enterprise directly, which has a significance to the students working for the enterprise in the future. In this mode, the target of cultivation is not so clear at first, but is screened when taking practice in companies.

Postdoctoral intellectual property strategy base is established by the cooperation between Wuhan University of Technology and SAIC-GM-Wuling Group in Liuzhou.

The idea is to build a training base, and Wuhan University of Technology carries out the practical teaching and research professionals in the use of al-Qaeda, building a specialized intellectual property protection strategy for SAIC-GM-Wuling Group, and achieving many systematic and reasonable aspects from the registration of trademark to the intellectual property protection of product structure. By the efforts in seven or eight years, Postdoctoral intellectual property strategy base is systematization. With the practice and research in the enterprise every year, our talents provide a comprehensive guide for SAIC-GM- Wuling Group's own research and development of IPR protection system. Meanwhile, the more outstanding students are retained by the enterprises in the internship process, which forms a harmonious situation.

3.2 Business as the Applicant

3.2.1 Training Talent
In the training phase, business still pays attention to technical training, while study of the theory can enhance the practical technology as a supportive measure. In order to achieve the purpose of servicing company as soon as possible, you can use short-term mode .When Training is completed, the school sent to a recognized certificate of completion, while they do not have to configure a special diploma.

To legal profession, we should first set up our own professional guidance committee, which is composed by experts from enterprises and institutions. According to the different needs of enterprises, we set up different subjects. For example, in Automotive Legal Training Program, because it is a training mode, the arrangements of training time should not be too long.The theory do not have to cover all legal knowledge, it just only need to served as a training course. (Showed in below).

3.2.2 Undergraduate Cultivation Degree

Companies usually need a higher overall quality of talent: a solid theoretical basis and strong practical ability. So we can rely on "double degree" mode, to go comprehensive training in personnel. This stage of the teaching time can be controlled in three years. Practice training is necessary as well as the law of professional knowledge.

For example, for automobile production units, the legal personnel with professional knowledge in addition, if automotive product quality disputes, traffic accidents and liability, they will work in the business environment will adapt faster. Therefore, schools should set up business management, automotive theory, marketing strategy and other courses. So it can be cultured for subsequent corporate culture, quality consciousness and competitiveness of enterprises and have a complete understanding of consciousness and feelings.

3.2.3 Postgraduate Training of Personnel

First, the training period can be divided into 2-year and 3-year. It can take 2-year mode if it is the commissioned research Graduate school. It have enough time to finish the relative course and credit. Students who need to get knowledge and higher research can choose 3-year, enjoying the learning and practice opportunities supplied by school.

Second, in the training courses, for the purpose of taking teaching and research combined. They can strengthen their theoretical knowledge through public courses, What is more , they can get a lot of valuable information from legal cathedra and case analysis. The legal ability in practice could be improved in this mode.

Again, the mentoring and coaching focus on "mentoring" as the core. Instructors lead the students to participate in scientific research, to finish the originality of the high level of academic papers. Tutors should organize students to take part in the relevant co-operation with research institutes ,broaden their knowledge of drawing on existing social achievements.

Finally, the students enter and explore the projects related with co-operation companies. In process of combination of production and research, students proficiency and academic abilities are strengthened. The practice of students to participate in the actual operating and business papers not only help improve the quality of observation, but also help to identify forward-looking business problems of capitalization of knowledge.

3.3 The Interface between the Culture Stages

Higher education of china after the completion of the nine-year compulsory education can be divided into two types: one is the general higher education; the other is of higher vocational education. General higher education is directly from high school, Dr. level, through in the end. And higher vocational education refers to secondary, vocational (equivalent to college level) education. If the educated want to continue their studies, they can be achieved by means of social examination. [4]

First of all, the College should be noted in the curriculum at all stages of the design of the interface. Training stage is true with technology-based and PhD in preparation for the study. Bachelor of Law Master degree courses mostly attend the courses under the corresponding categories. For example Wuhan University of Technology and Law in the undergraduate curriculum arrange for civil law, commercial law.

Secondly, College can be taken as "2+3" training mode, specializing in early stages. If the learners who would like to further enhance their academic qualifications, you can phase in Training half of the period after the creation of undergraduate directly related disciplines, to cut short Undergraduate teaching time.It can be completed in the stages original Training ,and the time of training probably be cut short from 6 years to 4 or 5 years. Learned the same gold content, enterprises need to shorten the training time. For the MA degree Commissioned by enterprise, it can test their knowledge through examination, thus can allow them exemption related courses, and speed up the training.

4   Research Strategy Will Be Applied to the Legal Significance of Innovation and Personnel Training System

4.1   Pairs of Schools Is an Advanced School System, and That Make Teaching and Research Full of Vitality

Firstly, there is a wealth of practical experience. It help to develop comprehensive educations. Strengthening the law faculty's ability is a key part of the work of personnel training.

Secondly, the combination of production and research achieved a number of scientific research. Through the many years of practice, a lots of professional teachers in the accumulation of a solid theoretical foundation acquired a wealth of practical experience. The students carry out specific research, which not only addresses people's real problems, but also enhance the theoretical level and publish many related papers.

[5] For example is the text of Law and the Liuzhou Wuling SAIC-GM IP Strategy Group to establish the station, Dr. No, for many years the co-operation strategy of the College of intellectual property research formed a special order papers, monographs, academic reports and other forms. Scientific strategic goals of Hubei development and implementation also provides a valuable experience.

4.2   In Terms of Students, It Is an Advanced Practice Mode, They Can Find Research Subjects from the Community

The combination widely enhance the students' work ability. Students can improve the overall quality. College students should have a basic theoretical knowledge of the practice and have a strong operational skills .As a result ,students can adapt quickly to positions of work, and have a certain capacity for sustainable development, Which may be the maximum effiectiveness for students in production and research strategy. Through the cooperation with enterprises to increase the proportion of practical teaching, it can help to train and improve their vocational skills. The school's moot court, legal debates, provide a simulation environment for the students under the preliminary practice of comprehensive skills training. The training base on the real environment to provide students with comprehensive skills under the practical training. Students not only can develop a strong ability to analyze and solve practical problems, but also can be trained to master new situations. [6] At the same time, high-quality graduates can help to improve the current employment rate.

4.3 To Company, It Is an Advanced Management Mode, Technology and Talent to Ensure More Thing

Faced with the economic development of companies and enterprises to the serious impact and great challenges, all sectors of society are taking various measures to cope with full speed up the transformation of scientific and technological achievements. As an important and effective measures, that have been implemented in which the positive. Meanwhile, industrial restructuring and upgrading down are important to speed up the efficiency of the production and research projects. Enterprises can provide free legal talent to direct the College practice. The law students can explore the combination of cutting-edge issues and innovations in the theory and practice. To strengthen research cooperation, we should enhance innovation capacity constantly, promote sustainable development of enterprises. So that it can reduces the input costs of enterprises, and technology and human resources have effectively guaranteed.

Acknowledgement. This work was supported by the level teaching and research projects in Colleges and Universities in Hubei Province in 2009. The establishment No is 2009095. Thanks to the school of arts and law. WHUT for strong support and helpful discussion.

References

1. MATERIA. Combination higher vocational and technical education to promote the comprehensive development. Tongren Vocational and Technical College (3) (2005)
2. Zhao, G.: Combination Research and Education Experience. Zhengde University (12) (2008)
3. Xu, G.R., Wu, J.: Production and research integration model, to enable enterprises, universities benefit both. Shanghai University of Engineering Education (3) (2001)
4. Yao, H.: Added on the convergence of higher education in the United States at all levels and types of communication. Ningbo University (Educational Science Edition) (2) (2009)
5. Huang, H.: In the form of cooperative education and practice. Higher Education Research (3) (2000)
6. Zhang, E., Yang, P.L.: Domestic and foreign colleges and universities of cooperative education model. Teaching and Research (3) (2006)
7. Chen, J.: The new situation Research Strategy Innovation and Development Alliance. China Renmin University Press, Beijing (2008)

Thinking of High-Quality Courses Construction

Dongping Yang, Juanjuan Li, Zhengyan Li, and Guoqiang Sun

The Aviation University of Air Force, Aeronautical Control Engineering Department, China
yangdongping6688@sohu.com

Abstract. This paper mainly introduces the thinking of the high quality course construction. First the specific purpose of the high quality course construction is introduced. Then the content of high quality course construction is provided. At last several issues on building high-quality curriculum are introduced.

Keywords: high-quality courses, innovate, construction.

1 Introduction

Course teaching is a primary means of training of talented persons. Course teaching quality is directly affecting the quality of personnel training core essential factor. Construction quality program is to promote modern teaching engineering project teaching content, is also a critical step and an important element in the modernizing of teaching.

2 The Object and Request of Building High-Quality Curriculum

High-quality courses Construction is guided by cultivating new military talented person, who have all-round diathesis, knowledge structure, comprehensive abilities, creative spirit and innovation capability. With advanced teaching thought the teaching idea for the pilot and In order to improve the comprehensive quality of students, the high quality course, which is to training students' innovative spirit and practice ability as the key point, is created harmony to impart knowledge and cultivating ability raise quality for the integration of the course teaching mode by using of scientific teaching methods, advanced teaching means and interactive teaching. The high quality course construction, which is the core of the course content and teaching methods in the innovation system construction, affect many contents, which are included construction of teachers' team, innovation of teaching means, optimization of teaching process and evaluation reform of teaching. So the human-oriented ideas and advanced teaching ideas are to embody during teaching; Content of courses, which including latest information on subjects, should be to adapt to the requirements of new military revolution and close to modernization of the army and the military preparation; It should be constantly updated also; Content of course should to have the own special features and should meet the needs of quality education; lots of good textbooks should be including; qualified teachers should be needed.[1]

Y. Wang (Ed.): Education Management, Education Theory & Education Application, AISC 109, pp. 355–359.
springerlink.com   © Springer-Verlag Berlin Heidelberg 2011

3 The Main Content of the Course Construction

The content of high quality course construction includes six aspects.

3.1 The Content Construction of Teaching

Not only the latest information on subjects, new military revolution, modernization of the army and the need of military preparation but also education, scientific, advancement, pertinence and practicability should be embodied in building high-quality curriculum. The changes of theory, knowledge, technology and experience in teaching should be contained in course construction. We continued to combine construction of teaching content of high-quality courses with teaching reform of this subject, and handle curriculum development about one or a series correctly.

3.2 Building the Teacher Troops

The quality of teaching can improve by high-quality teachers. The lecturer, who is the course of the teacher, has been granted professor, who has much experience in teaching or attend research activities, or lecturer, who has much experience in teaching too, is responsible for the quality of the course design and development.[2]

Every excellent course must build a reasonable structure and high level business teachers. Without one, effective teaching experience and teaching achievement can not be continued. So the important task of excellent course construction is to build excellent teachers.

At the same time it also used as a measure of the excellent course construction achievements of evaluation important scale.

3.3 The Reform of Teaching Methods

Scientific and reasonable advanced teaching method is to improve the teaching quality. It is the bridge between the knowledge and student's quality and ability. Key to the construction of teaching methods is to construct distinctive and flexible methods, which is followed by thoughts of Quality Education and Creative Education. So students are the center of the class. The teachers' task is to deal with impart knowledge and ability training. Teachers can teach students knowledge methods, the research thinking, critical spirit and creative thinking. Study enthusiasm of student, meanwhile, can be played.

3.4 The Construction of Teaching Material

The construction of teaching material constitutes an important component of the high quality course construction. The high quality course requires a set of advanced and applicable first-class teaching materials. The lecture teachers should write actively materials or should use the high level of excellent teaching materials at home and abroad or other of excellent teaching material in the army. Teaching material aid on excellent course, Experimental instruction and foreign books can match material of excellent course. Teachers guide student to read a lot of reference materials. So it can train student's ability of mastering the latest development and develop subject knowledge.

3.5 Teaching Measure Construction

Modern educating technique that is using of Internet technologies or computer multimedia technology must be fully used. If so, education efficiency and teaching effect can be improved. In order to adapt to the development of modern teaching, integrating education resources, adjusting contents and discuss the relationship between teachers and students of the teaching mode of modern information are depended on conditions and characteristics of modern education technology.

3.6 The Experiment Content Construction

The experiment and practice is an important mean of training students' innovation ability and practice ability of the high quality course. Reform of teaching contents and forms is based on the tracking application to the professional fields to the development of science and technology progress. Students' independent working ability and the innovation ability should be fostered. We should Increase the proportion of comprehensive designing experiments.

Students can learn to division in the field of professional new technology and new theories and new knowledge by visiting apprentice skills practice links such as comprehensive practice.

Application innovation ability and Knowledge of skills can be fostered and improved.

4   Several Issues on Building High-Quality Curriculum

4.1   Establishing Impetus Mechanism and Encouraging Famous Teachers to Participate in the High-Quality Curriculum Building

Giving lectures by famous teachers is a very important guarantee for building high-quality curriculum. In order to ensure the quality of building high-quality curriculum, universities should demand and encourage famous teachers to shoulder responsibilities of building high-quality curriculum actively, and pay special attention to do a good job in building reasonable echelon of teaching faculty, which take famous teachers as the core and good instructors as the backbone. Meanwhile, to eliminate the obstacles on system and mechanism, practical measures on input and mechanism should be taken. [3]For instance, Universities should budget special expenditure for the high-quality Curriculum building; There should be corresponding mechanism of stimulation to reward and commend these teachers who participate in building high-quality curriculum; Flexible mechanism of personnel and instructor accountability system should be established; Besides, these mechanisms of supervision and evaluation, such as the students' assessment of teaching and the supervision to teaching came from other teachers should also be further improved .

4.2 Attach Importance to the Reform and Innovation of Teaching Method and Teaching Means

It is a key sign of building high-quality curriculum to optimize and integrate the curricular system and teaching content , strengthen the teaching , on which are based basic theory practice and the frontier of cross-disciplinary science and form new teaching content and more smart teaching methods according to the need of teaching. There are two aspects to reform the teaching content and teaching method. First, we should solve the relation between imparting knowledge and cultivating competence of students. Curriculum teaching must emphasize both the impartment of basic theory and knowledge and the training in innovation ability and practice ability of students. Second, it is necessary to integrate teaching and research to promote the translation from research achievements to teaching content. Third, lay emphasis on the using of advanced teaching method and teaching means. Active use of flexible teaching methods, such as Modern information technology and network technology, play an important role in reforming the ideological concept and teaching methodology of traditional teaching. Posting the syllabus, teaching plans, exercises, experiments, teaching documents and references etc of high-quality courses on the site, which means full development and utilization of the web-based educational resources of universities, realize the sharing of quality teaching resources.

4.3 Maintain a Strict Standard for Curriculum Selection to Guarantee the Quality of the High-Quality Curriculum Building

To ensure the building keeping high quality, the selection of curriculum must be highly organized and meticulously planned to achieve a high-quality result that will accept the final acceptance of experts. First, the declaration of curriculum and review process should be strict. We should formulate relevant standards , clear declaration condition and review process to make sure every project declared by teachers can be reviewed under the precondition of a Open , impartial and equal environment . The panel, formed under an arm of universities, should conduct rigorous reviews of projects declared, show them publicly and hear the Views of all parties. Second, carry out the high-quality curriculum building in a planned and efficient manner. The panel should be formed regularly to inspect, supervise and coordinate the progress of the high-quality curriculum building and have comments from students by means of class assessment to improve the content of the high-quality curriculum building continuously. Third, the panel must keep the final review strict on the basis of the content and standard of the high quality curriculum building. During the process, the affect that curricula act on the change of teaching ideas and personal cultivating modes and the promoting in educational reform must be emphasized. In addition, the innovation, progressiveness, systematicness, practical applicability, scientific and demonstration affect of teaching content also should be attached importance.

References

1. Gao, C., Wang, P.: Exploration and Practice of Excellent Course Construction (2011)
2. Cheng, H., Gao, C.: Thinking and Practice of Excellent Course Construction (2000)
3. Lu, X.: Promote quality education course teaching research resources sharing (2008)

Transformation of Education on the Course of Control Theory

HaiSen Ke, Dong Wei, Min Xie, EnHui Zheng, and XiuYing Zhou

College of Mechanical and Electrical Engineering,
China Jiliang University, Hangzhou, Zhejiang, China
hske@cjlu.edu.cn

Abstract. According to the characteristics of the Control Theory, combined with the advantage of case method teaching itself and the actual situation of the students, cases teaching methods is proposed. But we should also note that the past decade year between teachers and students is dwindling communication space of ten years. So, how to fully arouse the enthusiasm of the students and use all kinds of teaching methods, in the completion of the teaching task and the training of the students' self learning ability, innovation ability, become the focus of attention of the teachers. Based on the experience on the course, the paper aims at increasing the quality and effect through the analysis of student's condition and developing the method correspondingly.

Keywords: Case teaching, Works, Education Reform.

1   Introduction

Automation major is the national characteristics of China JiLiang University, so the professional training of the students' creative ability should also be considered throughout the courses of any teaching. The control theory is one of the backbone courses which the automation major students need to major in, and it covers extensive knowledge, involves content more, emphasizes the theory and practice of the course.

The case teaching method, which selects typical instructive cases under certain teaching purpose and organizes students to analyze these cases with providing some context, is kind of interactive teaching method that can cultivate overall capability of students[1-2].

The development of network technology has been paid much attention to by the people in the past decades years and now is very similar to nearly everyone. The network, which has the features of peculiar information transfer fast, extensive coverage and man-machine interactive etc, has been used everywhere. The students who born nearly 80/90s like the Internet and use it to communication especially. Therefore, taking the network as the tool or teaching platform for the teaching mode is necessary. Someone[3-5] strive to accommodate the characteristics of the network teaching and introduce it into students' teaching in class and extracurricular study, which can let the students to learn the necessary useful knowledge in self actualization, and at the same time learn some expand knowledge in the interest or outside pressure.

Y. Wang (Ed.): Education Management, Education Theory & Education Application, AISC 109, pp. 361–365.

During the actual teaching courses, how the process of teaching and practicing can be individualized to achieve the purpose of personal training is one of the most important considerations in changing of traditional methods. In the paper, we focus on seeking help to improve students' ability to study independently, train students to use knowledge to solve practical problems in the ability of the complicated target control. At the same time, with the interactive teaching condition of the website, some interesting or relatively complex examples were shown for students online communication, which can improve the students' learning enthusiasm and initiative, expand students' knowledge, improve the teaching effect of the course, and provide strong support for the graduation design and future employment[6].

2   Analysis of the Status of Study and Teaching

2.1   Current Problem in Teaching

The development of science and technology has brought convenience to education, and at the same time, produced the negative effects on education. The past decade year between teachers and students is dwindling communication space of ten years, which in turn reduces the teaching fun. Most teachers who teach more knowledge with the help of PPT found that cannot bring the happiness of study as expected, instead, may cause the rebellious attitude of the students, and even disgusted at learning. The main cause is that the students can not effectively take notes, effectively preview, unable to keep up with the rhythm of the class. It should not be desirable that we leave the convenience of the technology progress to the teachers and the troubles of the technology progress to the students. The pattern will only aggravate the opposition between teachers and students and can not be lasting.

2.2   Current Status in Teaching

According to the summary in recent years, the feedback of graduated students, and the communication with senior students, it can be found that most students have no satisfactory of theory knowledge level. Even those students who have better test scores lack the systematic understanding and mastery about the basic science. Most of them have less self-learning ability and cannot be interested in professional curriculums' learning. Besides, they have the weaker skill in the actual technology.

The students who are the 80 or early 90 have growing up in a happy and lonely environment, which would allow them to have some different from our characteristics. Therefore, we must take those characteristics that they grow up with mobile phones and the development of the network into consideration in the teaching.

2.3   Current Demands of Society

By investigating a large number of engineering enterprises, especially those enterprises related with Automation, we conclude that following problems existing in university graduates.

Don't want to start from the most common;

Lack of understanding of first practical environment;

Lack of suffering of spirit;

Don't understand social, and blend in the society as soon as possible;

Poor communication skills;

Lack of comprehensive talents.

3 Design of Teaching

3.1 Reform of Teaching Methods

First, we will rebuild the communication between teachers and students and listen to the voice of the students. With the development of technology, the mobile phone and network have been everywhere, which are the common equipments for the college students to communicate and learn. It can't solve the problem of better learning to limit the students to use the Internet or mobile phone. On the other hand, the best solution is to fully respect the students' choice, oneself also added to the ranks of the students, using the Internet and mobile phones with students the way of teaching and communication. We should share the convenience brought by the technological progress with the students, otherwise the classroom teaching or experimental teaching will only become such patterns dominated by the teachers and which will only intensify antagonism between teachers and students, not lasting.

Secondly, in the classroom teaching, the multimedia is only the auxiliary teaching tools but not the dominant. The use of the traditional teaching methods, such as the blackboard writing, the blackboard drawing, the body language teaching, are not only adjust teaching rhythm but also give students to think about rhythm of the time. Besides, we also note the interaction between teachers and students, improve the students' attention by asking questions and practice, discussion etc, increase the learning interest of the students by stimulating students' thinking.

Finally, in the process of teaching, case teaching was introduced which breaks the principle of imparting knowledge to students based on complete discipline system, and it claims that curricula should pay attention to those inspiring contents selected from the numerous and widely materials, which can facilitate students understand the nature of things. The case teaching process can be divided into case analysis, discussion and conclusion, but it should be adjusted according to the specific situation during the actual implementation, such as teaching progress, students' reaction. Although the methods can effectively stimulate the enthusiasm of students, and should be welcomed by students, there are also many problems in the actual teaching process.

3.2 Increase the Extracurricular Experiment Teaching

Laboratory is the base of teaching and the scientific research which were paid more attention to nowadays. Students can consolidate the theoretical knowledge learned in the classroom and apply the study of the theory to practice process through participating in the experiments. The purpose of this reform is not changing the experiment equipment conditions that most experimental into designing experiment, i.e. students are required to complete the design of the experimental process, experimental requirements outside class.

In the experiment, in order to avoid falling into repeating validation, and at the same time, can realize the smooth transition by the validation to design thinking, we design the experiment using the accumulation mistakes found in practice of teaching students, and reform continuously experiment content combined with theoretical interpretation, try our best to make students from passive to active. On the other hand, we should also pay attention to the experimenter is only just began to learn, so the design work can't be put forward according to the imaginary which need the guidance and the related knowledge must be considered.

3.3 Reform Assessment Methods

In order to guarantee the implementation of education process, we should give evaluation on the students' classroom questioning, attendance, quizzes, experiment, etc. As usual combined results, those results will be in certain proportion included in the final results. Develop a "daily practice record" system for student to record student achievement and performance. Adopt flexible practice approaches, for example, continuous practice and discrete practice.

4 Conclusions

Our classroom teaching reform and practice have achieved some results after several years of effort to control theory. From the past 2 years (8 class), classroom teaching and practice, the communication between teachers and students are constantly improved, students' feedback results are more and more better. Visible teaching reform for the improvement of students' all-round development, innovation consciousness and practical ability training is very important, students learn to also show a higher enthusiasm. We achieve the expected teaching goal that "To me study" turns into "I want to study".

Acknowledgement. This paper has been supported by the ZheJiang Province Level Excellent Course((2011)9); The Teaching Reform Project of China JiLiang University (HEX201007, HEX2011006); The Teaching Reform Project of College of Mechanical and Electrical Engineering, China Jiliang University.

References

1. Zou, L.: Investigation of education value and teaching process about case methods of teaching. Journal of Wuxi Education College 4(2), 39–41 (2004)
2. Zhang, J., Jin, Y.: On the essence and Characteristics of case study-based teaching. Journal of Chinese Society of Education 1, 48–52, 62 (2004)
3. ABET. Criteria for Accrediting Engineering Technology Programs, 11 (2003), http://www.abet.org

4. Chen, Q.: Web-based autonomy learning in college English teaching. Foreign Language World 6, 125 (2006)
5. Cen, C., Cai, Y.: China Public Service Development Report 2006. Press of China Social Sciences (2007)
6. Ke, H., Wang, W., Wei, D.: Transformation of education on the course of programmable logic controller. In: 2010 International Conference on Education and Sports Education, July 17-18, pp. 205–206 (2010)

The Study of Specialized Courses Using the PDCA Cycle

Chi BaoQuan, Huang ZhenHai, Zheng EnHui, and Wang GuiRong

China Jiliang University Mechatronic Engineering; Hangzhou 310018
{bqchi,hzh,ehzheng,lilygrwang}@cjlu.edu.cn

Abstract. In order to improve the level of the university education, the focus of educational reform has been on the specialized courses in China's universities and colleges. Now, the study model of specialized courses is discussed using the PDCA (plan-do-check-act) cycle. It provides the feasibility and the concrete applications using the PDCA cycle, and indicates that it a good method to study the specialized cousese. Although it is a management approach being successfully applied in the quality improvement, it can strengthen the awareness and be interested of college students.

Keywords: PDCA cycle, specialized courses, quality of the education.

1 The Origin of PDCA Cycle and the Relation of between PDCA Cycle and University Education

PDCA cycle is a mode of quality improvement, coming from Dr. Edward Deming which is renowned as the guru of quality management[1]. PDCA cycle is constituted by the first letters respectively of four English words: Do, Check, Plan and Action. It indicates four different stages, that the fist stage is Plan: including making policy and objective and setting procedure and plan; the second stage is Do: according to the content of the plan, the concrete implementation is arranged; the third stage is Check: the result of the concrete implementation is inspected in accordance with the target set and standard; the last stage is Action: according to the inspection and feedback of the information obtained, affirmation and doing are given for the good result and running another PDCA cycle is used to solve problems still in the practice[1, 2]. The characteristics of the model are nothing less of four stages; big ring with little ring; quality level rising like as the ladder[1, 2].

Since the implementation of total quality management in our country, the quality of product and service are made great improvement. As an effective tool of a total quality management, PDCA cycle is widely applied to the practice, and obtains obvious results in quality management and quality improvement[2, 3].

The course of college students studying in China's universities and colleges can be divided into two kinds, one is basic course and another is specialized course. The specialized course is deeply study on technical knowledge based on the learning of basic course, which can form the technology base for the further work. The practice of the specialized course is the relation between theory and practice, therefore learning specialized course has chief significance. For the technical and seriousness, the model of studying specialized course is boring. In order to better teaching and learning, the

Y. Wang (Ed.): Education Management, Education Theory & Education Application, AISC 109, pp. 367–370.
springerlink.com   © Springer-Verlag Berlin Heidelberg 2011

practical, vivid and interesting is the development trend for the education model of specialized course, which can give full play to teachers and college students's initiative, and make college studentss to play a guiding role, and train the ability for practical use of specialized course, especially the practice ability, ensure that the college students's professional technical level has risen steadily and it can suit the needs of entering into work .

2   The Feasibility and the Need for the PDCA Cycle in the Specialized Courses Teaching

Alough PDCA cycle is initially using in quality management, its application range is gradually enlarged with more and more research on it. It can be used in many fields like transportation, medical bills and english teaching etc[5、 6]. The model can be also used in specialized course. Due to the experimental condition and practice base of the restrictions, fundamental technology is morly emphasised by the traditional specialized course education, and practice ability is weak for solving problems using the learing. Therefore PDCA cycle can betterly arouse college students' interest in the professional skills and train the professional awareness.

(1) PDCA cycle can arouse college students' interest in the professional skills, so as to enhance college students as guiding role in the process of the education. Interest is refers to a kind of psychological tendency for person to contact with certain things and to perform an activity. If students are interested in a things, work and activities, they can devote themself to learning more for solving the problem in the field, such as electronic competition and model ships ect. Because specialized course has more technical knowledge and more boring than basic course, some college students are difficult to study it. The reason is that they are not interest in learning it, in addition to the basic knowledge is not mastered familiarly. PDCA cycle can give a working routine: the first step is planning, and the second step is doing, and the third step is checking and the last step is action. The study of specialized course is arranged as PDCA cycle for college students, therefore they can know where their shortcomings, and sovle the problem in themselves, and at last they are interested in do it.

(2) Using PDCA cycle can construct the training mode of specialized courses. Now more the traditional specialized courses focus mainly on the basic principle of knowledge, and therefore the ability of solving problems is not enough for college students. College students coming from that environment, they can not begin to do some work based on what their learning. The reason is that they lack the exercise to solve the practical problem. Therefore to solve specific problems and master the methods to solve the problems should be trained in specialized courses education. Dealing with specific problems, plan should be consided, and then do and check and action. If there are some problems still, another PDCA cycle should be done. Where time is needed, it erects a bridge across the gap between the theory and engineering practice.

(3) Using PDCA cycle can cultivate the professional awareness, and it is the foundation of forming the enterprise culture consciousness for the future. Cultural awareness is very important, it will affect the person's behavior and habits. As we can see, enterprise culture is heavily influenced on the cohesion of the enterprise. Using of

PDCA cycle can promote students to learn and maste more technology knowledge in the professional field, for later work it can lay a foundation. Professional culture consciousness is gradually formed in the mind of the students and also literacies are formed, with continually running PDCA cycle. That are foundation to accept and integrate in enterprise culture for students' later work.

3   The Concrete Applications Using the PDCA Cycle in Studying the Specialized Courses

In the study, to learn one course can be regarded as one PDCA cycle, and to sovle a problem also can be consided as one PDCA cycle. For example, first, We need to develop a plan: a specific design scheme should be completed in one day or two days; second, actual operation is done: students should do it as the providing from the first step; third, the second day or another time, students should check the result of the second step; last, based on the result of the check, it will be passed or be running another PDCA. Such as, this one problem is solved, students can do another like above. By doing this as many times as necessary, the studying specialized course can be consided as the cycle of spiraling up, and the quality of learning must be improved.

PDCA cycle is given as a good method for students. Doing as scientific work running, a good habit must be formed in the study, and the professional awareness can be cultivated. In return, the professional awareness can decide the behavior and habits of students. Therefore, using PDCA cycle in study is like to lay the foundation forward the successful road.

In the current society, with the rapid development of science and technology, new knowledge and new technology of the corresponding professional field are growing ever stronger, therefore the college students can not master all knowledge of the corresponding professional field within a few years. Thus, teaching the method of studying and training the corresponding interest and the professional awareness can impove the ability of sustainable development of college student, that can make them adapt to the new economic. Therefore PDCA cycle as a model of the impovement, using it in specialized course must become a necessity.

Acknowledgments. The research is supported by the projects of teaching reform of China Jiliang University(HEX201006,HEX2011006).

References

1. Janet, H.C.: Small-Scale Study Using the PDCA Cycle. Today's Management Methods, 209–222 (1996)
2. Foster, S.T., Ogden, J.: On differences in how operations and supply chain managers approach quality management. International Journal of Production Research 46(24), 6945–6961 (2008)
3. Geng, J.H., Gao, Q.S., Fang, A.L.: Quality Improvement:a Typical Complex Adaptive System. Commercial Research 9, 52–54 (2007)

4. Saxena, S., Ramer, L., Shulman, I.A.: A comprehensive assessment program to improve blood-administering practices using the FOCUS–PDCA model. Transfusion 44(9), 1350–1356 (2004)
5. Qing, J.: Appling PDCA method to manage the road traffic. Shanxi Architecture 34(20), 247–248 (2004)
6. Jiang, Z.L.: Improving College Students English Writing Ability by Applying PDCA Models. Journal of Chongqing University of Arts and Sciences(Social Sciences Edition) 27(4), 97–99 (2008)

Discussion on Experiment Course of Signal and System Reform

Huang Xiaohui, Jia Zhenhong, and Li Xingang

College of Information Science & Engineering, Xinjiang University, Urumqi 830046, China
hxhdemail@sohu.com

Abstract. Signal and system is a basic theoretical courses for communication majors, and experiment teaching, which play an increasingly important role in cultivating students' creative spirit and training their practical ability, is the basic links and an important way for students to understanding and application this theory. This paper introduces the status and insufficient of our school on signal and system experiment, and put forward some thinking of experiment teaching reform.

Keywords: signal and system experiment, status and insufficient, experiment teaching reform.

1 Introduction

For communication majors, the signal and system theory is the foundation to master this professional course, It is essential to do a certain number of experiments. The students have to master the professional theoretical knowledge systematically, which requires students to do relevant experiments when learn theoretical knowledge. Experimental teaching is the basic links and an important way of teaching. The importance of experimental teaching is that a student can learned more from the experiment than from the theory, and more intuitive. Experimental teaching play an irreplaceable role in the whole education with their independent status, which is the essence of experimental teaching. So I think that we have to reform and innovation experiment teaching to adapt the development of the information age. Experiment will play an increasingly important role in training students in the innovative spirit and practical ability.

The experiment on Signals and Systems has played an important part in promoting the curriculum teaching effect and developing students' practical ability since its establishment. but there are still many deficiencies.

2 The Status and Insufficient of Signal and System Experiment Teaching

2.1 A Single Experimental Teaching Methods, the Experimental Content Is Old

Experimental teaching still remain in the traditional teaching model. Students who are guided by the teachers or experimental guiding books do experiments through the operation steps in the class as a unit fixed at the same time, but them do not know why to do so , even do not know what operating results show. Students' assessment of the experiment eourse is mainly based on experimental report, or teachers' impression on students. This experimental teaching is inefficient, place students on a state of passive learning and mechanical operation, leading to teachers and students feel boring, stop the student's logical thinking and creative imagination, and it is not conducive to develop of experimental ability and research capacity of students, it also does not meet the education modern needs.

2.2   The Experiment and Theory Can Not Be Joined Organically

In the history of scientific development, experiment and theory interdependence each other, the theory is a summary of experimental facts, the basis of experiment is a scientific theory, experiment is also to determine and correct mistakes of theory. From the view of experimental role in the scientific development, experimental teaching should be attented by teaching managements, educators and students. However, the current teaching on theory and experimental of signal and system can not be synchronized. Theoretical knowledge, mainly in the form of mathematical expression, whereas the experimental conducted in the circuit, a lot of theories are not verificated by experimental timely, and experimental methods, steps, and the experimental phenomena has not been guided and reasonable explanation by theory. So on one hand theoretical teaching is not a good guide to the experimental teaching, on the other hand, laboratory teaching can not serve well to the theory teaching.

2.3   The Experiment Is Too Formal

Signals and Systems experiment equipment is a experimental box which is integrated and modular, students in the experiment only need to choice experiments module requiremented, connect some wires, no need specific choice of circuit components to build the circuit, so that the students only completed experiment formally, them can not grasp the installation method of the circuit, commissioning skills, is not conducive to develope students hands-on and innovation. Some students can not use basic laboratory equipment proficiently, resulting in the experiment did not go well, and even error results are due to operators' mistakes.

3   Some Suggestions

3.1   The Reform of Experiment Syllabus and Experimental Teaching Plan

Syllabus is the plan of based on the teaching, every practice course should have a standardized practice course syllabus and specific quality standards. With the education reform and the development of science and technology continuously, the teaching programs and curricula have to be revised, and gradually increase the independent experimental course, including some experiment based Matlab software. It may be helpful to increase experiment hours, so that fully reflect the theory teaching and experimental teaching is different but equal in the whole educational system, try to change people's old ideas than experiment teaching is subordinate to the theory teaching. Meanwhile, the examination of the experimental and assessment methods should be reformed, increasing the credit score of experimental course

3.2 Independent Course, Separate Evaluation

Independent experimet course from theory is helphul to improve the status of the experimental teaching, teaching in the traditional experiment, the content of experiment course on signal and system theory are based on the progress and content of theory teaching arrangements, completely dependent on theory, students do not pay attention on experiments in general, many of them are not interested in this kind of experimet. In order to improve students' operational capacity and ability to analyze problems, and change the view of ' Emphasis on theoretical and practical light ' , we should made signal and system experiment into independent curriculum, self-contained, focuse on training, and individual assessment. Independent experiment course is expected to improve independent status of the teaching experiment, to ensure that teaching time, and enhance the systemic experiment, it is benefit of reasonable course arrangements and unified management, it is beneficial for students to appraise, to stimulate enthusiasm for learning. So that experimental teaching and theory teaching as interrelated aspects of the two independent teaching.

3.3 More Software Experiment

The establishment of more software experiment, It is difficult to achieve with hardware on a lot of signal and system theory content, but very easy to experiment with the software, on top of, software experiment is intuitive, can be a good complement the lack of hardware experiments.

3.4 Suggestions on How to Do That

According to the experimental curriculum and training objectives, we can divided the experiment of signal and system into three parts: hardware verification experiment, software experiment, design experiments base on hardware and software.

The contents of hardware verification experiment include the equipment operation, electrical parameter tests, the verification of basic theory, the kind of experiments base on hardware. So that students can understand the signals and systems theory more depthly, and to develop students' practical skills. The contents of software experiment include contect which can not be achieved on the hardware circuit, ideal circuit simulation, the experiment using programming and simulation method. It can enhance students understanding of the knowledge and improve the analytical capacities of students.

The design experiments base on hardware and software, focuses on design and debugging of small-scale experimental circuit, this methods combining simulation and physical experimental. First, computing, simulation analysis, revised design parameters on software. Second, making hardware circuit. Finally, testing. Comprehensive training students to apply the knowledge to analyze and solve problems, to enable students to integrate theory with practice, master the design and development skills, improve their overall quality, the creativity of students.

4 Concluding Remarks

Signal and system experiment teaching is the basic links and an important way of the theory teaching which is difficult for students to learn. We have to reform and innovation experiment teaching to adapt the development of the information age .

References

1. Zhang, L.-J.: Reseach On Enhance of Signals and Systems Experiment Teaching. Science & Technology Information 23, 460–489 (2008)
2. Zhang, L., Zhou, Z.: The university signal and the system educational reform discuss. Journal of Zhoukou Normal University 27(5), 69–71 (2010)
3. Lv, Z., Ding, G.: Experiment System Design of Signals and Systems. Experiment Science & Engineering 8(1), 51–52 (2010)
4. Lu, Y.: Reform ing experimental teach ingmethods and building a new experimental platform. Laboratory Science 4, 23–24 (2009)

Survey on the Concept of Shaping Behavior and Cognition of Female Flight Attendants

Song Faming[1], Shu Jianping[2], and Wen Xiaoyuan[2]

[1] Civil Aviation flight university of China, Guanghan, Sichuan, China
[2] Chengdu Sport University, Chengdu, Sichuan, China
{afmsung,bxbjk98}@163.com, cwxysfm@126.com

Abstract. The concept of shaping behavior and cognition of female flight attendants were investigated and analyzed, the results showed that professional female flight attendants, body mass index (BMI) in normal, 34.9% of the people in a slim state; 60.2% of respondents have been shaping, body sculpting is 21.1%; 85.6% body sculpting movements, regulate diet accounted for 67.1%, 56.8% body sculpting diet; in shaping reasons not of their own body satisfaction accounted for 72.8%, 51.4% love beautiful; 10.3% of regular exercise, 79% sometimes sports, 10.7% do not exercise; present and future options exercise shaping methods are in first place is yoga; control diet and exercise remain the first choice to prevent weight gain.

Keywords: female flight attendants, shaping behavior, Cognition.

1 Introduction

Flight attendants have beautiful, dignified, elegant appearance and give people a good image features, because they bear important aircraft cabin safety and service duties and represent the airline's image, personal qualities will affect the airline's image and effectiveness. Excellent flight attendant must have good physical, skilled professional skills, quality of service, elegant appearance manners, good psychological quality and hard-working spirit and other qualities. Flight attendant's personal qualities not have congenital, but need be trained and accumulated over a long period (especially physical). During the time, they need evolve and improve gradually. To become a qualified flight attendant, then they gradually must build and shape the external appearance instruments in their daily life. In the past, if people's physical fitness was to live healthier and live longer, yet today, people have keen on shaping the physical form of exercise, which will be affected by a higher level of demand driven, that is, the pursuit of the body external morphology and body posture perfect. Physical beauty and wholesome beauty are complementary, physical beauty is inseparable from health, and vice versa. However, in order to maintain the physical beauty, some female flight attendants severely affect their health because of body shaping blindly. Thus, by means of analysis of the concept of body shaping behavior and cognition, it designed to make female flight attendants girls have a correct understanding of body shaping, and then put forward feasible suggestions.

Y. Wang (Ed.): Education Management, Education Theory & Education Application, AISC 109, pp. 375–382.

2   Research Subjects and Methods

2.1   Research Subjects

Randomly select 332 female flight attendants from China Civil Aviation Flight University.

2.2   Research Methods

The study used literature and data, questionnaires, interviews, statistical methods and analyzed the implications of data collection. According to the study design questionnaire, a total of 11 sub-problems, 332 questionnaires were returned of 321 copies, 299 valid questionnaires, 93.1% efficiency.

3   Result

3.1   Basic General

In the 86 respondents, 168 female freshman, sophomore 131 girls, mean age (19.2 ± 0.8) years, minimum 18, maximum 21 years old. BMI mean 18.78kg / m², of which less than 18.5kg / m² 34.9%, greater than 18.5kg / m² and less than 22.9kg / m² accounted for 65.1%. BMI is from the immunological point of view of data to drawn up to determine the basis of nutritional status through a variety of body composition assessment, it is the internationally recognized standard to judge the human body. (Body mass index) BMI = weight (kg) / height (m) ², the World Health Organization has been to determine the human body as an important indicator of the degree. According to the 1998 WHO recommended standards of the Asia-Pacific, BMI <18.5 is too light weight, 18.5 ≤ BMI <23.0 for the standard body weight, BMI ≥ 23.0 as overweight. According to the study population for our body to determine criteria: BMI <20 as underweight, 20 ≤ BMI <24 as normal weight, 24 ≤ BMI <26.5 as overweight, BMI ≥ 26.5 as obesity. This standard survey indicate the mean BMI between the normal range, but only above the lower limit of normal (18.5kg / m²) ,explain that respondents belong to slim-type, indicating that respondents are more symmetrical shape, good posture.

3.2   The Analysis of Body Sculpting Experience, Self-perception of Body

Table 1 shows, has tried shaping and being body shaping are the number of 180 and 63 respectively, accounting for 81.3% of the total survey. Has been tried in the shaping of students, sophomore higher than the proportion of freshman; but is body sculpting, freshman higher than the proportion of sophomore. In General, it has not obvious difference between freshman and sophomore students, 81% and 81.7% respectively. It shows that in ordinary life, students have high self-demanding from aspects of maintaining good body and intense body shaping behavior.

Table 1. Body shaping of students from different grades

	Tried	being	no
freshman	88	48	32
sophomore	92	15	24
total	180	63	56
percentage	60.2	21.1	18.7

Can be seen from Table 2, in less than 18.5kg / m² persons, great, good, fair, bad numbers are 14,69,35,3;in greater than 18.5kg / m² and smaller than 22.9kg / m² persons, great, good, fair, bad numbers are 5,53,112,7. It shows body mass index affect the student's self-perception of body.

Table 2. Body mass index and self-awareness of table (number)

	great	good	common	no-good
<18.5kg/m²	14	69	35	3
≥18.5kg/m²且 ≤22.9kg/m²	5	53	112	7

Table 3 shows that in less than 18.5kg / m² 121 students, 25 people are body shaping, accounting for 20.7%, and in which 25 people, 23 were dissatisfied with their body, accounting for 92%; in more than 18.5kg / m² and less than 22.9kg / m² 178 students, 38 people are body sculpting, accounting for 21.3%, and in which 38 people, 63.2% were dissatisfied with their body. It indicates that their body awareness is heavier than the truth, so that the girls tend to body shaping. This reflects many girls are now in pursuit of fashion, "the beautiful backbone ", the blind pursuit of slim, so as to take a series of shaping behavior.

Table 3. BMI and shaping with the crowd of table (number)

	be shaping	Dissatisfied with their body
≤18.5kg/m²	25	23
≥18.5kg/m²且≤22.9kg/m²	38	24

3.3 The Analysis of Cognitive Fitness Motivation and Sources of Knowledge

Can be seen from Table 1 and Table 4, in 243 girls, in descending order reasons for downsizing were not satisfied with their body, love beautiful, increased self-confidence, forced to future employment pressure, physical fitness, and other respectively. Love pretty in second place that it is not only an individual manifestation of the pursuit of beauty, but also to meet the psychological needs of social groups, so-called "beauty of the heart, in everyone." Dissatisfied with their body in first, the body means more plastic, focusing on the external image of the shape, in order to meet their requirements through body shaping. Therefore, tend to body sculpting has become a common phenomenon. Although respondents face a severe employment situation, but only 28.4% of people think that the purpose of shaping is the future of the employment pressure. Through the data obtained, the purpose of shaping body is to maintain good body. only 27.2% people choose body shaping in order to make themselves more healthy. Overall, the awareness of body sculpting of respondents is still relatively narrow. Table 5 shows, in descending order body of knowledge sources networks, newspapers and magazines, friends and classmates, television, books.

Table 4. Cognitive fitness motivation（N=243）

	Dissatisfied with their body	Love beautiful	increased self-confidence	physical health	forced by the pressure of future employment	other
Frequency	177	125	114	66	69	11
Proportion	72.8%	51.4%	46.9%	27.2%	28.4%	4.5%

Table 5. Source of shaping knowledge（N=243）

	newspapers and magazines	books	television	friends and classmates	networks
frequency	159	97	108	137	183
proportion	65.4%	39.9%	44.4%	56.3%	75.3%

3.4 The Analysis of the Shaping Behavior Method

Obtained from Table 6, the 243 shaping students, body shaping in the first row of the chosen method is to choose movement with 208 people, accounting for 85.6%; in second is to choose a diet adjustment with 163 people, accounting for 67.1%; in third is to choose a diet with 138, 56.8%; came in fourth place with 19 body shaping underwear, accounting for 7.8%; the drug of choice body sculpting, body shaping, very few instruments. It indicates that the respondents have higher levels of education, have a strong ability to judge things, so that the choice of body shaping method is also more rational, does not blind.

Table 6. The selected method of body shaping（N=243）

	diet	drugs	Regulate diet	exercise	surgery	Body shaping underwear
frequency	138	9	163	208	5	19
proportion	56.8%	3.7%	67.1%	85.6%	2.1%	7.8%

Can be seen from Table 7, in descending order, the selected method of body shaping was introduction by friends, by other means, advertising, introduction by doctors. Introduction by doctors in last place indicates that in the shaping process, the respondents lack in expertise and body sculpting expert guidance.

Table 7. Source of body shaping method （N=243）

	Friends.	Advertising.	By doctors	other
frequency	191	76	35	117
proportion	78.6%	31.3%	14.4%	48.1%

Derived from Table 8, in descending order, presenting symptoms in the process of body shaping was fatigue, dizziness, anxiety, anorexia, difficulty concentrating, insomnia, low self-esteem. It shows that nearly half of body shaping will bring fatigue and other physiological effects, and bring to a small part of people adverse effects. Therefore, in the shaping process, students should strengthen the fitness science education of flight attendants, rather than simply improve their own image for the pursuit of fitness.

Table 8. Body shaping manifestations （N=243）

	fatigue	dizziness	anxiety	insomnia	self-esteem	anorexia	Cannot concentrate
tried	97	66	56	48	17	31	42
shaping	7	10	7	11	7	4	14
frequency	104	76	63	59	24	35	56
proportion	42.8%	31.3%	25.9%	24.3%	9.9%	14.3%	23.1%

3.5 The Analysis of Body Shaping and Sports

Sports sociological theory indicates that the dominant philosophy of consciousness, ideas led behavior. So, the correct level of fitness concepts and cognitive has decided student's body shaping methods and behavior, plays an important role in health. Simultaneously, cognitive affects behavior. The Many students concerned about their individual size, the more power to take shaping behavior.

Table 9. Movement situation （N=243）

	regular exercise	sometimes exercise	do not exercise
tried	21	146	20
shaping	4	46	6
number	25	192	26
proportion	10.3%	79%	10.7%

Table 9 shows, 25 were regular exercise, 10.3%; 192 sometimes exercise, 79%; 26 people do not exercise, 10.7%. It reflects the students' participating in sports is not high emotion. This body shaping with the selected method has inconsistent with Table 6 ranked first body shaping movement, indicating that flight attendants can recognize the students the importance of body shaping movements, but in the specific implementation process they always can not adhere to, rather can achieve the body shaping effect by regulating diet, dieting. It reflects the female flight attendants have the cognitive understanding of body shaping, but behavior is inconsistent with the cognitive.

Table 10. Selection method of sports fitness at present and in future （N=243）

	running	aerobics	ball	swimming	yoga	other
present	62	24	17	28	97	10
future	49	63	41	62	159	18

Table 10 shows, in descending order, sports body shaping methods are selected yoga, running, swimming, aerobics, ball games, other. The future exercise shaping mode in descending order are selected yoga, aerobics, swimming, running, ball games, other. Yoga has been described in depth among the college as students of fashion movement. Because yoga can control breath and regulation, so as to make the body heat, speed up the metabolism, after a period of practice, flexibility and body shape will be greatly improved. Fat content reduced, and the body does not let up. Among them, stretching can consume a large energy. Aerobics choice now in fourth place, either in the next selection came in second place, Prevalent in society, a wide range of aerobics fitness due to the school environment and school conditions and other restrictions, affect them body shaping.

3.6 Prevention of Weight Gain Methods

Table 11 shows, the choice of measures to prevent weight gain in descending order were controlled diet, exercise, reducing sleep, the other. It shows the control diet and exercise is still the best ways and means to prevent weight gain.

Table 11. Prevention of weight gain methods （N=243）

	Control diet	exercise	reduce sleep	other
frequency	191	186	42	17
proportion	78.6%	76.5%	17.3%	7.0%

4 Conclusions and Recommendations

4.1 Conclusions

4.1.1 Most of the female flight attendants can understand shaping cognitive, but behavior and cognition are contrary, did not properly understand the relationship between fitness and health. BMI index within the range of slim, but they still not satisfied with their own body, still the body shaping. This is a relationship of their professional requirements.

4.1.2 Fitness motivation is not clear. Many students in dealing with beauty think that the slim, "bean sprouts" type is beautiful. In fact, the pursuit of "Healthy America" is our modern goal. A variety of college life and modern material life are filled with the diary life of students, and people who exercise is very few for the target of physical fitness.

4.1.3 Lack of proper body sculpting guide. Diet and exercise are best way to prevent weight gain or weight loss. Before or in the process of body shaping, who are few with weight loss expert and body shaping doctor.

4.1.4 Stylish sports - yoga has been the first choice of body shaping for college girls. By mimicking nature in all things, and find the natural self, away from turbidity, light-weight fitness.

4.2 Recommendations

Female flight attendants have misunderstanding in the shaping of cognitive and behavioral aspects, the following recommendations:

4.2.1 Schools should pay attention to women's health issues, so that they define the relationship between fitness and health.

Fight attendants are likely to be a good civil aviation employee with good body, because good body is vital for security and. Through various channels, they learn about fitness knowledge to guide their scientific fitness, draw up reasonable plan and diet plan. Such as: the development of body shaping exercise program, using the American College of Sports Medicine proposed in recent years continued twenty to thirty minutes of jogging (aerobic exercise), heart rate control in 150 times / min, for weight reduction and control is the best way and exercise heart. Diet should be low-fat, high protein, high in vitamins and low sugar. Fruits help solve hunger, but also add snacks. Salt and water balance during body shaping should be prohibited unreasonable dehydration (service diaphoretic, diuretic, laxative) and control water. Body shaping during the day in 2000ml water intake should be controlled about the way and be taken a few times to the appropriate supplement.

4.2.2 Students should strengthen the training of modern mass aerobics and dance physical training.

Aerobics is a modern public financial gymnastics, music, dance in one, by hand or use the fitness equipment, sports to fitness, which is a typical aerobic exercise and more appropriate for girls. Particularly the use of fitness exercise equipment to achieve the strength and beauty of unity. Dance form is based on the theory of human science, blend with sports, music and dance, through a variety of physical training to achieve better health, beautify the body, the purpose of a more upright posture ideal training activities. Reducing caloric intake (diet), increasing the body's energy expenditure (exercise), result in a negative calorie balance, so that the body consumes stored body fat.

References

1. Lu, Y.-Z.: Sociology of Sport. Higher Education Press, Beijing (2001)
2. Chunjiao, S., Chen, J.: The research on weight loss concept and knowledge for female college students. Youth Research (11) (2000)
3. Sun, J., Su, H.: The investigation on weight-loss behavior of female college students. Journal of Physical Education (5) (2005)

4. Lei, L., Wang, J.: Beijing female students tend to weight the impact of factors. China Health Journal 19(3) (2005)
5. Dingshi, Y., Xu, Y.G.: The survey on college girl weight conscious awareness and behavior. Donghua Institute of Technology (2) (2007)
6. Wang, M.: Henan University. The survey on student awareness of sports and physical exercise. Shanxi Normal University Institute of Physical Education (2) (2005)

The Observable Characteristics of Effective Teaching in Professional Experimental Courses

Hongmei Ai, Baomin Wang, Junying Bai, and Puguang Lu

School of Civil Engineering, Dalian University of Technology, Dalian, 116024, China
{ahmei}@dlut.edu.cn

Abstract. The main problems in the experiment teaching of professional course were analyzed in the paper. And the basic idea of effective teaching was elaborated. Three aspects meanings of "effectiveness" were analyzed detailed, namely, teaching efficiency, teaching effect and teaching benefits. The view that the effectiveness of experiment teaching should be investigated in the teaching effect and long-term teaching benefits was put forward. According to the characteristics of the professional experimental courses, observable characteristics of effective experiment teaching in teaching objective, teaching resource, teaching preparation, classroom model and evaluation system were indicated clearly.

Keywords: experiment teaching, effectiveness, characteristics.

1 Introduction

As an important part of the practice teaching, experiment courses can help students to improve their basic experiment skill, comprehensive quality and the awareness of innovation. Experiment courses in engineering courses provides lots of opportunities for college students to contact with practice of production and research, to learn how to solve the practical engineering problems, to prepare well for the research work or technical work later after they graduate from the university. However, for a long run, education in China pays more attention to theory and neglects the practices. Experiment courses are always set as accessory course, which leads to the lack of relative theoretical research on it, the weakness of teaching stuff, the obsolescence of the teaching contents and the unscientific evaluation [1]. In these years, in order to turn out interdisciplinary talents, many universities provide much more time and space for students' personally development by reducing the classroom teaching time, class hour and credit hour. Under this situation, to enrich the content of experiment teaching and improve the effect of experiment teaching by increasing the lab-class hour is hard to achieve because of many factors involved. Thus, revolution of experiment teaching in universities usually focused on the aspects of style, method, content and system of teaching. A significant topic in the revolution of experiment teaching is how to improve the efficiency of experiment teaching under the condition of existing class hour.

Y. Wang (Ed.): Education Management, Education Theory & Education Application, AISC 109, pp. 383–387.
springerlink.com

2   What Is the Effective Teaching

The pay by teachers does not equal to the quality of teaching, because individual professional quality, ability level and environmental factors will all influence the teaching effects. In the same way, studying quality of college students does not really rest with their efforts, but with the individual situations, the ability of understanding as well as creating and their acknowledgment to the course. Therefore, the teaching investment isn't equal to the quality of teaching. There exists a problem of the 'effectiveness' of teaching.

Effective teaching is a kind of educational ideas came from the scientific teaching movement that originated from the first half of 20th century. The term can be defined as follow: to reach the prospective teaching goal effectively by scientific teaching methods and ways, under the guidance of the scientific educational value [2].

From the definition above, three meanings are included in the 'effectiveness' of teaching: teaching efficiency, teaching effect and teaching benefits [3][4][5]. Teaching efficiency reflects the accomplishment of the teaching and studying by both the academic and students in the quantitative teaching movements; teaching effect is the directive result of teaching by examination or evaluation; teaching benefits are the expression of the contribution of teaching activities to the development of individual abilities, which are measured by the social requirements.

Attention should be paid here that 'effective teaching' can not be taken as 'teach or learn most by least time' simply. Teaching efficiency is just a method to improve the 'effectiveness' of teaching; in the same way, 'effectiveness' of teaching should not be judged by the score acquired by students because the teaching effects are short-term effects and the teaching benefits are the terminal target of the 'effectiveness' of teaching.

In modern teaching theory, the only index of the teaching benefits is whether the students get their improvements or developments. Therefore, whether the teaching is 'effective' or not is decided by the profit of students, which focus more on the long-term teaching benefits, namely, the improvement of their learn ability including the ability to solve the problems, ability to research and create and the ability to learn in all their lives, ie[6].

3   The Observable Characteristics of Effective Teaching in Professional Experimental Courses

The purpose of experiment teaching is to strengthen students' understanding to abstract theory knowledge, train students' applying ability of theory knowledge and to cultivate students' innovation and exploration spirit by self-operation or demonstration [7].The final start of experimental teaching is to promote students studying useful knowledge, developing useful ability and cultivating good quality of non-intelligence, which are benefits of effective experimental teaching. So judging a way of experimental teaching effective or not, the best method is to inspect the specific conduct of experiment teaching, and estimate if it has those characteristics that contribute to objectives mentioned.

Scholars hold different viewpoints on characteristics of effective teaching, because of the different investigation emphasis. Many viewpoints tend to grasp the essence feature of effective teaching from the whole teaching. For example, Qiuqian Song proposed the characteristics of effective teaching are as follow, teaching objective is based on student development, dialectical unity of preinstall and generation, teaching effective and more knowledge amount, ecological balance in teaching, teachers conducts are based on student development[8].Some viewpoints tend to actual teaching circumstances, such as teachers' conducts. For example, Hong Wei proposed that enhancing students' capacity by teaching, teaching expression should be clearly, teachers should have their own style and features, teaching should be seriously and extrude the emphasis, teachers should incentive students' initiative. According the regression analysis on teaching behavior characteristics to totality evaluate, result show that these features can explain teachers' teaching effect of 93.8%[9].

From feature of professional experiment course, the whole process of effective teaching should possess observable characteristics as following:

3.1 Teaching Objective of Multilevel and Continue Progressing

Teaching objective is the materialization of teaching aims, it is also the primary standard of judging if a way of teaching is effective. The final purpose of teaching of professional experiment course is to enhance students' capacity of practice and innovation. However, the existence of individual difference results in different quality of teaching, although the students received the same teaching, every student obtains different development[10].Therefore, combining different specialty, different student, teachers must formulate multilevel and continue progressing teaching objective.

3.2 Impeccable Instructional Resources for Experiment

Instructional resources includes two aspects, they are hardware and software. Hardware means experiment teaching place, teaching facilities, instruments and so on, hardware are significant objective guarantee for experiment teaching. Insufficient experiment teaching fund will result in laboratory cannot reach the demand, teaching facilities old-fashioned, limited instruments, enhancing the experimental quality will be empty talk. Software means quality of teachers, teaching materials for experiment course, courseware and so on. Teachers specialty level is a important criterion to judge the teaching is effective or not, high efficiency or not[11]. Without adequate and high quality teachers, effective experiment teaching is an empty talk; if teaching materials and courseware are not match the course, teaching effect will be influenced.

3.3 Full Preparations of Experiment Courses Beforehand

The preparations before experiment courses include the situation about both teacher's and students'. For the aspects of teaching and learning, having perfect preparations before class is the important guarantee to realize effective experiment teaching.

Teacher's preparation before class is full or not, which directly reflect the teacher's teaching attitude and ability. Modern science and technology have been developing rapidly, and the new technology appear constantly, the specialty experiment course should be in the dynamic development process accordingly. Thus the teacher should increase the amount of effective knowledge during class by enriching himself and updating the teaching contents constantly.

On the other hand, the students should master the experiment's principle, method and procedure by reviewing relevant professional knowledge and previewing experimental instruction, which is help to finish experiment efficiently under teacher's guidance and improve the learning effect.

3.4 Interactive Classroom Mode

In the traditional experiment courses, teacher always being as main body and adopting compulsive teaching method. Making the students in a step-by-step passive learning state, which limiting their ability of innovation, curiosity, thirst for knowledge and the development of creative thinking[12]. Actually, the outstanding feature of experiment should be practicalness, if the teacher spent a lot of time on teaching, there are little time can be used for practicing. Therefore, the professional experiment courses should adopt interaction as main form, teaching and leaning crossly. In the whole process, the teacher should be help students to resolve problems that being in the course of experiment, which strengthen the teaching quality effectively.

3.5 Scientific and Effective Evaluation System

The evaluation system includes two aspects that are both students' learning effect and teacher's teaching quality. Scientific and effective evaluation system is the important feature of professional experiment courses, which is avail to find and solve problem both in teaching and learning sides, so that improve academic record or improve teaching quality.

In all kind of university, the appraisal of experiment courses is commonly depend on the experiment report book and lack of the investigation of the process. This kind of performance appraisal way cannot fully and truly reflects the improvement degree of students' relevant knowledge, ability and quality, which make negative effect on the student's study enthusiasm.

Nowadays, the foundation of evaluation system about teacher is basis on the mode of theory course. Evaluation system have a poor operability and the feature of experiment unconspicuous, teacher's teaching quality is cannot have scientific and justice evaluation.

4 Conclusion

Professional experiment is not only the important channel for students to take pare in the engineering application, but also has great significance for cultivating the applied talents with innovative consciousness and high quality. But at present for the deviation of education concept and the restriction of the experiment resources, many national colleges tend to verify the basis knowledge and cultivate the basic skills instead of the innovation and overall quality. It does not suit the demand of the modern society. Basing on the effective teaching concept, professional experiment teaching must be done with a view to improvement of the students' abilities. It should be made to realize the maximization of the teaching efficiency and give full play to the unique functions of experimental teaching through the improving of teaching resources, teaching mode, the evaluation system and so like.

References

1. Guo, D.: Discussion on the Effectiveness of Experimental Teaching. The Theory Monthly 10, 111–112 (2002)
2. Cui, Y.: Effective Teaching: the Concept and Strategy. People's Education 6, 46–47 (2001) (in Chinese)
3. Cheng, H., Zhang, T.: Discussion on the Validity of the Teaching and the Improving Strategy. JIANGXI Educational Research 5, 34–36 (1998) (in Chinese)
4. Yao, L.: The Elementary Approach to the Implications of Effective Teaching. Modern University Education 5, 10–13 (2004) (in Chinese)
5. Mou, H.: The Summary of the Study of Effective Teaching's Implication and Features in China. Reading and Write Periodical 10, 36–37 (2008) (in Chinese)
6. Song, Q.: On Teaching Effectiveness. Curriculum, Teaching Material and Method 10, 25–29 (2004) (in Chinese)
7. Zhang, T., Fan, H.: Analysis on Experiment Teaching Purpose with Philosophy. China Modern Educational Equipment 8, 156–157 (2008) (in Chinese)
8. Song, Q.: The Implication and Features of Effective Teaching. Exploring Education Development 1A, 39–42 (2007) (in Chinese)
9. Wei, H., Shen, J.: Study on Teaching Characteristics of Effective University Teacher. Journal of Southwest China Normal University (Humanities and Social Sciences Edition) 5, 33–36 (2002) (in Chinese)
10. Ai, H., Wang, L.: Research on the New Teaching Model and Evaluation System for Teaching and Learning of Building Materials Course. Journal of Architectural Education in Institutions of Higher Learning 19(2), 81–85 (2010) (in Chinese)
11. Long, B., Cheng, X.: Conceptual Reconstruction and Theoretical Thinking of Effective Teaching. Journal of Educational Science of Hunan Normal University 7, 39–43 (2005) (in Chinese)
12. Xu, Y., Yin, Y., Bai, X.: Discussion on the Innovation Reform of Experiment Teaching in Building Materials Courses. Journal of Yellow River Conservancy Technical Institute 1, 74–75 (2004) (in Chinese)

Optimal Design of High-Radix Router's Switching Fabrics Based on Tile

Xian-Wen Wu[1], An-Hua Mo[2], and Li-Quan Xiao[3]

[1] Hunan Vocational and Technical College of Railway, Zhuzhou Hu'nan 412001
[2] Army Representative Office of Dengzhou District, He'nan 474000
[3] School of Computer Science, National University of Defense Technology, Changsha 410073
wxw_4221xh@126.com

Abstract. In this paper, we first analyze the applications of low-radix router, put forward the need and possibility to design high-radix router, summarize advantages and disadvantages of various switching fabrics on the basis of low-radix router. Then in-depth we study the router's internal traffic of interconnection network which uses fat-tree topology. A high-radix switching chip is implemented by a comparison on different port allocation policies of a 32-port switching chip, and that tile is appropriate to be used in the high-radix router design is pointed out.

Keywords: High-radix router, Tile, Switching fabrics, Network structure.

1 Introduction

The traditional parallel communication is widely used in high-performance computers by the advantages of technology matured, interface achieve easily, high bandwidth, low delay and low cost. However, its further performance enhancement will cost expensive due to wiring difficulty, clock skew, power consume, large area, limited distance [1]. Therefore, further improvement on the parallel communication bandwidth is facing "electronic" bottleneck. Serial communication is one of the key technologies to overcome these problems. It solves the problem of clock skew, high bandwidth, long distance, high interconnection density, anti-interference ability and so on. Moreover, the link bandwidth of serial communication improves more easily, because the bandwidth of a single serial communication can range from 2.5Gbps to 10Gbps. Furthermore, InfiniBand technology which supports the use of DDR on a single serial link, supports 4×, 8× and several other serial links used in parallel, to further improve the bandwidth based on Coarse Wave Division Multiplexing (CWDM) and Dense Wavelength Division Multiplexing (DWDM) etc.

  With the continuous development of the serial transmission technology, high-performance routing switches in the interconnection network play an important role, for example, the Cray designed the YARC chip with 64-port×3-link [2], Myrinet designed 32-port×4-link [3] switch chip, Mallanox designed the InfiniBand switch chip [4] with 24-port×4-link, SGI company is designing 16-port×12-link chip [5].

Y. Wang (Ed.): Education Management, Education Theory & Education Application, AISC 109, pp. 389–395.

Through research and analysis, this paper describes an optimal design of a 32-port chip of high-radix router.

2 The Design Basis of High-Radix Router

With the maturity of the serial communication technology and the progress of ASIC design process, the time to design high-radix router has come. The basis of designing high-radix can be described as below First, the use of serial transmission technology can greatly reduce the number of pins in switch chip, thus it is possible to integrate more interconnect ports on a chip. One disadvantage of the serial communication technology is long delay. To overcome this shortcoming, we can reduce the diameter and delay of the interconnection network, for example, measures such as high-radix router can be taken. Second. according to the derivation of William J. Dally and John Kim [6], the relationship among optimal router radix(k), the total bandwidth of a router(B), the delay of a packet's header traverse a router(t_r), the number of interconnection network nodes(N), packet length(L) can be expressed by expression (1).

$$klog^2k = Bt_rlogN) \ /L \tag{1}$$

Despite of the decreasing of t_r as technology development, in general, $(Bt_rlogN)/L$ continued to show a growing trends, so the radix increasing over time remains. Third, with the upgrading bandwidth of a single pin in serial communication, higher bandwidth of port can be achieved with fewer pins. Lastly, the enhancement of ASIC design and production process make high-radix router realization possible. In theory, with the number of ASIC chip logic unit grows in the speed of linear or square, and ASIC design and production technology level from 1.3um to 65nm to 45nm is on the rise, so the density of logic units within a single ASIC chip increases rapidly. Therefore, the chip logic resources will no longer be the limiting factor in the design of high-radix router switching fabrics.

3 Applications of Low-Radix Router

Low-level router (low-radix router) has been widely used in the traditional parallel interconnection networks. Its transmission medium is electric, with a number of multi-pin parallel ports, which may cause a lot of signal loss.

Crossbar switch routers usually use low-radix fabric in which the input cells are exchanged to the corresponding output-ports through a centralized controller. Centralized controller first claims a path according to the head of the destination address of a cell and the routing algorithm, then determines whether the path is idle or not. When it is idle, the cell is sent to the corresponding output port. The specific application is shown in table 1.

Table 1. The specific application

Switching fabric	Abbreviation	Shortcomings	Improvements
Switching fabric with input buffer	IQ	head blocking, cell throughput only 58.6%[12]	VQ can solve the head blocking, appropriate algorithms can achieve 100% data throughput, but when N increased the delay became larger, and the exchange rate is constraint, which becomes not practical
Switching fabric with output buffer	OQ	Concentrated in the packet destination address, request an infinite buffer queue length, packet loss can not meet	Requires N times acceleration within the chip, under the condition of large scale and high wire speed, the acceleration becomes impractical[7]
Switching fabric with input and output buffer	CIOQ	The delay of a single packet not be considered	Behavior to match, the same input model is applied to OQ and CIOQ

As the development of serial technology, the bandwidth of a single pin in chip ports has been greatly improved. By the condition of the same port number, the router bandwidth may increase significantly. Conversely, if the total bandwidth keep unchanged, the total pin number may reduce, so that the available chip signal can be increased. The specific circumstance is shown in table 2.

Table 2. The corresponding table with the relationship between technology and the number of ports

Technology	Requirement	Port number	Pin number	Remarks
Parallel interconnect link	32bit, 400MHz DDR link, Bidirectional bandwidth	8 bidirectional port	(32bit data+1bit clock)×2(LVDS) ×2=132bit 132×8=1056	Further add port number is very difficult
Serial link	6.25Gbps×8Lanes	8 bidirectional port	40	The same pin scale, port number can surpass 24

Analyzed from Table 2, if the message packet size remains unchanged, the parallel port can increase the number of transmission links, so network will develop from traditional links with low number and high bandwidth[2] to high number and low-bandwidth links. This made foundation for the use of high-radix router with multi-port and low-bandwidth.

4 The Feature of Tile

The structure of tile and the relationship between its input and output are very neat, simple, and implemented easily. Extension of switches based on tile can be easily realized by the copy of tile. A single tile structure consists of input buffer (IB), the routing computer unit(router), the line bus(row bus), the line buffer(RB), 8×8 sub-switch, the column buffer(CB), the column channel(CC), the output multiplexer, which is shown as Figure 1.

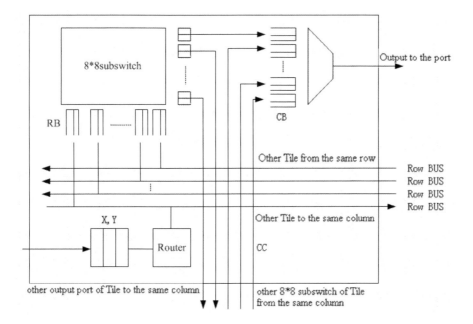

Fig. 1. The basic structure of tile

The YARC router used in BlackWidow multiprocessor system with 64×64 switch chip is made up of an 8×8 array of tiles where each tile contains logic and buffer needed by input and output port.

As the increase of router radix(k), if a centralized approach is still used to allocate the exchanging resource, the connection complexity, chip area, exchange delays will become unacceptable. A conclusion can be drawn from Fig.1, a tile is a single unit, when there is a demand expansion of units, it only need to make a copy of a separate tile because of its high symmetry and easy expansion which simplify the wiring within chip.

5 Optimal Design of High-Radix Router Based on Tile

In development process of high-radix router, YARC tile was first proposed and adopted in the design of 64-port high-radix router. On this basis, the paper will optimize the design of the implementation of tile switching fabric. In order to reduce the complexity, only a high-radix 32-port switch fabric is described in this paper.

5.1 Design Principle of the Tile Structure

In addition to the general characteristics, we mainly consider two questions in tile structure design. The first one is the number of ports in switch chip, which determines how to choose the number of Columns and Rows. On the other hand, the ports, column and row in tile structure are all identified, so the allocating policy of column output ports also should be considered. The following design principles are required to think when design row and column. First, the size of a single sub-crossbar can neither be too small, nor be too big, usually controlled under radix 8. Second, the number of sub-crossbars should be controlled so as to reduce the difficulty of wiring in chip. As the number get larger, sub-crossbars will become smaller, result in wiring difficult. Third, in the switch fabrics based on tile, buffers are allocated at each I/O port, and buffer number is equal to port number×(R+C), so the design of crossbar is related to the buffer resources. For example, designing of a 32-port switch needs 32 crossbars with 384 buffers and 8 rows and 4 columns when the 4×8 crossbar structure is adopted. In contrast, if using 8×8crossbar structure, only 4 rows, 4 columns, 16 crossbars, 256 buffers are needed. The latter's complexity is smaller and use fewer buffer resources.

5.2 Flow Strategic Design

Topologies such as Mesh [8], Torus [9], Fat-tree [10], Clos [11] are usually adopted in the design of high performance interconnection networks. There are many advantages to use Fat-tree according to the results of design and experience. The diameter of a Fat-tree is small, and its bandwidth is half-bandwidth and end nodes are identical, there will not be net bottleneck and provide a wealth of redundant paths. Therefore, design topologies using Fat-tree can solve the traffic imbalance problem. The mapping between input ports and output ports will affect the distribution of the exchange structure, the actual flow situation of 32 switch chip shown in Table 3.

Table 3. The actual flow situation of 32 switch chip

Input port	Efficient output port
0,1,2,3	[0,1…..15]16,17,18,19
4,5,6,7	[0,1…..15]20,21,22,23
8,9,10,11	[0,1…..15]24,25,26,27
12,13,14,15	[0,1…..15]28,29,30,31
16,20,24,28	0,1,2,3
17,21,25,29	4,5,6,7
18,22,26,30	8,9,10,11
19,23,27,31	12,13,14,15

In the crossbar design process, all relationships should be designed regardless of full flow design not network nodes. According to the optimal routing strategy of Fat-tree, the ports utilization will become lower as router radix get bigger. When the radix tends to infinity, we can only obtain 25% utilization rate. Therefore, it is necessary to improve the efficiency of the exchange structure.

We firstly analyze a continuous distribution of the ports and part of a continuous distribution manner used by YARC. The specific analysis is shown in Table 4.

Table 4. The relationship between port utilization and manner of port allocation

Manner	Samples				utilization
	Row	Input port	column	output port	100%
Continuous allocation	R0	0,1,2,3,4,5,6,7	C0	0,1,2,3,4,5,6,7	50% ,
	R1	8,9,10,11,12,13,14,15	C1	8,9,10,11,12,13,14,15	25% ,
	R2	16,17,18,19,20,21,22,23	C2	16,17,18,19,20,21,22,23	0,not
	R3	24,25,26,27,28,29,30,31	C3	24,25,26,27,28,29,30,31	balance
Part of continuous allocation of YARC	row	Input port	column	output port	
	R0	0,1,2,3,4,5,6,7	C0	0,4,8,12,16,20,24,28	62.5% ,
	R1	8,9,10,11,12,13,14,15	C1	1,5,9,13,17,21,25,29	12.5%, not
	R2	16,17,18,19,20,21,22,23	C2	2,6,10,14,18,22,26,30	balance
	R3	24,25,26,27,28,29,30,31	C3	3,7,11,15,19,23,27,31	

We can see from the table, neither part of a continuous distribution of YARC nor the continuous distribution can reach a balanced distribution of all the sub-crossbar traffic. Parts of the flow are very large and busy, while the other parts are idle most of time. It can be further optimized.

On the basis of the existing, the paper explores the discrete distribution manner on row and column, ports assignment is shown in Table 5.

Table 5. The discrete distribution manner on row and column

Row	Input port	column	Output port
R0	0,4,8,12,16,20,24,28	C0	0,4,8,12,16,20,24,28
R1	1,5,9,13,17,21,25,29	C1	1,5,9,13,17,21,25,29
R2	2,6,10,14,18,22,26,30	C2	2,6,10,14,18,22,26,30
R3	3,7,11,15,19,23,27,31	C3	3,7,11,15,19,23,27,31

Rows and columns of Table 5 arranged in combination may constitute 16 crossbar. Take any sub-crossbar to analyze, such as the output port (1,5,9,13,17,21,25,29) and the input port (2,6,10,14,18,22,26,30), the relationship of crossbar can be drawn based on actual traffic flow is shown in Figure 2. The rest are to follow the flow relationship, so the actual flow can be distributed evenly to all the crossbars, and achieve efficient use of resources.

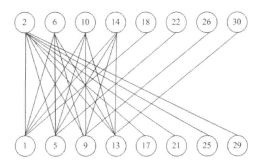

Fig. 2. Crossbar flow relationship

6   Conclusion

In this paper, we first analyze the feature and technology of low-radix router, then in depth study the influence from the router's internal traffic under optimal routing in fat-tree network, and make a comparison on different port allocation policies of a 32-port switch chip. Finally, we proposed an optimal design that can balance the internal traffic and improve switching performance of the chip.

References

1. Lee, K., Lee, S.-J., Yoo, H.-J.: SILENT Serialized Low Energy Transmission Coding for On-Chip Interconnection Networks. In: ICCAD, pp. 448–451 (2004)
2. Scott, S., Abts, D., Kim, J., Dally, W.J.: The Black Widow High-Radix Clos Network. In: Proceedings of the 33rd International Symposium on Computer Architecture, pp. 16–28 (2006)
3. Boden, N.J., et al.: Myrinet A Gigabit-per-Sencond Local Area Network. IEEE Micro 15, 29–36 (1995)
4. InfiniBand Association. InfiniBand Architecture Specification Release 1.0 (2001)
5. http://www.sgi.com
6. Kim, J., Dally, W.J., Towlesl, B., Gupta, A.K.: Microarchitecture of a High-Radix Router. ACM SIGARCH Computer Architecture News 33, 420–431 (2005)
7. Leonardi, E., Mellia, M., Neri, F., Marsan, M.A.: On the Stability of Input-Queued Switches with Speed-up. IEEE/ACM Transactions on Networking 9, 104–118 (2001)
8. Adve, V.S., Vernon, M.K.: Performance Analysis of Mesh Interconnection Networks with Deterministic Routing. IEEE Transactions on Parallel and Distributed System-s 5, 225–246 (1994)
9. Alcover, Lopez, R., Duato, P., et al.: Interconnection network design: a statistical analysis of interactions between factors. Parallel and Distributed Processing, 211–218 (1996)
10. Leiserson, C.E.: Fat-Trees: Universal Networks for Hardware-Efficient Supercomputing. IEEE Transactions on Computers 34, 892–901 (1985)
11. Clos, C.: A Study of Non-Blocking Switching Networks. The Bell System Technical Jounal 32(2), 416–424 (1953)
12. Zheng, Y.: Queuing theory and its applications in modern communications. People's Posts & Telecom Press, Beijing (2007)

Forecast of Employment Situation and Countermeasure Research on Twelfth "Five-Year Plan" Period in Hebei Province of China

Xiaomei Shang[1], Zheng Li[1], Shituan Zhou[2], and Guilan Song[3]

[1] School of Electrical Engineering and Information Science,
Hebei University of Science and Technology, Shijiazhuang 050018, China
[2] Political Department of People's Liberation Army No. 256 Hospital,
Shijiazhuang 050800, China
[3] Department of Personnel, Hebei University of Economics and Business,
Shijiazhuang 050061, China
lzhfgd@163.com

Abstract. Hebei province is with a large amount of people and the labor resource is very rich, and the unemployment by corporations' reforming and adjusting economy frame work is intertwined with farmers without arable land, resulting in serious providing and demanding contradiction. In this paper, prediction and analysis about the tendency of labor force supply and demand in Hebei Province during the twelfth "Five-Year Plan" period are implemented. Some relevant countermeasures and suggestions are presented to improve the employment situation.

Keywords: Employment, Situation, Forecast, Countermeasure.

1 Introduction

The important measures of our country's the twelfth "Five-Year Plan" is to realize GDP further improve in the socialist market economy system; At the same time not only emphasized "rich", but also "enriching people"; pursue this is a new China 60 years later, then on starting to achieve the comprehensive construction well-off society grand goal to be an important meaning. The central further the expansion of domestic demand, promoting economic growth point of the major initiatives making Bohai Sea area rapid rise in the region has become a new economic growth point in Hebei province, and the geographical advantage of Hebei is speeding up converting into the development advantage will form a new talent areas [1, 2]. In the 12th "Five-year Plan" period, Hebei province will remain in the important period of strategic opportunities. Also is in the key years of rapid economic development and economic reform of further deepening. The opportunity is unprecedented, also is the challenges. As the nation's sixth populous province, during accomplishing these tasks, Hebei province will meet one issue cannot be got around, that the huge population of labor in employment problems. How to solve this problem is directly related to social stability and the effective utilization of labor resources, which contain the speed and

Y. Wang (Ed.): Education Management, Education Theory & Education Application, AISC 109, pp. 397–404.
springerlink.com   © Springer-Verlag Berlin Heidelberg 2011

process on the reform and opening up and economic construction. Employment problems at present and in the future in our province, is one of the most important problems in our social economic life during quite a long period.

2 Prediction of Contradiction on Labor Supply and Demands in the Future Five Years

2.1 Forecasts of Labor Supply

Too large a population and a low quality of people is the biggest difficulty which faced us during our social and economic development. Since the overall planning, the population growth rate in Hebei has dropped significantly, especially after 1990, the interval time of every 10 million population's growth in Hebei province is obvious, the wide and fast population growth momentum were controlled effectively. But as the big population base, too high a proportion of child-bearing age, the role of inertia of the population growth, on March 31, 2009, total population reached 70 million in Hebei province. In established population cases, there are three aspects which lead to the basic factors affect labor supply: labor participation, transfer of agricultural labor to non-agricultural industries rich and towns' release of the rich people.

1) Labor participation of new working age population
Labor law stipulates that all 16 May participate in social work. So we could make 16 years the criterion to define new working age population. According to Hebei bureau on population and social workers of historical data, new working age population in the labor participation in 1978 ~ 1990 has been wander up and down 65% .After 1996 it has remained at 90%, compared the "eleventh five-year plan" period labor participation rate, it is a little lower. This is because, on the one hand, in recent years, due to higher education's transfer from elite education to mass education, the students in school increased and respected decreased the people's participation in economic activities [3-5]. On the other hand, because some of the lost workers can not find work and then become "passive unemployed man" in a long time, or part of the employee who not satisfied at their jobs and have other economic sources, and gradually lose desire to find work, at last become "the conceptual unemployed man" and go out from economic activity population.

According to the trend of the decline of the labor participation, We could assume that in the next 5 years, the labor participation rate of the new working age population were 85%, 87.5% and 90%. The forecast results see table 1.

In the three prediction table 1 listed indicated that, during 2011 ~ 2015, new labor supply in our province would remain at 180-about 1.9 million, with an average new to nearly 400000 annually, according to the "2009 statistics yearbook" relevant data calculation in Hebei province. It can be seen that in the next five years, new labor employment pressure is still relatively large.

2) The shift of agriculture rich labor to non-agricultural industry
As a big agricultural province, in the process of industrialization, our province accompanied by transfer from agricultural labor to non-agricultural industries, this is

Table 1. Forecasts of 2011 ~2015 new labor supply (unit: ten thousand persons)

Year	Labor participation rate %		
	85.0 %	87.5 %	90.0 %
2011	41.65	42.875	44.1
2012	39.95	41.125	42.3
2013	34.85	35.875	36.9
2014	37.4	38.5	39.6
2015	24.75	25.88	27.0
Total	179	185	190

not rebel against the objective law. It is a kind of the labor force resources' own optimization during the employment in the structure.

According to 2009 statistics yearbook in Hebei province, in 2008 per capita GDP in Hebei province is $1537.5, according to the library where the "standard of a structure" contrast methods' measure, when GDP per capita more than $1000,then the first industry employment labor force is 15.9%, but in 2008 the first industry employment population in Hebei province of the provincial total employment population is 39.8%, and library times where compared to the "standard", the first industry structure of labor obtain employment is 23.9% much more [6]. When according to the total computation employees in Hebei province in 2008 and don't consider the new increase of the labor force, we should make the first industry labor force in 5.9238 million, and should make sure the labor force transfer of about 8.8899 million , the year average transfer to nearly 1.8 million people.

3) The release of surplus employees in town

The town in surplus employees in our province is mainly refers to the unit laid-off workers from public ownership of enterprise or business. "Laid-off workers" in China is a new word since the reform and opening up, employees that laid off in the market economy, market competition, are the inevitable result of the adjustment of industrial structure. In the town of labor employment, the whole province in 2001 applied the unified the unemployment registers card system, but for varies of reasons, unemployment that registered accounts only 75% of the actual unemployment, therefore, it is a need modify the actual unemployed amount. According to social and the labor safeguard in Hebei province, the data released by the statistics are as shown in table 2 [7-8].

As is shown in the picture, the layoffs in our province are rising annually. In the next five years, as the financial crisis of the impact will not be stop in a short period, the economy in Hebei province would be suffer a serious case. Some enterprises suffered in the mass in the financial crisis, many enterprises have to reduce costs through the economic crisis, leading to the rising number of laid-off workers and it will continue to increase. Moreover, with the speed up of the technology advances, and enterprise capital organic component will also be constantly improved, these may give rise to new surplus person. They will go to social labor market, forming a new labor supply. At the base of the unemployment person in town during 2003-2008 is 286000 in average, if over the next five years that unemployment is for 300000 in average a year, the actual unemployment after correction for that is about 400000.

Table 2. The statistics of unemployed in town during 2003~2008

Year	Registered unemployed person	Registered unemployed rate	Actual unemployed person (ten thousand people) (%) (the modified data /ten thousand)
2003	25.7	3.9	34.27
2004	28	4.0	37.33
2005	27.8	3.9	37.07
2006	29.0	3.8	38.67
2007	29.3	3.8	39.07
2008	32.2	4.0	42.93

4) Big number increase of college graduates

Since 2003, the province in high expansion in the first university graduate after enrolment, the employment model of university graduate will have major changes, new employment pattern is about to be formed, with enrolment expansion continues to expand, the number of graduates of a successive incremental trend at present.

Table 3. The increased university undergraduate students

Year	2005	2006	2007	2008	2009	2010	2011
Register number (ten thousand)	47.36	55.76	56.18	57.48	55.9	50.3	48.5
Recruit number (ten thousand)	28.5	27.30	29.89	31.1	33.34	36.92	37.34
The acceptance rate (%)	60.18	48.96	53.2	59.6	73.40	76.99	54

From the table can see, the enrolment has increased every year during 2005-2008, and 2008 was a turning point, after that the enrolment is reduced year by year. Even though the enrolment is reduced after 2008, the plan to recruit students is still in increase, and slightly more, so made the acceptance rate 70% in 2010 and 2011. In the next five years the number of college entrance examination in Hebei province will drop slightly, but wise plan will have increased slowly and therefore our province will improve the acceptance rate of the university entrance. From the table to predict from the annual recruit students number during 2007~2011, the next five years the number of graduates of Hebei province would be in an average of 330000, according to the national average employment rate 70% a year, there will be 99000 people cannot normal employed, these people will inevitably come to join the team in employment, as a social personnel to join the ranks of employment, this increased the employment pressure of society.

The huge pressure of employment: there are 1.8 million shifted surplus labor force in an average year in agricultural fields; The town will set 400000 surplus staff and workers, and 400000 or so new Labour⬜ the 99000 graduates who can not find work entering the labour force. Don't consider the town's population, the public registration unemployment combined the three above every year close to 2.5 million people who need to obtain employment, so the heavy task you can imagine.

2.2 Forecasts of Labor Demand

Influence factors of labor demand is nothing more than two kinds: one kind is economic growth factors; another kind is economic structural change factors.

1) The influence of Economic growth factors on the demand for labor
Economic growth factors, strictly speaking are point to the growth of economic output. The main measure is gross domestic product, or GDP growth rate. In recent years, domestic and foreign complex economic situation has spread to many industry development and to talented person's demand, formed the demand on the talents going soft, thus appeared the growth of GDP , high input on capital is not always bring higher employment growth, even the drops of employment rate, also won't automatically into employment opportunities' expand. Job growth also depends on elastic index of the economic growth, namely employment GDP growth of 1% each, creating jobs in the percentage of growth.

From the 1990s GDP growth of 1% in our province before every job growth above 0.3%, after 1991 years it has been falling, in recent years, it has dropped to 0.1% to the left. In our province the cause caused GDP growth employment elastic there are two points: one is in the economic development process of labor productivity production decreased the number of labor unit of GDP; another is the original of the employment of the surplus workers are laid off again that they obtain employment labor post which created by GDP growth.

The next five years in GDP growth is to be 7%, 8% or 9%, the employment elastic index were to be 0.11, 0.13 and 0.14, according to the labor force in 2009 Hebei statistics yearbook, the employees has a 37.2566 million calculation in 2008. In the next five years, it is necessary to make the following measuring labor requirements (see table 2). According to the forecast, in our province, the next five years on average every year the amount of new labor demand will be between 286900 ~ 469400. Even if labor demand to achieve a high prediction; the labor supply will also be greatly more than the labor demand scale.

Table 4. Forecasting during 2011 ~ 2015 of labor demand (unit: ten thousand)

Employment elasticity coefficient	GDP growth		
	7 %	8 %	9 %
0.11	28.96	32.79	36.88
0.13	33.90	38.75	42.59
0.14	36.51	41.43	46.94

2) The influence on demand factors by economic structure changes

Through the analysis about the 2009 Hebei economic yearbook data, for nearly 20 years to three big industry in GDP respectively in the proportion of the workforce in three big industry and the distribution situation to carry on the analysis, the first industry employment of 1978 years for 16.2161 million, by 1999 to 16.5325 million, an increase of 316400 people in the national economy, the proportion from 76.88% in 1978 to 49.76% in 1999, down by 27%; From 2000 to the end of 2008, the first industry professionals number increased from 33.8571 million to 37.2566 million, increased by 3.3995 million, the proportion in the national economy has fallen from 49.56% to 39.76%, a decrease by 10. From the data above, it is easy to find that although the first industry production value of its absolute value is increased, it in its GDP its proportion was declined year by year. The declining of the proportion of the labor force in the first industry indicated that the first industry employment basic saturation was unable to absorb redundant labor force.

From analysis of the second industry data, from 1978 to 1986 the number of employees risen from 2.9285 million up to 6.0742 million, both have a big development in the national economy, from 13.88% up to 23.13%, risen about 3.1457 million and nine percent respectively, both have large development; from 1987 to 1992, between 1993 and 2001 the proportion in the national economy is 23% and 27% respectively in nearby concussion, slightly up small first again came and slightly rose slightly born. After 2001 and later to 2008 increased. The second industry has a great contribution to the economic and the increase on people's income in our province, but with the constant deepening of reform, the readjustment of industrial structure, the labor absorption capacity of the second industry will gradually abate.

From the third industry structure data we can see, from 1978 to 2008 by leaps and bounds, all the employees from 1.9495 million to 10.7423 million or nearly six times, proportion in the national economy from 9.254% to 28.83%, increased 20%. This shows that no matter from employment or proportion, the third industry presents a rapid growth, explaining to the third industry contribution to employment, and obviously has a huge potential for development. Even so, the proportion of the third industry in our province also meet very far from the "Eleventh Five-Year" plan industrial upgrading adjusted target 35%, if press to realize this aim, the next five years society will provide around 400000 jobs every year.

In all, all the analysis above shows that, in the future five years the labor demand every year in our province would be about 900000, compared to 2.5 million, the employment situation would be very serious, and the employment pressure will be exist for a long time.

3 Countermeasures and Suggestions

During "Five-Year Plan" Period, our province in the field is still had to face the employment is a serious labor surplus supply of conflict. The basic form of the employment of decided to our goal is not to destroy the unemployment, but as far as possible will moderate unemployment in unemployment control of the area. To realize the goal of moderate unemployment should be mainly to take the following measures:

3.1 Keeping a Rapid Economic Growth

To solve the employment problem, fundamentally, depends on the size of the economy and the speed of the growth. Only the scale of the economy continues to expand, can the society continues to create new jobs. Restrict factors of economic growth in our province, recently most due to the weak market demand, for a long-term main due to the insufficient of the technical innovation ability. Therefore, to keep a high economic growth rate , we should make great efforts by creating effective demand and improving the technical innovation ability.

3.2 Make Full Development of the Third Industry; Speed Up Construction of All over the City, Especially Small and Medium-Sized Town

Make full development of the third industry, speed up construction of all over the city, especially small and medium-sized town, and expand the surplus labor force employment space. Arrange the laid-off workers positively and the shifts of surplus rural labor force to the third industry, and give some policy support. Since the reform and opening up, although our province has a quick development in some businesses which close to people's life such as transport food service industry. But high quality services, such as finance, economic information and technical consultation, and legal, accounting, all the industry and quality of these businesses needs to be expanded and improved.

3.3 Ensure the Full Employment of Graduates

The graduates with higher education are a group of intellectuals with knowledge, cultural. As long as more guidance, the maximum employment of the graduates can be made sure. Strengthen the education of employment of graduates, make sure that education graduates can fully aware of the current employment of the severity of the situation, should make more efforts to adapt to the development of the society, and constantly looking for opportunity in this industry, and understand the needs of the society at present and its trend of development. Proficient in an skills or expertise, making themselves as enterprise need talent during the university, all these will increase the employment of the core competitive power. For example, the software engineering theory is very wide, the employment market is big, and the social demand is also the biggest of all kinds of professional. General requirements for computer, and information related professionals, have years of relevant experience in software development. At present, the society about JAVA/J2EE, VC+ +, net, database such these technologies' require is in dominant. In addition, familiar with Windows, Linux kernel development application platform and development on software, all these would be a must. For software developers, familiar with data structure, algorithm design and software engineering idea is very necessary. And hardware positions need experience, and the more abundant experience the easier to find the right job.

3.4 Establishing a Rural Surplus Labor Force Is Orderly Circulation Mechanism, to Realize the Agricultural Industrialization and the Scale

About the problem of the move out of labor force of agriculture, establishing a rural surplus labor force is orderly circulation mechanism, to realize the agricultural industrialization and the scale. To give full play of the role of rural labor repertory, and control the scale and speed of shift of the labor force. We should make big efforts to develop modern agriculture, strengthen the construction of new rural and the reconstruction project in urban village, make full use of the rural labor reservoir,　use relevant policy to control the rural workforce flow to the employment in the cities with big pressure.

4 Conclusions

The predictions on labor supply and employment are meaningful and help to make scientific policies and future judgement on population and economic situations. Based on the static datum and analysis, the countermeasures and suggestions for the labor problem are presented with detail. As a province with large mount of people, Hebei province must make a full and proper application on the labor sources to implement coordinate harmonic development on economic and society. It is expected that this research can be helpful to the reforms and policies on the development of Hebei province of China.

Acknowledgments. This work is supported by the Planned Project of Department of Science and Technology of Hebei Province and the Teaching Research Project of Polytechnic College of Hebei University of Science and Technology.

References

1. Li, J.: Reports on Human Development in Hebei Province. Hebei People's Publishing house, Shijiazhuang (2007)
2. People's Government of Hebei Province. Hebei Economic Yearbook. China Statistics Press, Beijing (2008)
3. Wang, N.: Mathematical Modeling of our country urban employment. Logistics and Management 12, 117–119 (2009)
4. Zhang, L., Zhang, D.: The Comparation and Study of Forecast Model with Obtaining Employment in Sichuan. Journal of Shangqiu Teachers College 9, 49–51 (2008)
5. Liu, H.: Employment Situation in China the Next Decade and Some Measures. Nankai Economics Studies 4, 41–47 (2000)
6. Yong, S.: Middle and Long Terms of Economic and Social Development and Human Resources Forecast Analysis in Hebei province. Economic Forum 18, 62–63 (2009)
7. Wang, J., Lin, L.: Analysis on Labor Force Participation Rate and Labor Supply in the Future in China. Population Journal 4, 19–24 (2006)
8. Xue, F.: Prediction and Analysis on Labor Force Employment Situation in China. Population Journal 2, 53–55 (2008)

Analysis of Employment Situation of Shandong Province Based on Factor Analysis

Ma Jun[1] and Han Xin[2]

[1] MPA Center, Shandong University of Finance, Jinan, Shandong Province, China
jamesma007@sina.com
[2] Postgraduate student of Shandong Universtity of Finance Jinan,
Shandong Province, China
hanxin2o1o@163.com

Abstract. Based on the analysis of the level of Shandong's economic development, in this thesis, the factors that influence the employment are divided into six aspects such as the level of social and economic development, laborer quality, the level of urban and rural development, living standards of laborers, working environment and so on, having 30 indexes in total. Three major employment impact factors— social and economic development factor, industrial structure factor and laborer quality factor — are obtained from the factor analysis of the 30 indexes. And a linear regression model of employment impact factors to the employment rate of Shandong Province is set up by taking advantage of data of Shandong statistical yearbooks between 1978 and 2008.

Keywords: Factor analysis, employment impact factor, linear regression.

1   Introduction

Today, the growth of GDP is no more regard as the only method to judge the national economic level as well as the regional economic condition. All the aspects shall be taken into consideration for measuring the economic level, and guarantee the GDP growth and ensure the quality of whole macro-economy in the meanwhile. As one of core indexes of macro-economy, employment is always the social focus, and is also one of the most important factors in national economy.

With the further advance of state-owned economic system reform, adjustment of industrial structure, urbanization speed and the shift of surplus rural labor force, the severe situation of employment in China becomes more outstanding. The employment problem raises high concerns of government departments and numerous scholars, and they discuss the existing forms, influence and solutions of unemployment in China from different visual angles. Professor Shi Jiwei (People's Publishing House, 2006) discusses the concept of full employment, features of full employment policy and policy choice; Professor Li Xiaomei describes the major factors that influence the laborer employment in details from four aspects such as the economic growth slowing down, adjustment of industrial structure and technological innovation, laborer quality barrier and barrier of laborer market system; Professor Xu Zhangyong (Hubei Social Sciences, 2007) summarizes the profound reasons for the enormous employment

Y. Wang (Ed.): Education Management, Education Theory & Education Application, AISC 109, pp. 405–417.

pressure in China at present stage; Professor Gao Jie (People's Publishing House, 2007) makes an empirical analysis on the basis of theoretical discussion of the relationship between government's R&D investment and employment with the sample data; Professor Jiang Yuefeng (2002) from Northwest Normal University studies the quantitative relation between economic development and employment increase; Professor Zheng Haiyan (Economy and Management, 2003) studies on the basis of the indexes such as the growth rate of gross national product (GNP), growth rate of national consumption, growth rate of staff salaries, laborer population index, etc., through the regression model analysis, and comes to the conclusions that the growth rates of GNP and national consumption are in inverse proportion to the unemployment rate, and growth rate of staff salaries is in direct proportion to the unemployment rate; Professor Wang Chunlei (Research on Financial and Economic Issues, 2007) from Dongbei University of Finance and Economics brings up the following tax policies such as decreasing the macro tax burden to improve the economic growth, adjusting tax policy to improve the labor-intensive industries development, and strengthening preferential tax policy to improve the tertiary industry development, improve the development of medium and small-sized enterprises and the non- public sectors of the economy and establish urbanization strategies, so as to promote employment. [1] Professor Li Yining [2] analyzes the systematic and structural reasons for the employment pressure in China and brings up countermeasures in the financial crisis.

There are many achievements on the study of the prediction of employment amount (i.e. quantity of employment) in the country. Commonly two methods are employed: the first method is to calculate employment elasticity in the direct method and integrate with qualitative analysis to give out the rough estimated value of employment elasticity thus to predict the quantity of employment (Guo Qingsong, 2004; Yuan Zhigang, 1998). This method is easier to operate, but more subjective factors are added into the estimation on the employment elasticity, thus prediction on the quantity of employment becomes inexact. The second method is to set up regression model for prediction; it is common to set up a single - element regression model between the quantity of employment and GDP (Yu Zhangsheng, 2001; Wu Xiaoping, 2000; Jin Xiaona, etc., 2003); some scholars also add other impact factors such as capital, time trend, etc., and this predicting method seizes the principal contradiction and corresponds to the economic significance of employment elasticity. In addition to the two common methods above, there are other predicting methods such as setting up random effect model with the panel data to predict the demand elasticity and employment demand (Zhang Chewei, 2005) and predicting the quantity of employment by improving neural network model (Su Bianping, etc., 2006). Each of the methods above has its merits, but there is still room for improvement.

Since reform and opening up, Shandong's economy has made remarkable progress. From 1978 to 2009, the local growth rate of link relative ratio is over 10% every year; especially after entering the 21st century, the growth rate of Shandong's local gross product is over 15%. However, in this overall environment of economy rapid developing, Shandong's employment rate still fluctuates at 2%, and even negative growth in some years.*

In recent years, Shandong Province regards enlarging employment as a major goal of economic and social development in the whole province, and it makes efforts to improve the pulling ability of the economic growth to the employment. Now a new situation of calling for the worker to find a job on his own, enterprises able to hire employees independently, the government to promote job creation, two-way selection between employees and employers and market competition in employment has been created. From the analysis on the data of statistical yearbooks during "Eleventh Five-Year Plan" period, the economic growth speed of Shandong is constantly quickened and growth speed of quantity of employment is quickly increased, but the employment rate increases slowly. There are two major reasons: (1) the laborer supply increases at a high-speed, but the demand is relatively insufficient, and the contradiction between aggregate supply and aggregate demand will remain for a long time; (2) the structural unemployment issue is outstanding. During "Eleventh Five-Year Plan" period, through the economic structure optimization in the whole province, the imbalance of laborer supply and demand among different industries is eased to some extent, but the proportion of employed population in the primary industry still fluctuates at 40%, larger than proportions of the secondary and tertiary industry. The structural contradiction of the laborer quality unsuitable to post demand is becoming increasingly conspicuous. Although the ratio of the number of college graduates each year and the annual quantity of employment increases, the proportion is still less than 1% by 2008. Furthermore, the employment rate of college graduates is very low. At present, Shandong Province is in an important economic and social transitional period, there are still many inadequacies in employment structure, policy and serious situation. A series of social problems may be raised if handled improperly, which will directly affect the overall interests of reform, development and stability of that province. [3] Therefore, it is crucial to industrial structure optimization and job enlargement.

In this thesis, factor analysis is employed to comprehensively analyze the six aspects which influence the employment as the level of social and economic development, laborer quality, the level of urban and rural development, living standards of laborers and working environment, and three impact factors — economic and social development factor, industrial structure factor and laborer quality factor — that influence the employment are obtained. A comprehensive analysis is made on the overall employment level, the employment level in each industry and the quantity of employment of Shandong Province, by setting up a linear regression model of employment impact factors to the employment rate of Shandong Province; the final result indicates that the three factors have larger influence on the primary industry and tertiary industry, which means changing of the three factors can result in reasonable adjustment of industrial structure of Shandong Province, increase of quantity of employment and improvement of employment rate.

2 Analysis of Influencing Factors of Employment

2.1 Decomposition of the Factors Influencing Employment and the Construction of Index System

With the development of the society, although the income level is still the main reference standard for people to choose employment area, social factors and environmental factors have been increasingly affecting the people's employment choice.

Zhu Jiaming, Wang Li,[4] etc. have constructed 18 indexes which influence employment in Main Influencing Factors Analysis and Future Forecast of Quantity of Employment of China, including quantity of employment, GDP, fixed assets investment, fiscal revenue, fiscal expenditure, net export, money supply, revenue, expenditure rate of scientific research, marketization index, industrial structural adjustment level, natural population growth rate, urbanization level, laborer resource rate, resident consumption level, Engel's coefficient, human capital, urban-rural income disparity. In The Application of Principal Component Analysis Method to Economic Forecast by Neural Networks, Zhang Xinghui, Du Shengzhi,[5] etc., through the application of principal component analysis method, synthesize 23 factors that influence unemployment as GDP, primary industry added value, secondary industry added value, tertiary industry added value, per capita income of urban residents, aggregate investment in capital construction, total energy production, electric energy production, production of steel, production of cement, total retail sales of consumer goods, total import, money supply M1, money supply M2, retail price index, present price of industrial added value, unemployment rate and quantity of unemployment into 3 main components and use them as the input of the neural network to forecast the unemployment.

The index selection of this paper starts with level of social and economic development, laborer quality, level of urban and rural development, living standards of the laborers and working conditions, and each factor includes the following indexes respectively (as shown in the table 1).

Table 1. Index construction

Level of social and economic development	Local total output value GDP
	Total output value of primary industry
	Total output value of secondary industry
	Total output value of tertiary industry
	Total output value proportion of primary industry
	Total output value proportion of secondary industry
	Total output value proportion of tertiary industry
	Total industrial output value
	Total export
Laborer quality	Natural growth rate
	Birth rate
	Mortality rate
	Number of college graduates
	Operating expenses for culture, education and science
Development level of urban and rural	Urban construction investment
	Rural construction investment
	Consumption of urban residents
	Consumption of rural residents
	Disposable income of urban residents

Table 1. *(continued)*

			Rural per capita net income
			Residents consumption disparities
			Urban consumer price index
			Rural consumer price index
standards of the laborers	Living		Personal income
			Personal expenditure
			Consumer price index
			Per capita housing area
			Number of hospital beds per ten thousand people
			Number of doctors per ten thousand people
ions condit	ng condit	Worki	Dust emission
			Industrial waste water emission
			SO$_2$ emission

2.2 Selection of Employment Impact Factor

1. Principle of factor analysis

The basic idea [6] of factor analysis is to divide the original variables into groups according to the significance of correlativity, and to make the variables in the same group have higher correlativity and the variables from the different groups have lower correlativity. Each group of variables represents one basic structure and is indicated with an unobservable integrative variable, and this basic structure is called common factor. The general factor analysis model is given below.

Assume there are n samples and each sample observes p indexes, and there is a higher correlativity between the p indexes. For the convenience of analysis and eliminating the influence caused by the differences of the observation dimension and order of magnitude, standardize the observation data of the sample, and regard the mean value of the variables as 0 and the variance as 1 after standardization. For the sake of convenience, the original variables and the variable vectors after standardization are all indicated by X, and the common factors after standardization are indicated by $F_1, F_2, \cdots, F_m (m < p)$. Supposing that:

(1) $X = (X_1, X_2, \cdots, X_p)'$ is observable random vector and mean vector $E(X) = 0$, covariance $\mathrm{cov}(X) = \Sigma$, and covariance matrix Σ is equal to correlation coefficient R;

(2) $F = (F_1, F_2, \cdots, F_m)'(m < p)$ is unobservable variable and the mean vector $E(F) = 0$, covariance matrix $\mathrm{cov}(F) = I$, which means various components of vector F are mutual independent;

(3) $\varepsilon = (\varepsilon_1, \varepsilon_2, \cdots, \varepsilon_p)'$ and F are mutual independent, and $E(\varepsilon) = 0$, and Σ_ε, the covariance matrix of ε, is diagonal matrix

$$\mathrm{cov}(\varepsilon) = \Sigma_\varepsilon = \begin{bmatrix} \sigma_{11}^2 & & & 0 \\ & \sigma_{22}^2 & & \\ & & \ddots & \\ 0 & & & \sigma_{pp}^2 \end{bmatrix}$$

Which means the various components from ε are mutual independent, and then model

$$\begin{cases} X_1 = a_{11}F_1 + a_{12}F_2 + \cdots + a_{1m}F_m + \varepsilon_1 \\ X_2 = a_{21}F_1 + a_{22}F_2 + \cdots + a_{2m}F_m + \varepsilon_2 \\ \cdots \\ X_p = a_{p1}F_1 + a_{p2}F_2 + \cdots + a_{pm}F_m + \varepsilon_p \end{cases}$$

is called factor model, and the matrix form of the model is:

$$X = AF + \varepsilon$$

Among which

$$A = \begin{bmatrix} a_{11} & a_{12} & \cdots & a_{1m} \\ a_{21} & a_{22} & \cdots & a_{2m} \\ \vdots & \vdots & & \vdots \\ a_{p1} & a_{p2} & \cdots & a_{pm} \end{bmatrix}$$

2. Selection of employment impact factor based on factor analysis

Table 2.

Total Variance Explained

Component	Initial Eigenvalues			Rotation Sums of Squared Loadings		
	Total	% of Variance	Cumulative %	Total	% of Variance	Cumulative %
1	23.313	77.712	77.712	19.091	63.637	63.637
2	4.371	14.571	92.283	5.866	19.552	83.189
3	1.294	4.314	96.597	4.022	13.408	96.597

Extraction Method: Principal Component Analysis.

Total Variance Explained

Component	Initial Eigenvalues			Rotation Sums of Squared Loadings		
	Total	% of Variance	Cumulative %	Total	% of Variance	Cumulative %
1	23.313	77.712	77.712	19.091	63.637	63.637
2	4.371	14.571	92.283	5.866	19.552	83.189
3	1.294	4.314	96.597	4.022	13.408	96.597

Based on the data of Shandong statistical yearbooks between 1978 and 2008, using the statistics analysis software, SPSS, to standardize the 35 index data and perform the factor analysis to the data after standardization, and from SPSS output variance explained table (Table 2) and Scree Plot (Figure 1), we can see that the first three eigenvalues are higher and other eigenvalues are all lower. The cumulative contribution of the first three common factors to the sample variance is 95.977% and 3 common factors are selected in this paper. So there are three impact factors and contribution rates of each factor are as follows (as shown in the table 2).

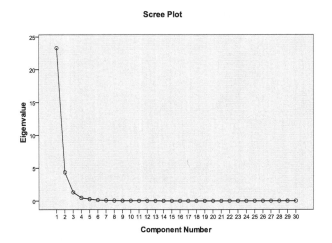

Fig. 1.

Estimates of factor loading are as follows (Table 3):

Table 3.

Rotated Component Matrix[a]

	Component		
	1	2	3
X1: Local total output value (hundred million Yuan)	.957	.276	.071
X2: Total industrial output value (hundred million Yuan)	.969	.225	.065
X3: Total export (hundred million dollars)	.981	.124	.067
X4: Resident consumption of Shandong Province	.932	.352	.075
X5: Rural resident consumption	.885	.449	.099
X6: Urban resident consumption	.936	.343	.069
X7: Consumer price index of the province	.619	.465	.141
X8: Urban and rural resident consumption disparities	.953	.290	.054
X9: Disposable income of urban residents	.909	.408	.080
X10: Rural per capita net income	.875	.468	.100
X11: Disparities in urban and rural income levels	.924	.373	.068
X12: Total output value proportion of primary industry (%)	-.642	-.733	-.155
X13: Total output value proportion of secondary industry (%)	.759	.393	-.329
X14: Total output value proportion of tertiary industry (%)	.341	.797	.474
X15: SO_2 emission (ton)	-.008	.698	.619
X16: Number of college graduates (person)	.971	.104	.105
X17: Operating expenses for culture, education, science and public health services (ten thousand Yuan)	.960	.268	.070
X18: Number of hospital beds per ten thousand people	.824	.495	.235
X19: Number of doctors per ten thousand people	.558	.540	.307
X20: Per capita housing area (m^2)	.878	.350	.160

Table 3. *(continued)*

X21: Industrial waste water emission (ten thousand tons)	.910	.202	.205
X22: Dust emission (ton)	-.735	-.048	.507
X23: Total output value of primary industry	.870	.470	.126
X24: Total output value of secondary industry	.967	.233	.064
X25: Total output value of tertiary industry	.953	.290	.067
X26: Birth rate (%)	.103	.191	.963
X27: Mortality rate (%)	.159	.384	.890
X28: Natural growth rate (%)	.060	.050	.972
X29: Rural consumer price index	.635	.650	.153
X30: Urban consumer price index	.933	.351	.055

The factor analysis model of indexes system of employment situation can be indicated from the table above. And from the factor loading matrix after rotation, we can see the common factor F_1 belongs to the social economy indexes, thus F_1 mainly reflects the overall social and economic development level of Shandong Province; the common factor F_2 has larger load on X_{12} (total output value proportion of primary industry) and X14 (total output value proportion of tertiary industry), which reflects the industrial structures of the primary and tertiary industry of Shandong Province; the common factor F_3 has larger load on X_{26} (birth rate), X_{27} (mortality rate) and X_{28} (natural growth rate), which reflects the basic quality of the laborers of Shandong Province, and is called laborer quality factor.

3 Employment Situation Analysis of Shandong Province

3.1 Linear Regression Model of Employment Impact Factors to the Employment Rate of Shandong Province

Based on the data of Shandong statistical yearbooks between 1978 and 2008, and using the Eveiws to establish the linear regression model of employment impact factors to employment rate of Shandong Province. The interval of this sample is from 1978 to 2008, and the model result is estimated by least square method as follows:

$$R = 5.334F_1 + 5.277F_2 + 1.872F_3 + 53.468 \tag{1}$$

The results are shown as Table 4

Dependent Variable: R

Method: Least Squares

Date: 01/11/11   Time: 12:41

Sample: 1 31

Included observations: 31

Table 4.

Variable	Coefficient	Std. Error	t-Statistic	Prob.
F1	5.333841	0.376318	14.17376	0.0000
F2	5.277459	0.376318	14.02394	0.0000
F3	1.871652	0.376318	4.973590	0.0000
C	53.46799	0.370198	144.4306	0.0000

R-squared	0.939907	Mean dependent var	53.46798
Adjusted R-squared	0.933230	S.D. dependent var	7.976724
S.E. of regression	2.061178	Akaike info criterion	4.404346
Sum squared resid	114.7083	Schwarz criterion	4.589377
Log likelihood	-64.26737	Hannan-Quinn criter.	4.464662
F-statistic	140.7677	Durbin-Watson stat	1.172230
Prob(F-statistic)	0.000000		

The formula (1) and Table 4 show the changes of common factors F_1 (social and economic development factor) and F_2 (industrial structure factors of the primary and tertiary industry) have obvious effect to employment rate, and common factor F_3 (laborer quality) also has certain influence on employment rate. Therefore, the employment rate can be enhanced from the following three aspects specifically.

(1) Increase the government expenditure, expand the total social demand and boost the economic growth. Improve the pulling capacity of economic growth to employment, keep a high economic growth speed, create jobs and increase total employment [7]. "Sustaining economic growth, ensuring people's wellbeing and maintaining stability" is the foundation in the current circumstances.

(2) Adjust the industrial structure, expand employment and enhance the employment rate. The elasticity of employment of the primary industry is greater, and the labor force of the primary industry will increase when the capability of absorbing labor force of the secondary and tertiary industry is inadequate; as an emerging industry, the tertiary industry has advantages in absorbing labor force, and with the development of economy and society, the employment proportion of the tertiary industry is continual rising and the elasticity of employment of the tertiary industry is usually higher than the primary and secondary industry, which means the development of the tertiary industry has greatest pulling capacity to employment. With the reduction of the capability of absorbing labor force of the secondary industry and the rapid development of the tertiary industry, the tertiary industry will become the main industry that spurs employment growth all over the world and it will naturally play the role of absorbing surplus labor [8]. Shandong is populous province with agricultural population of 58.6 million. The problems that Shandong Province faces are that the primary industry is overpopulated and the tertiary industry is underdeveloped. Therefore, adjusting the industrial structure and enhancing the pulling function of the tertiary industry to employment are quite necessary for enhancing the quantity of employment and the employment rate of Shandong Province.

(3) Improving the laborer quality has certain impact on enhancing the employment rate.

3.2 Analysis of Influence of Industrial Structure Adjustment on Employment

Based on the proportion data of quantity of employment of three major industries of Shandong Statistical Yearbooks between 1978 and 2008, and using Eveiws to establish the linear regression model of employment impact factors to the quantity of employment of three major industries of Shandong Province. The interval of this sample is from 1978 to 2008, and the model result is estimated by least square method.

The proportion model of the quantity of employment of the primary industry is:

$$R_1 = -9.34F_1 - 7.689F_2 - 4.316F_3 + 59.592 \tag{2}$$

Among which R_1 represents the proportion of the quantity of employment of the primary industry.

Dependent Variable: R1

Method: Least Squares

Date: 01/11/11   Time: 14:25

Sample: 1 31

Included observations: 31

Variable	Coefficient	Std. Error	t-Statistic	Prob.
F1	-9.340404	0.341159	-27.37842	0.0000
F2	-7.689420	0.341159	-22.53910	0.0000
F3	-4.316499	0.341159	-12.65244	0.0000
C	59.59193	0.335611	177.5623	0.0000
R-squared	0.981311	Mean dependent var		59.59194
Adjusted R-squared	0.979234	S.D. dependent var		12.96707
S.E. of regression	1.868605	Akaike info criterion		4.208176
Sum squared resid	94.27550	Schwarz criterion		4.393206
Log likelihood	-61.22672	Hannan-Quinn criter.		4.268491
F-statistic	472.5583	Durbin-Watson stat		1.287820
Prob(F-statistic)	0.000000			

The proportion model of the quantity of employment of the secondary industry is:

$$R_2 = 3.679F_1 + 3.284F_2 + 2.932F_3 + 22.32 \tag{3}$$

The proportion model of the quantity of employment of the tertiary industry is:

$$R_3 = 5.651F_1 + 4.391F_2 + 1.382F_3 + 18.104 \tag{4}$$

The comparison of the coefficients of the equations (2), (3) and (4) above indicates the three common factors have a negative influence on the proportion of the quantity of employment of the primary industry and have the biggest impact, which means improving the level of the three common factors could decrease the proportion of the quantity of employment of the primary industry and transfer the surplus labor of the primary industry; the three common factors also have a great influence on the proportion of the quantity of employment of the tertiary industry, and improving the level of the three common factors could increase the proportion of the quantity of employment of the tertiary industry and make the tertiary industry absorb more laborer, to expand the employment; the three common factors have a general influence on the proportion of the quantity of employment of the secondary industry. Although improving the level of the three common factors could increase the proportion of the quantity of employment of the secondary industry, the increase is less than the tertiary industry.

4 Summary and Prospects

Based on the factor analysis on the 30 indexes which influence the employment situation of Shandong Province, social and economic development factor, industrial structure factor and laborer quality factor which influence the employment situation of Shandong Province are obtained. And linear regression model which influence the employment situation of Shandong Province is obtained after conducting linear regression to the employment rate of Shandong Province by using the three factors. Obtaining the three factors by aiming at the industrial structure of Shandong Province and adopting factor analysis, and conducting linear regression to the three major industries of Shandong Province respectively, we can see that the three factors have greater influence on the primary and tertiary industry from the comparison of coefficients, which means the industrial structure of Shandong Province can be reasonably adjusted, the quantity of employment can be increased and the employment rate can be enhanced by changing the three factors.

References

1. Zhu, F.: Study of Employment Impact Factors of China (December 2009)
2. Li, Y., Han, Z. (eds.): The Cross-century Themes and Problems of Chinese Economy. Economic Science Press, Beijing (1999)
3. The "Eleventh Five-year" Plan about Employment Issued by Shandong Provincial Department of Labor and Social Security. China Labor Consultation Network, http://www.51Labour.com/
4. Zhu, J., Wang, L., Tong, J., Guo, Y.: Main Influencing Factors Analysis and Future Forecast of Quantity of Employment of China. Mathematics in Practice and Theory (August 2010)
5. Zhang, X., Du, S., Chen, Z., Yuan, Z., Mo, R.: The Application of Principal Components Analysis to Economic Forecast by Neural Networks. The Journal of Quantitative and Technical Economics (4) (2002)

Research on the Discipline Construction of Medical Information Retrieval Based on the Evidence-Based Decision-Making

Wei-Li Chen

Library of HeBei United University, Tangshan, Hebei, China
lilychen2008@126.com

Abstract. Evidence-based decision-making is the implementation of evidence-based medicine model. To conduct evidence-based decision-making, firstly, policy-makers must systematically collect and evaluate of existing research evidences, and based on the best available evidence, to take into account the amount of available resources and resource allocation values, finally make reasonable decisions. Therefore, the information retrieval is an essential component of evidence-based decision-making. If there is no access to medical information retrieval technology as protection of evidence, evidence-based decision-making is simply impossible to achieve, and the correctness of information retrieval determines the success or failure of evidence-based decision-making. In order to improve doctors' awareness and ability of evidence-based medical decision-making, it should take medical information retrieval knowledge management theory and technology as guidance, and take measures to improve the academic level of the construction and service, providing evidence for the evidence-based decision support services.

Keywords: Evidence-Based medicine, Evidence-based decision-making, Medical information retrieval, discipline construction, Knowledge service.

The emergence of evidence-based medicine in the late 20th century led to a significant change in the medical profession, which was seen as a major way of improving the quality and efficiency of health services and controlling the rising health care costs of the world in the 21st century. Evidence-based medicine requires the implementation of evidence-based decision-making, which provides development opportunities for the medical information retrieval discipline and presents new challenges. Medical information retrieval discipline should be based on features of evidence-based decision making, guided of knowledge management theory, made efforts to strengthen the discipline construction and vigorously carry out knowledge services to provide evidence for evidence-based decision-making and protect access to their own development at the same time.

1 The Rise of Evidence-Based Decision-Making

Evidence-based decision-making refers to the decision-making based on the evidence; decision-making model is evidence-based medicine. Evidence-based medicine was put forward by Sackett Group of Canada in the late 90s of 20th century, the core idea is that

"Doctor's careful, accurate and sensible application of the best evidence currently available to, Make decisions when they face the specific treatment of patients" With the development of evidence-based medicine, the current evidence-based medicine in clinical practice is no longer limited to the diagnosis and treatment of individual patients, but also to health care regulations and policies, public health and the development of prevention strategies, health service organizations and Management, health technology access, approval of new drugs, health care planning, clinical guidelines and uniform-service processes to determine the patient's choice of services, medical malpractice lawsuits and medical and health services and all related activities and behavior. The current accepted definition of evidence-based medicine is "a kind of science of which taking into account economic efficiency and value to carry out the medical practice based on the best available evidence"

The rise of evidence-based medicine, the decision marked the way to start the practice of medicine from the traditional style based on subjective experience of the modern evidence-based research into evidence-based decision making. Traditional medicine is experience medicine-based, that is based on physician experience, intuition, or pathological and physiological principles to the handling of patients, a result of their practice: some truly effective therapies to the public because they do not understand the long-term is not used; some practical invalid and even harmful therapy has been widely used in the long-term that may be effective deduce from the theory,. The modern medical model is based on the experience medical, emphasizing evidence-based decision-making, that is handling of patients according to the scientific research, carefully collected on the basis of medical history and physical examination, requiring clinicians to effective information retrieval, formal methods used to evaluate the clinical literature, to found the most relevant and correct information, the most effective application of the literature (ie, evidence), evidence-based deal with clinical problems, develop disease prevention and treatment measures.

Effective implementation of evidence-based decision-making will accelerate the application of low-cost interventions, the practice of medicine out of the existing measures are not valid, prevent newly invalid measures entry to medical practice, and thus increase the proportion of effective measures to improve the quality and efficiency of health services and make full use of limited health care resources.

Evidence-based decision-making required to provide medical services to patients based on the current basis of evidence available. Evidence is scientific discovery in accordance with disease etiology, diagnosis, treatment and outcome of the general discipline. In some sense, medicine practice has always been the decision-making according to evidence, the key difference between the traditional medical practice and evidence-based medicine is the definition and location of evidence. The doctor's medical experience, high qualification doctor's guidance, textbooks and medical journals scattered study regarded as the main evidence of clinical practice by experience medicine, and evidence-based medicine believe that the evidences associated with medical practice and decision-making is multi-level, and some evidence is reliable, some are not reliable; some are directly related, some are indirectly related. The evidence can be used directly to guide the practice of medicine from medical observation and scientific research is based on research of human units about human disease and health of the general discipline, such research methodology is epidemiology. The results of basic research is scientific evidence, such as biochemical studies, the results of molecular biology studies and animal studies are one source of new ideas and new methods of medical practice, to a certain extent for health care decision-making , but such evidence can not be directly used to guide medical practice, needed verification from epidemiological studies.

The implementation of evidence-based decision-making, benefit from the methodological results of epidemiological studies, and technical support is dependent on the computer and the network as the main subject of modern information retrieval theory and technology. Because being evidence-based decision-making, policy-makers must first systematic collection and evaluation of existing research evidence, based on the best available evidence, taking into account the amount of available resources and resource allocation values, and make reasonable decisions. Therefore, the information retrieval is an essential component of evidence-based decision-making, and it's impossible to achieve without medical information retrieval technology as protection of evidence. Information retrieval is correct or not determines the success or failure of evidence-based decision-making.

2   The Requirements of the Evidence-Based Decision-Making to Information Retrieval Discipline

All decision-making of evidence-based health care should be affected by three factors: the research evidence, the available health resources and resource allocation in value orientation. As the main practice body of evidence-based decision-making, clinicians have a decisive impact on evidence-based decision-making whether they have the awareness, aspirations and capabilities. The implementation of evidence-based decision-making, not only put forward new demands to the traditional medical education, clinical management, but also to the construction and service of medical information retrieval disciplines.

2.1   To Raise Physician's Awareness for Evidence-Based Decision-Making

Clinician's awareness and desire for evidence-based decision-making is the primary condition to implement evidence-based medicine. Affected with the traditional clinics model, doctors are accustomed to their own experience-based clinical practice, most doctors awareness of evidence-based decision-making is still relatively weak, many studies proved by the low-cost medical methods are not to extensive; many methods also prove harmful routine clinical use. In the process of evidence-based medicine popularity, medical information retrieval services for doctors undertaking the task of evidence-based medical training, training the primary objective should be to raise awareness of evidence-based medicine doctors, allow clinicians to realize the significance of evidence-based decision making and at same time to improve the initiative and awareness of implementation of evidence-based decision-making.

2.2   To Improve the Doctors' Information Management Capacity

Information management capacity of doctors is basis of implement evidence-based decision making, including information retrieval capabilities, evidence of assessment of capacity and management of personal knowledge. In the environment of evidence-based decision-making, policy makers need to be experts of collected, retrieval, assessment and management information .Only with a strong information management capability, clinicians can obtain the latest and reliable information from a larger range of clinical studies of others, to guide their own treatment decisions, also to accumulate clinical experience to be able to improve the judgments Capacity of the disease. Therefore, in the teaching and training of medical information retrieval discipline, should be targeted to the content of teaching and experimental design, so as to continuously improve the information management capacity of doctors.

2.3 To Improve the Level of Evidence Protection

Follow the evidence is the essence of evidence-based medicine. The most convincing evidence comes from a large scale of randomized controlled clinical study which by internationally recognized (Randomized Controlled Trial, RCT), but because of the relatively small study for RCT, in the medical literature, other types of clinical studies can also be used as evidence. However, the vast medical literature today, more than 200 million annual articles published on the current biomedical journals of the world more than 22,000 kinds, that could say that its large number and rapid aging. In addition, medical textbooks and monographs on average half-life of 7 years, the medical journal literature half-life of 5 Years. As busy physicians, together with information retrieval by its own capacity constraints, is not always possible to obtain the necessary independent evidence. Medical information retrieval discipline should play to their advantages, through a variety of services to meet the medical needs of the medical evidence.

3 Strategies of Medical Information Retrieval Discipline

Evidence-based medicine is a revolution that transforms knowledge into the quality of health services and efficiency. The emergence of evidence-based medicine means new opportunities for medical information retrieval discipline but also presented new challenges. Discipline should be based on medical information retrieval knowledge management theory and technology as guidance, to take measures to improve the subject construction and service levels.

3.1 Knowledge Management Theory as a Guide for Discipline Construction of Medical Information Retrieval

Knowledge management is the advanced stage of information management, is the new management theory that requirements of the economic development of knowledge. Information retrieval is the sub-disciplines of information management discipline, of course, but also to knowledge management as a guide. Traditional information management is an automated idea that believes computer technology is all-powerful information management factors and neglects the subjective factor in the role of knowledge and innovation, so information management limited to the traditional management of explicit knowledge or data Management. Compared with the theory of information management, knowledge management, information technology, both the emphasis on the role of innovation in knowledge, but also attach importance to people's initiative and creativity to play a role, with particular attention the role of people's tacit knowledge. Inspiration of Knowledge management theory to medical information retrieval discipline construction is the following aspects.

First, need to optimize the knowledge structure of subject research, teaching and service personnel of medical information retrieval. First of all, personnel of information retrieval should have a certain medical background of biomedical disciplines, so that they can have the tacit knowledge related to the angle from the user to think and services. Secondly, through a variety of ways to improve medical information retrieval staff interests, work experience and skills and dedication and so the level of tacit knowledge, tacit knowledge which is related to the work of medical information retrieval makes the decisive factor.

Second, need to conduct the medical information retrieval research based on knowledge management. Extensive integration of knowledge management philosophy, psychology, social science, information science and technology, knowledge and science, library science, information science, literature, communication, acceptance and management of school subjects such as science and technology principles and methods, is the product of multi-disciplinary cross-cutting, multidisciplinary contribute to this knowledge innovation play in the various subjects factors. For medical information retrieval discipline it should be based on knowledge management as a guide to locate and research disciplines: medical information retrieval should goal at support evidence-based decision-making, by strengthening medical research on biological medicine, management, decision sciences, epidemiology, and computer Science, and information resources management in particular medical clinical practice for the service of evidence-based decision-making.

Third, need to strengthen the fundamental theories of medical information retrieval research. With the development of computer science, network environment and digital hospital, the traditional disciplines of medical information retrieval theory is based on evidence-based decision-making has not completely meet the requirements of information retrieval. Discipline should incorporate medical information retrieval based medicine, grid, meta-data theory, artificial intelligence and medical informatics and other disciplines of the latest research to explore the theory that adapt to the needs of medical information retrieval.

3.2 Knowledge Management Theory as a Guide for Teaching and Training of Medical Information Retrieval

In 1992, evidence-based medicine first appeared in the "Journal of American Medical Association," it was called "new model of medical education", its core is to develop the ability of doctors interpret medical literature. Therefore, the implementation of evidence-based decision-making put new and higher requirements for the education of various medical disciplines including medical information retrieval discipline. In the medical education of information retrieval based on evidence-based decision making, knowledge management theory can provide very useful insight:

First, medical information retrieval should enhance the ability of evidence-based decision-making of user as the goal of teaching. Knowledge management theory holds that people's ability to form mainly tacit knowledge, and tacit knowledge, including technical factors, cognitive factors, experience elements, emotion factors and faith elements, which requires mainly should be case teaching and practice teaching in he teaching of medical information retrieval, mainly through the continuing reform of teaching methods to strengthen the academic links with evidence-based decision-making, so improve the user's intuition, inspiration, skills, and tacit knowledge level of evidence-based decision.

Second, medical information retrieval teaching should pay attention to doctor's knowledge innovation ability. Evidence-based decision-making is the decision-making based on evidence, the evidence itself is knowledge, has led to some differences in knowledge due to differences in study methods and rigor. Knowledge is derived from the data conversion. In the process of the data into knowledge, wisdom plays a key role. Wisdom is mainly tacit knowledge, including questions, critical, thinking, cognition and decision-making capabilities. Therefore, to enhance the user's knowledge of evidence-based decision-making ability and innovation ability, teaching in medical information retrieval, we should abandon the rote, simple teaching approach, students should be active participation by the absorption of the teaching process to improve student critical spirit and independent thinking ability of knowledge innovation.

3.3 To Provide Knowledge-Based Services for Evidence-Based Decision-Making

The so-called knowledge-based services, is the process of implementing of the evidence-based decision-making, medical information retrieval services to the entire process of participation in the services provided for the user deeper, more focused, more practical help physicians solve clinical problems. Mainly from the following aspects:

First, newly ideas, involve in the whole process of information services. In order to provide the whole process of evidence-based decision-making services (including literature, identification , screening, evaluation, synthesis, etc.), information retrieval personnel should regularly in-depth to the first line of clinical, teaching and research, and consult medical experts with an open mind to understand and master the specific Clinical problems, in order to provide the best evidence initiatively and purposely for clinical staff to retrieve.

Second, take part in the literature review or meta-analysis of the system. Achieving technical basis of evidence-based medicine is the randomized controlled trials and its review or meta-analysis of the system, including analysis, literature search, data sorting and other work that are important part of meta-analysis. Information retrieval should extensive collection of published and unpublished research results about the problem, assess for them, and finally write out the high-quality systematic reviews or meta-analysis literature under the participation of the medical experts, epidemiologists, statisticians together.

Third, provide services initiatively to clinical practice of evidence-based decision. In addition to provide accurate source of evidence for the evidence-based medicine, it should also take a variety of ways to conduct some training for medical staff on evidence-based medicine, clinical science and medical literature databases to retrieve evidence and evidence-based medicine in database search methods so that they can retrieve clinical problems closely related with the best scientific evidence timely and accurate.

Fourth, conduct research and practice of evidence security services based on the hospital information system, Hospital information system is a major tool for physicians, medical evidence of the system will be integrated into the hospital information system, will greatly improve the efficiency of look and apply for medical evidence, is an important trend in medical information research. To this end, medical information retrieval discipline should be strengthened research and practice of evidence security services based on the hospital information system, particularly electronic medical records and medical decision support systems, allowing physicians will be able to easily access all the required Kinds of evidence when they use these systems in service for patients, realize the goal pose of knowledge management theory of "Provide the right knowledge to the right people at the right time ".

References

1. Muir, G., Tang, J.: Evidence-based medicine and Evidence-based health care decision-making. Peking University Medical Press, Beijing (November 2004)
2. Bai, C.: Evidence-based medicine and Medical Information Services. Medical Information 20(9) (2007)

The Development of Scientific Activities among University Students Majoring in Logistics

QiuHong Zhang[*], Lianying Cao, Yujian Lv, and Qi Qian

Traffic and Logistics Engineering Department of Shandong Jiaotong University,
Jinan, China 250023
zhangqiuhong6849@sina.com

Abstract. Contemporary college students are the future talents of the reserve army, whether they have the practical ability of science and technology is directly related to the development of China's comprehensive national strength. Academic and technological activities are effective means to improve students' academic and practical ability of innovation. To the prevailing lack of practical ability among university students for the current status quo, this paper represents the current situation of academic and scientific activities among university students majoring in logistics as well as the merits and demerits . Furthermore, it puts forward a feasible solution to the basis mechanism and cooperation to the development of logistics contest .

Keywords: Logistics, Scientific competition, Technological innovation, Practice teaching,

For a long time, because of the ideological constraints of traditional education and exam-oriented education system of personnel selection, many domestic colleges and universities one-sided emphasis on higher education in the teaching of theoretical knowledge and regard classroom teaching as the main way of'preaching', giving priority to the teachers' teaching .In order to cram up the examination, most of the students passively accept the instilling knowledge, out of production, out of the society. This situation forces the students to pay more attention to knowledge while ignoring ability, overlooking the overall quality of students and innovation ability. As a result, the structure of students becomes so rigid that they have to experience a long period to adapt to the society after graduation, leading to poor practical work. This situation forms a sharp contrast to the growing international competition for creative talents in the 21st century, which requires undergraduates to have a certain amount of scientific research capacity and creative spirit.

In recent years, the school is continually innovating personnel training mode, optimizing personnel training program, putting the training program curriculum theory and practice of teaching system optimization in the first place of revision, continually building the practical teaching system of practice and training, graduate internships, social surveys, extra-curricular academic and scientific activities, scientific research

---

[*] Corresponding author.

project composed of undergraduates. University of the training methods and means are continually reforming and developing, introducing science and technology activities into the process of teaching students, organizing more and more students to throw themselves into scientific research activities so that they can exercise actually from scientific research when in the school.

As a popular field, the development of high-tech logistics is imperative to China's logistics industry. Science develops hand in hand with the team's research in universities, holding logistics contests among university students not only improve their practical ability, but also the achievement is a remarkable reflect of the younger generation's collective intelligence. Additionally, it adds the crowning touch to the development of logistics industry.

1   The Status of Scientific Contests in Logistics

National contest on logistics design by university students is drawing increasing attention these years, logistics Class Steering Committee and China Federation of Logistics & Purchasing offer a variety of measures to encourage students to be innovative in science and technology, through these activities. It drives domestic universities to carry out a variety of logistics contest. There are an army of students proposed valuable ideas when participating in the contest, which not only promoted the development of education, but also do a great help to the development of China's logistics industry. As sponsors, logistics companies not only carry out their own propaganda, but also provide students the opportunity to study and practice.

During the scientific contests in logistics, universities do well in supporting and encouraging technological innovation, and the forms are also extensive. Domestic universities have given strong support to the contests both in terms of human and financial resources, well-known experts in logistics enterprises have always been invited to give lectures in some universities, based on this platform, students learn a lot about cutting-edge technology in the developing of logistics. Many universities hold various activities, such as the creative competitions in logistics, academic experience-sharing sessions to develop interests and raise the awareness of science and technology innovation among lower grade students. During participating in the competition, they have mastered certain knowledge and skills and extended the learned knowledge to some degree; therefore, their own ability has been gradually improved.

2   Factors Impeding the Scientific Activities in Logistics

2.1   Lack Understanding of the Logistics Enterprises among Universities

As the students' science and technology practice is booming, there still exists many unfavorable factors blocking its development. Due to the lack of accumulation in corresponding knowledge and ideas during the students' usual time, especially among freshmen and sophomores, there are not enough opportunities to exercise in enterprises, thus they always feel a great loss when facing this kind of activity which are aiming at improving practical ability, and they can not find the area they are good at. In the meantime, they have not sufficient time to think and research, and ultimately makes the logistics science and technology activities at a low level hovering. It does great harm to form the students' awareness of science and technology innovation.

2.2 The Logistics Practical Ability of Students Needs to Be Improved

Logistics design competition is the study of applying theoretical knowledge into practice, so we should strengthen theoretical knowledge and accumulate professional and systematic logistics knowledge. Theoretical knowledge is the base of competition, while it is practice that makes the program feasible. As a result, we should be engaged fully in practice. In the process of competition, students realized that they had committed a "book worship" error, which is piling too much advanced technique into the program. Apparently, it seemed that the overall operational efficiency was greatly improved and the advancement in technology was realized. However, from the perspective of the operators, it ran contrarily to the original intention of the company's operation and it could not make a reasonable balance between cost and efficiency. In short, it lacks practical value.

3   The Way Students Develop Scientific and Technological Activities

3.1   Enhance the Learning of the Fundamental Knowledge and Strengthen the Mastery of Theories

Professional technology contest is a test of using theoretical knowledge to solve the practical problems. Its basis is the knowledge which is learned in the classroom. Therefore, solid professional basis is fundamental, and temporary learning must bring a unsatisfactory result. Students have to strengthen theory attainment independently and accumulate basic knowledge of systematic logistics.

3.2   Promote the School-Enterprise Cooperation and Increase the Experiment and Practice Skills

Through scientific activities we can increase practical ability. Strengthening cooperation and exchanges with the enterprise, and increasing the exercise out of school make the students participate in science and technology activities with both examples and practice opportunities and make campus training and external practice closely, They can not only enhance the students' theoretical practical level, but also improve the levels of technology of enterprise. Students touch the tendency of enterprise and society as soon as possible. For schools, not only does it improve the utilization rate of various teaching resources, but also it can promote the consummation of practice teaching system and flourish academic atmosphere, closing the relationship between teachers and students.

3.3   Strengthen the Exchange between Departments

In the process of participating in the game, team members should be various. The major is the students who major in logistics and it is necessary to absorb the students who major in computer science, marketing and management in the term. A logistics solution design involves many areas, and various disciplines crisscross. Therefore, on the basis of the students who major in logistics, a term which is consisted of different students is beneficial for a comprehensive design and strengthens the power of itself.

Each scientific activity is closely related to the major under each department. In order to take full advantage of each major, our university plays more attention to the popularity of these activities, thus enhancing the level of professional cooperation and exchanges. When participating in the scientific and technological activities, only with a more reasonable professional team strengths, in-depth cooperation, a clear division of labor, from a higher point of view to look at the program, can a more open design ideas be formed upon the program so that it will maximize team effectiveness greatly.

3.4 Increase input and Provide Better Material Guarantee for College Students' Science and Technology Activities

Material guarantee is the base for the good development of college students' science and technology activities .We should set up the foundation for logistics activities ,provide funds for science activities which are set up after screening or takes part in important science campaign and meet the students need in science and technology activities for necessities such as raw materials, experimental tools. Secondly, we should set up logistics technology activities practice base, which depends on science and technology activities management department of school, science and technology communities or important scientific activities team, and provide experimental site and the necessary equipment for college science and technology activities. Lastly, in order to take advantage of teaching and research facilities which we have to necessarily support the college science and technology activities ,we should strengthen the openness of laboratory and logistics engineering training center .Aiming at providing time and space where students can learned by themselves , mobilizing the students' subjective initiative and strengthening students' ability to innovate, we can develop open experimental teaching in logistics and allow the students who are audited to do independent experiment utilizing the facilities that we have.

3.5 Learn the Advanced Experience of Foreign Countries, and Establish a Rational Mechanism of Technology Application

President Kurt • Rainer • Ku Cile of Berlin University believes that technological innovation and application of knowledge is the root reason of Germany's becoming strong and powerful. University is not only the discoverer of scientific knowledge, but also the participant in conversing of scientific and technological achievements. There are three most common ways for technology products of Berlin University flowing into the market. The first way is the direct talk among universities, economic sectors, industry sectors and consumers, discussing the problems about the curriculum, subject content and improvement. Based on this, new program of undergraduate and graduate education and teaching plan can be established. The second way is that universities can join the Berlin Technology Foundation. Through the network of the foundation, the cooperation between universities and the private sectors or the public sectors can be promoted, thus realizing technology transfer. Berlin Technology Foundation was established to link research activities with industry sectors, building a technology transfer platform for both industry professionals and researchers. The last way is to establish technology transfer institutions. In the 70s of the 20th century, Technical University of Berlin set up the technology transfer agency to enable researchers to transfer technology by means of applying for patents and licenses.

In the field of industrial, academic and research cooperation with enterprises, researches have been gradually done thorough by college, thus some good results have been achieved. In the design competition in the logistics, logistics enterprises' active participation in scientific and technological activities can greatly contribute to the win-win situation between businesses and universities. The wisdom of the students is like a scientist, while universities, businesses and society are just like experimental plots. Scientific and technological progress will increase the effectiveness of the experimental fields. Improved efficiency has a higher demand on society, thus a virtuous cycle of industrial chain is created. If the chain of the industry is well-managed, scientific and technological activities can have a better innovation. To further integrate technology and society and to maximize the benefits of scientific and technological activities are the effective way to develop scientific and technological activities. The combination of students' technology practice works and the demand of society can maximize the significance of scientific and technological practice and promote the advance of society. " Useful innovation is the best innovation.," said Kaifu Li.

Nowadays, technological innovation activities of China's college students are facing a rare historical opportunity for development. Although problems are multifaceted among those activities, causes of these problems are varied as well. Therefore, we should give a high priority to not only these problems but also the causes of them and analyze them in depth, creating a new situation in scientific and technological innovation activities. If only we regard the technological innovation activities of college students in a developing sight, they will grow a healthy, thorough, lasting growth.

References

1. Meng, T.: Improving the Quality of Graduate Design and Training Creative Talents (6) (2006)
2. Wang, D., Lin, H., Xu, W.: Some Ways to Set Up Innovative Experiments in the Undergraduate (12) (2008)
3. Zhang, S.: Development of Students' Scientific Activities—in the Case of Beihang University Students' Innovation Activities. Science and Technology Communication (11) (2010)
4. Yi, H.: The Problems and Suggestions of College Students' Science and Technology. The Science Education Article Collects (2010) (Ten—day on)

The Review and Prospect of Physical Education under China's Accession to the WTO within Last 10 Years

Jiasi Luo

Shenyang Sport University, china
luojiasi@yahoo.com.cn

Abstract. Since China joined the WTO 10 years ago, competition with different countries has been continued among different kinds of fields. Physical education, as a special area, has also got many opportunities and challenges at the same time. During the new period of history, we should think of the future movement of physical education, take precautions before it is too late, and come up with new plans positively to figure out a development model of physical education which is more suitable for China.

Keywords: Education, WTO, Review, Prospect.

1 Introduction

On December 11th 2001, China became the 143rd member of WTO after 15 years hard negotiation. Since then, China has entered into the mainstream society as a more open posture, and the Chinese people have also become tighter to the world.

Until now, China has been joining the WTO for ten years. During the past ten years, huge amount of imposts, service and capital have entered Chinese market, and Chinese products, service and capital have also exported to the world. The rapid growth of the economy in China makes it competitive in many fields of the world. Obviously, the entering of WTO for China has far-reaching effect among different kinds of fields, including Physical Education (PE). For the ten years period of entering WTO, our Chinese Physical Education has caught many new opportunities and faced many challenges as well; it is essential for us to review and prospect the Physical Education under the new environment of joining WTO.

2 The Enter of WTO Brings More Opportunities and Changed to the Physical Education in China

2.1 Diversity Is Taken on among the School-Running Subjects and the Resource of Operating Expenses

For the last ten years, the present PE school-running subjects have no longer only been controlled by the country, the region, the PE system or the individuals, other types, such

Y. Wang (Ed.): Education Management, Education Theory & Education Application, AISC 109, pp. 433–437.
springerlink.com   © Springer-Verlag Berlin Heidelberg 2011

as foreign PE schools, co-operation systems have been played important roles which improve the diversity of school-running subjects. At the same time, the resource of operating expenses has been increased on both quality and quantity. This would solve the contradictory between operating expenses and high quality physical education, and provide better opportunities and choices to the students who are interested in PE.

2.2 Internationalization Is Taken on among the Teaching Contents and Teaching Methods

For the last ten years, more and more advanced courses, text books and training equipments have been introduced to the country through formal channels, and new concepts and methods have widen our vision, especially on the application of IT to the physical education, which breaks the mode of traditional teaching methods and brings wider room for the improvement of PE.

2.3 Scientification Is Taken on among Training Modes and Methods

For the last tem years, quantities of foreign PE training equipments, excellent coaches, athletes and relative researchers have been introduced to our country, which results the revolution of traditional training modes and methods. The training modes, methods and equipments has become more scientific, more reasonable and more advanced, so the training could stay consistent with the characteristics of motion and individuals, and the concept of lifelong physical education' has been set up gradually.

2.4 Legalization Has Been Taken on among Education Management

The most characteristic factor that distinguishes market economy from planned economy is the legalization of market economy. For the last ten years, everything has been in accordance with world standards. Under the competition of physical education among the global environment, the more normative a country's PE is, the more energetic its PE system will be, the more market shares it will take and vice verse.

2.5 Providing the Learning Opportunity to Physical Educators for Improving Fundamental Qualities

21st century, after all, is the century for intellectual competition. During the 10 years after China joint WTO, we have had more chances to communicate directly with foreign educators and to understand their ways of thinking and acting. In this way, we are learning their advantages to improve ourselves so that we can become compound physical educators and meet the needs of modern society.

3   The Challenges of PE in China Face after the Joining of WTO

3.1   The Challenges to Perceptions of PE

The main ideas in Chinese PE follow the former Soviet Union. The features of planned economy are also obvious. Although the reform and openness has been practiced in China for over 30 years, the old features of planned economy show up from time to time. After the acceding WTO, the new ideas, perceptions, training format, and forward-looking PE theories have been spread to China. This has caused a challenge to traditional PE in China. Confronting these challenges, we need to alter traditional ideas in order to adapt to the tide of international PE development. To give an example, we need to form the awareness of marketizing PE and providing service to other industries so that to expand the market of PE; we need to set up the concept of advanced, follow-up and lifelong service of PE, etc. We need to be flexible and proactive, rather than waiting for the doom.

3.2   The Challenge to the Defects of Chinese PE

Over the period of reform and openness, the last fort of planned economy–education, including the physical education, has been gradually adapting to the needs of market economy. However, some parts still can not keep the pace of the development of market economy and democracy. PE has the regular pattern as other education; also, it has its own features. Many issues in PE, especially in the area of 'market admittance', need to be solved through reform.

3.3   The Challenge to the Quality of Physical Educators

With the rise of comprehensive strength of different countries, many countries have increased the investment in PE.This has stimulated the development of modern physical education. On the other hand, the different concepts, ideas, approaches and patterns in different countries have been communicated and coalesced. As a result, a physical educator should be able to use English to communicate with foreign investors. At the same time, he or she should also be equipped with research and development skills, teaching skills, training skills. Moreover, he or she should be familiarwith Chinese PE development and also be ethical. Only in this way, can an physical educator face the challenges and deal with them successfully.

3.4   The Challenge to Culture of PE

Due to the fact that there are different cultures in the east and in the west, problems exist in importing and exporting education concepts including the PE concepts in different countries. The cultural exchange will last not only a decade or two decades; we need to prepare for the long-term challenge.

4   Concerns about the Future Development of PE

4.1   How to Deal with the Problems about Protecting Domestic PE after the WTO Application

During the ten years after WTO, the market of PE in China has been further developed and opened to the world, which includes the following several aspects: Firstly, the general limitation to the physical educators has been reduced. For example, some foreign PE institutes set up branches in China during the recent years, and even, made investments to the domestic physical colleges and institutes, which lead to the joint physical colleges and the joint PE development and training institutes. Secondly, the area of PE has been broadened. After the entrance of some foreign physical educators, physical development and training institutes, and physical management institutes to the Chinese market, the condition of the current physical trainees has been influenced a lot. Thirdly, the scale of PE operation has been broadened. The legal scale of the operation for foreign physical investors and educators will be further increased. As stated above, PE is a public or sub-public industry, but not a private industry. In addition to its commercial function, social, cultural, moral and physical functions have also been undertaken by PE. The market economy does not mean to marketize all. Even for the western developed countries, with their high level of marketization, their education still needs the support of the public funds. Although most of the non-compulsory education is undertaken by the private institutes, it is mainly impacted by the management and control of the government. Therefore, to be the precondition of connecting the domestic PE market with the international market, sufficient rules and legislation are necessary to protect the internal interests. In addition, we should learn from the competition among internal, external and joint institutes, and improve our level of development, education and training by exploring our own market features. What is more, we should make our scope and view of PE, and our system of PE known to the whole world, and improve with the world under our legislation and the rules of WTO.

4.2 How to Deal with the PE Industrialization and Globalization after the WTO Application

As we know, during the recent years the international education market tends to be more and more competitive. Large numbers of Chinese students choose to study in foreign educating entities, which lead to the outflow of resources of student marketing. Seen from the current condition of PE in China, the competition is not so intense, but it is certain that the PE market will have to face the problem in the future. Currently, studying abroad is very popular around the world and has become an industry. It is supposed to be a good method to spread the sense of worth and attract talented people abroad, but now, it has been used by some countries to improve the domestic economy. It is predictable that Chinese PE will join with the world in the near future, and the amount of joint education entities, cooperative education entities and single proprietorship entities will increase, and the market will be more competitive. PE, as a special and professional major, is limited by stadiums, equipments, instruments and many other factors, in which area Chinese PE entities still need further improvement compared with the developed countries. Through joint activities and cooperative activities, and learning from the experience overseas, both PE and relative industries will be better off.

In conclusion, WTO has provided Chinese PE entities with much more opportunities, but it is also a challenge to the Chinese market. We should prepare sufficiently to make full use of the opportunities. In addition to education and training, another aim of PE in China is to present the importance of PE and give a true and fair view of what is PE to the public. Due to this aim of PE, development to the PE market should be considered twice before actions are taken. The market economy does not mean to marketize all. There are many ways to open Chinese PE market to the world, and we should choose the most suitable one based on the features and condition of the Chinese market.

References

1. Lu, Y.: The Effect of Joining WTO on China Sports Industry. Journal of Physical Education (5) (2000)
2. Dong, J.: The Opportunities and Challenges Confronted with by Sports Industry. Sports & Science (3) (2001)
3. Wang, C.: The Research of the Relationship between Technology Economy and Athletics Industry. Sports & Science (6) (2001)
4. Yao, H., Wu, L.: Research on Theoretical Connotation of School Sports Industrialization and Characteristics. Sports & Science (7) (2001)
5. Zhu, X.: Research on the Development Countermeasures of China Sports Lottery. Sport Science (3) (2002)
6. Fang, Q.: Discussion on the Rise of Sport-Star Advertisement and its Development Countermeasure. Sport Science (3) (2002)

Financial Incentive Mechanism Effects of Low-Renting Public Housing with BOT Mode

Fangqiang Liu[1,2]

[1] College of Construction Management and Real Estate, Chongqing University, Chongqing 400044, P.R. China
[2] Chongqing Traffic & Tourism Investment Group Co., Ltd, Chongqing 400021, P.R. China
lfqcq@163.com

Abstract. The reason why private capitals are unwilling to invest Low-renting Public Housing project is low return on investment. Therefore, in order to attract private capital into this project through BOT mode, investment incentives must be proposed. Because financial incentive is the most commonly used investment incentive method, this paper makes quantitative analysis on incentive effects with financial stimulation policies based on Low-renting Public Housing profit function, so as to propose corresponding policy recommendations for our Low-renting Public Housing construction.

Keywords: Low-renting Public Housing, BOT, finance, incentive.

1 Introduction

In the Second and Third Session of the Eleventh National People's Congress, two years in row Premier Wen Jiabao has clearly put forward: actively promote Low-renting Public Housing [1,2]. But if Low-renting Public Housing still adopts the Low-rent Housing Construction mode (which is only by public finance investment), it may lead to insufficient capital investment [3,4] and its effect is not very ideal for Low-renting Public Housing construction in China. In allusion to this situation, many scholars propose to attract private capitals into Low-renting Public Housing construction and operation through BOT mode [3,5], and government has also issued polices to promote it [6]. However the rental and sale ratio is very low in most cities, commercial residential buildings are unable to make profit with rent, while the rent of Low-renting Public Housing is even lower than commodity houses, it is impossible to make a cost-recovery basis, let along gaining financial returns; while the prerequisites of private capital investment is to recover cost and get reasonable profit, so if government wants to successfully introduce BOT mode into Low-renting Public Housing construction, investment incentive mechanism is a must to meet private capital profit needs and get a win-win situation between government and investors.

Nowadays, financial incentive is one of commonly used incentive methods among countries. Some scholars and organizations have researched financial incentive methods and tools [7], financial incentive mechanism in American Low-renting Public Housing development [8,9], financial incentive policies in venture capital of developed countries [10], financial incentives promote the development of Beijing

Y. Wang (Ed.): Education Management, Education Theory & Education Application, AISC 109, pp. 439–444.
springerlink.com © Springer-Verlag Berlin Heidelberg 2011

biomedical industry [11] and so on. But there are still no literatures on the excitation effect of financial incentives in Low-renting Public Housing investment, therefore we need to build up project profit function based on Low-renting Public Housing characteristics, and get the financial incentive effect from it in order to set up a more scientific, reasonable and applicable financial incentive mechanism for developing Low-renting Public Housing with BOT mode.

2 The Profit Function of Low-Renting Public Housing project with BOT Mode

2.1 The Profit Function of Low-Renting Public Housing project

The profit function of Low-renting Public Housing project can be shown as:

$$\pi = \sum_{t=T_j+1}^{T_y}\left(\frac{P_t\alpha Q - b_t Q + \delta C_t N}{(1+r)^t} - \frac{f_t}{(1+r)^t}\right)$$
$$-\sum_{t=T_j+1}^{T^r}\frac{Di}{(1+r)^t} - \frac{D}{(1+r)^{T^r}} - \sum_{t=T_0+1}^{T_j}(\frac{I-D}{T_j(1+r)^{t-1}} + \frac{tDi}{T_j(1+r)^t})$$

function(1)

among which : T_0 is the start time of Low-renting Public Housing construction; T_j for operation start time; T_y for transfer time; T_e for project scrap time; P_t for the rental price at time t ; Q for construction area of Low-renting Public Housing(assume it is irrelevant with time, the area of Low-renting Public Housing remains unchanged within the operation period,); α for the rental ratio of Low-renting Public Housing; C_t for rental price of parking space; N for number of parking space; δ rental ratio of parking space; b_t for maintenance cost of keeping a normal operation of Low-renting Public Housing project; f_t for other cost(including taxation cost) apart from financial cost and maintenance cost within t period; I for total investment on Low-renting Public Housing; r Return On Investment Capital, that is the maximum revenue if investors put capital in other projects;

In order to simplify the discussion, we assume that investor gets loan from bank at the initial period of the Low-renting Public Housing project, the total amount of loan is D, during the construction period the bank gives the allowance of credit is $\frac{D}{T_j}$ at the beginning of each year, the term of loan is T^r, payment methods: payback loan interest every year within the loan period, and repay the principal in full at the end of loan period(this assumption will not have significant impact on the result) .

According to function (1), profit function π is the profit which has taken project unit opportunity cost into the prerequisites, but not the usual accounting profit. $\pi > 0$ means investors make better profit in Low-renting Public Housing; while $\pi = 0$ means investors just cover the opportunity cost in Low-renting Public Housing project, but there is no difference whether participating in the project or not.

So from above analysis we can see that in order to attract private capitals into Low-renting Public Housing construction, there must be $\pi \geq 0$.

2.2 Numeric Example

Take use of function (1) and get the profit of investor participating in Low-renting Public Housing construction. The total investment of one Low-renting Public Housing project is 2.693 billion, and total area: 954,700 square meters, rent: 11yuan/ month. Square, annual rate of rent growth is 3%; 3575 parking space, parking fees 300 yuan/ month·one, annual rate of parking fee growth is 4%, 80% of investment is loan from bank, annual interest rate is 6.4%. Assume franchise operation period is 30 years, unit maintenance cost is 1yuan/ square, unit management cost is 1 yuan/square, taxation cost in the operation period takes 9.5% of rent, investors require that the return on investment capital should be 11.60% the 2008 average rate of profit [12]. Based on these data, the maximum profit of investors is π = -0.79984 billion, which is far smaller than zero. If private capitals participate in Low-renting Public Housing construction, it would be massive losses.

3 Preferential Rate Incentives

3.1 Interest Rate and Preferential Rate Policy

Rate (also known as interest rate) is the ratio between interest amount and principle within a period of time, usually expressed by percentage, interest rate computing formula: interest rate= interest amount÷ principle÷time×100%.

Preferential rate refers to central bank formulates low interest rate for certain state-prioritized economic departments, industries or products in order to promote its activity and adjust industry and product structure.

Nowadays, Low-renting Public Housing is national key low-income housing projects, which is welfare-oriented, so Low-renting Public Housing is in accord with preferential rate requirement and government should provide loans on favorable terms for investors who participate in Low-renting Public Housing construction.

3.2 Incentive Intensity of Preferential Rate Policy

Assume government provides favorable interest rate for investors who participate in Low-renting Public Housing project as i_0, and $i_0 < i$. With other conditions unchanged, investors profit is:

$$\pi = \sum_{t=T_j+1}^{T_y} \left(\frac{P_t \alpha Q - b_t Q + \delta C_t N}{(1+r)^t} - \frac{f_t}{(1+r)^t} \right)$$

$$- \sum_{t=T_j+1}^{T^r} \frac{D i_0}{(1+r)^t} - \frac{D}{(1+r)^{T^r}} - \sum_{t=T_0+1}^{T_j} \left(\frac{I-D}{T_j(1+r)^{t-1}} + \frac{t D i_0}{T_j(1+r)^t} \right)$$

Function (2)

Derivation of i_0 in function (2) and get the incentive intensity of this measure:

$$\frac{\partial \pi}{\partial i_0} = -\sum_{t=T_j+1}^{T^r} \frac{D}{(1+r)^t} - \sum_{t=T_0+1}^{T_j} (\frac{tD}{T_j(1+r)^t}) \prec 0$$

Function (3)

because $i_0 < i$,   then : $\pi|_{i_0} > \pi$.

From function (3) we can see, project profit function is inverse with interest rate, which means the lower of interest rate, the higher of profit function; while raising interest rate leads to decrease of profit function.

3.3 Preferential Interest Rate Incentive Mechanism Analysis

With other conditions unchanged, preferential rate policy will decrease project total expense and it has no influence on project total investment and total revenue. With project total investment and revenue unchanged, the decrease of project total expense will raise Return on Investment. Therefore preferential rate gives a positive incentive for private capital participating in Low-renting Public Housing construction.

The preferential loan rate for investors can be granted through policy financial institutions, housing accumulation fund loan or commercial loan through commercial financial institutions with government interest subsidy (notes: this paper takes subsidizing interest payment of loans into preferential policies.)

Even though preferential rate could effectively raise the rate of return for project investors, it still restraint by Commercial banks 'cost of funds, the preferential loan rate cannot be lower than the deposit rate at the same period(except for discount government loans). In fact, preferential rate margin is restricted by many factors, it is even so for housing accumulation fund loan, and so only with this measurement cannot guarantee a positive profit for project investors, therefore other incentive methods have to be considered.

3.4 Empirical Analysis

Whit the Low-renting Public Housing case 1, in order to reflect the incentive effect of preferential policy, we assume that government grant preferential rate as 4.0、4.4、4.8、5.2、5.6、6.0、6.4(unit is %), then calculate investor profit (table 1).

Table 1 shows the relation between Low-renting Public Housing preferential rate and investor profit, with other variables unchanged, every 0.4% loan interest decrease, project investor profit increase 660 million yuan.

Table 1. The incentive effect of preferential rate measurement

SN	Lending rate(%)	Income(million)	Notes
1	4.0	-38342.40	
2	4.4	-45282.69	
3	4.8	-52222.99	
4	5.2	-59163.28	
5	5.6	-66103.58	
6	6.0	-73043.87	
7	6.4	-79984.17	

4 Incentive of Financing Guarantee

4.1 Definition of Financing Guarantee

Financing guarantee means that guarantor stand for vouchee to provide beneficiary with a guarantee payment of principle and interest, financing measurements include loan, issuing marketable debt security (stocks excluded), overdrafts, deferred payment, credit limit granted by banks and so on. Usually bank loan requires guarantee: government guaranteed loans, enterprise guaranteed loans, and secured natural-persons guaranteed loans and so on.

4.2 Incentive Mechanism Analysis of Government Guaranteed Financing Policy

Financing guarantee can increase investor profit from many aspects. First government guarantee can reduce the financing cost of project investors, such as get preferential interest rate or reduce the loan-related cost of raising funds to increase project investor's profit. Second, speeds up the procedure of getting loan with government guarantee to reduce project construction period, which indirectly increases project investor's profit (due to $\frac{\partial \pi}{\partial T_j} < 0$). Third, lifts the line of credit through government guarantee, reduces own-funds pressure and improves its efficiency. Project investors can get more loans from financing institutions with financing guarantee so as to optimize fund structure.

4.3 Empirical Analysis

With the Low-renting Public Housing case 1, in order to reflect the excitation effect of government financing guarantee, we assume the fund structure of Low-renting Public Housing(proportion of loan fund) is 40%、50%、60%、70% and 80%, then calculate investor profit(table 2).

Table 2. The incentive effect of preferential interest rate measurement

SN	Lending rate(%)	Income(million)	Notes
1	40.0	-130248.80	
2	50.0	-117682.64	
3	60.0	-105116.48	
4	70.0	-92550.33	
5	80.0	-79984.17	
6	90.0	-67418.01	

Table 2 shows the relation between Low-renting Public Housing funds structure and project investor profit, with other variables unchanged, every 10% increase of loan fund proportion, project investors' profits increase about 125.6616 million yuan.

5 Conclusion

It's can be found from above theoretical and empirical analysis that Low-renting Public Housing is a guaranteed program aims to low-income residents, its rent is restricted by residents' consumption capacity, even with financing incentive policies like preferential rate, financing guarantee and so on, it is still difficult for project investors to get a reasonable return with effective operation methods, which cause a lack of activity for private capital participation. Therefore it is suggested while government providing financing incentive policies through BOT mode, other investment incentive measurements should be taken into consideration, such as favorable taxation, land assignment, adjusting concession period, subsidies and so on to promote the activity for private capitals investing Low-renting Public Housing construction , so as to solve the fund "bottleneck" finally during the development process.

Acknowledgment. This work is supported by The Chongqing Soft Science Research Program (CSTC,2010CE0120).

References

1. State Council: Report on the Work of the Government at NPC and CPPCC, [EB / OL] (March 05, 2009), http://news.qq.com/a/20090305/001675.htm
2. State Council: 2010, Report on the Work of the Government at NPC and CPPCC [EB/OL] (March 17, 2011), http://www.most.gov.cn/yw/201103/t20110317_85455.htm
3. Li, D.: On the construction of Low-renting Public Housing with BOT finance. Economic and Social Development 8, 8–11 (2010)
4. China International Capital Corporation Limited. On 10 million low-income housing construction program 2011. China International Capital Corporation Limited, Beijing (2010)
5. Wang, Q.: Low-renting Public Housing takes the leads and makes break through [EB/OL] (July 30, 2010), http://www.gxnews.com.cn/staticpages/20100730/ newgx4c52ee80-3146195.shtml
6. State Council. Some opinions on Encouraging and guiding private investment with healthy development [EB/OL] (May 13, 2010), http://www.gov.cn/zwgk/2010-5/13/content_1605218.htm
7. World Investment Report. "Investment, Tradeand International Policy Arrangements", United Nations Conference on Trade and Development, United Nations, New York and Geneva, 178–180 (1996)
8. Yan, Y.: Low-income housing tax preference certificate: financial incentive mechanism on low-cost housing development in the U.S. Beijing Planning and Construction 4, 52–54 (2007)
9. Solomon, R.: The 2006 Public Houmng Finance update. Journal of Housing and Community Development 63(5), 24–30 (2006)
10. Su, Q.: Venture capital taxation and financial incentive policy analysis of Developed countries. Foreign Economy and Management 24(6), 18–22 (2002)
11. Wu, J., Zhao, Q.: Financial incentives boost the development of Beijing biomedical industry [EB/OL] , http://www.financialnews.com.cn/yh/txt/201006/17/ content_293033.htm (June 17, 2010)
12. National Bureau of Statistics, China Index Research Institute. 2009 Statistical Yearbook of China Real Estate. China Statistics Press, Beijing (2010)

Research on Psychological Health of Poor College Students

Qiuhong Zhang[*], Xiaoli Ni, Jian Liu, and Junni Zhao

Department of Transportation and Logistics,Shandong Jiaotong University
Jinan, 250023
zhangqiuhong6849@sina.com

Abstract. Due to the multiple reasons of economic, academic, lack of communication, poor college students more easily show some obvious psychological symptoms, such as anxiety, low self-esteem and depression. Besides to providing the economic assistance to poor students, university teachers should also guide them to a correct their view of poverty, open psychological lectures, systematically construct the mental health education. Through acceptance of the psychological explanations, universal mental health knowledge, and understanding of the characteristics of adolescent psychological development and the law, poor students will improve their self-awareness and self-improvement skills.

Keywords: university students, pressure adjustment, mental health.

With the reform of China's higher education system, tuition for university students has been raised. It brought new problems to those students grown under difficult economic situation. 'Poor students' has been a special group at colleges and universities. It was shown that poor students accounted for 7% to 10% in total university students. Although the members of this group change every year, a series of troubles caused by financial difficulties are the same. There are many common features, arousing widespread concern. The government and the universities are taking a series of measures to alleviate their plight. A number of groups from various channels lend their helping hand. But it can not fundamentally change the existing situation of poor students. Most of them can not get rid of the image of 'poor students'. With more mental pressure than non-poor students, these poor students run into poverty. It is a new phenomenon -psychological poverty.

1   Mental Health Status of Poor Students

80 poor students, randomly selected from two grades, were investigated in this research which includes 69 boys and 11 girls. 75 students are from rural areas and 5 are from urban places. In contrast, 74 students were randomly selected as a group of non-poor students, including 57 boys and 17 girls, 32 rural students, 42urban students.

---

[*] Corresponding author.

Y. Wang (Ed.): Education Management, Education Theory & Education Application, AISC 109, pp. 445–449.
springerlink.com   © Springer-Verlag Berlin Heidelberg 2011

The results from the test comparison, there are significant differences in the indicators, reflecting the poor mental health was significantly worse than non-poor students. On the personality characteristics of poor students, and the perseverance of high tension, excitement and fantasy of the low; showed restraint in behavior, serious, perseverance, responsible, careful and thoughtful, conscientious, realistic, practical, intense distress, serious, prudent , cautious, introspective, follow the conventions, for proper and reasonable; in the mental health status, summarizations, obsessive compulsive, interpersonal sensitivity, hostility, psychotic 5 students scored higher than non-poor, their low level of mental health easily suffer from various mental illnesses, personality and mental health are caught in a 'psychological poverty' at home. Moreover, women are more prone to poor students in psychology of poverty. China Foundation for Poverty Alleviation from the recently published '*A Survey of Poor Female Students Report*' point of view, 84.7% of the poor female college students from poor families suffer from the pressure.

2 Causes of Psychology Poverty

Psychology of poverty is due to economic poverty, poor students, and produces a series of personality characteristics and mental health changes. Most poor students do not want to say their own, is understandable. Because they go to college, to the big city, living environment has undergone great changes, this contrast large beyond their imagination.

Poor students because of economic constraints, the busy rush of life that is, without economic strength, or the energy to carry out more interpersonal relationships, participation in campus cultural life. This not only makes their behavior runs counter to the current campus culture, but also make them a psychological self-inductance shabby, the inferiority complex wells, and produce defensive reactions, obsessive-compulsive symptoms and some hostility, impulsive behavior will be demonstrated.

3 Measures of Mental Health Education

Youth mental health is the most important resource for the growth and development of students.Mental education is not only for preventing mental disorders, but also for promoting growth and development of students, which is more important. Guidance and consultant are the two fundamental measures of mental health education.

3.1 Guidance

3.1.1 Correct Understanding of Poverty and Scientific Self-positioning

Different people have different understanding of stress, their coping styles and the impact of stressors on them are different. Studies have shown that the face of any potential stressors who regard them as a threat rather than a challenge to people have more negative reactions. To alleviate the psychological pressure of poor college students, college teachers should on the one hand to encourage students to accept the reality and establish a "poverty is not my fault", "poverty is temporary," "poverty can be changed" and "poverty is not terrible "and rational knowledge; On the other hand, teachers should guide students objectively, comprehensively and correctly understanding of themselves, based on its own terms, energy and ability to identify themselves as necessary, formulate a reasonable development goals, and guard against the blindly comparisons, to avoid unnecessary adding to their stress.

3.1.2 Providing Guidance of Self-decompression

To reduce the economic difficulties of impoverished students, college teachers should encourage students to self-improving, to ensure the basic learning time, based on the full use of free time in a variety of part-time job. More attentions should be paid to guide students to actively self-decompress. First of all, to guide them to learn to talk, let them talk to the students, families and teachers of their inner pain, translate the others' understanding, comfort and help into a huge spiritual force, and build them into their own psychological support system. Second, to guide them to open themselves ; humbly ask others to obtain the necessary advice and help. Finally, teachers should guide the poor students participant in sports activities, transfer or eliminate stress. Subject to various factors, many poor students' free time is relatively little, they generally spend the limited time on learning or part-time work, rarely participate in school sports activities. The accelerated pace of life, physical and mental state is not conducive to stress relief. Teachers can guide and supervise them often participate in sports activities, not only allow them to do too much work, but also can release their stress and anxiety, mitigation, transfer and eliminate their feeling of pressure, so as to improve their learning efficiency and quality of life and promote the healthy development of physical and mental.

3.2 Counseling

In the counseling process, the first lectures, seminars after the group counseling, and individual counseling for some students, and finally psychological training activities.

3.2.1 Open Psychological Lectures

Mental set up a seminar every month, a total of 5 lectures, "Introduction of Mental Health Standards," "psychological conflict and psychological adjustment", "personality traits and personality test", "emotional awareness and regulation," "environmental coordination and interpersonal relationships "and a focus on health education, personality and emotion regulation. Mental health lectures to help students update the concept, the concept of personality and quality of ideas, the importance of mental health awareness, understanding personality and emotional health is an important indicator of mental health, to help poor students to form a correct understanding of poverty and health psychology, students the healthy personality, enhance their sense of participation, self-esteem, self respect, self-improvement, improving emotion regulation ability.

3.2.2 Group Counseling

The theme of group counseling is growth. 32 participation of the voluntary composed of four support groups received by the two "group counseling training" as the head teacher. First, develop group counseling plan, the development group name, nature, objectives, time, design basis, group structure, and the principle of confidentiality and effectiveness evaluation. Growth of support groups is a living, homogeneous, dynamic groups, group counseling process is built on mutual respect between members of the group, shared, sincere and accept the above atmosphere, free chat, experience sharing, resource sharing.

Growth guidance is divided into four stages: the founding group activities, self-understanding, and growth experience and growth objectives. Beginning to explore support groups, self-awareness of the basic point, recalled growing up, growth targets for the end to rebuild, with new insights on the growth and establish personal development goals. Through group counseling, a group dynamic will form, which may play an important role on individual growth.

3.2.3 Individual Counseling

Advisory teachers in individual counseling expressed their understanding and support of the poor students. The emotional guidance, cognitive restructuring and behavioral counseling guidance help students to adapt to the environment and ease pressure, and improve self-awareness, self-acceptance and self-appreciation, get support from their mental world and the spiritual world, and strengthen self-esteem, self-reliance, self-confidence, active pursuit of knowledge.

3.2.4 Interpersonal Communication Training Team

The training of interpersonal skills is mainly to teach knowledge, method of communication. Interpersonal communication training aims at helping students by the implementation of the program. Intensive training methods, and training for cooperation, autonomy, responsibility and so on can help the students master the knowledge and skills to communicate with the others, and enhance the ability of communicating. It also helps to create a successful communication of psychological quality, and establish collaborative relationships.

Interpersonal communication training can be divided into four phases: the training gives a good first impression, language training, body language training, and open self-training. The entire training process of a training activity by a component, each activity includes the creation of scenarios, collaborative learning process, the students with the theme of the introduction of a basic content-related scenarios, playing different roles to play from different angles to get experience, understanding of the role based on this exchange, consultation, debate, share and deepen the understanding of activities to clarify the idea of irrational, resulting in a leap in understanding. Training in interpersonal communication is through a series of activities will be cognitive, behavioral, emotional, mental process the ability to harmoniously integrate the training.

4 Summary

Mental health education cares about the growth of poor students and the ability to help themselves in the process of education, and strives to take the psychological characteristics of poor students in account. It also creates a very warm learning environment where they can be respected and be trusted. Mental health education acts as a guide to encourage the students who have difficulties in economy, and mobilizes the capacity of self-education students in the education process.

Mental health education is a comprehensive mode including education, student growth and development-oriented education model. In this mode, the task of the mental health is the creation of psychological courses and psychological counseling. Education of mental health aims to build systematic psychology education programs. By holding psychological lectures, opening mental health education courses and psychological courses, and popularizing the psychological knowledge, we can understand the psychological characteristics and laws in the growth of adolescent. These can help the student to have self consciousness and the ability to improve themselves. The curriculum should pay attention to the role of psychological training which helps to improve their speculative power, autonomy and to grow their accumulated experience and methods. Psychological counseling is an integral part of health education. The school psychological counseling mainly focuses on the guides. And it helps students relieving mental confusion, correcting psychological disorder. Psychological counseling should be based on different situations of individual counseling and group counseling services to meet the needs of growth and development of students. It should also do good to promote their learning, living, interpersonal relationships and the environment and play an inherent potential to adapt to mature and successful.

Mental Health Education has brqad prospects, all aspects of students are involved in psychological problems and mental health. Mental health education will be more prominent , and it will become increasingly important.

References

1. Chang, C.-H.: Modern psychology. Shanghai People's Publishing House, Shanghai (1994)
2. Pang, L., Hu, F., Han, R.: Attention to poor students: Problems, Causes and Solutions. Peking University Education Review 2(2), 39–42, 49 (2004)
3. Zhang, L.: Poor College Students between Anxiety and Life Events Research. Shaanxi Normal University (Social Science Edition) (4), 155–157 (1998)
4. Chen, Y., Wei, C., Li, B.: Psychological Status of Poor College Students and Countermeasures. Liaoning Education Research (2), 62–63 (2004)
5. Yang, Z.: On the Economic Poverty of the Impact of Mental Health. Higher Education (2), 115–119 (2004)
6. Zhang, Y.: Psychiatric Rating Scale Manual, pp.17–27, 35–42, 154–160. Hunan Science and Technology Press, Changsha (1996)
7. Psychological Science 25(5) (2002)

Reform and Practice of Talents Training Model of Electronic Information Engineering Majors

Xia Zhelei, Xiao Binggang, and Wang Xiumin

College of Information Engineering
China Jiliang University
Hangzhou, China
bgxiao@cjlu.edu.cn

Abstract. The Establishment of Personnel Training Model provide an important guarantee for the quality of personnel training, The domestic and international research status of electronic and information engineering teaching is analyzed. The reforms from different aspects and angles of personnel training have been carried out by many colleges and universities , but less involved in the testing area. Based on the specific situation in education of our school, several key issues of the process of cultivating talents who are equipped with the creativity and characteristics of electronic measurement is researched.

Keywords: Teaching Reform, Electronic Information, Creative Talent, Electronic Measurement Characteristics.

1 Introduction

In recent years, with the development of informationization in the society, electronic information industry has become a industry which is the strategic mainstay of our national economic development and the leading industries. The number of colleges and universities which found electronic and information engineering major is increasing fast, but the education quality are unavoidably unevenness. Theory and practice is combined by the way of reforming the practice of teaching for the talents of the electronic information major in many domestic colleges and universities. The designing concept and practice of the training program for electronic information major such as the orientation and objectives for talent training, the design of teaching process, teaching contents and design of curriculum system, practice teaching settings have been introduced by Tang Chaojing etc who is in the National University of Defense Technology. To build a multi-practice teaching system for the practical teaching system which is divided into four levels as follows: the skills base level, the design level, integrated application development level and technology innovation level is put forward by Xiao Yingwang etc who are in South China normal university. A kind of a core two basic points of educational philosophy of electronic information engineering education for undergraduate training, which is focusing on improving teaching quality, and make a series of explorations and practice from the basic theory and engineering practice is proposed by Luo Zhengxiang etc who are in Electronic Science and Technology University. To build an innovation talents training system through the

Y. Wang (Ed.): Education Management, Education Theory & Education Application, AISC 109, pp. 451–455.

effectively combination of professional construction and discipline construction by relying on the advantage of discipline superiority, reform the practice teaching system by making use of the results of discipline-building, exerting scientific advantages, integrate the inside and outside of school and class, form the resultant force for training practical talents with innovation ability is proposed by Tan Qingguo etc who are in Electronic Science and Technology University. The practice teaching system in foreign universities for the electronic information major is mainly reflected the concept of "student-centered", such as study-discuss curriculum are set up and the basis of profession and comprehensive practice ability is combined in electronics and information major in the famous university Stanford University and MIT, students are engaged in the practice and research by enter the business and research institutions.

Although reform for training mode for electronic and information engineering major is conducted in many colleges and universities, but the electronic measurement, standard education, bilingual teaching and other aspects are referenced too little, and the reference curriculums are almost not existing. Contrast with the major characteristics of China Jiliang University, innovation of personnel training model with characteristics of electronic measurement for electronic and information major is imminent, how to improve the ability of students for electronic measurements and standards in electronic information major is a urgent research and study subject in current. On the basis of the analysis of these issues, the training of applied innovation talents which start from the society demand and electronic information as the target, combine the characteristics of China Jiliang University after analysis of the teaching model in current, conduct the research and study for the innovation of personnel training model with characteristics of electronic measurement for electronic and information major is put forward.

2 Training Orientation

According to the educational philosophy of China Jiliang University:" Measurement made school, Standard made people, Quality made enterprise ", the problem of orientation setting for undergraduate education in electronic and information major is solved and the problem that engineering and technical personnel lack of the ability for engineering practice, innovation and quality is dealt by teaching students on the basis of the social need and electronic and information engineering major, the applied and innovation talents are trained as the target, the training guidelines and principles for cultivating applied innovative personnel by "solidify the base of the discipline, optimize the curriculum setting, outstand the characteristic of measurement and standard, strength the innovation ability, pay attention to the engineering practice create the measuring culture" is established.

3 Setting of Theory Teaching Mode

3.1 Optimization Curriculum System, Enhancing the Setting of Featured Curriculum

The featured curriculum of measurement and standard is relied on, the core of curriculum system is regarded as the engineering theory and engineering practice education, the curriculums system is optimized, the characteristics of the curriculum setting is strengthened, the training mode for electronic and information talents with characteristic of measurement and standard is formed. The talent training scheme is revised timely, keeping pace with the times of talent training scheme and in line with the requirements of the applied innovative talent is ensured by making requirement analysis and social graduates follow-up through the way of the experts of enterprise industry and professional guidance committee are closely relied on. The specific contents are as follows: (1) based on the quality curriculum of province-- "Signals and Systems", " Electromagnetic theory and microwave technology" and the quality curriculum of school--"Communication Circuits", "Analog Circuits "and so on ;"Radio Measurement and Testing", "Electronic Measurement Technology " and other featured curriculums is added to Electronic and Information Engineering curriculum system; the teaching content of measurements for digital media parameter to "Digital Video and Audio Technology ","Digital Signal Processing ", "Digital Image Processing" is added and the teaching content of RF testing for mobile communication parameters to "Communication circuit", " Digital Mobile Communications " is added; (2) Construction of the curriculum for standard education: "Standardization and Intellectual Property"," management system certification" and other standard curriculums is added to the curriculum system, a database platform with quality standard literature resources which is based on "quality inspection characteristic literature database" is created to provide standards, verification regulation, documents for rules and regulations, quality inspection books as the main contents of the comprehensive literature resources, the education for standard in electronic information field is added into the related specialized curriculums , a training talent mode with the characteristics of standard education is construct; (3) Bilingual education in the basis professional curriculums is carried out as follows: "Algorithms and Data Structures", "Modern Logic Design", "Data Communication and Network" and other basic curriculums.

3.2 Construction of Teaching Materials

Curriculum construction is focused on the construction of teaching materials , breaking the limitation of the traditional subject system, closely combining the need for electronic information and measurement major, outstanding the engineering, application ,innovation and developing electronic information materials with high quality and characteristics. The research of specific construction of teaching material content are as follows: (1) The professional teaching material "digital electronic technology" which is already existing is continued to construction. (2) The characteristic teaching material is compiled and pressed: the "Electronic Measurement Technology" which is of measurement characteristics is compiled, it is used to the professional teaching of the curriculum -- "Electronic Measurement Technology", related press is contacted in order to public it and let it as the professional textbooks for other schools.

4 Setting of Practice Teaching Mode

The provincial finance and the school-enterprise cooperation laboratories which are already existing is strengthened by the construction, the practical teaching role for electronic measurement technology of the provincial Laboratory demonstration center -- "RF2000 RF circuit test" is given full play; a central government laboratory -- "Wireless Communication comprehensive laboratory measurement techniques " is build, some corresponding teaching experiments is carried out to make students improving their engineering capabilities and ability of creativity and innovation; The laboratory opening dimension to the outside is expanded and students are encouraged to participate in variety kinds of academic innovation competitions and opening topics in the laboratories, an open and research-based practical teaching platform for more students is set up; The experiment teaching and engineering practice are combined.

5 Conclusion

Teaching reform is a continuous process of exploration, summarizing and improvement , much things still remains to be done, how to combine the characteristic of our school with cultivating innovative talents is a discussion which is worth pondering and serious. Establishment of personnel training program is the theoretic basis for teaching. The issues of how to teach in a better way is a long-term question that each college teachers is required to explore, the reform should be continuously updated teaching ideas by breaking the shackles of traditional thought in the teaching, teachers should good at finding and solving potential problems in the process of teaching, the courage to reform is owned, and work hard to cultivate more high-quality innovation talents for country.

Acknowledgment. The authors gratefully acknowledge the financial support from the 2011 key teaching reform project of China Jiliang University "Innovation of Personnel Training Model with characteristics of electronic measurement for electronic and information major", the project of the operation mechanism of Zhejiang Province Philosophy and Social Sciences（09CGYD054YB）and the 2010 teaching reform project of China Jiliang University "Research and Practice of Case Study Approach Used in the information professional curriculums".

References

1. Tang, C., Mao, J., Du, G.: Electronics and Information Professionals cultivation plan and design. Higher Education Research (2009)
2. Xiao, Y.: Electronic and Information Engineering and Applied Personnel Practice Teaching System. Changchun University of Technology (2009)
3. Luo, Z., Yu, Y.: Quality as the core, based on both theory and engineering practice, and enhance the quality of personnel training and Practice. University of Electronic Science and Technology Social Sciences Edition (2005)

4. Tan, Q., Yang, Y., Jiang, N.: Relying on electronic information subject Advantage Building Innovative Training System. University of Electronic Science and Technology (2010)
5. Yang, X.: American Research Universities Commentary on the Characteristics of Undergraduate Education Curriculum Reform. Foreign Education (2003)
6. He, Z., Yang, G.: MIT Undergraduate Education Features and Implications. Comparative Education (2003)

The Exploration and Practice in the Training Model for Interdisciplinary Science Professionals

Baohua Tan[1], Fei Yang[2], Chuyun Huang[1], and Guowang Xu[1]

[1] School of Science, Hubei University of Technology, Wuhan, China, 430068
[2] School of Foreign Languages, Hubei University of Technology, Wuhan, China, 430068

Abstract. Cultivating versatile and innovative professionals is an important direction in the reform of higher education, and the training of interdisciplinary professionals is an important measure of creating versatile and innovative professionals. This paper discusses the training model for interdisciplinary science professionals, and taking the training of interdisciplinary science professionals at the Hubei University of Technology as an example, it expounds the mechanism and model of the training of interdisciplinary professionals.

Keywords: professionals training model, higher education reform, interdisciplinary training, innovation.

1 Introduction

With the rapid development of economy, culture, science and technology at home and abroad, and the striking advance of modern science and technology and culture, science and technology have become increasingly significant in promoting economic development. Social demand for talents has taken on a diversified and multi-level development tendency, and there has been a high degree of differentiation and synthesis of disciplines. In the new situation, the traditional model of education is greatly challenged, exposing more and more limitations, especially with the professional division too small, and the scope too narrow, so that after graduation students with poor adaptability have difficulty in possessing a strong comprehensive ability and innovative capacity [1].

"National Medium and Long-term Educational Reform and Development Plan (2010-2020)" and "National Program for Medium and Long-term Talent Development (2010-2020)" requires that a large number of various high-quality engineering and technical professionals with powerful creative ability and adaptation to economic and social development should be trained in order to serve the strategy of the state to take a new road of industrialization, and build an innovative country [2,3]. Building an innovative country is a major strategy for us to face the future, and the core of building an innovative nation is technological innovation. For institutions of higher education, the broad knowledge base is a source of innovation, while interdisciplinary cross-training model is an effective way to develop versatile innovative professionals. The interdisciplinary cross-training model, a new model different from the traditional single-discipline professional development, refers to the intersectional training between different subjects in different schools and different

Y. Wang (Ed.): Education Management, Education Theory & Education Application, AISC 109, pp. 457–463.

professionals, in which the knowledge is more extensive, the structure is more rational, the adaptability is stronger and the innovative possibilities are greater [4].

2 Interdisciplinary Professionals Training Model

In order to meet the requirements of higher education reform and cultivate versatile and innovative professionals, many colleges and universities have tried and explored the ways to carry out the interdisciplinary professionals cross-training, including: the curriculum-style cross-training model, the instructors' joint mentoring model, the professional cross-training model and the joint school-running model [5]. These 4 interdisciplinary training models each have their own advantages and disadvantages.

2.1 Interdisciplinary Professionals Training Model

The curriculum-style cross-training model
The curriculum-style cross-training model emphasizes the concept of diluting professionals, the use of arts and science intersection, polytechnic intersection, and basis intersection, etc. Given the discipline-based course and the professional-based course according to subject categories, the junior students are trained in a wide-ranged way, while the senior students mainly focus on a particular area of the expertise-depth study, indicating characteristics of the profession [6]. Now there are universities which are exploring and practicing the combination of two or more intersectional disciplinary courses into one course to learn, for example bio-materials technology, automotive electronics technology, etc.

The instructors' joint mentoring model
In our current education system, head teachers and counselors can generally help students with difficulties in life and ideological issues, but it is difficult for them to give their students substantive help and guidance in their professional studies. Establishing specialized instructors can effectively solve this problem, but the instructors' knowledge and energy are also limited, so to realize the interdisciplinary professional training needs different specialized instructors to instruct together, namely implementing the instructors' joint mentoring system to solve the problem of interdisciplinary knowledge fusion.

The professionals cross-training model
Professional intersection refers to the establishment of interdisciplinary and intersectional new specialty, the formulation of interdisciplinary training plan and curriculum system, and the interdisciplinary and synthetic teaching will help the students to obtain comparatively comprehensive education. For instance, Oxford University requests that the creation of all professions should include in a certain degree intersectional and professional content in order to better cultivating interdisciplinary professionals [7].

The joint school-running model

The joint school-running refers to the integration of the high-quality teaching practice resources of different schools, and the cooperation in cultivating students. Generally, it can be conducted through domestic joint school-running and international joint school-running, etc. [8]. At home the joint school-running often refers to the collaboration conducted in the same city between universities by taking mutual recognition of credits, minor second major, degree, etc. For example, the "Ten Hubei University Joint Second Profession Minor" activities concerning the 2008 undergraduates are currently carried out, which is a typical exploration of the joint school-running training model. But the international joint school-running often refers to 2 to 3 years of studying at home, and then 1 to 2 years of studying abroad, resulting in the gaining of the undergraduate school record issued and acknowledged by the two sides of the international joint school-running.

2.2   Analysis of Advantages and Disadvantages of Various Models

The curriculum-style cross-training model is conducive to fully developing scientific literacy of students, and enhancing their flexibility in professional employment. However, the current higher education system, in essence, is a kind of education for training professionals, in which the curriculum structure and the credit structure are obviously imbalanced. This has made these intersectional disciplinary courses mutually independent with a self-contained "collage style", which is lack of organic links between courses, and there are few interdisciplinary courses in a real sense.

Although the instructors' joint mentoring system is a good interdisciplinary training program, there are some difficulties in the execution and it is very difficult for the education system to guarantee a smooth implementation. This kind of training model requires high-quality teachers on the one hand, and, on the other hand, it requires the establishment of new professions for the training in different intersectional models, which is obviously not realistic.

Currently, the joint school-running model is more likely to be carried out in the 211 and 985 schools, while the provincial colleges and universities are also carrying on positive and effective exploration and practice. However, the ongoing mutual recognition of credits and a minor in the second profession, degree, etc. is not mature, and it is still in the initial stage of research and exploration. Meanwhile, in the international joint school-running model, only a few students have the opportunity to study abroad; and domestic joint school-running, in fact, cannot escape from the fixed school-based constraints.

Therefore, in general, in order to improve the quality of interdisciplinary professionals training, the most fundamental issue is raising the level of academic faculty, and the key to raising the level of teaching staff is carrying on high-level interdisciplinary scientific research.

3   Interdisciplinary Scientific Research Promoting Interdisciplinary Professionals Training

The Hubei University of Technology has a history of more than 50 years and it is an engineering-based and multidisciplinary key provincial university. The Electronic

Science and Technology disciplines of College of Science, Hubei University of Technology, mainly relies on two undergraduate programs in Electronic Information Science and Technology and Optoelectronics Information Science and Technology. In 2003 the college started the enrollment of undergraduate students majoring Electronic Information Science and Technology, and in 2004 it started the enrollment of undergraduate students majoring in Optoelectronics Information Science and Technology. As a result, in 2010 the college obtained the first-class enrollment eligibility for "Optical Engineering" masters. The college strictly trains the students to guarantee the quality of the disciplines, relying on an interdisciplinary training model, leveraging cross-disciplinary research and as a result, it has turned out academic graduates of excellent professional quality and strong adaptability, with the fame of "wide professional interface, strong practical hands-on ability, learning fast, and being easy to retain". Hence it has received each employer's universal high praise, and has obtained a good social reputation.

3.1 Interdisciplinary Faculty Training

The Electronic Science and Technology discipline is a intersectional professional discipline of Hubei University of Technology, including electronics and photonics technology with a number of related disciplines, like electronic technology, electronic engineering, information technology, computer application, optical engineering, optoelectronic technology and so on. The tasks of all disciplines are closely linked; in addition to acquaintance with their own professions teachers should also understand the relevant disciplines. For example, teachers of photonics technology discipline should be familiar with electronic technology, information technology, computer applications and other disciplines. There are 2 ways in which teachers can gain cross-disciplinary knowledge: one is to have access to relevant information, and the other is to have discussion with teachers of relevant disciplines, through which interdisciplinary knowledge will soon be gained in a comprehensive way. This can help develop a group of high-level cross-disciplinary faculty, and lay a good foundation for the interdisciplinary professionals training.

3.2 The Establishment of Interdisciplinary Curriculum

The most promising breakthrough in the 21st century focuses on cross-disciplinary disciplines. Therefore, in the establishment of curriculum, modeled on the curriculum-style cross-training model and profession-style cross-training model [9], we set the intersection of courses and professions, for example, we add such courses as SCM theory and application, theory and application of sensors, and EDA to the establishment of courses of photonics information and science professions; while adding photovoltaic technology, optoelectronics and other optical technology information and up-to-date science lectures to the establishment of electronic information and science professions in order to broaden the professional caliber and develop the adaptability of students. Meanwhile, in order to coordinate the credit system reform, each profession gives students a greater choice and autonomy in the establishment and choice of the curriculum.

3.3   The Training of Students' Interdisciplinary Research Capacity

Undergraduate students are the basis of academic personnel training, while graduate students are an important research force in the research and development of disciplines [10]. In the research work on disciplines, focus should be laid on attracting students to participate, whether they are directly or indirectly involved in the project. Part of the graduate students is directly involved in the project and has mastered the interdisciplinary knowledge and research capabilities. Undergraduate students and most graduate students not involved in the project can also acquire the relevant interdisciplinary knowledge through the teacher's explanation and experimental demonstration. In order to complete their research tasks students involved in the project must work on the knowledge in the relevant fields, take elective courses or study the related curriculum independently. Students working at different disciplines often have discussions together, accept guidance from teachers of their own courses and project-related teachers, and they are influenced by what they see and hear in the research process, with their knowledge quickly widened and ,as a result, they have mastered the interdisciplinary knowledge and research capabilities. Thus, applying the interdisciplinary research and training model, students have improved their professional quality constantly, have strengthened their professional capacity continuously, and have enhanced their professional standards unceasingly.

3.4   The Construction of Interdisciplinary Experimental Platform and Innovative Laboratory

The basic idea for the construction of interdisciplinary experimental platform and innovative laboratory is that we should take the students as the center, adopt open management, take creative thinking and innovative ability training as the core, and regard cultivating high-quality, interdisciplinary professonals as the purpose; we should also aim at improving students' professional quality and ability and take developing student's innovative ability, team spirit and school tradition of studying theory linked with practice as the fundamental content.

University experimental teaching cannot merely be verification of theoretical courses. It should be more likely to be the primary means of cultivating students' innovative ability and become an important place to stimulate innovation among students [11]. According to the needs of economic and social development for high-quality personnel training in the new century, we have combined the construction of curriculum closely with the construction of the key laboratory, adjusting the laboratory setting. We have synthesized the experiments at all levels, and based on this we have established a new experimental curriculum teaching system in which a scientific mutual connections have formed and the fusion, the penetration and the seepage of related contents have been established. We have set up a brand-new experimental teaching demonstration center in order to assist and promote students' awareness of innovation and innovative ability. Meanwhile, interdisciplinary innovation laboratory provides a fundamental platform for teachers and graduate students to do scientific research and therefore it can better serve the discipline construction.

4   Conclusion

General Secretary Hu Jintao said at Qinghua University Centennial Conference, "Quality is the lifeline of colleges and universities." The talent fostering is the primary responsibility of institutions of higher education, and developing versatile and innovative interdisciplinary professionals is the core task.

This article discusses the interdisciplinary professionals training model and, taking the interdisciplinary professionals training at Hubei University of Technology as an example, describes the mechanism and model of scientific research promoting professionals training. In teaching and research work we will actively study and comprehend the essence of "National Medium and Long-term Educational Reform and Development Plan(2010-2020)" and " National Program for Medium and Long-term Talent Development (2010-2020)" to further explore models and methods of the professionals training and to develop more creative and high-level professionals able to meet the needs of society.

Acknowledgment. This work is supported by Teaching Research Fund of the Hubei Provincial Department of Education (Project No: 2009224), and it is also supported by three Teaching Research Fund Projects of the Hubei University of Technology (Project No.: 2006015, 2010013, 2010014).

References

1. Xu, Q.: Interdisciplinary Talent-training and Innovation in Discipline organization. Liaoning Education Research 1, 21–23 (2004) (In Chinese)
2. National Medium and Long-term Educational Reform and Development Plan (2010-2020) issued by PRC State Council (May 2010) (in Chinese)
3. National Program for Medium and Long-term Talent Development (2010-2020) issued by PRC State Council (May 2010) (in Chinese)
4. Xu, G.: Primary Research on the Advantage and Design of Inter- discipline Training Pattern. Journal of Guangdong Institute of Public Administration 14(3), 90–92 (2002)
5. Li, P., Wang, S.: Analysis of Cases in Interdisciplinary Talen-training Model. National Administration College of Education 1, 91–95 (2004) (in Chinese)
6. Wu, X., Yu, H., Chen, C.: Comparison of Interdisciplinary Talent-training Models and Enlightenment. Journal of Zhejiang University of Technology (Social Science Edition) 7(4), 396–399, 425 (2008) (in Chinese)
7. Liu, B.: Comparison of Interdisciplinary Education and Research at Domestic and Foreign Universities. Journal of Chongqing Institute of Technology (Social Science Edition) 4, 175–177 (2009) (in Chinese)
8. Zheng, Y., Zhao, Y., Li, J.: Primary Analysis of Talent-training Models in the Construction of Interdisciplinary Cross-professions. Science and Technology Imformation 9, 12–60 (2008) (in Chinese)

9. Zhang, Y., Ren, L., He, J., Zhang, H., Liu, X.: Interdisciplinary Disciplines and Cross-disciplinary Talent-training. China Medical Engineering 6, 109–111 (2002) (in Chinese)
10. Liu, Y., Hu, J.: Exploration of Constraints of Interdisciplinary Talent-training. China Higher Education Research 3, 59–61 (2004) (in Chinese)
11. Peng, B., Cui, Y.: Construction and Practice of Interdisciplinary Experimental Teaching Demonstration Center at Engineering Colleges. Professional Space 10, 209–210 (2008) (in Chinese)

Enabling Nursing Students' Critical Thinking with Mindtools

Chin-Yuan Lai[1], Sheng-Mei Chen[1], and Cheng-Chih Wu[2]

[1] National Taichung Nursing College, 193, Sanmin Road, Section 1, Taichung, Taiwan
{yuan,csm}@ntcnc.edu.tw
[2] National Taiwan Normal University, 162, Heping East Road, Section 1, Taipei, Taiwan
chihwu@ntnu.edu.tw

Abstract. This paper reports our implementation of a web-based learning support system to foster nursing students' clinical reasoning skills. Computer-based mindtools were implemented in the system to help students engage in critical thinking within a nursing process framework. The findings showed that the majority of students considered the application tools helpful to them in doing clinical reasoning. However, a few students experienced difficulties in using the tools and suggested ways to simplify and unify the application tools.

Keywords: Critical thinking, Nursing education, Mindtools.

1 Introduction

The theory-practice gap in clinical nursing education has long been documented in nursing literature as a critical area for improvement [1], [2]. Facione and Facione [3] indicated that critical thinking could provide an engine to enhance nursing students' knowledge and then clinical decision making. Paul and Heaslip [4] argued that critical thinking not only could assist nurses by applying their knowledge in clinical practice, but also adjusting themselves to their patients' demands. Many studies developed strategies to promote students' critical thinking skills. Kuiper [5] used self-regulated learning strategies to enhance metacognitive critical thinking abilities as 32 new graduate nurses reflected during 8-week preceptorship programs. Murphy [6] adopted an evidence-based approach, including focused reflection and articulation via journal writing and post-conferences, to promote students' clinical reasoning. In recent years, critical thinking in nursing has become synonymous with the widely adopted nursing process model of practice. In this model, clinical judgment is viewed as a problem-solving activity, beginning with assessment and nursing diagnosis, proceeding with planning and implementing nursing interventions directed toward the resolution of the diagnosed problems, and finally culminating in the evaluation of the effectiveness of the interventions [7]. Brunt's [8] review of critical thinking in nursing suggested that additional research studies are needed to determine how the process of nursing practice can nurture and develop critical thinking skills, and which strategies are most effective in developing critical thinking. Duron, Limbach, and Waugh [9] have identified a five-step framework that can be implemented in virtually any teaching or training setting to effectively move learners toward critical thinking. These five steps include: 1) determine learning objectives; 2) teach through questioning; 3) practice before you assess; 4) review, refine, and improve; and 5) provide feedback and an assessment of learning

Computer technologies could be used in all subject domains as tools for engaging learners in reflective, critical thinking about the ideas they are studying. Using computers as mindtools, including semantic organization, dynamic modeling, information interpretation, knowledge construction, hypermedia, or conversation tools, will facilitate understanding more readily and more completely [10]. Among various mindtools, concept maps defined as semantic representations have frequently been used in nursing education to support students' critical thinking. Wilgis and McConnell [11] found that concept mapping was a valuable teaching and evaluation strategy for novice graduate nurses that could be used by nursing educators to improve critical thinking and identify and correct areas of theoretical and clinical deficiency. Vacek [12] used both a concept map and a conceptual approach to teach the concept of psychosis instead of focusing on content and found students could identify and implement nursing actions for patients with psychosis regardless of the etiology or health care setting.

Although many disciplines have used computer technology as mindtools to promote students' gains in critical thinking, integrating mindtools with critical thinking teaching is still relatively new to nursing. The purpose of this study is to implement a web-based learning system and provide mindtools to support nursing college students' learning in a "Clinical Nursing Practice" course. The course is an essential component of a nursing education program. It aims to help students link theory with practice at the final stage of the training program. In the study, students used our system and mindtools to develop their understanding of the nursing process in clinical settings. This paper reports our implementation of a system and the effects on the students' clinical learning.

2   Mindtools to Support Critical Thinking

The nursing process involves five major steps [13]: A-Assessment, D-Diagnosis, P-Plan, I-Implementation, and E-Evaluation. Below we describe how mindtools are incorporated in each step.

2.1   Assessment (What Data Is Collected?)

Assessment is a crucial step in the nursing process. It is the phase where the nurse systematically gathers comprehensive, relevant, valid, reliable and complete information from the patient and other credible sources in order to identify health problems, make a diagnosis, and then plan, implement and evaluate the given nursing care. During this phase, students used Scales to collect patients' data and then used Spreadsheets to analyze the quantitative information to consider the implication of patients' conditions. We collected various symptom assessment scales in Scales to support the gathering and processing of patients' information.

2.2 Diagnosis (What Is the Problem?)

Diagnosis is the second phase of the nursing process which usually entails making a statement of the patients' health problems. It involves an intellectual activity wherein students use the diagnostic reasoning process to draw conclusions about the patient's health status and decide whether nursing intervention is needed. According to the standard nursing diagnosis format, students first present the nursing diagnosis and then include the problems and original statements [14].

To assist students to implement this decision-making process, we developed a *Patient Problem Retrieval System* to allow them to search patients' health problems. Like an intelligent information search engine, found on the World Wide Web, when a student inputs a keyword such as "anxiety", the screen displays the defining characteristics (i.e. symptoms) about the "anxiety" of psychiatric patients according to five aspects: a physical dimension, a mental dimension, a social dimension, an intellectual dimension, and a spiritual dimension. After students have a complete understanding of patients' health problems and their relative definitions, they use semantic networking tools to draw a concept map to present the relationship between the nursing diagnosis and the health problems (i.e. *Clinical Reasoning Web* [15]).

2.3 Plan (How to Manage the Problem)

In this phase, students determine problems that need individually developed plans and those that can be addressed by routine interventions contained in standards of care, model care plans, protocols and other forms of preplanned, routine care [14]. To attain this aim, students first must seek one central problem which contributed to related ones from the *Clinical Reasoning Web*. They would then refer to *Resources* to formulate an individualized nursing plan. We located relevant resources of nursing intervention and health education in *Resources*. Students could use it as mindtools to help them access and process information. For example, if a student infers that a patient's main problem is "impaired social interaction", he or she will draft a nursing intervention plan about improving this problem such as encouraging patients to participate in morning exercises, team therapeutic activities, etc.

2.4 Implementation (Putting the Plan into Action)

Implementation is the fourth phase of the nursing process in which students apply the plan and put their hypothesis to the test. During this phase, except for the hospital's routine nursing intervention, the instructor guides students by designing planned team activities for all patients. Each day, during the three weeks of nursing practicum, one student planned and arranged one kind of team activity to serve as the host of the activity. The other students provided assistance while the activity was proceeding. To make these daily activities successful, students had to work together while still being occupied with their daily duties. Students, therefore, frequently used *Forum* as a discussion platform.

2.5   Evaluation (Did the Plan Work?)

Evaluation is ongoing throughout the various phases of the nursing process. The student evaluates whether sufficient assessment data have been obtained to allow a nursing diagnosis to be made. The diagnosis is in turn evaluated for accuracy and appropriateness to the patient's health problems. The nurse evaluates whether the expected outcome and interventions are realistic and achievable. If not, an alternative plan should be formulated [14]. At this time, students may check back to the mindtools (*Scales, Clinical Reasoning Web* or *Patient Problem Retrieval System*) to analyze the problems which caused the plan to fail and again access the material in *Resources* to reframe another nursing plan.

3   The Learning Support System

A web-based learning system, along with the above mentioned mindtools, was implemented to support students' clinical practicum sessions. The system contains nine sections: *Environment Introduction, Practicum Guide, Bulletin, Forum, Resources, Performance, Scales, Performance Sharing*, and *Self-evaluation*. These functions were designated in accordance with a five-step framework of critical thinking teaching [9].

The *Forum, Resources, and Scales* "functions" correspond to the *Forum, Resources, and Scales* "mindtools," respectively. Other functions, such as the *Environment Introduction* and the *Practicum Guide,* were to direct the students to quickly enter into clinical learning. The *Bulletin* shows course information and the instructor's message to students. In the *Performanc*e section, students were asked to submit a series of assignments which could present their clinical learning and performance, including Reflective Journal Writing, Nursing Process Reflection, Clinical Reasoning Web, OPT Worksheet, Activity Design, etc. *Performance Sharing* encouraged students to analyze and learn from each other's work. *Self-evaluation* is to help students learn how to assess their own performance.

Among the various assignments, the Reflective Journal Writing, the Nursing Process Reflection and the OPT Worksheet were designed to allow students to frequently practice critical thinking. Firstly, Reflective Journal Writing allows students to better assess their needs, and at the same time, allows their instructor to provide them with necessary guidance and support. It was designed following Johns Model of Reflection [16], which is based on five cue questions which enable students to analyze their experiences and reflect on the process and outcomes. Furthermore, the Nursing Process Reflection enables students to reflect on the nursing process. We adopted Nielsen, Stragnell, and Jester's [17] guide of Reflection on Clinical Reasoning in designing this assignment. Finally, the OPT worksheet proposed by Pesut and Herman [15] provides students with a framework of clinical reasoning to evaluate the "circulating nursing process." When the students submit these three assignments to the system or revise and submit them again, they quickly get the instructor's feedback on-line.

4   Results and Discussion

We used the web-based system to support a three-week nursing clinical practicum session. Eight female students enrolled in the practicum session participated in this study. Students were each provided a Netbook (ASUS Super 10A) to use throughout the practicum. The practicum session was designed to provide students with clinical experiences in the area of psychiatric nursing. The research instrument used in this study was a post-activity questionnaire with ten open-ended questions. The students were asked if and why they thought the applications, especially those designated for psychiatric practice or engaging their critical thinking, were helpful in facilitating their clinical reasoning to provide any additional comments about the use and design of tools for their future clinical learning. That is, we did not ask students about all applications because some were already extensively used (such as spreadsheets and a discussion forum). To analyze the data generated from the open-ended questions, the content analysis method was used.

Table 2 is the summary of the number of students reporting on each application. Among the seven applications, six (Numbers 1 to 6) were reported as useful by almost all students. One application (Number 7) was thought not so useful: four students agreed that it was helpful; four students didn't. Below we describe the benefits, problems, or suggestions which the students reported on each application.

Table 2. Number of students reporting helpfulness, unhelpfulness and suggestions or problems on the applications

Applications		helpful	not helpful	suggestions or problems
1.	Scales	8	0	0
2.	Patient Problem Retrieval System	8	0	0
3.	Resources	8	0	3
4.	Clinical Reasoning Web	7	1	2
5.	Reflective Journal Writing	7	1	0
6.	Nursing Process Reflection	8	0	2
7.	OPT worksheet	4	4	5

All students indicated that *Scales* was useful in their gathering of patients' information. Their opinions included statements, such as "It helps me a lot –I have more understanding of my patient's condition"; "It allows me to collect patients' data quickly, clearly and "grasping" the weight-bearing point, I didn't need to collect additional information"; "The Scales allows me to apply the collected data to finish other assignments such as my Nursing Process Reflection and Clinical Reasoning Web."

All eight students agreed that the *Patient Problem Retrieval System* and *Resources* were useful in clinical judgments. The students said: "The retrieved information is coherent." "The data are complete." "The information is appropriate and helpful to me. It helps me when making a clinical judgment." However, despite the above comments, three students strongly suggested that more information was needed in

Resources to allow them to retrieve relevant information more conveniently and profoundly.

Seven students thought the *Clinical Reasoning Web* was helpful to clinical judgment; one student said its benefit was limited for her because it took time. The benefits which the students reported were that it deepened their understanding of nursing problems and symptoms and their causal relationship, and the relevance with clinical judgment. One student duly noted: "It makes me more attentive to my patient".

Seven students said that the *Reflection Journal Writing* was helpful to their clinical learning. As students indicated, their Reflection Journal Writing not only promoted their self-awareness, but also enhanced their professional development. They commented: "Due to the instructor's suggestion in my Reflection Journal Writing, I could understand my drawbacks and lack of information, and thus get the direction for further learning." "It helps me recall what happened today- to process and reflect on how I could improve." "While reflecting, I learned how to reduce my anxiety." One student, however, indicated that keeping a journal was unhelpful because it took time.

All students agreed that *Nursing Process Reflection* benefited their learning. Their reported benefits included: "having more understanding of patients' background information and collected data...", "better recognition of patients' various symptoms...", "better understanding of the relationship between this assignment and clinical judgment". Two students, however, indicated they had problems in writing this assignment. One student said she had some difficulty in understanding the meaning of some guiding words in the Nursing Process Reflection Assignment; the other student reported that she was not yet accustomed to using it for clinical refection.

The effect of the *OPT Worksheet* is neutral: half of the students indicated that it was helpful; the other half thought it was not. The reasons why it was unhelpful included that its features repeated those of other tools, or it was too complicated and took too much time to finish. One student commented, "I could use Scales to make clinical judgments without an OPT Worksheet." Another student said, "The framework of OPT is excellent for clinical judgment, but the written words in OPT are not easily understood, I must query the instructor or other students". Maybe in future re-designing, we could simplify this worksheet or integrate it with other applications.

Besides the various opinions on different applications, we asked them how this clinical learning experience could affect their future nursing work. Most students posited that the experience was very different from previous practicum sessions, and on the whole it was constructive, although there were many assignments to be finished. They indicated that they gained substantially not only in self-growth, but also in professional development. Most importantly, they said they will be willing to use these application tools to help them collect patients' information and make clinical judgments in the future. Two students, however, reported that using the computer was not convenient for them in the hospital because it was difficult to find a seat to access the system.

5 Conclusions

The results show that students felt the application tools of the system on the whole were helpful in facilitating their clinical reasoning. Most students opined that they are ready to use them in future nursing work. However, a few students pointed out several issues, such as lack of time to finish the assignments, having difficulties in understanding the functionalities of application tools, and inconvenience in using a computer in the clinical setting. Further simplification and unification of the application tools may alleviate some of the problems. Moreover, as some students suggested, with more detailed information about patient care being added into the system, the effectiveness of the tools would become even more obvious.

References

1. McCaugherty, D.: The Theory-Practice Gap in Nurse Education: Its Causes and Possible Solutions. Findings from an Action Research Study. Journal of Advanced Nursing 16, 1055–1061 (1991)
2. Landers, M.G.: The Theory-Practice Gap in Nursing: The Role of the Nurse Teacher. Journal of Advanced Nursing 32(6), 1550–1556 (2000)
3. Facione, N.C., Facione, P.A.: Holistic Critical Thinking Scoring Rubric. The California Academic Press, Millbrae (1994)
4. Paul, R.W., Heaslip, P.: Critical thinking and Intuitive Nursing Practice. Journal of Advanced Nursing 22, 40–47 (1995)
5. Kuiper, R.: Enhancing Metacognition through the Reflective Use of Self-regulated Learning Strategies. The Journal of Continuing Education in Nursing 33, 78–87 (2002)
6. Murphy, J.I.: Using Focused Reflection and Articulation to Promote Clinical Reasoning: An Evidence-Based Teaching Strategy. Nursing Education Perspectives 25, 226–231 (2004)
7. Tanner, C.A.: Thinking like a Nurse: A Research-Based Model of Clinical Judgment in Nursing. Journal of Nursing Education 45(6), 204–211 (2006)
8. Brunt, B.A.: Models, Measurement, and Strategies in Developing Critical Thinking Skills. The Journal of Continuing Education in Nursing 36(6), 255–262 (2005)
9. Duron, R., Limbach, B., Waugh, W.: Critical Thinking Framework for Any Discipline. International Journal of Teaching and Learning in Higher Education 17(1), 160–166 (2006)
10. Jonassen, D.H., Carr, C., Yueh, H.-P.: Computers as Mindtools for Engaging Learners in Critical Thinking. Tech Trends 43(2), 24–32 (1998)
11. Wilgis, M., McConnell, J.: Concept Mapping: An Educational Strategy to Improve Graduate Nurses Critical Thinking Skills During a Hospital Orientation Program. Journal of Continuing Education in Nursing 39, 119–126 (2008)
12. Vacek, J.: Using a Conceptual Approach with a Concept Map of Psychosis as an Exemplar to Promote Critical Thinking. Journal of Nursing Education 48(1), 49–53 (2009)
13. Nursing Process, http://en.wikipedia.org/wiki/Nursing_process
14. Chabeli, M.M.: Facilitating Critical Thinking within the Nursing Process Framework: A Literature Review. Health SA Gesondheid 12(4), 69–90 (2007)

15. Pesut, D.J., Herman, J.: Clinical Reasoning: The Art and Science of Critical Creative Thinking. CENGAGE Delmar Learning (1998)
16. Johns, C.: Framing Learning through Reflection within Carper's Fundamental Ways of Knowing in Nursing. Journal of Advanced Nursing 22(2), 226–234 (1995)
17. Nielsen, A., Stragnell, S., Jester, P.: Guide for Reflection Using the Clinical Judgment Model. Journal of Nursing Education 46(1), 513–516 (2007)

Canonical Chinese Syntax Awareness Facilitated by an e–Learning Program*

C.C. Lu[1], C.H. Lu[2], M.M. Lu[3], C.H. Hue[4], and W.L. Hsu[2]

[1] National Hsinchu Univ of Education
[2] Academia Sinica
[3] Natl Univ of Tainan
[4] Natl Taiwan Univ

Abstract. Being aware of syntactic structures is important in learning a new language. However, the syntactic concepts could be very elusive and vague and learning syntactic structures could be very dry and tedious.

In this study, based on our analysis of Chinese regulated verses in Tang Dynasty, we designed an e-learning program. The aim of the program is to make the language learners familiarized with the frequent grammatical structures of the language, alone with learning the vocabulary in the same semantic field.

The canonical Chinese syntactic patterns emerged from our corpus after delicate analysis, and a database which contained various word classes was created. Several steps were set up for helping the Chinese learners to make up one line in an antithetical couplet. First, the learner was asked to choose the topic, and then to choose the semantic frame and the canonical syntactic tablet. Second, one line in an antithetical couplet selected from our database was shown on the screen, and the learner's task is to make up another line by choosing words from a fixed set offered by the program. Matched with the screening criteria coded in our program, several candidates were suggested for each slot of the couplet. The learner was asked to choose the words for his own preference and finished the poem. After that, the learner could start another new trial.

In order to examine the learning effect, a syntactic awareness test was constructed and given to the Chinese learners before and after the training session. In addition, some testing items were presented to the learners during the training session and their responses were recorded and graded, in order to assess their formative learning effect in using this e-learning program.

1 Introduction

Learning Chinese as a second language has been prevalent these days. Many learning tools and software, focusing mainly on the pronunciation or vocabulary building, are developed for helping the Chinese learners. However, as Ravelli (2004:123) mentioned, it is not just access to the organizing vocabulary that is important, but also the "control of the supporting colligation patterns [that] is also crucial." Therefore, in this study, we try to provide a learning program to promote Chinese learners' awareness in Chinese sentence structures.

* This study was supported by NSC 99-2631-S-001-001.

As suggested in Hoey (2005) , different types of input are hypothesised as having different degrees of impact on the primings of language use. With particularly valued input, like literary or religious texts, or the words of a close friend, Hoey elaborated in his argument, the primings could be particularly salient. The input of interesting e-learning programs, we postulated, could also be one with large effect.

2 Program Development

The motivation for our software program is trying to help Chinese learners to be aware of the usage of the canonical word order patterns which could be very useful to compose and make sentences in speaking and writing Chinese, and to parse and comprehend most sentences in listening and reading Chinese.

2.1 The Materials

The Materials used in this program were the Chinese Regulated Verses. The Regulated Verses became an established verse form at the beginning of the Tang dynasty (618-907 AD). It was also known as the Modern Style in contrast to the Ancient Style. There were strict metrical rules followed by the poets for composing the regulated verses. A typical regulated verse consisted of 8 lines, with 5 or 7 syllables in each line. The syllables in the end of the 2nd, 4th, 6th, 8th lines were rhymed. The tone for each syllable in a line was chosen to fit a fixed pattern of tonal categories. And it usually contained two antithetical couplets in the poem: the four lines in the middle, that is, lines 3 & 4, and line 5 & 6. The two lines in an antithetical couplet contrasted each other in meanings, tonal categories, as well as in grammatical structures. Nonetheless, due to the sound change as time went by, the same character might be read differently nowadays. Therefore, in our program, the learning focus was on the antithetical part rather than the other phonological regulations.

Even though, the corpus of Tang poems was adopted with three reasons. The first reason is that their constructions reflected Chinese highly-used syntactic patterns, still frequently used in modern society. The second reason is their descriptions were beautiful. Basically, they were pleasant cultural heritages to enjoy. And what is more is that the frequent word orders embedded in the regular texts would enable the learners to focus on the analogical practices, in that the retrieval of word order patterns in random texts with various numbers of words is much more difficult than those in regular expressions.

However, not all data of Tang poems was suitable for designing e-learning. One of the important reasons was that in proposing the candidates of the same semantic prosody, the lyrics which shared similar semantic frames would offer more choices than the lyrics with divergent themes. Therefore, the database was selected for the pilot program, in order to meet the ends of our curriculum. After analyzing the works of numerous Tang poets, 152 poems describing the natural scenes written by Wei Wang were adopted.

2.2 Canonical Word Order Extracted from Our Corpus

The design rationale in this study is based on colligation, the grammatical combination of words or a word category (Firth, 1957; Hoey, 2005; Hunston and Francis, 2000; Stubbs, 2002). Colligation defines the grammatical company and interaction of words as well as their preferable position in a sentence. For Instance, grammatical pattern [verb+to-infinitive] is an example of colligation.

Both Sinclair (1991) and Hoey (2005) contended that co-occurrence frequencies indicate principles of linguistic patterning across traditional levels of analysis, including grammatical patterns and patterns of discourse. And Hunston and Francis (2000) have built on Sinclair's work to propose a description of language in terms of patterns.

It is shown in Hoey (2000) that how corpus analysis reveals previously unnoted types of regularity. Thus, we started to work with corpora to abstract the canonical word order patterns. The word order patterns here we mean the abstract indicated by the co-occurrence frequencies (Sinclair, 1991, p. 103). In this sense, as we shall see below, the term "patterns" has much in common with the construction grammar model proposed by Goldberg (1995).

The analysis and classification of the corpus is based on our working experiences of The Grammatical Knowledge-base of Contemporary Chinese in doing Chinese Information Processing researches from 1995. Through the continuous development in the past years, our lexicon has extended to 73,000 entries based on the classification of seventy thousands words. We did the segmentation and parsed the words with the GKB (The Grammatical Knowledge-base of Contemporary Chinese) dictionary. Then we double-checked the results and abstracted the canonical word order patterns. Five canonical word orders were extracted from our analysis in the pilot work. Please see the details in the Appendix.

2.3 The Scenario of the Program

Our program was designed for use in the classroom or for self-learning. The level was set as intermediate, suitable to the Chinese learners who had some basic knowledge of Chinese characters. Encouraged step by step, the learners could observe the pattern of the word order information provided in the program, and build up the grammatical concepts through statistical learning (Li, Farkas, & MacWhinney, 2004). A scenario is shown as Fig 1.

The users followed the steps shown on the screen:

1.The five frequently-used canonical syntactic patterns used in the Antithetical Couplets in Chinese Regulated Verse were shown on the screen. The users had to choose one from them.

2.The first line of some certain antithetical couplet was shown on the screen.

3.The choices for the first part of the antithetical couplet were suggested by the computer. The user had to make a choice from the candidates.

4.The choices for another part of the antithetical couplet were suggested by the computer. The user had to make a choice from the candidates.

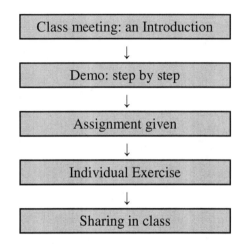

Fig. 1. The Scenario of the program

For instance, after the user had chosen a syntactic pattern, the first line of some couplets which matched the pattern was shown, as:

江流	天地 外 ，
Topic	Comment
NP	NP+locative

1. The first line of a couplet pair was shown on the screen;
2. The learner was asked to choose the contrasting NP for the
   Topic part, from the choices suggested, and finished the Topic part;
3. Then the learner had to choose an NP for the Comment part;
4. Then the learner had to chooses a locative for the previous NP, and finished
   the Comment part.

The learners might implicitly acquire the canonical word order patterns as they picked up the words suggested during the exercises. However, the teachers might also provide an explicit focus during their learning, and pinpoint the relations among the words in a salient pattern, in order to facilitate the learner's metalinguistic awareness for Chinese grammars. That would be very helpful for the patterns which learners appeared to have difficulties in acquiring. Beside of that, this program might also make the learners familiarized with the synonyms and antonyms in the same semantic field, as well as the lexical items which were associated within a semantic schema.

3   Conclusion

The program has some significant advantages. On the top of them is to help the language learners to identify those patterns which are syntactically regular because of their high frequencies of occurrence from samples of text far too large for the learners to handle. The awareness of canonical word orders in Chinese may guide the learner' attention in other authentic situations and may help them to become a fluent user of Chinese later on.

References

1. Firth, J.R.: Papers in Linguistics,1934-1951. Oxford University Press, London (1957)
2. Goldberg, A.: Constructions: a construction grammar approach to argument structure. University of Chicago Press, Chicago (1995)
3. Hoey, M.: The hidden lexical clues of textual organization: a preliminary investigation into an unusual text from a corpus perspective. In: Burnard, L., McEnery, T. (eds.) Rethinking Language Pedagogy from a Corpus Perspective, pp. 31–42. Peter Lang, New York (2000)
4. Hoey, M.: Lexical priming: A new theory of words and language. Routledge, London (2005)
5. Hunston, S., Francis, G.: Pattern Grammar: a corpus-driven approach tothe lexical grammar of English. John Benjamins, Amsterdam (2000)
6. Li, P., Farkas, I., MacWhinney, B.: Early lexical development in a self-organizing neural network. Neural Networks 17, 1345–1362 (2004)
7. Ravelli, L.J.: Signaling the organization of written texts: Hyper-Themes in management and history. In: Ravelli, L., Ellis, R. (eds.) Analyzing Academic Writing: Contextualized Frameworks, Continuum, London, pp. 105–130 (2004)
8. Sinclair, J.M.: Corpus, Concordance, Collocation. Oxford University Press, Oxford (1991)
9. Stubbs, M.: Words and Phrases: Corpus Studies of Lexical Semantics. Blackwell, Oxford (2002)

Appendix

Pattern 1

Topic	+	Comment
↓		↓
NP		Locative Phrase
		↓
		N/V + Locative Marker

Examples:

江流	天地	外，
River	World	Outside

山色	有無	中。
Mountain View	Exist+Not exist	In between

This pattern consists of a topic and a comment. The topic is consisted of a disyllabic NP, originally might come from the compounding of two monosyllabic nouns, and a Locative Phrase, whose head is a locative marker. The locative marker is a special part of speech in Chinese. It behaves as a head in a head-final position. So its position is different from the proposition in English. The component in the Locative Phrase before the head locative marker could be a compound disyllabic N or a compound disyllabic V.

Pattern 2

Topic   +   Comment
↓   ↓
NP   VP
↓   ↓
Adj+N   Locative Phrase   +   V
                                                     ↓
                                      N + Locative Marker

Examples:

明+月   松 + 間   照 ，
Bright+Moon   Pine + Between
Illuminate

清+泉   石 + 上   流 。
Clean+Spring   Rock + Upon   Flow

This pattern consists of a topic and a comment. The topic is consisted of an NP, consisting of a monosyllabic adjective and a monosyllabic noun. The monosyllabic noun is the head of the NP and the monosyllabic adjective is its modifier. The comment consists of a Locative Phrase and a monosyllabic verb. The monosyllabic verb is the head of the VP and behaves as a predicate. The Locative Phrase before the head verb is its modifier. The Locative Phrase consists of a monosyllabic head locative and a monosyllabic noun before it.

Pattern 3

Topic   +   Comment
↓   ↓
Locative Phrase   Predicate Phrase
↓   ↓
N + Locative Marker   NP   +   V/Adj
                                      ↓
                                     N+N

Examples:
屋 + 上   春 + 鳩   鳴 ，
House +Upon   Spring+ Dove   Twitter

村 + 邊   杏 + 花   白 。
Village + Beside   Apricot + Flower   White

This pattern consists of a topic and a comment. The topic is consisted of an Locative Phrase, which consists of two parts: a monosyllabic noun and a locative marker, which is head of the Locative Phrase. The comment is a Predicate Phrase, consisting of a head predicate and a subject of the head. The head predicate head in a Chinese sentence could be either a verb or an adjective. The subject is a NP, with a compounding structure of two monosyllabic nouns. The position of the subject NP is before the predicate.

Pattern 4

	Topic		+	Comment
	↓			↓
Adj	NP			Adj
	↓			
	Adj +N			

Examples:

皎潔	明 星		高 ，
Pure	Bright+Star		High

蒼茫	遠 天		曙 。
Vast	Far + Sky		Sunrise

This pattern consists of a topic and a comment. The topic is consisted of an NP, which consists of two parts: a disyllabic adjective and a disyllabic NP, which consists of a monosyllabic adjective and a head noun. The disyllabic adjective is the modifier of the following NP. The comment consists of a monosyllabic adjective, which behaves as the predicate for the topic. But in Chinese sentence construction, the predicate adjective could be immediately used as a comment without a BE verb.

Pattern 5

Topic	+	Comment		
↓		↓		
NP		V	+	NP
↓		↓		↓
Adj/Vi+N		V		Adj/N + N

Exampless:

荒 + 城		臨	古 + 渡 ，
Desolate+Town		Face	Old+ Ford

落 + 日		滿	山 + 秋 。
Setting+Sun		Fill	Mountain+Autumn

This pattern consists of a topic and a comment. The topic is consisted of an NP, which has a head monosyllabic noun and its modifier, which could be an adjective or a verb showing the status or the action of the head noun. The comment consists of a VP, which has a monosyllabic head verb and its object disyllabic NP, consisting of a monosyllabic head noun and its modifier, which could be an adjective or a noun.

Multilayer Fundamental Physics Curriculum-Group-Construction with Opening and Mutual Learning in Selective Instruction and Discussion

Shi-jun Xu[1], Xiao-ling Ren[2], and Jia-qing Cui[1]

[1] School of Science, Xi'an Technological University, Xi'an 710032, China
xushijun000@sina.com, cuijiaqing@xatu.edu.cn
[2] School of Computer, Xi'an Polytechnic University, Xi'an 710048, China
renxiaoling2@163.com

Abstract. In common polytechnic universities of China, the curriculum-group-construction of fundamental physics has many difficulties, such as few teaching hours, less interaction between teachers and students, and bad effect etc. A high effective curriculum construction mode was proposed, which core was development of excellent physics course. Based on the excellent course, multilayer fundamental physics curriculum-group-construction mode with opening and mutual learning in selective instruction and discussion was established. In the four-level-course practices in common polytechnic university, we set up a multidimensional physical teaching materials system, an online interactive learning system, a selective instruction mode in class, a discussion-teaching mode, a comprehensive examination pattern, a multi-role demonstration experiment system, and an optimization of modern education technology and multiple teaching resources and means. The multilayer fundamental physics curriculum-group-construction mode has improved learning interest, learning ability and innovation ability for most students. Because of many distinct characters and remarkable achievements, this mode should be popular with other majors or courses.

Keywords: curriculum construction, multilayer course, fundamental physics, discussion, interaction, science and technology platform.

1 Introduction

In the stage of popular higher education, the fundamental physics teaching in common polytechnic universities (especially in the west) is facing with tremendous pressures. The pressures show no excellent teachers, few hours (about 100), large-scale students in class, weak learning foundation, no advanced teaching methods and modern educational technology which is far away from the requirements of information development, insufficient laboratory equipment and demonstration experiments, single assessment or examination mode which is difficult to develop good teaching, bad studying initiative, weak studying interest, and less teacher-student interaction. Meanwhile, the curriculum construction theory of fundamental physics[1-2] and the guiding thought is not scientific, no prominent construction achievements and

Y. Wang (Ed.): Education Management, Education Theory & Education Application, AISC 109, pp. 481–488.

characteristics lead to low effectiveness. Accordingly the theory cannot play a fundamental role in the subsequent courses teaching and developing students, physics-quality and science accomplishment.

During the curriculum construction in the past, based on the teaching real status of the common polytechnic universities and lots of various teaching research projects or funds, we had established a curriculum-group-construction mode which set excellent course as core course. The construction achievements and characteristics of fundamental physics were prominent and distinguished. Under the fundamental physics teaching guiding thought, which was selective instruction and discussion, we had constructed a multilayer fundamental physics course-group system with opening and mutual learning characters. We had improved learning interest, consequently, developed students learning ability, and ultimately enhanced teaching effectiveness. It had supported the teaching fundamental physics in large-scale enrollment and the multidisciplinary development of universities. The course construction practice and theory was awarded the Outstanding Teaching Award of Shaanxi Province.

2   Basic Facts of Fundamental Physics Curriculum-Group-Construction

2.1   Curriculum-Group-Construction Model and Guiding Thought

We found that the excellent course construction and the curriculum-group-construction are encouraged and relied on each other [3-4]. The domain of curriculum-group-construction is larger than course construction. Because the excellent course is constructed on the curriculum-group, many excellent teachers and teaching resources can be concentrated to build carefully one or two courses in the excellent course construction. At last, we have formed a new course construction mode: "the key course-group of the school" to "the key course-group of the university" to "provincial or national level excellent course". This curriculum-group-construction mode, which set excellent course as core, can not only build excellent course on the basis of the curriculum-group but also promote the curriculum-group-construction with the help of the excellent course construction. This "excellence" plays a strong demonstration effect on the teaching reform of the other courses, and plays a leading role on the overall teaching effects, meanwhile it is easy to break the "bottlenecks" of poor teaching condition.

Therefore, the guiding thought of fundamental physics curriculum-group-construction is as follows. Based on the university-running orientation and characteristics of in common polytechnic universities, which are the emphasizing on engineering practice and the pushing manufacturing technology forward, we have optimized the teaching contents and teaching methods with the guiding thought of selective instruction and discussion. Finally, we have constructed a multilayer curriculum-group system in fundamental physics with "selective instruction & discussion & open & mutual learning" and a teaching mode in fundamental physics with "open and interactive study". All teachers must strengthen the monitoring for teaching research and teaching process, rationalize the relation of course-group and excellent course, highlight the university physics excellent course construction and ultimately improve the quality of teaching.

2.2 Main Ideas and Achievements of Curriculum-Group-Construction

The constructing ideas of teaching contents are the enhancing classic physics, separating modern physics, linking with engineering & physics history. Based on the teaching conditions of common polytechnic universities, we require some teaching materials of "reinforcing the fundamental contents and linking with engineering contents", and optimization or integration of multiple teaching resources and teaching means. We must carry out the comprehensive examination pattern and promote excellent course construction by the teaching research projects.

In recent years, based on the key curriculum-group-construction platform, with the carrier of teaching and research projects, we have made significant progresses on the training excellent teacher, the integration of course contents, the optimization of teaching methods and means, the construction of multidimensional and systematical teaching materials [5], and the comprehensive assessment or examination. The teaching quality can be keep at a higher level. The characters of the basic course platform and engineering and opening and interaction are highlighted. The student interests in learning are improved. The learning abilities are cultivated. The students' physics-quality and science accomplishment are promoted. Now our College Physics course is award the title of Shaanxi Province Excellent Courses.

3 Contents of Fundamental Physics Curriculum-Group-Construction

3.1 Training Excellent Teachers

The teacher team is developed effectively by the extensive and in-depth teaching research activities, possessing a series of reasonable makeup on title of technical post, age, education background and major. They have great potentials, rich teaching experiences, a spirit of collaborative and innovative.

In a word, the now teaching level is much higher than five years ago. We have 16 teachers, including 3 professors, 5 associate professors, 6 instructors, 6 young teachers under the age of 35, one teacher over the age of 50. The proportion of master is 90% and the proportion of doctor is 31%. Their majors are distributed throughout condensed matter physics, optics, optical engineering, physical electronics and optoelectronics etc. We have obtained a series of fruitful achievements of course construction and reform. For example, awarded 25 teaching prizes in the last five years, included five provincial prizes, undertaking 2 national research projects and 5 provincial or ministerial research projects, 18 SCI or EI papers. In addition, the teachers of curriculum-group are also committed to the construction of applied physics undergraduate major and physics postgraduate major. The achievements have served effectively the university-running idea "making the engineering and technology as the centre, making arts and science as the support, the harmonious development of multi-disciplinary".

3.2 Optimization and Integration of Teaching Contents

According to the fundamental physics teaching thought, we have established a multilayer fundamental physics content system for common polytechnic universities(as shown in figure1).

Considering the constructing ideas of teaching contents, which are the enhancing classic physics, separating modern physics, linking with engineering & physics history, we have

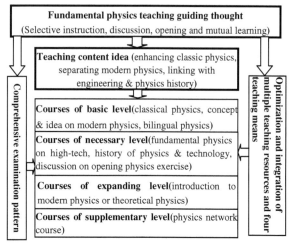

Fig. 1. Thought and structure of teaching contents

created a four-level-course system by effectively expanding a single course to series of courses. The four-level-course system includes the courses of basic level, necessary level, expanding level, and supplementary level. These courses of basic level are core and main courses, which are almost required. The courses of necessary level (e.g. three or five elective courses) are supplement and feature courses. The courses of expanding level are wide extension and sublimation. All each level subjects are prioritized and complemented each other, thus the optimal integration to course group is achieved. This reform idea of teaching content system is so distinguished compared with other domestic universities.

For the practical teaching contents, we have mainly completed the observation and analysis of physical phenomena through opening experiment, demonstration experiment, video courseware showing, computer simulation and other modern education technologies.

3.3 The Construction of Teaching Conditions

We have built a multidimensional teaching material system of fundamental physics (as shown in figure2).

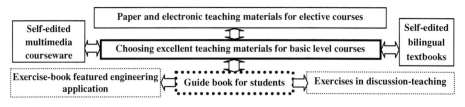

Fig. 2. Multidimensional teaching materials of fundamental physics

The system includes the choosing excellent domestic teaching materials, which selection rules are "applicability first" and "taking excellence into account". The breakthrough of fundamental physics multidimensional teaching materials is suitable supplementary teaching materials (the guide book for students, the exercises in discussion-teaching, the exercise-book featured engineering application). The self-edited bilingual textbooks and multimedia courseware are supplement of teaching materials. The paper and electronic teaching materials for elective courses are expansion of public elective course textbooks and electronic materials. This multidimensional teaching materials system is interrelated and promoted each other. It is suitable for the university-running orientation and characteristics of in common polytechnic universities. This study emphasizes development of learning ability, using modern education technology, integrating various media and resources.

The online physics teaching system including physics theory instruction, physics experiments, physics exercise, physics world and physics garden, provides the students an interactive learning environment, shares lots of teaching resources, enlarges the channels of learning physics knowledge for students, realizes the self-learning dream of students.

3.4 Teaching Method and Mean and Examination Mode

Many achievements in teaching methods and means and examination model are developed by the extensive and in-depth teaching research activities. These are the selective instruction mode in class, the discussion-teaching mode to exercise [6], the comprehensive examination pattern, the multi-role and opening demonstration laboratory with research functions, and the optimization of modern education technology and multiple teaching resources and means a variety of teaching methods. Especially, we had proposed a small class exercise teaching model with opening and interactive discussion, which was affirmed by the Physics and Astronomy Teaching Advisory Committee of Ministry of Education of China (PATACMEC)!

Selective instruction mode in class. In common polytechnic universities, combining the teaching conditions with the university-running orientation and characteristics for developing advanced talents, we had studied a selective instruction mode in class. The model emphasizes on comprehensive and correct understanding to the basic physics concepts and laws, basic applied ability at a specific direction, an overall understanding of the physics framework system, scientific methods and physical history. The mode characters are "instructing selective core contents" and "the teaching tendency in engineering", The combination of selective or intensive instruction and active learning is widely used in class teaching. The instruction focuses on the basic ideas and concepts and laws and methods and the knowledge structure, as well as, the physics history, the scientific method, the tendency of engineering applications. At the same time of the elective instruction in class, the discussion teaching with interaction and opening is applied in the four-level courses.

Discussion-teaching to exercise. The discussion-teaching to exercise is sure a supplement to the above selective instruction mode in class. In the practices, we had explored many various-flexible and rich-content teaching methods, such as "one student discussion, other students re-correction and correct", "random panel discussion", "student representative debate", "different classes with different discussion questions" and "non-average score", "speech score and innovation score" and so on.

Comprehensive examination pattern. The comprehensive examination mode or pattern is consistent to the teaching guiding thought of selective instruction and discussion. It can promote the interaction of teaching-learning, accurately get the unity of the teaching process and examination objectives. We turned "the only one exam" to "the multiple and multi-form assessment", made the subjective ability as the assessment center. The final scores are made up of four parts: exercise class scores, demonstration design results, small paper scores and last exam scores. Final examination is often related to the basic concepts, basic rules and physics ideas, avoiding excessive skills and applied knowledge, For example, we increased some discussion questions and subjective questions, focused on developing thinking ability of student.

Multi-role and opening demonstration laboratory with research functions. For the fundamental physics series courses in common university, it is required establishing the multilayer teaching models with opening and mutual learning characteristics. In the models, a important content is establishing versatile and opening laboratory, which has demonstration and research functions, and will promote the laboratory level. Based on demonstration experiments teaching, we had explored a teaching reformation plan aimed at integrating theoretical course teaching and experiment course teaching. Obviously the multi-role and opening demonstration laboratory is key[7]. To building versatile high level science and technology activities platform for teachers and students, the laboratory was integrated and constructed through integrating existing experimental resources and techniques and funds. Above versatile functions had included the entity demonstration, the CAI demonstration and manufacture, the CAI on physics, the opening and typical experiment, the interesting physics exposition, the groping interaction experiment of teacher and student, etc.

The running of this laboratory is so efficient and unhindered that it has brought about a striking teaching effect and promoted scientific research capability and administration level of teachers.

Optimization of modern education technology and multiple teaching resources and means. To the mass-enrolling in the common university, we have proposed a new teaching model of multilayer fundamental physics curriculum-group, which features are opening and interaction. In the model, exploiting and optimizing multiple teaching resources and means are essential conditions raising teaching effect. Four teaching means are optimized and integrated, which include traditional teaching method, multimedia teaching means, demonstration experiment teaching means, and opening-discussion

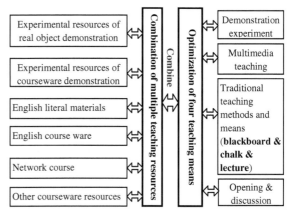

Fig. 3. Optimization of multiple teaching resources and means

teaching. Multiple teaching resources are also collocated and optimized, which are real object or courseware demonstration experiment, English literal or courseware material, network courses, and etc. According to different course types and layers, many kinds of resources are dissimilarly carried on allocation, as well as, refined-chosen instruction contents and the best combination point of traditional-software teaching are respectively designed and found (as shown in figure3).

Through lots of teaching practice, almost students' study interest and ability have been promoted with the new teaching model. The ultimate teaching achievements are favorable and remarkable.

4 Characteristic and Merits and Achievements of Physics Curriculum-Group-Construction

The fundamental standard of success or failure of curriculum-group-construction is characteristics and achievements. The research has made outstanding achievements, and condensed the distinct character.

The establishment of four-level curriculum system with characters: "setting excellent course as the core course", "clear priorities and complement each other and promotion each other", not only played the overall effect on the course-group, but also realized the important construction for excellent courses.

The constructing ideas are consistent to the teaching idea "selective instruction and discussion", which are the enhancing classic physics, separating modern physics, linking with engineering & physics history. The research highlight the engineering and opening and interaction, embody the combination of theory and practice, embody the combination of science and technology, embody the combination of science knowledge and historical materialism. The students' learning interest, learning ability and innovation ability are improved. The students' physics-quality and science accomplishment are promoted.

We have got the comprehensive achievements in teaching reform & research and so much excellent teaching results and papers published. The teaching methods and teaching resources are full utilized and optimized. We establish a structural teaching material construction model, and a perfecting the comprehensive assessment model. The construction teacher team by the extensive and in-depth teaching research activities is a low-cost teacher construction program with high efficiency in common polytechnic universities.

In the past six years, the representative achievements of curriculum-group-construction were "university physics" awarded the title of provincial excellent courses. The teacher team had undertaken 16 teaching research projects, nearly 13 academic research projects, won 30 various teaching awards (over 10 high-level teaching awards), published over 90 teaching and academic research papers(over 20 key papers) and 6 well-known textbooks including two foreign lecture notes, built a structural multidimensional teaching materials construction mode for common university, built a online rich resources learning system for college physics.

In recent years, all teachers, teaching effects were good grades from the experts and the students. The instruction quality including new teacher, was higher than average level of our university. The small class exercise teaching model with opening and interactive discussion was praised by three commissioners of the PATACMEC from Tsinghua University. Total student opinion surveys show that the teaching interaction enhances significantly the teaching and learning interests.

5 Conclusions and Prospect

We have studied the curriculum-group-construction mode, which core is physics excellent course development. Based on the excellent course construction, we have established the multilayer fundamental physics curriculum-group system with opening and mutual learning in selective instruction and discussion. Through this teaching study and practices, the learning interest, the learning capacity and the innovation capacity for the most students are all improved. Because of many distinct characters and remarkable results, this mode should be propagated to broad disciplines and courses in many common polytechnic universities.

Many students show some low learning interest, which is the main problem in the current teaching. Therefore, our next studies include optimization of teaching contents and methods, strengthening multidimensional teaching materials and teaching resource construction, establishing a suit of scientific running method on the comprehensive examination, and heightening teacher level.

Acknowledgment. This research is partially supported by Natural Teacher Scientific Research Fund of China under Grant GJJKY2011121, Ministry of Education in Humanities and Social Sciences Planning Fund (No.08JA790039), and Educational Reform Research Fund from Xi'an Technological University (No.10JGY26 & 11JGY27).

References

1. Li, Y.F.: Problems on Excellent Course Construction in University. China Higher Education Research 1, 91–93 (2007)
2. Huang, B.Y., Xiang, G.X.: Analysis and Consideration on Current Natural Excellent Course Construction. China Higher Education Research (9), 72–75 (2007) (in Chinese)
3. Xu, S.J., Ren, X.L.: Multilayer Basic Physics Course-group-construction with Selective Instruction and Discussion and Opening and Mutual Learning. Modern Education Science(Higher Education Research) (6), 105–106 (2009) (in Chinese)
4. Qian, Y.: Consideration to Curriculum-group-construction in Quality-project Background. Modern Education Science(Higher Education Research) (6), 144–145 (2008) (in Chinese)
5. Li, X.J.: Construction and Evaluation of Teaching Material in University. China Electric Power Education (10), 73–74 (2011) (in Chinese)
6. Xu, S.J., Ren, X.L., Zhou, J.: Classification and Comparison of the College Physics Exercises Based on Education Informatization. In: Proceedings of 2011 Second ETP/IITA Conference on Telecommunication and Information (TEIN 2011), Phuket, Thailand, vol. 4, pp. 277–280 (2011)
7. Xu, S.J., Ren, X.L.: Construction of a Demonstration and Multi-functional Opening Laboratory of Basic Physics. Research and Exploration in Laboratory 24(8), 112–114 (2005)

Research on Automation Specialty Application Talents Training*

Rongmin Cao, Denghua Li, Zhong Su, and Yingnian Wu

School of Automation, Beijing Information Science & Technology
University, Beijing 100192, China
rongmin_cao@yahoo.com.cn

Abstract. With the social production and continuous improvement in the degree of automation, the demand for application talents in automation major areas is also growing, at the same time the demand on the comprehensive ability and innovative ability of application talents higher and higher, which gives automatic professional education and the training of talents to a higher standard. Automation, as a characteristic major which reflects the close combination of information and industrialization, is not only faced with the "information technology to stimulate industrialization", "take the road of new industrialization," the historical development opportunities, but also faces the huge impact of competitions between majors. How to cultivate the application talents with strong innovation ability to fit the needs of social development is an important issue for the education and research of automation major, which should be seriously considered. Based on University of Illinois at Chicago of the United States(UIC) visit, UIC and our school automation professional education plans were compared, the United States application talents training is worth learning.

Keywords: application talents, automation specialty, practice teaching systems, innovation ability.

1 Introduction

The automation school of Beijing Information Science & Technology University comprises 3 departments and 1 center. They are: Control Engineering Department, Electrical Engineering & Automation Department, Intelligence Science & Technology and Electrical Practice Center. There are 25 labs, such as Automation Control Theory Lab, Computer Control Lab, Motion Control Lab, Artificial Intelligence Lab, Intelligence Network Lab, and Electronic Power & Electronics Labs.

The present total faculty number in the School is about 70, among them with titles of full and associate professor is 35. The automation school has established academic relationship and cooperation with Mid-Sweden University. The Automation School pays great importance in scientific research. After several years' construction and

* Supported by Key Project Higher Education Reserch of Beijing Information Science & Technology University (2009GJZD11) and Automation Characteristic Specialty Construction of Beijing.

development we have been concentrating its efforts in the research of inertial technology, advanced control theory and application, motion control, measuring technology and robotic control. These technologies have been applied in the field of national defense and lots of industry.

2 Education Scheme of Automation Program of Beijing Information Science and Technology University

2.1 Education Object and Characteristics of Automation Program

There is developing in an all-round of morally, intelligence, physical and aesthetics. There is coordinate unity in quality, ability, knowledge. Automation graduates can be in the scientific research institution, national defence, higher institute , departments such as the enterprises and institutions, etc. They are engaged in the industrial process control, automatic detection and test, network analysis, system design, scientific and technology development and they can work in fields such as embedded controlling, network controlling and navigating and guidance, etc.

1) Sturdy natural science basic knowledge, management science foundation.
2) Excellent humane social science foundation and foreign language capability.
3) The engineering foundations and certain professional knowledge of more broad fields such as automation control, electronic technology, embedded controlling, information measuring and dealing with, the computer technology and network technology, etc.

2.2 Length of Schooling and Degree

1) The professional basic length of schooling was 4, implement the elastic length of schooling, the length of schooling was 3- 6.

2) Graduate that accord with the regulations of " degree regulations ", award engineering bachelor's degree.

2.3 Automation Professional Main Subjects

Beijing information science & technology university automation professional rely on subjects as: control science and project, electrical engineering, computer science and technology.

2.4 Course Structure and Schedule

Course structure of automation professional includes: Public basic course, professional foundation courses, specialized courses. Required credits 67.5credits, 51credits and 13credits respectively, as shown in Table 1 to Table 3.

3 Education Scheme of Electrical Engineering on UIC

The Electrical Engineering curriculum is concerned with analysis and design of modern electronic systems, devices, and signals for a broad range of applications such as wireless or network communication, electrical power and control, and multimedia information technology. The curriculum provides a wide background in the fundamental theory of electrical engineering and in the mathematical and scientific tools necessary for an electrical engineer to meet the current and future challenges of a professional career. The field of electrical engineering is currently evolving at a rapid pace since it has a major role in the accelerated growth of the technological world. This requires the modern electrical engineer not only to have a sound basis in the fundamental principles but also to have the capacity to learn and assimilate novel advances as soon as they materialize. These qualities are anticipated in the curriculum, which includes not only a sound theoretical background but also offers a variety of courses that develop the student's ability to gain knowledge autonomously and to combine it with contemporary design techniques. Courses are in diverse areas such as signal processing, power electronics, communications, optical and electromagnetic technologies, control systems, integrated circuits, multimedia networks, and image analysis.

Table 1. Public Basic Course structure of automation program

Course title	Credits
Cultivation of ideology and morality and Elementary knowledge of law	3
The Conspectus of Marxism Basic Principles	3
The Course of Essentials of Chinese Modern History	2
The Survey of Mao Ze Dong Thought、Deng Xiao Ping Theory and "Three Representation" Important Thought.	4
College English（1）（2）（3）（4）	16
Advanced Mathematics （1）（2）	11
Linear Algebra	3
Probability and statistics	3
University Physics（1）（2）	6.5
College physical training（1）（2）（3）（4）	8
Military affairs theory	2
The C Programming	3
The Simulation of Automatic Control System	3

Table 2. Professional foundation course structure of automation program

Course title	Credits
Engineering Cartography	2.5
Circuits Analysis	4
Mathematics for Control Engineering	3
Analog Electronic Technique	4
Digital Electronic Technique	4
Electrical Machines and Drives	4
Fundamentals of Computer Software	4
Principle of Automatic Control	5.5
Modern Control Theory	2.5
Principles of Microcomputer & Interface Technology	4
Power Electronics	3.5
Measuring Technology & Instrument	3.5
Computer Control System	3.5
The Simulation of Automatic Control System	3

Table 3. Professional foundation course structure of automation program

Course title	Credits
Inertial Technology	2.5
Process Control	3.5
Motion Control System	3.5
Intelligent Control in Networked Control Systems	2.5

The educational objectives of the Electrical Engineering undergraduate program are for its graduates to:

Have knowledge of fundamental principles in electrical engineering and fundamental scientific principles and tools to design and develop products and practical solutions for problems in public and private sectors.

Demonstrate an ability to function independently and in multidisciplinary teams with the communication skills and ethical conduct necessary for professional success;

Demonstrate an understanding of the need for life-long learning, acquiring new knowledge, and mastering emerging technologies and new tools and methods;

Have knowledge necessary to pursue graduate/professional education and/or engineering practice.

Opportunities are available to participate in the activities of the student chapter of the Institute of Electrical and Electronic Engineers (IEEE) and Eta Kappa Nu, the honor society of electrical engineering. An interest in robotics can be pursued by joining the Engineering Design Team, a College of Engineering student group.

3.1 Degree Requirements—Electrical Engineering

To earn a Bachelor of Science in Electrical Engineering degree from UIC, students need to complete University, college, and department degree requirements. The Department of Electrical and Computer Engineering degree requirements are outlined below. Students should consult the College of Engineering section for additional degree requirements and college academic policies.

3.2 BS(Bachelor of Sciences) in Electrical Engineering Degree Requirements Credits

Nonengineering and General Education Requirements 50
Required in the College of Engineering 55
Technical Electives 17
Additional Mathematics Requirement 3
Electives outside the Major Rubric 3

3.3 Total Hours—BS in Electrical Engineering 128

Nonengineering and General Education Requirements:
ENGL 160—Academic Writing I: Writing for Academic and Public Contexts 3
ENGL 161—Academic Writing II: Writing for Inquiry and Research 3
Exploring World Cultures coursea 3
Understanding the Creative Arts coursea 3
Understanding the Past coursea 3
Understanding the Individual and Society coursea 3

Understanding U.S. Society coursea 3
MATH 180—Calculus Ib 5.
MATH 181—Calculus IIb 5.
MATH 210—Calculus IIIb 3.
MATH 220—Introduction to Differential Equations I 3.
PHYS 141—General Physics I (Mechanics)b 4.

PHYS 142 —General Physics II (Electricity and Magnetism)b 4.
CHEM 112—General College Chemistry Ib 5.

3.4 Total Hours—Nonengineering and General Education Requirements 50

Required in the College of Engineering
Electrical Engineering Core Courses .
ENGR 100—Orientationa 0a
One of the following courses: 3
CHE 201—Introduction to Thermodynamics (3) .
ME 205—Introduction to Thermodynamics (3)
CS 107—Introduction to Computing and Programming 4.
ECE 115—Introduction to Electrical and Computer Engineering 4.
ECE 225—Circuit Analysis 4.
ECE 265—Introduction to Logic Design 4.
ECE 267—Computer Organization I 3.
ECE 310—Discrete and Continuous Signals and Systems 3.
ECE 322—Communication Electromagnetics 3.
ECE 340—Electronics I 4.
ECE 341—Probability and Random Process for Engineers 3.
ECE 346—Solid-State Device Theory 4.
ECE 396—Senior Design I 2.
ECE 397—Senior Design II 2.

3.5 Electrical Engineering Advanced Core Courses

Three of the following courses, each with a laboratory: 12
ECE 311—Communication Engineering (4) .
ECE 317—Digital Signal Processing I (4) .
ECE 342—Electronics II (4) .
ECE 350—Principles of Automatic Control (4) .
ECE 367—Microprocessor-Based Design (4) .
ECE 424—RF and Microwave-Guided Propagation (4) .

3.6 Total Hours—Required in the College of Engineering 55

Technical Electives Courses Hours. Seventeen hours chosen from the following list.
Those courses not used to meet the advanced electrical engineering core requirement
can be used as technical electives.

However, no more than a total of two courses below the 400-level may be used to
meet the technical elective requirement. Also, no more than one course from outside
of the Electrical and Computer Engineering Department may be used to meet the
technical electives requirement. 17.

PHYS 244—General Physics III (Modern Physics) (3) .
CS 385—Operating Systems Concepts and Design (4)a .
ECE 333—Computer Communication Networks I (4) .
ECE 347—Integrated Circuit Engineering (3) .

Electrical and Computer as.
ECE 366—Computer Organization II (4) .
ECE 368—CAD-Based Digital Design (4) .
ECE 401—Quasi-static Electric and Magnetic Fields (3) .
ECE 407—Pattern Recognition I (3) .
ECE 410—Network Analysis (3) .
ECE 412—Introduction to Filter Synthesis (3) .
ECE 415—Image Analysis and Computer Vision I (3) .
ECE 417—Digital Signal Processing II (4) .
ECE 418—Statistical Digital Signal Processing (3) .
ECE 421— Introduction to Antennas and Wireless Propagation (3) .
ECE 423—Electromagnetic Compatibility (3) .
ECE 427—Modern Linear Optics (3) .
ECE 431—Analog Communication Circuits (4) .
ECE 432—Digital Communications (3) .
ECE 434—Multimedia Systems (3) .
ECE 436—Computer Communication Networks II (3) .
ECE 437—Wireless Communications (3) .
ECE 442—Power Semiconductor Devices and Integrated Circuits (4).
ECE 445—Analysis and Design of Power Electronic Circuits (4) .
ECE 448—Transistors (3) .
ECE 449—Microdevices and Micromachining Technology (4) .
ECE 451—Control Engineering (3) .
ECE 452—Robotics: Algorithms and Control (3) .
ECE 458—Electromechanical Energy Conversion (3) .
ECE 465—Digital Systems Design (3) .
ECE 466—Computer Architecture (3) .
ECE 467—Introduction to VLSI Design (4) .
ECE 468—Analog and Mixed-Signal VLSI Design (4) .
ECE 469—Computer Systems Design (3) .
MCS 425—Coding and Cryptography (3) .

3.7 Total Hours—Technical Electives 17

Additional Mathematics Requirement Courses Hours.
One of the following courses: 3
MATH 310—Applied Linear Algebra (3) .
MATH 410—Advanced Calculus I (3) .
MATH 417—Complex Analysis with Applications (3) .
MCS 471—Numerical Analysis (3) .
MATH 481—Applied Partial Differential Equations (3) .

3.8 Total Hours—Additional Mathematics Requirement 3

Electives outside the Major Rubric.
Three hours from outside the ECE rubric 3.

3.9   Total Hours—Electives Outside the Major Rubric 3

Students preparing for the Fundamentals of Engineering Examination, which leads to becoming a Licensed Professional Engineer, are advised to use these hours to take CME 201—Statics and one course from the following courses: CME 203—Strength of Materials, CME 260—Properties of Materials, or ME 211—Fluid Mechanics I.

4   Conclusion

Obtained through visit research: goal is training applied talents in the united states; personnel training mode is centered closely around the school and enterprise combined with the center of attention to practical teaching; in talent evaluation, and have established a strict examination system, while the introduction of social of evaluation mechanisms. the United States application talents training is worth learning.

References

1. Hong, W.Y.: Robotics Education and Innovative Abilities. Electrical & Electronic Education 8, 6–9 (2005)
2. Zhang, G., Zhu, H.: Students and Innovative Ability and Mechanism. China University (6), 11–12 (2003)
3. Yang, C., Liu, J.: Foreign Training for Undergraduates comparative study. Nanjing Institute of Technology (Social Science Edition) 7(3), 25–28 (2007)
4. Sun, G.: Asking, "Applied University". Pui Ching University in Guangdong 7(2), 7–9 (2007)
5. Tong, W., Duan, Z.: Automation Practice Teaching and Creative Talent. In: China Automation Education Annual Conference Proceedings. Mechanical Industry Press, Beijing (2007)
6. University of Illinois at Chicago of the United States(UIC) Undergraduate Catalog (2009-2011)

On the Developmental Stages and Cultivation of Academic Awareness for Graduate Students in China

Kaige Ren[1] and Guonian Wang[2]

[1] Marxism Institute, China University of Geosciences, Wuhan, P.R. China
[2] Scholl of Foreign Languages, China University of Geosciences, Wuhan, P.R. China
pxjlh@163.com

Abstract. Graduate students' awareness of academic norms is analyzed in terms of its composition and developmental stages. The features in the formation of students' academic consciousness are also summarized. As to how to cultivate and shape their academic decency, suggestions and discussions are put forward in such aspects as educational philosophy, training patterns and ways of teaching.

Keywords: Graduate students, Academic awareness, Developmental stages, Cultivation.

Academic norms serve to establish a set of general principles and main guidelines for academic research conducts. They aim at promoting academic advances, and should turn out to be the intrinsic consciousness in scholars of all research fields. In other words, academic norms and academic awareness shall exert a subtle influence on researchers and their academic endeavors in an invisible but omnipresent way. (Yang, 2004:13) The academic awareness in graduate students is the consciousness of observing academic regulations and discipline criteria; it is manifested in the ways of value judgment, rational perception, willpower and practice while they approach the subject, content and method of their research projects. Such awareness is closely related to students' academic decency, and they combine to constitute the base of creative capacity for graduates. Upon admission, they shall keep a rigidly serious attitude toward experiments, researches and paper writing; they shall also cultivate the desirable academic consciousness and form absolute honesty in academic researches. This is of great importance to their present academic pursuit, their future academic research and their desired career.

1 Composition of Graduates' Academic Awareness

The system of academic awareness consist the subsystem of cognition, of knowledge and of experience. The knowledge subsystem functions as reflection of academic conducts and as raw materials for the other two subsystems. It is through this subsystem that graduate students perceive academic behaviors, seek research methodology and rules, and understand academic norms defined by academic communities. As the basis of graduate academic consciousness, the knowledge

Y. Wang (Ed.): Education Management, Education Theory & Education Application, AISC 109, pp. 497–502.
springerlink.com

system is acquired partly dependent on graduate cognitive capacity and intelligence levels, and partly dependent on the completeness of the academic knowledge itself, as well as on how it is acquired. The development and fulfillment of the knowledge system represent the width and depth in which graduates perceive academic norms.

The process of cognizing academic norms is one that, in the scope of the norms, turns academic research logic into the researcher's own thinking logic. The cognition system of academic awareness is a series of reflections on academic conducts based on academic performances. It assists to optimize and perfect the knowledge system. Meanwhile, it serves to guide and evaluate academic researches. Any academic research graduate students are engaged in is virtually an act that aims to realize certain values or findings. Therefore, cognition is key to the formation of graduates' judgment of academic values.

The experience system is the understanding and comprehension generalized and verified through the practice of academic researches. Locke remarks that all our (human) knowledge is based on experience, and most fundamentally it comes from experience. (Jin, 2010:99) The experience system unites the freedom and necessity of academic awareness, modifies and controls various research practices, thus avoiding unnecessary repeated research efforts, and generating a strong continuity and creativity for academic researches. The enrichment and mutual modification of the knowledge system and cognitive system can help promote the experience system. The experience system plays a significant guiding role for the awareness system, and it influences and determines the awareness system.

The academic consciousness in graduates is a complex and unique system of thoughts. Of the system constituents, the cognitive system is the core, without which the academic research would lose its momentum, and the experience and knowledge would also be impossible. The knowledge system is the basis, without which there would be no cognition or experience. Without experience system, research could not be conducted, and knowledge and cognition would not be generated. The three factors are closely intertwined and mutually generative, and combine to exert a decisive effect on the academic awareness for graduate students.

2 Developmental Stages of Graduates' Academic Awareness

Academic awareness is formed on the basis of practice and performance graduates conduct on a certain subject in a specialized and systematic manner, and it develops in two ways: vertical and horizontal. The vertical direction means graduates' cognitive capacity and academic ethics develop to higher level; and the horizontal direction means their logical cognition of academic knowledge evolves to social cognition. Such development accords with the general rules of human awareness development in the following order: Spontaneity—Growth—Maturity—Consciousness—Freedom.

The Spontaneous Stage. In such stage graduates know nothing or little about research contents or methods; much less do they know about the certainty and regularity of academic researches. Their understanding of academic norms comes from their own thesis writing experience upon college graduation, or from the influence of family nurture, school education and social interaction. They may have bits and pieces of academic knowledge, but is unfamiliar with the solemn, scientific and systematic features it has. Therefore, students at this stage are likely to borrow or copy other researchers' ideas, methods, or words. Plagiarism and transplantation characterize graduate academic practices at this state.

The Growing Stage. Students at present are familiar with academic instruments and standards as well as their values, but as to the nature of academic norms they are still puzzled. They observe academic authorities and depend on academic teamwork. They observe academic regulations reluctantly, not voluntarily. They are mandated by norms, not disciplined on their own. They acquire these academic rules mainly from team workers and their own observation, and have not laid desirable emphasis on the norms.

The Mature Stage. As their academic levels improve, they now have an objective and rational understanding of academic norms, to which they voluntarily subject themselves. They, however, need further studies on the concrete requirement of academic norms. At this stage, graduate students regard the norms as the commonly observed agreement among the community. They know that there is a common belief, value and orientation for the academic realm. Furthermore, they can offer their preliminary judgment and evaluation to academic conducts and performances.

The Conscious Stage. "In the perspective of morality, consciousness is first of all recognition of heteronomous morality, a mastery of social morals, and a self-decided selection under free will."(Ma, 2009:53) Students at this stage are good at academic research and have reached considerable achievements with a profound comprehension of the values and significance of academic norms. They are willing and ready to observe the principles and rules stipulated or agreed, and they are able to make systematic and scientific analysis, reasoning, judgment, induction and deduction in terms of scientific ethics and academic values.

The Free Stage. Students are fully aware of the necessity and certainty of academic norms. They have established a truthful attitude to academic researches, and understood what academic norms mean to life. They now deem academic norms as part of their life and research, as a means to realize their aim and perfect themselves. They take the regulations for granted, not just for certain. The norms have now turned to an instinct for them, as well as a passion, obligation and belief.

3   Features of Graduates' Academic Awareness

3.1   Mobility of the Formation of Academic Awareness

The academic awareness system that consists of knowledge, cognition and experience is an open system with self-organization, self-adaptation and self-construction. Once the system is constructed, its contents are not constant but changeable. New information comes into the system, making it develop and fulfilled. When the information input reach certain degrees, the cognition, experience and knowledge subsystems will all change, breaking the original balance of each subsystem. A new set of academic norms will subsequently come into being. The academic norms usually develop from order to disorder and to order again, and evolve from lower levels to higher levels. The understanding of academic norms is likely to fluctuate due to the integrity, source and feedback manners of incoming information.

Diversity of the Orientation of Academic Awareness

Academic norms originate from the rules for academic conducts in academic practices that are generally acknowledged by academic communities. In terms of logic, fact judgment and value judgment are premises for norm judgment, and are universal indications or indication systems to coordinate human behaviors. (Ma, 2009:53) The norms involve both fact cognition and value judgment, and thus are basically value judgments. Due to the diversity of social structures and value orientation, the same norm contents may mean different things to specific social members. As knowledge researchers and academic practitioners, graduate students may differ and vary in their value orientation and career choice. These multiple choices unavoidably bring about diversity in the understanding of values and significance of academic norms.

3.2 Passivity of Cultivation of Academic Awareness

In the cultivation of academic norms, graduate advisors play a key and unmatchable role. A supervisor's academic achievements and quality will directly influence students in their cognition and observance of academic norms. "Individuals habitually follow authorities and copy their peers in developing a self-governed moral system."(Qian, 2000:59-68)

Due to the above-mentioned features, the formation of graduate students' academic awareness also relies on the process of awareness cultivation.

4 Cultivation of Graduates' Academic Awareness

The cultivation of academic awareness is a comprehensive and systematic process that shall go with individual personality and their cognitive levels. Desirable academic consciousness could only be formed by flexible education and formal training.

4.1 Individual Orientation

Researches show that the instruction of academic norms is not necessarily related to the formation of individual academic ethics. Graduates have their own learning experience, values, beliefs and career orientation, a fact that makes it hard to achieve a universally adaptable academic standard or academic cognition. For the same reason, academic norms shall be multi-oriented so they can fit the vast majority of graduate students bearing different morals, wills and values. It is therefore unwise to apply the same education methods and academic contents to all students who are psychologically and logically mature, and are personally unique. Instructors of academic norms are expected to familiarize themselves with students as regards their academic foundations, moral levels and norm cognition. They shall employ a flexible instructive strategy while informing students of academic norms. Only in this way could graduate students raise their understanding of academic norms.

4.2 Teamwork Orientation

"When norms function, they are related to inner consciousness and are not purposefully observed. Legal conducts are habitual and internalized."(Yang, 2008:5-14) Norms may vary due to different subject, fields and manners they apply to. Graduates can only achieve proper moral judgments and conduct selection through continual academic performances. Academic norms are the unanimous selection of all members in an academic community with the same pursuit of values and the same profit orientation. It is proved that within an academic team that shares the common belief, its members are able to observe the requirements of academic norms in a conscious way, and improve their own moral levels and self-discipline. As newcomers and successors of an academic group, graduate students would and could develop a decent research habit and an efficient academic awareness through long-term, repetitive and normalized academic training, as well as the influence of the group's scientific spirit.

4.3 Content Orientation

Supervisors are academic leaders and are responsible for the cultivation of student academic norms. "In most cases, supervisors are mentors to graduates and will pose a life-long influence on them."(Deng, 2006) Supervisors as role models in observing academic norms can guarantee a correct recognition and serious observance of the rules for their students. The acquisition of the whole set of academic knowledge serves as the basis for students' academic moral selection. Without real-life understanding and judgment of case studies, they would fail to make proper judgment in lien with the norm contents, and were likely to break the confinement of morals and norms. Thus, it is advised that in the instruction of academic norms and in the direction of academic researches, supervisors shall both lead and limit their students. They shall not only tell students what they "should" do and how they "should" do them, but also warm them what they "should NOT" do and in which way they "should NOT" conduct them. Scientific spirits shall be advocated and academic misconducts shall be despised. Their immune system could only be established in this way.

Graduate students are the cornerstones and successors of the nation's sci-tech undertakings. Their academic creativity, academic awareness and academic accomplishments are decisive for the nation's future development. If these would-be scholars fail to lay a solid and rigid academic foundation at the initial stage, it would be depressive. (Zhou, 2004:8) Their decent academic awareness and habits are possible, given that their personalities and individualities are respected, the team work is emphasized, and proper contents are informed. Only when their elegant academic consciousness is cultivated, will they be able to understand and related to different ideas and concepts stipulated in the academic norms; to obtain sound cognition, accurate judgment and correct selection of the norm contents; to effectively raise their academic morals; and to tap their creative potentials.

References

1. Deng, Z.L., Deng, X.M.: An Open Letter to Oppose Academic Corruption and Academic Misconducts [EB/0LI] (March 22, 2006), http://www.acriticism.com
2. Jin, S.: Possibility of Rational Morality: The Premise of Locke Moral Philosophy. Journal of Southwest University (Social Sciences Edition) 07, 99 (2010)
3. Ma, Y.Q.: Unscrambling the Characteristics of Moral Self-discipline. Studies in Ethics 05, 53 (2009)
4. Qian, W.L.: The Individual Generative Mechanism of Moral Consciousness. Social Sciences in China 04, 59–68 (2000)
5. Xu, M.Q.: How to Perform Norms. Journal of Academics 07, 56–60 (2002)
6. Yang, G.R.: On Regulations and Norms. Journal of Academics 03, 5–14 (2008)
7. Yang, Y.S., Zhang, B.S.: Introduction of Academic Norms, p. 13. Higher Education Press, Beijing (2004)
8. Zhou, L.W.: On the Role of Graduate Supervisors. Academic Degrees & Graduate Education 08, 8 (2004)

Exploration on Construction of Scientific Research Network Platform in Colleges and Universities

Fanmei Liu and Yunxiang Liu[*]

Shanghai Institute of Technology
yxliu@sit.edu.cn

Abstract. The rapid development of internet has exerted a profound influence on education and scientific research. In recent years, in order to excavate excellent teacher's recessiveness knowledge and share them together, enhance the ability of teaching and reaching, the developments of research platform for teaching based on internet is gained more and more attention. Now, there are many such platforms for teaching, almost every educational department has their research and teaching platform about every disciplines and fields. As shown by the investigation, they are somewhat alike and quite tedious in the style of design. Especially, the content about one disciplines are not comprehensive enough, at the same time, there are lack of innovation. This thesis is aim at how to work out these problems and how to construct the research and teaching platform for special disciplines. The special scientific research platform for teaching of Chinese have been designed and constructed in this paper.

Keywords: teaching and research platform, information, network technology.

1 Research Background and Current Situation

1.1 Thoughts on Improving Teachers' Scientific Research Level

Scientific research is the way to cultivate a teacher with good qualification and innovation ability. It is also an important part to promote educational development and revolution. Only by actively participating in scientific research, can a teacher improve his own quality and research ability, shifting from an experience-based teacher to a research-based one[1].

1.2 Background on Building Scientific Research Platform

With the fast development and rapid spread of network technology, network information has greatly influenced people's way of life and working, even the way in education and scientific research. Nowadays, Internet has become another crucial platform in economic and educational development, which signals human social progress[2]. The openness, interactivity and timeliness of the network technology will change the monotonicity in the process of educational research.

---

[*] Corresponding author.

Y. Wang (Ed.): Education Management, Education Theory & Education Application, AISC 109, pp. 503–509.
springerlink.com © Springer-Verlag Berlin Heidelberg 2011

1.3 The Current Situation of Development of Network Platform at Home and Abroad

It is not until the middle of the 1990s that China began to carry out network education, which is relatively laggard but develops very fast. So far, the Internet of education and scientific research has connected over 500 colleges and universities with rapid expansion to big and medium qualified cities. Our country has so far established a basic platform of network education and people can study via the network. Jonathan D.Leavey, an expert in online education market research, predicted that "the market of network education in China will boom dramatically in a few years and there is opportunity now for leapfrog development in China's network education market[3]." In the U.S. the Clinton Government declared to establish Internet-based education in the country where the Internet technology is the most advanced. The government educational departments, research institutes, colleges and universities shall all be connected through Internet. In 1998, England had implemented the Network Education plan. Every classroom and other learning places have been connected with internet to obtain learning materials and resources by the year 2002 in Singapore.

1.4 The System Development of Network Platform at Home and Abroad

Currently, there have existed some internet-based scientific research platforms or learning platforms and technologies related to platform development have made great progress theoretically and practically. For example, the distinctive teaching and research, learning network platforms overseas include WEB CT platform, Virtual-U platform, WISH platform (WEB Instructional Services Headquarters) and LUVIT⟨ Lund University Virtual Interactive Tool) etc. In China, the well developed platforms include network multimedia teaching and research software platform in Jinhaihang New Digital School, the school network explored by Founder Group, the Virtual Learning Area system developed by educational science and technology department of Capital Normal University, the self-learning platform school network by Kelihua, etc.

2 Main Problems

2.1 The Platforms for Special Disciplines Are Much Rarer, and the Resources Are Insufficient

Also the current scientific research platforms are nearly identical, The knowledge for specific discipline supplied is insufficient and far from abundance, The resources from one current Platform are confused and desultory. Especially, there is little change in the way of the resources were provided, "Books copy phenomenon " are so seriously that make the current platforms not adapted to the current situation[4].

2.2 Hardware and Software Facilities Are Not Perfect

The development of scientific research website is uneven, in economically developed area, the construction of websites in provincial college is perfect relatively, and is well used and achieve good results. However, in some under-developed areas, especially in the West, funds shortage and hardware insufficient phenomena are popular. They can't meet the needs of the new requirements of current information education.

2.3   The Support Databases for Scientific Research Platform Need to Be Improve

Since the database construction is mixed and disorderly, and lack of pertinence for specific subjects in most scientific research platforms, for the teacher, it is difficult to use and the document retrieval, literature search are inefficient.

3   Solutions to the Problems

3.1   Strengthen the Distinctive Discipline Construction

The platform should fully reflect the characteristics of the disciplines, break through the traditional scientific research network with general and mixed disciplines and realize simplified management. The platform construction should also be combined with teachers' self-learning to meet their needs.

3.2   Improve the Database and Discipline Resources Construction of the Platform[5]

The platform construction should be combined with the trend of education informatization, the mechanism of information sharing and the principle of resource sharing, attach importance to the learning resource development, design and storage model in order to standardize the resource construction. In addition, we should enhance the search engine capability so that teachers can find the sources they need timely.

3.3   Offer a Relatively Sound Interactive Function

The platform provides teachers with several collaboration tools in research activities, and support synchronous and remote information communication. We can upload or download references and other people's research findings on the platform, and at the same time share our learning results and materials [6].

3.4   Simple and Convenient in Use

The major function of network scientific research platform is to provide teachers with learning sources in a convenient way. Therefore, it should be simple and convenient enough in the platform construction for most teachers to download and read papers and documents [7].

4   Designing and Programing of Main Function Modules of the Scientific Research Platform

4.1   Development of Platform Management Interface

The management interface includes function modules such as user management and resource management. Its major functions are modifying username and code, establish normal users and register one's own username, etc. The structure of backstage database which reserve users' information is reflected in Table (1):

Table 1. Backend Database

Id	Field Name	Data Type	description
1	UserId	Int	ID Number for users
2	UserName	Varchar(50)	User name
3	UserPwd	Varchar(50)	password
4	Ename	Varchar(50)	User name
5	Email	Varchar(50)	E-mail

4.2 Development of Message Board Function

The message board offers teachers the chance of giving their opinions exactly and timely. While the opinions will be kept in the database of the server for anyone visiting the website to read. The message board is mainly developed by setting DSN, establishing database connection, making web page displaying the current message, web pages where visitor can leave his own message and web pages which show detailed information of a certain message [8].

In order to realize the functions such as online message, checking messages and backstage management, user's message information including name of the user, time, contact address, the theme and content of the message must be reserved in the database. The relevant information of the database in reflected in Table(2):

Table 2. Message board database design

Description	Field Name	Data Type	Field Size	Requied	Allow Zero Length	Indexed
Theme Numbers	ID	AutoNumber	Long Integer			Y
Spokesman name	Username	Text	20	Y	N	N
Message theme	Subject	Text	100	Y	N	N
Time/Date	DateTime	Date/Time		Y	N	N
Email Address	Email	Text	50	N	Y	N
Message Contents	Content	Memo		N	Y	N

4.3 Development of Discussion Room in Education and Scientific Research Platform

The discussion room aims to communicate and study teaching methods and theories for several professional teachers at the same time. In fact it is a tool for information exchange. A professional teacher can send messages online to other teachers and they could reply and give their own opinions.

The backstage database adopts Access Database named chat.mdb, which contains three tables--alluser, onlineuser, chatlog. Alluser is used to reserve all the registered users' basic information of the discussion room. Onlineuser reserves all the discussion information of the discussion room like usernames and the text format of the message, etc. Chatlog is used to reserve information data of the online users so that one can quickly find the online user. The relations between the tables are as Figure (1).

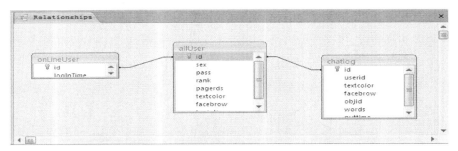

Fig. 1. The figure shows the relationship of the database tables, it is one-to-one relationship between Table Onlineuser and Alluser. While it is one-to-many relationship between Table Alluser and Chatlog.

5   Practical Study on Education and Scientific Research

This paper explores disciplinary faculty's recognition and the usage of the network education and scientific research with the methods of questionnaire, follow-up survey and interviews. By analysing the data of the survey with EXCEL tools, it proves the relations between the usage of network platform and scientific research performance.

5.1   The Attitude to the Use of Education and Scientific Research Platform

According to the graph which analyses the usage information of the platform by surveying 77 Chinese teachers in colleges and universities, it looks into nine aspects including the participation level, flexibility, use frequency of the platform, sharing degree of teaching resource pool, the improvement of the level of education and scientific research, interactivity, the construction and perfection of resource pool, the satisfaction degree of message board function, practicability. The Figure (2) is the result of statistical analysis with EXCEL tools.

The second graph shows that so far some disciplinary teachers still do not know much about the platform. While for those who have used the platform they are quite satisfied with the capability and interactivity. Most teachers think that the resource sharing ability is qualified, through which they could find relative materials and information timely and rapidly.

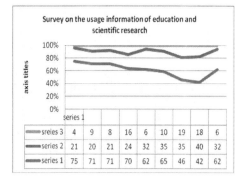

Fig. 2. The figure is the Survey on the usage information of education and scientific research, The number one to nine represents nine aspects and series one, two and three means the number of participants in each aspect.

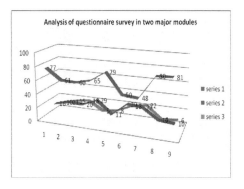

Fig. 3. The number one to nine represents nine aspects and series one, two and three show the number of participants in each aspect in the figure.

5.2 Questionnaire Survey on Online Communication and Message Board Function Modules

Online communication and message board modules as the major function modules enable teachers to communication without the limit of time and space. Therefore, the construction of these two modules should not be neglected. We survey on nine aspects including teachers' satisfaction degree of the two modules, the convenience degree of online communication, the degree of promotion in scientific research, the utilization of resources, the gains in communicating with teaching expert, the rate of replies, the classification of backstage database, the function of discussion room and the effects of information distribution. Figure (3) is pictorial diagram with Excel software,

The survey displays that there is a higher degree of participation in this module and it has provided teachers with great convenience and is popular among them. By communication they have solved the problems they had in teaching and scientific research. We can also tell from the graph that some functions of the message board module are not good enough, for example the design of message content database classification is not clear, and it is inconvenient to seek information. All these need improvement.

6 Conclusion and Expectation of Subsequent Work

This paper only has stage achievement and there are still many problems to be studied and explored in the process of combining theory and practice in the research. It includes the follows:

6.1 Further discussion on how to perfect the construction of message board and online communication resource pool and how to improve the design of message content database classification.

6.2 Further study on how to reasonably build teaching background information module and how to classify database and improve the speed of inquiry.

6.3 How to apply the education and scientific research platform to other disciplines except the Chinese subject and what modules to be added to realize these functions.

All in all, combined with new course revolution of college Chinese, this paper has preliminarily studied on the education and scientific research of teachers in the network circumstance. We believe that with the constant improvement of education and scientific research level and development of network technology, there will be more qualified education and scientific research platforms.

References

1. Su, Z.: From experience to research transformation. GanSu Education (14) (2006)
2. Li, x.: The Impact of development of information technology to the modern life. Science and technology information (24) (2008)
3. Cai, j.: To strengthen the construction of the informatization of education web site promoting. Journal of Hunan Business College 06 (2005)
4. Yang, l.: Construction and Application of Long-distance Education System of the School of Chinese Communist Party. ShangDong Normal University (May 2004)
5. Liu, f.: The Design & Educational Application of Informationization Campus Network in Colleges. Central China Normal University (2004)
6. Gao, y.: The network teaching platform design how to develop skills. The Science Education Article Collects (01) (2009)
7. Ding, x., Huang, j., Yang, c.: Design and Exploit a Generating System (TAGS) of Network Teaching Activities. Modern Educational Technology (2) (2010)
8. Chen, y.: Dreamweaver Chinese version from approaches to master, p. 357. China Machine Industry Press (2009)

The Entrepreneurial Motivations and Barriers for Technical University Students in Taiwan

Su-Chang Chen[1], Hsi-Chi Hsiao[2], Chin-Pin Chen[3],
Chun-Mei Chou[4], Jen-Chia Chang[5], and Chien-Hua Shen[6]

[1] Associate Professor, National Penghu University of Science and Technology, Taiwan
csc@gms.npu.edu.tw
[2] Chair Professor, Cheng Shiu University, Taiwan
hchsiao@cc.ncue.edu.tw
[3] Associate Professor, National Changhua University of Education, Taiwan
iechencp@cc.ncue.edu.tw
[4] Associate Professor,, National Yunlin University of Science and Technology, Taiwan
choucm@yuntech.edu.tw
[5] Associate Professor, National Taipei University of Technology, Taiwan
jc5839@ntut.edu.tw
[6] Professor, Transworld University, Taiwan
shen17@ms51.hinet.net

Abstract. This study aims to investigate perceptions of entrepreneurial motivation and barriers for technical university students in Taiwan. Questionnaire survey was conducted in this research. A total of 1360 questionnaires were distributed to 34 technical universities in Taiwan. After eliminating the questionnaires with incomplete answers and invalid ones, there were 847 effective samples from 22 technical universities, with a valid return rate of 64.16%. The results indicated that the top three motivation of start a business for technical students (in descending order of mean scores) is "creating something of my own", "the chance to implement my own ideas" and "personal independence". The top three barriers of start a business for technical students (in descending order of mean scores) is "lack of initial capital", "lack of knowledge of the business world and the market", and "lack of a high level of entrepreneurial competence". The motivations of start a new business for Taiwan's technical students are similar with the US students. But in entrepreneurial barriers is little difference.

Keywords: Entrepreneurial Motivation, Entrepreneurial Barriers, Start a Business, Technical University.

1 Introduction

In April, 2011, the unemployment rate in Taiwan is 4.29%. Unemployment rate of people with educational level of and above university is 4.80% and it increases by 1.52% in comparison to 3.32% in 2001. Because start a new business can reduce

Y. Wang (Ed.): Education Management, Education Theory & Education Application, AISC 109, pp. 511–517.
springerlink.com © Springer-Verlag Berlin Heidelberg 2011

unemployed rate and make economic growth, many nations take entrepreneurship education as the valid means that solves social jobless problem [1].

In the 1990s, 55% of university students and 66.9% of high school students in the USA were interested in entrepreneurship. In the 1980s, there were 34.3% students interested to startup a new business in the UK. In Singapore, about 5.3% university students were self-employed in the 1990s [2]. According to the statistics of the U.S. Small Business Administration (SBA), two thirds of university students included entrepreneurship in their career planning in 2001 [3]. In a survey on the entrepreneurship intent of university students in Taiwan, Chen & Song found that the percentage was at medium level [4]. Gong & Hsieh discovered that 51.8% students of Chongqing University in China were interested in entrepreneurship [5].

Small and medium businesses play an important role in Taiwan economy. According to the 2010 statistics presented by the Taiwan Small and Medium Enterprise Administration, Ministry of Economic Affairs indicate there are approximately 1.24 million small and medium firms in Taiwan that (a) represent 97.7% of all employer firms; (b) The employment number reaches to 8.19 million, that is 78.06% of all employee, it increase 0.23 million comparison in 2009 [6].

The nature of technical university is focused on the workplace in the training of human resources, not on academic research it emphasizes with highly practical [7]. Due to the education can influence on the attitudes and aspirations of youth. Universities need to teach them to understand how to develop potential enterprise even while they are still students in school [8]. The entrepreneurs are not inherent, the knowledge and skills of entrepreneurship can be taught [9]. It is found that graduates with entrepreneurship program training were more likely to start a new business and had stronger entrepreneurial motivations than other graduates [10]. Therefore, it appears that technological universities should be preparing their students for entrepreneurship as to work for oneself, not only to work for others [3]. However, enhancing the students' understanding of the processes involved in the developing a new business was therefore an important objective in university education [11]. So, the Ministry of Education (MOE) of Taiwan launched the U-Star Plan in 2009 to provide university students with early contacts entrepreneurship and related entrepreneurship services, including entrepreneurship guidance services [4].

To find out what motivation and barriers for entrepreneurship of technical university students in Taiwan is the motivation of this study. In other words, the aim of this study is to investigate perceptions of entrepreneurial motivation and barriers for technical university students in Taiwan.

2 Literature Review

There are numerous definitions of entrepreneurship. However, the term of entrepreneurship refers its origin to the morph entrepreneur, which in turn originates from the 12th century French compound verb *entreprendre* (*entre + prendre*). In short, it means "to undertake" in modern English [12]. Simply speaking, entrepreneurship is creating a new enterprise [13].

Rasmussen and Sorheim said that universities could play a role to contribute to entrepreneurship both indirectly and directly by commercialization of research and by being the incubation for new business [10]. Entrepreneurship education could provide a valid learning curriculum program, and to promote entrepreneur intension and ability for graduated students [14]. According to Bechanrd and Toulouse study, entrepreneurship education was a series of training courses for educating students with the formation concepts and development of business [15]. Wang and Wong indicated that entrepreneurship programs can provide students with training to take calculated business risks [3]. Shepherde mentioned that entrepreneurship education can reduce the risk of entrepreneurship failures [16]. That means that the entrepreneurship education will seek to better prepare their students for a changing labor market [2]. Gottleib and Ross considered that entrepreneurship education played as referring to the education of applying creative thoughts and reforms to enterprise [17]. Smith indicated that entrepreneurship education has become increasingly important [18]. However, promoting the students' understanding of the processes involved in the developing a new enterprise was therefore an important objective in entrepreneurship education [11].

Entrepreneurial education has been thriving in Taiwan in recent years. In terms of the number of schools opening entrepreneurship courses, there were eighteen universities opening related entrepreneurship courses in 2003, and increased to eighty nine in 2007. In terms of course variety, there were 102 entrepreneurship courses in 2005 [19]. The variety expanded to 550 related entrepreneurship courses in 2009, including nine universities opening degree programs and thirty three universities credit programs [20].

Shinnar, Pruett and Tonet found that the faculty and students were in agreement regarding the rank and importance of the top two motives of entrepreneurship: the chance to implement my own ideas and personal independence. In addition, money—although important—was not crucial. Both groups included only one financial motive in their top five: the opportunity to be financially independent. Likewise, wanting to make more money than by working for wages received a low rank and importance. They also found that the top five barriers as ranked by students referred to risk, lack of capital, current economic situation, competence, and knowledge. Faculty agreed with the importance of risk, competence, and capital as barriers to starting businesses. Last, both groups ranked procedural and operational issues as relatively unimportant [2].

3 Method

This paper is to investigate perceptions of entrepreneurial motivations and barriers to startup a business for technical university students in Taiwan. It was conducted with a questionnaire. The survey instrument was original developed at the University of Alicante, Spain, and translated into English by Shinnar et al [2]. This study translated English version into Chinese.

The survey consisted of 5 Likert-type scale questions. The questionnaire was distributed to the samples selected from 34 technical universities through the department chairperson. The department chairperson delivered the questionnaire to 40 students in the day school each university. A total of 1360 questionnaires were distributed. There were 22 universalities back. After eliminating those with incomplete answers and the invalid copies, a total of 847 copies were valid. The valid return rate is 64.16%.

There are 286 samples that have intention to start a business in future, but 561 samples are not.

4 Results

For 286 have entrepreneurial intention students, the top third motivation of start a business for them is "creating something of my own", "the chance to implement my own ideas", "personal independence". These results are similar with Shinnar et al research [2]. In addition, money—although important—was not top three crucial items. The "building personal wealth" is only fourth important items.

However, the last five motivation of start a business for them is "following a family tradition", "the difficulty of finding the right job", "gaining high social status", "having more free time", "making more money than by working for wages".

It shows in Table 1.

Table 1. Motivations for start a business

Motivations	M	SD	Raking
Creating something of my own	4.346	0.672	1
The chance to implement my own ideas	4.252	0.680	2
Personal independence	4.231	0.693	3
Building personal wealth	4.056	0.873	4
Being at the head of an organization	4.045	0.817	5
The opportunity to be financial independent	4.028	0.837	6
Improving my quality of life	3.804	0.848	7
Dissatisfaction in a professional occupation	3.629	1.013	8
Receiving fair compensation	3.626	0.888	9
Managing people	3.591	1.078	10
Making more money than by working for wages	3.584	0.983	11
Having more free time	3.570	1.124	12
Gaining high social status	3.479	0.958	13
The difficulty of finding the right job	3.476	0.939	14
Following a family tradition	3.304	0.907	15

For 561 not have intention to entrepreneurship students, the top five barriers of start a business for them is "lack of initial capital", "lack of knowledge of the business world and the market", "lack of a high level of entrepreneurial competence", "lack of experience in management and accounting", and "lack of ideas regarding what business to start". This is little difference with Shinnar et al research [2]. The rank of their study is "excessively risky", "lack of capital", "current economic situation", "lack of competence, and knowledge".

The last five barriers of start a business for them is "having to work too many hours", "lack of support from family , friends", "doubts about personal abilities", "fear of failure", and "irregular income".

It shows in Table 2.

Table 2. Barriers for start a business

Barriers	M	SD	Raking
Lack of initial capital	3.657	0.956	1
Lack of knowledge of the business world and the market	3.426	0.970	2
Lack of a high level of entrepreneurial competence	3.423	1.008	3
Lack of experience in management and accounting	3.371	0.999	4
Lack of ideas regarding what business to start	3.290	1.000	5
Current economic situation	3.287	0.992	6
Excessively risky	3.276	0.961	7
Irregular income	3.094	1.010	8
Fear of failure	2.909	1.092	9
Doubts about personal abilities	2.902	1.107	10
Lack of support from family , friends	2.825	1.014	11
Having to work too many hours	2.706	0.979	12

Students ranked the lack of knowledge of the business world, lack of a high level of entrepreneurial competence, lack of experience in management and accounting, and lack of ideas regarding what business to start as a more important barrier, possibly indicating a need for accounting, leadership, management and business word courses in entrepreneurship programs in Taiwan.

5 Conclusions, Recommendations and Limitations

5.1 Conclusions

This study produced three key findings. The first conclusion of this study is that top three motivation of start a business for technical students in Taiwan is "creating something of my own", "the chance to implement my own ideas" and "personal independence".

Secondly, the top three barriers of start a business for technical students in Taiwan is "lack of initial capital", "lack of knowledge of the business world and the market", and "lack of a high level of entrepreneurial competence".

Thirdly, the motivations of start a new business for Taiwan's technical students are similar with the US students. However, their entrepreneurial barriers are little difference with the US students.

5.2 Recommendations

This study just focuses on student's viewpoint; it can collect teacher's opinions and to compare both difference. It can find the gap between teachers and students on entrepreneurship as Shinnar et al research [2]. Then, it can improve the effectiveness of entrepreneurship education in Taiwan.

Secondly, the result indicate that students ranked the lack of knowledge of the business world, lack of a high level of entrepreneurial competence, lack of experience in management and accounting, and lack of ideas regarding what business to start as a more important barrier. It is suggested that the schools can provide complete entrepreneurial programs, and allow university students to have the opportunity to learn organizations of firms, market analysis, product development, fund raising and corporate operation before they enter the society.

Thirdly, this study only focuses on technical university. One could also replicate the same process in academic universities or junior colleges.

5.3 Limitations

There are a number of limitation in the study. First, the data for this study were collected through questionnaire. Samples provided opinions about perceived entrepreneurial motivations and barriers. All data were reported by the same group of respondents. Therefore, the observed relations may be, in part, a result of common-method effect. However, this limitation is consistent with the limitations of most survey research.

Secondly, samples of this study were selected from students of marketing- and logistics-management-related departments. Therefore, the observed relations may be, in part, a result of common-method effect. However, the graduates of these departments are more likely to start a new business in Taiwan. Therefore, the conclusions are acceptable in this respect.

Acknowledgments. The authors would like to acknowledge the support of the National Science Council of Taiwan for this research under the project number of NSC NSC 99-2511-S-346-001-MY3, 99-2511-S-230-001-MY3.

References

1. Chen, S.C., Hsiao, H.C., Chou, C.M., Chang, J.C., Shen, C.H., Liang, R.D.: Developing a Research Framework of Entrepreneurship Education Curriculum for Department of Marketing and Logistics Management in Technological Institutes and Universities. In: Pro. 2010 Inter. Conf. on Manage. and Serv. Sci., vol. 1. Wuhan, China (2010)
2. Shinnar, R., Pruett, M., Tonet, B.: Entrepreneurship Education: Attitudes Across Campus. J. Edu. for Bus., 151–158 (January, February 2009)
3. Wang, C.K., Wong, P.K.: Entrepreneurial Interest of University Students in Singapore. Technovation 24, 163–172 (2004)
4. Chen, S.C., Song, M.H.: The Entrepreneurial Intention for University Students. Leisure Industry Res. 9(1), 47–60 (2011)
5. Gong, L., Hsieh, L.I.: The Entrepreneurship Mental and Counterplan of University Students. Youth Studies 3, 68–72 (2009)
6. Small and Medium Enterprise Administration, Ministry of Economic Affairs: White Book of Small Business,
   http://www.moeasmea.gov.tw/ct.asp?xItem=9504&ctNode=689&mp=1
7. Gurel, E., Altinary, L., Daniele, R.: Tourism Students' Entrepreneurial Intentions. Annals of Tourism Res. 37(3), 646–669 (2010)
8. Luthje, C., Prügl, R.: Preparing Business Students for Co-operation in Multi-disciplinary New Venture Teams: Empirical Insights from a Business-Planning Course. Technovation 26, 211–219 (2006)
9. Okudana, G.E., Rzasa, S.E.: A Project-Based Approach to Entrepreneurial Leadership Education. Technovation 26, 195–210 (2006)
10. Rasmussen, A., Sorheim, R.: Action-Based Entrepreneurship Education. Technovation 26, 185–194 (2006)
11. Katz, J.A.: The Chronology and Intellectual Trajectory of American Entrepreneurship Education: 1876-1999. J.l of Bus. Venturing 18, 283–300 (2003)
12. Hoang, B.P., Huang, H.S.: An Empirical Study of Business Entrepreneurship. J. Entrepreneurship Res. 3(3), 1–27 (2008)
13. Low, M.B., MacMillan, L.C.: Entrepreneurship: past Research and Future Challenges. J. Manage 14, 139–161 (1988)
14. Chou, C.M.: Entrepreneurship Education Promoting Commercial Education Students' Employments. Ming Dao J. 1(1), 15–30 (2005)
15. Bechard, J.P., Toulouse, J.M.: Validation of a Didactic Model for the Analysis of Training Objectives in Entrepreneurship. J. Bus. Venturing 13(4), 317–332 (1998)
16. Shepherde, R.P.: Academic Resources for Entrepreneurship Education. Simulation and Gaming 27(3), 365–374 (2000)
17. Gottleib, E., Ross, J.A.: Made not Born: HBS Courses and Entrepreneurial Management. Harvard Bus. School Bulletin 73, 41–45 (1997)
18. Smith, M.O.: Teaching Basic Business: An Entrepreneurial Perspective. Bus. Edu. Forum 58, 23–25 (2003)
19. Liao, M.H.: The Current Status of Entrepreneurship Education for Higher Education in Taiwan. Master's thesis of Graduate Institute of Education, National Dong-Hwa University (2008)
20. Lai, Y.C.: Business Plan writing Skills. Workshop of Business Plan Writing Skills and Practice. Far East University, Tainan (2010)

The Development of Architectural Design Management System Based on Petri Nets

Wen Ding

Hunan mechanical & electrical polytechnic,
410151 Changsha, China
706901219@qq.com

Abstract. Petri nets is a kind of graphical modeling tools, its intuitive and preciseness can describe system and process between the logic relation of the complex activities, so are widely used in the software development process. In this paper, by using Colored Petri Net on the software development process modeling and software demand in the detailed design and realize the detailed modeling, design model and the C# program code the automatic conversion, and to improve the efficiency of software development.

Keywords: Petri, Software development, building design management system, Colored Petri Net.

1 Introduction

Affect large-scale distributed software development processes of many developed a high degree of complexity, leading to large-scale software development success rate is very low. Limited resource constraints in the development of the software successfully, not only to strengthen the software development process management, but also the use of advanced software development techniques and tools, the introduction of advanced development ideas. Petri nets as a formal graphical modeling tools can be applied to the software development process needs analysis, architecture design and coding and implementation phases, to validate the various stages of design results in the development of early software design problems found to improve the success rate of software development.

Architectural Design Architectural Design Management Systems will be people, and processes for integrated management of cash. Institute of management to the project as a carrier, through the introduction of Internet-based collaborative management system, to distributed architecture of remote collaborative design. Its main contents include: contract management, quality management, schedule management, cost management, human resources management, remote collaborative design. Architectural Design projects include: planning and design, architectural design, construction design, structural design, electrical design, water supply and drainage design, HVAC design, intelligent building design. Petri net application because of its simple, strong and has a description of visual modeling capabilities,

Y. Wang (Ed.): Education Management, Education Theory & Education Application, AISC 109, pp. 519–526.

which identify the Architectural Design Management Systems for the development of modeling. In this paper, colored Petri net modeling tool CPN TOOLS.

2 Petri net Description

From 1962, Dr. Carl Adam Petri of Petri net since, Petri net theory has been widespread concern and development [1]. From 1980 to convene the first Petri net theory and application of international seminars, the annual international seminar once continuous, Petri net theory and applications are constantly enrich and improve. Basic Petri net is defined as follows [2]

$$PN = (S, T; F) \tag{1}$$

Where $S \cap T = \varnothing$, $S \cup T \neq \varnothing$, $F \subseteq S \times T \cup T \times S$, $dom(F) \cup cod(F) = S \cup T$, $dom(F) = \{x | \exists y : (x, y) \in F\}$, $cod(F) = \{y | \exists x : (x, y) \in F\}$ .S and T are known as PN's Place collection of transition set, F is the flow relation. Place and they are called transition elements and T_ S_ elements. $X = S \cup T$ is called set of elements of PN.

Site because, based on the base can be defined on the P / T net (Place-Transition Nets, PTN) system as follows [3]

$$\sum = (S, T; F, K, W, M_0) \tag{2}$$

$N = (S, T; F)$ Constitute the basic network, K On the capacity of the function of N, W For the weight function, M_0 For the initial marking.

3 Colored Petri Net

With the Petri net application in practice, revealed two shortcomings: no data and hierarchical concept, it is not possible by having well-defined interfaces to build a large model of the subnet [4]. Colored Petri net as a class the most well-known high-level Petri net, its development goal is to overcome these serious shortcomings. To this end, the introduction of a hierarchical decomposition of data structures and concepts, and programming language combining the power of PTN, PTN to describe the synchronization of concurrent processes provide a basis for the programming language was defined data types and data provide a basis for action. Unlike Petri nets only one token)and an integer to describe the location by the state, colored Petri net is attached to each mark as a mark on the color data values. In colored Petri net, for a given location, all tags must have the same type of markers belonging to a particular color, the location of all these types constitute the set of colors.

CPN is defined as follows [5,6]

$$CPN = (\sum, P, T, A, N, C, G, E, I)$$

(3)

Where $\sum$ The color set, P is the place's collection, T is the set of changes, A is the set of arcs, $P \cap T \cap A = \varnothing$ .N is the node function, $N \in \{A \rightarrow P \times T \cup T \times P\}$ .C is the color function, $C \in \{P \rightarrow \sum C\}$ .GIs a guard function, $G : T \rightarrow Exprs$,And satisfies $\forall t \in T : [Type(G(t)) = Bool \wedge Type(Var(G(t))) \subseteq \sum]$ .E is the arc expression function, $E : A \rightarrow Exprs$, And satisfies $\forall a \in A : [Type(E(a)) = C(p(a))_{MS} \wedge Type(Var(E(a))) \subseteq \sum]$ .I is Initial function, $I : P \rightarrow Exprs$,And satisfies $\forall p \in P : [Type(I(p)) = C(p)_{MS} \wedge Var(I(p)) = \varnothing]$.

4 Colored Petri Net-Based Needs Analysis Modeling

Requirements analysis is the basis for the entire software development, software systems development to ensure the success of an accurate grasp of the software requirements analysis software development process is an important step, is a decisive step. Requirements development process is generally divided into four stages: requirements elicitation, requirements analysis, preparation of specifications and requirements verification. Currently, demand modeling There are many ways, including: structured analysis, object-oriented analysis and formal methods. The introduction of formal requirements analysis in the language can reduce ambiguity, improve accuracy, you can lay the foundation for the verification, but also allows for reasoning and implementation requirements.

CPN model used to build demand, not only to clear, intuitive description of the software requirements specification, to express the relationship between functions, and can demand dynamic simulation model to analyze the performance of the demand model, demand model accuracy, consistency and accessibility and other aspects of verification.

4.1 Management Process Analysis

Architectural design management system include: access management, program information, program design, program approval, program changes and other functions, the specific processes shown in Figure 1.

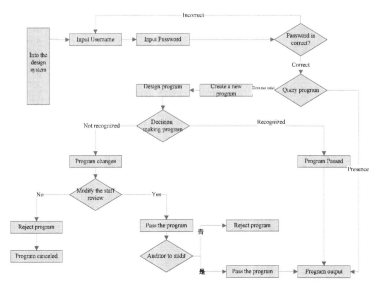

Fig. 1. Architectural design management process

4.2 Colored Petri Net Modeling

Building design according to the user on the system's functional requirements for approval, the use of CPN theoretical design process modeling system functions and establish the program based on Colored Petri Net model of the design approval system needs. First, define the color set of requirements models and variables are as follows :

```
val n=5;
colset DESIGNER=index d with 1..n;
var p:DESIGNER;
colset SCHEME=index s with 1..n;
fun schm(d(i))=1`s(i);
var sch,sch1,sch2:SCHEME;
```

Colored Petri Net-based demand model shown in Figure 2.

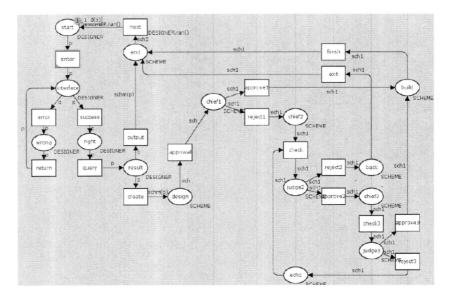

Fig. 2. The design based on system requirements CPN model

5   Colored Petri Net-Based Software Design Modeling

Software design is not specifically programmed, but refined into a prone process drawings, the main task is to design each module of the algorithm to determine the local data structure. The phase is not only logically correct to achieve the functions of each module, and the design process should be clear, readable and efficient.

Detailed design of the basic tasks include algorithm design, data structure design. Using the algorithm proposed in this paper process modeling and CPN CPN model into C # code.

5.1   Modeling Algorithms

In the design system, the audit program by the program stored in the database, the ordinary user can check the output of these programs. The algorithm idea is to record the output into the array, and then with the output. Model used in the color sets, functions and variables defined as follows.

```
val n=10;
colset UNIT=unit timed;
var u:UNIT;
colset RECORD=index id with 1..n;
var r:RECORD;
colset NUM=int;
var k,j:NUM;
```

```
colset AT=int;

colset RECORDTIME=product RECORD*AT;

var p:RECORDTIME;

colset ARRAY=list RECORDTIME timed;

colset ARRAYTIME=product ARRAY*AT;

var a,a1:ARRAY;

fun expTime;

fun intTime()=IntInf.toInt(time());
```

Color sets the color integer NUM is set to output statistics is the number of records into the array. ARRAY is a timed color sets RECORDTIME list of color sets, indicated by the output record array. Color set ARRAYTIME and AT is the Cartesian product ARRAY color set, representing the current output record contained in the array and the current simulation time.

Algorithm design model shown in Figure 3.

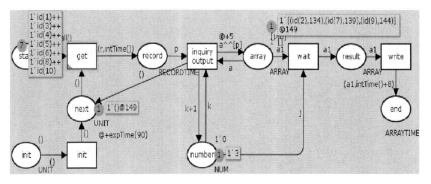

Fig. 3. Output algorithm model program

5.2 Model into Code

CPN model and color of the functions declared set directly into the programming language of functions and data types. CPN model in order, concurrency, choice, loop four structures, can also be easily converted into a programming language structures.

Conversion rules are as follows:

(1) sequence structure: the action took place between the expression context. In the programming language code, the order can change the structure where each function corresponds to a statement, such as assignment, calculation, comparison, input and output.

(2) complicated structure: CPN model of the concurrent structure of the language code into a program that a variable used for many operations.

(3) Select the structure: there is only one action can occur, the choice of model structure corresponding to the program code execution, such as if, case statements.

(4) loop structure: sometimes it is necessary to perform an action repeated several times. Model loop structure corresponding to the code in the loop, such as for, while and until statements, etc.

Based on the above CPN model structure and code mapping rules, the algorithm in Figure 3 models can easily generate C # program, sample code is as follows:

```
Public void readData(DataSet ds)

{

        DataTable dt=ds.Table[0];

        int count;

        Integer[][] tomac=new
Integer[dt.Rows.Count,dt.Fields.Count];

        Object oExcel,oBook;

        oExcel = new Excel.Application();

        oBook=oExcel.Workbooks.Add();

        for(int i=0;i<dt.Count;i++)

        {

          for(j=0;j<dt.Fields.Count;j++)

          {

            tomac[i,j]=Convert.toInt32(dt.Row[i][j]);

          }

        }

        oSheet.Range("A1").Resize[dt.Rows.Count,dt.Fie
lds.Count].Value=ds;

        ds=null;

}
```

CPN combines the advantages of programming language, provides a definition of data types and data values of the basic unit of operation. Using CPN for algorithm design, coding can easily establish the detailed design of the algorithm model of personnel into the program code.

6 Summary

This paper proposes a CPN-based architectural design management system development process model approach. Using CPN needs analysis phase of software development to establish the system functional requirements model validation process is correct; the detailed design stage with the CPN to describe the data structure and algorithm design, build algorithm model. This model also gives the rules to the code conversion, the CPN model into a more convenient program code. But the CPN into the program code when there are issues such as accuracy, is the next contents of the study.

References

[1] Carl Adam Petri. Communication With Automata. Tech. Rep. RADCTR-65-377, Rome Air Dev. Center. NewYork (1966)
[2] Yuan, C.: Petri net theory. Electronic Industry Press (1998)
[3] Cai, L.: Software Scenario Testing based on CPN State space. Computer Applications and Software 27(9), 37–40 (2010)
[4] Jensen, K., Michael Kristensen, L., Wells, L.: Coloured Petri Nets and CPN Tools for Modelling and Validation of Concurrent Systems. Int. J. Softw. Tools Technol. Transfer 9, 213–254 (2007)
[5] Jensen, K.: An Introduction to the Theoretical Aspect of Colored Petri Nets. In: de Bakker, J.W., de Roever, W.-P., Rozenberg, G. (eds.) REX 1993. LNCS, vol. 803, pp. 230–272. Springer, Heidelberg (1994)
[6] Jensen, K.: An Introduction to the Practical Use of Colored Petri Nets. LNCS, pp. 237–292. Springer, Heidelberg (1998)

Inquiry-Based Education and Its Implication to Education in China[*]

Tu Huiwen[1] and Xie Feng[2]

[1] School of Economics & Management, Zhejiang Sci-Tec University
[2] School of Finance & Management, Zhejiang Finance & Economics Institute
huiwentt@yahoo.com.cn

Abstract. As scientific literacy becomes an important goal in education nowadays. Inquiry-based teaching has also become more and more important method in education process. In this article we analysis the nature and character of inquiry-based teaching method, then we compare the characteristics of inquiry-based teaching to traditional teaching method, after these, we further explain inquiry-based teaching from the view of constructivism theory. At last, the paper proposed a few suggestions for education in China.

Keywords: scientific literacy, inquiry-based teaching, traditional teaching method, constructivism.

1   Introduction

With the rapid development of global economy, the human society has stepped into a new era. At the mean time, everyone has to face with great change in his every day. He has to deal with a great deal of problems and to make decisions. The ability to solve problem becomes more and more important in our daily life. So, how to develop people's ability in the new era? Education plays the essential role in this circumstance. Then the problem is: what kind of knowledge and ability should be taught to the students? How to teach this knowledge and ability?

Traditionally, education is focused on mastery of content, with less emphasis on the development of skills and the nurturing of inquiring attitudes. Teachers focused on giving out information about "what is known". Students are the receivers of information, and the teacher is the dispenser. But in the modern society, students need opportunities not just to hear explanations which were given by experts like teachers, books, films, and the Internet, but they also need to practice using the ideas themselves to gain confidence in their use, and through this process develop a familiarity with, and understanding of science as its practiced among scientists in the world outside of the classroom. Facts can easily be obtained from any text book, memorized and reiterated. However knowledge must be constructed through dialogue and made meaningful. Students must understand this difference between knowing the facts or the information and having knowledge of a particular scientific phenomenon. As Siegel pointed that,

---

[*] This study is supported by Zhejiang Natural Science Funding Project (No.Y6090471).

Y. Wang (Ed.): Education Management, Education Theory & Education Application, AISC 109, pp. 527–534.
springerlink.com   © Springer-Verlag Berlin Heidelberg 2011

"education has as its fundamental task both the development of students' reasoning ability, and also the fostering of a complex of attitudes, habits of mind, dispositions and character traits, such that students are not only able to reason well; they also care about reasons, and organize their beliefs, judgments and actions in accordance with the deliverance of the reasoned evaluation of reasons." [1] Science educators, government sponsored agencies, curriculum developers, and other stakeholders of science education have all called for a conscious effort to be made to teach "authentic" ways of conducting science which focuses on the goal of develop students' scientific ability in general, and on understanding of the nature of science in particular[2].

To gain the status of lifelong learning ability, it is not enough to have reading and writing skills as before. Science and technology have become so important in modern life that the ability of citizens to understand and use science can spell the difference between prosperity and decline. Helping students to develop into scientifically literate citizens is a vital objective proposed in recent science education reform initiatives.

Scientific ability is often recognized as the ability to apply that knowledge and understandings in everyday situations, and an understanding of the characteristics of science and its interactions with society and personal life. Scientific ability addresses the understandings and habits of mind that enable people to grasp what those enterprises are up to, to make some sense of how the natural and designed worlds work, to think critically and independently, to recognize and weigh alternative explanations of events[3]

Science educators have long recommended that learning with inquiry be placed at the core of science instruction to actively engage learners in the processes of science. Under this background, many science teachers strive to help the students understand and apply scientific concepts, participate in scientific inquiry and understand the nature of science. And many countries are trying to reform their traditional education model to teach students scientific ability. Among these countries, education reform in America is a typical case.

There has been a lot of education reform documents been signed by America governments to promote science teaching in these years. Major reform documents in science education, such as Science for All Americans from Project 2061 of the American Association for the Advancement of Science and National Science Education Standards from the National Research Council[4], have stressed the importance of inquiry based learning and understanding the nature of science for improving the science literacy in students. In general, these reform initiatives have recommended an inquiry-based approach to science education, real-world problem solving for students.

Many researchers have also indicated the importance of inquiry based constructivist approach to science learning. They proposed that, through inquiry-based science teaching and learning, students come to understand the values and assumptions of the knowledge and the process by which the knowledge is created. Scientific literacy has become a critical issue for all citizens of the United States.

So inquiry-based teaching has become an important education method nowadays. In this article we analysis the nature and character of this method, then we compare the characteristics of inquiry-based teaching and traditional teaching method. after these, we further explain inquiry-based teaching from theory of constructivism view. At last, the paper proposed a few suggestions for education in China.

2   The Nature of Science Inquiry

2.1   The Content of Scientific Inquiry

Generally the nature of science refers to the values and assumptions inherent to science, scientific knowledge, and the development of scientific knowledge. Scientific inquiry is defined as a systemic and investigative activity that scientists employ in an attempt to provide an explanation of the world and to discover and describe relationships between objects and events based on the evidence they collected[5]. Inquiry is the basic method through which the unknown of nature are revealed. Students follow the same inquiry process in their classrooms as scientists do can develop the skills and method as scientist do. Scientific inquiry involves, asking questions, seeking evidence, making observations, gathering data, formulating explanations, testing explanations, evaluating explanations, communicating and justifying proposed explanations, interpreting and responding to criticisms, formulating appropriate criticisms, engaging in critical argumentation, and reflecting on alternative explanations. In addition to these, Wu & Hsieh have identified seven phases of scientific inquiry: asking and deciding questions, searching for information, designing investigations, carrying out investigations, analyzing data and making conclusions, creating hypothesis and verified, and sharing and communicating findings[6]. These phases are very similar to the traditional scientific method.

Through scientific inquiry, the epistemology of science, science as a way of knowing, or the values and beliefs that are inherent in the development of scientific knowledge can be developed. Developing an understanding for the key aspects of the nature of science, tentative and empirical nature of science, the role of imagination, subjectivity, and creativity in science, understanding the distinction between observation and inference, theories and laws, and the social, cultural, political and economical embeddedness of science, and scientific inquiry are consistent with the goals of science education reform documents Benchmarks for Science Literacy, National Science Education Standards and the Michigan Curriculum Framework Science Benchmarks[7].

2.2   The Process of Inquiry Based Teaching

Teaching through "inquiry" involves engaging students in the research process with instructor support and coaching at a level appropriate to their starting skills. Students learn discipline specific content but in doing so, engage and refine their inquiry skills[8].

2.2.1   Selecting a Problem

This model begins with the teacher selecting a puzzling situation or problem that is genuinely interesting and stimulating to the learner. It may be a scientific problem or a puzzling event. It may be a scene from a play or a story that requires the students to formulate an out come. And this problem is the start of the inquiry based teaching. It encourages students' questions and analysis and discussion about this phenomenon.

2.2.2 Presenting the Rules

Before beginning the inquiry lesson, the teacher explains the process to the class; in this model, the entire class can participate. The teacher is the main source of data and will respond only to questions that can be answered yes or no, thus placing the burden of framing the question on the learner. The teacher may choose to add additional information or guide the questioning, but the responsibility for hypothesizing must remain with the students; the teacher is in control of the process but not in control of the out come.

2.2.3 Gathering Data

In most cases, students ask questions that require the teacher to do the thinking. Each question must be asked as a tentative hypothesis. The student cannot ask the questions which would require the teacher to give the information. Rather, the question must be phrased so that the teacher can respond with a yes or no answer. The teacher may decide to add information or expand on the problem at any time; it is important to let the students experience some frustration as they question. There is a temptation for the teacher to rephrase the question.

The data should be recorded on the board or on data sheets kept by each student.

2.2.4 Developing a Theory

When a student poses a theoretical question that seems to be an answer to the problem, the question is stated as a theory and written on the board in a special area reserved for this purpose. Students may ask to discuss the information and frame hypothetical questions they will ask the teacher. Depending on the nature of the problem, the teacher directs the students to other sources of information. Students are encouraged to ask hypothetical questions at this point.

If the students reach a point where the theory they have posed seems to be verified, then the class accepts the theory as a solution and moves to the next stage.

2.2.5 Explaining the Theory

In this step, students are asked to explain the theory accepted as a tentative solution and state the rules associated with that theory. In addition, they must determine how the theory could be tested to see if the rules can be generalized to other situations.

Through the process of inquiry, individuals construct much of their understanding of the natural and human-designed worlds. Inquiry is not seeking the right answer -- because often there is none -- but rather seeking appropriate resolutions to questions and issues. For educators, inquiry implies emphasis on the development of inquiry skills and the nurturing of inquiring attitudes that will enable individuals to continue the quest for knowledge throughout life.

2.3 Difference between Inquiry-Based Teaching and Traditional Teaching

There are mainly five essential features of inquiry-based teaching, they are:

- Learners are engaged by scientifically oriented questions.
- Learners give priority to evidence, which allows them to develop and evaluate explanations that address scientifically oriented questions.

• Learners formulate explanations from evidence to address scientifically oriented questions.

• Learners evaluate their explanations in light of alternate explanations, particularly those reflecting scientific understanding.

• Learners communicate and justify their proposed explanations. [4] ---

These techniques seek to engage students with meaningful questions about everyday experiences, emphasize using a method of investigation to evaluate some form of evidence critically, and engage learners in a social discourse to promote the knowledge-construction process.

The main differences between inquiry-based teaching and more traditional teaching are as the table below:

Traditional classroom	Inquiry based classroom
Curriculum begins with the parts of the whole. Emphasizes the basic skills	Curriculum emphasizes big concepts, beginning with the whole and expanding to include the parts
Materials are primarily textbooks and workbooks	Materials include primary sources of material and manipulative materials
Strict adherence to fixed curriculum is highly valued	Pursuit of students questions and interest is valued
Learning is based on repetition	Learning is interactive, building on what the student already knows
Teachers disseminate information to students; students are recipients of knowledge	Teachers have a dialogue with students, helping students construct their own knowledge
Teacher's role is directive, rooted in authority	Teacher's role is interactive, rooted in negotiation
Knowledge is seen as inert	Knowledge is seen as dynamic, ever changing with our experiences
Students work primarily alone	Students work primarily in groups

3   Inquiry Based Education from Constructivist View

3.1   Constructivist View of Learning

The origin of constructivist theory of learning is from Jean Piaget's theory on how people learn as a result of his study on children's stages of cognitive development. And Piaget's ideas contributed greatly to the direction, meaning and understanding of constructivism. In Piaget's theory, there are two important ideas: one is that learners construct their own knowledge schemes in relation to, and filtered through, previous and current experiences; the other is that referring to the interaction with the environment, the student assimilates complementary components of the external world into his existing cognitive structures or schemata. Learners need to construct knowledge by themselves in order to make it meaningful[9]. So a lot of researchers regard that constructivism is a "meaning-making" theory, and a learner –centered theory of learning.

Thus constructivism is a theory based on observation and scientific study. And it is about how people learn. Constructivism takes a cognitive approach to look at how people learn. It regards knowledge is built or constructed by the learners according to their prior experiences and the understanding they bring to the learning place, instead of knowledge transferred directly from the teacher to the learners. From the view of constructivists, learning means constructing, inventing, and developing learner's own knowledge. To learn is not to get knowledge by information or knowledge transmitting by outside teachers, but on promoting learning through learner's intellectual activity such as questioning, problem generating, investigating and problem solving[9].

3.2 Inquiry Based Education as a Constructivist Learning Process

As we can see from the discussion above, there are five essential components of inquiry-based science instruction. These components include: (1) developing science problem-solving skills, (2) communicating science ideas, (3) developing science laboratory skills, (4) developing data analysis skills, and (5) having frequent hands-on activities. The process of inquiry is modeled on the scientist's method of discovery, viewing science as a constructed set of theories and ideas based on the physical world, rather than as a collection fo irrefutable, disconnected facts(National Science Foundation, 1997). Inquiry teaching leads learners through the experience of scientific inquiry. Learners develop their understanding of fundamental scientific ideas by directly experiencing materials, consulting resources that include experts, and dicussing among themselves. In the inquiry teaching process, learners are required to organize their prior knowledge with new science knowledge and skills via their innate qualities of curiosity, openness and skepticism[10].

In a science inquiry class, the educator provides the learners with the critical elements of the scientific process. It promotes learners to make observations, construct hypotheses, conduct experiments, report results, and derive and communicate conclusions. Through this process, learners would form concepts of what they observed, what they investigated, and what they repeated tested[11].

4   Implication to Education in China

With the great challenge stimulated by rapid global change to the people nowadays, scientific literacy has become a vital ability for people to deal with this challenge effectively. Many countries have sponsored education reform to teach and encourage people's scientific literacy. Among these education reform affairs, inquiry-based education act as an very important method. The discussion of inquiry-based education in this paper has practical significance for improving the quality of teaching and learning. They provide ideas for policy makers as well as teachers to design educational programs to enhance student learning outcomes. It offers great implication to the education reform in China.

4.1 Emphasis Training of Students' Scientific Literacy

Education is the most valuable asset that an individual of today can possess. However, the primary purpose of education is to develop discernment of knowledge and correct behavior. Learning should be meaningful, not merely an attempt to stuff information into young minds. It should prepare students for the future, and focus on meaningful understanding, make students be realistic in its expectations. These guidelines provide a useful criterion against which any educational reform or method may be measured.

Traditional education methods are far from providing students with meaningful methods and provide them with useful scientific skills. Inquiry-based instruction can be a important supplement to our education system. Students may benefit greatly if teachers are encouraged to add inquiry-based activities to their classrooms[12]. Such encouragement would be consistent with suggested educational reforms that promote the use of inquiry-based science instructional methods in Chinese schools and a departure from repeated use of traditional science instructional methods.

4.2 Emphasis the Important Role of Inquiry-Based Method in Teaching

As a useful teaching method, inquiry-based teaching has become a important method in many countries education system. And it gives a useful reference on how to execute china higher education reform.

Traditionally, education approach in Chinese higher school is focused on mastery of content, with less emphasis on the development of skills and scientific literacy. The current system of education is teacher centered, with the teacher focused on giving out information about "what is known." Students are the receivers of information, and the teacher is the dispenser. Much of the assessment of the learner is focused on the importance of "one right answer." Traditional education is more concerned with preparation for the next grade level and in-school success than with helping a student learn to learn throughout life.

As discussed in this paper that inquiry-based teaching is a useful tool of which can develop students' scientific ability and lifelong learning. It's urgently for China to adopt this teaching method in our education system, especially in our higher education. We should pay great emphasis on the important role of inquiry-based teaching in our education system. At the mean time we should explore effective way of inquiry-based teaching, and fostering development of Chinese higher education course.

4.3 Integrate Inquiry-Based Teaching and Traditional Teaching in Education

Although inquiry-based teaching is a important education tool in science teaching, it is not the whole substitute of traditional teaching methods. Its usage is also with conditional. When we adopt inquiry-based teaching method as a useful education tool, we must also pay great emphases to traditional teaching method, such as teacher-centred instruction, Instruction based on textbooks, individual written assignments, direct instruction and lectures, seatwork, etc. When we adopt various teaching method, we should choose the appropriate method according to the goal of periodical learning era. Traditional methods contribute a lot when the goal is require students to grasp some basic theory and knowledge. It is just some less contribution when it is used in develop students scientific skills. From this point of view, whether traditional teaching methods or inquiry-based methods are all useful tool in the education system. What we should do is integrating these teaching methods effectively in the whole education period.

4.4　Enhancing the Scientific Ability of Teachers

In the inquiry based teaching process, the science acquirement of students is largely determined by the course design of teachers. Teachers play a vital role in the inquiry based teaching process. They are essential in promoting student learning effect. So teachers with high lever science knowledge and teaching skills are of great important to inquiry based teaching.

In order to build an effective teacher system, administrator and stakeholders of schools should pay great attention to the building process of teachers. At first, schools can step up recruitment efforts to hire teacher candidates who have strong academic credentials and who have completed a rigorous teacher preparation program. Secondly, schools should provide teachers with training programs. It will provide science teachers with the time essential to learn science content, practice effective teaching methods. Teachers should understand clearly that not only the concepts students are expected to learn, but also they should develop their own scientific ability in the teaching process.

References

1. Siegel, H.: Why should educators care about argumentation. Informal Logic 17(2), 159–176 (1995)
2. Solomon, J., Scott, L., Duveen, J.: Large-scale exploration of pupils' understanding of the nature of science. Science Education 80, 493–508 (1996)
3. McComas, W., Almazroa, H., Clough, M.P.: The Nature of Science in Science Education: An Introduction. Science and Education 7(6), 511–532 (1998)
4. National Research Council, National science education standards. National Academic Press, Washington, DC (1996)
5. Martin-Hansen, L.: Defining inquiry: Exploring the many types of inquiry in the science classroom. The Science Teacher 69(1), 34–37 (2002)
6. Wu, H., Hsieh, C.: Developing sixth graders' inquiry skills to construct explanations in inquiry-based learning environments. International Journal of Science Education 28(11), 1289–1313 (2006)
7. American Association for the Advancement of Science, A, Benchmarks for science literacy: A project, report. Oxford University Press, New York (1993)
8. Von Seeker, C.E.: Effects of inquiry-based teacher practices on science excellence and equity. The Journal of Educational Research 95(3), 151–160 (2002)
9. Pajares, M.F.: Teacher beliefs and educational research: Cleaning up a messy construct. Review of Educational Research 62, 307–332 (1992)
10. Edwards, C.H.: Promoting student inquiry. The Science Teacher 64(7), 18–21 (1997)
11. Richardson, V.: In constructivist teacher education: building a world of new understandings. Falmer Press, Washington, DC (1997)
12. Bodzin, A.M., Cates, W.M.: Enhancing Preservice Teachers' Understanding of Web-based Scientific Inquiry. Journal of Science Teacher Education 14(4), 237–257 (2003)

Study on Construction and Management of Innovational GIS Laboratory

Lifeng Yuan[*] and Xingfei Liu

College of Geography & Biological Information Nanjing University of Posts and
Telecommunications, Nanjing 210003, China
yuanlifeng7833@126.com

Abstract. GIS is a technological and comprehensive interdisciplinary. With vigorous development of education on GIS specialty in China, construction of innovational GIS laboratory has become necessary road when multiattribute GIS people with special skills are trained for market demand. This paper take Nanjing University of Posts and Telecommunications for a case study, surrounding a objective of talent training of our school and GIS specialty feature, putting forward a way of construction thinking with taking "four platforms" as main body of innovational GIS laboratory , and expatiating on its contents, and including objective and principle of construction, and professional laboratory construction, teaching team construction, innovation platform construction and laboratory management construction.

Keywords: GIS specialty, laboratory construction, GIS talent training.

1 Introduction

The Geographic Information System (GIS) is a computer technology discipline which established for geographical research and decision service. Based on the geographical spatial database and the support of the computer software and hardware, GIS can select, manage, operate, analyze, simulate and display the spatial related data and provide many kinds of spatial and dynamic geographic information with the analysis method of geographical model. The specialty of Geographical Information System (GIS) has enrolled students since 1999, and 151 universities have recruited GIS specialty undergraduate throughout the country by the year 2009 (data from "China Education and Research Network"). The specialized education of GIS has showed the flourishing trend. The geographic information system is a discipline with higher practicality which emphasis on technology and practice. In order to grasp the technology of GIS, the students need enough time to understand all kinds of GIS knowledge and operate the all functions of GIS. GIS professional teaching is not only to help student learn the theory and technology, but also to develop the students' ability of acquirement, processing and analysis the spatial information using high

---

[*] Supported by the School Teaching Reform Project (JG01611JX17) and the laboratory construction and equipment management research project (No. 2011XSG05、2011XSG06) of Nanjing University of Posts and Telecommunications.

technology through a lot of practices, it is significant to prepare for future work to students and this is what the particularity of specialized education of GIS.

The academician ZHU Qingshi said, the president of the South Science and Technology University, "No first-class laboratory, No first-class university." Therefore, the GIS laboratory with modern equipment and advanced management mode is a cradle for innovative GIS talent training. In the paper, on the basis of educational reformative achievement on aspects such as personnel training and construction of GIS specialty in the domestic and foreign universities in recent years, combining the actual conditions of GIS specialty of Nanjing University of the Post and Telecommunications, centered on the personnel training objectives and characteristics of GIS specialty, the construction and management of innovational GIS laboratory are discussed in detail.

2 Target and Feature of Professional Talents Training

The GIS specialty is a new developing comprehensive and technological interdisciplinary. According to the basic requirements of specialty personnel training, through fully researching domestic and international GIS specialty teaching and curriculum construction, depending on long-term education base of communications and IT talent and the advantages of information technology for the national economic construction and development needs, facing to the requirements of GIS specialty in enterprises and institutions, especially in the ministry of information industry, and promoting the field of GIS applications in communications and the crossing and integration between GIS technology and communication technology, our school has been promoting positively the popularization of the application of GIS in the field of communications and the crossing and integration of GIS techniques and communications techniques, and training senior specialized personnel who acquire rudimentary knowledge and basic skills such as computer techniques, GIS techniques, RS techniques, communications techniques, spatial location techniques, etc., and which can be engaged in work related to GIS such as the application study, technological development, production management, in the fields of education, scientific research, communication, survey and drawing, the municipal administration, resource, travel and land.

GIS specialty characteristics of our school are to train students in GIS research, design, and development and applications ability in the communications industry. GIS specialty is relying on the background of computers and communications disciplines in our university, aiming to the direction of the mobile and intelligent geographic information mainly, taking communication and information processing as the main application areas, and using the integration between spatial information technology and communications technology. Students will be learning: the basic theory, knowledge and technology about geographic information systems, computer, communications, remote sensing, and global positioning systems (GPS). They will be trained in basic research, technological development of scientific thinking and experimental training in order to master GIS research, design and development of basic skills and the initial teaching, research, development and management capabilities.

3 Construction Objective and Principle of Innovative GIS Laboratory

On the support of modern software and hardware equipment, the management mode of advanced education, the concept of "people-oriented", the overall construction objective of innovation GIS laboratory is to provide the open experimental environment, to cultivate the innovative atmosphere especially to train the students' innovative ability on basic of strengthening students' ability of application, design development and integration. According to the cultivation objective and feature of GIS professional talent, innovative GIS lab construction should comply with the following principles:

- Principle of goal. Take the cultivation of students' design capacity, development ability, and innovation ability and integration capability as the target of the construction of innovative GIS laboratory and establish the general frame of innovative GIS laboratory, keeping in good tune with the training objective and features of GIS specialty.
- Principle of advance. Configure advanced instruments and equipment for GIS laboratory and establish experiment instrument booking, using platform and open laboratory management platform based on C/S and B/S structure.
- Principle of integration. Strengthen the organic combination of theoretical teaching and practice teaching. To teach with visual image, the course teaching should follow the requirement of integration if the condition allows, combining classrooms and labs together and "teach, learn and do "together organically.
- Principle of open. Open laboratory, assure enough open time, provide autonomy learning conveniences and loose research environment, encourage exploration, tolerate failure.

4 Innovative GIS Lab Platform Inner Construction

According to the principles above, taking the cultivation of students' comprehensive quality as the core, taking the cultivation of the ability of technical application as the major aspect, taking the cultivation of the development ability of GIS software as the key point, taking the combination of production, learning and research as the route, the construction of innovative GIS laboratory have been designed systematically with the main body of " four platforms plus four modules ", the basic frame of which is shown in Fig 1.

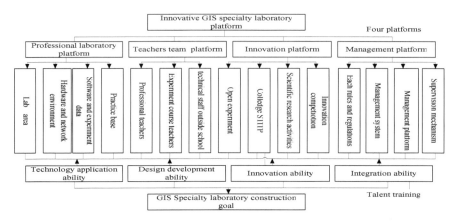

Fig. 1. The basic frame of construction of GIS specialty laboratory

4.1   The Professional Laboratory Construction

As the main battlefield of practice teaching, the laboratory is the vital guarantee for the personnel training and scientific research. Since we set up GIS in 2005, we gradually establish and perfect the software and hardware infrastructure and teaching environment. GIS professional laboratory is invested and constructed by the school and the project of "Central authorities and locality together-build the featured advantaged discipline laboratory of the university". At present the investment is about 5 million yuan. The professional laboratory platform construction includes GIS laboratory, GIS hardware and network environment, GIS software and data center and practice base.

(1) GIS laboratory
The GIS laboratory is the main battlefield where students make experiment, take practice, and train their ability. Our university GIS laboratory covering an area of about 120 m2 and including mating teaching and perfect network infrastructure is a comprehensive open experimental teaching laboratory based on computer technology, network communication technology and the multimedia technology.

(2) GIS software and data center
At present, our GIS laboratory have softwares such as : ArcGIS 9.3（Engine 9.3）、Erdas 8.7、digital photogrammetric system VirtuoZo、ENVI 4.7、AutoCAD 2006、SuperMap 6、MapInfo 9.0、MapX、southern surveying and mapping CASS 7.0、SQL Server 2005. Spatial database is the GIS core. Now there are Nanjing TM/ETM remote sensing images, 1:100000 vector foundation geographical data, 1:50000 topographic map data, land use data and soil data, etc, and some teachers' scientific research project data and data support with books.

The above software and hardware equipment and spatial data provide a strong material guarantee for student to make experiment and take part in teaching practice. In addition, in order to constantly perfect the laboratory construction, we are actively considering buying other hardware and software to meet teaching requirement and the scientific research.

(3) practice base

As the important components of teaching and research work and the base of students' knowledge innovation and technological development, practice bases include the practice bases inside school and the practice bases outside school. The practice bases inside school rely mainly on Students' Scientific and Technical Innovation Activity Centre. The practice bases outside school have signed cooperative agreements with Jiangsu Provincial Bureau of Surveying and Mapping , East China Mineral Exploration and development Bureau , Suzhou Yulong Scientific Education Co., LTD enterprises and been exchanging regularly, progressively establish the combination mechanism of production, teaching, research closely combined with teaching and social production; engage experts and scholars of the practice bases outside school to join our school GIS professional guidance committee, to participate in programming the specialized training, and to conduct reform and construction of courses, construction of practice bases and the like.

(4) GIS hardware and network environment

GIS is a system including data collection, input, data management and analysis, data output and transmission, so it need corresponding hardware facilities and other auxiliary facilities. The concrete main hardware equipment is shown in Table 1.

Tabel 1. The hardware experiment equipment of laboratory platform

ID	Equipment name	Type	Quantity
05010105	microcomputer	DELL VOSTRO 200	75
05010105	notebook PC	DELL VOSTRO 1400	8
05031000	Embedded GIS experiment box	EELIOD 270	35
05010104	server	DELL 1430	4
03230243	Navigation engineering experimental platform	GPS-portable	6
03191212	GPS static intelligent equipment	HD 8200X	7
05031000	GPS receiver	V-700	1
03040270	Total speed Intelligent measuring devices	NFS-3S2	2
05010105	palm computer	HP5965	6
03190300	oscilloscope	TDS1012B	2
03210129	Spectroradiometer	HH	2
06040201	graphic plotter	HPDJ500+A1	1
05010501	color laser printer	HP2605	1
06020200	scanistor	SX36	1
03040225	total station	NTS-302B	8
05010549	laser printer	HP1008	13
03170300	digital camera	DSC-W170	2

4.2 Teaching Body Construction

Teaching body construction is the core content of GIS lab construction. It has put forward higher requirement for the teachers that GIS is a specialty of high practicality and the renewal of professional knowledge is very fast. We have laboratory professional teachers and experimental teachers. Now 100% of all the GIS professional teachers have master's degree or higher academic degree, 60% have doctor degree and most of them are young doctor teachers employed in recent years. Encourage teachers to participate in the scientific research project and international academic activities, cultivate their innovation abilities, and engage the technical staff of the practice base outside school as part-time practice teachers.

4.3 Innovation Platform Construction

Innovation platform construction includes Open experiment, Science and Technology Innovation Training Program, Scientific research and Innovation competition and so on. Open experiment mainly is that teachers according to their research project or research interests , draw up the topic and the students sigh up ,then after divergent selection the teachers make experiment scheme and guide students separately or collaboratively to finish. In addition, students can also according to their interests put forward questions and design scheme through teachers' examination and guiding, then students grope for experimental procedure to complete the experiment and write experiment report or paper.

The purpose of STITP provides a science and technology innovation opportunity for undergraduates, cultivates students' innovation consciousness and makes them contact and understand academic front, clear the academic dynamic development, strengthen students' team cooperation spirit and communication skills, make use of the teaching resources and personnel superiority, teaches students in accordance with their aptitude, improve teaching quality and managerial benefit. The school provides the research fund for STITP project, the student carry out the project, the teacher is responsible for guiding, inspecting and supervising students to complete the project, and guiding students to how to reply on the project and how to make the summary and promotion of the research achievements.

Scientific research activities are that students can participate in scientific research project of teachers, train their abilities of theory with practice and scientific research innovation practice and team cooperation, "Research" promotes "learning" and contact closely with "production". Innovation contest by encouraging and guiding students to taking part in the national GIS design contest, improve the students' GIS application, design and development ability. In addition, through setting famous teachers lectures and holding regular research report, we cultivate the passion of scientific research and innovation consciousness of the students; strengthen the innovation spirit and innovation ability of the students.

4.4 Laboratory Management Construction

Laboratory management platform is the basic guarantee for experiment and practice teaching activities. Laboratory management platform construction mainly includes as below:

(1) Establish and perfect the regulations. Such as the post responsibility system, laboratory use regulations, operating rules, valuable instruments use and management way, etc. Use the advanced management method and standardized management program.

(2) Definitude mangers and their responsibility. Take three levels management system those are target decomposition, right and censure discernible (vice President of the college for managing experiment --- lab director---lab professional manager and experiment course teachers). Managers are responsible for the registration of laboratory instruments and equipment and experiment teaching management work, the technicians are responsible for laboratory equipment use, maintenance and the students' experiment guidance and so on.

(3) Modern equipment management platform. Our laboratory instruments and teaching management platform are based on C/S and B/S model that can improve management efficiency.

(4) Computer system management. Install unified several computers software by installing the networking computer software distribution system; ensure that computer system is a pure system after every restart by installing computer protest card.

(5) Open laboratory. Our lab is open for students that provide learning environment and innovation atmosphere.

(6) Supervision and sampling. For supervising lab operation and management, the supervision groups regularly or irregularly supervise and inspect the operation condition of laboratory.

5 Conclusion

The construction of innovative GIS laboratory is the key to the cultivation of GIS talents, and the main carrier of promoting students' practice ability. Centre on training objective and features of GIS specialty, taking the cultivation of students' comprehensive quality as the core, taking the cultivation of the ability of technical application as the major aspect, taking the cultivation of the development ability of GIS software as the key point, taking the close combination of production, learning and research as the route, establish the advance-with-the-time cultivation modes of specialized personnel of GIS specialty practical education. Through the construction of innovative GIS laboratory, students trained by our school GIS specialty will be able to become GIS talents that are of strong practice skills and innovation ability and meet economic development and social need.

Acknowledgments. The work is supported by the School Teaching Reform Project (JG01611JX17) and the laboratory construction and equipment management research project (No. 2011XSG05、2011XSG06) of Nanjing University of Posts and Telecommunications.

References

1. Liu, X., Zou, Y.: On Construction of Innovative GIS Laboratory. Research and Exploration in Laboratory 27(1), 134–138 (2008)
2. Zou, Y., Liu, X.: The Construction of Innovational GIS Laboratory for Training Qualified Personnel in GIS Field. Geomatics World 2, 25–30 (2007)

3. Yuan, L., Ma, M., Zhang, H.: A Study of the Talent Training Mode and Curriculum System Construction of GIS. Higher Education of Sciences 1, 40–43 (2010)
4. Kuang, Y.: Exploration of GIS Laboratory System Composition and Management Model. Experimental Technology and Management 20(4), 133–135 (2003)
5. Yuan, L., Jiang, C.: A Study of GIS Specialty Practice Teaching System Construction. China Education Innovation Herald 34, 70–71 (2009)

Analysis of Professional Skill and the Teaching of Effective Interface Ordnance N.C.O.

Zhu Tian-yu[1], Jiang Zhong-bao[1], Liu Yun[1], and Duan Shao-li[2]

[1] Wuhan Mechanical Technology College, Luoyu east road.1038,
430075 Wuhan, China
[2] Wuhan Technical College of Communication,
430000 Wuhan, China
liuyun111700@tom.com

Abstract. August 2010, the PLA units associated together human resources and social security men made a "People's Liberation Army soldiers on active service professional skill requirements", marks the professional skill army soldiers entered a new stage of development. Soldiers for the full implementation of professional skill, this combination of armaments to protect the professional soldiers of the actual situation of professional skill, Analysis of the professional skill and armaments to protect the professional noncommissioned officer education and teaching of articulation.

Keywords: Professional skill, N.C.O.

1 Introduction

Troops professional skill is required in accordance with national occupational skills standards and qualification criteria, combined with the needs of military posts through the identification of military administration of the examination, the soldier's skill level objective and impartial, scientific and standardized evaluation and certification activities. Soldiers, professional skill makes explicit the ability of professional soldiers, is the main basis for the employer forces. It is becoming a force specification professionals, soldiers, professional conduct standards. Will also become non-commissioned officers to measure and evaluate education and teaching institutions are important to adapt to the needs of military construction measure.

Noncommissioned officer academies training goal is to serve the front line troops of technical talents, therefore, to highlight the skills of noncommissioned officer education, training, vocational skills training should be assessing the level of NCO academies as an important indicator of the quality of education. Noncommissioned officer education and professional skill of the effective convergence is to establish a competency-based education quality noncommissioned officer academies, and establishing a force development needs of the noncommissioned officer education and training system and promote the development of education required for non-commissioned officers. And also to improve the quality of soldiers, to promote job access system, non-commissioned officers to enhance students' ability to practice and innovation, and promote the development of non-commissioned officers closely with the military education requirements.

2 According to Ordnance Characteristics of the Teaching Institutions, Non-Commissioned Officers, Professional Skill to Build a Long-Term Operating Mechanism

Strengthen vocational and technical appraisal of the management mechanism and organization and management system is to achieve the professional skill and educational institutions, non-commissioned officers an important guarantee for effective convergence conditions. The establishment of relevant institutions in the non-commissioned officers of professional soldiers Occupational Skill Testing, non-commissioned officers and students of vocational skills appraisal of the normal institutions of learning into the non-commissioned officers training, vocational skills identified by the soldiers stand on the professional skill of the specific work plan, organize and coordination.

Soldiers, noncommissioned officer academies for professional skill, to meet the requirements of military training program, strict implementation of national occupational skills certification rules and regulations, according to "People's Liberation Army soldiers on active service professional skill requirements", from the appraisal team, reporting conditions, assessment methods, test services organization, performance statistics, certificate management, establish the corresponding six quality control system. Strict accordance with the professional skill level outline strict assessment criteria and identification of active troops to maintain professional skill assessment of the seriousness and authority of national vocational qualification certificate.

3 Deepening the Reform of Curriculum, and Professional Skill to Effectively Link

To ensure the professional skill work carried out, non-commissioned officers in the development of professional institutions, training programs should be vocational skills and professional identification of the corresponding content into one, starting from the training objectives, job requirements and work to build process-oriented professional competency model curriculum, vocational skills assessment test will be incorporated in classroom teaching, and professional skill and effective convergence and promote curriculum development.

According to "People's Liberation Army military training and assessment framework", and "military armaments to protect the professional soldiers and identification of occupational skills standards outline level" requirements, continuously improve teaching standards. First, the course content system architecture, and change the vertical to horizontal, that is vertical at disciplines, sub-categories of courses, into a horizontal position to the ability of the premise, the work process in accordance with the needs of the re-organization of an integrated curriculum of course; Second, teaching is organized and professional qualification certification system integration, emphasizing the integration of knowledge and ability. Note that according to course content and job requirements, the most advanced and latest knowledge rapidly incorporated into the curriculum.

Protection of professional soldiers under the armaments requirements of job competency standards, selection of materials. Non-commissioned officers should be based on their respective branches of the military academies and professional orientation, preparation of related materials in their own teaching materials to enhance professional while on the basis of partial self-compiled teaching materials, the choice of professional skill designated national resource materials. Development of teaching materials to introduce troops positions standard, required for professional positions related to new knowledge, new technologies, new means reflected in the textbooks to improve the teaching of advanced and practical.

Highlighting the practical teaching, the practice of teaching as a teaching center of the link, the direction of educating people to achieve educational and professional skill-building needs and the forces of "Zero." To work and study as a starting point, it is recommended as the core task-based project-oriented teaching reform, and "learning to do one" practice-oriented curriculum system. Around work and study, to create and develop the quality of jobs, job training capacity consistent teaching and learning environment, job content and work to achieve the teaching, teaching methods and ways of working, teaching and learning environment and working atmosphere of consistency

4   Exploration of Convergence with the Professional Skill of Teaching Methods and Examination, Assessment Methods

Basis for the different characteristics of students, professional features, and actively explore different levels of teaching, project teaching, case teaching, field teaching diverse teaching methods, students develop the ability to apply effective. In specific teaching practice, the different characteristics of the various types of work, expertise and relevant skills will be divided into basic skills training, professional skills training, skill training, military training four parts, in the skills training process to exacting standards by simple to complex, from single to comprehensive, step by step.

Actively explore the examination, assessment methods of reform. In the assessment, the practical ability to increase, the proportion of ability assessment by examination in the form of reform, and gradually establish a competency-based assessment, routine exam and skills test consistency of the examination, assessment methodology, and with the national vocational qualification certificate system integration .

5   Members of Professional Skill Training Evaluation Team Building "Double" Teachers

Assessor is to achieve the professional skill and educational institutions, non-commissioned officers of the key conditions for effective convergence. Therefore, non-commissioned officers institutions to focus on "dual" faculty ranks.

New circumstances, institutions, non-commissioned officers is to train qualified teachers of key military personnel, noncommissioned officers education faculty, the designers are not only teaching the theory and execution, or practice of practitioners and mentors. Improving the quality of teacher education is to achieve the goal of non-commissioned officers, military personnel to promote the sustainable development of the fundamental guarantee.

The assessor is the key to the implementation of professional skill of the conditions, institutions, non-commissioned officers training staff is to achieve the fundamental guarantee. Therefore, professional skill evaluation in accordance with the relevant standards staff, faculty and state institutions, non-commissioned officers to make professional skill combination of evaluation staff in order to better achieve the professional skill and education institutions, non-commissioned officers effective interface.

Construction and optimization of "dual" teaching staff. "Double" teachers for professional non-commissioned officers set up a very strong practical significance of institutions is no trivial matter, this part of the teachers directly determine the characteristics of these institutions, non-commissioned officers, therefore, non-commissioned officers institutions should increase efforts to build a variety of ways, "Double type "faculty team.

References

1. Zhang, C.-Y.: Study on countermeasures of implementing "double-certificate" in higher vocational colleges. Journal of Minxi Vocational and Technical College 9(3) (September 2007)
2. Ruan, L.: Improving Identification Law of Professional Skills To Promote the Development of Higher Vocational Education. Journal of Tianjin Vocational Institutes 9(6) (November 2007)
3. Mao, B.: Discussion on the Connection of the Evaluation System of Vocational Education and the Identification of Vocational Skill. Vocational and Techical Education (Education Science) 26(1), General No.419 (2005)
4. Xu, D.: Practice Exploration on the Development of Professional Skills Identificantion in Independent College, Vocational And Techical Education (Education Science) 27(25), General No.479 (2006)
5. Yang, S.: On the professional skill and vocational technical education. Henan Social Sciences 12(2) (March 2004)

Research on the Construction of
the Clothing Video Database

Jianping Liu, Lu Chang, and Huilan Chen

Donghua University, Shanghai 201620, China
ljp2006@dhu.edu.cn

Abstract. Donghua University is a national key university specializing in textile and clothing. The Clothing Video Database developed by the library of Donghua University has largely enriched the digital resources of the university and improved the library's capacity and level of service. This article describes the methods and procedures of establishing the Database, introduces the idea of presenting the videos and their explanation documents on the same interface, and elaborates the feasibility and importance of providing users with a multi-angle combination of resources. We also discuss problems appeared in the development of the Database, their corresponding solutions, and our future plans. The Clothing Video Database can serve as a reference in establishing video databases, particularly for other university libraries.

Keywords: clothing, database, video, digital library, web design, university library.

1 Introduction

As a leading university in textile and clothing industries in China, Donghua University has developed its library into the center of textile resources in China. It has also finished the construction of Clothing Video Database (CVD), which is a sub-project of the constructing Contemporary textile information reference platform, a 10th-five-year subject characteristic project [1]. The key advantage of the database is that it accomplishes the goal of posting video resources and closely-related data on the same page. The combined resources from multiple aspects make the database more user-friendly. Thus, it is highly rated by the users and experts after launching. This article introduces the research and practice of constructing database, illustrates the importance and feasibility of providing users with resources that combine multi-dimensions and multi-perspectives.

Many universities have their own key majors and education features. Over the years, each should have its own special collection of certain subjects [2]. A considerable number of universities around the world have their video database, e.g., The Image Collection [3] of Harvard University, Williams Memory Project [4] of Williams College, audio and visual database of Tsinghua University etc. The information in these video databases [5] is of great value for communication and scientific research from historic, social and cultural perspectives. However, many

Y. Wang (Ed.): Education Management, Education Theory & Education Application, AISC 109, pp. 547–554.

users find it extremely difficult to extract useful information from those videos by themselves. We believe that users can obtain much more useful information if these databases have background documents attached to videos.

2   Procedures to Establish the CVD

2.1   Choosing Topics for the Database

Clothing products possess both practical functions and motion beauty. When people choose clothing products, they usually pay much attention to the dynamic visual impression, especially how they look like when wearing the clothes in action. Thus, dynamic visual impression is a very important discipline in teaching, researching and product designing of clothing subject. Based on our analysis on the library's book collections, key subjects setting, user demand and other aspects, we have decided to establish a CVD to further meet the requirements of textile subject development.

The goal of the project is to offer a visual dynamic experience when users search for information related to clothing. Moreover, while reading the explanation documents of videos, users have a clearer and better understanding of clothing products' technology, art, and its social culture. Although there exists a number of CVDs in the Internet, most of them are purely video databases, which cannot meet users' needs of clothing teaching and research.

Our database displays videos as well as their relevant information on the same interface. This will improve user efficiency and enhance pleasure. In this way, the CVD can help users to achieve their goals more efficiently and conveniently. The establishment of the CVD is very important to the development of university digital library with subject characteristics as well as to the advancement of clothing and textile research. Fig. 1 shows the system structure of the CVD.

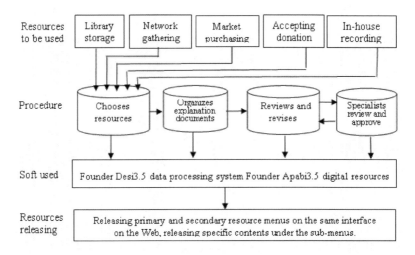

Fig. 1. System structure of the CVD

2.2 Functions and Releasing Plans of the Database

Based on users' needs, we include both searching and browsing function in the Database. We used content-based retrieval and multi-field search method for the following two reasons. The first reason is that content-based retrieval is convenient and fast, and the second reason is that all video resources in our database can be easily described in words, which provides the possibility of using the content-based retrieval method. This project is a sub-project of CALIS characteristic subject database. Fig. 2 shows the "CVD" metadata design.

Therefore, it adopts the source data standard and corresponding recording rules of CALIS characteristic subject database [6]. We followed "source data rules of video information" and "source data recording rules of video information" with slight modifications according to users' feedbacks. After trials and corrections, the final structure of source data can concisely reflect main ideas of the resources and elegantly illustrate the characteristics of the resource.

Undoubtedly, visual effect is essential to designing and purchasing of the clothing products. However, many people would like to know more about the social and cultural background and other relevant information when they views images of clothing products. Nevertheless, most existing video databases on the topic of clothing collect video resources only, which cannot meet users' needs in researching and teaching.

To resolve this problem, our CVD displays videos and their explanation documents on the same interface, with video information as primary resource and its explanation documents as secondary resources. For example, the interface titled "Christian Dior 2004 spring and summer fashion show" includes three video documents (we divided the video into three separate documents due to the requirement of the file size) and four explanation documents. Each explanation document has their own theme, including brand development of Christian Dior, brand culture, and characteristics of its famous designers, etc.

3 Gathering and Organizing Resources of the Database

3.1 Selecting Types of Resource

The library of Donghua University has a notable collection of videos on the topic of clothing. There are a large number of video resources in teaching and research, and they are stored in many different places, including libraries, reference rooms of the departments, as well as private collections by faculty and students. This makes it difficult for users to search for and to make use of these resources, and pulls down the utilization ratio of these clothing video resources. Based on the answers of a questionnaires distributed to users, our project team compared various clothing video resources. In our first development period, we decided to collect video resources of famous clothing brand, fashion shows of the top designers, and important national and international clothing design competitions.

3.2 Collecting Resources

(1) Searching for resources from our library. We selected the video resources according to the requirements of database development. In particular, we used DVDs attached in clothing periodicals and a small number of individually purchased fashion show DVDS from the library collection.

(2) Searching for resources from the Internet. We searched clothing videos by Google and baidu for the video resources released from the professional clothing websites. We compared these Web resources, and then archived the qualified ones into the resource database.

(3) Searching for resources from professors and specialists of clothing. After years of service, the librarians have established good relationship with professors and specialists of clothing in the university. Our project received a lot of help from these professors and specialists, in that some of them provided us with their personal collection of clothing videos that they collected over years without hesitation. Moreover, they offered many helpful advices regarding the releasing, searching, browsing, and explanation document writing.

3.3 Processing Resources

We established the processing standard, including the standard of the size, form, naming of the document and rules of placing the document, so as to ensure the qualification of resources.

(1) Standardized processing of video resources. Limited by the conditions of the database construction software, we standardized the size and content of all video resources. First, we transformed videos into MPG form, which is a well accepted international standard of compression algorithm of motion images. Second, we divided the content in the DVDs into multiple parts, each corresponding to one video resource data to be put into the database. We deleted all content in the original videos that are unrelated to our releasing title. After making sure that all finished resources are no bigger than 50M, we named them by their order name to be released and saved them into a document file called "video names".

(2) Writing explanation documents. Prior to our project, we cannot find any service of clothing video resources with explanation documents. The main reason is that the explanation documents need to be written and edited by subject specialists. However, specialists usually cannot realize the importance of writing explanation documents for digitized resources, not to mention that the total number of specialists is small to begin with. The library of Donghua University has more than 60 years of experience with the construction and service of characteristic resources in the subjects of textile and clothing. Our reference librarians have the following advantages: a. Some have research experiences in the topic of textile and clothing; b. They developed the skills of quickly finding out the needs of users. c. They have previous experience with constructing digitalized library. In our project, all the explanation documents were written and compiled by these librarians, and these documents received highly positive feedbacks from experts as well as general users.

Because of the characteristics of the video resources, explanation documents should describe the content of the video resources from multiple aspects. Thus, our librarians referred to various resources and asked specialists for advice in the process of writing these documents. After the draft was completed, it was proofread and revised by specialists. All explanation documents were titled by the themes being expressed, and we created explanation document files for each video resource.

(3) Naming and verifying documents. We created the document file named "video resources to be released", and saved the files with their descriptive titles. We then indexed all object data, including videos and their explanation documents in the "video resources to be released" document, and sent them to team members who were responsible for data verification. The main tasks of these team members were to check whether the file size and format of the documents were standardized and whether the description documents looked comprehensive. After the first round, the data were sent to specialists for another round of verification. The main tasks of the specialists were to check the accuracy and comprehensiveness of the explanation of description documents and to offer detailed suggestions. We improved the documents according to these suggestions, and asked advice from specialists and general users before we put them into the database.

Fig. 2. "CVD" metadata design   **Fig. 3.** Data processing and verification

4 Processing and Releasing Data

In our project, we used the software platform developed by Beijing Founder Electronic Company affiliated to Beijing University (Apabi3.5 digital resource platform, Desi3.5 processing system). This software platform is able to manage documents, images, and multi-media files and has the function of full-text search services, including searching, browsing, and downloading. This soft platform was authorized by CALIS project of characteristic database.

First of all, we created a database called the "CVD" in the platform. Please refer to Fig. 2 for the structure managements of the source data of the Database.

Second, we processed resources in Founder Desi3.5 processing system. We opened the "CVD" just we just created, and clicked the menu called "producing and uploading". We uploaded all resources with multiple formats under one topic title to one data file, so that we could release each video with its explanation documents on the same interface. According to the characteristics of individual video resources, we selected the appropriate processing method from the following fours provided by the platform – indexing, classifying, menu creating and multi-media linking. In the initial period of platform testing, we used double checking to ensure the quality of our database. Following the completion of data processing, these data were sent to the project members who were responsible for checking the standardization and accuracy of the data. Please refer to Fig. 3 for the procedures of data processing and verification.

5   Problems and Solutions in the Development of the Database

5.1   Resource Collection for the Database

One problem is that the resources for the Database are rather fragmented. The project members need to search public online information, purchase ready information from providers, and inquire provision from users. To develop a successful Database, the existing resources of our library can only serve as the basis; a large amount of supplementary resources should be collected through multiple channels. Please refer to Fig. 1 for the data collection channels. Although several instructors in Fashion College and staffs and users of the College Data Room own a considerable amount of resources, they lacked the motivation to contribute to this project due to the following two reasons. First, they are unwilling to share their personal collection of featured resources related to the subject with the public. Second, they have concerns about possible loss of their collection. The project team members spent a lot of time and efforts to persuade these resource owners to share their resources, yet the results were still unsatisfying. On the other hand, though Fashion College, located in the fashion center of China—Shanghai, often holds or organizes events such as fashion releases and international fashion forums, the university library were only allowed to record a limited number of such events due to various constraints. Thus, there is a long way to go before we can obtain sufficient video resources to meet the needs brought up by the development of the subject.

Below we provide three solutions to resolve the problems that we met during the collection of video resources. (1) We can strive for the support policy at the university level for digital library resource development. The support from the university will not only increase the input from the library administrators, but also makes it a lot easier to obtain resources from faculty and staffs from the university. (2)We can improve the service of our library, in that those users who are satisfied with our subject-specific services are more likely to contribute to the development of our digital library. (3)We should set up a fashion video resource collection team for regular subject-specific video collection. In addition, we should carefully make short-term and long-term planning for the development of digital resource.

5.2   Writing Explanation Documents

While culture creates clothing, clothing carries the essence of culture. In other words, fashion is the combination of our material and spiritual creation [7]. Therefore, it was challenging to write explanation documents for every video in the database. The two most difficult parts were to decide on the themes of the explanation documents for each video based on its content and to write the explanation objectively, accurately, and completely.

To overcome these difficulties, our project team selected one librarian with a comprehensive background in multiple subjects (including experience in teaching and research of fashion courses, library retrieval service and digitalization) to come up with the themes for videos and draft the table of content of their explanation documents. Her draft was then reviewed and refined first by the project review team and then by professors in Fashion College. Once the table of content was finalized, this librarian wrote the full content of the explanation document based on the refined table of content. Notably, the proper table of content and search phrases can lay the foundation of an objective, accurate and complete explanation document, with the latter task being more demanding. After the draft of explanation documents were finished, they were again reviewed and refined by the project team and professors in Fashion College. After several reiterations, the completed explanation documents received many positive feedbacks from the database users. Our case illustrated the importance of having librarians with backgrounds in multiple subjects in the database development team.

5.3 Communication and Collaboration with the Software Provider

When we developed the Database, we were able to upload, process, and publish files with multiple formats attached to each video, which was unusual. This was achieved through frequent communication with the software provider. However, we still had problems with automatic format conversion of the uploaded files as well as limited of file size, despite our repeated suggestions to the software provider. In addition, we did not achieve the goal of developing our own software and upgrading its functions. The development of databases requires continuous communication with the software provider, because the communication can improve mutual understanding and bring out better results.

Due to the constraints faced by the internal system of the software provider, they pay more attention to solving specific problems in software application. However, as for the needs of extended development and function upgrade, neither could the provider's representatives have enough patience to listen to our needs, nor could they accurately pass them to the developers. This problem has delayed the digitalization and service improvement of our library. During the process, we also realize that it is important to redevelopment the library application platform and that it is necessary to have professional staffs other than the software provider in our team to achieve this goal.

6 Conclusion

According to user statistics gathered for its beta and go-live versions, the CVD has high search and download frequency. In the Digital Library Development Sub-project inspection of 211 Project, the review experts give the Database very positive comments. Notably, the ideas of composing explanation documents for videos from multiple perspectives and displaying primary and secondary data of different types in one interface are innovative and meaningful to the development of other video databases.

The completion of this project suggests that university libraries are able to accomplish resource sorting and processing and to use featured video resources as basis to create multi-dimensional database that can greatly improve users' information utilization efficiency and satisfaction. The methodology used in this project can serve as a reference for other universities. As more universities get involved in the collective development and sharing of such database and relevant services, different methodologies would emerge, thus advancing the database development and improving user experience. As such, the university library and users would be in a win-win situation.

The abundance of information in the 21th century provides us with diverse ways of obtaining information. How can university library, as a professional information service organization, maintain its unique position in such environment? The answer is to constantly improve our service to help users achieve their goals conveniently and efficiently. Furthermore, the collaboration and information shared by university libraries all over the world are likely to give the users the most enjoyable experience.

The development of library video database can, at the university level, better meet the needs of subject development. Meanwhile, at the library level, it can promote innovative design, improve library service, and advance the collection and research of the featured subjects. In the new information era, this practice can help university libraries return to the main stream of information service, thus promoting their overall development.

Acknowledgements. This work was supported by Chinese CALIS project and Donghua University of Arts Research project. We thank Professor Bao Mingxin and Liu Xiaogang of Donghua University for their constructive suggestion.

References

1. http://www.lib.whu.edu.cn/calis4/area_spec_intro.asp (June 21, 2011)
2. Xu, R., Li, C.: Discussing the building of database of special topics in agricultural machine Photograph. New Technology of Library and Information Service 4, 78–79 (2004)
3. http://arboretum.harvard.edu/eastern_asia/galleries/baskets/index.html (June 21, 2011)
4. http://drm.williams.edu/cdm4/item_viewer.php?CISOROOT=/collections&CISOPTR=10&CISOBOX=1&REC=8 (June 21, 2011)
5. http://www.lib.tsinghua.edu.cn/find/find_media.html (June 21, 2011)
6. http://cdls2.nstl.gov.cn/2003/SpcMetadata/ (June 21, 2011)
7. Ning, J., Ma, Q.: Research on clothing consumption culture. Journal of Textile Research 7, 117–120 (2006)

Strengthening Cooperation with IT Enterprise, Promote the Practical Teaching of Information Specialty

Wang Liejun and Jia Zhenhong

College of Information Science and Engineering
Xin Jiang University
Urumqi, China
wlj@mail.xjtu.edu.cn, jzhh@xju.edu.cn

Abstract. This paper deeply analyzed the main training problem of the current applied practical talents in information specialty and discussed strengthening cooperation with IT enterprise can promote information specialty's practical teaching reformation. Cooperation with IT enterprise also can improve the effectiveness of practical teaching and college student's professional standards. It explored a new way to improve the innovation capability and overall professional quality of the information specialty college students.

Keywords: IT enterprise, information specialty, practical teaching, teaching reformation.

1 Introduction

With the fast development of information technology, enterprises have developed an increasing demand for the applicable specialists of information major who bear practical creativity. The fact is on the one hand such professionals are hard to find, and on the other, the employment of integrated university graduates majored in information has become more prominent. In recent years, College of Information Science and Engineering, Xinjiang University has actively collaborated with well-known enterprises, carrying out exploration of practical teaching reform of its major by establishing laboratory, training base and professional competition collaboratively.

2 Problems Exist in Nurturing Informational Specialists

2.1 Professional Teaching Paid Little Attention to Practice

Experimental facilities for information majors of our university are mostly experimental boxes, which cannot provide related materials detailed enough for students. Moreover, the limited experimental classes made it difficult for students to comprehend the content completely. The usual case is that experiments end with the validation of the results. The role of students in the experiment is just to plug or draw

Y. Wang (Ed.): Education Management, Education Theory & Education Application, AISC 109, pp. 555–559.
springerlink.com   © Springer-Verlag Berlin Heidelberg 2011

leads and do validation. In traditional teaching, tests of students mainly focus on written ones, resulting in the neglect of practice.

2.2   Experimental Teaching Lagged behind the Development of Industry

Experimental teaching of universities lagged behind usually the latest technological development of electronic information industry, making it hard for students to be employed since they cannot meet the needs of the enterprises.

2.3   Lack Prefessional Teachers with Practical Teaching Experience

Many professional teachers who seldom participate in engineering practical projects know little of the enterprises' needs, which definitely affects the depth and width of their professional experimental classes.

3   Strengthing the Cooperation between Unvisities and Enterprises will Enhance Teaching Reform

3.1   Professional Teaching Paid Little Attention to Practice

The cooperation between universities and enterprises serves as a bridge for universities to adjust their experimental classes and build related professional courses based on their knowledge of market needs. Universities can arrange their teaching plan to meet the needs of enterprises and dynamically improve related teaching plans and teaching contents.

3.2   Help to Nurture Professional Teachers in Universities

The cooperation will do good to the direct communication between enterprises and professional teachers who can promptly find the technological needs of enterprises and the development orientation so as to renew their knowledge to meet the needs of experimental teaching of related courses.

3.3   Help to Improve Students' Practical Capacity and Interes

The cooperation enables students to come to know the latest informational technology quickly. Students can apply the latest electronic equipments and software. Their way of doing experiments is no longer confined to the rigid experimental boxes. They come to realize the aim and significance of practice and their interest in improving capacity of doing practical works increase.

4   Measures Taken in Cooperation between Universities and Enterprises

In recent years, college of information science and engineering of Xinjiang Universities intensifies the cooperation with IT enterprises in various aspects by taking the following measures.

4.1   Establish Laboratory with Well-Known IT Enterprises Collaborately

In recent years, the college has built college students collaborative laboratory with American TI, Altera, the world-famous semiconducting chips manufacturer and college students innovative laboratory with Beijing Dasheng, a domestically well-known producer of teaching facilities. These enterprises supply the laboratories with the latest equipments, related soft wares and materials free of charge annually.

4.2   Establish Training Base with Well-Known IT Enterprises Collaborately

The college has set up a united training base with Zhongxing communication corporation, a famous manufacturer of communication equipments at home and a training base out of school with Lenovo Group, Xinjiang Branch.

4.3   Improve Experimental Teaching Plan, Teaching Content and Textbooks through Communication with IT Enterprises

In recent years, teachers of related specialties have been sent out to be trained in enterprises to enhance the application of experimental teaching with communication with IT enterprises. Based on the social needs, the experimental teaching plan has been remodified to decrease checking experiments and increase comprehensive and designing experiments. Experimental materials have been renewed based on the new teaching plan and the new content to make it more practical.

4.4   Nurture Teachers by Selecting Them to Receive Training in IT Enterprises

Every term, a group of excellent young teachers of different specialties has been sent out to be trained in IT enterprises to enlarge their knowledge width and improve their capacity in practical innovation. Meanwhile, senior engineers from the IT enterprises has been invited to deliver lectures, perform training to help teachers improve their knowledge about market's demand and practical ability.

4.5   Participate in the Professional Competitions and Innovation Project Organized by IT Enterprises

Every year students would send to take part in the professional competitions and innovation project organized by IT enterprises, during the process of which students can experience the pleasure of practical manual work and get to appreciate the process; the teamwork awareness and spirit could also be cultivated.

5   Effectiveness of the Cooperation with IT Enterprises

A series of measures taken in cooperation with the IT enterprises obviously exert a positive effect on the improvement of the creativity of information-majored students of college of information science and engineering in Xinjiang University. The collaborative establishment of laboratory and training base saves the insufficient experimental fund, offers students an opportunity to access the latest information technology and shortens the gap between study in school and work in enterprises.The renewed teaching plan and contents are highly welcomed by students who are no more confined to rigid experimental boxes and repeating validating experiments. Based on the features of learning experimental courses, the original experimental textbooks have been renewed, many validating experiments deleted, comprehensive and devising experiments added and times for experimental classes adjusted. Two books out of the teaching materials have been published and the rest will be officially published successively in future.Eight young outstanding teachers have received professional training in the cooperation and intercourse with IT enterprises. Teachers of the college have organized a team to explore experimental teaching equipments and professional practice equipment suitable for their own students and gained patent. Students respond favorably to those equipments in years of application. Students of the college have done well job in the professional competitions. Eight pieces of work have been awarded the second prize in the national college students' electronic devising competition, the 16th best work and excellent work in the national college students' robot competition, the second prize in the national college students challenge cup technological out of school competition and the third prize in the Altera Asia innovation devising competition. Students participated in the collaborative innovation laboratory has hosted four national college students innovation projects and awarded Excellent when they are completed. According to the follow-up statistics, the number of students passed graduate exams has increased and the employment has obviously improved.

6 Conclusion

It will effectively enhance the experimental teaching reform of informational profession in universities to intensify the collaboration between universities and enterprises. The new teaching mode has diminished the separation of theory and practice in teaching and created an engineering environment to connect theory to practice for students to improve their capacity in practice. The students has been inspired to be active and creative; both the teaching effect and the professional level of students have been improved, which prepares a new way to nurture the practical innovation and professional all-around development of information-specialized students.

Acknowledgments. The paper was sponsored by second "21 Century Education and Teaching Reform Project" of Xinjiang University.

References

1. Chen, N., Yi, Y.: Contesting to promote electronic information Curriculum Reformation. Journal of Electrical & Electronic Education 23(01), 13–14 (2001) (in Chinese)
2. Ma, J., Zeng, X., He, C., Liang, X.: The Exploration of University Enterprise Cooperation in Promoting the Teaching Reform. Journal of Electrical & Electronic Education 31(02), 17–18 (2009) (in Chinese)

3. Pan, L., Zhu, C.: Research and Realization on the Reform of Electronic Practice Course. Laboratory Research and Exploration 20(01), 117–118 (2001) (in Chinese)
4. Jia, Z.: National Undergraduate Electronic Design Contest and Promoting Teaching Quality of Electric and Electronic Information Specialties in Universities in Border Area. Journal of Electrical & Electronic Education 29(s1), 37–39 (2007) (in Chinese)

A Laboratory Measurement Method of Antenna Radiation Pattern

Hui Xie, Yujun Liang, and Qin Wang

Electronics Engineering College, Navy University of engineering,
Wuhan, Hubei, China

Abstract. The radiation pattern is the most demanding in measurement steps and most difficult to interpret of all antenna measurements. This paper introduced a laboratory measurement method of antenna radiation pattern, and demonstrated this experiment in detailed steps such as connection of the laboratory measurement devices, tuning at a correct frequency, points for attention, etc. A quarter-wave omnidirectional antenna was shown as a model, and the measurement result is fully consistent with theoretical value.

Keywords: far field, radiation pattern, laboratory measurements.

1 Introduction

Of all antenna measurements considered, the radiation pattern is the most demanding in measurement steps and most difficult to interpret. Any antenna radiates to some degree in all directions into the space surrounding it. Therefore, the radiation pattern of an antenna is a three-dimensional representation of the magnitude, phase and polarization of the electromagnetic field. In most cases the radiation in one particular plane is of interest [1, 2, 3], usually the plane corresponding to that of the earth's surface, regardless of the polarization of the antenna [4, 5]. Measurements of radiation pattern should therefore be made in a plane nearly parallel to the earth's surface.

2 Tuning at a Correct Frequency of an Antenna

To check if an antenna is tuned at the correct frequency, we can use a Directional Coupler and a Spectrum Analyzer. The signal is internally generated by the Tracking Generator of the Spectrum Analyzer, which is connected to the input port of the Directional Coupler. The antenna is connected to the output port of the Directional Coupler.

The signal reflected by the antenna is sampled at the reflected port of the Directional Coupler and displayed on the Spectrum Analyzer. We can then check if the minimum amount of reflected power corresponds to the correct frequency value, and if the amount of reflected power is low enough. Be careful to correctly terminate the other ports with 50Ω dummy loads if the Directional Coupler is of the

Y. Wang (Ed.): Education Management, Education Theory & Education Application, AISC 109, pp. 561–565.

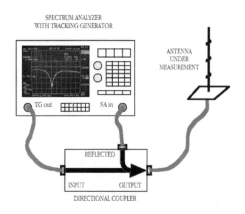

Fig. 1. Connection of the laboratory measurement devices

bidirectional type and to keep the cables as short as possible to avoid any resonance effect. The best electric length for cables is an N times the half wavelength [6].

To easily read the value of the reflected power, a calibration of the Spectrum Analyzer is required. In a first step, the reflected power when no antenna is connected to the output port of the Directional Coupler should be measured, and this value should be set as the reference value. In a second step, the power reflected by the antenna under test should be measured. In this way, we measure how much lower the reflected power is, in comparison to the case in which all the power is reflected.

For our quarter-wave omnidirectional antenna we got the following graph on the Spectrum Analyzer's screen:

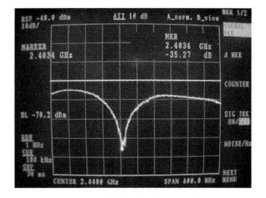

Fig. 2. Measurement result of a quarter-wave omnidirectional antenna

We can see that the antenna is correctly tuned at 2.44 GHz, and the reflected power is about 35dB less than the case without the antenna. This is a good result.

3 Measuring the Antenna Radiation Pattern

The antenna under test is connected to a transmitter, which can be a Signal Generator, and the other one is connected to a receiver, which can be a Spectrum Analyzer or a Power Meter. In our case, the receiver will be a Spectrum Analyzer. The antennas are mounted over tripods at fixed positions. For the antenna under test, a suitable mount is required which can be rotated in the horizontal plane with some degree of accuracy in terms of azimuth angle positioning. In the photo, one of such mounts is shown.

Fig. 3. Antennas are mounted over tripod at fixed positions

The distance between the tripods should be more than a couple of meters to measure the far field. It is assumed that the antennas have been carefully matched to the appropriate impedance and accurately calibrated and matched devices are being used.

To prepare the measurement, switch on the Signal Generator and the Spectrum Analyzer well in advance and let them stabilize. Set the frequency of the Signal Generator to 2.44 GHz, with no modulation and disable the RF output until you connect the antenna. Set then the Spectrum Analyzer for a center frequency of 2.44 GHz and a frequency span of 20 MHz. Align the geometrical axis of the antenna under test so that it points to the reference antenna. Set the zero of the azimuth scale. The elevation angle of the antenna under test should be also zero, and the two antennas should be at the same height. Connect the antennas, switch on the RF output of the Signal Generator and set its level high enough so that on the Spectrum Analyzer you can see the peak of the signal well over the noise floor. Record the value you read on the Spectrum Analyzer's screen on paper. This will be your Reference Level.

Without changing the elevation setting, carefully rotate the antenna in azimuth in small steps of 5 degrees. You can alternatively rotate the antenna to permit signal-level readout of 3 dB per step. These points of signal level corresponding with an azimuth angle are recorded and then plotted either manually on polar coordinate paper or printed with the use of a computer. On the polar paper, the measured points are marked with an X and a continuous line is then drawn by hand, since the pattern is a continuous curve.

Following the described procedure, we measured the radiation pattern of the cantenna. We aligned the antenna with the receiving one, which in our case was a 4 dBi omnidirectional one. We reset the azimuth angle to zero. The value read on the Spectrum Analyzer, which was used as Reference Level, was -60 dBm. In the table 1, there are the values read rotating the antenna of 5 degrees at a time in both directions and the hand-made graph of the radiation pattern. From the radiation pattern graph, the 3dB beamwidth can be estimated in the order of 70 degrees [6]. Due to obstacles near the area of measurement, the angle range was limited to 180 degrees.

4   Laboratory Measurement Results

We got the laboratory measurement results shown in this table:

Table 1. The laboratory measurement results of the quarter-wave omnidirectional antenna.

Angle (degrees)	Measured Level (dBm)	Rel. Gain (dB)
o	-60	0
5	-60	0
10	-60	0
15	-60.25	0.25
…	…	…
80	-78.75	18.75
90	-82	22
…	…	…
-90	-82	22

The radiation pattern can be drew as fig.4, and it is fully consistent with theoretical value.

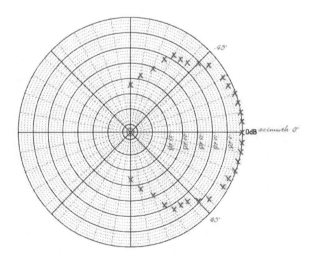

Fig. 4. Radiation pattern of the quarter-wave omnidirectional antenna

References

1. Volakis, J.L.: Antenna Engineering Handbook, pp. 17–21. McGraw-Hill Professional studio (2007)
2. Balanis, C.A.: Antenna Theory Analysis and Design, pp. 27–38. A John Wiley and Sons, Ltd, Publication (2005)
3. Zheng, S., Jianhua, Z., Ye, H.: Antenna and radio wave propagations, pp. 6–10. Xidian university Press (2004)
4. Li, L.: Antenna and radio wave propagations, pp. 26–34. Science Press (2009)
5. Yi, H., Kevin, B.: Antennas from Theory to Practice, pp. 263–272. A John Wiley and Sons, Ltd, Publication (2008)
6. Zennaro, M., Fonda, C.: Radio Laboratory Handbook of the ICTP School on Digital Radio Communications for Research and Training in Developing Countries. Cables and Antennas 1, 120–130 (2004)

On the Model of Postgraduate Student's Self-management from the Perspective of Self-organization Theory

Wang Jijun and Zhao Long

Liaoning Dalian Dalian University Graduate College
wangjijun@dlu.edu.cn

Abstract. The research on the postgraduate student's self-management is an important topic and this paper analyzes the current situation of the postgraduate student's educational management in Dalian University in order to display the postgraduate's student's subjectivity thoroughly and improve the postgraduate student's comprehensive quality. This paper explores and realizes a new training model of the postgraduate student's self-management and helps to educate and manage the postgraduate student in Dalian University.

Keywords: Organization Theory, postgraduate student, training model, self-management.

1   The Current Situation and Key Issue of the Postgraduate Student's Self-management

The postgraduate student education develops at very fast speed and the total number increases quickly in Dalian University with the transformation from "teaching university" to "teaching-research university" of Dalian University. The total number of the full-time postgraduate student is 800 until the end of 2010. The postgraduate is a kind of high-level talent with a high-education background. They are the backbones of our country and play a very important role in the socialist modernization process. It is a key and urgent issue on how to apply scientific outlook on development thoroughly in the new period and strengthen the political education work and self-management of the postgraduate student.

1.1   The Problems on the Postgraduate Student's Political Education

The postgraduate student's education and management work of Dalian University develops gradually. Although it makes great progress, there are still some problems.

(1) It is insufficient for the postgraduate to pay attention. Some postgraduates believe that the politics is far away from their life and major and job and they do not care about it in mind. They also do not study it carefully.

(2) The study organization system is too loose. It always adopts the system of tutor-responsible for the postgraduate student. And it is looser than the college student. There is no regular learning organization for the postgraduate student.

(3) It is difficult to make unified arrangement for the postgraduate student. There is a great age difference among the postgraduate students and they experience different things. The sciences and engineering graduate students often go out for experiments,

Y. Wang (Ed.): Education Management, Education Theory & Education Application, AISC 109, pp. 567–574.

and they stay at school not as long as the other postgraduate student. As a result, this kind of life style also brings some difficulty for the administrator of a school to make schedule on the political education work.

(4) There is a little restriction on the mechanics of learning. The postgraduate students only need to get some compulsory scores. There are few detailed rules and management measures about the political education work, which lead the postgraduate student not to care about the political education.

1.2 The Main Problems on the Postgraduate Student's Self-management

The feature of the postgraduate education is research quality, self-management quality and self-program quality. Therefore, the self-management is very important. The main problems on self-management of the postgraduate student are followed:

(1) They do not have a clear goal for the postgraduate study. Many students don't know their future career. Some students choose different majors, also called interdisciplinary, when they take part in the entrance exams for postgraduate schools. The problem is what they have learned do not match with what they will learn. The other reason is that many postgraduate students do not know how to choose a right job, because they are not clear about their own features and their future career.

(2) The postgraduate students are often indifferent. Each postgraduate student has their own tutor to instruct their study and research. The science and engineering postgraduate student usually does experiments in laboratory, and they talk less with other postgraduate students. In addition, the postgraduate students are older than undergraduate student, and they are maturer and stronger. They consider things more realistically and do not pay much attention to the interpersonal relationship. Some postgraduate student is even in a full-closed or half-closed state.

(3) The postgraduate students do not have a strong sense of time. Many postgraduate students own lots of free time. They spend too much time to get entertainments when they are free. It is a kind of waste.

(4) The postgraduate student can not work under high pressure. Most of them were born in 1980s, and they are the only child in a family. They experience a little stress and frustration. Once they face it, they may feel down in the dumps for a long time and are not interested in any thing, which is bad for their mind and body.

As a result, the author analyzes the political education work of the postgraduate student based on the self-organization theory and proposes a new model of the self-management on the postgraduate student.

2 Analyzing on the Necessity of Postgraduate's Self Management

The current situation of postgraduates' ideological and political education and the absence of their self management help us realize that no matter to consider the characteristics of postgraduate group or the social factors of the current education of postgraduates, the postgraduates' self management education needs to strengthen in postgraduate education period to promote the formation of self management ideas and concepts, to improve postgraduates' non-intellectual factor and emotional intelligence capabilities, to enhance the quality of postgraduate education.

2.1 The Characteristics of Postgraduate Group

With the characteristics of high level of knowledge, perfect personality and strong independence, the enrolled number of postgraduates is less than the number of undergraduate student. Moreover, the significant regional differences result in the difficulty of forming the traditional concept of "group". For postgraduates have more self-control time, it is not sufficient just to rely on the traditional management method to meet the needs of contemporary students. It demands us to combine with the situation of postgraduate education with the guidance of relevant departments to play the advantage of student organization for self management.

2.2 Social Factors

At present, the employment brings great pressure to school students with little social experience. Few contemporary postgraduates experienced frustration with weak ability to resist the pressure. The society tends to employ the individual with practical ability. Therefore, the postgraduates in school should catch the practice opportunity on campus to enhance the ability of communication, cooperation, expression, interpersonal communication, resistance to stress and other abilities to improve their overall quality. We first have to learn to survive and self management to improve the skill of surviving. To do more practice in school will enrich ourselves, enhance self management and improve our overall quality.

3 Analyzing on Postgraduate Self Management Mode Based on Self Organization Theory

3.1 Introduction of Self Organization Theory

Self organization theory is a systematic theory formed and developed in the end of 1960s. Its main research objects are the formation of complex self organization system (including natural systems, social systems) and the development mechanism problem. From this, the essential content of self-organization theory can be seen: the openness of the system is the Prerequisites for system to produce self-organization behavior; the nonlinear mechanism is the fundamental basis of organizational behavior; away from equilibrium is necessary condition to produce self-organizing behaviors; fluctuation is an important opportunity to adjust the behavior of self-organization.

3.2 The Applicability of Self-organization Theory to Self-management Mode

3.2.1 Self-organization—An Open System

"Open" means the system continuously exchange material, energy and information with the external environment. The student organization based on self-management mode is an open system. A large number of students join in the system and parts of them leave the campus and the system every year. Elections are carried out every year in the internal of student organization. In addition, self-management system has relation with other departments and organization both on campus and out of campus.

They exchange kinds of information, such as employment information exchange. Therefore, self-management mode is an open system.

3.2.2 The Non-Linear Interaction Effect among the Elements in the System of Self-management Mode

Linearity refers to the function relation between two variables; a factor change will lead another factor change. Self-management consists of a number of sub-elements, but their relation is not a simple superposition relationship, between them there is a complex non-linear effect. The change and development of any element is not the simple effect of another element but the comprehensive effect of each element [5]. For example, self-management model should adapt to the school's goal of educating people, to carry out a variety of extracurricular activities, and supervised by the majority of students and the higher authorities of the management. In the process, funds have to put into the activities and effect by the previous activities which has a certain tradition. Therefore, there are nonlinear interactions between the various elements in self-management system.

3.2.3 Self-management Mode—Far Away from Balance System

Away from the balance is the power for development, the only way to exchange materials energy, information with the external environment, from the disordered state to ordered state, for example, to carry out academic activities is a weak link (far from balance) in this system, the system will establish academic department, using a variety of measures for the academic activities, forms to plan and invite experts to conduct lectures and other forms of activity to compensate for this deficiency, to reach a balance state and the system also grow up. Therefore, the self-management mode is away from balance system.

3.2.4 The Fluctuation in the Internal System of Self-management Model

"Fluctuation" is a concept of statistical physics, which referring to the random changes of the characterization of the nature of a physical system in the vicinity of its mean[6].For a system, its features are always fluctuating in reality, so they generally only consider the average. These parameters are called "fluctuation" in the real value of the average deviation. Prigogine pointed out that self-organization mechanism is the 'order through fluctuations'[7]. Fluctuations are ordered builders. If the "up" is a positive concept of self-management, the "fall" is not consistent with the system. Such fluctuations can promote postgraduate self-management to form a more orderly system structure. Obviously, there are fluctuations in self-management system. Therefore, the self-management mode of postgraduates is an organic systems with self-organizing properties.

4   The Construction of Postgraduates' Self-management Model Based on Self-organization Theory

4.1   The External Environment of Postgraduates' Self-organization System

The school-education of postgraduates is a large system with its goals and plans to foster talents. Outside the system, there are leadership and surveillance from both the school degree committee and department degree subcommittee, the software and hardware environment support from the postgraduates management system, as well as the qualified teachers.as shown in Figure 1:

1. School degree committee, department degree subcommittees
2. The software and hardware environment of postgraduate's management system
3. Teachers to cultivate postgraduates. 4. Postgraduates' self-organization system

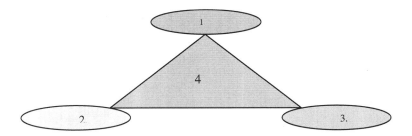

Fig. 1. The external environment of postgraduates' self-organization system

From the above figure, we can obviously find the connections between each system outside and inside the self-organizing system. There are material, energy and information exchanges between the external and internal systems. The external systems provide the funds, teaching equipments with the postgraduates self-organizing system , and also train goal for it.

4.2 The Importance Put on the Overall Development of Students by Higher Authorities

A real talent should not only have the ability to learn, but also have good morality, physical health, healthy personality and stress resistance ability, which are both explicit and potential capabilities. Thus, the postgraduate school administrators should pay attention to the overall development of students. They should support and guide the overall development of students, such as encouraging students to participate in sports activities, social activities, volunteer activities to increase opportunities for students to train themselves and provide a platform for students to develop more comprehensively. Otherwise, the self-organizing mechanism will be failed if it just support and restraint the big systems.

4.3 The Construction of Postgraduates Organizational System Based on the Self-organization Theory

Dissipative structure theory tells us that through continuous exchanges in material, energy and information ascepts with the outside world, a far-from-equilibrium opening system may spontaneously change from disorder into order state, when   the external conditions reach a certain threshold. The synergetics shows that its power comes from the synergtic effects of the competition and cooperation among a large number of subsystems within the system, and also the resulting order parameter control. In short, competition and collaboration is the fundamental driving force of the evolution of self-organization. For postgraduates self-management system, the competition and collaboration within the various subsystems (each departments and sub-students unions) and elements is the driving force of the evolution of the system.

4.3.1 The Collaboration and Competition among the Postgraduates Organizations

Within the self-management system, the student organizations should play their respective roles. The existing organizations are mainly postgraduates union, party branches as well as other postgraduate students organizational forums. The postgraduate union puts even more stress on student's study and living services. The party branches care more about student's thought tendency, develop the students who are positive to join the Communist Party and train him to be qualified member. Other organizations and forums will focus on raising student's interests and hobbies outside the classroom. The postgraduate organizations must collaborate to serves for students' overall development, and do not have the unnecessary loss and waste resources of self-control large-scale system because of the competition, as is shown in Figure 2:

1. Graduate party branch, Youth League branch .2. Graduate student union.
3. Other Graduate Organization Forum. 4. Postgraduates' self-organization system

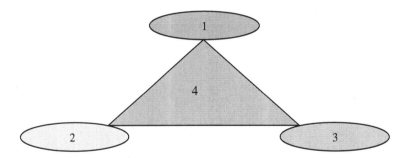

Fig. 2. Graduate evolution of self-organization system and circulation

4.3.2 The Construction of the Postgraduate Union

(1) Rules and Regulations. Any organizations should have set up their own rules and regulations to restrain their members' behaviors. For one thing, the union has to adhere to the basic codes of conduct; for the other thing, it has to provide credible proofs for its decision-making. For example, the proofs of financial currency, rewards and punishment mechanism, and working procedures of each departments. Rules and regulations can help to decrease human's influence on programes so as to make the system operating orderly and enhance the Union's efficiency and binding force in organizing activities.

(2) Organization Design. The graduate Union should set President, vice President, the Union Office, Academic Department, Practice Department, Information Department, Propaganda Department and Sports Departments etc. Any activities should be held in the charge o g the vice President, and guided and supervised by the

relevant departments. Each department should shoulder some corresponding responsibilities. Each department has its own function, existed independently and the cooperation and coordination in departments is the foundation of a virtuous operation system. Each department should also keep contact with the Union's subsidiaries which is helpful for the flow of information, energy and materials.

(3) Team Member Motivation. Each year there would be some students who could get the rewards from the Graduate School. Either be awarded the Excellent Student Leader or the scholarship. Evaluating the outstanding students Union's branches will help to keep a balance in competition. In addition, the members of the Union will be give certain priority opportunity in joining the party.

(4) Increase the Democratic Supervision. The graduate Union is an organization providing service for the general students. The mass students' opinions are given a prior consideration. It has to strengthen its communication with the mass students. By setting up the suggestion box to increase the transparency of our work. Any reasonable and constructive suggestions will be accepted in order to serve the students. It is also a good way to strengthen the information flow in the self management system of graduate student.

(5) Strengthen communication and collaboration with other brother institutions. The Graduate Students Union should carry out activities not only according to its actual conditions, but also aiming at promoting communication and collaboration with other brother institutions to make the best of the both sides. What's more, communication and communication can also to provide the students opportunities to exchange and learn from each other, making information, energy and materials' interaction with the outside world, and providing a powerful driving force for system to move forward.

4.3.3 The Individual Cooperation of the Graduates with Their Own Efforts

The self-management of a graduate depends not only on the work of school administrators and student organization, more importantly, on the graduate's own effort. Under the guidance of the school administrators and student organization, graduates themselves should make corresponding efforts.

  1. Moral Character   2.Intelligence Learning Development
  3. Sports Fitness Exercise 4.Individual Graduate Student Self Organization Unit

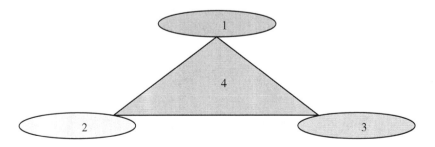

Fig. 3. The Self Organization Mode of the Individual Graduate Student

They should make their own life planning, accomplish their goals down-to-earth and allocate their time reasonably. They should, in the school life, enrich your life through mastering the practical professional skills and comprehensive executive ability, in order to prepare for the future society. They should take the advantage of their free time to actively take part in social practice activities to accumulate social experience. Based on this they should strengthen the physical exercise and have a strong body. The graduate, as an element of this system, each should have a positive enterprising heart and strive to become an all-round developed person with a good personality who learns well, work well and does well in innovating.

5 Conclusion

In the graduate student organization system, since each level of the system is an all-closed-circulation system, each has a specific target (order parameter) as a guide. It is implemented by each department and organization, and inspected and then accepted by the supervision authorities. In such a closed system, we can timely summarize some problems emerging in the work, analyze the causes, find the countermeasures and make up for the deficiencies, so that this system can continuously self-expand and make progress to higher levels.

By constructing graduate students self management mode, we are more clear to our training targets, more specific to master the training process, and more distinct to the relationships that we should handle in the process. We find and solve the problems existing in the cooperation and communication between each organization in the graduate student self-organization system. We also strengthen the department of each level between competition mechanisms, and reasonably make each department in concert to realize the common goal of graduates' all-round development. The model makes the material, energy and information of the system reasonably and orderly flow. The supervision function of graduate students from individual organizational units can play better. The establishment of the graduate students' self-organization system has far-reaching significance and value management practice to improve graduates' moral training, scientific research and academic ability, and the physical qualities.

References

1. Xue, T.: Graduate Education. Guangxi Normal University Press, Guilin (2003)
2. Xue, T.: Graduate Etudents Education and Management. Guangxi Normal University Press, Guilin (2004)
3. Hu, X.: College Graduate Student Research Training Mode in Our Country—From a Single to Dual Mode. East China Normal University, Shanghai (2004)
4. Lu, P.: Chinnese Postgraduates' Training Mode study. Wuhan University of Technology, Wuhan (2006)
5. Liao, W.: The Professional Degree Graduate System Structure Research in Our Country, vol. 136. South China University of Technology, Guangzhou (2010)
6. Hu, H., Lou, H.: The self Organization Theory and Social Development Research. Shanghai Science and Technology Education Press, Shanghai (2002)
7. Sun, Z.: The self Organization of Social Evolution Theory- the Method and Model, vol. 55. China Social Science Press, Beijing (2004)

An Fast Max-Min Ant Colony Optimization Algorithm for Solving the Static Combinational Optimization Problems

Zeng Lingguo

Zhejiang Normal University,
Jinhua, Zhejiang 321004. China

Abstract. Response to the MMAS's characteristic of solving Static combinatorial optimization, we was studying the update rule of pheromone of solving Static combinatorial optimization problems. The new rules will run on the total information, avoiding the MMAS operation to calculate the total information and check whether the pheromone cross-border operation in each iteration step. The experimental results on the traveling salesman problem show that using this technology can significantly speed up the MMAS algorithm to reduce consumption of time.

Keywords: Computational Intelligence, Initial Heuristic Algorithm, Ant Colony Optimization, MMAS, Travel Salesman Problem.

1 Introduction

In terms of ACO arithmetic, The more solution we obtained in the local area including optimal solution, the greater possibility to obtain optimal solution. But The ACO algorithm running time was usually limited by us, so we always hope that they can obtained the optimal solution in a relatively short period of time [1]. Therefore, reducing the computation of every ant on each iteration step, the ant colony can explore more solution at the same time. To obtain better quality solutions. In this paper we try to seek such rapid ant colony optimization. so, we study the MMAS base on Travel Salesman Problem—TSP.

So-called TSP is to point to a person start from one city and visit the other city except outside the source station based on exactly once according to certain path, finally to return to start city, the path is shortest. Can use all the connection of Weighted without directed graph $G = (V, E)$ to diagram TSP, V is the graph's vertex set, Corresponding to n cities, E Is the graph edge set, the weight of edge e_{ij} is the distance d_{ij} between city i and city j. in where $d_{ij} = d_{ji}$, $i, j \in V$。 TSP is typical of a NP-hard problem of combinatorial optimization problem, by many of the new algorithm used as test problem [2].

2 Use MMAS Solving TSP

Using MMAS solving TSP, the first important thing is modeling by the pheromone. using a real τ_{ij} to correspond e_{ij}。 The real τ_{ij} was called a pheromone values, denotes the ants strong degree when they chose edge e_{ij}. All the pheromone value set constitute pheromone matrix $\tau = (\tau_{ij})_{n \times n}$ 。before the Algorithm, Assigned to each pheromone a maximum value of positive Numbers.

2.1 The Structure of Solutions

Specific solving process is as follows: First, randomly selected a departure vertex for m ants, and set a taboo list at vertex. so when the ant select the last city, they must avoid the city with taboo list. Therefore, Ants tectonic solution meets TSP constraints. Supposing ant k now is in city i, and Randomly choose other city j to arrive with tern. Choose rules as follows:

$$
p_{ij}^k(t) = \begin{cases} \dfrac{[\tau_{ij}(t)]^\alpha [\eta_{ij}]^\beta}{\sum_{h \in Alw(k)} [\tau_{ih}(t)]^\alpha [\eta_{ih}]^\beta}, & \text{if } j \in Alw(k) \\ 0, & \text{otherwise} \end{cases} \tag{1}
$$

Where : $p_{ij}^k(t)$—the probability when the ant k in the city i and in the iteration step k;

$\eta_{ij}\underline{\quad} = 1/d_{ij}$, Heuristic information;

$Alw(k)$—the city set that ant k have not been visited;

α—the parameters of the importance of controlling pheromone information

β—the parameters of the importance of controlling heuristic information ;

$\tau_{ij}(t)$—of the pheromone values in Iteration step k.

2.2 Pheromones Update

After all the ants have constructed a complete solution, they calculate the path length $L^k(t)$, $k = 1, \cdots, m$ based on the Taboo list. The Pheromone matrix simulates the Pheromones volatile according to this rule:

$$
\tau_{ij}(t+1) = (1 - \rho)\tau_{ij}, \ i, j \in V \tag{2}
$$

Where: ρ—Pheromones volatile rate. So far, the optimal solution T^{gb} or the current optimal solution $T^{ib}(t)$ of the iteration step has been allowed to release pheromone in its path. Under normal circumstances, these two solutions need cross use according to certain scheme. Using $T^{ib}(t)$ is good for that ant colony explore and obtains the more optimal solution area, and prevent search stagnated in a suboptimal subspace, Using T^{gb} is good for that ant colony focus on searching in the optimal solution neighborhood, hat makes search into better solution is more likely. Scheduling optimal solution T^{gb} and $T^{ib}(t)$ solution needs to consider the size of searching space, the large the space, the more using T^{gb}.Because only using a best solution to release pheromone, so

$$
\tau_{ij}(t+1) = \tau_{ij}(t+1) + \frac{1}{L^b(t)}, \ \text{if edge}(i,j) \in T^b(t) \tag{3}
$$

Where : $T^b(t)$—the shortest path of the optimal solution $T^{ib}(t)$ or T^{ib}; $L^b(t)$—the path length of $T^b(t)$.

In hyper-Cube Frame, we need to limit pheromones value within [0, 1], so we changed the formula 3 into formula 4:

$$\tau_{ij}(t+1) = \tau_{ij}(t+1) + \rho, \text{ if edge}(i,j) \in T^b(t) \tag{4}$$

The difference of the two kinds of pheromone release rules is that the formula 3 release pheromone increment includes the information of Path length on the path of allowing solution. the difference of this path length is that ant system(AS) show the key effect in Release rules, it's power that AS seek better solutions. But in the MMAS, Because only allowing the best solution update pheromone, so the pheromone solution component may be the factors of Constituting the optimal solution. AS rely on the path length to distinguish the good solutions components and not good components, But MMAS isn't like this AS. The other different is the slightly less amount of calculation by formula 4. And the maximum value of any pheromone does not exceed 1.

2.3 Pheromones Restrictions

In the references [4], the maximum value $\tau_{\max}$ of pheromone is the optimal solution function of the question examples. But the optimal solution of the problem examples is not understood before solving actually., From algorithm beginning to now we can get an optimal solution, and use it to replace the question examples optimal solution. So the pheromone's maximum value $\tau_{\max}$ is a variable. The pheromone's minimum value is $\tau_{\min} = \tau_{\max}/a$, in which, a is a parameters. The pheromone's value be limited in $[\tau_{\min}, \tau_{\max}]$, so the probability p_{ij} that the ant in a city i select a city j must accord $0 < p_{\min} \le p_{ij} \le p_{\max} \le 1$ .Because in the HCF, the pheromones value is automatically restrictions on [0, 1] interval. Therefore, in order to prevent some pheromone value reduced to 0, We set the pheromone's maximum value $\tau_{\max}$=0.999, and set pheromone's minimum value $\tau_{\min}$=0.001.

2.4 Pheromones Initialization

In order to better explore the solution space of question examples in the early stages, MMAS initialized the pheromone value to maximum $\tau_{\max}$.Because the pheromone volatile rate ρ is a relatively small quantity. So, the pheromone on each edge slowly appear difference, this enhanced the ability that MMAS explore more solution space.

When the ant search stagnated in a subspace, in order to enhance the explored ability, MMAS would initialize the pheromone value at the proper time. It would begin to Pheromones initialization, when the appointed iteration step is over or Search is stagnation

3 MMAS Algorithm

Algorithm1 had shown MMAS solving TSP process.

Input   TSP question example
Output   the obtained optimal solution T^{gb} from the algorithm begin
S01 Calculating the distance d_{ij} between the city i and city j, Heuristic information $\eta_{ij}^{\beta}, i, j = 1, \cdots, n;$
S02 each element of initialization pheromone matrix τ of each element is maximum value
S03 while didn't meet the end conditions do

S04 Heuristic information multiplied by pheromone information $\tau_{ij}\eta_{ij}^{\beta};$
S05 for each ant $k = 1, \cdots, m$ do
S06 Randomly put the ant k in a city $i_0;$
S07 Put ^{i_0} into Taboo list $Tabu_k$, and the current city pointer $i \leftarrow i_0;$
S08 while there is dis-visited city by the ant k do
S09 Randomly selects a city to visit $j \in (V - Tabu_k)$ base on formula 1
S10 $Tabu_k \leftarrow Tabu_k + \{j\}, i \leftarrow j;$
S11 end while
S12 end for
S13 obtain the best solution T^{ib}, and update $T^{gb};$
S14 Update all the pheromone Update (τ), base on formula 2 and formula 3
S15 Checking whether the pheromone value is limited in $[\tau_{\min}, \tau_{\max}]$ scope
S16 end while.

4 Total Information Updated Rules

Although MMAS is the most commonly used one of the best performance ACO algorithm, but it has deficiency. In solving the static combinatorial optimization problem, the step 04 and step 05 in the algorithm 1 is needless.their Computational complexity are $O(n^2)$, and each iteration step must run, so this greatly increased the amount calculation of the whole algorithm .we can rearrange MMAS's pheromone matrix updating rules to avoid those Unnecessary calculation. This can't reduce the whole algorithm complexity, but can reduce algorithms running time in the case of constructing the same solution or can increase the algorithm tectonic solution number in the case of the given total run time. Like this we can obtain more solution space and improve search more optimal solution opportunity.

In the formula 1 of MMAS algorithm solution rules, the α been generally appointed 1. The heuristic information η is static. Therefore, the heuristic information $[\eta_{ij}]^{\beta}$ would be calculated and saved before the algorithms 2 enter into outside loop (S03--S16). These provided conditions for us to rewrite the MMAS pheromone update rules. For convenient statement, we first provide solutions of tectonic rules and omit the parameter α, as follows:

$$p_{ij}^k(t) = \begin{cases} \dfrac{\tau_{ij}(t)[\eta_{ij}]^\beta}{\sum_{h \in Alw(k)} \tau_{ih}(t)[\eta_{ih}]^\beta}, & \text{if } j \in Alw(k) \\ 0, & \text{otherwise} \end{cases}$$

We called $\tau_{ij}(t)[\eta_{ij}]^\beta$ the total information corresponding the edge e_{ij} of the iteration step t, Notes for $TI_{ij}(t)$. The solutions tectonic rules can be written as formula 5 in total information form.

$$p_{ij}^k(t) = \begin{cases} \dfrac{TI_{ij}(t)}{\sum_{h \in Alw(k)} TI_{ih}(t)}, & \text{if } j \in Alw(k) \\ 0, & \text{otherwise} \end{cases} \tag{5}$$

At the both sides of formula 2, multiply by $[\eta_{ij}]^\beta$ at the same time, can obtain:

$$\tau_{ij}(t+1)[\eta_{ij}]^\beta = (1-\rho)\tau_{ij}(t)[\eta_{ij}]^\beta, \ i,j \in V$$

So total information volatile rule is as followed

$$TI_{ij}(t+1) = (1-\rho)TI_{ij}(t), \ i,j \in V \tag{6}$$

At the both sides of formula 3, multiply by $[\eta_{ij}]^\beta$ at the same time, can obtain:

$$\tau_{ij}(t+1)[\eta_{ij}]^\beta = \tau_{ij}(t+1)[\eta_{ij}]^\beta + \frac{1}{L^b(t)}[\eta_{ij}]^\beta, \text{ if edge}(i,j) \in T^b(t)$$

So total information volatile rule is as followed

$$TI_{ij}(t+1) = TI_{ij}(t+1) + \frac{1}{L^b(t)}[\eta_{ij}]^\beta, \text{ if edge}(i,j) \in T^b(t) \tag{7}$$

At the both sides of formula 4, multiply by $[\eta_{ij}]^\beta$ at the same time, can obtain:

$$\tau_{ij}(t+1)[\eta_{ij}]^\beta = \tau_{ij}(t+1)[\eta_{ij}]^\beta + \rho[\eta_{ij}]^\beta, \text{ if edge}(i,j) \in T^b(t)$$

And then we can obtain the total information release rules in the HCF:

$$TI_{ij}(t+1) = TI_{ij}(t+1) + \rho[\eta_{ij}]^\beta, \text{ if edge}(i,j) \in T^b(t) \tag{8}$$

The purpose of the formula 6 and formula 7 is directly update pheromone value and directly obtain new total information in the total information matrix, and the purpose of the formula 6 and formula 8 is too. So, the step 04 of the algorithm 2 could be canceled. On the basis of solution tectonic rules formula 5,the effect of the new pheromone update rules and the original rules is exactly same, but reduce the calculated amount.however, how to check whether pheromone value cross-border is a new problem. The local pheromone update rule of the ant Colony System—ACS given us an inspirer. Using pheromone volatile rules formula 9 instead of formula 2:

$$\tau_{ij}(t+1) = (1-\rho)\tau_{ij}^k(t) + \rho\tau_{\min} \tag{9}$$

Suppose the τ_{ij} wasn't enhanced from algorithm begin, so $\tau_{ij}(t+1)$ mould be expressed as follows:

$$\begin{aligned}
\tau_{ij}(t+1) &= (1-\rho)^{t+1}\tau_{ij}(0) + \rho\tau_{\min}[(1-\rho)^t + \cdots + (1-\rho)^0] \\
&= (1-\rho)^{t+1}\tau_{ij}(0) + \tau_{\min}[1-(1-\rho)^{t+1}]
\end{aligned}$$

When t was enough large, the $(1-\rho)^{t+1}\tau_{ij}(0)$ and $(1-\rho)^{t+1}$ come near 0, $\tau_{ij}(t+1)$ come near $\tau_{\min}$ .In other word, when $\tau_{ij} < \tau_{\min}$, The revised pheromone value volatile rules automatically enhanced the value of τ_{ij} ; when $\tau_{ij} > \tau_{\min}$, it could reduce the value of τ_{ij}; In a word, pheromone $\tau_{ij} \geq \tau_{\min}$. On the other hand, the maximum value of τ_{ij} has been automatically limited by the formula 3. In fact, supposing that τ_{ij} was enhanced at each iteration step, so we could obtain as follows:

$$\begin{aligned}
\tau_{ij}(t+1) &= (1-\rho)\tau_{ij}(t) + \rho\tau_{\min} + \frac{1}{L^b(t)}, \text{ if } edge(i,j) \in T^b(t) \\
&= (1-\rho)[(1-\rho)\tau_{ij}(t-1) + \rho\tau_{\min} + \frac{1}{L^b(t-1)}] + \rho\tau_{\min} + \frac{1}{L^b(t)} \\
&\vdots \qquad \vdots \\
&\leq (1-\rho)^{t+1}\tau_{ij}(0) + [\tau_{\min} + \frac{1}{\rho L^b}][1-(1-\rho)^{t+1}]
\end{aligned}$$

where, $L^b = \min\{L^b(1), \cdots, L^b(t)\}$ 。 When t was enough large, the $(1-\rho)^{t+1}\tau_{ij}(0)$ and $(1-\rho)^{t+1}$ come near 0, $\tau_{ij}(t+1)$ come near $\tau_{\min} + \frac{1}{b}$. in other words, $\tau_{ij}(t+1) \leq \tau_{\min} + \frac{1}{\rho L^b}$。 At the both sides of formula 10, multiply by $[\eta_{ij}]^\beta$ we can obtain:

$$\tau_{ij}(t+1)[\eta_{ij}]^\beta = (1-\rho)\tau_{ij}(t)[\eta_{ij}]^\beta + \rho\tau_{\min}[\eta_{ij}]^\beta, \ i,j \in V$$

The total information's volatile rule to limit the lower value of pheromone is as follows:

$$TI_{ij}(t+1) = (1-\rho)TI_{ij}(t) + \rho\tau_{\min}[\eta_{ij}]^\beta, \ i,j \in V \qquad (10)$$

The algorithms 2 given solving TSP process based on total information updated pheromone value.

Algorithms 2 FMMAS algorithms of TS

Input TSP question example

Output the given optimal solution T^{gb} from the algorithm begin to run

S01 Calculating the distance d_{ij} between the city i and city j, Heuristic information $\eta_{ij}^\beta, i, j = 1, \cdots, n$;

S02 each element of initialization pheromone matrix τ of each element is maximum value

S03 while didn't meet the end conditions do

S05 for each ant $k = 1, \cdots, m$ do

S06 Randomly put the ant k in a city i_0;

S07 Put i_0 into Taboo list $Tabu_k$, and the current city pointer $i \leftarrow i_0$;

S08 While there is dis-visited city by the ant k do

S09 Randomly selects a city to visit $j \in (V - Tabu_k)$ based on formula 1

S10 $Tabu_k \leftarrow Tabu_k + \{j\}, i \leftarrow j$;

S11 end while

S12 end for

S13 select the best solution T^{ib}, and update T^{gb};

S14 Update all the pheromone Update (τ) based on formula 7 and formula 10

S16 end while

5 Conclusions

Using the characteristic of solving Static combination optimization problem, we improved pheromone update rule, so the pheromone update was run at the total information, and canceled the total information calculation and canceled the operation crossing the border check pheromone value. The calculation experiments in TSP standard test examples show that: the new algorithms by us given would greatly reduce the time of the MMAS, in the case of finishing the same solution structure, and improve the ability of MMAS search optimal solution.

References

1. Dorigo, M., Caro, G.D.: The Ant Colony Optimization Meta-heuristic. In: Corne, D., Dorigo, M., Glover, F. (eds.) New Ideas in Optimization, pp. 11–32. McGraw Hill, London (1999)
2. Rosati, L., Berioli, M., Reali, G.: On Ant Routing Algorithms in Ad Hoc Networks with Critical Connectivity. Ad. Hoc. Networks 6(6), 827–859 (2008)
3. Dorigo, M., Stützle, T.: Ant Colony Optimization, pp. 20–89. Mit Press (2004)
4. Stützle, T., Hoos, H.H.: MAX-MIN Ant System. Future Generation Computer Systems 16(9), 889–914 (2000)
5. Dorigo, M., Gambardella, L.M.: Ant Colony System: A Cooperative Learning Approach to the Traveling salesman Problem. IEEE, Transactions on Evolutionary Computation 1(1), 53–66 (1997)
6. Blum, C., Dorigo, M.: The Hyper-cube Framework for Ant Colony Optimization. IEEE Transactions on Systems, Man and Cybernetics, Part B 34(2), 1161–1161 (2004)

Practice Study on Integrating Teaching and Research in a Graduate Course

Xiao-Qun Dai

National Engineering Laboratory for Modern Silk,
Soochow University, Suzhou 215123, China
daixqsz@suda.edu.cn

Abstract. The relationship between teaching and research in the modern university is of international concern. A strong teaching and research nexus will enhance the quality of university teaching and research, enhance the quality of research by allowing the cross fertilisation of ideas and learning between academics, and deliver potential benefits to students and staff. In this study, we tried to integrate teaching and research through a graduate course teaching practice in terms of course content design and pedagogy. The course profile was completed, including of objectives, content, class activities and assessment. In each aspect, research was integrated to cultivate students' various abilities and skills. Through practice, the course was improved and commented positively.

Keywords: Course design, teaching, research.

1 Introduction

Entering 21st century, as the information techniques developing very rapidly, high education faces many challenges. Since various university evaluations put emphasis on research, in many high educations, research become focus, teaching is of less importance[1]. On the other hand, since students can get information easily by many ways, especially through the internet, spoon-feeding teaching in class becomes even unacceptable. Such one way communication is very boring and easy to lose attention.

As western nations move towards information-based economies, they find themselves competing in higher education in global markets. This is placing a premium on the preparation of students for competitive workforces; and universities and their staff are finding it increasingly necessary to orient their work, especially their teaching, to these objectives. Academic leaders and managers need to understand the conditions teaching and those that inhibit connections and seek to compartmentalize academic activities. Likewise, governments are increasingly interested in and committed to funding research that has direct benefits for the economy. To the extent that universities can successfully combine these two activities, they are likely to be rewarded. These rewards will take various forms including sustained student enrolments, industry support and diverse research funding.

Coaldrake and Stedman note that most academics are expected to be involved in both research and teaching, and that 'the interplay between the two is widely held by academics to be a necessary part of ensuring quality.' The relationship between

Y. Wang (Ed.): Education Management, Education Theory & Education Application, AISC 109, pp. 583–587.
springerlink.com   © Springer-Verlag Berlin Heidelberg 2011

teaching and research in the modern university is one of international concern. A strong teaching/research nexus will potentially serve universities better as they move into an uncertain future, enhance the quality of university teaching and research. A strong nexus through scholarships of application and integration allows universities to relate more closely to new and important clienteles. A strong nexus also enhances the quality of research by allowing the cross fertilisation of ideas and learning between academics, students, industry, professional associations and other stakeholders, and delivers potential benefits to students and staff[2].

From many aspects, such as research institutional policy, departmental strategy, curriculum and programme design, many efforts are made to encourage linking teaching and research[3,4]. Oxford Brookes University, in the context of the move to semesters, required all courses to identify explicitly how they 'deliver' teaching-research links. At Earlham College (USA), applications for research grants have to identify how the proposed research will benefit students through producing a 'pedagogic impact statement'. McMaster University (Canada) has established an inquiry project, to develop inquiry based courses across the university, starting in year one. Individual academics applying for promotion at the University of Auckland (New Zealand) need to show how their research and teaching are linked. For course design, workplace-oriented learning, practice regarding course, inquiry based courses, research-led course, research-based course, many ways are tried[5].

In China, we also face the same challenge, the mode teaching only or research-intensive does not meet the demand of students cultivation in high education[6]. Through granting various teaching research projects, such as teaching innovation, curriculum design and pedagogic research and etc., we are also trying to improve teaching. Postgraduate education and training are a particularly important part of the university's academic profile. The focus of this study is graduate course, the purpose is on two aspects: to build a research-led course and do pedagogic practice.

2   Methodology

According to UK standard framework standard 3 designed for experienced staff who have an established track record in promoting and mentoring colleagues in learning and teaching to enhance the student learning experience, a research-led course was designed to support and promote student learning in all areas of activity, core knowledge and professional values through mentoring and leading individuals and/or teams; and incorporate research, scholarship and/or professional practice[7].

The course was improved through a couple of teaching cycles in terms of context and teaching ways.

3   Research-Led Course Design

3.1   Course Objectives

Clothing provides human body a small and portable environment, it should meet the demands from human body psychologically and physiologically, and have some other particular functions. In the course "science of clothing environment", knowledges from many areas, such as textiles, clothing, human physiology, materials and ergonomics are involved in the course, it is many disciplines crossed.

The objectives of this course were set as: to gain the scientific knowledge for clothing environment and know the latest research progress in related areas; and to draw out the requirements for clothing design. Through this course study, students have to learn to apply the scholarship in practice; to know the terms in English in the area; and to develop skills, such as independent-learning, application of knowledge, abilities to solve problems, scientific writing and presentation.

3.2 Course Content

The course is divided into several theme modules: human thermal physiology and clothing environment, relation between garment construction and materials mechanical properties, garment pressure and movement comfort, clothing hygiene, new materials and functional clothing. Each module includes theoretical background, application in the industry, related measurement methods, latest research progress. Through searching for research information, studying research papers, students also contribute to the course context.

Courseware is in PPT format and is written in English to help students to know the terminology in English. Since the clothing science research is well advanced in Japan and we may have to read many research papers in Japanese, some terms in Japanese were also introduced.

4 Pedagogic Study

4.1 Class Activities

The ideal classroom for the 21st century is: thinking; task-focused; students are engaged in meaningful learning tasks, challenged to ask questions, think, discuss, apply and evaluate their new understanding and skills; teamwork; transcdendence. Teach more by teaching less: the class usually starts with a problem, students are led to find the answer by themselves.

Students will be motivated to study and learn more outside class if they understand the subject matter and feel interested in it. Therefore, the course is set as guide study, the teacher gives lectures on theoretical parts, students have to participate in many ways, reading research papers and case study.

Usually, a class is started with a brief summary of the previous class session, and a reminder of where we are in the topic currently working on. If there are any problems arising from the homework, or any questions from previous class, then discuss them during this section. Then the lecture is given on new material. The lecture is often well structured with a brief outline, a list of objectives, main content, and a summary. To arouse interests and motivate students, application examples or demonstrations are always included during the lecture.

Many activities are also incorporated in class, such as quiz, group discussion and presentations. Students can learn by investigating issues, framing questions, solving problems, testing and interpreting evidence, by critical engagement with and application of research literature and reports. This simulates the student research process and is often delivered in group and project work.

4.2 Assessment

Students need to be pushed to learn more, so through the whole course study, many assessment tasks are assigned, reading research papers, writing literature review essays, doing some experiments, writing reports, and so on. As final coursework, students are required to find a research topic, design a research plan and carry out it using what they've learned from this course as well as what they've learned previously. They are asked to write a research report and present their research achievements in class. All these, regular assignments, performance in class, projects, exams, contribute to the summative subject evaluation according to a fixed percentage.

4.3 Evaluation of Teaching

After several classes, a feedback form is assigned for students to submit suggestions and comments on the course. Our department review team also come to watch the class randomly, their comments also help to improve teaching. Sometimes, comments and advises from experienced colleagues are also collected. Such reviews can clarify where research skills are taught, practiced and assessed, and a more integrated delivery of research and scholarship within all modules or units can be achieved.

Students are often happy to complete something meaningful and they are eager to experience challenge. It is believed that the learning process is most effective through doing something with a clear purpose in mind. We just try to include more activities into class to get students involved and provide them opportunities to develop various skills they may need in the future career. Fortunately, students are always lenient if only you show your enthusiasm and efforts. Now both the students and colleagues begin to comment positively about this course.

5 Conclusion

In this study, we tried to integrate teaching and research through a graduate course teaching practice in terms of course content design and pedagogy. The course profile was completed, including of objectives, content, class activities and assessment. In each aspect, research was integrated to cultivate students' various ability and skills. Through practice, the course was improved and commented positively.

Acknowledgement. A project funded by the Priority Academic Program Development of Jiangsu higher education institutions(PAPD). This work was also funded through a graduate curriculum project entitled "Science of clothing environment", granted by the Graduate Department of Soochow University.

References

1. Lucas, L.: Linking reasearch and teaching: the significance of research funding and evaluation policies. Research and teaching: closing the divide? In: An International Colloquium Marwell Conference Centre, Winchester (March 2004)
2. Zubrick, A., Reid, I., Rossiter, P.: Strengthening the nexus between teaching and research. Department of Education, Training and Youth Affairs, Commonwealth of Australia (2001)
3. Healey, M.A., Jenkins, A.: Linking teaching and research in disciplines and departments. The Higher Education Academy, York (2007)
4. Healey, M.A., Jenkins, A.: Developing undergraduate research and inquiry. The Higher Education Academy, York (2009)
5. Gibbs, G.: Institutional strategies for linking research and teaching. Exchange (3) (2002)
6. Jenkins, A., Healey, M.: Institutional strategies to link teaching and research. The Higher Education Academy, York (2005)
7. UK professional standards framework fro teaching and supporting learning in HE. The Higher Education Academy (2006)

The Development and Application of Virtual Instrument Technology in the Experimental Teaching

Jianqiang Liu, Xingqi Fu, Xingcheng Zhang, and Jianye Song

School of Physics, Shandong University, Jinan, Shandong, 250100, China
jqliu@edu.sdu.cn, fu10501@163.com, zhangxingcheng@gmail.com,
jenemy110@163.com

Abstract. Virtual instrument is a new type of testing instruments and a great breakthrough of the traditional instrument. Now it is a hotspot of the test system research. This paper reviews the application of virtual instrument technology in experimental teaching, introduces the concept and basic structure of virtual instrument, analyzes its developing status and trend, and then, take the development of virtual oscilloscope for example to introduce how to use the Lab VIEW to develop experimental platform.

Keywords: virtual instrument, lab view, experimental teaching, experimental platform.

1 Introduction

With the development of the computer technology, information network technology and communication technology, the experimental method continuously updated and new measuring theory constantly appear, the demands of teaching research work are more and more high on the experiment. Experiment content need to update to match the high-speed development of the science and technology; meanwhile the experiment is more and more complicated with large data increase exponentially, which more measuring need instrument to measure and control. In this process, because of time-consuming on the establish of various instruments, the uncertainty of stable operation on large system, the limitation on the data reading and the pretty big purchase expenses of instruments, the development of experimental work has been greatly restricted. In order to meet the demands of the education teaching and the development of the industrial area, national instrument company put forward the concept of virtual instrument technology firstly in 1986, which created a powerful new generation test instrument and brought great changes on measuring measurement and control technology.

The introduction of the virtual instrument technology makes the whole experiment process much simplified, if you write different data processing program with different experimental needs, it can form different instruments function. As long as there is a computer system with virtual instrument software and a DAQ (Data-Acquisition) device, you can isolate needed measuring data and start to make a measurement with

Y. Wang (Ed.): Education Management, Education Theory & Education Application, AISC 109, pp. 589–594.
springerlink.com   © Springer-Verlag Berlin Heidelberg 2011

simple connection. There is no need to change measuring devices when replaced the measuring objects, only need to write a software according to the test requirement, so this strengthen repeated serviceability of the virtual instrument and DAQ equipment and saved much purchase expenses of equipment [1, 2]. To virtual instruments, the hardware can just solve the problem of signal input and output, the key of the whole apparatus is the software. From the earliest Lab VIEW and Lab Windows/CVI software to the appearance of Lab VIEW 8.5, the functionality and intellectualization of virtual instrument is further expansion. Virtual instrument can simulate various tests phenomenon effectively with simple operation and easy-studying-easy-using, so students can understand all kinds of abstract physical theory.

2   The Development of Virtual Instrument Technology

With the development of science and technology, traditional instruments can't satisfy the requirement of industrial production and experimental research. Virtual instrument can be widely used in communication, electronic measurement, automation, aerospace, military engineering, automobile, power engineering, mechanical engineering, construction, railway transportation, geological exploration, biological medical and many situations which require scientific analysis with high-performance tt&c equipment for its intelligent, automation, flexibility. Developments of virtual instrument depends on the computer, software, high quality A/D acquisition card and regulate amplifiers, so computer hardware and software make virtual instrument mixed with high-tech together closely. Using constantly perfect computer and network technology, we can write powerful software (from the earliest Lab VIEW software and Lab Windows/CVI software to LabVIEW8.5) to make the function of virtual instrument more powerful and broader prospect of its application.

Virtual instrument not only widely used in industry, but also in experimental teaching for short development and maintenance time, easy extension and integration, which has been a new means of teaching. It became a developing tendency to deduce physics experiment together with the multimedia teaching and virtual instrument technology, and walked into the laboratory and electronic technology class slowly. Students can complete various electronic instrument operations by the computer and network assistance to improve learning efficiency and need not worry about damaging experimental instruments. It also can bring students broad vision and high-tech learning atmosphere [3]. At present, virtual instrument mainly developed to the following directions: Hanging-outside virtual instrument, PXI-type integrated virtual instrument, networked virtual instrument [4]. Hanging-outside virtual instrument could effectively prevent measured signal and computer form mutual interference to make test result accurate and reliable. PXI-type integrated virtual instrument could simply integrate different test equipments into the same system because of its expandability and compatibility. Networked virtual instrument made test message and result sharable and even could control experiment facility in a ling distance, so it was hopefully used to solve problems of experiment in distance education. In addition, Energy Savings, compact and flexible were another development direction for virtual instrument. Following this way portable virtual instrument would be born.

3   The Basic Characteristics and Composition of Virtual Instrument

The virtual instrument technology is the software based on computer, that is to say, the software is loaded into a computer rely on computer hardware and operating system, the measured data is collected to virtual instrument software of PC by data acquisition device and carried on the corresponding data analysis and processing to realize the function of various measurement analysis instrument. Its essence is fully combined the traditional instruments hardware and the latest computer software technology to realize and expand the function of traditional instruments. The core technology of virtual instrument is the software. In order to meet the different needs, users can construct test instrument (system) flexibly to achieve completely different function of measurements testing to realize "the instrument becoming the software", "The software is the instrument". Compared the virtual instruments with the traditional instruments, the virtual instruments have more advantages.

Virtual instrument is an instrument based on computer measuring, it simulated the appearance of traditional instruments and the switch, buttons, display unit of manipulate instrument in the computer screen on the way of various graphics through the software. Its operation is realized by computer mouse or keyboard. It collects data through the general underlying data acquisition module and processes data by computer software. We can write different data processing software to realize different instruments function with different test requirements. The emergence of virtual instrument truly reflects a new conception is that "the software is the instrument", "use the software to do the instrument". A typical virtual instrument measuring system is shown in Fig. 1.

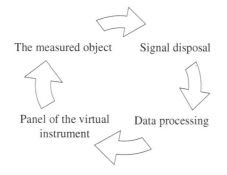

Fig. 1. A typical virtual instrument measuring system

The tested object is converted to electrical signals by various sensors and the measured useful signal is removed from mixed signals after disposed (such as filtering, amplification, etc.) through the signal disposal module, which is put into the data acquisition card. The computer collects measured data according to the virtual instrument software. Users can easily finish collection, analysis, display and storage of the tested amounts in this way.

4   Build Experimental Platform with Virtual Instrument Technology

The characteristics of virtual instrument are the software combined with the hardware to solve practical problems. The hardware collects various signals by general data collector match with various sensors and signal regulates circuit. The software realizes the data analysis and processing by writing programming with graphical programming software Lab VIEW (Laboratory Virtual Instrument Engineering Workbench).

Lab VIEW is a programming language developmental environment and professional software which using graphical programming language - G language to produce massive program massive program for specialized test measurement and signal processing. The hardware has been greatly degree of reuse in soft and hard combination way. Under the condition of invariable in the hardware, researchers can focus more on innovation but not the construction of test system by changing the software to realize various complex test measurement and signal processing scheme.

Oscilloscope has been widely used in signal testing, signal comparison, logic analysis and other areas and is a necessary experimental instrument in the electronics field. Here take the experiment of construction of oscilloscope as an example to illustrate how to quickly build a virtual instrument testing platform.

Before you start designing virtual instrument, you should determine the design demand, for example, if you need to design an oscillograph with basic function, firstly it can collect time-domain waveform signal and bias or gain control the collecting signal, and then it can display some basic information such as effective value and peak value [8]. You can do prototype simulation which is using the functions of simulation arbitrary signals to generate signal after demand confirmation, then use waveform display control to display signal, in addition to display, you should measure time signal use the waveform measurement tools provided by Lab VIEW. The measured peak, root mean square and other information displayed through numerical control, then use specified gain (on zoom) and multiple bias (specified offset add operation) calculation method to finish gain and bias function design. After Prototype simulation, you can connect the system to the data acquisition card and put the collected data to the virtual oscillograph to make actual signal collection and analysis [9].

The design of virtual oscilloscope made a reference to the function traditional oscilloscope and combined the characteristics of virtual instrument and powerful information processing capability of computer to realize the waveform display, storage, and printing, generate reports, time domain, frequency domain parameters of automatic measurement, display, query and automatic calculation function of phase difference. From the demand to prototype design and the programming on designing of the virtual oscilloscope, the whole process is so simple and clear and has very short time-consuming that we can quickly develop all kinds of experiments. The virtual instrument technology can also be used in electrical experiments of filter experiment, basic circuit virtual experiment and other series of university basic experimental tutorial.

5 Conclusions

Virtual instrument phased out and replaced the traditional instruments; its testing technology is a new milestone. It is more intelligent, changeable combined with high-tech technology in modern measurement and control technology; it will become the mainstream of the future testing instruments and represent the new direction of the development of instrument, which provides convenience for industrial production and experimental teaching [10]. An important feature of virtual instrument is that it can simulate various physical phenomenons. If it is introduced into the experiment teaching, and then you can use the Lab VIEW graphical programming environment and data acquisition hardware to achieve various experimental function easily. For example, we can use the Lab VIEW to write simulation experiment procedure for emulation teaching with no hardware. If the hardware is not invariable, we can easily change and increase or decrease the experimental function through changing the program. The main reason is that Lab VIEW is graphical programming language and easy to use, students can quickly grasp its usage and better understand the principle and function of the instrument in the process of writing virtual instrument programmed by them, which have a very good teaching effect.

Acknowledgments. The authors thank the sub-project of the National Teaching and Research Key Program of China during the 10th Five-Year Plan Period (AIA050007-l4), the Teaching Research Foundation of the Commission of College Physics Teaching Guidance (WJZW-2009-13-hd, WJZW-2010-33-hd and WJZW-2010-21-hd), the Teaching Research Foundation of Shandong Province (2009123) and the Teaching Research Foundation of Shandong University for financial support.

References

1. Zhang, J., Han, X.: LABVIEW Graphical Programming and Application Examples. China Railway Publishing House, Beijing (2005)
2. Dai, P., Wang, S.: Testing Engineering and Application of LABVEIW. Publishing House and Electronics Industry, Beijing (2006)
3. Li, W., Luo, Q., Li, Z.: The application of virtual instrument technology to demonstration experiment teaching of science and engineering. IEEE International Symposium on IT in Medicine and Education 1, 592–595 (2009)

4. Weide, Q.: The Virtual Instrument Technology and Its Application and Prospect. Electrical Drive Automation 29(4), 6–9 (2007)
5. Liu, P., Cao, H., Qiu, P.: The Development Process and Application of Virtual Instrument. Shandong Science 22(1), 80–83 (2009)
6. Zhu, Y., Guo, H., Xu, J.: The circuit and component test system based on virtual instrument. Petroleum Instrument 24(5), 85–86 (2010)
7. Zeng, T., Hou, J.: The Study on Application of Virtual Instrument with oscillography/frequency spectrum display function in Teaching Experiment. J. Electric. Electron. Edu., 64–66 (2001)
8. National Instruments: LABVIEW. Professional Software of Testing and Automation, 7–14 (2001)
9. Zhang, X., Hui, W.: Experiment teaching based upon virtual instrument. Electronic Instrumentation Customer (2011), doi:10.3969/j.issn.1671-1041
10. Olteanu, R.L., Dumitrescu, C., Gorghiu, G.: Related aspects to the impact of virtual instruments implementation in the teaching process. Procedia Social and Behavioral Science 1(1), 780–784 (2009)

Changes and Characteristics of Vocational Curriculum in Taiwan: 1964-2010

Chuan-Yuan Shin[1], Kung-Huang Lin[2], and Hung-Min Lin[3]

No.69, Lane 485 , Dapu Road , Changhua City,Taiwan(R.O.C)
eyrir@ymail.com

Abstract. Although vocational education in Taiwan began more than 150 years ago, no solid foundation has been established to ensure its further development. It is because it was simply designed to fulfill the need for production human resources (HR) of businesses in the absence of a curriculum. Also, despite a set of regulations announced in 1902 (the 28th year of Emperor Guangsui's Regime, Qing Dynast), they were simple rules without specifying the contents of any curriculum.

The light of vocational education eventually rose until the restoration of Taiwan along with the participation of the US Advisory Group. With the concerted efforts of the government, schools and industries, the first official curriculum tailored for vocational education was announced in 1964 to put vocational education on the right track and ensure its stable future development.

Whether it is the unit trade training curriculum, cluster-based ladder curriculum and course group curriculum, the efforts in curriculum planning for vocational education are witnessed. This also allows the localization of the vocational curriculum in industrialized countries, such as Europe, the USA, and Japan, in Taiwan and paves way for the healthy development of vocational education in Taiwan.

Keywords: vocational school curriculum, changes and characteristics, Curriculum Standards.

1 Introduction

The trend of education reforms seems to have swept across the globe following the rises and falls of new and old businesses, the launch and replacement of new workplace skills, and the changes in and impacts of new education concepts. It is the same in either developing or developed countries. Since the government moved to Taiwan, it has placed tremendous investments and made huge efforts to promote vocational education, and large quantities of base-level industrial and business HR were thus fostered. Apart from making undeniable contributions to national economic constructions, this has given an opportunity for young students interested in learning skills to develop confidence and diligence for the future.

On top of an important tool to accomplish vocational education, the vocational curriculum plays a decisive role in the overall development of vocational schools because it is

Y. Wang (Ed.): Education Management, Education Theory & Education Application, AISC 109, pp. 595–606.

(1) the core of school administration;
(2) the blueprint of teaching;
(3) the re-construction of student's confidence; and
(4) the consensus for developing characteristics.

Given that curriculum revision, industrial structure and HR demand are interrelated, and such revisions must be able to foster better quality HR, curriculum revisions are thus important and difficult. This also explains why the planning and design of vocational curriculum must go through the following long and rigorous processes prior to announcement, including study, discussions, drafting, revisions, public hearings etc, conducted by and with course experts, psychologists, business representatives, school representatives, parent representatives and the relevant administrative personnel. Therefore, it is a huge and painstaking process.

Changes have been made to the educational goal, curriculum structure and scope and content of instruction to a certain extent in the "curriculum standard" (later called "curriculum guide or framework") over time as a result of the time-specific political, economic, social and educational policies. The aim of this study is to conclude the trend of evolution of vocational curricula over time by analyzing the content differences among these curricula and investigating the time-specific circumstantial conditions, intellectual backgrounds and their characteristics. In doing so, it is hoped that the finds of this study can provide a reference for reviewing and improving the revision of the vocational curriculum in the future.

2 Description of Vocational Curricula over Time

When drafting the curriculum, the Ministry of Education (MOE) must consider the time-specific educational system, government financial conditions, industrial structure, social structure and HR demand. Therefore, the educational goals, curriculum structure, taught courses and length (hours) of professional practicum of the curricula planned and designed in different times vary.

2.1 1964-1972: High Development of Unit Trade Training Curriculum—Vocational School Curriculum Standards of 1964

Given that to meet the industrial needs is the central of unit trade training curriculum, the course contents, skill level and practice equipment should made to meet the workplace requirements. Therefore, in order to enhance the employment rate of vocational school graduates increased and fulfill the need for national economic development, the revision of the curriculum for industrial and vocational schools began in September 1962 (Department of Secondary Education, MOE, 1962). The Curriculum Standards for Industrial and Vocational Schools were also announced on 10 October 1964. The major revisions included (Shih, 1990):

(1) Increase of course hours for practical courses: In the curriculum standards announced in 1952, the proportion of course hours among general, professional (including related courses) and practical courses was 35%-33%-32%. These standards were revised to 30.7%-30.7%-38.6% in the

revision of 1964 to increase the course hour of practical courses by 6.6% to reinforce vocational skill training.

(2) Classification of vocational schools into 6 categories: agriculture, industry, business, fisheries, nursing & midwifery, and domestic science.

(3) Except the business and domestic science categories, all other categories were subdivided into related sections, particularly the industry category which contained up to 53 sections.

(4) Vocational schools were divided into two levels: senior high and junior high vocational schools, and the study period of each level was 3 years.

2.2 1973-1984: Prime Time of Unit Trade Training Curriculum— Vocational School Curriculum Standards of 1973

Junior high vocational schools came to an end with the rise of the 9-year compulsory education in 1968. The number of vocational schools increased from 111 schools in 1962 to 171 schools in 1972. In addition to increasing the student ratio of senior high vocational schools to senior high schools to 7:3, the government revised again the course hour of practical courses for vocational schools in the Curriculum Standards for Senior High Vocational Schools 1974 to cope with the HR demand for export expansion and the Ten National Construction Projects. The major revisions included (Shih, 1990):

(1) Abolition of the curriculum standards for junior high vocational schools: When the 9-year compulsory education was implemented in 1968, the student recruitment of junior high vocational schools was immediately suspended, and the curriculum standards for junior high vocational schools were also abolished in academic year 1972.

(2) The nursing & midwifery and fisheries categories in the original 6 categories (industry, business, agriculture, nursing & midwifery, fisheries and domestic science) were changed into the medical technology and marine technology categories.

(3) The sections in individual categories were adjusted and increased to 61 sections.

(4) The course hour of practical courses was also adjusted. The proportion of course hours of general, professional and practical courses in the industry category was adjusted to 31.6%-26.3%-42.1% to cope with the HR demand as a result of the industrial structure transformation from labor-intensive to technology-intensive.

2.3 1985-1995: Curriculum Transformation—Rise of Cluster Curriculum and Vocational School Curriculum Standards of 1986

The aim of Taiwan's 10-Year Economic Development Plan (1980-89) drawn in 1981 was to develop technology-intensive and knowledge-intensive industries with low energy density but high added value; i.e. the development of mechanical, IT, electronics, electrical and vehicle strategic industries (Lee, 1989). To cope with such an impact, various reforms were made to the vocational education in the 1970s, including

(1) The promotion of the competency-based instruction in 1982.

(2) The implementation of the Industrial and Vocational Education Reform Phase II in 1982.

(3) The implementation of the pilot project for Extension of the Vocational Education Focus Compulsory Education in 1983.

(4) The review and improvement of the defects in the "unit trade training curriculum", adoption of the cluster curriculum, and announcement of the Curriculum Standards for Senior High Vocational Schools 1986.

(5) Dual-rail practice of Cat A and Cat B in the curriculum standards.

The major revisions of the Curriculum Standards for Senior High Vocational Schools 1986 (Cat A) included:

(1) Classification of vocational schools into 7 categories, including industry, business, agriculture, domestic science, marine technology & fisheries, medical technology, and fine arts; and only the medical technology category was not revised.

(2) Planning and design of cluster-based sections and curriculum frameworks to cope with the rapid changes in industrial structure and increase the opportunities for changing careers. The industry category was classified into 5 clusters: mechanical engineering, electrical engineering, civil engineering, chemical engineering and craft.

(3) The general courses and their weekly course hours of each category were very similar. For example, there were 4 lessons for Chinese Language, 2 lessons for English Language, 2 lessons for The Three Principles of the People, and 2 lessons for the Introduction to Social Science, and they were the same to all sections and categories.

(4) Reinforcement of computer skill training to cope with the need for industry automation and computerization: in addition to making the Introduction to Computers as a compulsory course to all sections and categories, IT was also integrated to various professional courses.

(5) Reduction of course hours for practicum and increase of professional theory courses: The general-to-specialized curricular framework was applied to provide adaptive education for students to meet the employment and change of employment needs.

(6) About 5-10% course hours for optional courses were added for schools to arrange courses based on local differences.

(7) The proportion of course hours of general, professional and practical courses was about 35%-30%-25%, and the remaining time was left for activity-based courses, military education and nursing.

2.4 1996-2005: Vocational School Curriculum Standards of 1998 and Promotion of an Integrated Curriculum for Vocational Education

The call for educational reform started in 1995 was irresistible. The Department of Vocational Education adopted the second national highway expansion policy (擴展第二條國道政策). Academies asked for promoting to institutes of technology or universities of science and technology. As a result, the opportunities for further studies of vocational school graduates escalated. Vocational education aiming to foster base-level technical HR was challenged by both the subjective and objective circumstances, and most vocational school students had a strong desire for further studies. In order to avoid a waste of educational resources, the connection of the curriculum of vocational schools, 4-year institutes and 2-year junior colleges became an important issue in vocational education curriculum. The Vocational Education Integrated Curriculum Planning Taskforce was formed with experts and representatives of vocational schools, 4-year institutes and 2-year junior colleges to study and revise the curriculum for the following objectives (MOE, 1998):

(1) To maintain a balance between employment and further study of students.
(2) To cope with the implementation of the credit-year system.
(3) To simplify taught courses in order to reduce the work stress of students.
(4) To increase activity-based courses in order to release the physical and mental stress of students.
(5) To simplify curriculum structures in order to provide all sections and categories with greater flexibility for course planning and design.
(6) To horizontally and vertically integrate the curriculum of vocational schools, 4-year institutes and 2-year junior colleges.
(7) To increase the proportion of courses determined by schools in order to give schools more spaces for operations.
(8) To plan and design general courses for cultural literary, liberal arts and nationalism development.

In order to cope with the trend of educational reform for more senior high schools and universities, the MOE piloted the comprehensive senior school in 1996, hoping to reduce the number of vocational school students every year and maintain the proportion of senior high and senior high vocational school students at 50%-50%. This period can be considered as the "decline of vocational schools" when the functions of vocational schools were questioned. The major revisions of the Curriculum Standards for Senior High Vocational Schools 1998 included:

(1) Change from the course hour to credit year system: Students were required to complete courses of at least 22 credit hours but not more than 34 credit hours every semester (not including activity-based courses).
(2) Schools could freely arrange the "school courses" to increase the flexibility of course planning and design.
(3) Free selection of school optional courses for students to provide students with more space of learning.
(4) Simplification of curriculum framework: The original curriculum framework (general courses, professional fundamental courses, professional courses, optional courses and common activity) was simplified into general courses, professional courses (including practicum) and activity-based courses.
(5) Field-specific design for general courses: Including Chinese Language and Literature, Foreign Language and Literature, Social Studies, Mathematics,

Natural Sciences and Fine Arts. Courses in each field were opened as necessary.

(6) Increase of activity-based courses: At least 4 lessons a week should be arranged, compulsory, no credit (originally 2 lessons a week.

(7) Reduction of the variety and lessons of taught courses: 30 to 34 lessons a week (including PE, ME&N, activity-based courses and flexible teaching time); and 10 to 12 courses a semester.

(8) Cancellation of the cluster curriculum and Cat A and Cat B curricula.

(9) General courses were about 40-50&%, practical courses were about 20%, school courses were about 15-35%, including general, professional and practical courses.

2.5 2006-2010: Drafting of Common Cores for Late Secondary Education and the Temporary Curriculum Guide for Course Groups 2006 and Curriculum Guide for Course Group of Vocational Schools 2010

Although senior high schools and senior high vocational schools both fall in the late secondary education, the former is the preparatory education for university education and the latter the preparatory education for base-level technical HR for different fields. Influenced by the scholar supremacy tradition in Chinese culture (all except scholars are at inferior ranks), general senior high schools are always the first choice of most junior high school graduates wishing to further their studies, and the 5-year colleges or senior high vocational schools are the choices when there is no other choice. While both the curriculum and teaching materials of general senior high schools and senior high vocational schools are considerably different, the MOE thus announced the Common Core Curriculum Guide for Late Secondary Education in 2004 in order to standardize the quality of late secondary education. According to this Common Core Curriculum Guide, the requirements of the following courses are: 8 credit hours for Chinese Language and Literature, 8 credit hours for Foreign Language and Literature, 6-8 credit hours for Mathematics, 4-6 credit hours for Integrated Science, 6-10 credit hours for Social Studies, 4 credit hours for daily-life-related courses, 4 credit hours for art-related courses, 4 credit hours for PE, totally 48 credit hours of compulsory courses. While such a guide has imposed severe restrictions on the time for practical courses of vocational schools and the syllabus of some courses (Chinese Language and Literature and History), disputes arose all over. As a result, the vocational school curriculum that was supposed to implement in academic year 2006 was announced as the "Temporary Curriculum Guide", which was also called the "Temporary Guide 2006", to collect the opinions from different parts of society. The MOE announced the Course Group Curriculum Guide for Vocational Schools in 2009 which was implemented in the academic year 2010. It was also called the "Curriculum Guide 2010".

Vocational schools have long been classified by category and section. In the Temporary Curriculum Guide for Vocational Schools 2005, they were classified by group and section into 15 groups, whereby homogeneous sections in each category were categorized into the same group (e.g. Art & Craft and Interior Design sections were categorized into the Design Group). The major revisions of the Temporary Guide 2006 and Curriculum Guide 2010 included

(1) The common cores of 48 credit hours in 7 fields in the late secondary education were all categorized as compulsory courses in the Temporary Guide 2006. Although flexibility was granted in the Curriculum Guide 2010, it was a minor adjustment aiming to standardize the cultural literacy of the late secondary education.

(2) In both the Temporary Guide 2006 and Curriculum Guide 2010, professional and practical courses of 15-30 credit hours in each group were classified as the MOE compulsory courses. They were the core courses of each group aiming to develop the core competency of students.

(3) The categories and sections of vocational schools were categorized into 15 groups, i.e. mechanical engineering, powered machinery, electrical and electronic engineering, chemical engineering, civil engineering and construction, business and management, agriculture, domestic science, food processing, restaurant and hospitality, marine technology, fisheries, fine arts, design, and foreign languages; aiming to link and integrate the sections and departments of vocational schools, institutes and colleges, and universities of science and technology by attribute.

(4) Vocational schools should offer courses of 192 credit hours within 3 years (not including activity-based courses), and the required credits for graduation of vocational school students were 150 credits. Schools should offer professional and practical courses of no less than 60 credit hours, including no less than 30 credit hours of practical courses.

(5) Only the MOE-required general, professional and practical courses were specified in the MOE curriculum guide, and schools should plan and design courses for the remaining credit hours (both compulsory and optional) according to student needs and school conditions. School courses should be planned and designed according to the defined administrative procedure and reported to the competent authorities for review and approval.

3   Characteristics of Curriculum Change over Time

The vocational curriculum was announced in the form of "curriculum standards" for the first time in 1964. Although the proportion of course hours and total weekly course hours of general, professional and practical courses; and the title and course hour of general courses were specified, the contents of neither the professional nor practical courses were defined. Vocational schools planned and designed these courses by modeling relevant courses in industrialized countries, such as Europe, the USA and Japan; or according to the actual needs. In the absence of a consistent syllabus for such courses, the standard of graduates from different schools varied. After years of study, discussions and revisions, the Curriculum Standards for Vocational Schools were eventually announced in 1964. Until the Course Group Curriculum Guide for Vocational Schools announced in 2009, the MOE has announced five set of formal curriculum standards, and the characteristics of each set are described below.

3.1 Curricular Standards and Equipment Standards for Vocational Schools 1964

3.1.1 Unit Trade Training Focus
Due to the success of the unit trade training for equipping students with the professional skills of a particular occupation was piloted by 8 senior high vocational schools in 1955, the idea of preparatory education for workplace skills became the focus of the curricular standards announced in 1964 in order to provide base-level technical HR for different industries.

3.1.2 Planning Curriculum in the American Way
In the beginning of Taiwan restoration and the government's retreat to Taiwan, the vocational curriculum was planned and designed according to the curriculum practiced in the Meiji Period of Japan. With the suggestions for curriculum planning and the support of equipment and teaching materials for practical courses from the USAID in 1955, the vocational curriculum was planned and designed in the American way.

3.1.3 From Theory-Based to Practice-Based Curriculum
The course hours of practical courses were comparatively less due to equipment and space inadequacy or the practical skill inadequacy of instructors. As a result, what students have learned may not be practical at work. After the success of the unit trade training pilot project, the importance of practical courses eventually caught the public attention.

3.1.4 Curricular Standards and Equipment Standards for Senior High Vocational Schools 1974

3.1.5 Continuous Expansion of Practical Course Proportion
As the industrial structure has changed from labor-intensive to technology-intensive, the demand for technical HR escalated. Therefore, the educational goals of vocational schools also changed into provision of technicians for industries. This made practical courses more important and the proportion of course hours thus increased again from 38.6% to 42.1%.

3.1.6 Increase of Categories and Sections to Cope with the Division of Labor in Industries
The division of labor became more delicate in industries as the economic development prospered. As the original section categorization of vocational schools was unable to meet the modern demand, the number of sections increased to 61.

3.2 Curricular Standards and Equipment Standards for Senior High Vocational Schools 1985

3.2.1 Replacing Unit Trade Training Curriculum with Cluster-Based Ladder Curriculum
The industrial structure changed again from technology-intensive to capital- and knowledge-intensive with the rise of automation and computerization. The demand for unit trade professional skills gradually declined, and the need for changing employment

of workers increased. The cluster-based ladder curriculum was planned and designed to fulfill the comprehensive professional skill demand in workplaces in order to facilitate workers to change their job.

3.2.2 From Practical Course Focus to the Balance of Theory and Practice

The proportion of weekly course hours of practical courses reduced from 42.1% to about 25%, marking a significant change.

3.2.3 Co-existence of Cat A and Cat B Curricula

The huge reduction of the course hours for practical courses was questioned by vocational schools and their faculty members. Even the industry was worried about the skill level of graduates. After heated debates, both the Cat A (theory-focus) and Cat B (practice-focus) curricula were announced at the same time for vocational schools to choose what fitted their schools most.

3.2.4 Planning Curriculum from General to Specialization

Courses began with the comprehensive general knowledge of a particular industry and gradually toward the special knowledge and skills in workplaces.

3.2.5 Adding IT-Related Courses for the Advent of the Information Era

The Introduction to Computers, Application of Computers, Word Processing and other computer-related courses were added to the vocational curriculum for the first time. Also, many professional courses were instructed in the assistance of computers (CAI), such as accounting IT systems, computer-assisted drawing (CAD) etc, marking a big change in the vocational curriculum.

3.3 Curricular Standards and Equipment Standards for Senior High Vocational Schools 1998

3.3.1 Lifelong Education Emphasizing Fundamental Courses

Chinese, English and Mathematics are the 3 root courses. However, their importance was often second to professional and practical courses in the vocational curriculum. With the rise of lifelong education concept, their importance eventually regained. Therefore, the course hours of these 3 root courses were largely increased in the vocational curriculum where the educational goals also changed to meeting the employment and lifelong education needs of students.

3.3.2 Increase of Space for Autonomous Operations of Schools by Increasing School Courses

The proportion of school-determined courses increased from 5-10% to 15-35%. Apart from meeting the special needs of local industries, this has given more space for autonomous operations of schools, thus forming the school-based curriculum.

3.3.3 Valuing Mental and Physical Development of Students by Increasing Activity-Based Courses

The number of PE lessons increased from one lesson a week to 2 lessons a week, and activity-based courses (including class meeting and week assembly) were increased from 2 lessons a week to 4 lessons a week, hoping to release the work stress and promote the healthy mental and physical development of students.

3.3.4 From Course Hour to Credit Hour and Abolition of the Repeat System

Traditionally, students were required to repeat the same year if they failed a certain amount of courses in the academic year. Apart from wasting educational resources, this hurt the emotion of students. Therefore, the course hour system was changed into the credit hour system that students only need to repeat the courses they failed.

3.4 Course Group Curricular Guide for Vocational Schools 2010 (Curriculum Guide 2010)

3.4.1 Never-Ending Disputes and Disintegrated Opinions

Different parts of society hold different and even diverse views on the proportion of Classical Chinese and Modern Chinese works in the Chinese Language and Literature course, the contents of Chinese history in the History course, and the impact on the late secondary education of the common cores of 48 credit hours curriculum planning for vocational schools. The curriculum guide that was supposed to be implemented in academic year 2006 was thus announced in a "temporary" form, hoping to make more time to settle these disputes. In the Curriculum Guide 2010 announced in 2009, however, the syllabus for the Chinese Language and Literature and History remained unchanged.

3.4.2 Replacing Category with Group to Determine Group Core Competency and MOE-Required Professional and Practical Courses (15-30 Credit Hours)

Vocational schools were originally classified into 7 categories, including industry, agriculture, business, domestic science, marine technology and fisheries, fine arts, and medical and nursing. Apart from promoting the medical and nursing to the institute level, the original sections in each category were re-organized into 15 groups. The required courses and their credit hours were determined according to the core competency of each group.

3.4.3 School-Determined Courses for Reference to Emphasize Space of School Operations

The proportion of school-determined courses was maintained at about 25% in the curriculum standards announced over time. In the Curriculum Guide 2010, however, the proportion increased to 40-50%, thus giving much more space for schools to develop the school-based curriculum.

3.4.4 Change of Military Education and Nursing and Reduction of the Number of Lessons

The title of Military Education that has been used for many years was first changed into General Education on National Defense (Temporary Guide 2006) and later All-out Defense Education. Also, the course hour which was compulsory for 3 years (2 lessons a week) was also changed to compulsory for one year (1 lesson a week). The Nursing course was also changed into the Health and Nursing, one lesson a week, a compulsory for both boys and girls.

4 Conclusions

The development of the vocational curriculum is in effect a history of industrial structure development; a history of the economic, political, social and cultural changes of a nation. Education evolves along with the industrial and social development. In return, industries obtain high-quality HR in the support of education. Both are thus complementary and inseparable. By reviewing the vocational curriculum evolution in Taiwan, the following conclusions are made.

4.1 Gradual Formation of a Localized Curriculum Framework

The vocational curriculum announced during 1964 to 1973 was unit trade training focus. In the curriculum announced in 1986, the focus was changed into the cluster-based ladder type. Nevertheless, both models imported from the USA. It was not until the curriculum announced in 1998 and 2009 that the light of localization was eventually seen as the ethnicity of Taiwan, student traits and school cultural were considered in both curricula. In the future, how to develop a vocational curriculum that is tailored for students of Taiwan according to the social structure, economic development, political culture and educational characteristics of Taiwan will be prime mission.

4.2 Formation of the Goal for Integrating Lifelong Education with Vocational Education

The goal of vocational education was to prepare base-level HR for industries, and the pursuit of further studies at 2-year junior colleges, institutes of technology and universities of science and technology was considered as going astray from the goal. Today, institutes of technology and universities of science and technology increase like mushrooms after rain. Along with the improvement of the family financial power and the academic-qualification preference of industries, the pursuit of higher academic achievements has become a common trend in vocational school graduates. Therefore, it is necessary for vocational schools to set their educational goals that meet the student's need for higher academic achievements in order to cope with the actual needs.

4.3 Horizontal and Vertical Connection and Integration of Curricula become a Central Issue in Curriculum Development

Although the common core curriculum for late secondary education, the planning of the integrated vocational curriculum and the skill education reform plan were independently planned curricula, the connection and integration of these curricula have become a central issue, and the effectiveness of the proper use of educational resources has been seen.

4.4 Distinction of Formative Education and Preparatory Education for Career

Formative education emphasizes the acquisition of practical and specialized work skills in order to quickly supply the HR desperately required by industries. Therefore, the faculty, equipment and course contents of formative education should resemble to that of the designated workplace. Preparatory education aims to equip students with the comprehensive and fundamental industry-specific professional knowledge. Therefore, students will need professional training at the designated workplace prior to employment. This favors the rises and falls of an industry and avoids functional unemployment from causing social problems. Both types of vocational education are functionally different, and how to effectively and suitably differentiate and use them will directly affect the planning of curriculum contents. Therefore, vocational educators should carefully consider the curriculum for vocational education.

References

1. Taiwan Province Archives, Education annuals, 5, Annuals of Taiwan. Taipei: Taiwan Province Archives (1970)
2. Jiang, W.S.: Teaching materials and methods for vocational categories and sections. Shita Books, Taipei (2000)
3. Lee, D.W.: Curriculum development for vocational education: theory and practice (1989)
4. Shih, S.Q.: Curriculum development for vocational categories and sections. In: Teaching Materials and Methods for Vocational Categories and Sections, pp. 129–161. Shita Books, Taipei (1990)
5. Gao, M.M.: Curriculum for industrial and vocational education. Hualien Senior High Vocational School, Hualien (1968)
6. Department of Secondary Education, MOE, References for the revision of the curriculum standards for vocational schools. Department of Secondary Education, Taipei, MOE (1962)
7. Ministry of Education, Curriculum standards and equipment standards for senior high vocational schools. Ministry of Education, Taipei (1998)
8. The Third Chinese Education Yearbook, Chapter 6: Vocational Education. Cheng Chung Bookstore, Taipei (1957)
9. Yang, C.X.: Technical and vocational education: theory and practice. San Min Bookstore, Taipei (1985)
10. Jiang, Q.B.: Unit trade training and industrial and vocational education. In: Education and Culture, vol. 142, Ministry of Education, Taipei (1957)
11. Gu, B.Y.: A Dictionary of vocational education. Chinese Society of Industrial and Vocational Education, Taipei (1958)
12. Hsiao, H.C., Ho, S.C., Shih, S.C., Huang, L.T.: Consultation and Guidance for New Curriculum Development in Vocational High Schools. Department of Technological and Vocational Education, Ministry of Education, Taiwan (2003)
13. Hsiao, H.C., Chen, M.N., Yang, H.S.: Leadership of vocational high school principals in curriculum reform: A case study in Taiwan. International Journal of Educational Development 28, 669–686 (2008)

Development Analysis of Featured Industrial Base in Shijiazhuang Based on SWOT Analysis

Wei-li Shi

Department of Science and Research
Hebei University of Science and Technology
Shijiazhuang, China
swlsjz@hebust.edu.cn

Abstract. As a form of economic organization, featured industrial base is an effective vehicle for regional innovation. It also has played a crucial role on the local economic development. In order to grasp the development of featured industrial base in Shijiazhuang, and to enhance the competitive strenge of the base, the SWOT analysis is used to analyze the internal strengthes and weaknesses, and external opportunities and threats. Based on the analysis, the decision-making basis on formulating the development strategies will be provided to improve the efficiency of technological innovation in Shijiazhuang featured industrial base.

Keywords: Featured Industrial base, SWOT Analysis, Industrial chain, Competitive forces, Innovation, Strategy.

1 Introduction

Because of the rapidly rise of the private economy in our country, the specialized division of the work become more and more delicate. Economic development has completely getted rid of the age of individual development. The competitive and cooperation idea of "Symbiosis," "win-win" or even " multi-win" has become an important part of developing enterprise. Any organization, regional and national development must be built on the basis of competitive advantage, distinctive industrial base was born and grown up under this environment. It concentrated the talent, technology, capital and other advantages of resource, then established the featured industrial base relying on a number of distinctive industries, higher correlation of upstream and downstream industries, completely industrial chain, high level of techanical skills, and closely relationship between key enterprises of strong innovation ability with the technology-based SMEs.

Featured industrial base in Shijiazhuang was established after developing the high-tech industrial development zones, it was built by the government in order to accelerate the development of high-tech industries around the traditional industries and industries with competitive advantages in Shijiazhuang. After several years of rapid development, featured industrial base in Shijiazhuang has played an active role in guiding the development of high-tech industries, cultivating characteristic high-tech enterprises, developing the regional advantage of high-tech industries. Currently, the base covers the electronic information industry, materials, optical communication, textile and leather, biomedicine, electric power automation, advanced manufacturing equipment and other fields, forms the industry cluster with professional characteristics, which promote the aggregation of regional competitive industries. In order to take full advantage of the development of featured industrial base in Shijiazhuang, we use the method of SWOT analysis to analyze its internal strengths and weaknesses, external environment, opportunities and threats ,which could evolve the develop strategy of featured industrial base in Shijiazhuang.

2 Overview of SWOT Analysis

SWOT analysis method, also known as trend analysis or self-diagnosis, was proposed by professor Weihrich (H. Weihfich) from University of San Francisco in the United States at the early 80s in the 20th century, The four letters of SWOT analysis represent: strength, weakness, opportunity and threat. SWOT analysis is the study of internal strengths, weaknesses, opportunities and threats of the object, enumerate in the form of matrix through the surveys, systematic analysis of thought and analyze after matching the various factors. The conclusion of SWOT analysis usually plays a role with a decision-making[1]. Using this method, you can have a comprehensive, systematic and accurate study about the scenarios of the subjects. Appropriate development strategies, plans and countermeasures can come out from this result. The method of SWOT analysis is commonly used for strategic management, which could used to make the group development strategy and analysis the situation of competitors, the guiding idea is fully grasp the main competition's strengths and weaknesses, and external environment whose opportunities and threats to make the suitable future development strategy. It can help to make full use of the advantages, overcome the weaknesses, seize opportunities and meet challenges. S, W is the internal factors, O, T is the external factors. According to the complete concept of competitive strategy, strategy should be an organic combination of the business "can do" (ie, organizational strengths and weaknesses), and "may do" (it is opportunities and threats of the environment). The analysis of advantages and disadvantages is main focus on the comparison between competitor's strength and its main competitors. The analysis of opportunities and threats will focus on the changement of external environment and its possible impact on competitors. The organizations with different resources and capabilities face to the same changement of the external environment, its opportunities and threats may be totally different, which can integrate the four aspects, competition is the connection link between them[2].

3 SWOT Analysis about Development of Featured Industrial Base in Shijiazhuang

3.1 Basic Characteristics of Featured Industrial Base in Shijiazhuang

Till the end of 2009, the operating income of featured industrial base in Shijiazhuang above billion reached the number of 69, range from 1 to 4. 9 billion reached the number of 21, range from 5 to 9.9 billion reached the number of 10, above 10 billion reached the number of 2.The largest scale is Pingshan Metallurgy (operating income reached 15.8 billion), Jinzhou Textiles have the largest number of employees (60,185 peoples), Xinji leather is the largest export industry (Export sales 45.6 billion), Zhaoxian starch is the largest leading enterprises (Hebei Jingye Group), the largest number of enterprises is Luquan building materials (76). Overall featured industrial base in Shijiazhuang (2009) was shown in table1.

Table 1. Overall of featured industrial base in Shijiazhuang in 2009

Administrative county (Excluding municipal districts)	Featured industrial base total (a)	Total operating income(billion)	The proportion of the city's private economy (%)	The percentage of total city's GDP
17	69	1499	30%	12%

3.2 Using SWOT Analysis Method to Analyze the Develop Strategy of Featured Industrial Base in Shijiazhuang.

Above-scale featured industrial base in Shijiazhuang, there are Xinji Leather, Wuji Wood Line, Gaoyi Pottery, Zhengding Plates, Zhengding Food and Feed, Zhaoxian Starch, Jingxing Calcium and Magnesium, Luquan Cement, Pingshan Metallurgy,Etc.In all the bases, The first four industry of above is the key industry clusters, which made tremendous contributions of the economic development and technological innovation to Shijiazhuang. Now using the SWOT analysis to analyze and compare the feature industrial base in Shijiazhuang, the result of this survey is shown in Table2 [3].

1) Advantage Analysis
a) Advantages:
It has an unique geographical base, resources, popularity and industrial advantages in the feature industrial base in Shijiazhuang, its type includes traditional industries, government training industries, the emerging dominant industries,etc. There are historical formation of the traditional industries such as Xinji, Wuji leather industry, Zhaoxian starch industry, Jingxing Calcium and Magnesium industry, Luquan Cement industry; After years of cultivating the characteristics of the formation of competitive industries such as Zhengding, Gaocheng sheet metal furniture industry, Gao Yi Pottery industry, mining, Pingshan metallurgy industry.In recent years, the rising competitive industries such as Jinzhou textile and garment and decoration materials industry, Shenze soap and detergent industry, Lingshou, Pingshan stone industry, Luancheng pharmaceutical industry, Xinle carving industry, Zhengding, gaoyi textile printing and dyeing industry. The features and benefits of the industrial base in Shijiazhuang is unique and more prominent relative to other places. It is also the important manifestation of the core competitiveness of the economy in Shijiazhuang.

Table 2. SWOT analysis of the feature industrial base in Shijiazhuang

Strengthes	Opportunities
Unique geographical, resources, popularity and industrial advantages, cement, stone, ceramic, leather, starch, textile, sheet metal, pharmaceutical and other special industries covering most of the leading industries and traditional industries.	The tenth of two five-Years Plan of national economic and social development in Hebei province, national policy support, economic restructuring and regional economic development, development of surrounding areas, industrial structure adjustment and the market transformation in Beijing and Tianjin.
Weaknesses	threats
Ability of the government's macroeconomic management and control needs to be improved, small scale industry chain is short, the advantages of cluster is not obvious, dependence on resource, software and hardware construction is not support, lack of talent people.	Opening, investment environment and business environment, Establish and perfect the system of all levels and types of service system, modern logistics, the development of Beijing and Tianjin, other featured industrial base impact.

b) Geographical advantage and environmental benefits:

Shijiazhuang industrial base,who is at the central and southern North China Plain, north of Beijing and Tianjin,south to lead Zhengzhou and Central Plains hinterland, east adjacent to Bohai Sea and the North China Oil Field, West to depend Taihang Mountains,and it is the neighbor with the national coal and heavy chemical industry base in Shanxi Province. It is the goods distribution center of Shanxi and Hebei provinces in the history. The basic conditions are developed in capacity of road, rail, shipping and transport, while to attract foreign investment or maximize the business. Because of the unique geographical location and transportation advantages, it has the advantage of certain market radiation. The radiation of most industrial bases is for more than half of our country. In particular, Xinji leather base is depending to East Asia, Europe and the Americas.

2) Factors of disadvantage:

a) To be further improved of the government's macroeconomic management and regulatory capacity

Compared with municipalities and coastal areas,there is a certain "gap updating " in ideology and institution, there are still many problems in function transformation, management system and so on. It need to improve management tools and management methods.The guidance and support is necessary in development of industrial clusters.

b) Small-scale, shortened industry chain, some constraints in the development of the base

Compared with advanced provinces, the industrial base is small-scale and short-chain in Shijiazhuang, capacity of market-operating is limited, links' connection is not closely. The dominated large-scale industrial clusters are less who have a greater impact in the country. Many enterprises have similar products, specialization is not clear, the product correlation is low. There don't form the industrial chain that many enterprises arrange the work and ooperate with each other. Development of the bases was main in the traditional way, the modern circulation types such as Chain Operation take up smaller proportion.

Market system is not connected with the development of leading industries, the market level is not quite reasonable, function of markets is not perfect, service system around markets is imperfect, funds/talents and other markets are behind the market.All these factors are limitted the expand market in a certain extent .

c) A certain gap in human resources

In the personnel structure and personnel training, there is still a gap to develop "multi-billion dollar class industrial (industry) gathering area". It's a lack of talent with mastering modern knowledgeand practical experience in commerce management and trade circulation.The real thing is still relatively difficult for all kinds of talents in the economies of scale and accelerate the bases.

d) It will promote the development of industrial base in Shijiazhuang who is relying on the development of the surrounding areas in Shijiazhuang and the market transformation of industrial structure adjustment in Beijing and Tianjin.

In recent years, the pace of economic development under the jurisdiction of counties in Shijiazhuang and its surrounding areas is gradually accelerating who created favorable conditions for the development of industrial base scale and proposed new requirements for speeding up the industrial base in Shijiazhuang. Most important characteristics such as national drug city, the textile industrial base and agricultural base will promote the development to a new phase in special industries of Shijiazhuang. Relationing with Beijing and Tianjin, more development opportunities will come to the industrial base in Shijiazhuang whose industrial structure will be restructuring and market will be transformating from Beijing and Tianjin.

3) Opportunities:

a) Opportunities will come from implementing"The twelfth five-year-plan outline for national economic and social development of Hebei province"for the development of industrial base in Shijiazhuang.

In "The twelfth five-year-plan outline for national economic and social development of Hebei province",it submitted to the Economic Circle around Central Capital and to expand the economic uplift of the coastal zone,who will foster a number of industrial (industry) gathering area, zone, and large enterprise groups above one hundred billion. Under the guidance of the purpose, there will signify for development opportunities of industry base in Shijiazhuang: First of all, it will help to improve the investment environment; The second, it is conducive to promote the structural adjustment of business enterprise; The third, there will help to regulate the order of circulation;The fourth, it will help to speed up the pace from traditional commerce to modern business.

b) Promoting the development of industrial base by the strategic adjustment of economic structure and regional economic development.

With the strategic readjustment of economic structure,there will also have new developments and changes in regional economy; In particular, It is prominent of the comprehensive economic strength and industrial advantage in Shijiazhuang,while bringing significant opportunities for the development of industrial base with the strategic adjustment of economic structure. As a regional central city of Shijiazhuang in north China, it will get more opportunities in development of the base and even the whole economy [4].

4) Possible challenge:

a) the challenge of opening to the outside world

After implementing "The twelfth five-year-plan outline for national economic and social development of Hebei province", the pace of opening to the outside world is speeding up, who challenges the construction of the industry base in Shijiazhuang. In this case✚ Most enterprises of foreign commercial enterprises, especially the "aircraft carrier " enterprises will seize and divide up market sharing. On this basis, they will increase competition within the industry and lead to brain drain.

b) New demands from the development of modern logistics in industrial base.

Based on the rapid development and progress from modern logistics of transportation and information technology, that puts forward the new requirements to the industrial base in Shijiazhuang.The rapid development of other industrial base also impact the industry of Shijiazhuang. In the next period of time, Beijing and Tianjin as two economic centers around the Bohai Sea,it will continue to expand difference between Beijing and Tianjin and other areas by "polarization effect " from Beijing and Tianjin. Meanwhile, the development of industrial base of characteristic industry base surrounding areas of provinces and cities from North China have a certain challenge to the development of the industrial base in Shijiazhuang [5].

4 Conclusion

Through the analysis including the strengths, weaknesses, opportunities and challenges of Shijiazhuang industrial base, we have the following conclusions:

1) Strategy of advantages – opportunity. We must construct characteristic and industrial bases in from of the principle of high standard, deepen the reform of opening-up; We will actively carry out economic restructure and improve the proportion about services in economic growth; Enhance the competitiveness of industrial base, in order to provide financial and technical support for construction of the base.

2) Strategy of advantage - threat. We must accelerate institutional innovation, deepen the commercial system, establish large business groups, and speed up enterprises of small and medium businesses; Following the scale of development, growing the economic power,then to focus on building regional characteristics, and cultivating its own brand.

3) Strategy of weakness-opportunity. Effectively change the functions of government, create a good environment for the characteristics of industrial bases; Studying and formulating market rules, then to strengthen the market system, to distribute the reasonable commercial distribution, to increase investment in business infrastructure.

4) Strategy of disadvantage - threat. Enhance the comprehensive strength and competitiveness; Learn advanced experiences, and promote industry upgrading; Focus on the development of modern logistics industry; Accelerate the pace of technological innovation, Promote the buildings of circulating information; Pay attention to the training; Establish and improve the incentives of distributing companies.

The specific implementate strategies of four key industrial bases about Xinji leather, Wuji Wood Line, Gaoyi Pottery, Zhengding Plates include:

1) Speed up Xinji leather market reforming and pace improving , promote the development and prosperity of leather industry and relevant industries . Under the thought of laying equal stress on   development and improving, pursuing both intension and extension development, Speed up Xinji leather market reforming and pace improving, and consummate relevant preferential policy and measures, gradually, develop Xinji leather market to a domestic top grade and international well-known leather production& trade centre.

2) Depending on Zhengding Panel Market, develop the markets of panel, furniture and other building, decorating materials. For Zhengding panel market, we should enlarge the scale and region, improve the level, raise the profile. Meantime, Speed up the construction of decorating market and modern furniture market, to make a pattern of mutual supporting and mutual promotion.

3) Create Wuji special wood moulding market,with a new change, base on Hebei province, and face the domestic markets, construct a industry base of Wuji characteristics,and gradually, make it develop to the important supporter and relier of local economy.

4) Explore actively the development methods of ceramics market. Looking forward to the future, improve ceramics' creating ability,reduce promptly the gap between internal and international advanced technology. Improve comprehensively the development of ceramics industry, including core technology, material, process and so on.

References

1. Jin, Z.-M.: Strategic Management, pp. 11–18. Tsinghua University Press, Beijing (2004)
2. Wu, Y.-G., Deng, B.: Discussion on the Development Strategy for Characteristic Industries in Chongqing City Based on SWOT Analysis. SCI-Tech Information Development & Economy 20(24), 132–134 (2010)
3. Lu, X.: The Role of Featured industrial bases in local government. Productivity Research (17), 140–142 (2009)
4. Sun, C., Wang, Z.-B.: Analysis and Countermeasures in China's Equipment Manufacturing Industry Based on SWOT Analysis. Science and Technology Management Research (5), 163–165 (2007)
5. Gao, J.-S., Zheng, Y.-L.: Research on the Development Strategy of Hebei Information Service Industry Based on the SWOT. Journal of Information (6), 97–99 (2008)

Coordination Development Prospects of Rural Education and Community in China

Jing Tian[1], Ling Wang[1], and Zongling Zheng[2]

[1] College of Educational Science and Management, Yunnan Normal University,
Kunming 650500, China
[2] College of Economics and Management, Yunnan Normal University,
Kunming 650500, China
jingtian2003003@163.com, wlyn@263.net, zling75@yahoo.com.cn

Abstract. The city-oriented rural education in China now, both guiding ideology and teaching content, has not well adapted the development of national social economic, and can't meet the needs to build a harmonious society. Using the theory and methods of the coordinated development of systems science, combining with the policy guidance of China's economic social development, and education conditions, this paper proposes a diversification development idea of rural education to help the rural population achieve more development path, which is local advantages resources should be integrated into the rural education system. This study provides a coordination development picture of education and community; it is a useful supplement of rural education theory in the process of urban-rural integration.

Keywords: Rural education, community development, the integration of rural and urban education, educational diversity.

1 Introduction

Since reform and opening-up, China has experienced great changes dramatically. China's GDP has increased over 9.8% in average annual. National per capita income increased by nearly 50 times; and about 5 million people alleviated their poverty. China's human development has made great progress, but the imbalance of China's economic development, the human development' gap between regional and urban-rural is increasing. However, many poor communities are not covered in economic growth, especially in remote ethnic areas. Poverty is more severe in the mountain areas which natural conditions are relatively harsh. To internal development inequality between urban and rural areas, human capital endowment is an important reason, and education is determine the improvement of living conditions of the rural population. With China's new rural construction, the police from Rural Support City turn to City Financed Rural and the integrated urban and rural development. In this context, issues to be addressed of rural education and rural economic, social and cultural development have been undergoing a qualitative change in many key areas, the changing means that we have to find more effective idea of rural education development.

Y. Wang. (Ed.): Education Management, Education Theory & Education Application, AISC 109, pp. 615–621.
springerlink.com © Springer-Verlag Berlin Heidelberg 2011

This paper described the rural education problems of rural development currently in China. Then, using the theory and method of coordinated development, it demonstrate an association picture of the school and the social, economic, natural in rural development, based on rethink the theory and practice of local development, the paper attempt to present an new ideas of coordinated development of rural education and rural community development.

2   The Issue of Rural Education

Science and technology are productive forces. Education is the medium and bridges which convert the science and technology to the productivity. Thereby, Education establishes the legitimacy services for the economic and social. School education is the most critical and most important stage in human life. School education has changed its original rural communities, when education police of the country was implemented. Rural schools and communities link into the individual, family and community different relationships.

2.1   Rural Education and Economic Development

The relationship between education and the economy is combined closely by the people (workers), Education can improve the quality of workers, and the qualities of workers determine the level of labor productivity, thereby affecting economic growth. And economic development determines the necessary material basis and conditions of educational development. Practice has proved that level of economic development determines the scale and speed of education development [1].

Before "the two frees of charge and one allowance police" [1], most of the funding of compulsory education was undertook by local governments. The investment of rural education reflects the country's economic structure inequality. In the poor remote mountainous rural areas, financial investment in basic education can only guarantee teachers' salaries and it can't improve educational conditions. In this particular historical period, the villagers in addition to commitment to education surcharges, taxes, student tuition and fees, but also offer both personally work for the school, the obligation to offer materials for school.

In many poor mountainous areas, the average lands of villagers are only two acres. Economic model is mainly alpine dry potatoes, corn. There is little irrigation system to the main dependent on the weather. The cash income of villagers is Meager. In addition to potatoes and corn, the villagers are lack of circulation capital and goods changing with the outside world, and lack of capital accumulation conditions too. The "lag-type rural communities" [2] are fall into the modern economic development to catch up with demand for universal basic education resources.

Moreover, "the two frees of charge and one allowance police" is only for the nine-year compulsory education, many villages or town is only set to junior high school. If students want to enter high school, vocational school, college, they must leave their homes, and burdened tuition, beard travel expenses and living expenses of more than urban households. The investment in education reflects the structure of inequality between urban and rural areas. It is an education spending which villagers' rural income bear the urban economic level. A few of students can afford the cost of education, and have ability to enter college; they generally stay in the city when they graduated. It can be said in the pyramid of the education system, the funds of rural flow into the cities in the form of education investment. Through the path of higher education, rural best intellectual resources are out of rural areas too.

2.2 Rural Education and Community Development

Rural education is bond for rural development and social heritage. Its main function are socialization, social control, selection and distribution of various social roles, assimilation newer and maintenance sub-culture, and preparing for innovation and change of rural community. Along with the popularization of compulsory education, basic education is converging to various regions of China, The development of rural education should promote local development, but based on the local position, rural education showed a contrary side to rural community development. Local knowledge in rural include agricultural knowledge, folk, embroidery and other ethnic knowledge, knowledge of plant and animal classification, customs, rituals Etc. But knowledge and norms of rural school education is city-oriented. The knowledge is prepared mainly for the students to for further education, and emphasis on language, mathematics and other major subjects. From students into the school, they began to feel from another world values. Knowledge of school education neglect local knowledge and wisdom of rural life accumulated. That the rural school education of city-oriented made advantages and disadvantages of city world become students' goals and copies. And the advantages and disadvantages of rural world is actually become lower-level value of the disadvantages [3]. Powerful rendering of foreign culture is almost annihilation perspective of the indigenous cultural values, and highlights the students. Rural students look at cities with different mentality enterprises. As a result, rural students depart from "their world" consciously and unconsciously, they give up the value from the culturally and ideology which rooted in rural life. Once these students grow up, and they escape the country inevitable. Mostly illiterate and semi-literate, or the sick and elderly can not be accepted by the city, they stay in the rural. The major force of rural development has become less and less. This trend is impact rural communities and the development of China society.

2.3 School Education and Personal Development

Education plays a leading role in person's integrated development. Education is a means of tools to human's growth and development. And human growth and development is a goal for education development. Education is also a symbol of hope for the future education becomes a kind of symbol of imagination, to promote the students' hard work, but also driven by their family members to make contribution to support their studies.

Education enables people to move between urban and rural areas mobile, and made personal to develop. However, there is a narrow channel for students entering a higher school in China. Even if the students is full of expectations to their future, and the reality is that they can not further their studies, or even they are admitted to the higher school, students may give up due to their inability to pay school tuition. Students who can't further their studies come back to the village, which they learned the knowledge culture is different knowledge of everyday life in villages. Label into different levels of education and culture community to form interpersonal invisible divide once again to promote these young people go out to work outside the village to find a way out.

An important function of education is empowerment which has played a larger role to rural population flowing into cities. In fact, the process of basic education is empowerment process, the school passed language, mathematics and other basic knowledge which is necessary for daily life. However, in everyday life, if the lack of the environment of application knowledge, the knowledge may have deserted from the "empowerment" to "loss of power." In short, the schools transmit cultural knowledge is the basis of other knowledge, but also a basic condition to get benefits. However, schools exists some problem which neglect agriculture knowledge, and separate from rural real-life conditions. Education has become a consumer and strengthens the marginalized groups in poor areas vulnerable situation, and marginalized groups have become particularly poor, particularly weak.

3 The Coordination Development Relationship of Rural Education and Rural Social Economic

Systems theory is a scientific theory, the basic point is: "Any large system is complicated by the large number of subsystems. The coordination between subsystems and subsystems ensure a large system's presence of organic. According to systems theory, integration and coordination between systems is a symbiotic relationship. And only when the Internal, external relations of system is coordination and unity, the system can fully play the system's overall function, to ensure the smooth operation of the system. The coordinated development of rural education and rural communities is adaptation, integration and facilitate of rural education and rural political, economic and cultural systems of the relationship [4]. Build a rural education framework to meet requirement of the economic and social development need seek education coordination of external and internal coordination in essence.

Rural education is a subsystem of the rural socio-economic system, and which is a sub-subsystems of the national socio-economic system. National socio-economic system provides funds, managers, teachers and the corresponding educational requirements to rural education development. Rural socio-economic system provides sites, students and other supporting conditions for rural education, and makes requests to training rural people. Rural education provides basic education for rural children, and furnishes cultural and technical training for the villagers, but also supplies labors and personnel for national social and economic systems. (See Fig.1).

Rural education system is a large complex system, including the management system of rural education, personnel systems, the system of implementation, structural system, etc.. Various education subsystems are also sub-open, dynamic, and their interactions affect the coordination development of rural education. The relationship and the organic link between the subsystems in the rural education system are basic to achieve rural education system functionality. That the rural education system itself has the purpose, integrity, relevance, and dynamic balance and other characteristics determine coordination, multi-track and universal of the rural education.

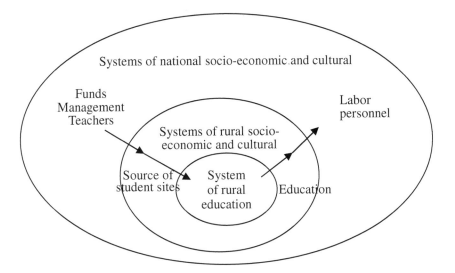

Fig. 1. The system relationship of rural education, economic social and national economic social

4   Strategy of Rural Education Development in the Urban-Rural Integrated

The seventeenth Congress of the Communist Party of China in 2007 first proposed the "urban-rural integration". It will strengthen the position of agriculture as the foundation of the national economy, take a path of agricultural modernization with Chinese characteristics, set up a permanent mechanism of industry promoting agriculture and urban areas helping rural areas, and form a new pattern that integrates economic and social development in urban and rural areas [5].

In 2008, Third Plenary Session of the Seventeenth Central Committee of the Communist Party of China described 'urban-rural integration', the view that "China has entered an important period which efforts to break into the urban-rural dual structure, and format a new pattern of integration of urban and rural economic and social development." [6]. Urban-rural integration is means that the resources, technology, capital, labor and other factors of production are flowed rationally, cooperated mutually, complement each other between urban and rural areas, to achieve a goal which resources are allocated optimally and economic and social are sustained and coordinated development in urban and rural areas. Its meaning in the following areas: "First, the integration of urban and rural space, which means two geographical of rural and urban areas entity into one regional complex of unified continuous, network-like, multi-node, and permeable. Second, complementary functional structure, the integration is not means to the disappearance of the center - external economy, only means that the two interdependent. Third, the share of urban and rural infrastructure, the mean is efficient transport and information networks to link urban and rural production and living closely. Four urban-rural populations, capital, technology, information and other factors are flowed highly smooth. Five coordinated urban and rural environment, urban and rural living standards are improved together. Urban-rural integration is not means the same of urban and rural areas, but changes the urban and rural opposition; rethink the idea of Emphasis on urban, rural depression, from urban, rural respective small cycles, small systems to large urban and rural unified cycle, large-scale systems, and establish urban and rural general idea, share the resources through playing the city radiating advantage and the association between urban and rural edge, and ultimately win and prosperity, achieve win ultimately and common prosperity.

An echo with the urban-rural integration is the integration of urban and rural education, 2010 through the "National education reform and development of long-term planning programs (2010-2020)" "built development mechanism of urban and rural educational integration" was wrote in document officially. The proposition "Integration of urban and rural education" was put forward, with profound policy implication and active policy-oriented. Integration of urban and rural education is through 'urban and rural development', through two-way communication, interaction of urban and rural education to achieve balanced development of urban and rural education, to narrow the gap between urban and rural education [7].

The integration of urban-rural education claims that rural education has to service to rural and urban, its target is mutuality symbiosis of rural and urban. The integration development claims also that rural has to selection an education mode according to itself social economy and culture which the whole benefit of rural education will be obtained furthest. The management and operation mechanism of education adaptive rural development should be structured and perfected to get sustained economic development and social progress. Therefore the rural education investment should be increased by states, and the structure of rural education is improved, vocation education, continuing education and all forms of education at all levels in rural become booming to give a well conditions for rural residents.

For the compulsory education, the integration of urban-rural education implements the unification and standardized in education content and process, however, special circumstances of rural education should be regarded, for example, the urban oriented education mode in rural education should be modified, and rural elements should be added into the education and teaching process, so that the students could study more knowledge about rural development, and the combination between the school education and rural culture could be highlighted, which rural children are more easy to accept the school education. The reason for doing this is that the education process is just a kind of choice and distribution process, in which the distribution has not only economic and social factors, but also has intellectual level, non-intellectual factors. Thus, the village school education should not only make a few students on the "entrance list", but also make other rural students having other way by training their labor skills and attitudinal. Although, Pareto defined the generalized elite as those achieved outstanding achievements in every field of human activity, for rural children, it is not only way to upper society by the "jump out country", they can also become an elite in the country. Therefore, the school education in the country should advocate the diversification of education to realize flow up of the rural population by infusing city culture and propagating rural subculture.

In process of urban-rural integration, the vocation education and continuing education in rural will be planed based on the needs of rural education and the national requirements to promote urban-rural and regional cooperation, to strengthen the overall coordination of all kinds of education resources and comprehensive utilization. The development of rural education is uneven in different areas and regional in China, it decides the vocational education and continuing education in China have no a unified mode, and should follow these principle "adjust measures to local conditions, classification guidance, the gradient promoting ". In short, the talents trained in country education is or isn't adapt to the needs of the construction of urban-rural integration, can or can't promote economic and social development, it is the final standard that rural education is success or failure.

5 Conclusions

The essence of urban and rural education integration is comply fair education, so that each people can enjoy equal national educational resources, and achieve the goal of balanced development of urban and rural education, but urban and rural education have different development way. However, there are many obstacles. To achieve this goal of urban and rural education integration, the education system must conduct in-depth and thorough reform; acknowledged the gap between urban and rural areas, complementary advantages; innovative educational management; perfect mode of investment in education; improve the quality of urban and rural teachers; strengthen inter-school counterpart support, so as to achieve the true integration of urban and rural education.

Acknowledgments. The research reported here is financially supported by the Humanity and Social Science Funds of Education Ministry of China (Grant: 10YJC880112). And part works is financially supported by the Science Research Foundation of Yunnan Education Department (Grant: 2010Z081).

References

1. Wang, F.: Education Planning Research, pp. 16–17. Southwest China Normal University Press, Chongqing (2007)
2. Lu, K., Zhu, Q.: Studies on Development Power of the Post-developed Rural Community in the New Rural Construction. Issues in Agricultural Economy (8), 46–49 (2008)
3. Liu, T.: Problems and Prospects of Rural Education, Reading, vol. (2), pp. 19–14 (2001)
4. Sun, L., Sun, F.: Studies on Concerted Relations between Economic and Social Development with Rural Education, p. 138. China Agriculture Press, Beijing (2007)
5. Hold High the Great Banner of Socialism with Chinese Characteristics and Strive for New Victories in Building a Moderately Prosperous Society in All Respects. Speech by Hu Jintao to the 17th National Congress (2007),
   http://www.en8848.com.cn/yingyu/28/n-64328-6.html
6. CPC Central Committee on A Decision about Major Issues of Rural Reform and Development, http://www.hbwy.com.cn
7. Shao, Z.: Concept Change and Institutional Innovation: From Balanced Urban and Rural Education to Integration of Urban and Rural Education. Fudan Education Forum 8(5), 14–19 (2010)

The Three Basic Working Attitudes That College Teachers Should Highly Emphasize

Shouhui Chen, Zheng Guo, and Yanjie Zhang

College of Textile, Zhongyuan University of Technology, Zhengzhou, Henan,
450007, PR China
Henan Key Laboratory of Functional Textiles Material, Zhengzhou, Henan,
450007, PR China
island0410@gmail.com

Abstract. Nowadays, college teachers, especially the teachers with little educational experience, should try to constraint themselves before they could teach students. The three basic working attitudes that the college teachers should highly emphasize are being steady, being seriously and being responsible. The main character of the modern young students is being blundering. How could they can changed by education? This is the duty of the college teachers should bear.

Keywords: College teachers, responsible, seriously, steady.

1 Introduction

Teachers are praised as the soul's engineer of human being [1]. The college teachers are the last tutors before the students leave for the society. For the college students, the campus is playing a role of the small society. In this small society, the effect of the attitudes, which the college teachers treat with people and things, on the college students would not be estimated [2, 3]. With years of working experience as a college teaching, I have realized that for the Chinese college teachers the three most needed and basis working attitudes are being steady, seriously and responsible.

2 Being Steady

Being steady means doing our own work cautiously and conscientiously, with more activity and less pomposity. The most common word that the college teachers use to comment the students is non-steady. However, what about ourselves? With all kinds of titles, achievements and rewards, it is not easy to measure which parts worth the value. Especially, in the modern society, the words such as "academic corruption" and "science paper corruption" are popular in the campus [4]. In this kind of campus atmosphere, how could the college teachers ask the students being steady?

There is a saying that the best education is exemplary teaching [5, 6]. Education is composed by the explanation of outlook on life and on the world and the exemplary teaching in the teaching practice. To the students, the exemplary teaching is much

Y. Wang (Ed.): Education Management, Education Theory & Education Application, AISC 109, pp. 623–625.
springerlink.com

more important and effective. If every college teacher could be steady when they are doing their teaching or researching work. Then the students could learn from them. In all kinds of competitions and activities, the student would pay more attention to what they are doing and what they could learn from that, rather than the prize itself. And then with the steady atmosphere in the campus, when the students graduate, they could grow and change from being above their business to being steady. The steady working attitude could be transited to the students if every college teacher could insist on being steady during their daily work. With the effect, when the college students enter into society, they would work steadily. If things go on like this, the modern young would not seek only for the quick benefits.

For the college teachers, being steady means a lot to themselves. Nowadays, the college teachers are busy applying all kinds of awards and prizes, however, they are suffering from how they can improve their research work to another level. Actually, since the college teachers have spent a plenty of time on the surface working, more and more energy has been distributed. Finally, the lack of energy has seriously restricted our achievements. If the college teachers could spend more time on their actual work, especially on research work, the working basis could be improved heavily. Therefore, in the future, the great achievements could be expected.

3   Being Seriously

The second working attitude is being seriously. For the college teachers, being seriously means preparing every lesson carefully and treating every student as your own child. The teachers should remember that the students are watching and learning from them every time and everywhere. I always try to tell the students that when doing anything or any task, whatever important or not, they should try their best. Otherwise, they should quit from the beginning. It is because that we could not know what kind of things would play an important role in the future, and we could not which kind of things would help us to solve the problems in the future. What the students could do now is to try their best to do what they had met or been told. When doing these kinds of things, they should take care of every detail. There is a saying that success depends on details. When the students are trying to earn some title in competitions, they should pay more attention to details than other things.

There are a lot of things, which seems like simple. However it is very difficult to finish them perfectly. Teaching students is a kind of thing like this. For the college teachers, there is ever-lasting when trying to teach students. When the college teachers could treat every thing seriously, they could learn more. Finally, they would find that there is nothing trouble but achievements.

There is another kind of understanding about being seriously, which tells that do not trying to refuse anyone readily when someone asks for help. It is possible that a chance of finding a new continent would slide away when we turn down others [7]. Either colleague or students, when we are trying to help them, new methods to solve the same problems could be found. At the same time, it is possible for us to find new questions. However there is a precondition that is we should treat every people and every thing seriously.

4 Being Responsible

For the college teachers, the third working attitude is being responsible. To a great extend, the teaching job could be done better or not, which depends on our conscience in education. Our responsibility could decide the way we behavior when treating with students [8]. Could we pay more attention to the students' future? Or we could just finish it quickly. We should try to tell if our honor is more important than transiting the useful and practical knowledge to the students. As college teachers, we should think all time what we could do to make their college more meaningful. We should try to supervise them to study foreign language harder, to learn how to search literature, to be familiar with the office software and to write an excellent literature review. These basic skills could benefit their whole life. There is no one to force us to do these kinds of work. The strong responsibility would drive us to do more and more for the students.

5 Summary

Being steady, seriously and responsible, is not only a kind of working attitude, but also a kind of habit, which the college teachers should have. If the college teachers could put this kind of habit into practice, the students and the teachers would benefit a lot.

References

1. Watson, B.: What is Education? The Inhibiting Effect of Three Agendas in Schooling. J. of Beliefs & Values-Studies In Religion & Education 30, 133–144 (2009)
2. Forrest, M., Keener, T., Harkins, M.J.: Understanding Narrative Relations in Teacher Education. Asia-Pacific Journal of Teacher Education 38, 87–101 (2010)
3. Erzikova, E.: University Teachers' Perceptions and Evaluations of Ethics Instruction in the Public Relations Curriculum. Public Relations Review 36, 316–318 (2010)
4. Sui, G.H.: Academic Corruption and Education in Teachers' Professional Ethics. J. of Chongqing University (Social Sciences Edition) 8, 26–31 (2002)
5. Carr, D.: Character in Teaching. British Journal of Educational Studies 55, 369–389 (2007)
6. Carr, D.: Revisiting the Liberal and Vocational Dimensions of University Education. British Journal of Educational Studies 57, 1–17 (2009)
7. Carr, D.: On The Contribution of Literature and the Arts to the Educational Cultivation of Moral Virtue, Feeling and Emotion. J. of Moral Education 34, 137–142 (2005)
8. Winch, C.: What Do Teachers Need to Know about Teaching? A Critical Examination of the Occupational Knowledge of Teachers. British Journal of Educational Studies 52, 180–185 (2004)

Application of Project Teaching Method in Higher Project Business Website Development

Songjie Gong

Zhejiang Business Technology Institute, Ningbo 315012, P.R. China
gsjei2011@163.com

Abstract. In teaching practice of business website development course, application of project teaching method combining theory teaching and practice teaching together closely, makes the best of creativity potentiality of students, improves the capability of students on analyzing and solving the practical problems. Introducing project teaching method, asking students to make a site plan with the value before studying during the teaching process, setting up a site designing project team under the teacher's guidance, learning while producing, to enable students put the knowledge they learned in use of the practical problems with value around directly, completing the designing production and publish. Teaching building on the site, the main use of the project method, according to website building process, the formation of project teams to carry out site planning, design, programming, etc. The project teaching method in business website development, divides the teaching content into a number of operational project, and the implementation of the project in which students operate mainly, training the students' vocational ability, in order to make full preparations for the students' employment.

Keywords: Higher vocational education, project teaching method, business website development.

1 Introduction

Apply knowledge in teaching courses in line with the principle of the use of project-based method, a typical practical project training content to web content for teaching, by asking questions, analyze problems and solve the problem, knowledge of the various parts of the series from the point by simple to complex layers of progressive, practical skills of students and application capabilities [1,2,3].

Project-based teaching is a professional activity-oriented, practical project as a carrier, the actual work will be teaching a combination of content and teaching organization and implementation of the project is consistent [4,5], so that students truly feel real practical needs, fully stimulate student motivation for learning, outstanding vocational skills training for students as a teaching method for learning a variety of practical and strong operational knowledge and skills. From the management perspective, the project is that as a management object, according to the limited time, budget and quality standards to complete the one-time task.

2 Project Teaching Method

Project-oriented teaching methods, teachers and students through the implementation of integrated project co-workers carried out teaching activities [6,7,8]. This project refers to the production of a specific, practical value of the product for the purpose of the task. The project reaching method is shown as figure 1.

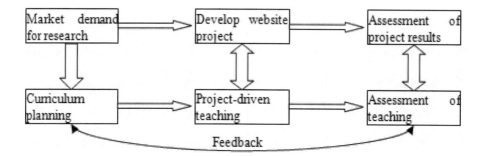

Fig. 1. Project teaching method

3 Implement of Business Website Development Course Based on Project Teaching Method

This stage is the practice of teaching, and the teaching purpose is to set up a project team to complete a comprehensive website design and development, learning, consolidation of technologies required for website building and training to improve their professional skills.

In the construction site before the project, teachers needed for the building according to the website first technology division of roles, and then according to the students for the technical characteristics and interests form the project team. When the project team set up to consider the ability of the members of the group balance problems, such as each team must have a strong organizational skills of students as a project manager, students with a strong programming skills as a programmer, female students divided equally to each group as a web designer.

3.1 The Project Programs

Project team members develop project implementation plans that the total project schedule, the schedule listed in the requests for a draft plan, art design, program design, process integration, system testing site specific completion time and the division of labor members of the group table.

Team members according to their role within the group as the knowledge needed to develop self-study program, and teachers by the project manager responsible for supervision and inspection of the members of the learning progress and learning.

3.2 Website Planning

According to members of the group's interests, discuss the group to build the site's theme. After determining the theme of the site, members of related topics through the Internet access the website, specifically about the content of these functions and structure. Collect relevant information, such information is stored in the project manager in the group of teachers to create machine folder. Project manager of the team members to compare the data collected, analyzed, selected from the basic to meet the requirements of the data and classify storage. The information collected to complete work, the preparation of a simple web site plan, and the planning document requirements include the following: the goal of building sites, site work analysis and recommendations, the overall site structure, the main part and site management features description, website building progress and implementation process.

3.3 Website Design

Web site design is including site design and the physical structure of the logical structure of web design. The physical structure of the website design requirements:

1) Do not store all the files in the root directory;
2) Click "Site Plan" sections set to design the site's directory structure;
3) Do not use directory names Chinese name, Each directory is an independent images subdirectory;
4) Execute the executable file and can not be separated files, the site's database file separate place;
5) Directory level not too much, not more than three layers.

Logical structure of the web site's link structure is designed based on website content richness. There are three basic ways: linear structure, hierarchy, the network structure.

3.4 Web Interface Design

According to the site plan site planning the overall structure of the book on the web page, two columns, feature pages and content page layout, and then use graphic design software interface design, we recommend using Photoshop CS software, web design images into PSD format.

3.5 Website System Commissioning

Website development system function modules to be good after the website and graphic design renderings to integrate, and then each module integrated into a system. In the integration process in general, there will be many uncoordinated issues, such as after the function can not be achieved, web link errors, the page image is stretch, ultra-wide web forms, etc. This will require constant testing to find the problem, modify the code, until the realization of the final site functionality and the implementation of the results page.

4 Conclusions

In teaching practice of business website development course, application of project teaching method combining theory teaching and practice teaching together closely, makes the best of creativity potentiality of students, improves the capability of students on analyzing and solving the practical problems. Introducing project teaching method, asking students to make a site plan with the value before studying during the teaching process, setting up a site designing project team under the teacher's guidance, learning while producing, to enable students put the knowledge they learned in use of the practical problems with value around directly, completing the designing production and publish. Teaching building on the site, the main use of the project method, according to website building process, the formation of project teams to carry out site planning, design, programming, etc. The project teaching method in business website development, divides the teaching content into a number of operational project, and the implementation of the project in which students operate mainly, training the students' vocational ability, in order to make full preparations for the students' employment.

References

1. Yu, M., Jiang, Y.: The application of project-driven method in Information System Department professional practical teaching. In: ICEIT 2010, vol. 2, pp. V2395–V2398 (2010)
2. Yu, J.: Innovation teaching method researching on embedded system course based on robot project-driven idea. In: EDT 2010, vol. 2, pp. 249–252 (2010)
3. Xu, H., Zhang, L., Zhang, Q., Yu, B.: Exploring research of "Project-driven" teaching methods of electronic technology courses. In: EDT 2010, vol. 2, pp. 273–276 (2010)
4. Pan, H., Wang, A., Yan, J.: Projected teaching method for XML technology course design. In: ETCS 2010, vol. 1, p 390-393, 2010.
5. Zhang, Y., Wang, X.: Practice of project teaching method in computer course. In: ICCSE 2010, pp. 855–858 (2010)
6. Ou, J., Huang, Z.-F., You, J.-S., Dong, Y.-N.: A project-oriented teaching method in digital learning environments. In: ICACTE 2009, vol. 1, pp. 311–317 (2009)
7. Xin, M., Qin, Z., Wang, L.: Research on project-driven teaching method innovation on post-graduate professional courses. In: IFCSTA 2009, vol. 3, pp. 127–130 (2009)
8. Wang, A., Li, L.: The research of project teaching method in technique and application of database teaching. In: EDT 2010, vol. 2, pp. 506–508 (2010)

Exploration and Practice on Project Curriculum of Business Website Development

Songjie Gong

Zhejiang Business Technology Institute, Ningbo 315012, P.R. China
gsjei2011@163.com

Abstract. In vocational education, project-based curriculum reform climax has come. Although in the process of curriculum reform, there are many sub-squeaks, and there are many difficulties, but most of the vocational educators recognize that the traditional mode of teaching in the discipline system of vocational education on the road to nowhere. The project of curriculum reform is the only way to improve the quality of vocational education and project curriculum is currently the major reform trend of higher education courses. The paper analyses the current process of teaching and existing problems, points out using project curriculum as carrier. Then it puts forward to the curriculum development ideas from these aspects of project features, implementation procedures, work process and the situation of the teaching of design.

Keywords: Higher vocational education, project curriculum, business website development.

1 Project Curriculum

1.1 Definition

Vocational education program is defined as task-centered selection organizes course content and to complete the task as the main approach to learning curriculum model [1,2,3]. The aim is to enhance the curriculum and the correlation between the work to integrate theory and practice, to improve the efficiency of students' professional ability [4,5]. Project curriculum is designed to change the traditional curriculum model; students create a professional capacity to vocational courses.

1.2 Character

Project course is characterized by:

(1) By a professional capacity to formulate curriculum objectives and focus on what students can do, not know;
(2) To task for teaching content and focus is to teach students how to complete the tasks, knowledge, skills, learning to carry out the process with task completion;
(3) Focus on the tasks of learning needs, the typical design of a product or service as a carrier "learning project", the organization of teaching.

Y. Wang. (Ed.): Education Management, Education Theory & Education Application, AISC 109, pp. 631–634.
springerlink.com

Teaching in order to expand according to the project schedule, each project was to learn the tasks can overlap and duplication, that is cross-learning project design tasks, as long as service in the task of learning on the line, not necessarily tied to a logical sequence of tasks. In the design project course, we should carefully understand the three basic characteristics.

1.3 Project Scheduling

In accordance with principles of project courses, students must learn the appropriate tasks to complete, and professional ability to achieve certain requirements need to select a typical product or service as a carrier organization of teaching. Project scheduling has the following basic pattern.

(1) Progressive, that is according to level of difficulty from low to high order items;
(2) Parallel, that is, the complexity of the project, there is neither the difference between, there was no obvious correlation, but arranged according to a parallel relationship;
(3) Process-type, that is, according to the work process in order of priority projects is completed, these projects together is a bigger project.
The project scheduling is shown as figure 1.

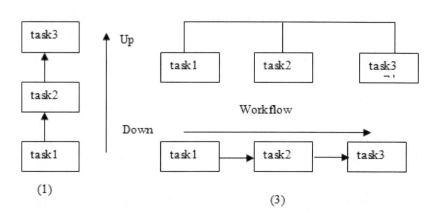

Fig. 1. Project scheduling

2   Practice of Business Website Development

Business website development is the professional website development computer applications vocational students in the direction of post-employment in one of the main jobs. This course is to train students to master the functional business website needs analysis, database design, page layout, the background to achieve the ability to deploy and distribute. These are the most important job site development and basic capabilities, so this course in computer application in a very important position, should be as a professional core courses and required courses [6,7,8,9].

This course is based on practical capacity-building, the curriculum content of the selection criteria made fundamental reforms to break the transfer of knowledge as the main feature of the traditional curriculum model, into a task-centered curriculum and teaching organization for students in specific projects to complete the process to build a theory of knowledge, and develop professional competence.

After industry experts in-depth, detailed, systematic analysis of the final course identified the following tasks: Business web site needs analysis, database design, page development, website testing, configuration and deployment site. The study is based on business website development projects for clues to the design. Highlight the course of professional competence training for students, closely around the theoretical knowledge of the selection task needs to be done, while fully taking into account higher vocational education of theoretical knowledge and learning needs, and integration of the relevant professional qualifications of knowledge, skills and attitude requirements.

The idea of business website development is shown as figure 2.

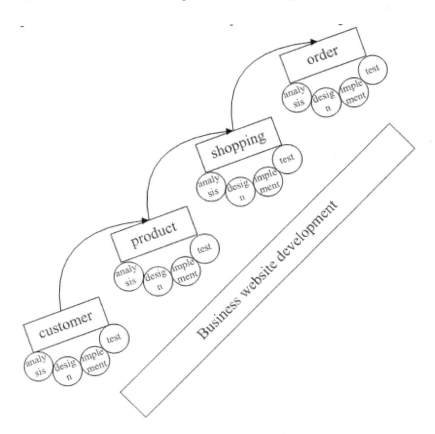

Fig. 2. The idea of business website development

3 Conclusions

In vocational education, project-based curriculum reform climax has come. Although in the process of curriculum reform, there are many sub-squeaks, and there are many difficulties, but most of the vocational educators recognize that the traditional mode of teaching in the discipline system of vocational education on the road to nowhere. The project of curriculum reform is the only way to improve the quality of vocational education and project curriculum is currently the major reform trend of higher education courses. The paper analyses the current process of teaching and existing problems, points out using project curriculum as carrier. Then it puts forward to the curriculum development ideas from these aspects of project features, implementation procedures, work process and the situation of the teaching of design.

References

1. Scribner, S.: Knowledge at Work. In: McCormick, R., Paechter, C. (eds.) Knowledge and Learning. Paul Chapman Publishing Ltd, Great Britain (1999)
2. Knoll, M.: The Project Method:Its Vocational Education Origin and International Development. Journal of Industrial Teacher Education 34(3) (1997)
3. Pepin, Y.: Practical knowledge and school knowledge:a constructivist representation of education. In: Larochelle, M., et al. (eds.) Constructivism and Education, p. 178. Cambridge University Press,
4. Willliams, P.J.: Design: the Only Methodology of Technology? Journal of Technology Education 11(2) (2000)
5. Greinert, W.-D.: European Vocational Training Systems:the Theoretical Context of Historical Development. In: Cedefop (ed.) Towards a History of Vocational Education and Training(VET) in Europe in a Comparative Perspective, Office for Official Publications of the European Communities, Luxembourg (2004)
6. Chappell, C.: New knowledge and the construction of vocational education and training practitioners. In: Proceedings of the International Conference Working Knowledge, Productive Learning at Work, Sydney, Australia, December 10-13 (2000)
7. Reetz, L.: Handlung, Wissen und Kompetenz als strukturbildende Merkmal von Lemfeldern. In: Bader, R. Sloane, P.F.E. (ed.) Lernen in Lernfeld, pp. S141–150. Eusl-Verlag, Markt Schwaben (2000)
8. Billett, S.: Learning in the workplace: strategies for effective practice, p. 34. CMO Image Printing Enterprise, Singapore (2001)
9. Rauner, F.: Berufliehe Kompetenzentwicklung-vom Novizen zum Experten. In: Dehnbostel, P., Elsholz, J., Meister, J. (eds.) J Alternative Positionen sigma, p. 117. Vemetzte Kompetenzentwiehklung: zur Weiterbildung, Berlin (2002)

On the Training Mode of Professional Printing Talents Based on the Social Demand and Employment-Orientation

Hong He, Haiyan Wang, Yulong Yu, and Jieyue Yu

Hangzhou Dianzi University, Hangzhou 310018
{hehong,why73,yylong}@hdu.edu.cn, 61982218@163.com

Abstract. Regarding the market demand and employment-orientation as the breakthrough point, we analyzed the current demand of printing industry as well as the present situation of personnel training, and put forward the standpoint that the personal training demand should be adapted to social needs in order to make the professional training conform to the market development and actual requirements of personnel training, thus training the compound talents with practical ability, employability, entrepreneurship, innovative capability.

Keywords: Printing, market demand, talents training mode.

1 Introduction

With the rapid development of information scientific technologies and economy, informationization, datumization, networking, and digitization have become a developmental trend of current printing industry, thus the standards of printing professionals must be correspondingly improved [1]. To exploit markets for printing professionals as well as develop their talents to their "factory", what we should do is to regard the market demand and employment-orientation as the breakthrough point, and then conduct the in-depth innovation on training mode of the printing professionals. Only in this way, the positive circulation----"Colleges and universities - Students - Business - Colleges and universities "can be formed, the talents will be accepted by this society, the enterprise can flourish, and colleges and universities can grow with vigor. However, numerous printing colleges and universities cannot rapidly and efficiently adapt the abrupt change in current society as well as the market, which may lead to a fact that graduates cannot meet the needs of both the community and enterprises. Although such a situation may be caused by so many factors, it is inseparable with the unbalance between current training mode of printing professionals in universities and the actual needs of society.

2 Demand Analysis on Technical Personnel of the Printing Industry

During the past 20 years, with the rapid economic development, the large increase in both printing corporate scale and the number of employees has led to a surge

Y. Wang (Ed.): Education Management, Education Theory & Education Application, AISC 109, pp. 635–639.

demand on a variety of technical personnel and management personnel. Besides, the wide application in modern information technology, CTP computer plate-making technology, and JDF technology, etc, promotes a shift of the traditional printing industry to media processing services. However, nowadays the technology, personnel, equipment of many printing companies still cannot satisfy the requirement of products' update. Moreover, these companies own enough funds to purchase advanced equipment, but the key problem lies in the lack of comprehensive and skilled personnel integrated the technique of operation in pre-printing, printing, after-printing process with maintenance.

Currently, most professional personnel in printing industry can be classified into 2 types: the first type is talents graduating from colleges and majored in the printing; the second type is common mechanic who learned the technique through vocational, technical school or two-year short-term training courses, so they operate the general press, however, due to the lack of theoretical foundation, they cannot operate the complexly advanced printing equipment. Actually, what those companies really need are highly skilled personnel with both theoretical knowledge and practical experience.

According to recruitment survey, the existing professional personnel in printing companies include three types, first is the talents graduating from colleges and major in the printing, and they are capable of basic theoretical knowledge of printing, but lack of practical experience; the second is the students graduating from ordinary technical schools, and they have some practical experience and skills, but lack of theoretical knowledge which limits their development in the companies. Moreover, to pursue profit, companies are not willing to spend time and money developing them. The third type is common workers who are less educated, so they can only engage in general laborers, and their limited knowledge structure restricts them to accepting and understanding advanced technologies, thus it is impossible for them to solve the technical problems.

In order to meet the requirements of the printing industry and to train large number of qualified personnel, we should make full use of the advanced teaching conditions to train some "knowledge-based and skills-based" talents who not only have the practical ability to participate in the actual operation, but also the strong professional knowledge, that is, they can integrate the practical experience with theoretical knowledge, thus welcomed and recognized by the companies [3]. Therefore, through the research on personnel training mode majoring in printing, universities' advantage of self-discipline, professional characteristics as well as the opportunities and situations of the rapid development of current printing industry can be used to explore a new and unique way for printing professionals' training. We should spare no effort to adapt the national economic restructuring, job market demand and the implementation of the "eight-eight strategy", to meet the requirement of national economic and social development as well as the requirement of journalism, publishing, printing companies for talents, for the purpose of training the technical personnel capable of technology and management. Local economic development requires and emphasizes the talents with solid basis, practical experience and high quality, which has very important theoretical and practical significance for speeding up the construction and teaching reform of printing engineering, as well as the cultivation of high-quality printing professionals.

3 The Current Situation of the Printing Talents Training

3.1 Inadequate Qualified Teachers as Well as Sufficient Faculties in Numerous Printing Institutions

In recent years, the rapid expansion of college enrollment only expands the scale of college and universities rather than enhance the quality. The key factor lies in the inadequate teaching staff, which directly leads to the decline in teaching quality [2].

3.2 Old Fashioned Teaching Plan, Unreasonable Supporting Materials, and Insufficient Innovation in Teaching Methods

With the development of technology, printing equipment is high-tech product integrated computers, electronics, automation, mechanical, optical and other disciplines in one. Thus, it is necessary for students to explore these high-new technologies such as CTP Technology, electronic shaft driving, digital printing, digital workflow, etc. However, nowadays many colleges and universities only emphasizes theoretical knowledge and analysis abilities rather than the practical ability; meanwhile, teaching materials are too old-fashioned which cannot reflect the social development and technology updates; for classroom teaching, in most cases universities in our country adopt the infusion method but lack of classroom communication and interaction. Moreover, printing requires strong practical ability, thus the simple theoretical teaching cannot stimulate students' initiative and creativity, resulting in the situation of lack of self-learning ability as well as practical ability.

3.3 The Mismatching of Theory and Practice Caused by the Unscientific Ratio of Theoretical and Practical Lessons

As an applied processing industry, printing particular stresses on experience and practice, and its manufacturing procedures such as film developing, computer-to-plate, press operation and maintenance, inspection of presswork quality, and printed process all need excellent practical ability. In addition, many professional courses are highly independent and not closely jointed. At present, the practical teaching hours are set fewer and unreasonable in many universities, ignoring the intrinsic link of all the courses and staying in the stage that practice for the sake of practice. This kind of arrangement is obviously unreasonable, for it cannot develop compound talents integrated theory with practice.

4 Regulation Strategies of Printing Professional Training

4.1 Printing Teaching Reform Starting from Social Needs and Employment Orientation

At present, China's education reform of printing should focus on the improvement of teaching idea, teaching content, and teaching methods. We should learn from foreign experience, emphasize the common education of school and enterprises, and give permissions to enterprises to participate in designing teaching programs and professional courses, while training personnel according to the pulse of the current industrial and technological development. As a result, to reduce the disconnection of the future courses or professional technology with actual production, and to better transform theoretical knowledge into practice.

Meanwhile, to speed up the reform of teaching methods and the gradual transition from the traditional "indoctrination" teaching transition to "production training projects" as the main body and professional program of "Combination of study and work" as the main line. In addition, the practice conditions of school can be used to carry out teaching on the spot and build creative classes. For example, "printing quality inspection and control" can be taught directly at the printing quality department, allowing students more time and exposure to actual production and digestion of knowledge, which improves their practical ability and motivation of studying and develops their areas of interest.

4.2 To Construct a "Compound" Talent as the Personnel Training Objective, and Develop a More Comprehensive Teaching Plan

The current printing industry has the characteristics of digitalization, networking, and informatization, and it is inter-crossed with other disciplines. Considering from its characteristics and integrated application in the aspects of high technology and equipment, we should determine the property and professional direction of printing subjects according to the core of the printing industry and technology, while using the surrounding area as an auxiliary discipline to develop a more comprehensive education plan.

4.2.1 Lay Solid Foundation, and Highlight the Basic Ability

Printing major is an integrated applied discipline based on light, electricity, machinery, automatic control, chemistry, and physics. Therefore, not only do we pay attention to basic education courses, but also add related content of basic knowledge and application in arrangements of professional courses, for lying a foundation for students to have a wide areas of employment.

4.2.2 Optimize Curriculum, and Improve Teaching Efficiency

If we take into consideration all aspects to set curriculum, then it will lead to an excessive repetition of teaching content and furthermore a sense of fatigue among students. Repeated courses also aggravate the conflicts of arranging the theory and practice teaching, resulting in insufficient hours for students to acquire the necessary knowledge and skills, and therefore reducing the effectiveness of teaching. Meanwhile, some overlapping and inter-related course contents are taught by different teachers. This will lead to lack of effective association of various teaching contents, so it is worth merging relevant courses [4].

4.3 Strengthen the Cooperation between Universities and Enterprise for the Establishment of Training Base

At present, the cooperation between printing universities and enterprises to establish internship training base not only can raise awareness and influence of the enterprises by virtue of universities, but also can solve the problem that universities lack of internship base. However, in light of current trends, the majority of training base is a mere formality [5]. If we put the training practice into effect, students can accept vocational training in both schools and enterprises, thus they may also have the opportunity to put the theoretical knowledge learned in school into practice. For the problems existing in the learning process, practice can help them digest, absorb and consolidate. Only in this way, can students master a strong practical ability and realize the truly organic integration between theory and practice. For example, we can arrange vocational internship for students, graduation internship and graduation design can also be carried out at the base established by both schools and enterprises, while enterprises should also be allowed to participate in the practical training, job arrangements and workload formulation to strengthen their cooperation in talents training, thus greatly integrating teaching, research, production. It is an inevitable measure to train "high-level, refined, and professional" talents for the engineering colleges in the new century.

5 Conclusions

Talents training should advance with the times and actively adapt to market changes. However, blind conformism for the market is inadvisable, that is, we must get rid of the restriction of negative factors of the market. The current printing colleges and universities have cultivated so many talents, which are worthy of learning [6]. Nowadays, the market has required the printing talents to improve their own quality, thus as the main force of talents training, institutions specialized in printing should cover the existing shortage in education to successfully achieve this complex project of talent training. Only in this way, can students master some practical knowledge, including both the solid theoretical knowledge that is conductive for their continuous study as well as the practical knowledge that can help them meet the requirements of the society.

Acknowledgment. Fund project: Funded by the Education Department in Zhejiang province(Y201120623), Natural Science Foundation in Zhejiang province(Y1100771), Educational Science Planning in Zhejiang province in 2011(SCG96).

References

1. Fan, S., Yang, S.: On the adjustment of knowledge structure of printing professionals. Journal of Zhuzhou Institute of Technology 06, 10–13 (2006)
2. Jing, C.: Diversity of talents training mode in college and universities: interpretation and response, pp. 115–117. Technology University Press, BeiJing (2001)
3. Cai, D.: Improve the teaching quality and enhance the teaching management. J. China's Higher Education 07, 32–34 (2006)
4. Yu, J.: Innovation of teaching experiment in the printing equipment and printing technology curriculum. Print World 15, 25–27 (2009)
5. Xue, M., Wang, A.: On the industry demand-oriented graduation internship conducted by schools and enterprises. China Education Development 15, 32–33 (2009)
6. Zeng, K., Hong, X.: The establishment of the new training mode of compound talents majored in packaging and printing. Journal of Hunan University of Technology 21, 86–88 (2007)

Think in Higher Education Administration Based on Quality Project

Hou Xianjun and Peng Wuliang

Institute for Higher Education, Shenyang Ligong University,
110159 Shenyang, China
Houxj1970@163.com

Abstract. Educational administration is of great importance in university management. This work studies the quality project used in higher education administration, and classifies the main stakeholders into two roles in experience: one is to supervise the teaching system, and the other is to promote the basic construction of teaching. In order to carry out the construction of quality project and gain more achievements and advance the college, administrators should seize the opportunity and combine the infrastructure and long-term goals with the implementation of quality project.

Keywords: Qulity project, higher eduction adminstration, education and teaching.

1   Introduction

Modern teaching management is to build high performance management organization for the preconditions [1,2]. Educational management is the organization of university teaching and management center [3]. Fully implement China's educational policy to ensure that the teaching of the normal direction to achieve the aims of education and training of qualified personnel are dependent on the scientific, academic and careful management.

At present, with rapid social and economic development, university education have been expanding, to further increase the number of students, schools have been expanding, to the Educational Administration raised new challenges and higher demand. University management, as an important part of Educational Administration, should maintain the normal order of teaching the school's hub [4]. The level of educational administration, a reflection of the level of school management is related to a university in the current competitive society in the survival, development and status.

2   The Current New Situation of Educational Administration Work

To fully implement the concept of scientific development and put the importance of higher education on teaching quality, Ministry of Education and Ministry of Finance of china jointly issued two important documents in January, 2007, which are "The

Y. Wang (Ed.): Education Management, Education Theory & Education Application, AISC 109, pp. 641–647.

opinion on the implementation of quality of undergraduate teaching and teaching reform project by Ministry of Education and Ministry of Finance" and "Ministry of Education on Further Deepening the Reform of Undergraduate Education to improve teaching quality and number of observations". Government started the relevant meets and decided to implement the "quality of undergraduate teaching and teaching reform project".

Implementation of quality engineering has very important significance. In recent years, rapidly increasing scale of higher education, quality has been greatly improved for China's rapid economic and social, health and sustainable development and the reform and development of higher education itself has made a great contribution. However, the quality of higher education can not fully meet the needs of economic and social development. Therefore, the quality of implementation of the project, to further deepen the reform of higher education teaching, improve the capacity and level of training to better meet the economic and social development of high-quality innovative talents, has important historical and practical significance. Meanwhile, the implementation of quality engineering is to uphold the scientific concept of development, full implementation of the State Council to deploy an important strategic decisions and initiatives to implement the strategy and this strategy through science and education an important part.

Since 1994, China began a pilot assessment of teaching universities and colleges, the Ministry of Education in 2003 formally established the periodic teaching evaluation system. Now more than 500 have accepted the assessment. Moreover, the provincial education department responsible for carrying out the vocational training institutions to assess the level has also been assessed 587, coming to an end the first round of assessment. Quality of implementation of the project, whether it is the Ministry of Education has accepted the majority of undergraduate teaching level evaluation of schools, or about to meet the assessment of schools, is a development opportunity and challenge. In carrying out the quality of engineering practice, the school if the "Quality" Construction and assessment of rectification, the results consolidate the combined efforts to promote the deepening of the reform of education, will greatly promote the overall educational level of school improvement and quality of the training improve and promote the continued strengthening of the teaching infrastructure for schools to further deepen reform, strengthen the construction, improve quality, and create favorable conditions, to lay a solid foundation.

3   The Current New Situation of Educational Administration Work

The implementation of the quality project is a system and a large project is to deepen the educational reform is an important part of covering the major part of undergraduate teaching. How to implement "quality engineering" as the starting point, to deepen the educational reform, promote the teaching of school infrastructure, quality of success of the project is to implement one of the key factors.

Smooth implementation of quality engineering, involving the school's many teaching units and functions. The academic management of university teaching is to organize and manage the center. Educational Administration Department on the one hand responsible for teaching the basic order and stability, security and orderly operation of the normal teaching of important functions. Also shoulder the training, education, teaching reform, teaching responsibilities and other important infrastructure, such as the training model of reform, professional development, curriculum development, teaching practice, materials selection and construction, teaching quality monitoring and many other content and links. If the educational administration department can seize the opportunity, through the implementation of quality engineering "wind." Implementation of the project and the quality of school infrastructure and long-term development goals closer together, will be able to promote the school to get better and faster development. Therefore, the university educational administration departments should attach great importance to the quality of project implementation; the implementation of quality engineering is the development of the schools as new opportunities and challenges. The work should be executed according to the following principles:

(1) Scientific decision-making and co-ordination arrangements. Decision theory to action decisions made before the action is called decision-making. Decision is the choice of action, behavior is the implementation of the decision-making. Correct behavior from the correct decision. Educational Administration departments should do a solid job of quality engineering, implementation of the preparatory work;

As an educational administration department should first carefully study and convey the spirit of the document the Ministry of Education and related work background in the academic management team and all the teaching units to do collective mobilization. School educational administration departments should be combined with their own reality, from the training model to promote reform, and strengthen professional development, promoting the teaching materials and promote curriculum development, teacher selection and teaching team teaching, strengthen the practice teaching, etc., from school level to the full implementation of the Quality scientific, rational planning and decision making, develop the quality of implementation of the project schools, advice and implementation.

After completion of the whole argument, in order to ensure quality construction and the depth and comprehensive implementation of the smooth realization of the goal of building the school of educational administration department should establish the appropriate quality of implementation of the project leadership and organization structure, a clear focal point of the working groups, divide the project work group of duties and tasks specifically related to the teaching units and functions as a charge related to the implementation of quality projects and contacts, the formation of the full implementation of quality engineering organization and support system. Meanwhile, the department of educational administration should be in school year budget, the establishment of a Quality engineering construction funds, the establishment of "quality construction projects funds' capital accounts, as the quality of the construction start-up capital, so as to effectively guarantee the quality of the project successfully carried out.

Following demonstration and deployment in the whole school completion, educational administration departments should further the teaching units, each unit with the party leadership, teaching and research director, teacher, student representative to analyze the strengths and weaknesses that currently exist, tap potential, clear goals, from a professional development, curriculum development, practice teaching, teachers teaching, team teaching and other aspects of in-depth analysis and research, combined with the overall quality of the school project implementation arrangements, the full proof, all from the reality, to develop the teaching quality of construction of the project implementation unit planning, and implementation of specific projects and construction methods. Implementation of quality scientific and engineering of the specific program must be in line with objective laws. In developing the implementation of quality engineering program, plans to gradually organized and implemented in accordance with the quality of engineering and construction projects related to project work.

(b) Find out the rules and accelerate the development. Quality engineering projects mainly in the form of project in construction, so the project should be targeted for management by objectives. Quality of project implementation in order to accomplish something, there should be planning objectives. Quality starts on the basis of academic management should be phased, step by step step by step implementation of quality projects. Quality of the construction projects will be a ladder-like development, such as to become a national construction projects, construction projects must be at the provincial level; to become a provincial construction projects, construction projects must be a field grade; to become a colonel-level construction projects must is a hospital-level construction projects, and so on. Therefore, the quality of implementation of the project must be down to earth, everything from reality, from colonel to carry out the quality of the construction project selection staff. Find out the laws, to accelerate development.

Ministry of Education, Ministry of Finance officially launched the quality of works, many colleges and universities have been carried out with university level courses, university level textbooks and other family planning projects related to selection. Moreover, some provinces and cities have been carried out, such as courses, projects such as building characteristics, professional judging. After starting the quality of engineering, school-related projects should be rapidly integrated, targeted to make the existing quality of engineering construction projects and construction projects consistent; the quality of new construction projects, school construction projects should be related to the prompt start of the level of field judging. Colonel in the quality of projects started to carry out named colonel-level series, while the school work closely with provincial and national projects related to selection of the work process, well-organized, positive declaration, and constantly improve the national, provincial, school three quality engineering systems.

(c) Adequate safeguards and normalized work. In fact, the quality of implementation of the project is not the school have access to many items of the project, but rather access to national, provincial, university series quality project after project, how to further strengthen the project construction, project objectives, expected results, so as to enhance educational level to improve educational quality. This is the fundamental purpose of the implementation of quality projects, and also the key to the success of the implementation of quality projects.

On the one hand, the school educational administration departments to take effective measures to ensure the quality series of smooth implementation of the project to strengthen the conditions for the implementation of quality engineering construction and protection. Implementation of the "Quality" is a comprehensive, system engineering, quality engineering to promote the further implementation. The school must strengthen the conditions for building multi-party efforts to guarantee the quality of project-related construction projects to achieve the desired effect. Set up a special account for the quality of construction projects to use. Must be based on scientific planning for special consideration, first tilt the principle, according to Reform of personnel training, professional development, curriculum development, practice teaching reform project needs a unified plan to strengthen the quality of engineering construction projects related to the conditions of support. Engineering projects such as quality reporting within the school to focus on the human, material, financial and other aspects of the integration of resources, which requires co-ordination arrangements for academic administration, the heart to a thought, an effort to make, the overall situation. Meanwhile, the educational administration department should enlist their support, to mobilize governments, schools and social aspects of power, the development of higher education initiative directed to the quality of implementation of the project up.

On the other hand, the school educational administration department should strengthen the quality of project monitoring and protection, not only to ensure, the smooth progress of construction projects at all levels, but also to focus on the normal development of construction projects. The so-called normal development, is to ensure quality construction of the project is relatively stable, standardized, with planning, predictability and common sense, the quality of schools in a timely manner for projects in the construction project, do not ask off the monitor, adjust the policies and measures which can ensure the smooth operation of the quality of health of the project. To ensure the smooth progress of the construction project, the school educational administration departments should enhance the inspection of construction projects. For example, the use of professional assessment, program evaluation, teaching quality monitoring, special inspections and other means, according to the project planning, the quality of the project to track the progress of construction, inspection and guidance to ensure the smooth progress of construction projects . Also, be pooled colonel, hospital-level teaching staff supervision and teaching units on the further implementation of school quality engineering and related projects of progress in the overall supervision and guidance to promote the further implementation of quality engineering. School educational administration departments must strengthen and improve the data collection and reporting system. To strengthen the teaching quality control, to guide the teaching units, teachers will further enhance the "Quality Project" energy inputs, the school should actively explore the implementation of teaching basic state data collection and reporting system for the various teaching units in the characteristics of professional, quality courses, teaching materials construction, the quality of teaching achievement awards and other engineering and related construction projects to be informed of the situation, while working to establish a campus Web site, teaching conferences, teaching and other forms of communication and the basic state of the carrier-based teaching timely data reporting system to ensure the "quality project "information transmission and communication, but also promote healthy competition in the formation of the school building, to create a competing construction units, striving to develop a good situation, and thus effectively promote the further implementation of quality engineering, quality of project implementation to ensure the normal development.

(d) Deepen the reform and focus on quality. Quality project in-depth implementation of education for the school staff and students updated ideas, deepen the educational reform to play a good lead role, in order to further deepen the educational reform to strengthen the teaching infrastructure, improve the quality of the training laid a solid foundation.

In practice, educational administration departments should actively deepen the quality of teaching reform. In carrying out the selection of specific projects at the same time, carry out reform, and efforts to integrate the various related projects, the overall system to promote quality construction projects. Ministry of Education clearly states: "attaches great importance to practical aspects, to enhance students practical skills. Necessary to strengthen the experiment, practice, practice, and the Graduation Project (Thesis) and practice teaching, in particular, to enhance professional practice and graduate internships and other important aspects for the inclusion of teaching program of practice teaching accumulated credits (hours), humanities and social sciences majors should generally not be less than the total credits (hours) 15%, professional engineering, agriculture and medicine should not generally be less than the total credits (hours) 25%. Propulsion Laboratory and experimental content of the Reform and innovation, to develop students practical ability to analyze problems and problem-solving ability. to strengthen the research in close cooperation, broaden the students off-campus practice channels, and society, industry and enterprises to build practice, practice teaching base to take all effective measures to ensure that students graduate and professional practice and quality practice time, to promote education with productive labor and social practice closely. Thus, the quality of works on the practice of teaching has a clear higher requirements, schools should take this requirement as a starting point to reform-based training model, combined with practical, carefully revised undergraduate training programs, promote the practice of teaching contents and curriculum reform, to ensure the humanities and social sciences, engineering, agriculture and health professional and specialty teaching programs in all practice teaching accumulated credits (hours) adjusted to not less than the total credits (hours) 25% of the requirements are carried out to optimize practical teaching system reform and improve the practice of teaching programs, and comprehensively promote the practice of teaching reform.

At the same time, schools should actively promote the educational administration department and the school system two teaching quality assessment and monitoring system, through the teaching evaluation, expert evaluation, peer review and other means to establish and improve the school's teaching quality monitoring system, so check the quality of university teaching institutionalized and normalized. Meanwhile, the inspector should also strengthen the role of supervision of experts, to strengthen the reform of undergraduate education and teaching management, teaching activities, guidance; construction supervision team of experts, scientific and efficient explore ways and means of supervision, development and improvement of teaching quality external evaluation and supervision mechanisms.

4   Conclusion

After a year of construction and practice, some schools the quality of the engineering system has been formed for the continued development of these schools and the high level of development laid a solid foundation. Practice shows that the university departments of educational administration and implementation of quality in the implementation of projects should first have a thorough understanding, to make a scientific decision-making, co-ordination arrangements, focusing on school integration and full use of resources; find out the quality construction of the development of the basic law, with the implementation "quality "in the east, to speed up educational reform. Meanwhile, the educational administration departments to take effective measures to protect the quality of construction projects normal development deepen reform, focus on quality. Only in this way, the implementation of quality project can achieve more significant results.

References

1. Andreas, H., Andreas, H.: Quality assurance in UK higher education: Issues of trust, control, professional autonomy and accountability. Higher Education 51, 541–563 (2006)
2. Lawrence, S., Sharma, U.: Commodification of education and academic labour. Critical Perspectives on Accounting 13, 661–677 (2002)
3. Cutright, M.: Chaos Theory and Higher Education: Leadership, Planning, and Policy. P. Lang, New York (2001)
4. Harvey, L.: The new collegialism: Improvement with accountability. Tertiary Education and Management 2, 153–160 (1995)

Comparison of Statistical Clustering Techniques for Correction Analysis of Achievements of the College Entrance Examination

Hu Xifeng

Department of Experiment Teaching administration,
Zhanjiang Normal College, Guangdong, 524048 P.R. China
HUXFXX@163.com

Abstract. In this study, three different statistical clustering methods: a hierarchical, k-means and an artificial neural network (Self-Organizing Maps, SOM) technique were applied to analyze the achievements of the college entrance examination. A comparison of the methods for correction analysis was attempted. The research results indicate that distance function has an important effect on the correction among the courses, and that the performance of SOM was similar to the cluster result obtained by other two clustering algorithms. The empirical results also show that there is a better correction between Chinese, Science basis and Selective course.

Keywords: Clustering, Correlation, Achievements, College entrance examination.

1   Introduction

Cluster analysis is a multivariate statistical technique that tries to group data based on a similarity or dissimilarity measure. It is a rather objective classification method since there are techniques and criteria to examining the minimization of the distance within a cluster and maximization of the distance between clusters, for finding the optimal number of clusters. Many different fields of study, such as engineering, zoology, medicine, linguistics, anthropology, psychology, and marketing, have contributed to the development of clustering techniques and the application of such techniques [1-3].

There are two main traditional methods of statistical clustering: the hierarchical and the non-hierarchical k-means [1,2]. In recent years, artificial neural network techniques gain interest and have increasingly been recognized as a useful statistical technique for the classification [3-5]. The aim of the present study is to apply several statistical clustering techniques (hierarchical, non-hierarchical k-means and Self-Organizing Maps (SOM)) to analyze the achievements of the college entrance examination, to test the effect of the clustering algorithms, and to compare the results from the different clustering techniques.

Y. Wang (Ed.): Education Management, Education Theory & Education Application, AISC 109, pp. 649–653.
springerlink.com   © Springer-Verlag Berlin Heidelberg 2011

2 Methods

The data to be clustered are the achievements or scores of 343 science students from Aizhou Middle School in Zhanjiang, Guangdong, China, in the college entrance examination in 2009, containing the scores of Chinese, Mathematics, English, Science basis and Selective course (that is, one of Physics, Chemistry or Biology), which are represented with object (or course) number of 1, 2, 3, 4, and 5, respectively, in this work. The hierarchical, non-hierarchical k-means and artificial neural network SOM are employed.

Hierarchical Clustering. Hierarchical clustering is a basic way to investigate grouping in data sets, and follows below a series of steps. (1) After the data are normalized(if necessary), the distance between every pair of objects in data sets can be calculated so as to obtain a distance or dissimilarity matrix. (2) This is carried out for all possible pairs of clusters so that the pair with the minimum distance is merged, then for these newly formed clusters, the process continues similarly so as to create bigger clusters until all the objects in the original data set are linked together into a binary, hierarchical tree. (3) For a given hierarchical tree, "cutting" the tree at a given level will yield the corresponding cluster with certain precision.

k-means Algorithm. k-means is a non-hierarchical clustering algorithm widely used in many applications that require partition clustering. Unlike the hierarchical clustering algorithm, k-means algorithms determine iteratively all clusters in one step, and do not create a tree structure, so k-means is more appropriate for clustering large datasets.

The main concept behind k-means is the utilization of centroids, one per cluster. Initially, the centroids are placed randomly, and the positions of the centroids are then re-estimated and located in the centre of the cluster they correspond to. This procedure is repeated until the k centroids are fixed.

Self-Organizing Maps. The self-organizing map (SOM) is a subtype of artificial neural networks and is considered an advanced approach of clustering that can produce reliable segregation even in difficult cases. This makes SOM especially appropriate for clustering high-dimensional data.

A SOM network contains two layers: (a) input layer consisting of data sets, and (b) the output layer consisting of k nodes for k clusters. The weights of the connections from the input neurons to a single neuron in the competitive layer are interpreted as a reference vector in the input space. Therefore, once a SOM is trained well, classification occurs when a output vectors is assigned as output node, which makes it possible to partition all input vectors into a certain number of clusters.

3   Empirical Results and Discussion

3.1   Classification Using Hierarchical Clustering Techniques

In order to carry out the hierarchical clustering, the distances between every pair of objects in data sets must first be calculated to obtain distance matrix. Then the hierarchical cluster trees for various distance functions were obtained, and were shown in Fig. 1. In the Fig. 1, the numbers along the horizontal axis represent the indices of the course or object, and the height of the vertical axis indicates the distance between the courses (i.e. height of a link). If a link is approximately the same height as neighboring links, there are similarities among these courses, so these courses could be grouped into a cluster. It could be found from Fig. 1 that the data sets can be grouped into three clusters using cosine function.

The cophenetic correlation coefficient can also be applied to describe the correlation between these achievement series (see Table 1). The closer the value of the cophenetic correlation coefficient is to 1, the better the clustering result[2]. As shown in Table 1, there appears the best clustering result using cosine distance, and the achievement series could be clustered into three clusters. Of course, another better cluster result could be obtained when correction distance function is used.

Table 1. Cophenetic correlation coefficient for different Distance functions

Euclidean	cityblock	cosine	correlation
0.6797	0.6823	0.9191	0.9139

The empirical results from the hierarchical cluster also shows that expect the clusters with only one variable or course, a cluster containing course 1, 4 and 5 appears using cosine distance, suggesting that there is the best correction between Chinese, Science basis and Selective course when cosine distance function is used. If correction distance function is used, another better correction could also be obtained between Mathematics, Science basis, and Selective course. The observation shows that distance function has an important effect on the clustering result between courses. That is, different distance function could create different clustering results in hierarchical clustering.

3.2   Clustering by k-Means Techniques

k-means uses an iterative algorithm that minimizes the sum of distances from each object to its cluster centroid, over all clusters. The average silhouette value(ASV) was often used as a measure of cluster quality at various cluster numbers[1], and shown in Table 2. For a given distance function, great ASV suggests a better cluster result. As shown in Table 2, ASV increases with the number of clusters k. When k equals 2, there are no distance functions using whichever achievement data sets. For cosine and squared Euclidean distance function, the k-means algorithms with k of 3 or 4 can cluster the achievement series very well, while using other two distance functions, a slightly less good clustering result could be obtained only at k=4.

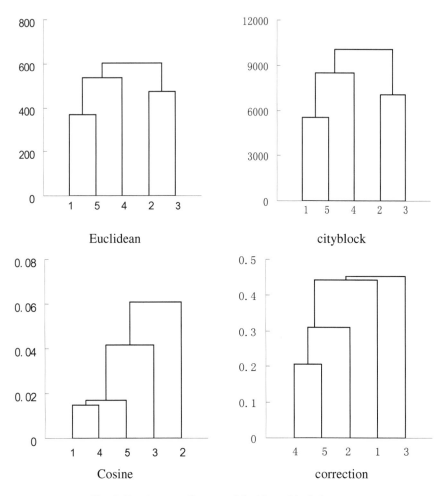

Fig. 1. Dendrogram diagram of the hierarchical cluster tree

Table 2. The average silhouette value for different Distance functions

k	sq Euclidean	correlation	cosine	cityblock
2	0.6427	0.3892	0.4384	0.4441
3	0.7420	0.6527	0.7614	0.4849
4	0.8109	0.7523	0.7758	0.7423

Like hierarchical clustering, the empirical studies from the k-means clustering algorithm also shows that the best clustering result also appears in following courses: Chinese, Science basis and Selective course when cosine and squared Euclidean distance function are used. The observation is consistent with what obtained using hierarchical clustering with cosine distance function.

3.3 Correction by SOM Techniques

For further investigation, SOM algorithm was often used to create a set of clusters of the datasets[3,5], and the Euclidean function was used as the distance function, and the hextop as the topology function with the learning rate of 0.02. After 1000 trials were conducted, the better clustering result could be obtained. SOM study shows that there are a better correction between Chinese, Science basis and Selective course.

It should be pointed out that when the hierarchical, k-means algorithm and artificial neural network SOM with appropriate distance function are employed, the same clustering result could be applied. In this sense, the selection of the appropriate distance function is more important for the better clustering results. In addition, the conclusion that there is a better correction between Chinese, Science basis and Selective course could be drawn from above three clustering algorithms, which implies that there exist common features in the three courses. Science basis course contains much more knowledge which is easier. In Selective course, about 80% students select the Chemistry or Biology course as examination courses. The students with good memory and solid professional basis would get a very good score in Chinese, Science basis and Selective course. Therefore, it is not difficult to accept the fact that high correlation exists between these three courses.

References

1. Grubesic, T.H.: On the Application of Fuzzy Clustering for Crime. Hot Spot Detection 22, 77–105 (2006)
2. Katsiotis1, A., Hagidimitriou, M., Drossou, A., Pontikis, C., Loukas, M.: Genetic Relationships among Species and Cultivars of Pistacia Using RAPDs and AFLPs. Euphytica 132, 279–286 (2003)
3. Bryan, B.A.: Synergistic Techniques for Better Understanding and Classifying the Environmental Structure of Landscapes. Environ. Mana. 37, 126–140 (2006)
4. Tsimboukakis, N., Tambouratzis, G.: A Comparative Study on Authorship Attribution Classification Tasks Using both Neural Network and Statistical Methods. Neural Comput. & Appl. 19, 573–582 (2010)
5. Tan, M.P., Broach, J.R., Floudas, C.A.: A Novel Clustering Approach and Prediction of Optimal Number of Clusters: Global Optimum Search with Enhanced Positioning. J. Glob. Optim. 39, 323–346 (2007)

Literature Review on Research of Real Estate Taxation in China

Xie Feng

Jiang xi university of finance and economics, Ph d student; Lecturer in Zhejiang
university of finance and economics, Hangzhou 310018
xiefeng9020@yahoo.com.cn

Abstract. Taxation on real estate has become a more and more important question in China. Although a great deal of literature on taxation of real estate have been published, consensuses need to be made in many aspects. This paper sorted literature in this area from three aspects, namely position, function and affection of real estate taxation. At last, the paper discussed the influence of real estate taxation to China.

Keywords: literature review, real estate taxation, function, influence.

1 Introduction

In many developed countries, property tax has become a principal source of revenue for state and local governments. And it is also an important component of a series of planned tax reforms in China. In China, the real estate market grew rapidly in recent years. At the mean time, many domestic scholars paid great attention to the study about real estate taxation. A lot of papers focused on the mode of the real estate taxation and emphasized on designing the real estate tax system. However, with house prices dramatically rises in China in recent years, the focus of property research turned to the questions on how to make use of real estate taxation to control the price of real estate, which are mainly through Logical analysis. Others looked at the influence of real estate tax on the real estate industry, government behavior, house buyers, land utilization and taxes revenue gap among different areas, etc. In general, domestic researches of real estate taxation are mainly care about various countermeasures of the government on property taxation, fewer on theoretic exploration.

2 The Position of the Real Estate Tax in Property Taxes in China

Nowadays many domestic literatures defined current property taxation of China. But no consensuses has been made at present.

Generally there are two kinds of property taxation, namely in narrow sense and in board sense. The property taxation in narrow sense only study tax in real estate possession period which is levied on the base of the possession fact. Property taxation in the lots of foreign literatures is in the narrow sense. However Liu rong heng (2006) held a different idea from the point of theoretical view that china's property tax system

Y. Wang (Ed.): Education Management, Education Theory & Education Application, AISC 109, pp. 655–660.

includes Resource Tax, Tax on Using Urban Land, Real Estate Tax, Tax on till-land Occupation, Vehicle And Vessel Tax and Tonnage Dues, which is in the narrow sense. But in fact Chinese property taxation not only has tax from property possession, but also has estate transfer taxation and estate income taxation.

Whatever in narrow sense or in board sense, real estate taxation is considered the most important in property taxation.

3 Function of Real Estate Taxation

3.1 Argument on the Real Estate Tax's Function to Control House Price

The argument that whether real estate taxation can prevent house price rising or not mainly was made around property tax. Chinese scholars made lots of study on the mode of the property tax. Huyijian (2004) thought, we have two models. In the model one, land conveyance fees are replaced by levying property tax. Meanwhile, we combined current Real Estate Tax and Tax on Using Urban Land into property tax. In model two, land conveyance fees are maintained. But current real estate tax and Tax on Using Urban Land are combined into property tax. The main different point of two models is whether we should change the current land use model into annual-rent one. The dissension occurs in the domestic researchers. Huyijian (2004) insisted model one have great difficulties in china. These difficulties are uncoordinated supply-demand relationship of land, the land possession for free or not and the difficulties in design, etc. However this model is positive from the theoretical point. The model two can not only sustain the balance of interests but also can combine the lease in batches and the annual-rent. Huyijian supported model two. Many scholars shared the same idea, e.g. Chenzhenyong(2007) and Wangzhibo(2008). However, Xiachangjie(2004) and Caomingxing(2010) support model two. Whatever they supported, all their studies focused on how to make their idea come into true and how to take the comprehensive reform, instead of the land grant fee itself.

It is popular in the domestic literatures that property tax can control the house price. They thought the house possession cost will be raised by levying the property tax so that the demand for speculative investment on house can drop, demand and supply of house will be in the harmony and the house price will fall down. But more and more researchers raised a different opinion and demonstrated it. Antifu, jinliang (2010) thought high profit led to high house price. Therefore it was the most important to lower the profit level. Only one tax can't do it well, although it can effect the relation of supply and demand of house to some degree. They thought we should depend on all kinds of economic levers, such as credit, land strategy, taxation, etc. They suggested demand for house could be defined to two sorts. One was demand from consumer ,the other for demand from investor. Demand from consumer should be encouraged and demand from investor should be discouraged. Xuchangsheng (2010) presented the property tax has little affection on lowering high house price level from the component element of real estate industry. In his article, the property tax maybe can lower the house price in the short term but from the long term its affection is limited. Caoyingxue, zhanzaisheng, liaoli (2008) also prove the property tax can affect the house price in the short term and not in the long term by the rational bubble model. So they held it was not the main goal to limit the house speculative investment for the property tax reform.

Yanweihua (2008) discovered taxation strategy give weight to the business tax, the Land Value Appreciation Tax and neglect the personal income tax and property taxation. Considering tax administration lagged behind, the failure of controlling the house price is inevitable.

Duxuejun, Hangzhonhua and Wucifang (2009) gave the details on how the real estate taxation influence house price by analyzing panel data of the whole country and areas. Their discovery included the affection of the real estate taxation on house price is remarkably different among different areas. Meanwhile they found different taxes have different affection on house price, namely the real estate tax and contract tax promote house price and Tax on Using Urban Land, tax on till-land occupation, the Land Value Appreciation Tax can limit house price.

The Kuangweida (2009)'s article is imbedded in housing characteristics and illustrates the property tax will cause the house price fall down ceteris paribus through constructing consumer-developer model and investor-developer model respectively. Meanwhile the scholar uses the panel data of 30 provinces in china from 1996 to 2006 and conducts the empirical test upon its theoretical models. The empirical outcomes show that the property tax will lower house price. The conclusion is the effect of property tax upon house prices is insignificant in middle and western china and significant in the whole country and eastern china. The author find interest rate play greater role than taxation in controlling house price.

In brief, more researches on property tax influence on house price are made by qualitative analysis , which draw a conclusion that the property tax has almost no influence on house prices. If research is made by empirical analysis, the conclusion is opposite. That is the influence is different among different provinces.

3.2 Function of the Real Estate Taxation in China

To study function of the real estate taxation, we should study function of property taxation first. The function of property taxation closely changes with government function. When the government function is limited, the property tax is mainly to collect revenue. At the period of capitalism with developed industry and commerce, the distance of wealth distribution is enlarged. The whole society starts to pay attention to relocating the distribution. Property taxation play main role in relocating the wealth. It is at this time that general property tax and death tax are considered important. Shengyuping (2005) stated china should take advantage of the redistribution function of property taxation. Gaoyajun, Wangyinmei (2006) analyzed the function of property taxation and present the function of property taxation should be exerted according to development of china's economics, income redistribution and taxation policy. Huangguolong (2008) think property taxation has four kinds of function including collecting revenue, redistributing the unfair wealth, encouraging being rich by hardworking and not by idleness, having fair tax burden. In these four functions, the most important functions are to ensure fair tax burden which is inherent, and redistribute the wealth, which also is the goal of property taxation.

Domestic study on the function of property is mature and great agreement is made. But these researches only are based on microscopy, which is relatively general and not definite, and ignore analyzing links, stages and exact taxes.

4   Strategy of Implementing Property Taxation Effectivly

In domestic literature, we know the experience of the real estate tax of western countries. The normal tax rate of the real estate tax has also been analyzed. The scholar presented heavy tax burden in dealings and low tax burden in possession of the real estate. But this view is not proved by data. Therefore most domestic literature mainly pay attention to the design of the property tax and possession taxation to solve how to exert the function of real estate. It is lack to study the dealings and beneficiary phrase of real estate.

On studying possession of real estate, liuminghui and cuihuiyu (2007) point out the shortage of china's current taxation system. That is heavy tax burden in dealings and low tax burden in possession of the real estate. We should make standard reform of taxation system of real estate possession and set reasonable aims on the basis of low tax rates board tax base and simple taxation. Dingyun (2009) introduce the taxation of real estate possession in USA, UK and HK and design taxation system of china's taxation of real estate possession with coordinated reforms. Sunyundong and duyuntao (2008) hold the same view. However they research on taxation of real estate possession in more detail, e.g. tax-free item, the exact way to levy, tax rate,etc.

As for the transaction of real estate, wangyouhui, denghongqian, aijianguo (2006) think the second hand house transactions can be sorted into two kinds. One transaction is influenced by the business tax, the other is not influenced. After studying two sorts of transactions by game analysis, they conclude the business tax can't lower house price and the tax can encourage to raise house price contrarily. For VAT and income tax in transactions of real estates can't transform from one taxpayer to another, the better way to control house price is VAT and income tax. Therefore we should depend on income tax of real estate to lower house price, not the tax on real estate transactions.

According to real estate economics, Wuxudong and lijing (2006) classified real estate into three kinds including primary market, secondary market and third market. They analyzed the tax burden from the three markets and the proportion. They found few tax and fee in primary market, heavy tax burden and high fee in second market, almost no tax and fee in the third market. But in renting house, the large-scale market and small-scale tax revenue exist in concurrence. In the end these scholar put forward their suggestion on how to reform china's real estate taxation system.

On how to make these design come true, domestic experts didn't make further research on the target of real estate taxation as a whole. Accordingly the study on the way how to make the target come true is one-sided. Experts tended to parrot what others say.

5   The Result of Real Estate Taxation

Domestic research on the influence of real estate taxation more with empirical analysis. These studies paid particular emphasis on influence on literated parts, such as government, house buyer, etc. But influence on the transfer quantity and the Ratio of Supply- Demand of house are ignored.

5.1 Influence on China's Real Estate Industry

Changli (2007) study the influence on real estate industry by analyzing land supply model, land price, the primary market and the secondary market, demand-supply ratio of housing, real estate enterprises, etc. The conclusion is that taxation reform of real estate can make the real estate industry more standard.

5.2 Influence on Government Behavior

Gonggangming (2005) thought the property tax can't make definite influence on local government behavior. It will encourage local government to take short-term behavior when land grand fee is charged by annual rent. But if local government can get in debts(such as to mortgage the future of the property tax for a loan),or issue bonds, the property tax will not encourage short-term behaviors. Therefore what influence local governments is not to levy the property tax, but to strengthen budget, standardize government behavior, to set up perfect supervision to financial revenue and expenditure.

5.3 Influence on House Buyer

wangxiaoming,wuhuiming (2008)classified house buyers into two kinds, consumers and investors. We can also further classified consumers and investors respectively into two parts, one with houses and one without houses. They study the house possession with different-income families, the property tax burden with different-income people, the cost after and before levying the property tax, the proportion of different –income people with two houses, investment profit from real estate industry after and before the property tax. As result they found the property tax had no influence on house buyers with demand for living, small influence on the proportion of demand for living and investing, great influence on investors and speculators.

5.4 Influence on Land Utilization Ratio

Wangzhibo (2010) set up the model of the property tax and land utilization according to the characteristics of china's real estate market. They studied the influence of the property tax on utilization ratio by land comparative statics and numerical simulation. They drew a conclusion that the influence was not positive. But changli (2007) put forward opposite idea that land annual-rent model can protect arable land and raise the utilization.

5.5 Influence on Revenue Distribution of the Property Tax among Different Areas

Wangming (2010) thought land grand fee should be preserved and the new property tax can be levied instead of several current real estate taxes. He also calculate the Gini coefficient of tax revenue and the real estate tax, and coefficient of dispersion of main taxes revenue, from 1999 to 2008. he conclude that the property tax can enlarge the distance of revenue distribution among areas and has negative affection on coordinated development of regional economy. So when we levy the property tax, we should make revelant system more perfect, such as transfer payment, integratedly reform land grand fee system. such studies are lack in domestic literature.

References

1. Gao, Y., Wangyingmei: Study on Function of Property Taxation. Journal of Zhongnan University of Economics and Law 158, 70–72 (2006)
2. Huang, G.: Study on Improving Our Property Taxation on the Basis of the Function of Property Taxation. Journal of South China Normal University(social science edition) 4, 49–52 (2008)
3. Shen, Y.: On Reforming Property Taxes and its Policy Choices. Collected Essays on Finance and Economics 120, 24–30 (2005)
4. Li, C.: Study on Reform of the real estate taxation on its affection on real estate industry. XiBei University (2007)
5. Wang, Z.: The Influence of Property Tax on Land Utilization. Journal of Finance and Economics 36, 90–100 (2010)
6. Kuang W.: Housing Characteristics, Property Tax and House Prices. Journal of Economy 4,151-160,(2009)
7. Du, X.-j., Huang, Z.-h., Wu, c.-f.: Impacts of Real Estate Tax and Local Public Expenditure on House Price Based on National and Regional Panel Data Analysis. China Land Science 23, 9–139 (2009)
8. Ding, Y.: Learn from International and HongKong Regarding Real Estate Tenure Tax. Journal of Central University of Finance and Economics 1, 16–21 (2009)
9. Hu, Y.: Choice on Model of the Property Tax and its System Design. Journal of Taxation 232, 25–28 (2004)
10. Chang, X.: Study on Affection of Levying the Property Tax and its Time. Journal of Taxation 232, 22–24 (2004)
11. An, T., Jing, L.: On several theoretical questions of Levying the Property Tax. Journal of Taxation 301, 36–44 (2010)

The Time Effect of DNA Damage and Oxidative Stress on Mice Liver Cells Induced by Exercise Fatigue

Su Meihua

Department of Physical Education, Zhangzhou Normal University, Zhangzhou Fujian, China
sumh1234@163.com

Abstract. To discuss the effect of DNA damage and its mechanism on liver cells induced by exercise fatigue. 24 male Kunming mice were randomized into four groups: Control group (CG) and exercise group (EG) was subdivided to three subgroups respectively indicating as 0EG, 24EG, 48EG(N=6/group). We built up the exercise fatigue model for mice through the protocol of repeated exhaustive treadmill running and use the single cell gel electrophoresis (SCGE) to detect the DNA damage of liver cells in different groups. We also measured the changes of SOD, GSH and MDA on liver tissue. Our studies showed that DNA damage of 0EG and 24EG were significantly higher than CG($P < 0.001$). DNA damage reached the highest at 24h and dropped to the pre-exercise level at 48h ($P > 0.05$). The activity of SOD and the content of MDA on liver tissue of 0EG and 24EG significantly increased compared with CG ($P < 0.001$). Those changes in 24EG significantly decreased compared with 0EG ($P < 0.05$). GSH concentration in 0EG reduced significantly compared with the control group ($P < 0.001$). Either the 24EG or 48EG of GSH did not change obviously compared with CG ($P > 0.05$). Those results lead to conclusions that exercise fatigue induces the oxidative injury and DNA damage on liver cells which exists some characters in different time after exercise. Exercise induced oxidative stress is one of the mechanism of DNA damage on liver cells.

Keywords: exercise fatigue, liver cells, DNA damage, single cell gel electrophoresis(SCGE), oxidative stress.

1 Introduction

The alkaline single cell gel electrophoresis assay (SCGE), which also called comet assay, has been used in many in vitro applications to assess DNA damage in individualized mammalian cells. This assay sensitively detects DNA single- and double-strand breaks induced by chemical compounds [1]. Exercise is usually said to increase the generation of reactive oxygen species that are potentially harmful. Strenuous exercise may increase the production of reactive oxygen species (ROS), leading to a situation of oxidative stress that has been discerned by detection of oxidatively damaged DNA bases and lipid peroxidation products [2]. Data have been presented suggesting that oxidative stress plays a role in the muscle fatigue that accompanies some forms of exercise, although these roles remain contentious [3].

Y. Wang (Ed.): Education Management, Education Theory & Education Application, AISC 109, pp. 661–667.

Some studies report that an acute bout of exercise increases the activities of superoxide dismutase (SOD), glutathione peroxidase and glutathione reductase in skeletal muscle of rats [4]. In animals, exercise training has been reported the activities of various protective enzymes in liver [5]. Some studies have shown that exercise-induced lipid peroxidation in response to various modes of exercise, including endurance running [6]. Recently, there has been growing interest in exercise-induced DNA damage due to its potential involvement in various disease states, because when the body is repeatedly exposed to situations that could damage the DNA, the likelihood that some of these changes will not be repaired or will be incorrectly repaired [7]. The relationship between oxidative stress due to vigorous exercise and induction of DNA modifications is unknown and their long term effects on health have yet to be elucidated [8]. The purpose of this study just once again to clarify what extend does the DNA damage occurred in repeated exhaustive exercise mice and how long can it be repaired by body's repair system. What the relationship is between exhaustive exercise, DNA damage, and antioxidant defense mechanisms in the mice.

2   Materials and Methods

2.1   Exercise Protocol

The exercise protocol was done as Marra et al. [9], but repeats it for seven days, once per day. It consisted of an acute exhaustive exercise bout of treadmill running. Mice were run at 28 m min^{-1} on a 28 slope for 90 min following a brief (10 min) warm-up (Omni-max small rodent treadmill). Mice were run during the early part of their dark cycle. Running mice were given three short practice runs (15 min at 15 m min^{-1} and 20 min at 20 m min^{-1}) 24 and 48 h prior to the test. Exhaustion was determined by failure to run after continued prodding and splaying within the treadmill lane. Control group mice were exposed to the noise and vibration of the treadmill for the same duration as the exercised mice. All animals were respectively sacrificed immediately (0SG), 24hours (24SG), 48huors (48SG) after exposure to the treadmill.

2.2   Alkaline Comet Assay

The alkaline comet assay was done as previously described by Godard et al. [10]. After automatic delimitation of nucleus head and tail as well as elimination of background fluorescence and touching cells, different parameters are calculated, describing nucleus geometry (length, areas . . .) and intensity (%of DNA in the head of the comet, tail moment . . .). Tail moment was defined by the product of the distance between the two barycentres of the head and the tail by the proportion of fluorescence in the tail of the comet [11].

2.3   Alkaline Comet Assay

The livers were removed and frozen in liquid nitrogen, and subsequently homogenized in ice-cold Tris-HCl buffer (25 mmol/L Tris, 1 mmol/L EDTA, 10% glycerol, and 1 mmol/L DTT, pH 7.4) with a glass homogenizer. The homogenate was centrifuged at 10 000 ×g for 20 minutes at 4°C. The supernatant and sediment fractions were separated, and the supernatant was aliquoted and stored at -80°C [12].

The activity of SOD, GSH and MDA was determined spectrophotometrically according to the method of the Nanjing Jiancheng Bioengineering Institute(China) with a spectrometer.

2.4 Alkaline Comet Assay

Data were expressed as the mean ± S.D. for the number of experiments indicated. Statistical analyses of the data were analyzed by one-way analysis variance (ANOVA). Significance level was set at p < 0.05.

3 Results

3.1 DNA Damage of Liver Cells Induced by Repeated Exhaustive Exercise

In Fig1, we use the comet assay parameter tail moment to express the DNA damage of liver cell induced by repeated exhaustive exercise. The higher of this value, the greater the damage that has occurred to the nuclear DNA. Statistically significant differences (p<0.05) in DNA damage were found in the liver cells of mice in OEG as compared to control group(CG),the DNA damage of liver cells in 24SG were significant higher than control group(CG) (p<0.01). There were no significant differences between the control group and the post exercise 48 hours group (48EG). The control liver cells sustained the least background damage.

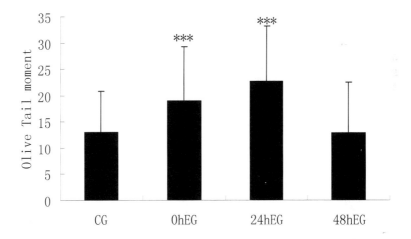

Fig. 1. Levels of DNA damage in liver cells of control vs. exercised mice. *P < 0.05 denotes a significant difference between control and exercised group. ***P < 0.01 denotes highly significant difference between control and exercised group.

3.2 Measurement of SOD, GSH and MDA

From the table1, we could see that the SOD activities in liver tissue of mice were largely altered in all post exercise groups. The SOD activity of post exercise immediately group (0EG) increased significantly compared with control group (P<0.001), and the post exercise 24 hours group (24EG) also has higher SOD activity than control group (P<0.05), and compared with the post exercise immediately group (0EG), the post exercise 24 hours group (24EG) had reduced the SOD activity and did not have statistical meaning (P<0.01). Also the SOD activity of the 48 hours group (48EG) did not have obvious difference compared with control group. GSH concentration in the post exercise immediately group (0EG) reduced significantly compared with the control group (P<0.01). Either the post exercise 24 hours group (24EG) or the post exercise 48 hours group (48EG)did not change obviously compared with control group (P>0.05). The post exercises immediately group (0EG) and the post exercise 24 hours group (24EG) both have higher MDA level than the control group (P<0.001), and the MDA level of post exercises immediately group (0EG) are significant higher than the post exercise 24 hours group (24EG),but the post exercise 48 hours group (48EG) still had higher value of MDA level compared with control group (P<0.05) and its MDA level was significant lower than the post exercise 24 hours group (24EG).

Table 1. The changing of SOD, GSH and MDA on liver tissue after repeated exhaustive exercise

Group	Activity of SOD , GSH and MDA		
	SOD (U/mgprot)	*GSH(mg/gprot)*	*MDA(nmol/ gmprot)*
CG	34.88±1.69	92.17±2.21	4.99±0.67
0hEG	53.56±2.76***	79.71±7.69**	13.73±0.85***
24hEG	37.75±2.91*##	88.73±7.65##	8.62±1.34***##
48hEG	36.45±3.04	83.99±6.77##	5.22±0.88##▲▲

Compared with CG, *P<0.05, **P<0.01, ***P<0.001; compared with 0hEG, # P<0.05, ##P<0.01, ###P<0.001, P<0.01, compared with 24hEG, ▲P<0.05, ▲▲P<0.01

4   Discussion

Single-cell gel electrophoresis (the comet assay) is considered a useful tool for investigating issues related to oxidative stress in human lymphocytes. Our results showed that DNA damage occurred post exercise immediately and 24 hours, the extend of DNA damage ranged differently, the post exercise 24 hours reached highest, DNA damage returned back as normal control post exercise 48 hours. It is in a good agreement with the following published results. Niess [13], Hartmann [14], [15] and Tsai [16] reported the DNA damage increase in comet tail moment 24 h after a maximal oxygen consumption test on a treadmill. All of these research results indicated that the extent of DNA damage and the DNA repair time can be an index to evaluate the situation of exercise fatigue on athletes.

In the current study, we investigated the hypothesis that mice exposed to repeat exhaustion exercise [17] would be subjected to oxidative stress, and that this would be manifested as either alterations in the status of oxidative defense mechanisms, or in the levels of DNA damage. Antioxidant status was assessed by measurement of two key antioxidants (GSH and SOD), and DNA damage by determination of DNA strand breaks using the Comet assays. The present study also investigates the status of lipid peroxidation. Malondialdehyde level is a marker of lipid oxidation. Antioxidants work together in human blood cells against toxic reactive oxygen species [18], [19]. Reactive oxygen species (ROS) cause lipid peroxidation and oxidation of some specific proteins, thus affecting many intra- and intercellular systems [20]. Since the protective action of GSH against oxidant injury is known to be due to its oxygen and other radical scavenging capacity [21], the depletion of GSH in liver tissue of exhaustively exercised rats is not surprising. So in our research, the data showed that GSH level reduced significantly post exercise immediately. The fact that, in the present study, a significant level of DNA damage was detected after exhaustive running in the mice probably shows that DNA might be a weak link in a cell's ability to tolerate oxygen free-radical attack. It is conceivable that the levels of exercise attained in our experiments could be associated with oxidative stress, and perhaps the deleterious effects associated with such stress. It is possible that a depression in the running performance of the mice could be attributed to disruption of the oxidant/antioxidant balance consequently resulting in oxidative stress[22]. SOD also plays a key antioxidant role, which protects against oxidative damage especially mediated by free radicals and lipid perioxidation[23]. According to our data, the liver cells were susceptible to the influence of elevated free radicals, because the SOD activities of post exercise immediately group (0hEG) increased significantly compared with control group, but the post exercise 24 hours group (24hEG) reduced significantly compared with (0hEG), and returned to the level of control group. So it may due to the activation of SOD to protect from the damage of elevated free radicals. We investigated the status of lipid peroxidation and found that MDA Levels were significantly elevated in post exercise immediately group (0hEG) and the post exercise 24 hours group, and the post exercise immediately group (0hEG) was the most highest among the four groups. The balance between oxidative stress and antioxidant defence mechanism may be impaired by depletion of enzymatic antioxidants and increased liver levels of MDA in exhaustive mice[24], [25]. The significant increase in DNA strand breaks in this study could partly be attributed to DNA enzyme repair activities, perhaps following the generation of oxidative lesions and base alterations[22], [26]. In conclusion, the current data indicate that strenuous exercise in the mice caused oxidative stress in these animals and this was associated with elevated DNA damage. In exhaustive exercise, DNA damage can be repaired by its own repair system, although the recovery time distinct from each other, and the exercise intensity was a critical factor in DNA damage[13], [14], [15].

References

1. Roy, M.D.: Approach for assessing total cellular DNA damage. Biotechniques 42(4), 425–429 (2007)
2. Aldred, S.: Oxidative and nitrative changes seen in lipoproteins following exercise. Atherosclerosis 192(1), 1–8 (2007)
3. Penkowa, M., Keller, P., Keller, C., et al.: Exercise-induced metallothionein expression in human skeletal muscle fibres. Experimental Physiology 90, 477–486 (2005)
4. Ji, L.: Antioxidant enzyme response to exercise and aging. Med. Sci. Sports Exerc. 25, 225–231 (1993)
5. Robertson, J.D., Maughan, R.J., Duthie, G.G., et al.: Increased blood antioxidant systems of runners in response to training load. Clin. Sci. 80, 611–618 (1991)
6. Mastaloudis, A., Leonard, S., Traber, M.: Oxidative stress in athletes during extreme endurance exercise. Free Radic. Biol. Med. 31, 911–922 (2001)
7. Weidner Maluf, S.: Monitoring DNA damage following radiation exposure using cytokinesis–block micronucleus method and alkaline single-cell gel electrophoresis. Clinica. Chimica. Acta 347, 15–24 (2004)
8. Hartmann, A., Niess, A.M.: Oxidative DNA damage in exercise. Handbook of Oxidants and Antioxidants in Exercise, 195–217 (2000)
9. Marra, S., Burnett, M., Hoffman-Goetz, L.: Intravenous catecholamine administration affects mouse intestinal lymphocyte number and apoptosis. Neuroimmunol 158, 76–85 (2005)
10. Godard, T., Fessard, V., Huet, S., et al.: Comparative in vitro and in vivo assessment of genotoxic effects of etoposide and chlorothalonil by the comet assay. Mutation Research 444, 103–116 (1999)
11. Mastaloudis, A., Yu, T.W., O'Donnell, R.P., et al.: Endurance exercise results in DNA damage as detected by the comet assay. Free Radical Biology and Medicine 36, 966–975 (2004)
12. Roy, M.D.: Approach for assessing total cellular DNA damage. Biotechniques 42(4), 425–429 (2007)
13. Niess, A., Hartmann, A., Grunert-Fuchs, M., et al.: DNA damage after exhaustive treadmill running in trained and untrained men. Int. J. Sport Med. 17, 397–403 (1996)
14. Hartmann, A., Niess, A., Grunert-Fuchs, M., et al.: Vitamin E prevents exercise-induced DNA damage. Mutat. Res. 346, 195–202 (1995)
15. Hartmann, A., Plappert, U., Raddatz, K., et al.: Does physical activity induce DNA damage? Mutagenesis 9, 269–272 (1994)
16. Tsai, K., Hsu, T.G., Hsu, K. M., et al.: Oxidative DNA damage in human peripheral leukocytes induced by massive aerobic exercise. Free Radic. Biol. Med. 31, 1465–1472 (2001)
17. Niess, A., Baumann, M., Roecker, K., et al.: Effects of intensive endurance exercise on DNA damage in leucocytes. Sports Med. Phys. Fitness 38, 111–115 (1998)
18. Hartmann, A., Pfuhler, S., Dennog, C., et al.: Exercise-induced DNA effects in human leukocytes are not accompanied by increased formation of 8-hydroxy-2V-deoxyguanosine or induction of micronuclei. Free Radic. Biol. Med. 24, 245–251 (1998)
19. Aldred, S.: Oxidative and nitrative changes seen in lipoproteins following exercise. Atherosclerosis 192, 1–8 (2007)

20. Child, R.B., Wilkinson, D.M., Fallowfield, J.L., et al.: Elevated serum antioxidant capacity and plasma malondialdehyde concentration in response to a simulated half-marathon run. Med. Sci. Sports Exerc. 30(11), 1603–1607 (1998)

21. Demirbag, R., Yilmaz, R., Güzel, S., et al.: Effects of treadmill exercise test on oxidative/antioxidative parameters and DNA damage. Anadolu Kardiyoloji Dergisi 6(2), 135–140 (2006)

22. Selman, C., McLaren, J.S., Collins, A.R., et al.: Antioxidant enzyme activities, lipid peroxidation, and DNA oxidative damage: the effects of short-term voluntary wheel running. Arch. Biochem. Biophys. 401(2), 255–261 (2002)

23. Lew, H., Pyke, S., Quintanilha, A.: Changes in the glutathione status of plasma, liver and muscle following exhaustive exercise in rats. FEBS LETTERS 185, 262–266 (1985)

24. Aniagu, S.O., Day, N., Chipman, J.K., et al.: Does Exhaustive ExerciseResult inOxidative Stress andAssociatedDNADamage in the Chub (Leuciscus cephalus)? Environmental and Molecular Mutagenesis 47, 616–623 (2006)

25. Bombail, V., Aw, D., Gordon, E., et al.: Application of the comet and micronucleus assays to butterfish (Pholis gunnellus) erythrocytes from the Firth of Forth. Scotland. Chemosphere 44, 383–392 (2001)

26. emmanouil, C., Smart, D.J., Hodges, N.J., et al.: Oxidative damage produced by Cr(VI) and repair in mussel (Mytilus edulis L.) gill. Mar. Environ. Res. 62(Suppl 1), 5292–5296 (2006)

The Application of "Functional Equivalence" in Trade Mark Translation

Zhuo Wang

Harbin University of Commerce, China, 150028
wangzhuo4648@163.com

Abstract. Trade mark translation, an intercultural communication, involves linguistic and cultural factors in both SL and TL. On the basis of the functional equivalence theory, the translated trade marks should attempt to achieve an ideal unity in terms of sound, meaning and culture, and the readers of the translated trade marks should be able to comprehend the translated trade mark to the point that the original readers of the trade marks must have understood and appreciated the original trade mark. This paper introduces the definition and characteristic of the trade marks, applies the principle of functional equivalence and discusses five approaches to the translation of trade mark in order to arrive at the functional equivalence.

Keywords: trade mark, trade mark translation, functional equivalence, cultural differences.

1 Introduction

With the increasing globalization of world economy and China's entry into the World Trade Organization, a great number of Chinese products flood into foreign market and foreign products into Chinese market. Then how to make their products or commodities better understood and accepted, and even welcomed by the consumers in the target market is, for all foreign-trade enterprises, the first and foremost matter to be taken into consideration. Consequently, it is self-evident for their goods to have a good trade mark and a good performance so to have the greater advertising effect and promote their sales.

For translating foreign trade marks into Chinese, the author would name only the following representative problems which should not be ignored or overlooked.

1.1 Chinese Characters Have Monosyllabic Pronunciations

Since there are a limited number of monosyllables, many Chinese characters share identical pronunciations. Thus it could cause uncertainty in choosing Chinese characters. A good case in point is that Psorales, the trade mark for a drug, was translated into破故纸when just coming into China. It is obvious that nobody in China was willing to buy such useless thing and thus this drug sold rather slowly in China.

1.2 Psychological Demands of the People between Countries Are Not the Same

So translated trade mark should satisfy or meet what the Chinese think. For instance, Catepillar Tractor Company in America chooses Catepillar as its trade mark. If it is put

Y. Wang (Ed.): Education Management, Education Theory & Education Application, AISC 109, pp. 669–674.

directly into Chinese, it will be毛毛虫，You can image what would the Chinese react at the first sight of the name.

So all those issues must be thought over under the specific situations when we go about the translation of foreign trade marks.

Considering the above-mentioned factors and problems, the author presents in the first place, the general information on the trade mark, such as its definitions, history. In the following Chapter, the author introduce a important translation theorists-Eugene A.Nida and his theory as functional equivalent. Guided by the principle and criteria, the writer descends the paper into the details of five different translating techniques and skills: literal translation, free translation, mixed translation, communicative translation, and transliteration. Furthermore, many first-hand examples are obtained and used to illustrate each concrete method. The author hopes such examples could give some guides or reference.

2 Functional Equivalence Theory and Techniques of Trademark Translation

2.1 Nida's "Functional Equivalence"

Functional equivalence is categorized into different levels of adequacy, with the minimal effectiveness and the maximal effectiveness at the two extremes. It is hard to achieve the maximal level of adequacy in translating while a translation under the standard of minimal level adequacy of equivalence can hardly be effective in communication. A maximal level of equivalence, according to Nida, is the case that "the readers of a translated text should be able to understand and appreciate it in essentially the same manner as the original readers did." And a minimal level is defined in the way the readers of a translated text should be able to comprehend it to the point that they can conceive how the original readers of the text must have understood and appreciated it. Of course, the closer to the maximal level translation is, the better it is.

2.2 The Application of Nida's "Functional Equivalence" in Trade Mark Translation

In the light of the Nida's translation theory，to fulfill the functions of trademark words and the ultimate purpose of inducing target consumers to buy the products, the approach with the target-culture orientation is widely taken as the basic translation strategy of the translation of trademark words. Under the guidance of the theory, five basic techniques (transliteration, literal translation, free translation, mixed translation and communicative transliteration) are adopted to translate trademark words.

2.2.1 Transliteration

Transliteration is one of the most commonly used methods in translating Chinese trade marks. Probably we, the Chinese, are familiar with the following illustrations, 美菱 (refrigerator), MeiLing;春兰(air-conditioner), ChunLan;轻骑(motorcycle), QingQi;长虹(TV set), ChangHong;海尔(refrigerator), Haier;红塔山(cigarettes), HongTaShan; 红豆 (shirt), HongDou; 健力宝 ((beverage), JianLiBao; 恒源祥 (sweater),

HengYuanXiang;步森(shirt), BuBen;森达(shoes), SenDa;拉芳(cosmetics), LaFang;泰山(cigarette), TaiShan;娃哈哈(milk), Wahaha and so forth.

From the above examples, we can see that transliteration is especially applicable in the translation of Chinese trade marks which are named after Chinese geographical names, places of interest or unique things with traditional Chinese styles in China. For instance, 黄鹤楼is better transliterated into HuangHe Lou rather than be literally rendered into Yellow Crane Tower which is too sluggish.

In translation practice, we find that foreign language trademarks are highly coherent in letters or words and can be pronounced easily in one breath, while the transliteration of Chinese trade mark diction are broken into independent words in accordance with the specific Chinese characters. So the transliterated version often lacks coherence. To avoid the disadvantage of transliteration in strict accordance with the standard Chinese pronunciation, we can use it freely, to some extent, such as the transliteration according to the pronouncing in local accent. The following are some examples,摩丽雅(washing power), MoRiYa;美而暖(clothes), MaLRan;立白(washing detergent), Liby;奇伟 (shoes oil), Kiwi;信婷(cosmetics), Sincin;凡士林(cosmetics), Fossillin;澳坷玛 (washing machine), Aucma;鄂尔多斯(sweater), Erdos;方太(cooker), Fotile;好迪 (spray), Houdy;哈德门(cigarette), HaTamen;格力(air-conditioner), Gree;青岛(beer), Tsingtao;格兰仕(micro-oven), Glanz;

2.2.2 Literal Translation

In translating Chinese trade marks, literal translation is to try to find English words, the same or similar in meaning to their corresponding Chinese originals. For examples, English version of王朝(grape wine) is Dynasty which impresses the consumers with the feeling of its long history. Moreover, it also implies that so long a history the wine has, the quality should be nice, as the way people's thinking works.永久((bicycle) is literally put into Forever that means the long durability of the product. Quite a few Chinese trade marks named after flowers, birds, animals and precious stones, etc, are rendered literally. For instances, 葵花(drug), Sunflower;月季(drug), China Rose;牡丹(cigarette), Peony;熊猫(TV),Panda;猴王(candy), Monkey King;狼牌(shoes), Wolf;狮王(toothpaste), Lion;小熊猫(cigarette), Lesser Panda;威龙(grape wine).

Nowadays, more and more trade marks made up of common nouns tend to be translated literally. For examples, Cigarettes:将军,General;东方,Orient;时代，Times;一只笔, A Pen;八喜，Eight Happiness;双喜，Double Happiness;红锡包，Red Flake.

Clothes, shoes:春竹，Spring Bamboo;雪中飞，Snow Flying;蓝冰，B1ueIce;乡村树，Village Tree;三枪，Three Gun;矫健，Vigor;双星，Double Star.

2.2.3 Free Translation

Free translation is rarely applied in translation of Chinese trade marks. However, if used properly, it can make the translated trade mark more easily to be understood and accepted by target language consumers. We would take a close look at the rendering of 杜康(wine). If transliterated, it should be put into DuKang which have no actual meaning and only are combination of letters to the target language consumers.

However, for Chinese consumers, at the first sight of the trade mark杜康，they would be inspired with the association to the inventor of the wine，杜康，and ancient Chinese culture and furthermore to the super quality of the wine. What target language version can make foreign consumers have the same or similar response? Final choice of Bacchus, God of wine in ancient Greek legend, undoubtedly is the best. Some good cases in point，创维(TV), Skyworth; 顺爽(shampoo), Hair Song;风影(shampoo), Dew;舒淇(cosmetics), Blue Seasky;温雅(cosmetics), Youngrace;统一(lubricating oil), Monarch;东盛(drug), TopSun.

2.2.4 Mixed Translation

In translating Chinese trade marks, semantic transliteration is to produce one English version which can convey and express both the meaning and the pronunciation of the source language trade marks. The rendering of回力(rubber-sole sports shoes) provides us with one very good illustration. In Chinese,回力means the great power to conquer the difficulty. Its English version, Warrior, denotes the courageous soldier. So they are similar in meaning. In addition，回力(Hui Li in Chinese Pinyin) pronounces quite close to Warrior. Since Warrior has the great power to conquer the difficulty, people wearing 回力shoes could also become the brave soldiers. More illustrations from cosmetics, foods, appliances, and computers and so on can give us a clearer explanation.

Cosmetics, shampoo, washing detergent:露美，Ruby;依波，Everbright; T透，Through;美媛春，Missmay;冰果，B-Cool;娇研，Jolly;诗乐氏，Swashes;诗碟，Shindy;贝芬，Befit;冷酸灵，Luclean;涤夫人，Deft;美露华，Mellowy.

Foods, drinks:乐百氏，Robust;非常可乐，Future Cola;百乐美，Belmerry;得利斯，Delicious;双汇，Shineway.

2.2.5 Communicative Translation

As is mentioned above, because of cultural differences, Chinese and the Western consumers' aesthetic conceptions are different. In domestic, through the shared language and culture, customers already know and accept the meanings of trade marks as well as their association. But in foreign markets, the situation is quite different. Some trade marks with nice meanings and favorable associations in source language culture may have different or even negative meanings in target language culture. So, trade mark translation should find out the functional and aesthetic equivalent between the two languages. And translated trade marks will get the same or similar, at least not opposite response to those from the source language customers. Otherwise, no matter how faithful to the original the translated trade mark is, from the view of pragmatics, it's a failure in translation.

Since marketing communication is usually designed to persuade the target audience to purchase products, the ultimate test of marketing communications is receiver's response-whether its mass communication is, in fact, received by the intended audience, understood in the intended way, and successful in achieving the intended objectives. From the view, communicative translation is one of the means to achieve marketing communication.

Communicative translation is customer-oriented translation. According to Newmark communicative translation attempts to render the exact contextual meaning of the original in such a way that both content and language are readily acceptable and comprehensible to the readership (Newmark, 47). It fully takes the advantages of target language and fulfills the needs of the target customers in the way that makes translated trade marks more idiomatic and easier to understand and remember. In other words, communicative translation can disculturalize the cultural words in target language.

For example, "East wind" is the English brand name for"东风" (truck). In Chinese " 东风"implies "favorable wind"-the key to success, because Zhu Geliang in Three Kingdoms succeeded in borrowing arrows owing to the east wind, which resulted in Chinese proverb:"万事俱备，只欠东风". In the other hand, the warm east wind can predict spring is approaching. In contrast, British is different from China in geographic location. So, east wind is a piercing wind coming south from the Arctic Ocean, which will never give the British people a good psychological response. In the contrary, west wind in England is warm and soft. Therefore, in Ode to west wind by Shelly is there the most famous last sentence: "If winter comes, can spring be far behind". As a result, when "East wind" (truck) is promoted abroad, it had better be translated into "west wind".

From the above mentioned examples, we have seen cultural components tend to be replaced by cultural equivalents in vocative text. In this regard, communicative translation, being set at the customer's level of language and knowledge, is more likely to create equivalent effect than other translating methods. As a result, it is often better than its original. But communicative translation concentrates on the message and the receiver's response and tends to undertranslate. At the same time, we should see that communicative translation doesn't suit some traditional Chinese brand translation, such as"世一堂"，"全聚德"，and"六必居"，because these haven't equivalent words in English.

3 Conclusion

Obviously, the first task for the translator is to produce a faithful and accurate translation. This is the elementary requirement but even at this linguistic level the translator may just as well come across many challenges, mainly caused by the flexible, creative and diversified language. Furthermore, translators could not always find correspondents in the text language for any word or expressions in the source language.

Secondly, in translating trade mark, a large amount of cultural information is hard to be transferred in equivalence between English and Chinese. Cultural problem in translating trade mark is so common that every translator just runs into it more often than not. Therefore, during the course of translating trade mark, culture brings about obstacles in one way or another and it is necessary to analyze the specific situations.

Finally, for a translator, the differences existing in the target language receptors should be kept in his mind. To tackle the problems arising from those restrictions mentioned above, making adjustments is one very useful translation method. Of course adaptation could not be made as one pleases.

The trade mark translation is a challenging, complicated and demanding task. The scope of discussion in this paper is far from comprehensive. Still new problems will keep coming up with the development of trade mark itself. However, the author of this paper strongly believes that with the growth of Chinese economy, this field of study will develop forward accordingly thanks to the painstaking efforts made by our translators and linguists.

References

1. Kramsch, C.: Language and Culture. Oxford University Press, Oxford (1998)
2. Morris, D.: The Naked Ape. Jonathan Cape, London (1994)
3. Newmark, P.: Approaches to Translation. Shanghai Foreign Language Education Press, Shanghai (2001)
4. NIda, E.: A Language Culture and Translating. Shanghai Foreign Language Education Press, Shanghai (1993)

A Probe into Image Shift in Translation

Zhuo Wang

Harbin University of Commerce, China, 150028
wangzhuo4648@163.com

Abstract. An article is linked with a line of images. Thus images should be paid enough attention to in translation. English and Chinese are two different languages in many aspects, so it is very important to deal with the problems of image shift. In order to prove the quality of translation, this thesis focuses on the image shift in three aspects: total image substitution, image substitution with image of the target language and image deleting. The thesis has three parts. Part one makes a brief introduction to the whole article. Part two analyses the factors affecting the meaning of image in English and Chinese. Part three focuses on the three ways of image shift in translation and other relative things.

Keywords: English to Chinese translation, image shift, source language, target language.

1 Images in English and Chinese

English and Chinese, which are both great in history and culture, abound in idioms, proverbs, sayings etc. All of these which have been used for quite a long time are peculiar not only in their grammatical constructions, but also in their specific meanings. When they are used, their structures and meanings can't be changed at random. All of them use concise, simple, vivid language, employ figurative images and most are quite well structured. Language, as a gem of a nation's culture, is characterized by its vivid images. So, the most distinctive feature of a language is the exotic cultural flavors expressed in it. A detailed analysis of the factors that affecting the meanings of images in both English and Chinese is meaningful in that it can not only display the peculiar culture of the two nations but also help translator a lot while handling the problem of image shift in translation study. So in this chapter, after the definition of images, more emphasis is put on the analysis of the factors that affecting the meaning of images.

2 Several Types of Image Shift in Translation

Generally speaking, most idioms, proverbs, sayings contain one or more images. These images have lost their superficial meanings with their stabilization. For example, the connotative meaning of the idiom "to eat the leek" has nothing to do with leek. Therefore, we shall begin with the analysis of image first, then probe into the relationship between the image and its connotative meaning and finally decide whether

Y. Wang (Ed.): Education Management, Education Theory & Education Application, AISC 109, pp. 675–682.
springerlink.com
© Springer-Verlag Berlin Heidelberg 2011

to keep, change or abandon the image. Comparing with shifting the concrete body which conveys the denotative meaning, the shift of the association which carries the connotative meaning is more difficult.

As an image is what makes our language vivid and expressive and the change of the source image is likely to add what is not there in the source text to the translated work and lose what is there in the source text, the source image should be preserved as long as it will not affect the smoothness of the text.

Whether the image can be properly shifted may directly affect the translation. This chapter mainly focuses on the modes of image shift in English to Chinese translation.

2.1 Preserving the Same Image in the Target Language

2.1.1 Reproducing the Same Image in the Target Language

Some images in one language culture can arouse much the same association and produce much the same esthetic pleasure among the readers in another language culture. The national coloring in one language is usually embedded in the image employed. When the national coloring does not affect the meaning and the understanding of the target language, we can reproduce the exact same image in the target language. For example:

heart and soul   全心全意   fish in troubled waters   混水摸鱼

gentleman's agreement   君子协定 castles in the air   空中楼阁

give somebody the green light   给某人开绿灯

In the above examples, the images "heart", "soul", ""fish", "waters", "gentleman", "agreement", "castles", "green light" are all conveyed by reproducing without any difficult understanding by the target readers because these images and their associations are the same in the target language. We know that the image which conveys the same association in the two languages can result in the same understanding of the two language readers. So under such a circumstance, if we reproduce the same image in the target language, both the superficial meaning and the connotative meaning in the source text can be passed into the target language.

2.1.2 Reproducing the Same Image plus Paraphrase

Sometimes an image in the source language has no corresponding image in the target language culture, or an image in the source language does have the corresponding image in the target language but may not convey a certain connotative meaning. In this case, simply reproducing the same image can not achieve the same esthetic and expressive effects in the translated text. So we may reproduce the same image with its connotative meaning added to the translated text. As a result, the target language readers can visualize the fresh image in their culture with the connotative meaning explained so that the receptors' horizons of expectations are broadened. The same esthetic feeling of the source language readers can be retained in the target language readers. For example:

(1) The planners were busy bypassing the Gordian knot.

计划制订者为躲避戈尔迪结—棘手问题;而忙得不可开交。

(2) Falstaff: what, is the old king dead?

Pistol: As nail in door. (Act V, Henry □)

福斯塔夫：什么!老国王死了吗?

毕斯托尔：死的直挺挺的，就像门上的钉子一般。(translated by Zhu Shenghao)

(3) I was not Pygmalion; I was Frankenstein.

我不是皮格马利翁一样的人，自己能享受自己创造的美;我是和弗兰肯斯坦一样，作茧自缚，自作自受。

(4) Bring owls to Athens.运猫头鹰到雅典，多此一举

(5) To carry coals to Newcastle.运煤到纽加索，多此一举。

In the above examples, Gordian knot in example one is an English allusion which may puzzle the Chinese readers if the image is directly reproduced. With the paraphrase of such an allusion, nothing is lost in the translation. Athens in the fourth example is a place rich in owl because according to the Greek Myths, the symbol of the city Athens is an owl. So it is unnecessary to bring an owl to Athens. Without the paraphrase, we Chinese readers may be confused about such an expression. Therefore, in such cases when the image in the source language is fresh to the target language readers, reproducing the same image with the connotative meaning added can not only produce the same esthetic and expressive effects, but also convey different cultural information so that the cultural exchange can be achieved and the receptors' horizons of expectations broadened as well.

Here we should pay attention that with the enlarging of the target readers' horizons of expectations and the development of cultural exchange these images which need paraphrasing in the target language today will become similar to them tomorrow.

2.1.3 Replacing Original Images with Familiar Target Language Images

Some images have their national colors. If such an image has no corresponding image in the target language culture, we may conduct the shift of the image by reproducing the same total image plus paraphrase. But if such an image has its corresponding total image in the target language yet different association is aroused in the two cultures, then we may replace the image with another one in the target language which has similar connotative meaning in the source language. For example: "to talk horse" and "吹牛", the total image of the two phrases are almost the same. If we do not change the image, it will cause misunderstanding and the stability of idioms in the target language will be affected. Therefore, we cannot translate "talk horse" into"吹马" (unless it is used to achieve some humorous effect) but into"吹牛". Similarly, we cannot translate "to spend money like water" into"挥金如水"but into"挥金如土". Another example: in English people say "love me, love my dog." We cannot simply translate it into"爱我，就要爱我的狗". Because in China dog represents the humblest, and always has a negative meaning, as many sayings show: "走狗" "狗仗人势"etc. In this case we should find another image in the target language. As in this case, we may translate it into "爱屋及乌"。 However, the stability of idioms will not be taken into consideration, if it is not translated into an idiomatic expression. The image in the source language could be rendered as"花钱像流水一样".This kind of idioms is also rare between two languages. The two idioms have slight difference. Undoubtedly, certain adaptation will have to be made in translating this kind of idioms, as can be seen through the following examples:

(1) Better be the head of a dog than the tail of a lion.

Translation A:宁为鸡头，勿为凤尾。

Translation B:宁当狗头，不当狮尾。

"Dog" represents the humblest life and "lion" the noblest. The connotative meaning of the idiom is that it is better to stay on a higher position in a lower class than stay on a lower position in an upper class. The Chinese phrase"宁为鸡头，勿为凤尾" means the same as the English one. The images in English phrase are "head of a dog" and "tail of a lion", while in Chinese they have become "head of a cock" and "tail of a phoenix". The choices of the images in two cultures clearly reflect their preferences to certain images and these two expressions. In this sense, the first translation version"宁为鸡头，勿为凤尾"is better than"宁当狗头，不当狮尾". If a Chinese reads the second version, he may be puzzled: why people tend to be the head of a dog?

(2) God's mill grinds slow but sure.天网恢恢，疏而不漏。

(3) He may always possess merits which make up for everything; if he loses on the swings, he may win on the roundabouts. (L. Strachey: Literary Essays)

他也许总是保持着弥补一切的优点，失之东隅的，他可以收之桑。

In the above examples, if the images in the source language are reproduced in the target language, it will misinterpret the communicative information. The familiar substitution images in the target language guarantee the comprehension of the connotative meaning of the source language text. Therefore, replacing image in the English phrase with the Chinese one not only conveys the same connotative meaning but also transmits the similar flavor. In the second example, "God's mill" is a religious image in Christianity and it refers to "punishment from God". The connotative meaning of the idiom is that one who has done evil things is sure to suffer punishment in the end. But the image is illegible to common Chinese readers and they cannot figure out what it refers to. So if the idiom is translated into"上帝的磨转得慢，但一定会转"，Chinese readers will get confused. While in the Chinese东隅(which means "morning") and桑榆 (which means "night") are used to compare to the two situations. The translator uses the familiar images to substitute the unfamiliar images so that the Chinese readers can have a better understanding.

2.2 Deleting the Original Image

Sometimes the image in the source text is hard to keep and even though it can be kept, the translated text would be lengthy in language form and difficult to understand. Or sometimes the image in the source language does not make sense to the target readers at all as the correlation between images and meaning itself is unclear or illogical in the source language and it is absurd or unimaginable for the receptors of the target language. For example:

(1)The teenagers don't invite Bob to their Parties because he is a wet blanket.

Translation A:少年们不邀请鲍勃参加他们的聚会因为他是一个令人扫兴的人。

Translation B:少年们不邀请鲍勃参加他们的聚会因为他是条湿毯子。

"A wet blanket", a blanket which is soaked with water, is converted to "a disappointment" in English. But in China, almost no one can understand such an expression. So in this sense, a translator should erase the original image, which seems to make the translated version lengthy and obscure.

(2) How could she remain as cool as a cucumber at such a critical moment?

在如此紧要关头，她怎么能保持如此冷静。

(3) Every family is said to have at least one skeleton in the cupboard.

据说家家户户都多少有点丑事。

(4) You'd better pull up your socks next term.下学期你最好加把油。

(5) By the winter of 1942 their resistance to Nazi terror had become only a shadow.

到了1942年冬季，他们对于纳粹恐怖统治的抵抗已经名存实亡。

In the above examples, all the source images are deleted. And we can see that it is much easier for Chinese to understand such expressions. Let's take example three for instance. What would a Chinese think when he reads "one skeleton in the cupboard"? Absolutely he should be confused, for no reason should a skeleton in the cupboard be related to "丑事". So in this sense, it is better to abandon the image and only preserve the connotative meaning or resort to other expressions with similar meaning in the target language.

2.3 Image Shift and the Context

In translation, image shift has a very close relationship with the context. The shift of certain image is usually influenced greatly by the context. So in translation the translator should take into consideration some unusual words which can arouse some special connotative meaning of the image. Different contexts of the same image may require different image shifts in the target language. Even if one expression has a very close corresponding expression in the target language, it cannot always be so translated in any circumstances.

Example: "a dog in the manger" is usually translated into"狗占马槽"， and it is similar in meaning with the Chinese phrase"雀巢鹤占".But according to the concrete context, it can be translated differently as follows:

(1) Don't be such a dog in the manger. Lend your bicycle to him since you will not go out this afternoon.
别那么不够朋友，既然你今天下午不出去就把自行车借给他用一用。

(2) Let me have the skates. You don't know how to skate. Don't be a dog in the manger.
把冰鞋给我，你不会滑冰，不要站着茅坑不拉屎。

(3) There you are the dog in the manger! You won't let him discuss you affairs and you are annoyed when he talks about his own.
你不干还不让别人干!不让他谈你的事，可他讲他自己的事你又恼了。

The above is a tentative analysis of the possible reactions a source language causes on the target reader. Nevertheless, the acceptability or non-acceptability of using any of the strategies mentioned above will depend on the context in which a given expression is translated.

2.4 Traps in Image Shift

Sometimes an image in the source language can find a corresponding total image in the target language, which seems as if they have the same connotative meaning. Actually it is not the case. The connotative meaning of the total image may vary greatly in the two cultures. The transplant of the image may lead to misunderstanding. More examples of this kind are given below:

(1) walking skeleton≠行尸走肉
The Chinese phrase refers to an utterly worthless person or one who is lack of capacity. The image in the phrase is to indicate a person. While the English phrase "walking skeleton" means a person who is lean as a rake or skin and bones, in which their figurative meaning are often ignored because"走"is equaled with "walking".

(2)lock the stable door after the horse has bolted/ is stolen≠亡羊补牢
The English phrase means it is too late to take precautions, emphasizing "it is too late". It is similar in meaning with the Chinese saying"船到江心补漏迟"， while the Chinese phrase originates from"见兔而顾犬，未为晚也；亡羊而补牢，未为迟也。"（《战国策•楚策四》） The phrase in its complete form is"亡羊补牢，犹未为晚"， which emphasizes "it is not late yet to take actions." Therefore it equals another English phrase "It is never too late to mend." So the replacement of the image in the target language will damage the connotative meaning of the source image and the target language readers can not get appropriate connotative meaning from the translated text.

(3) a big fish in a small pond≠小塘容不下大鱼，小笼装不住大鸟

The English phrase refers to "somebody who is important only among a small group." For example, "Just because Bill is treasurer of our Pub's Christmas savings club he thinks he is a financial wizard, but he is only a big fish in a small pond." So the phrase can be translated into "小池塘里的大鱼"or"村中无鸟雀为王"，while"小塘容不下大鱼"or"小笼装不住大鸟"means something is really too big and powerful to be contained in a small place.

From the above examples, we may find such traps really merit our attention. The replacement of the source image seems as if the same connotative meaning is conveyed. In reality, the connotative meaning is greatly damaged. Therefore, we should not take it for granted that these phrases in the two languages can correspond with each other. Needless to say, the replacement would mislead the target language readers.

3   Conclusion

It is widely acknowledged by Chinese and western scholars that the ways to shift images vary from translator to translator and the key to a successful shifting of expressions lies in the proper treatment of cultural difference in the hope of promoting cultural exchange in a significant way. Translation is by no means an easy job. Image shift is especially difficult. The translation of images often puts translator into an awkward situation, but a responsible translator will not stop in their effort to transfer cultural image in the hope of promoting cultural exchanges.

It can clearly have no existing "equivalence" in the target language: what is unique has no counterpart. Here the translator's bilingual competence is of help to him only in the negative sense of telling him that any "equivalence" in this case cannot be "found" but will have to be "created".

Translation is a creation-oriented activity. An image in a certain context may be transferred in different ways by different translators. But translators should bear in mind the following principle:

Translators should not rest content with the mere translation of the general meaning of the image. It is their obligation to transfer the cultural image as much as possible.

Immature and defective as it might be in many aspects, it is hoped that new insight into the image shift in the translation might be offered. The author firmly believes that the challenges and criticisms of this thesis will make it a better work and further research and whatever effort in this area will yield fruitful results.

References

1. Agnes, M.E.: Webster's New World Dictionary. Simon & Schuster Ltd (2003)
2. Homby, A.S., Cowie, A.P., Windsor Lewis, J.: Oxford Advanced Learner's Dictionary of Current English. Oxford University Press, London (2004)

3. Baldick, C.: Oxford Concise Dictionary of Literary Terms. Shanghai Foreign Language Education Press, Shanghai (2000)
4. Hatim, B., Ian, M.: Discourse and the Translator. Shanghai Foreign Language Education Press, Shanghai (2001)
5. Nida, E.A.: Language, Culture, and Translating. Shanghai Foreign Language Education Press, Shanghai (1993)

The Construction of Management System for Combination of Sports and Education

Shu Gang Li[1], Peng Feng Huo[2], and Hai Jun Wang[1]

[1] Department of PE, Hebei Normal University of Science and Technology,
Qinhuangdao, 066004, China
[2] Physical Education Department, Environmental Management College of China,
Qinhuangdao, 066044, China
lishugang126@163.com

Abstract. This paper finds that construction and development of college sports teams is facing difficult to combine with the school management system and irrational allocation of resources and related other issues with the continuous reform of education system through analyzing the current situation of combination of education with sport. This study suggests that university should establish management system of combination of education with sport matched with market services and education management under the guidance of combination of education with sport. And it is a main way to culture high-level athletes in college.

Keywords: combination of education with sport, resource distribution, management, college.

1 Introduction

Along with the education system reform, the development and construction of college sports teams faced and how the school management system, and how to make combined with social resources reasonable configuration, etc. Which college sports teams in the "combination of sports and education" policy guidance, and how to establish sustainable development education management in the market service matching integrated management system, is the present university to carry out the "combination of sports and education of training high level athletes the main issues.

2 Universities in the "Combination of Sports and Education of the Present Situation and the Existing Problems

2.1 Objective Orientation and Reality Gap Ability

According to the ministry of education and national sports bureau general higher school on further strengthening the construction of high level sport teams opinion "(education art [2005] no. 3) spirit, high level sports teams manager's goal is to cultivate all-round development of national high level sports talent, so it can bear the

Y. Wang (Ed.): Education Management, Education Theory & Education Application, AISC 109, pp. 683–688.
springerlink.com © Springer-Verlag Berlin Heidelberg 2011

world university games and the major international and domestic sports entries for the country, also task honor plan and competitive sports games contribute to sustainable development.

Sports department intends to gradually change the past level 3 network training system, widen the high cost of China's competitive sports mode of the talented reserves channels. However, after 20 years of development, the construction of college sports teams with the original target localization exist certain disparity. So far, college teams gradually become national athletes retire the -teach base. Efficient utilization recruit retired player and active athletes, swiftly high level sports teams, its purpose is to increase school visibility, improve school sports competition competitiveness. This hasting practice, caused the eager student status, hang in different training, free education outside of the special group. This phenomenon violated the "combination of sports and education", "Olympic plan" and "win honor for the sustainable development of competitive sports" for contributions original intention. Colleges and universities in China in the "combination of sports and education of practice, recruit only nominally does not participate in the actual training high-level practice, obviously didn't have driven team training level has the role of universities. Therefore, our country college as common from training "The Olympic Games is the object of honor" talent is far.

2.2 University Management and the "Combination of Sports and Education" Realization of Conflict

First, admissions imperfect system leads to matriculate significantly different. According to ministry of education college recruit about the provisions of the high-level athletes in Chinese universities, the enrollment of high level sport teams student has four basic ways: one is the cities they retired player; 2 it is not selected schools in various provinces such as they are the first team of students; Three is part; they are in service athletes Four is sports foundation good ordinary high school students. College admissions process, in practice because of its strength as well as by the master of the actual relative resources are quite different, each college enrollment condition, therefore enrollment situation excessive span appear, showing a polarizing individual colleges, and most universities demand to recruit to the high level of athletes. Leading to college admissions high level sports teams managed chaos, make the "order" cultivating combination of sports and education of high level sport generally echelon personnel ideas difficult to achieve.

Second, teaching" and "the body" that the proportion of contradiction aggravates. Imbalance The education and taking the contradiction between the MRT sports training in college sports has always been there. High level sports training water butt improve, depends on the relative concentration of the training time as a guarantee and unified planning of university of literacy class learning often and training the conflict, make training time faulting, no effective guarantee of consecutive, resulting in training not system. This phenomenon violated the "combination of sports and education of sport and education" original intention and feature. The "combination of sports and education of sport and education" enable college development tenet of high-level athletes, and it is also the development of competitive sports in ideal way. Universities can establish distinctive, with high level sports teams in corresponding teaching structure mode, will become hinder college in the "combination of sports and education of sport and education" of the key to success or failure.

Third, lack of scientific training hinder college high level sports teams managed training level. Coaches as sports training planners and organizers directly related to the level of scientific training and athletic performance improved. At present our country college coaches general lack of high level sports training in the theory and practice experience, not formed a high level of coach team. For example, in Hubei province on high level sports teams managed by survey gain coaches coach only 10% level title. Accept the professional training, received 28 per cent of the amateur training, and 3% of 69 percent haven't taken any form of training. Therefore, college coaches appeared formulating scientific training plan difficult, writing, scientific training syllabus and lesson plans and other scientific training problems. In addition, our country college athletic sports whether in coach's cultivation system or coaches training system lack of unified planning. Most high level sports coaches to the lack of necessary high level of training, management, experience of sports entries winning law cognition deficiency. Therefore, increasing college coaches team cultivation is also improve college in the "combination of sports and education of sport and education" of level of the key.

3   Constructing the "Combination of Sports and Education" Management System

The "combination of sports and education of sport and education" is not a simple combination of sports and education of two large systems, but a complex overall cohesion management project. In order to education system and the combination of sports and education of sport effectively, finish the national training high level sports talents, this paper will the task of college combination of sports and education management system for the target, divided into for establishing training plan, establish effective resource allocation, sports evaluation system, formulate performance and incentive feedback mechanism, establishing the logistics management guarantee system and expert consultation evaluation platform six aspects.

3.1   The Standard Set Consistent with Their Own Development Objectives

Set a high level sports colleges and, universities according to the development of school one characteristic can choose the school sports items and get the traditional advantage university widely biggish force project. In addition, combine the development of this area sport, targeted choose features for college sport project, "combination of uninterrupted supply of successful development of high level of athletes for university sports unit, the development of transport team lay a solid foundation.

3.2   Establish Tailored to the Characteristics of Its Own Training Plan

College sports teams according to their own development objectives and the school the real situation enrollment plan formulation explicit solutions, and according to recruit athletes competing plan and current team formulated to determine sports training plan. Sports training program should include training purpose, training objectives, training time, the training intensity, training effect, training objects, training methods, training means, etc. In addition, the development of college sports teams from sustainable development talented person echelon construction. Its contents should include training high-level echelon and high levels coaches' echelon develops. Reasonable college teams high level talents reserves is the universities in the "combination of the key to success.

3.3 Pioneering Diversified Allocation of Resources

In order to make the high level sports teams can reach the goals set by the school and obtain ideal achievements, colleges should input the appropriate resources, including physical capital input and exercise for human resources into two aspects. Along with the development of college sports market unceasingly thoroughly, social resources also unceasingly enter the universities of high level sport teams in construction. In actively expand market resources by hiring outside, sponsored the way national high level of college sports coaches guide. In the "combination of university practice in the process of mercerization business operation mode, the development of college sports team to say is like a double-edged sword, which promoted the rapid construction of college sports teams, but also influenced by the" combination of education for essential concept. Therefore, the development of social resources, the introduction of high level coaches, cultivates the high-quality management and training staff were required to college sports resources rationally.

3.4 Conduct Performance and Incentive Feedback Mechanism

First, set up information files. Sports information archives make the university in the "combination of management system, it is the core data part of this system performance feedback main support sources, it includes athlete information database, coaches information database, game information and achievement database. Second, f formulate appraisal target. In order to effectively assessment team sports training and promote the quality and training effect, deal with sports teams in training period, competitions, and establish the right target cycle of assessment, and constantly promote athletes, coaches, administrator's work efficiency, revising the development direction of the team. Detailed examination of the sports training goals should be including assessment target, game performance appraisal target, team construction development evaluation objects, etc. Third, formulate performance incentive system. Establish and perfect all kinds of incentive mechanism is a college of the orderly development of high level sport teams, also is an important guarantee of this system the core content of information feedback, it is reflective high level sports teams managed the practical work, is also the main indexes of the important way to adjust. Establish a complete set of incentive mechanism, should include achievement performance hook system, athletes comprehensive performance system, coaches performance system, management personnel performance system four aspects.

3.5 Perfect Logistics System

First, perfecting the system of recruit students. To establish a standardized reasonable university high-level athlete recruitment system, is the universities realize the "combination" echelon construction guarantee; it needs education department with sports departments jointly formulated connected, to send oh boy (xiao gao level sports reserve talented person cultivation admissions mechanism. For example, in establishing a fair, reasonable university high-level athlete recruitment quota system; In addition, can advance in secondary school students with athletic screening, and the establishment of related scholarship athletes, provide training funds from legal aspects, such as signing relevant contract to standardize and full college enrolling high level sports talents working. Secondly, formulate unique registration management mechanism. The purpose of the "combination is complete the teaching goals, and at the same time, and then use the advanced knowledge and methods of training, improve the competitive level of sports. At present, our country in the "combination" in the process of practice, the proportion of teaching and training and time schedules are contradictory boycott phenomenon, so "learning" and "training" didn't play a benign interaction. This paper argues that college explore the credit system in the "combination of student management pattern is to make the" combination "implement the fundamental into practice. Its pattern should include establishing separate teaching content, organization unique teaching materials and teaching plan, layout, formulate specific teaching schedule, establishing relatively independent performance management, etc.

3.6 To Establish an Expert Consulting Platform

As a scientific training of university college high level sports teams managed, a phase of the competition, training, management condition should be relevant experts assessment guide, it includes the annual review summarizing evaluation, evaluation, cycle assessment of the troop construction important contents. Through the annual summary evaluation, cycle of summarizing assessment has been completed, the work reasonable evaluation experts offer continued development of good advice, and actively improve limitations, and exert its advantages. To establish an expert consulting platform can be classified into regular hiring external experts and organization oneself experts two forms, the two forms of effective combination can will university practice in the "combination of normal evaluation and cycle assessment, effectively combines for university high level sports teams managed services.

4 Summary

University practice in the "combination of development of competitive sports in China is facing the world, is an important manifestation of sports development of national economic trend, as well as fully embodies sports to the people-centered development philosophy. University of high level sport teams in the construction of traditional training, management mode and method, but also a challenge for a new management model of exploration. Universities in the "combination of construction, national high level sports undertakes the mission, cultivation of talents education department also shoulder the task of campus culture construction, so it's development can't single itself in a separate management department, it should be as full systems have sports departments and the competent departments of education, form a tight integration of construction of new system. So as to truly become our country college high level sports training base of target service.

References

1. Zhou, Y.-c.: On The Combination of Sports and Education in the Perspective of Scientific Development Outlook. Journal of Beijing Sport University 01 (2008)
2. Fang, Q., Liu, Y.-z., Zhou, J.-s., et al.: A New Vision of "the Combination of Sports and Education" in the construction of high level sports teams for colleges and universities. Journal of Capital Institute of Physical Education 02 (2008)
3. Li, Y., Liu, C., Dong, Y. j.: Research on Some Question and Counter Measures of the Executing Strategy of the Sports and Education Combination. Journal of Huaibei Coal Industry Teachers College (Natural Science Edition) 2 (2009)
4. Dong, J.-f.: Research on the Trend of Combination of Sports with Education. Fujian Sports Science and Technology 3 (2009)
5. Tan, Z.-W.: An Analysis on the Implementation of School Sport Policy in High Vocational colleges. Journal of Jiangsu Radio & Television University 32 (2007)

Reflection on Golf Education Development in China under Leisure Sports Perspective

Xueyun Shao[1,*], Zhenming Mao[2], and Xiaorong Chen[3]

[1] Golf College, Shenzhen University, Shenzhen 518060, Guangdong, China
[2] College of P.E. and Sports, Beijing Normal University, Beijing 100875, China
[3] Department of P.E., Shenzhen University, Shenzhen 518060, China
sherryshao_1982@hotmail.com

Abstract. The purpose of this thesis is to discuss the practical issues of golf education, and come up with solutions to these problems through literature searching, interviewing and logical analyzing methods. In the backdrop of leisure sports education, golf education is exposed to problems such as shortage of professional golf educators and teaching materials, high tuition which provides high threshold to learners and lacking of education concepts. This thesis thinks deeply about the future of golf from theoretical side and put forwards solutions to the above problems such as combining golf materials and leisure sport materials, inserting golf education in the leisure sport education, adopting simple equipped practice range, cultivate golf professionals with leisure sport concept and developing indoor simulating golf.

Keywords: Leisure sports, Leisure sports education, Golf, Golf education.

1 Introduction

Golf education started late in our country due to many reasons. Experts have proposed universal problems such as shortage of teachers and teaching materials, lack of spare time, and unreasonable course offering. On the other hand, leisure sports education's entering China has aroused great interest among researchers.

Leisure sports developed vigorous in well-developed countries, China is also on its way to advance this. So people's recreational activities inseparable from leisure sports during times of its increasingly all-pervading today, therefore leisure sports education which represents persuading a high quality of life and training people to develop a lifelong habit of leisure sports may finally accepted. Leisure sports education has gradually become hot issues and some experts and scholars commence paying more attention to it.

Golf education will guide the golf games toward good direction, allowing golf management to be more professional, more systematic, thus providing better service to it. No doubt golf leisure sports education is the core of the development of golf, letting people receives good education to provide enough talents for the rapid development of the golf industry, and also provides academic guidance for the golf

---

[*] Room2702, Golf College, Wenke Building, Shenzhen University, No 3688, Avenue of Nanhai, Nanshan District, Shenzhen 518060, Guangdong, China.

Y. Wang (Ed.): Education Management, Education Theory & Education Application, AISC 109, pp. 689–695.
springerlink.com   © Springer-Verlag Berlin Heidelberg 2011

industry development. Including the construction of the education system, curriculum system, teaching theory, teaching training and constantly improve, past golf education has made giant achievement to the rapid development of the golf industry, approving a number of high-quality personnel who contributed a lot in the industry operation force. But the golf education starts late in our country, existing some shortcomings due to various reasons inevitably. Under such background as mentioned above, combining with some ideas of leisure sports education, this paper tries to find out the deficiencies and put forward some improved golf education proposals from the theoretical level.

2 Methodology

The purpose of this thesis is to discuss the practical issues of golf education, and come up with solutions to these problems through literature searching, interviewing and logical analyzing methods.

2.1 Literature

By reviewing recent years' literature of leisure sports, leisure sports education, golf education and reading lots of related scholars' research results, the author obtained the necessary thoughts and theory for understanding, and also searched related information through Internet as reference, combining leisure sports education and the education reality.

2.2 Interview

For the purpose of this study, the author communicated with tutors, related teachers and students about golf education issues, accessing relevant information, and then carried on further ponder and research.

2.3 Logical Analysis

The author obtained macro level of understanding and the related phenomenon by logical reasoning and analysis data or information coming after inspection or survey.

3 Results and Discussion

3.1 Leisure Sports and Its Development

What leisure sports means? With the rapid development of economy, people have more and more leisure time for leisure and entertainment, and leisure sports is the most popular choice, however what are the leisure sports? According to Baike's explanation by Baidu searching which shows: leisure sports is a component of social sports. Refers to the people in the leisure time to enhance physical and mental health, wealth and the creation of life, improve the self for the purpose of physical exercise activities [1].We can see it mainly refers to the leisure time and physical exercise activities. Scholar Maoxiang Xiong thinks that leisure sports is the activity that people want to meet their main purpose of mental and physical pleasure development needs in their spare time, it contains certain cultural elements inside [2]. Leisure sports is a freedom participate activities which is taken in spare times to meet the self actualization needs by physical exercise as the basic means under certain leisure state and environment. That is the expiation given by Guozhong Li, Xueqin Yang in their book" leisure sports"[3].

We can see from the above definitions that leisure sports understanding is inseparable from that of leisure time, leisure, sports and two categories, as shown in Figure 1, leisure sports can be seen as a unification of leisure and sports, take activities to under the leisure culture and leisure psychological experience, also access to enjoy the spirit besides the body exercise for health.

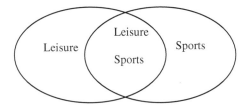

Fig. 1. The Interpretation of Leisure Sports

The trend of leisure sports future development. Nowadays people increasingly feels the fast pace of life, and also the increasing pressure, usually require appropriate leisure sports activities to relax keeping the living and working conditions. Because there is less and less time to exercise, healthy sports will still bring certain burden, so people started to take entertainment sports. Gradually, the leisure sports may become the mainstream methods. According to their own actual conditions such as the physical quality, consumption level, People choose different leisure sports such as walking, badminton, bowling, golf and so on. To participate in leisure sports is a kind of healthy lifestyle, health is the basic guarantee for people continuing to live and work, leisure sports is the essential part of outdoor keep us healthy, can give people to breathe fresh air, opportunities for exercise, but also can let a person away from tobacco and alcohol, to guarantee the health of some people.

There are many people who treat leisure sports as communication means, such as "prefer to play a round of golf instead of having dinner with you" has become an important way for people to meet new friends, to expand interpersonal relationship circle and business communication. After all, compared to other recreational activities, it can be widely accepted when some friends taking relaxing and health leisure sports. In addition, because of television, Internet and other media marketing force, people have more understanding of various sports activities, like golf is a common sport aboard, but in China there are still many people who are not very familiar with, and thus have a misunderstanding. Television and Internet media greatly accelerate people awareness of the penetration rate of leisure sports and aroused great interest of participation. Although unable to participate in the professional competitions, but it does not prevent people like it and participate in it. In view of the characteristics of leisure sports and influence, more and more people will definitely involved to participate in recreational sports.

Leisure sports education in colleges. Leisure physical education is introduced into our country recent years from developed countries such as Europe and America, in China is still an emerging discipline, leisure physical education is now started relatively late in our country, till at the beginning of the twenty-first Century some colleges offer relevant major, primary and secondary schools basically not involved. Through consulting relevant literature as well as the landing site to search out the institutions, the following colleges have related lessons, they are: School of sports science of South China Normal University, provided with a course called" Introduction to leisure sports" for the weekly classes; Guangdong Ocean University, opened the social sports specialty under the "Binhai Sports Leisure Management", with the help of the advantage of subject and marine features, based on Hospitality Management professional, sports science, management science, tourism science and sports leisure entertainment fitness of organic integration, the course emphasis on cultivation management and technical personnel; Guangzhou Institute of Sports offers major of leisure & management; Sports economy management college in Wuhan Institute of Physical Education also changed its name in 2005 having leisure sports specialty; Shenzhen Tourism College of Jinan University opened a golf and Leisure Management Specialty of Nanjing Institute of physical education; and the Occupation Technical College of Nanjing Sports Institute has fitness and recreation specialty and the Golf College of University Shenzhen opened" Leisure Sports Management" course.

Table 1. China's Leisure Sports Education in Universities/Colleges

No.	Universities/Colleges	Specialties/Majors
1	Guangdong Ocean University	Binhai Sports Leisure Management
2	Guangzhou Institute of Sports	Leisure Sports & Management
3	Wuhan Institute of P.E.	Sports Economy Management
4	SZ Tourism College Jinan Univ.	Golf and Leisure Management
5	Nanjing Institute of P.E.	Fitness and Recreation
6	Golf College, Shenzhen Univ.	Leisure Sports Management
7	South China Normal University	Introduction to Leisure Sports

3.2 Current Situation of Golf Education in China

Classification of Golf Education in China. The development of golf education in our country is not a long time, from personal development, golf education can teach person with golf skill, understand the culture of golf, and train students to become a professional golf talents; from the perspective of social development, golf education can delivery industry suitable talents, promote the development of golf game, guide the right direction for golf industry development.

Golf education can be divided into the golf specialty education and golf course in P.E., the golf professional education including golf diploma education and skills training. Golf diploma education can be divided into golf higher professional education and the golf specialized secondary education, while occupation skill training includes golf caddy training, course management personnel training as shown in Figure 2.

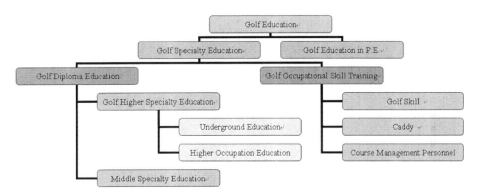

Fig. 2. Classification of China's Golf Education

Problems of China's golf education. Through consulting the literature and reading a lot of information, the author found some issue exists in China's golf education, such as shortage of teachers, lack of curriculum materials, not enough space and limit opportunity for golf practice and also not allowed in the specialty directory and so on. Undoubtedly these are the common issue raised by many scholars. In the leisure sports background, by searching of the relevant literature and reading a lot of information, the author still has the following considerations, combining with the leisure sports education, from the perspective of leisure sports education.

Golf education lack of leisure sports values. Under the leisure sports background, golf education nowadays lacks human-oriented values and leisure viewpoint of value. With the development of leisure sports education, health and sports leisure sports will eventually be replaced. In golf education, people also need leisure sports view guidance. The current golf education imparted to the students a lot of golf professional knowledge, cultivate them has the theoretical and practical ability to meet the needs of industrial development, but has ignored students' leisure sports education.

The idea of leisure sports education in the materials must be strengthened. From the existing golf related teaching materials can be found in these materials is to teach only golf professional knowledge, rarely involves the leisure physical education content. However, golf is a recreational sport, with its unique charm, if the golf teaching material learning into more leisure sports education, combining leisure sports education and golf education together can significantly improve the talent quality.

Shortage of teacher who has combined knowledge with leisure sports and golf. With the rapid development of golf, several universities open golf specialty or courses responsive to community needs, but in the lack of experienced golf teachers who has both theoretical knowledge and experience. Facing the same problem, leisure sports education also lack of teaching staff widespread. Therefore looking for teachers with the golf professional knowledge and leisure sports education knowledge at the same time is very difficult.

Golf education fees raise educational threshold. Golf in our country still belongs to the high consumption of sports, for example, golf clubs in the early years earned a lot. You have to pay 5000 RMB or more than 10,000 RMB for a set of foreign brand clubs; play a round of golf will take you 500 RMB to 1000RMB average. The higher prices of consuming golf also have some influence on the golf education fees.

3.3   Reflection on the Future Development of China's Golf Education

Golf education should leaning into the idea of leisure sports education. Unlike the previous way of life, leisure sports are a kind of new value pursuit. It not only means taking exercise in a relaxed pleasant atmosphere, but also means it is more important to enjoy the spirit in the process. As a recreational sport, golf have a lot in common in education content with other leisure physical education, thus referring to the leisure sports education concept, and integrating into the golf education, then students help students master the golf culture essence, deepen the golf leisure understanding in the learning process.

Cultivate golf personnel with thoughts of leisure sports education. The development of the industry is inseparable from the cultivation of talents, the golf industry as a leisure sports industry; enterprises need to have the leisure sports values talent. With the leisure sports concept talents are proactive, love, enjoy leisure sports spirit, suppose they entered the company will bring their own leisure sport idea to work, and also reflect in the way one gets along with people, customer service, not only enhance the self working and living level, also will own the concept of integration of leisure sports in the enterprise values and the enterprise culture, thus promote the rapid development of golf sport.

Combination with Golf teaching materials and leisure sports teaching materials. The quality of teaching level affects the quality of trained personnel quality, the cultivation of talents will be greatly discounted due to bad materials, so it is really a fantastic idea to consider put leisure sports education's contents into golf material with. Not let the content independent or forced to join, but to identify golf education and leisure physical education in common, then integration the two contents reasonable, in order to train golf talents with leisure sports knowledge.

4 Conclusion

This paper surveys and studies related content of leisure sports education and golf education, the development time of leisure sports education and golf education in our country is not so long, so it needs more practice and theoretical study. Under the background of leisure sports education, the development golf education has a new direction, However, we have to admit that there are several problems exists in the golf education, such as lacking of leisure sports education concept and content of in teaching, teaching material, course and so on. As a recreational sport, if it can in the integration of more leisure sports education, cultivating people leisure sports values, golf education can effectively promote the golf development in education.

References

1. Baike Baidu Searching LeisurSports (2011),
   http://baike.baidu.com/view/186152.htm#sub186152
2. Maoxiang Xiong, M.: Physical Education in University, vol. 98. Xiangtan University Publisher, Xiangtan (2008)
3. Li, G., Xueqin Yang, M.: Leisure Sports, p. 23. People Sports Publisher, Beijing (2007)
4. Kierstein, Resa, J.: Innovative Golf Programs. Parks and Recreation 45 (2010)
5. Hayward, Phil, J.: Supply, Demand, and the Real Future of Golf. Parks and Recreation 45 (2010)
6. Sherman, P.B., Jeff, M.: Golf Courses: An Asset for Communities. Parks and Recreation (2010)

On the Intellectual Property Right Protection Issue of Digitized Resources in a University Library

Tingrong Liu

Library, Shanghai University for Science and Technology, Shanghai 200093, P.R. China
shlgtsg@163.com

Abstract. The problem of intellectual property in various aspects during the constructing and applying of digitized resources in a university library is analyzed. The influences of relational law rules of china government in intellectual property protection to the digitizing process of a modern library are discussed. Some reasonable operating techniques are also put up for the digitized resources operation in the library administration.

Keywords: digitized resources, intellectual property, library, information service.

1 Introduction

With the wide application of computer technology and the development of the Internet, Digital type literatures which are published by CD and/or internet accounted for more than 50% of the world's total publications, and will take even greater proportion in the future affirmatively. literature resources system of a university library is also changing, many kinds of digital information resources is providing users with more convenient and quicker service, a luxuriant digital resources will become an important feature of modern library.

Library digital literature resources include the digitalized traditional literature documents and the digital resources published in the form of the electronic data medium or on a network, including online databases, electronic journals, digital books, computer software, video or audio materials. With the advancement step of the digital library, the conflicts between the sweeping public use of digital literature resource and the intellectual property protection become prominent increasingly in China, and related legal disputes are accumulating rapidly in recent years.

The contradictions between intellectual property protection and application of digital literature resources in university libraries will greatly affect the constructing process of the digital library in China. In this paper, based on a pertinent study of the relevant regulations of the National Copyright Law, a reasonable technique is put forward for the digital resources operating in library. It is propitious to a rapid and healthy development of digital library in a reasonable and lawful precondition and so has a positive meaning to the construction of China Digital Library.

Y. Wang (Ed.): Education Management, Education Theory & Education Application, AISC 109, pp. 697–704.

2 The Digitization of Traditional Library Collections

The digitizing of the traditional printed literature plays an important role in the digitization process of library resources. Generally those literatures are made into electronic documents for reader to access by scanning. In this process of digitalization image files are often used. However, text file can also be gained form images by OCR software. The advantage of image file is its reality for the original material and the process is relatively simple. But the image files need a large memory space so that to slowdown the transmission speed in a network. On the other hand, the text format files are more convenient for computer retrieval.

Almost all the traditional paper printed literatures in a university library can be digitized to provide more convenient serves for the reader. Digital document can be searched via a computer software and transfer through internet for reader to access other where, and digital copies also have the function to protect some precious documents. From a technical point of view, digitization of library collections is equal to the copy operation, so this process must abide by the reproduction right concerning national works copyright regulations.

Reproduction right has another name called refashion right, which refers to the right that makes documents into one or more copies by printing, photocopying, recording or video recording. Reproduction right is an important right among the copyright property rights in the world. Duplicating is an important means to make documents spread widely and serve more people, so the author's copyright is embodied mainly in the exertion of the reproduction right.

For those library collections which are protected under the intellectual property law at the present, the digitizing work should abide by regulations of intellectual property law. However, according to the 21st item of the China Copyright Law, in some cases, people can use a literature legally without the permission of the copyright owner and free for charge. This situation is called Reasonable Use in the law. Below are the 6th and 8th terms under this item.

(6) For classroom teaching or scientific research intention, to translate or replicate published work in a relative small amount, and used by teachers or scientific researchers, not for publication.

(8) Copying of the literature collection in library, archives, memorial hall, museum and art gallery etc. for the display or conservation need.

Form a careful analysis of the above two conditions, it can be concluded that a library digitization work of collection literature should accord with the purpose in 8th item for the strict restrictions which require the copied literature must belong to the library itself. In addition, the purpose must to be preservation, namely is to protect those precious or delicate collections, which are likely to damage under frequently use so an electronic alternative is needed to provide services instead of original one. The executors of document reproduction defined in 6th item should be direct users, such as teachers, students or researchers and for teaching or scientific research purposes. However, the library digitization is usually intended to provide information service for consumers, so its purpose does not conform to the regulation in strictly speaking. Although the difference between two purposes is certainly very difficult to distinguish, the object of digitization must belong to the library. For a literature comes form outside library collection, it contravene obviously the main requirements in the

8th term aforementioned, and the digitization does not belong to the Reasonable Use category.

To provide services under the environment of network is one of the most common features of a Digital Library. The digitization activity for protection printed literatures does not infringe the right of reproduction, but it is likely related to the deregulation of the right of information network dissemination in the China Copyright Law when a library provides its information serves via internet. This copyright protection item is just bringing forward recently to protect the author's rights in a network environment.

Regulations in the 47th item of the China Copyright Law define an unauthorized behavior to disseminate copyrighted works in public through information network as a tort. Only the copyright owner has the right to decide whether his works can be disseminated online, so it is a tort behavior predicatively to spreading digital literature resources without the copyright owner's authorization and a civil suit may be brought. According to the 47th item in the China Copyright Law the digital literature copyright is still owned by the original author even though it is digitized by library, and only the copyright owner has the rights of dissemination through the network. This right can not be infringed by anyone.

Therefore, after the digitization process of collection literatures in a university library is finished, the scope of the digital information spreading must be strictly limited in an extension of the library services. The library can only provide readers to brows inside its electronic reading room or via the network of the university. If the digital literature is spread to internet without permission of the author and other people outside the university could contact the literature content with no restriction, the library is beyond the Reasonable Use category and infringes the China Copyright Law even though it did not receive any economic benefits.

3   The Purchase and Use of Business of Commercial Digital Resources

The commercial digital resources of a university library generally refer to network databases, electronic journals, electronic books and other forms of electronic publications.

Because a digital literature is much easier to duplicated and disseminated than traditional printed document, the network database providers and publishing company always attempts to limit the scopes and modes of the digital resource sever in the contract of sale to their own economic interests. Therefore, the library should try to strive for more benefits for its reader according to the principle of Reasonable Use.

3.1   Network Literature Database

Network literature database has the advantages of quick updated of contents, high retrieval efficiency, large amount of information and it is the main part of library digital resources. Since most database provide free document retrieval on its website, full text database has become the most valuable resources of digital library, including domestic and foreign periodical k database, boo database, technical standard database, patent database and dissertations database. Equipped with those network literature database the university library can provide a convenient and efficient information service for the readers inside the university network coverage range legitimately. However, sometimes a problem of intellectual property rights also present by improper management of digital resources.

The network database different form traditional library resources and is actually a kind of virtual library. Library has no ownership of digital resource, only the right of use in a definite period of time. Library users cannot access the database resources when it is expired. That is to say, the network database can be used as a part of library collection resources only after being authorized and what the library purchase is only a right for the network database access, the ownership or copyright still belong to publishing company. In addition, either use the database directly through internet access to the database provider server or establish a mirror sites, publishers always try to attached strict limits on the use of the digital database in order to protect their own profit. When library and database provide sign the License Agreement the rights and obligations of both parties should be strictly defined, including scope, method and purpose of the network digital resources use, most content items reflects the attempt to protect the intellectual property rights of database publishing company.

The library and its legitimate user have the right of using the network database digital resource. There are many methods to define legitimate user technologically under the environment of network, such as IP address binding, the user password authentication and the concurrent access user number limitation. Many behaviors will be regarded as violations of intellectual property rights, such as unauthorized setting up a proxy to carry out database visit from unauthorized IP address, the using multi-threaded download software such as the "network ant" to download large amount of data balefully in a short time, or composing another databases using network database contents as its basic material.

3.2 Electronic Journals

Electronic journals are digital journals spread through the Internet and other electronic media such as CD. The application of academic electronic journals has exceeded the traditional printed periodicals in recent years. The quantity of electronic periodical purchased by university libraries in china also increased fast.

Electronic journals are a database according to academic subject which is built by database developer who has obtained relevant journals publishing distribution rights. User can get in to the homepage of electronic journals and searching, browsing and downloading papers. Now the electronic journal has a wide coverage in many subjects, but the time of publication often lagged behind than printed paper.

The application of electronic journals in university library should also abide to the copyright contract. It should be point out that if the subscription the network type electronic journals is stopped in some time, the library will lost the use right of all relative publication. In this case, a CD type journal should be required from publishers, or all purchased journals papers should be copied from the internet by the authorization of the publishers in order to maintain the benefits of library and the its readers.

3.3 Electronic Publications

Electronic publications is the multimedia information resources in various type of CD, magnetic disk, recorder tape or computer software, including the CD comes with books. Compared with the traditional paper publications, the electronic publications can contain a large amount of information, high reliability in use, stronger interaction capability. But its making and reading process requiring corresponding hardware and software supporting.

For university library the main problem of intellectual property right protection related to electronic publications is piracy, so the library should purchase legal electronic publications from the formal channels, providing network access and software of electronic reading for readers to use. Readers can perform a few copies of paper for the studying, researching, enjoying and other non-profit activities. Those activities are not beyond the scope of the Reasonable Use category defined by China Copyright Law. However, if the library made copies of electronic publications in mass by means such as CD writing or printing and carry out a paid service, it is a significant act of tort. In addition, without the authorization of the copyright holder, people can not edit their database by information copy form a commercial electronic publications, which activities also made a copyright infringement of the legitimate rights and interests. In recent years, some network database company and electronic periodical publishers agree that university library can save electronic copy as a backup in their license agreement, but the library should expressly restrict copyright.

4 The Resource Literature on the Internet

The computer network is an indispensable service tool to digital resources in library. Modern library provide readers with network service through its own website so the literature copyright problems in computer network should not be ignored.

The copyright in network specify a protection under the network environment. In the 10th item of the China Copyright Law, it is declared that traditional literature copyright is still effective in network electronic environment. The specific protected status of online literature copyright was also admitted officially in an important document "The Supreme People's Court on the trial of cases involving copyright disputes over computer network applicable legal interpretation". It is clearly explained as: "To spread literature to the public through the internet belongs to the methods of the literature application which is authorized by the copyright law. The copyright holders can use or permit others people use related literature by this method, and thus obtain reward.

Information resources on the internet are also under the intellectual property protection although it is open to the public professedly. However, browsing, copying, downloading resource on the Internet, if is purely for personal study, research, and appreciation and have non commercial benefit, it does not infringe the literature copyright. The library can not put unauthorized soft wares, music files, pictures, and movies on its own website. For valuable online resources, library must have the author's permission in advance and annotated the source when excerpt into websites, otherwise it will constitute an infringement of copyright. In addition, if the library put a intellectual resource links to webpage openly or download online information or software then edit to own digital resources, it also can cause the infringement of network author's intellectual property rights.

5   The Use of Digital Library and Management Strategy

5.1   Constructing and Arranging of Digital Resources

Digital library still needs to undertake resource constructing and arranging work in order to improve the efficiency of resource. Such as electronic journals catalog, information navigation system, special subject database, digital resource integration, network training of reader, personalized information service, which will be very helpful for the improvement of effective of digital resources application. It is needed to point out that all of these activities may involve the syllogistic rights of digital literatures, without the prior permission, it may also bring about the dispute of intellectual property.

The right of syllogism contains the copyright of changes the language form of literatures, includes adaptation, translation, annotation, organization and edition. A clear definition about the right of syllogism is stated in the 10th item of the China Copyright Law. Form this point of view some library activities very likely have violated the deduction right of the database provider such as classifying according to the resources disciplines, integrated of network database information, establishing a special literature database, or generating their own information service database according to the results of network database retrieval.

5.2   Digital Information Service

The main mode of library providing information services is document retrieval, science and technology original testify, document delivering and so on. The literature service may relate to the intellectual property rights, especially the document delivering form another library, if the processing methods or the intended usage is improper, exceeding the scope of the Reasonable Use; it also will constitute a tort. For example, universities libraries are generally carried out interlibrary borrow services in order to improve the utilization rate of electronic literature, replenish the shortage of the resources and increasing the possibility of the readers accessing to literature data he wanted. However, the development of interlibrary borrow operation will certainly reduce the number of database repeated purchase, lead to damaging the economic benefits of database providers in a certain extent. In order to prevent the spread of electronic information resources and illegal use, the database provider often ask to add regulation in digital database license agreement to limit the interlibrary loan service or totally prohibit. The library should comply with the relevant requirements in the license agreement.

5.3 Public Domain Literature Resource and Its Utilization

The protection of a given intellectual property has a limited period. Exceeding the statutory copyright protection period, the literature will belong to public domain, and the user not need the consent of the original copyright owner or pay the remuneration. In the China Copyright Law the general period of the copyright term is fifty years, the period of civil works begin to calculate from the death of the author, period of a legal person or organization works begin to calculate from the first published data. In addition, Documents of national and local laws, government regulations, resolutions of state departments, official decisions and orders, legislative, administrative or judicial files , and their official translations; news, calendar, tables of general public data, mathematic formulas are also not subject to copyright protection.

When take use of literatures which has entered the public domain, the library should also give intentions to respect of paternity right, right of revision as well we the protection the integrality.

6 Conclusion

At present, the protection of intellectual property rights is getting greater attention of the whole society. As a new type of digital resources of information resources to university library, the purchase, management and service of digital literature are inevitably facing the problem of intellectual property.

The development and construction of digital resource is the foundation of the digital library construction. Now the information technology and computer network is creating a new era of digitalization of university library. The network literature retrieval, online reading and download, network document delivering service, real-time update of the literature databases, electronic books and other new information technology and service means come in to use fast. The university library is facing more and more strictly restrictions of intellectual property during the process of digitalization. Therefore, the conflict between legitimate rights of the publication company and the benefits of readers should be handled correctly. In the premise of obeying the laws of intellectual property, the university library should make a full, reasonable and legitimate use of digital resources, providing readers with higher quality service. It is helpful for the enhancing the library management level and accelerating the development of digital library in china.

References

1. Witten, I.H., Bainbridge, D., Nichols, D.M.: How to build a digital library. Morgan Kaufmann Publisher, Burlington (2003)
2. Lin, y.: On the Network Environment and Intellectual Property Issues of Library. J. of Agricultural Library and Information Sciences 15(7)
3. McCray, A.T., Gallagher., M.E.: Principles for Digital Library Development. J. of Communication of the ACM 44(5)

4. China Copyright Law and Implementation Regulations. Founder Press of China (2003)
5. Hu, t., Hu, M.: On the Digital Library of Copyright Infringement. J. of Library Science of China (2001)
6. Koboldt, C.: Intellectual property and optimal copyright protection. J. of Cultural Economic 19(2)
7. Li, A.: The Introduction of Digital Resources of Academic and Management. Southeast University Press of China, Nanjing (2005)
8. Huang, B.: College Library Resources and Digital Resources Property Rights Protection. J. of Intelligence Theory and Practice of China 02 (2007)

Research and Exploration of Light Chemical Engineering Specialty Excellent Engineers School-Enterprise Cooperative Education

Lizheng Sha and Huifang Zhao

School of Light Industry, Zhejiang University of Science & Technology,
Hangzhou, 310023, China
slz9966@yahoo.com.cn

Abstract. "Excellent engineers education training plan" is the major project for China's higher engineering education reform, and the stage of enterprise learning is the key to the success of it. This paper introduces the ideas and implementing measures for training light chemcal engineering specialty excellent engineers through school-enterprise cooperative education in Zhejiang University Science & Technology. The key of school-enterprise cooperative education is to establish a workable engineering knowledge and engineering ability standard and to make a study plan which throughout four stages and five practice projects, thereby cultivating students' engineering awareness, engineering quality and engineering practice ability. Establishment of "order" educational mode can arouse the enthusiasm of enterprises to participate in talent training, and assures the success of school-enterprise cooperative education.

Keywords: light chemical engineering specialty, excellent engineers, school-enterprise cooperative education, engineering practice ability, enterprise learning.

1 Introduction

The developed countries are reforming their engineering education mode to meet the requirements of the development of modern engineering, for example, CDIO (Conception、 Design、 Implementation、 Operation) engineering education mode is the latest achievement in international engineering education reform in recent years[1]. Under the background of economic globalization, China actively promotes the reform of the system of engineers, and "Excellent engineers educating and training plan" is the major project for China's higher engineering education reform at the present stage. Through the implementation of "excellent engineers education training plan", we can cultivate and bring up a large number of high quality engineering and technical talents with innovation ability to meet the need of economic and social development[2].

The enterprise learning stage emphasizes the training of students' engineering consciousness, engineering quality and engineering practice ability and is the key to the success of "excellent engineers education training plan". For a long time, China's

Y. Wang (Ed.): Education Management, Education Theory & Education Application, AISC 109, pp. 705–710.

higher engineering education pays attention to theories, and makes light of practice, focuses mainly on knowledge but not the ability to apply this knowledge[3]. As a result, engineering design and practical education is seriously short, and it is hard to train high quality engineering talents. Zhejiang University of Science & Technology is one of the first pilot colleges implementing "excellent engineers education training plan", and light chemcal engineering specialty is one of the pilot specialty. In recent years, Zhejiang University of Science & Technology undertook active exploration on how to carry out excellent engineers school-enterprise cooperative education of light chemical engineering specialty.

2   Ideas and Thoughts of School-Enterprise Cooperative Education

Light chemial engineering specialty (pulp and paper engineering direction) has strong characteristics of engineering, the cultivation of students' awareness of engineering, engineering quality and engineering ability must face the engineering practice, and is achieved through training of engineering design, engineering operation, field production, technical management. In line with the teaching idea of "focus on ability training, face the engineering practice, enhance students' awareness of engineering, engineering quality and engineering ability", we research and explore the ways and details of school-enterprise cooperative education in the innovative education mode of "3+1". Taking the establishment of a perfect enterprise learning stage training scheme as the breakthrough point to attract enterprise to actively participate in engineering talents training reflects the thought of "learning by doing" in school-enterprise cooperative education [1,4], and make the graduates get engineering practical ability and good comprehensive quality. Graduates can take part in the pulp and paper production directly to achieve the training goal of "zero adjustment period".

3   Training Scheme of Enterprise Learning Stage

3.1   Knowledge and Ability Standards of Enterprise Learning Stage

Engineering knowledge and engineering ability training standards are the most important content of training scheme of enterprise learning stage, and the implementing system and evaluation system should be set up based on this training standard. According to the requirements of excellent engineers training, engineering knowledge and engineering ability can be divided into four modules: engineering basis knowledge, engineering expansion knowledge, basic engineering ability and comprehensive engineering ability. It is important to determine the knowledge points and ability points, to detail and perfect the content and requirements of these knowledge points and ability points, and to set up a series of operable knowledge and ability training standards.

3.2   Study Plan of Enterprise Learning Stage

In order to achieve the knowledge and ability standards required in enterprise learning stage, study plan should include five practice projects: cognition practice, on-the-spot teaching, field work, engineering practice and graduation design (thesis), and four stages: engineering cognitive ability training, basic skill training, engineering application ability training and engineering comprehensive ability training. Specific learning content and training should be set to cultivate students' engineering consciousness, engineering quality and engineering practice ability.

Stage for engineering cognitive ability training. In this stage, it is important to improve students' engineering consciousness and engineering cognitive ability through the implementation of cognition practice and other level 1 practice project. At the same time, as a support and improvement of level 1, on-the-spot teaching of specialized courses and other level 2 practice project must be emphasized to cultivate students' engineering consciousness and to improve their engineering cognitive ability and engineering learning ability.

Stage for basic skill training. This stage places great emphasis on training students' basic operation skills, and improving students' cognition to technology, equipment, control, installation, maintainance, product analysis and inspection, and these goals will be completed through the implementation of field work and other level 3 practice project.

Stage for engineering application ability training. This stage focuses on training students to master certain engineering method and to solve technical problems in the process of pulp and paper with knowledge and skills they have learned, thereby improving their engineering consciousness and engineering quality.

Stage for engineering comprehensive ability training. This stage focuses on training students' comprehensive application ability and further improving students' engineering quality and engineering practice ability. It can be completed through the implementation of graduation design (thesis), which is targeted at enterprise's actual engineering project, research products or technology.

3.3   Establishment of the Reliable Ways to Implement the Goals

Effective school-enterprise cooperative education relies on the reliable implementation ways. Specifically, it is necessary to create favorable conditions to cultivate engineering consciousness, engineering quality and engineering practice ability, and to organize the implementation.

Creation of cognition and operating conditions. According to the requirements of the level 1 practice project, engineering technicians in the enterprise explain the basic process of pulping and papermaking to students and show students around the production site, including the pulp and paper production line, auxiliary projects, papermaking wastewater treatment project, paper products testing process and so on.

According to the requirements of the level 2, part of the specialized courses are taught on the spot, that is, part of the classroom teaching is replaced by the on-the-spot teaching, for example, part of the course "papermaking principle and engineering" can be completed on production site,   teachers and the engineering technicians in enterprise explain the engineering knowledge to students, involving forming section, pressing section, drying section, calendering and finishing section. On-the-spot teaching can deepen the students' understanding to the complex engineering process.

According to the requirements of the level 3, students engage in field work in summer vacation. Through the field work, students can grasp the methods of equipment operation and equipment maintenance under the guidance of the enterprise technicians, meanwhile, they also learn how to communicate with others

Creation of production operation and management conditions. Students must take part in engineering practice according to the requirement of level 4, including production process and technology analysis, product quality control, raw materials and finished products inspection, production failure analysis, paper defects analysis, equipment installation and product design. Engineering practice greatly improves the students' the engineering practical ability and engineering quality.

Creation of engineering project and R&D conditions. Graduation design (thesis) is an important practice project for students to improve their comprehensive application ability. The topic of graduation design (thesis) comes from enterprise, it is genuine, challenging and innovative, and students can complete it with the help of teachers or enterprise technicians.

3.4   Establishment of Effective Assessment Methods

It is necessary to set up effective assessment methods to evaluate the students' engineering consciousness, engineering quality and engineering practical ability. For certain engineering practice, the assessment of engineering practical ability should be carried out by both school and enterprise[5]. Both teachers and enterprise technicians establish students' practice files, and guide and monitor the students' engineering practice. For students'operation skills, each enterprise has corresponding evaluation standard. The results of engineering design and products development can be given by teachers and enterprise technicians according to students' regular performance, on-the-spot operation, practice report, engineering design scheme or research achievements and thesis defense.

4   Establishment of Proper School-Enterprise Cooperation Educational Mode

Changing the enterprise's role from "object" to "subject" is very important and difficult during the implementation of "excellent engineers eduation training plan".

Whether the enterprises benefit from the cooperative education or not is the main factor for enterprises to determine their willingness to participate in cooperative education[6]. Excellent engineers' ability and quality win the favor of the enterprises so that enterprises are willing to participate in cooperative education. The "order" education mode let enterprises treat students as their own talents, thus enterprises become the "subject" of talents training.

4.1 Sign a Training "Order"

The career path of the graduates from light chemical engineering of our univerisity is pointing towards Zhejiang's papermaking enterprises, especially towards the main cooperative enterprises. So we can sign a school-enterprise cooperative training "order" with enterprises, which is based on the requirements of enterprises and students, and by way of two-way choice. School and enterprises sign a cooperative training contract and etablish students' training archives to ensure that the graduates give priority to serve the cooperative enterprises

4.2 Enterprise Actively Participate in Training Scheme Design

Under the "order" cooperative way, enterprises will actively participate in the design of training scheme, and some enterprises' talents employ ideas will be merged into the training scheme. In addition, the training standards and assessment methods will be more operable.

4.3 Establishment of the Training Base

Establishment of the "light chemical engineering specialty excellent engineers education training base" in cooperative enterprises will improve the social reputation and influence of the enterprises. It is necessary to establish "light chemical engineering specialty excellent engineers school-enterprise cooperative education committee", and invite the general manager of cooperative enterprises as main members, and invite enterprise engineers as advisors.

5 Conclusion

The key to implement light chemical engineering excellent engineers school-enterprise cooperative education is to establish operable engineering knowledge and engineering ability standards and to work out a study plan which throughout four stages and five practice projects. Establishment of "order " educational mode to ensure the graduates to give priority to serve the cooperative enterprises is the guarantee of success of school-enterprise cooperative education.

Acknowledgments. This project was funded by Department of Education of Zhejiang Province.(yb2010043)

References

1. Crawley, E.F., Malmqvist, J., Ostlund, S., Brodeur, D.R.: Rethinking Engineering Education–The CDIO Approach. Springer, New York (2007)
2. The Ministry of Education: The implementation of the ministry of education Excellent engineers education training plan,
   `http://www.moe.edu.cn/edoas/website18/24/info1277271957556124.htm`
3. Chang, e.F.: The main problems and countermeasures in China's higher engineering education. Forum on Contemporary Education: Macro Education Research 3, 61–62 (2009)
4. Zha, J.: On CDIO model under learning by doing strategy. Research in Higher Education of Engineering 3, 1–7 (2008)
5. Hu, Z., Liu, L., Ren, S.: Research on feedback line practice system basedon the capacity assessment. China University Teaching 4, 78–80 (2010)
6. Zhang, J., Tian, D., Cui, R.: An Empirical Research on Motives of Enterprises Participation in Cooperative Education. Research in Higher Education of Engineering 6, 23–27 (2008)

The Application of Fast Fourier Transform Algorithm in WiMAX Communications System

Zhiling Tang

Chongqing Technology and Business Institute

Abstract. We have proposed a memory based recursive FFT design which has much less gate counts, lower power consumption and higher speed. The proposed architecture has three main advantages (1) fewer butterfly iteration to reduce power consumption, (2) pipeline of radix-2^2 butterfly to speed up clock frequency, (3) even distribution of memory access to make utilization efficiency in SRAM ports. In summary, the speed performance of our design easily satisfies most application requirements of Fixed 802.16d and Mobile 802.16e WiMAX, which uses OFDMA modulated wireless communication system. Our design also occupies lesser area, hence lower cost and power consumption.

Keywords: FPGA, Fast Fourier Transform, Orthogonal Frequency, WiMAX Communications System.

1   Introduction

WiMAX – which stands for Worldwide Interoperability for Microwave Access is bringing the wireless and Internet revolutions to portable devices across the globe. Just as broadcast television in the 1940's and 1950's changed the world of entertainment, advertising, and our social fabric, WiMAX is poised to broadcast the Internet throughout the world, and the changes in our lives will be dramatic. WiMAX is providing the capabilities of the Internet, without any wires, to every living room, portable computer, phone, and handheld device. The WiMAX modules utilize the OFDMA scheme in their physical layer of communication.

OFDM exploits the frequency diversity of the multipath mobile broadband channel by coding and interleaving the information across the subcarriers prior to transmission. After organizing the time and frequency resources in an OFDMA system into resource blocks for allocation to the individual mobile stations, the coded and interleaved information bits of a specific mobile station are modulated onto the subcarriers of its resource blocks. Then OFDM modulation is cost effectively realized by the Inverse Fast Fourier Transform (IFFT) that enables the use of a large number of subcarriers—up to 1024 according to the Mobile WiMAX system profiles – to be accommodated within each OFDMA symbol. Prior to transmission, each OFDMA symbol is extended by its cyclic prefix followed by digital-to-analog (D/A) conversion at the transmitter. At the receiver end, after analog-to-digital (A/D) conversion, the cyclic prefix is discarded and OFDM demodulation is applied through the Fast Fourier Transform (FFT).

Y. Wang (Ed.): Education Management, Education Theory & Education Application, AISC 109, pp. 711–719.
springerlink.com © Springer-Verlag Berlin Heidelberg 2011

2   Architecture and Designmethodology

2.1   Radix-2^2 Decimation in Frequency FFT Algorithm

A useful state-of-the-art review of hardware architectures for FFTs was given by He et al. [6] and different approaches were put into functional blocks with unified terminology. From the definition of DFT of size N [7]:

$$X(k) = \sum_{n=0}^{N-1} x(n)W_N^{nk}, 0 \le k < N \text{ x + y = z}. \tag{1}$$

where W_N denotes the primitive N_{th} root of unity, with its exponent evaluated modulo N, x(n) is the input sequence and X(k) is the DFT. He [6] applied a 3-dimensional linear index map, and Common factor algorithm (CFA) to derive a set of 4 DFTs of length N/4 as,

$$n = \left(\frac{N}{2}n_1 + \frac{N}{4}n_2 + n_3\right)_N \tag{2}$$

$$k = \left(k_1 + 2k_2 + 4k_3\right)_N$$

$$X\left(k_1 + 2k_2 + 4k_4\right) = \sum_{n_3=0}^{\frac{N}{4}-1}\left[H\left(k_1, k_2, n_3\right)W_N^{n_3(k_1+2k_2)}\right]W_{\frac{N}{4}}^{n_3 k_3} \tag{3}$$

where n_1, n_2, n_3 are the index terms of the input sample n and k_1, k_2, k_3 are the index terms of the output sample k and where $H(k_1, k_2, k_3)$ is expressed in eqn (7).

$$H\left(k_1, k_2, n_3\right) = \left[x(n_3) + (-1)^{k_1} x\left(n_3 + \frac{N}{2}\right)\right] +$$

$$(-j)^{(k_1+2k_2)}\left[x\left(n_3 + \frac{N}{4}\right) + (-1)^{k_1} x\left(n_3 + \frac{3N}{4}\right)\right] \tag{4}$$

Eqn (4) represents the first two stages of butterflies with only trivial multiplications in the SFG, as BFI and BFII. Full multipliers are required after the two butterflies in order to compute the product of the decomposed twiddle factor $W_N^{n_3(k_1+2k_2)}$

(3). Note the order of the twiddle factors is different from that of radix-4 algorithm.

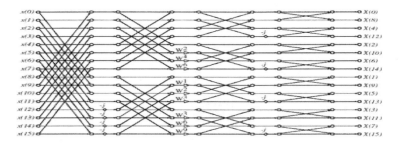

Fig. 1. Radix-2^2 DIF FFT flow graph for N=16

Applying this CFA procedure recursively to the remaining DFTs of length N/4 in eqn (3), the complete radix-2^2Decimation-in-frequency (DIF FFT) algorithm is obtained. The corresponding FFT flow graph for N=16 is shown in Fig. 1 where small diamonds represent trivial multiplication by $W_N^{N/4} = -j$ which involves only real-imaginary swapping and sign inversion [6].

2.2 Radix-2^2 FFT Architecture

Mapping radix-2^2 DIF FFT algorithm derived to the radix-2 SDF architecture, a new architecture of R2^{2}SDF approach is obtained [2]. Fig 2 outlines an implementation of the R2^{2}SDF architecture for N=1024, note the similarity of the data-path to R2SDF and the reduced number of multipliers. The implementation uses two types of butterflies; one identical to that in R2SDF, the other contains also the logic to implement the trivial twiddle factor multiplication, as shown in Fig 3 (i), (ii) respectively [2].

Due to the spatial regularity of Radix-2^2 algorithm, the synchronization control of the processor is very simple. A (log2N)-bit binary counter serves two purposes: synchronization controller and address counter for twiddle factor reading in each stage. With the help of the butterfly structures shown in Fig 3, the scheduled operation of the R2^{2}SDF processor in Fig 2 is as follows.

On first N/2 cycles, the 2-to-1 multiplexers in the first butterfly module switch to position "0", and the butterfly is idle. The input data from left is directed to the shift registers until they are filled. On next N/2 cycles, the multiplexers turn to position "1", the butterfly computes a 2-point DFT with incoming data and the data stored in the shift registers.

$$Z1(n) = x(n) + x\left(n + \frac{N}{2}\right)$$

$$Z1\left(n + \frac{N}{2}\right) = x(n) - x\left(n + \frac{N}{2}\right), 0 \le n < \frac{N}{2}$$

(5)

The butterfly outputs Z1(n) and Z1(n + N/2) are computed according to the equations given in eqn (5). Z1(n) is sent to apply the twiddle factors, and Z1(n + N/2) is sent back to the shift registers to be "multiplied" in still next N/2 cycles when the first half of the next frame of time sequence is loaded in. The operation of the second butterfly is similar to that of the first one, except the "distance" of butterfly input sequence are just N/4 and the trivial twiddle factor multiplication has been implemented by real-imaginary swapping with a commutater and controlled add/subtract operations, as in Fig 3(i) (ii), which requires two bit control signal from the synchronizing counter. The data then goes through a full complex multiplier, working at 75% utility, accomplishes the result of first level of radix-4 DFT word by word. Further processing repeats this pattern with the distance of the input data decreases by half at each consecutive butterfly stages. After N-1 clock cycles, the result of the complete DFT transform streams out to the right, in bit-reversed order. The next frame of transform can be computed without pausing due to the pipelined processing of each stage.

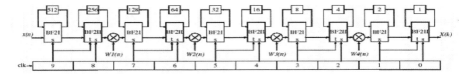

Fig. 2. R2^{2}SDF pipeline FFT architecture for N=1024

(i) BF2

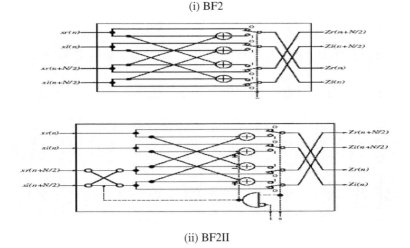

(ii) BF2II

Fig. 3. Butterfly structure for R2^{2}SDF FFT

3   Implementation in Verilog

VERILOG Hardware Description Language (VERILOG) was introduced by Gateway Design Automation in 1984 as a proprietary hardware description and simulation language [8, 9]. VERILOG synthesis tools can create logic-circuit structures directly from VERILOG behavioral descriptions, and target them to a selected technology for realization. Using VERILOG, we can design, simulate, and synthesize anything form a simple combinational circuit to a complete microprocessor system on a chip. VERILOG started out as documentation and modeling language, allowing the behavior of digital-system designs to be precisely specified and simulated [8, 9] and language specification allows multiple modules to be stored in a single text file. All these features of VERILOG will help better in simulation and synthesis of our proposed architecture.

The R2^{2}SDF presented above has been fully coded in VERILOG Hardware Description Language (VERILOG). Once the design is coded in VERILOG, the Modelsim XEIII 6.2c compiler [9] and the Xilinx Foundation ISA Environment 9.1i [10] generate a net-list for FPGA configuration. The net-list can then be downloaded into the FPGA using the same Xilinx tools and Texas Instruments prototyping board.

From the architecture of R2^{2}SDF in Fig 2, the butterfly blocks BF2I and BF2II are described as building blocks in VERILOG code. Booth multiplication algorithm for signed binary numbers is used for complex multipliers. Thus, the overall latency of the real implementation varies as the processing word length changes [2]. Look-up-table (LUT) based Random Access Memories (RAMs) and Flip-Flops are used to implement feedback memory of the very last stages where are the RAM blocks in the FPGA are used for the rest of the stages. Similarly, LUT-based Read Only Memories (ROMs) are used to implement twiddle ROMs of the very last stages whereas Block RAMs are used for the rest of stages [6].

Table 1(a). IMPLEMENTATION RESULT

Logic Utilization	Used	Available	Utilization
No. of Slices	3155	3584	88%
No. of Slice Flip flops	1514	7168	21%
No. of 4 input LUTs	5916	7168	82%
No. of bonded IOBs	32	97	32%
No. of Mult18x18s	16	16	100%
No. of GCLKs	1	8	12%

Table 1(b). TIMING SUMMARY

Minimum period	10.827ns (Maximum Frequency: 92.366MHz)
Minimum input arrival time before clock	5.406ns
Maximum output required time after clock	6.216ns

The implementation results after implementing in Xilinx Spartan3 FPGA [13] are listed in Table 1(a), 1(b). Table 1(a) shows the implementation results where as table 1(b) shows the timing summary. The resulting figures show that our implementation outperforms the other implementations of that kind. Its speed nearly matches that of the Xilinx core but its throughput is more than 3 times higher due to its pipeline nature.

4   Implementation of the Proposed FFT in OFDMcommunication System

The fundamental principle of the OFDM system is to decompose the high rate data stream (bandwidth=W) into N lower rate data streams and then to transmit them simultaneously over a large number of subcarriers. The IFFT and the FFT are used for, respectively, modulating and demodulating the data constellations on the orthogonal subcarriers [12].

In an OFDM system, the transmitter and receiver blocks contain the FFT modules as shown in Fig. 4 (a) and (b). The FFT processor must finish the transform within 312.5 ns to serve the purpose in the OFDM system. Our FFT architecture effectively fits into the system since it has a minimum required time period of 10.827 ns (Table 1 (b)).

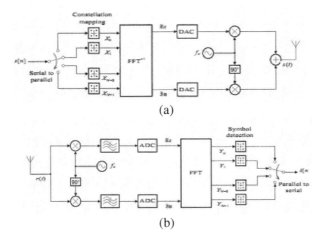

(a)

(b)

Fig. 4. OFDM Module: (a) OFDM Transmitter and (b) OFDM receiver

An OFDM carrier signal is the sum of a number of orthogonal sub-carriers, with baseband data on each sub-carrier being independently modulated commonly using some type of quadrature amplitude modulation (QAM) or phase-shift keying (PSK) This composite baseband signal is typically used to modulate a main RF carrier. s[n] is a serial stream of binary digits. By inverse multiplexing, these are first demultiplexed into N parallel streams, and each one mapped to a (possibly complex) symbol stream using some modulation constellation (QAM, PSK, etc.). Note that the constellations may be different, so some streams may carry a higher bit-rate than others.

The receiver picks up the signal r(t), which is then quadrature-mixed down to baseband using cosine and sine waves at the carrier frequency. This also creates signals centered on 2fc, so low-pass filters are used to reject these. The baseband signals are then sampled and digitized using analogue-to-digital converters (ADCs), and a forward FFT is used to convert back to the frequency domain. Orthogonal Frequency-Division Multiple Access (OFDMA) is a multi-user version of the popular Orthogonal frequency-division multiplexing (OFDM) digital modulation scheme. Multiple accesses are achieved in OFDMA by assigning subsets of subcarriers to individual users as shown in Fig 5. This allows simultaneous low data rate transmission from several users.

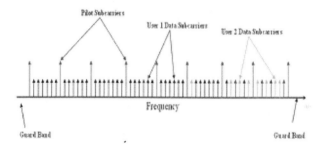

Fig. 5. ODMA Subcarriers Pattern

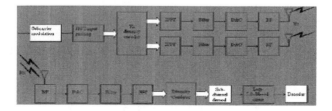

Fig. 6. OFDMA Trans-receiver

OFDMA can be seen as an alternative to combining OFDM with time division multiple access (TDMA) or time- domain statistical multiplexing, i.e. packet mode communication. Low-data-rate users can send continuously with low transmission power instead of using a "pulsed" high-power carrier. Constant delay, and shorter delay, can be achieved. OFDMA can also be described as a combination of frequency domain and time domain multiple access, where the resources are partitioned in the time-frequency space, and slots are assigned along the OFDM symbol index as well as OFDM sub-carrier index. OFDMA is considered as highly suitable for broadband wireless networks, due to advantages including scalability and MIMO-friendliness, and ability to take advantage of channel frequency selectivity. The trans-receiver structure of OFDMA is shown in Fig 6.

5   Conclusions

The FFT algorithm eliminates the redundant calculation which is needed in computing discrete Fourier transform (DFT) and is thus very suitable for efficient hardware implementation. A high level implementation of a high performance FFT for OFDM modulator and demodulator is presented in this work. The design has been coded in Verilog and targeted into Xilinx Spartan3 field programmable gate arrays. Radix-22algorithm is proposed and used for the OFDM communication system. The designed FFT is implemented and applied to Fixed WiMAX (IEEE 802.16d) and Mobile WiMAX (IEEE 802.16e) communication standards. The results are tabulated and the hardware parameters are compared.

References

1. Minallah, M.S., Raja, G.: Real Time FFT Processor Implementation. In: 2nd International Conference on Emerging Technologies, IEEE—ICET 2006, Peshawar, Pakistan, November 13-14, pp. 192–195 (2006)
2. Sukhsawas, S., Benkrid, K.: A High-level Implementation of a High Performance Pipeline FFT on Virtex-E FPGAs. In: Proc. of the IEEE Comp. Society Annual Symp. on VLSI Emerging Trends in Systems Design (ISVLSI 2004), pp. 229–232 (February 2004)
3. IEEE 802.16-2004, IEEE Standard for Local and Metropolitan Area Networks – Part 16: Air Interface for Fixed Broadband Wireless Access Systems (October 2004)
4. IEEE 802.16e-2005, IEEE Standard for Local and Metropolitan Area Networks – Part 16: Air Interface for Fixed and Mobile Broadband Wireless Access Systems Amendment 2: Physical and Medium Access Control Layers for Combined Fixed and Mobile Operation in Licensed Bands (February 2006)
5. WiMAX Forum, Fixed, nomadic, portable ad mobile applications for 802.16-2004 and 802.16e WiMAX networks (2005), http://www.wimaxforum.org
6. He, S., Torkelson, M.: A new approach to pipeline FFT processor. In: 10th Int. Parallel Processing Symp (IPPS 1996), pp. 766–770 (1996)
7. Proakis, J.G., Dimitris, K.: Digital Signal Processing: Principles, Algorithms and Applications. Prentice-Hall (1995)
8. Wakerly, J.F.: Digital Design Principles and Practices. Pearson Education (2006)

9. Modelsim manual, `http://support.xilinx.com`
10. Xilinx, Inc., `http://www.xilinx.com/`
11. Wold, E.H., Despain, A.M.: Pipeline and parallel-pipeline FFT processors for VLSI implementation. IEEE Trans. Comput. C-33(5), 414–426 (1984)
12. Bi, G., Jones, E.V.: A pipelined FFT processor for word sequential data. IEEE Trans. Acoust., Speech, Signal Processing 37(12), 1982–1985 (1989)
13. Xilinx, Inc. High-Performance 1024-Point Complex FFT/IFFT V2.0, `http://www.xilinx.com/ipcenter/`

The Study on Voltage Controlled Oscillator in Electronic Applications

Wu Wen

Chongqing Technology and Business Institute

Abstract. A low power and low phase-noise CMOS VCO Based on biasing of oversized MOS transistors is demonstrated. The LC-VCO used here dissipates 0.1114 mW power under 1.8 V supply voltage. The VCO frequency can be tuned from 11.6 to 12.5 GHz with a measured phase noise of 106 dBc/Hz at 400 kHz offset from 12.5 GHz oscillation frequency. Common figures of merit are 185.

Keywords: CMOS, Voltage Controlled Oscillator, low power.

1 Introduction

Bond wire inductor is being utilized by Ahrens and Lee for a 1.4GHz, 3mW CMOS LC low phase noise VCO [5]. Ham and Hajimiri, proposed an optimized LC-VCO with just 2GHz oscillating frequency [6]. 5.8 GHz fully integrated low power, low phase noise designed by Bhattacharjee et al [7]. Song and Yoon demonstrated a 1-V 5 GHz Low Phase Noise LC-VCO by Using Voltage-Dividing and Bias-Level Shifting Technique [8]. Moon et al used a Small VCO-Gain Variation to design 4.39–5.26 GHz LC- Voltage-Controlled Oscillator [9].

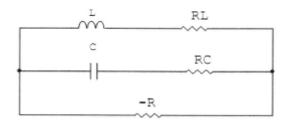

Fig. 1. Basic LC tank circuit

The oscillator requires a tank circuit. A parallel resonance tank comprises of inductance (L) and capacitance (C) is being considered. LC tank circuit acts as a resonance circuit in order to control the operation Voltage source is used. A general LC tank circuit is shown as in Fig. 1. L and C has some parasitic resistances associated which is RL and RC are shown respectively in Fig.1. Active components like MOS transistors are used to realize the negative resistance (-R) in order to compensate the losses due to RL and RC, shown in Fig. 2.

Y. Wang (Ed.): Education Management, Education Theory & Education Application, AISC 109, pp. 721–727.
springerlink.com

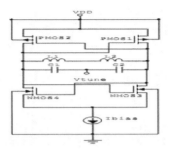

Fig. 2. Conventional LC-VCO Design

Fig. 2. shows a conventional LC-VCO core. L1, L2 and C1, C2 are the inductors and capacitors respectively. They form a tank and provide resonance to the core VCO. They are used in a pair to form a symmetrical design. So the differential output can be obtained in order to suppress noise. It also consist of two PMOS namely PMOS1 and PMOS2. Similarly two NMOS named NMOS1 and NMOS 2 are used. These MOSFETS are used in order to provide the feedback loop. Also the transconductance provided by them is necessary to start the oscillations. The current biasing Ibias is used in order to compensate variations due to active devices for process variations and temperature variations.

The oscillation condition for an oscillator is that transconductance of the oscillator and is the equivalent resistance of the tank at resonance must be equal. So the start-up condition for VCO is that negative resistance generated by the cross coupled pair transistors NMOS and PMOS must be enough to compensate the tank resistance given by following relation as follows as [1]

$$g_{mn} \geq \alpha_l g_{\tan k} \tag{1}$$

Once the VCO starts oscillating, the most important factor of interest in frequency of oscillation. The output frequency of a VCO is also known as center frequency. A standard VCO has a center frequency range typically from a few MHz to GHz [1].

$$f = \frac{1}{\tau} \tag{2}$$

Also the tuning depends on the variations due to various parameters of varactor and inductor. The tuning range shows the relationship between the VCO operating frequency and the tuning voltage applied. So the tuning range of the VCO is given by the relation [1-2]

$$L_{\tan k} C_{\tan k, \min} \leq \frac{1}{w_{\max}^2} \tag{3}$$

$$L_{\tan k} C_{\tan k, \min} \geq \frac{1}{w_{\max}^2} \tag{4}$$

Another important factor is power dissipated by the VCO. The power dissipated (P) is specified in mill-watts and it should as minimum as possible [1]. It is given the relation

$$P = I_{bias}V_{dd} \tag{5}$$

2   Circuit Design

The circuit schematic of the proposed LC-VCO is shown in Fig. 3. A simple complementary LC-VCO is considered as seen PMOS1 and PMOS2 are cross-coupled similarly the NMOS1 and NMOS2 are connected. The VCO core circuit requires the Biasing current in order to reduce the effect of device and process variations. It also results in the degradation in the phase noise. In this work, we have used a current mirror circuit in order to provide biasing to the circuit. LC-VCO starts oscillation when the transconductance of the circuit is greater then the tank transconductance. In order to achieve same and to operate device in low power region we have used two MOSFETs NMOS1 and NMOS2. These MOSFETS provide required transconductance in order to start the oscillations.

In order to suppress attenuations fully symmetric design has been considered. Also to operate the devices in the low power regime the MOSFET NMOS1 and NMOS2 are made wider. The width has been optimised in order to have low power. Here two pairs of MOSFET structure are used instead of a single pair structure i.e. complementary NMOS and PMOS cross-coupled differential structure. This facilitates lowering the supply voltage dependence. The values of the inductors, capacitor and the sizes of the MOSFETS PMOS, NMOS pairs are optimally chosen in accordance to the oscillation condition and frequency tuning range. The cross-coupled PMOS, NMOS pair are biased such that they get minimum voltage from the supply and hence dissipate minimum power.

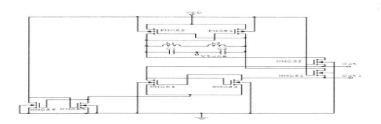

Fig. 3. Modified LC-VCO Design

Hence it is seen clearly that the LC-VCO design with above modifications is advantageous. As these provides following advancements as it removes dependency on external biasing and it operates the device with lower supply voltage. Hence the phase noise of a MOS oscillator in this design does not depend on the bias. It clearly provides advantage over the conventional LC-VCO design. It is inferred that from our work that VCO circuit has low power with low phase noise.

3 Results and Discussion

The modified VCO considered in this work has been simulated and its performance is analyzed. SPICE simulations are carried out for 180nm technology node for supply voltage 1.8V.

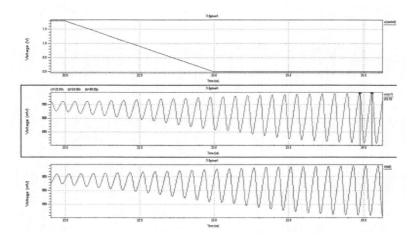

Fig. 4. Tuned Output Voltage variation with Time a) Tuning input, b) Tuned output, c) Tuned output-1

The oscillator is being kick-started using a transient pulse. The oscillations become stable after 25ns. Fig. 4. shows the input transient pulse and two differential outputs out and out1. Differential output has been considered in order to suppress noise. As the generated frequency is sinusoidal we are interested in the phase error produced called as phase noise. The phase noise of the LC-VCO has been analyzed. Phase noise is the frequency domain representation of rapid, short-term, random fluctuations in the phase of a waveform, caused by time domain instabilities. For this linear time noise model has been used to calculated noise ($L(\Delta f)$) [2].

$$L(\Delta f) = 10\log\left[\frac{2kT}{P_{sig}}\left(\frac{f_0}{2Q\Delta f}\right)^2\right] \qquad (6)$$

Fig. 5 shows phase noise using eq (6). The offset frequency f_0 been varied from 1Hz to 1kHz. The phase noise can be approximated at an offset of 1kHz as -117 dBc/Hz.

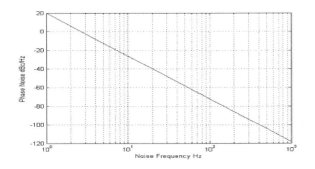

Fig. 5. Phase Noise of the Proposed LC-VCO

Figure of merit (FOM) takes all important VCO parameters like power, phase noise and oscillation frequency into account. FOM is given as [5]:

$$FOM = 20\log(f_0/\Delta f) - L(\Delta f) - 10\log P \qquad (7)$$

Using eq(7) in order to decide the roll-off between various improvement constraints the FOM factor has been evaluated for frequency of oscillation of 12.5 GHz with a power dissipation of 0.1114mW and a phase noise of -117 dBc/Hz at 1kHz offset. FOM is found out to be 175.

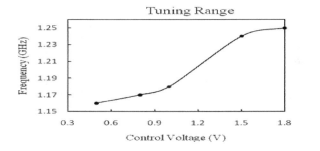

Fig. 6. Tuning Range of VCO

The tuning range of nearly 7.75% in the frequency of oscillation with control voltage varying from 0.5V to 1.8V [17] of LC-VCO is obtained as seen in Fig. 4.

Table 1. Summarized Comparison Results of This VCO and Previous Low Power CMOS VCOS [6]-[10]

Year	Ref.	Tech. (μm)	VDD (V)	Power (mW)	f_{osc} (GHz)	f_{tune} (%)	Phase Noise
2001	[6]	0.35	2.5	10	2	6	-117
2002	[7]	0.24	2.5	5	5.8	14	-112
2004	[8]	0.25	2.5	10	1.7	1	-134
2009	[9]	0.18	1.8	9.7	4.0	9.7	-113
2009	[10]	0.18	0.8	3.92	5.2	24.3	-116
2010	Present Work	0.18	1.8	0.1114	12.5	7.7	-117

From the table it can be seen that all the designs reported in literature have power in orders of milli-watts. However, in the present work the power is reduced by over an order. Hence it is analyzed that the present design is the best amongst the referred reported literature.

4 Conclusion

According to recent developments in the design of radio frequency (RF) front-end modules is implementation of low power, low phase noise voltage controlled oscillators (VCOs). So it is said that there is immense scope of research for design of LC-VCO [1-3]. Long has used NMOS as a biasing current source to design 2.4GHz Low-Power Low-Phase VCO [4].

References

1. Lee, T.H.: The Design of CMOS Radio-Frequency Integrated Circuits. Cambridge Univ. Press, Cambridge (1998)
2. Razavi, B.: Design of Analog CMOS Integrated Circuits (1998)
3. Kang, S.M., Leblebici, Y.: CMOS Digital Integrated Circuits - Analysis and Design, 3rd edn. TMH (2003)
4. Long, J., Foo, J.Y., Weber, R.J.: A 2.4GHz Low-Power Low-Phase-Noise CMOS LC VCO. In: IEEE Computer Society Annual Symposium on VLSI Emerging Trends in VLSI Systems Design (ISVLSI 2004), pp. 213–214 (February 2004)
5. Ahrens, T.I., Lee, T.H.: A 1.4GHz, 3mW CMOS LC low phase noise VCO using tapped bond wire inductances. In: International Symposium on Low Power Electronics and Design-ISLPED, pp.16–19 (August 1998)
6. Ham, D., Hajimiri, A.: Concepts and methods in optimization of integrated LC VCOs. IEEE J. Solid-State Circuits 36(6), 896–909 (2001)
7. Bhattacharjee, J., Mukherjee, D., Gebara, E., Nuttinck, S., Laskar, J.: A 5.8 GHz fully integrated low power low phase noise CMOS LC-VCO for WLA applications. In: IEEE Radio Frequency Integrated Circuit Symposium, pp. 475–478 (2002)

8. Song, T., Yoon, E.: A 1-V 5 GHz Low Phase Noise LC-VCO Using Voltage-Dividing and Bias-Level Shifting Technique. In: Topical Meeting on Silicon Monolithic Integrated Circuits in RF Systems, pp. 87–90 (September 2004)

9. Moon, Y.J., Roh, Y.S., Jeong, C.Y., Yoo, C.: A 4.39–5.26 GHz LC-Tank CMOS Voltage-Controlled Oscillator With Small VCO-Gain Variation. IEEE Microwave and Wireless Component Letters (August 2009)

10. Issa, D.B., Akacha, S., Kachouri, A., Samet, M.: Graphical Optimization of 4GHz CMOS LC-VCO. In: Design & Technology of Integrated Systems in Nanoscale Era-2009, pp. 33–37 (May 2009)

11. Chandel, R., Sarkar, S., Agarwal, R.P.: Delay and Power Management of Voltage-Scaled Repeaters for Long Interconnects. International Journal of Modelling & Simulation 27(4), 333–339 (2007)

12. Hajimiri, A., Lee, T.H.: Phase Noise in CMOS Differential LC-Oscillators. In: Symposium on VLSI Circuits Dig. Tech. Papers, pp. 48–51 (1998)

13. Hajimiri, A., Lee, T.H.: A General Theory of Phase Noise in Electrical Oscillators. IEEE Journal of Solid State Circuits 33, 179–194 (1998)

14. Dehghani, R., Atarodi, S.M.: Optimised analytic designed 2.5GHz CMOS VCO. IEEE Electronics Letters 39(16), 1160–1162 (2003)

15. Linten, D., et al.: Low-power voltage-controlled oscillators in 90-nm CMOS using high-quality thin-film post processed inductors. IEEE Journal of Solid State Circuits (JSSC) 40(9), 1922–1931 (2005)

16. MOSIS Service for HSPICE models, http://www.mosis.org

17. EDA tool used, http://www.tanner.com

Optimization Design Method of Mountain Tunnel Lining Based on Stress Mapping Return Arithmetic

Zelin Niu[1,2,*] and Zhanping Song[2]

[1] School of Highway, Chang'an University, Xi'an 710064, Shaanxi, P.R. China
[2] School of Civil Engineering, Xi'an University of Achitecture and Technology
710055, P.R. China
nz1109109@163.com

Abstract. Due to the construction of mountain tunnels which take the drilling and blasting method of construction, the lining structure parameter optimization is an important problem to be solved. The paper proposed an optimization design method of mountain tunnel lining based on stress mapping return arithmetic, can calculate the lining safety factor with nonlinear finite element method and D-P criterion. After the method theory is introduced, it is applied in Hei shansi lining of tunnel structure's analysis of Chang an lining of China. The numerical simulation experiment was conducted to study the typical cross-section of Chang an l tunnel lining thickness change on the stability of the tunnel states to discuss the correlation between the thickness of the lining of the tunnel structure and deformation and the internal forces. The result states the method is feasible and has guiding meaning for the tunnel construction.

Keywords: Tunnel engineering, lining thickness, optimum design, mapping return arithmetic.

1 Introduction

Mountain tunnel which uses drill-blasting method to excavate, on the influence of all kinds of construction factors, the variability of tunnel lining thickness is very great[1,2]. According to relevant data shows, the thickness of the second lining each section was content with all design requirements, only is about 50 percent of the tunnel total. The shortcomings of tunnel lining thickness weaken tunnel structure security reserve capacity, influence tunnel using life, endanger safety of the vehicles and personnel. therefore, to study the influence which tunnel lining thickness has on the reliability and security of tunnel structure, and to evaluate safety of lining structure become the current research hot issue in tunnel engineering fields[3-5].

Conventional tunnel lining calculation of e loading-structure just seems the surrounding rock as the loading, that is not accord to the actual state. The new austrian tunneling method renovates the rock –structure action together to guarantee the tunnel stability which means the loading-structure method has limit in the lining calculation. Aiming at that general method is difficult to calculate the lining safety factor

---

* Corresponding author. Tel.: +8615591859597.

considering the surrounding bearing effect, the paper proposed the simulation methods combining safety factor calculation and stress mapping return. The paper has studied the influence which lining structure thickness changes of Hei shansi tunnel and explored the relation between lining thickness and tunnel structure deformation and internal force, optimized and ensured tunnel lining safety thickness, has guided the design of Hei shansi tunnel lining.

2 The Safety Factor of Lining

In order to analyze the influence which lining thickness variation has on the tunnel structure security, we should do the numerical simulation analysis to tunnel construction, calculate internal forces of tunnel support structure, analysis safety coefficient of lining structure according to internal forces, examine the safety performance of the lining. Tunnel lining safety calculation is based on norm. In the analysis, according to the ultimate strength of lining concrete materials, the ultimate bearing capacity of eccentric loading of the lining is calculated, contrasting ultimate bearing capacity with numerical calculation of internal forces, the compressive (or tensile) strength safety coefficient of the lining different section is got. Namely:

$$N_{ultimate}/N \geq K code \tag{1}$$

In the equation, $N_{ultimate}$ is the ultimate bearing capacity of eccentric loading component, N is the numerical calculation of internal forces in the typical unit section of tunnel lining structure;

When the compressive strength controls, namely e = $M/N \leq 0.12h$, lining structure safety coefficient is.

$$N_{ultimate} = \phi \alpha R_\alpha bh \tag{2}$$

In the equation, φ is component longitudinal coefficient, as for tunnel lining, φ is 1; Ra is the compressive strength of concrete ultimate; α is eccentric effect coefficient of axial force, According to the type (2) sure; b is the cross-sectional widths, is 1m;h is the section thickness.

When the tensile strength controls, namely e=$M/N \geq 0.2h$

$$N_{ultimate} = \phi \frac{1.75R_l bh}{6e/h - 1} \tag{3}$$

In the equation: φ is component longitudinal coefficient, as for tunnel lining, φ is 1; R_l is concrete ultimate tensile strength; b is the cross-sectional width, and is 1;h is the section thickness.

3 Stress Mapping Return Arithmetic of Plasticity

It is well known that the rock and soil has the material nonlinear property which not be reflected by linear elastic constitutive model. Considering the FEM convergence speed and calculation accuracy, the paper introduced Druker—Prager yield criterion and stress mapping return arithmetic.

Druker—Prager yield criterion ie generalized Mises yield criterion, which is expressed as:

$$F(\sigma) = \overline{\sigma} + 3\alpha\sigma_m - k_d = 0 \tag{4}$$

The paper solved the constitutive equation based on two step of prediction and stress mapping return arithmetic. It is divided to the elastic prediction step and plastic adjusting step. It is a integral arithmetic that change a group of constitutive equations to a group of nonlinear algebraic equations and solving the algebraic equations[3]. The elastic prediction-plastic adjusting steps are as follows:

3.1 Elastic Prediction

While plastic factor $\Delta\gamma = 0$,

$$\varepsilon_{n+1}^{etrial} = \varepsilon_n^e + \Delta\varepsilon \tag{5}$$

$$\alpha_{n+1}^{trial} = \alpha_n \tag{6}$$

$$\sigma_{n+1}^{trail} = \overline{\rho}\frac{\partial\psi}{\partial\varepsilon^e}\bigg|_{n+1}^{trial} \quad A_{n+1}^{trail} = \overline{\rho}\frac{\partial\psi}{\partial\alpha}\bigg|_{n+1}^{trial} \tag{7}$$

In the formulas, $\gamma(t)$ is plastic factor, $\alpha(t)$ is inner variable, A(t) is harden parameter, $\Phi(t)$ is yield function, $\overline{\rho}$ is density, $\psi(t)$ is strain energy function. From formula (8) to (10), the result is elastic stress state, is not final solution of constitutive equation. If it is accord to formula (11), that is in the elastic scope or on the yield surface, the solution is also the final one.

$$\sigma_{n+1} = \sigma_{n+1}^{trial} \tag{8}$$

3.2 Plastic Amending

If the solution is not accord to (11), then the follow formulas, ie the mapping return equations should be calculated.

$$\varepsilon_{n+1}^e = \varepsilon_{n+1}^{etrial} - \Delta\gamma N(\sigma_{n+1}, A_{n+1}) \tag{9}$$

$$\alpha_{n+1}^e = \alpha_{n+1}^{etrial} - \Delta\gamma H(\sigma_{n+1}, A_{n+1}) \tag{10}$$

$$\Phi\left(\sigma_{n+1}, A_{n+1}\right) = 0 \qquad (11)$$

It is also accord to the condition of $\Delta\gamma > 0$, and the solution will be finally solved, the paper iterated above process adopting Newton-Rophson method which has good convergence.

4 Engineering Application

4.1 Numerical Model

Example Hei shansi tunnel is the secondary road in mountain hilly terrains, the tunnel length is 431m. The biggest excavation span is 12.54 m, excavation height is 9.07m. The tunnel is located in the hinterland of the loess plateau area ,which belongs to northern Shaanxi plateau's gully region. The lithology characters of tunnel level IV surrounding rock is Loess soil, and high content is in the surrounding rock is (moisture content 17% ~ 24%),the whole structure of surrounding rock is complete. According to the thickness variation and shortages of actual tunnel lining and surrounding rock types and buried deep size, etc, the finite element simulation calculation establishes 2d finite element model which simulates tunnel space excavation to analysis. The numerical calculation model was showed in Fig.1.

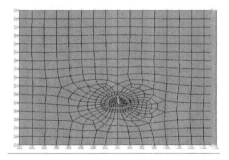

Fig. 1. The Numerical analysis model of the example tunnel

The x axis is the tunnel width direction, the y axis is the depth direction. The model adopts 8 node plane units to simulate the surrounding rock and surrounding rock reinforcement area (the surrounding rock. reinforcement area is used to simulate anchor role area and primary support); the beam element Beam2 is used to simulate the second lining. The parameters of numerical experiment parameter is shown in Table 1. According to geological investigation data, the ground stress field of example tunnel address gives priority to gravity stress field.

Table 1. The model parameters

name	Grain density g/cm³	Elasticity coefficient /GPa	Poisson's ratio μ	Cohesive force /kPa	Angle of friction
Adjacent rock mass	2.0	0.16	0.42	90	30°
Reinforces rock mass	2.0	0.16	0.42	90	30°
Two lining	2.3	30	0.17	40	34°

4.2 Calculation Results

The calculation of the safety coefficient of lining strength To analysis the variation of the supporting structure internal force aroused by the different thickness of the lining, the lining thickness is 40cm、45cm、50cm, the calculation results of supporting structure moment and axial force and shear and safety coefficient when four cases appear respectively are showed in figure 2 ~ figure 5. The figures show:

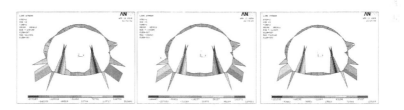

(a)lining thickness40cm   (b)Lining thickness 45cm   (c)lining thickness 50cm

Fig.2. The lining curved distance chart of different thickness for Hei shansi tunnel

(1)   Both of the most-positive, negative moment of tunnel lining take place near the arch feet, the yang arch has apparent apophysis trend, and the moment becomes greater as the Lining thickness increases, when the lining thickness is 50 cm, both of the positive and negative bending moments reach the maximum, on this time, the maximum positive moment is 125.31kN·m, the maximum negative moment is -121.04kN·m.

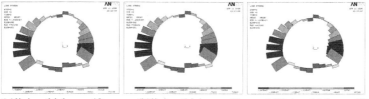

(a)lining thickness 40cm   (b)lining thickness 45cm   (c)lining thickness 50cm

Fig. 3. The lining axle strength chart of different thickness for Hei shansi tunnel

(2)   The Axial forces of the vaults, arch waist and arch feet of surrounding rock are mostly pressure, the lesser tension zone appears in vaults central and Arch supine central. The lining maximum pressurized axial forces are concentrated on both sides of the sidewalk, and the maximum axial forces grow greater as lining thickness increases, when the lining thickness is 50 cm, the maximum axial force reaches the greatest, and the greatest appears in the central of the inverted arch.

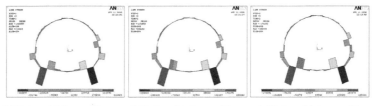

(a)lining thickness40cm   (b)lining thickness 45cm   (c)lining thickness 50cm

Fig. 4. The shear diagram of different thickness lining for Hei shansi tunnel

The maximum positive and negative shears of the lining all appear near arch feet, the shear appears obviously near arch feet, all the shears are not great in the other parts. At the same the results of numerical analysis shows, the most disadvantageous position of tunnel lining structural damage all appears in the tunnel arch feet, the tunnel destruction will start from the arch feet, and then make lining structure loss the wholeness, lead to the tunnel final instability. Therefore, in the analysis, the arch feet parts of tunnel support structure will be the key control parts in design and construction.

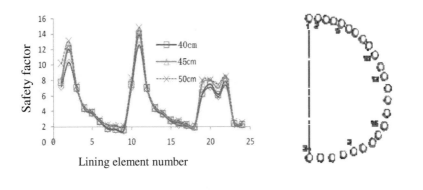

(a) safety coefficients   (b)Typical part Numbers

Fig. 5. Typical spot safety coefficient of different thickness lining

5 Conclusion

The paper proposed the simulation methods combining safety calculation factor and stress mapping return arithmetic in order to simulate the tunnel lining structure. Using the methods to simulate the Hei shansi tunnel typical sections the results show as follows:

The most disadvantageous position of tunnel lining structural damage all appeared in tunnel arch feet, the tunnel destruction started from arch feet, and then made the wholeness of lining structure loss, led to tunnel eventually instability. As to safety coefficient numerical, the arch feet and arch waist parts of the tunnel lining are structural stress concentration area, so they are also the danger zone of safety coefficients.

To the safety coefficient 2.0 of tunnel structure strength for standard, when the lining thickness increases to be 45 cm, the strength safety coefficient in lining key parts has all met the requirements. So as for Hei shansi tunnel level IV surrounding rock, when lining structure thickness exceeded 40 cm, to attempt to improve the tunnel structure security through increasing lining thickness is expensive, the tunnel lining structure of Hei shansi tunnel level IV surrounding rock parts is advised to be 40 cm.

It has proved that the stated the proposed method is feasible to optimize the mountain lining design parameters. The simulation results has guiding meaning in the tunnel engineering.

References

1. Guan, B.: Introduction to tunnel mechanics. Southwest Jiaotong University publishing house, Chengdu (1993)
2. She, J.: nfluence of Variability of Lining Thickness on Reliability of Tunnel Structure. Journal of Chongqing Jiaotong Institute 13(5), 83–90 (1994)
3. Wu, j., Qiu, W.: Regularity of the distribution of the thickness of tunnel linings and structure reliability analysis. Modern Tunnelling Technology 41(1), 22–25 (2004)
4. Gao, B., Lin, A., Zhao, W.: Research on Computation Methods for Structural Reliability Index of Tunnel Lining. Journal of Southwest Jiaotong University 31(6), 583–589 (1996)
5. Ted, B., Wing, K., Brian, M.: Nonlinear Finite Elements for Continua and Structures. Tsinghua University Press, Beijing (2002)

Pareto Analysis of Learning Needs about Adult Courses

Shuang Li and Ling Zhang

School of Science, Shandong Jianzhu University, Jinan 250101, China
{lish,zhangling052}@sdjzu.edu.cn

Abstract. To improve classroom teaching quality of adult education, teachers are supposed to teach for learning needs of students. In this paper, simple and intuitive Pareto statistical methods is employed for the analysis of courses and learning needs of adult learners with questionnaire and classification. The study puts forward some key factors for teaching contents and presents some analyses of balancing teaching contents, which is significant to flexible teaching and meaningful for teaching effectiveness.

Keywords: Adult course, Learning needs, Pareto analysis.

1 Introduction

If we regard adult learners as consumers, the curriculum and course content as a service of goods, the core of teaching adults is satisfying the course requirements of students. Usually, it is difficult for teaching to reach the target, because the teachers only pay attention to contents of course itself, with the expense of students' learning needs. Clearly, the needs determine attitude, and attitude determines learning's effects. Because adult students are of various backgrounds, a larger number of courses and learning is inevitably varied. As a teacher, it is impossible to meet all the needs of students, in particular, to guide and correct the negative of those unreasonable demands, which needs carefully designed questionnaires and careful analysis. Analysis of the point is to seize the principal contradiction, addressing key needs, Pareto analysis provides us with a simple and intuitive grasping the main issues of statistical analysis methods. Therefore, according to survey data, using the Pareto method, analysis of curriculum and adult learning needs of students is bound for teachers to adapt their teaching programs, which will play a positive role in teaching and learning.

2 Pareto Analysis

Pareto analysis is a simple and popular statistical method named after the Italian economist Pareto (Vilifiedo Pareto). Through a large number of statistics, Pareto found that 80% of the wealth is concentrated in the hands of about 20% of people[1].

Y. Wang (Ed.): Education Management, Education Theory & Education Application, AISC 109, pp. 737–740.
springerlink.com

Economists Kane (V E Kane) in his "fault prevention", also pointed out that a similar conclusion, for example, 80 percent of goods were sold to 20% of the customers, and 80% of a production problem focused on 20% of the links, etc. These examples focus on the concept of "significant minority and the majority which does not matter", "Pareto analysis pointed out that, in a specific problem", an important minority is the main reason for the problem, and "the majority does not matter "is derived from a variety of different factors. As a problem identification tool, Pareto analysis is widely used in various fields. People use it to identify the most important issues and causes of these problems, just because we understand the problem, we can prioritize and grasp the point.

Pareto analysis of the main conclusions is reflected in the Pareto chart. The Pareto chart is a factor as independent variable to elements of the frequency, relative frequency or probability of the dependent variable column, column axis from left to right in descending order by height accordingly, the highest in the left and the lowest ranking in right; this arrangement visually highlights the most important category, in all kinds of attention at a glance to identify the key elements, namely those with the greatest and greater frequency in the diagram on the left.

3   The Design of Questionnaire about Courses of Adult Learning Need

For the first class to face the higher number of complex components of adult classes combined together, the teachers always introduce the objective of course, objects and requirements. After students study the nature of the curriculum, they will have a rough idea, thus the teachers can make a simple program needs questionnaire. Through many discussions with students and questionnaires summary, we conclude that the needs of students focus on career, education, interests, and other four areas (which four areas?). As the answer to the questionnaire is simple and intuitive, we need a second classification of the above four categories, so the learning needs of adult learners get 10 categories: A) to get a new job or learning for employment ; B) for the current professional needs or the requirements of the school; C)for the promotion of job titles and learning; D) in order to obtain comprehensive system of vocational education and learning; E) for career development or further education and learning; F) for undergraduate or specialist qualifications and learning ; G)to obtain requirements or credit requirements for school and learning; H) learning for interest or professional interest in school; I) learning for course preferences; J) not to matter or other. For category E, it is indicated that students could either choose "not to matter" or fill out in accordance with their specific content of actual thought. Questionnaire requires students to make only one choice in order to ensure the authenticity of the results of the design, the test being conducted with secret ballot.

4   The Adult Learning Needs of the Pareto Statistics

In this paper, we take a case study of Probability theory attended by 4 classes combined, majoring in Engineering Management,2009 grade, Adult Education School of our university .To find out some main factors of learning needs, we organized four classes of 154 questions; the following table shows the questionnaire's results and the Pareto diagram is as follows:

No.	Item	frequency	probability
1	Employment to education	38	0.247
2	professional needs	28	0.182
3	Titles office professional	14	0.091
4	education courses	11	0.071
5	Career	10	0.065
6	educational required	23	0.149
7	credits required	12	0.078
8	Professional interests	8	0.052
9	courses preferred	4	0.026
10	Other	6	0.039

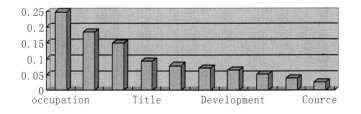

According to four classifications of demand components, the frequency of occupational factors is 101 with probability 0.656; the frequency of academic factors is 35 with probability 0.227; the frequency of hobbies factors is 12 with probability 0.078; other factors, the frequency is 6, the probability 0.039. Classification of the four elements of the Pareto diagram is showed as below:

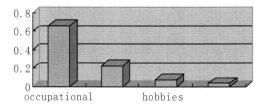

5   Explanations and the Conclusions of the Needs of Adult Learners

(1) Focus on demand. According to the first level of classification, from Pareto chart we know that in adult teaching and management, what teachers need to focus on and what is the most pressing issue, are the interest factors and occupational factors. Because they not only reach 73.4% of all the factors, but these two factors are also positive factors which are to promote students to learn the course. With those two elements solved, the vast majority of teaching and management issues will be solved. Therefore, one of the keys of teaching is to make the teaching curriculum and professional practice and professional development closely combined, and students of basic education, professional interests and expertise closely combined. Arousing the enthusiasm of the students, could not only solve the major problems of teaching, but also could also play a model role with efficiency.

(2) Balance the demand. Based on the statistical results of the second category, employment needs, job needs, education needs, job title needs, credit needs and educational needs of the system accounts for 81.8% of total demand. However, according to the first level of classification analysis, amongst these needs, four occupational factors are positive learning factors, but education and credit factors are neutral or negative learning factors. Therefore, teachers in the selection of course content, assignments and tests in the subject requirements as well as in the usual teaching management must pay attention to balance between the two requirements at the same time, pay attention to the requirements of students in the occupational factors in the selection of role models in order to play an exemplary role to promote the negative students for setting up their confidence in learning the course well.

(3) Pareto analysis does not mean that only the key factors are to be taken care of, turning a blind eye to other secondary factors. One goals of teaching is as much as possible paying attention to each student and can not be careless for any link. Pareto analysis suggests that we should focus on the focus areas, while, about other factors, you can use counseling, classification guidance, teaching and learning, race contest, etc. to promote the development of the overall teaching effectiveness.

(4) Students who focus on demand management factors vary with student's category, and occupational factors and elements of education, also vary from industry to industry, professional difference to professional differences. Therefore, the different types of groups and different projects have different classification levels and methods, which will produce a different analysis of the key factors of conclusion. Importantly, the simple Pareto analysis can give us a clear process to solve the problem of teaching and management of good advice and inspiration to optimize the distribution of course content.

Reference

1. Yuan, W., Pang, H.: Statistical Exercises and Cases, pp. 28–35. Chinese Higher Education Press, Beijing (2006)

Virtual Experimental Platform in the Network Database Application Development[*][**]

Xiaoyu Wu, Wei Dong, Guowei Tang, Huyong Yan, and Liquan Yang

College of Computer and Information Technology,
Northeast Petroleum University DaQing, 163318, China
{dongxiaoge105,crystalgoal}@163.com

Abstract. To develop the network virtual experiments education, is to adapt to the modern trend in education to improve the higher education level of the urgent need of modernization. This articles combination with the schools situation to analysis and research, realized virtual experimental platform in the network database application development. Virtual experimental platform, can make the client in do not need to install any database management system of cases, students and teachers can complete database application series of experiments in lesson and various auxiliary functions. In this respect this paper made many try, obtained the certain effect.

Keywords: network, data base, Virtual experimental platform.

1 Introduction

In modern education, experimental teaching occupied a very important position, for the students innovative quality education is an important means. At present, various universities in experiment teaching, mostly exist experimental form, content, demand is higher and higher and experimental software, device, equipment, site, funds safeguard relatively lagged development of contradictions, experimental security restrictions in a certain extent has affected the developing of the teaching and students' practical innovation ability. Virtual experimental platform[1,5] as a traditional experimental a necessary beneficial supplement, already can save yourself a lot of education funds, also make the experiment on time and space effectively outspread.

2 For Existing Network Virtual Experiment Research

Virtual laboratory concept is suggested by the university of Virginia William prof wulf at first in 1989, it describes network[6] communication technology and the multimedia simulation technology based on the virtual laboratory environment.

[*] Supported by foundation: Oil pipeline to weld intelligent identification system. No. GZD9A120.

[**] Supported by foundation: Welding operation technology detection analysis and diagnosis platform research. No. 115510106.

Y. Wang (Ed.): Education Management, Education Theory & Education Application, AISC 109, pp. 741–746.
springerlink.com   © Springer-Verlag Berlin Heidelberg 2011

First is MIT's WebLab, WebLab is an online laboratory, in 1988 development and put into use. Basic structure including equipment detector (a HP4155B semiconductor parameters analyzer) and a computer (realize double mission: equipment controller and Web server). The devices that want to test placed in the link the test equipment detector, through the GPIB interface communication between the device and your computer.

In autumn 2000, SMA (Singapore2MIT Alliance) microelectronics disciplines students, in Singapore (but lecturer and teaching assistants as far away as the Massachusetts institute of technology) can visit and use WebLab. Students simply click on "equipment" menu, can choose appropriate instruments and equipment . The students in experiments almost no difficulty, generally reflect good: "through the Internet, we should get the experimental data from MIT lab, is very interesting. In the information transfer process technical advantages are very strong "; "This training really well, we can put the real data compared with simulation data".

Another is the virtual laboratory[2,4] of the Carnegie Mellon university, it provides a unique links between the traditional laboratory platform and future experiment.

From these examples, this paper absorb experience, combine with the school actual condition to analysis, find the virtual reality technology defects, and took some coping principles.

3 Virtual Experimental Platform in the Network Database Application Development

3.1 Courseware and Practical Teaching Plan

Virtual experiment is to use computer and simulation software to the simulation the experiment environment and process, let the student through computer to do the experiment, thus to replace or strengthen the traditional real experiments. The teaching design of the virtual experiment system, from the view point of the skills of the students master to consider, let students understand and master what they learn, it is necessary to make the students more hands-on practice, fully reflects the subject status of students ; from the basic characteristics of the course to consider, the experiment system to design rich interactivity operation, in order to let the experimenter achieve more hands-on exercises to the purpose; judging from the basic content of courses, it contains a relational database principle, data structure principle, programming skills and the expression of algorithm thoughts and many other aspects of knowledge, virtual experiment system cannot cover all the knowledge, the design emphasis on the relevant principles and skills, and other related knowledge in content using the words, pictures, video and Flash teaching courseware to supplement. Network virtual laboratory with prominent advantages: (1) is completely based on network, the breakthrough temporal limitations (2) greatly reduced the cost (3) real-time interactive ability, is advantageous for the exchange (4) to facilitate resource sharing. The features of experimental course is many concept, practical strong, more wide range and therefore recommended in computer classroom (or computer multimedia classroom) using the way of flow guidance for the teaching form, lectures and computer should unification consideration.

3.2 Teaching Lesson Plans

This course content mainly includes: database introduction, relational database, relational database language SQL, relation system query optimization, relational data theory, database design, database recovery technology, concurrency control, database security, database integrity, etc. To improve the students applying the knowledge database to the practical of operation ability, and pay attention to design the plan of adapting to the network teaching environment, hope to provide certain guidance for students. Lesson plan mainly divided into the following four aspects, as shown in Fig 1.

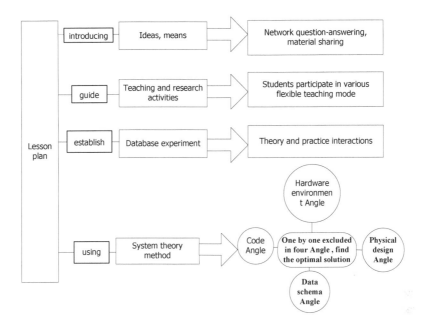

Fig. 1. Lesson plan figure

Through the establishment of database teaching website to answer questions online and correct school assignments, and share the teaching materials on the internet, sufficient use teaching video data other ways effectively play the modern information technology to the teaching process function. In the traditional "teach/discussion" way basis, increase "small project development, case, analysis open source, document retrieval, thesis speech" etc. flexible teaching methods, to realize the teaching , learning and research with each other and promote the interaction, forming the lively study atmosphere. Project of case teaching, raises the student systematic thinking, for example, in the database system design and system performance optimization experiment teaching process, in all kinds of affecting factors interactions. In the teaching experiments, through driving students to change the ideas for the lab lesson, formed the characteristic of database teaching, have already received preliminary results.

3.3 The Problems That Should Be Paid Attention to in the Platform Construction

Database principles courses teaching reform and innovation is a systematic project, which will not be accomplished overnight, in addition to the great efforts, but also pay attention to scientific methods and means, can be able to success. In teaching reform should notice the following problems:

（1） Compilation and selection experimental teaching material
（2） Improvement teaching method and testing method
（3） Teachers must have the ability of practical development database principles.

3.4 Strengthen the Teaching Experiment Depth, Develop Database System Design Skills

Strengthen the actual database system development, make students learn knowledge truly to apply their knowledge. People know now information system development is to use the development platform, and backstage to use the database to manage large amounts of data model. When students study database system application and development of basic principle, know how to establish a database, know integrity control of the database, safety control, concurrent control , the knowledge of database restore, but not equal to know how to use the database system developing platform to create the database system or carries on to data processing. while how to use the database system developing platform to create the database system or carries on the data processing, developing a practical miniature database system, is the students should really master. This requires should strengthen the teaching of course design in the education, assigned by the teacher or chooses to topic by student, to develop the actual database system.

The way of this paper is: in the beginning of the semester, select a group application system often to use in real life as the topic of curriculum design, issued to students in the form of commitments, student in groups to achieve. Topics include: worker archives management system, the system of managing student file, student performance management system, books data management system, warehouse management system, material management system, teachers' scientific research management information system, salary management system, etc.

3.5 The Variety of Teaching Methods

Auxiliary class teaching, based on classroom teaching, combining multimedia teaching means make classroom teaching vivid image. Combining with the teaching carrier, strengthening the experiment link, can active classroom atmosphere, improve the teaching effect. In this paper, emphasis on basic knowledge, at the same time, aim at the engineering nature and applicability of, strengthen the training of experimental link in the teaching, to enhance student team spirit and train the awareness of team. Computer experiment closely combining the basic knowledge and basic technologies, from the

R database installation to the matched experimental link of database system theory, to help students understand and digestion course content (such as query optimization, concurrency control, restore, etc.), lets the student in the actual database practical and laboratory, and submit standardized computer experiment report.

This theory with practice, book knowledge learning and actual systems experiment closely, not only helps students the knowledge of basic theory, and can make students more flexible more in-depth study knowledge, cultivating the ability of autonomous study and Extrapolate ability, still can stimulate the students' learning passion.

Database design and its application development, request the student in one R database, using development tools of appropriate (can optional) application system, to develop a small database application system for one department or unit (can choose the actual departments can also design a virtual department). Request 5 ~ 6 classmates form a development team, each student bear different roles, including: project administrator, system analyst, System designer, system developer, system testers, DBA. Through the practice, master the database design in bookish and the content of database programming chapters. Meanwhile, very important is to cultivate the team cooperation spirit, division of labor cooperation, and common complete database design and application development. For big assignment requires each group according to books design method and process given: detailed design reports in every stages of the database design, write the main functions of the system and instructions, submit running systems (must be really running), write harvesting and experience, has resolved and unresolved problems, further perfect ideas and suggestions. Finally each team conduct 60 minutes reports and reply, explain design scheme, and demonstrates the operation of the system and to report the division of labor and cooperation situation.

4   Main Research Achievements of Database of Virtual Experiment

Develop a network-based database application virtual experimental platform, in the case of the client does not need to install any database management system, students and teachers can complete every experiments and various auxiliary functions of a series of database application course, the experiment include 2 front-end database development tools principle and application experiment, 1 data middleware principle and application experiment, 2 database management system principle and use experiments, auxiliary function such as experiment guidance and demonstration, experiment report submit and reviewing, examine and evaluation, etc. The students produce experiment report after the computer practice, change the way in the past to ensure experimental achievement according to the experiment report, but pay attention to students' computer practice process, is also the students' experiment results mainly depends on the students of computer practice process, exactly what content that is the students independent operation, which is requires teachers to enhance the inspection dynamics, thus cancel the phenomenon of experiment report plagiarism, the experimental results as the primary way to measure students' usual performance.

5 Project Promotion Value

This project will establish an effective database application experimental teaching system, the integrated use of simulation technology, multimedia technology and network technology to construct virtualization database application series course experiment environment in the integration of experiment teaching, experimental operation and test report, realizing from auxiliary teaching, autonomously experiment to the experimental report submitted online and correcting the whole operation and management, to meet the needs of online experiment teaching. The results of researches on the next computer professional database application series course teaching has certain experiment guidance function, but also for related major to provide a complete database application series course experiment virtual platform.

6 Conclusion

Virtual experimental platform as a necessary and beneficial supplement of the traditional experiment, it is more paying attention to the cultivation of students' theoretical relation practice ability and practical capability, this to mobilize students' learning enthusiasm, improve the ability to solve problems is very good. Virtual experiment has advantage of usage convenience, low cost, less disturbance and easy to implement the repeated experiments under the same conditions, good sharing in the network environment, benefited range widely etc, and is the important content of network teaching. Anyhow, virtual experimental platform both can save a lot of education funds, also make the experiment extend effectively on time and space.

References

1. Yin, H., Chai, Y., Qu, J., Guo, M.: Computer hardware course virtual experimental platform development and design. The Modern Scientific Instruments 02 (2009)
2. Liu, X., Zhang, W., Cheng, H., Wang, B.: Virtual laboratory types and developing trend. Computer Application and Research (11) (2004)
3. Li, J., Sun, S., Jiang, P., Xia, X.: Construction teaching platform based on network virtual laboratory. College Age, B version (07)(2006)
4. Li, J., He, W.: Long-range virtual laboratory related technical research. In: Chinese Instrumentation Institute the Ninth Youth Academic Conference Proceedings (2007)
5. Yang, C.: Open virtual experimental platform design and development research - To education technology professional for example. Modern Education Technology (01) (2009)
6. Zhang, H.: The virtual laboratory based on the network. Contemporary and Long-Range Education Research (02)(2005)

An Empirical Analysis on the Interdependence Relation between Higher Education Tuition Expenditure and Urban Residents' Income—Based on ECM Model

Renjing Xu

Department of Foreign Language, Nanchang Institute of Technology, P.R. China, 330099
xubin9675@163.com

Abstract. Higher education is a quasi-public product, and the form of tuition standard of higher education should base on social benefits and market benefit. Our public college tuition standard is the embodiment of social benefit and market benefit. But, there also exist theoretic defect and practical difficulties in formulating college tuition standard, such as higher education tuition be high, people burden getting worse and not consider the economic bear ability of residents, etc. On account of the above-mentioned existing problems, in this paper, we firstly introduced the research background and research meaning of higher education tuition; Then, we carried out an empirical analysis on the relationship between Chinese higher education tuition and urban residents' income with error correction model of econometrics; Finally, we made analysis and explanation on the above-mentioned operation result.

Keywords: higher education tuition, Residents' income, ECM models.

1 Introduction

In the 1960s and 1970s, because higher education have played an important role of promoting the economic development of every country in the world, world countries, including developed countries and developing countries, have put huge resources into higher education, and higher education also got unprecedented development. But, after 1980s, most world countries appeared the coexist of economic growth slowdown and inflation, government revenue declining, cutting public education funding, which lead to higher education funding dropping and the strain of college financial. For instance, from 1980 to 1988, the annual growth ratio of college student number of low-income countries, low-and-middle income countries, middle-upper income countries and high-income countries were 8.8%, 6.6%, 6.1% and 4.3% respectively. But the annual decline ratio of corresponding public higher education funding of low-income countries, low-and-middle income countries and middle-upper income countries, except the high-income countries have 0.9 percent annual growth rate, were 12.3%,9.1% and 4.6% respectively, which resulted in the global financial crisis of higher education. In this context, the cost sharing theory proposed by John stone (1986) become the guidance for world countries raising fund. World countries all carried out the measure of students and their parents sharing education cost for mobilizing private

Y. Wang (Ed.): Education Management, Education Theory & Education Application, AISC 109, pp. 747–755.

resources to support the development of higher education, which became an important trend of world higher education finance.In view of the higher education charging status and higher education importance, domestic and foreign relevant scholars have made a large amount of research, the main achievements were as follows: Elchanan Cohn (1979) made an analysis on education's benefits, education's cost and the relationship between them, educational production and cost function, education finance and education plan. He put the education costs into two broad categories: direct costs and indirect costs. Direct costs mainly were the cost of education service provided by school, and indirect costs were mainly included the income forgone due to go schooling, tax deduction enjoying by higher schools, etc. Mark Bray (2000) pointed out that higher education charges not only because of the financial strain, but also because people have realized a free education in colleges, far from unfair, was unfair. The reason was that the likelihood of the youngsters coming from wealthier families entering higher schools was higher than that of poor families. CuiYuPing (2006) thought that tuition should based on the average cultivate cost to pricing and should not be based on the marginal training cost to pricing. Su Liangjun and Xun Bianxia(2006) made use of space autoregressive modeling and tuition data of Chinese universities undergraduates to carried out an analysis. At last, they found that college tuition not only was affected by school oneself, local economic development level and majors influence, but also was affected by college tuition levels of surrounding universities in the same region. Xie Zuoxu, Chen Xiaowei (2007) and   Pacharopouloset al (2008), through the university student sample survey of 10 provinces, city, district in Chinese mainland, made a analysis on the effect tuition on different social strata children. Took social advantage position class and basic class children as example, he discussed the relationship between social classes and tuition fees of higher education, college students possessed the attitude on the tuition fees of higher education and charge policy and higher education opportunities of different social strata, etc and got the conclusion that the tuition of higher education has an important influence on different classes children. Wen Jian, Lei Lijuan, Feng Dandan (2009), Hu Maobo and Shen Gong (2009) took the fees charged situation of ordinary higher education as the research object, refer to per capita disposable income of town family, per capita net income of rural families, each year state financial appropriation for ordinary university, per capita GDP and per capita training expenses as main factors to establish a the multiple regression model of econometrics. At last, he, according to the analysis result, put forward the corresponding proposal: First, to scientifically and reasonably sure universities charge standards; Second, to constantly increase national finance investing into ordinary universities; Third, to appropriately control the admissions scale of ordinary university, etc. Cui Shiquan and Yuan Liansheng (2010) thought that quasi-public product attributes and commercial properties of higher education could be located in users paying for higher education tuition. Tuition collection should base on private benefit and service providing cost, and the forming mechanism of tuition was based on the marketing pricing of private benefit section under government intervention. Wei An'duo (2010), in terms of the actual situation of China's universities charge, used spatial autoregressive models and combined with the some new characteristics of our economy in recent years to calculate an more reasonable actual amount of higher education tuition. Yuan Lei, Liu Huihong (2010) thought that, in private resources was invest into higher education background, college tuition should execute a difference pricing based on different universities, disciplines and individual consumers, etc. In short, there were a lot of scholars studying on the higher education tuition problem, and we can't introduce all of them. However, there was very little study on the interdependence relations between higher education tuition and residents' income. But even so, their researches and analysis have reference value and quotable significance for our in-depth studies on the higher education tuition problem. In this paper, we used the econometrics error correction model to study the relation between higher education tuition and urban residents' income level and hoped the study has some enlightenment for later scholars studying higher education tuition problem.

2   The Overview of Cointegration Theory and Error Correction Model

Cointegration theory and method was put forward by Engle Granger in 1987. In the economic field, economic time series are usually nonstationary. But our application econometric model exist dynamic stability hypothesis, and cointegration technique compensates for the lack of stable hypothesis. The basic thought of co-integration relationship was that although two or more variables sequence were non-stationary sequence, they combination may have stability and has a long-term stable relationship (co-integration relationship). If two variables all were single integrated variables and had same single integrated order number, there existed a possible co-integration relation. Cointegration meaning lies in revealing whether there has a long-term stable equilibrium relationship between variables. Cointegration concept established the theoretical basis of seeking for the equilibrium relationship between two or more non-stationary variables and setting up error correction model of variables existing cointegration relationship.

Cointegration test main steps were as follows:

The first step, if sequence y1t, y2t, y3t,...,Ykt all were single integrated, and the established regression equation was:

$$y_{it} = \beta_2 y + \beta y_{2t} + \cdots \beta_k y_{kt} + u_t \tag{1}$$

The estimated residuals was: $\hat{u}_t = y_{1t} - \hat{\beta}_2 y_{2t} - \hat{\beta}_3 y_{3t} - \cdots - \hat{\beta}_k y_{kt}$; (2)

The second step, to inspection whether residual sequence $\hat{u}_t$ were steady, that was to say, to judge whether residual sequence $\hat{u}_t$ contained unit root. Usually we use ADF test to determine whether the residual sequence were steady.

The third step, if residual sequence were steady, we can determine that there was a co-integration relationship among y_{1t}, y_{2t}, y_{3t},...,y_{kt} variables. Otherwise, there didn't exist a cointegration relation among y_{1t}, y_{2t}, y_{3t},...,y_{kt}. Error correction model was put forward by Hendry, Davidson, Srba and Yeo in 1978, so it called DHSY model. Error correction model is a specific forms of difference equation model, which reflect the "error correction mechanism" between long-term equilibrium and short-term fluctuations. The main content of the error correction model with double variables: to consider a autoregressive distributed lag model (ARDL (1, 1)) of two variables:

$$Y_t = \beta_0 + \beta_1 X_t + \beta_2 Y_{t-1} + \beta_3 X_{t-1} + \xi_t \tag{3}$$

We changed the model into single order difference form, which form was as follows:

$$\Delta Y_t = \beta_0 + \beta_1 \Delta X_t + (\beta_2 - 1)Y_{t-1} + (\beta_1 + \beta_3)X_{t-1} + \xi_t \tag{4}$$

$$= \beta_0 + \beta_1 \Delta X_t + (\beta_2 - 1)(Y_{t-1} - \frac{\beta_1 + \beta_3}{1 - \beta_2} X_{t-1}) + \xi_t$$

Set $ecm_t = Y_t - \dfrac{\beta_1 + \beta_3}{1 - \beta_2} X_t$, $\alpha = \beta_2 - 1$, and the model was turned into as follows:

$$\Delta Y_t = \beta_0 + \beta_1 \Delta X_t + \alpha ecm_{t-1} + \xi_t \tag{5}$$

Where, ΔY_t stands for the short-term volatility of explained variables, ΔX_t stands for the short-term volatility of explanatory variables, ecm_{t-1} stands for error correction items. α was called correction coefficient, which reflect the correction pace of Y deviating the equilibrium. Therefore, the short-term fluctuation of explained variables can be decomposed into two parts: one is the short-term fluctuations of explanatory variables, another is the adjustment effect of long-term equilibrium.

The most commonly used estimate method of ECM models is the two-step method put forward by Engle and Granger, which basic inspection procedure is as follows:

The first step, make cointegration regression for model (6), and got $\hat{k}_0$, $\hat{k}_1$ and residual sequences (7).

$$Y_t = k_0 + k_1 X_t + u_t, \qquad t = 1, 2, \cdots, T \tag{6}$$

$$\hat{u}_t = y_t - \hat{k}_0 - \hat{k}_1 x_t, \qquad t = 1, 2, \cdots, T \tag{7}$$

The second step, we use $\hat{u}_{t-1}$ substitute for $Ecm_t = Y_t - \dfrac{\beta_1 + \beta_3}{1 - \beta_2} X_t$, that is to say, we use OLS method to estimate the parameters of $\Delta Y_t = \beta_0 + \beta_1 \Delta X_t + \alpha ecm_{t-1} + \xi_t$.

3   The Description and Analysis of Indicators Data

In this section, we refer to 1995 ~ 2009 as samples period and took per capita disposable income of urban residents and per capita higher education tuition as variables to analysis the stable relationship between them. All the data are come from "China statistical yearbook". We according to the comparable price to calculate annual data, and took the data in 1995 as radix 100. Each index data was deal with consumer price index. In order to got easily stationary sequence, we made natural logarithms for each variable respectively. The transformation can eliminate heteroskedasticity between variables and do not change the cointegration relation between variables, so as to improve the reliability of the estimated. Meanwhile, in order to expediently survey the elasticity of consumer spending on the income, in this section, we took natural logarithm form of TUI and YC respectively. Then we took difference form of TUI and YC, which are written for LnYC, LnTUI, DLNYC and DLNTUI.

4 Integration Test

Most time series of economic indicators have non-stationary characteristics. If prior not consider the stationarity of time sequence and made linear regression on direct non-stationary data, there would appear "spurious regression" phenomenon. From the trend diagram of LnYC and LnTUI (see chart 1), we can found that the data of LnYC and LnTUI have an obvious rising trend. Therefore, before the cointegration analysis of variables, we first made Unit Root Test to inspection variables were stationary or not.

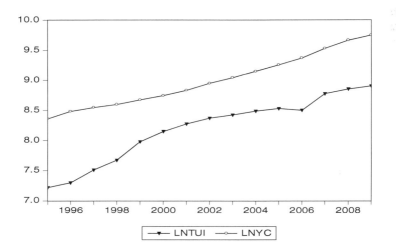

Fig. 1. The Change trend diagram of LNTUI and LNYC

In this paper, we adopted ADF (Augment Dickey - Fuller) test. ADF test results are showed in Table 1.

Table 1. The ADF unit root test statistics table of urban residents income and higher education tuition

Variable	Test Type	ADF value	$\triangle 1$	$\triangle 2$	$\triangle 3$	stability
LNTUI	C,0,1	-1.802	-4.00	-3.10	-2.69	No
LNTUI	C,T,1	-1.236	-4.80	-3.79	-3.34	No
LNTUI	0.0.1	4.469	-2.74	-1.97	-1.60	No
DLNTUI	C,T,1	-3.658	-4.89	-3.83	-3.36	Yes

Table 1. *(continued)*

LNYC	C,0,1	2.119	-4.00	-3.10	-2.69	No
LNYC	C,T,1	-0.673	-4.80	-3.79	-3.34	No
LNYC	0.0.1	13.585	-2.74	-1.97	-1.60	No
DLNYC	C,T,1	-3.466	-5.30	-4.01	-3.46	Yes

Note: LnYC, lnTUI and D stand for the logarithm of per capita urban resident's disposable income, the logarithm of per student of higher education tuition and 1 order difference respectively. In test type (C,T,K), C,T and K stand for Constant Term, trend variable and lagging orders respectively.

From table 2, we can found that the ADF statistical value of original time sequence (TUI and YC) are greater than 10% critical value. Therefore, the two sequences have unit root and they are nonstationary. However, the ADF value of (C, 0, 1) test forms of DLNTUI is less than 10% critical value, and the ADF value of (C, T, 1) test forms of DLNYC is less than 10% critical value. Thus, we can reject the null hypothesis, namely two first-order differential sequences don't exist unit root and are stationary. That is to say, LNTUI ~ I (1) and LNYC ~ I (1), and they can be undertake cointegration analysis.

5 Granger Test of Causality

Cointegration test indicators there is a long-term equilibrium relationship between variables. But, we still need further test whether there is a causality relationship between variables. If the variable X can help predict Y, namely when we can according to the past value of Y make regression and plus X past value, it can significantly enhance explanation capabilities of regression equation. Therefore, we can say X is the Granger-causes of Y, otherwise, X isn't the Granger-causes of Y. We used granger causality test to test the causality relationship between per capita urban residents' disposable income and per student higher education spending. Under 10% significant level test, the testing results are shown in table 2.

Table 2. The inspection result of Granger causality relationship between variables

variable	lagging orders	Null hypothesis	F-Value	P-Value
LNTUI, LNYC	2	LNYC isn't the causality of LNTUI	1.09	0.183
		LNTUI isn't the causality of LNYC	1.02	0.203
LNTUI, LNYC	3	LNYC isn't the causality of LNTUI	3.40	0.084
		LNTUI isn't the causality of LNYC	2.93	0.093

From table 3, we can found, under 10% significant level, there exists a two-way causality relationship between urban residents' income and higher education tuition.

6 Cointegration Test

In this paper, we adopted the E - G two-steps proposed by Engle and Granger (1987) to test whether there exist a cointegration relationship between LNTUI and LNYC.

First, we got the cointegration equation, which ia as follows:

$$LNTUI=-2.25529+1.162LNYC$$
$$(-2.127) \ (9.870)$$
$$R2=0.882, \ \bar{R}^2=0.873, \text{F-statistic}=97.42 \tag{10}$$

According to the significant test of above-mention model parameters and fitting indicators value, we can judge the regression equation yielded good results.

Second, To test the stationarity of residual sequence of regression equation.Test method is unit root test (ADF test), and test results are shown in table 3.

Table 3. The statistics table of ADF test of residual sequence

Variable	Test Type	ADF value	$\triangle 1$	$\triangle 2$	$\triangle 3$	stability
e_1	C,0,1	-4.014	-4.06	-3.12	-2.70	Yes
e_1	C,T,1	-3.877	-4.89	-3.83	-3.36	Yes
e_1	0.0.1	-4.143	-2.75	-1.97	-1.60	Yes

Note: $\triangle_1, \triangle_2$ and $\triangle_3$ stand for Critical value(1%), Critical value(5%) and Critical value(10%) respectively.

The critical value of ADF value of (0, 0, 1) test forms of residual sequence e1 is less than that of 5% significance level. Therefore, we thought residual sequence is stationarity, namely e1⊏ I (0). It shows that there exists a co-integration relationship between LNTUI and LNYC, namely there exists a long-term stable equilibrium relationship between them. From cointegration equation (10), we can see that the coefficient of urban residents' income is 1.162, which explain when, in terms of the long run, per capita town residents' income increase 1 percentage point, the higher education tuition spending will increase 1.162 percentage points. Thus, there is a significant positive correlation between higher education tuition expenses and urban resident's income.

7 The Establish of Error Correction Model

Since there exist a co-integration relationship between town residents' income and higher education spending, we can build the error correction model of DLNTUI and DLNYC (shown in equation 11).

$$DLNTUI_t = 0.182 + 0.596DLNYC_t - 0.313ECM\,(-1)$$

$$(1.695)\quad(2.895)\qquad\quad(-3.545)\qquad\qquad\qquad(11)$$

$$\bar{R}^2 = 0.875,\qquad F = 79.483$$

In the above-mentioned ECM models, error correction coefficient is negative and accord with reverse revision mechanism. From the equation (11), the short-term elastic of urban residents' income to higher education spending is 0.596, which indicates, in the short run, when urban residents' income increase 1 percentage point, higher education spending will increase 0.596 percentage points. The coefficient of Error correction item is - 0.313, which explains the adjustment strength of the short-term volatility to long-term equilibrium is 0.313. It shows the adjustment strength is significant.

8 Conclusion

Through the above-mentioned empirical analysis on the co-integration relationship between higher education spending and urban residents income, we can find there exists a long-run equilibrium relationship between urban residents' income and higher education spending. The augment of urban residents' income can promote higher education spending, and vice versa. In order to substantially improve urban residents' income level, we must take active measures to promote higher education consumption of urban residents.

References

1. Cohn, E.: Political Influences on State Policy:Higher-Tuition, Higher-Aid, and the Real World. Public Policy and Higher Education, 287–293 (1979)
2. Pacharopoulos, et al.: High Stekes on Tuition. The Chronicle of Higher Education (34), 209–236 (2008)
3. Bray, M.: The Value of the ability to deal with disequilibria. Journal of Economic Literature (13), 872–876 (2000)
4. Cui, Y.: University tuition discount and tuition pricing. Journal of Education Development Research 2, 27–30 (2006)
5. Su, L., Xun, B.: An analysis on factors affecting College tuition and their space correlation. Journal of Mathematical Statistics And Management (7), 400–406 (2006)
6. Xie, Z., Chen, X.: The influence of China college tuition on the children coming from different social strata–an empirical investigation and analysis. Journal Of Education and Economy (2),12–15 (2007)
7. Wen, J., Lei, L., Feng, D.: An empirical analysis on college tuition—based on an econometric. Journal of Management (10), 86–90 (2009)
8. Hu, M., Shen, H.: The Difficulties and countermeasures of College tuition standard. Journal of Education and Economic (1), 10–15 (2009)

9. Cui, S., Yuan, L.: College tuition properties: public service charges, price or users pay for education development. Journal of Education and Economic (11), 24–28 (2010)
10. Wei, A., et al.: The college tuition analysis—based on Space regression models. Journal of Liaoning Engineering Technology University (Natural Science Edition) (5), 182–184 (2010)
11. Yuan, L., Liu, H.: Discussion on price discrimination and education fair of college tuition. Journal of Education Exploration (11), 14–16 (2010)

Outdoor Sports and Teenagers' Moral Education

Tao Yuping

Chengdu Sports University
Chengdu, Sichuan Province, China 610041
Taoyuping8844@126.com

Abstract. The socio-economic development and growth in the living standard in our country not only creates the wonderful condition for the development of teenagers but also multiplies laziness. What's more, the negative impact brought by the examination-oriented education system fails to improve the teenagers' health and psychological quality. So, it is significant for us to try to use "tribulation" or "frustration" to exert an influence on the moral cultivation of the teenagers and the best way to do so is the outdoor sports. This article uses the method of comparative analysis and deductive reasoning to analyze the moral function and effect of the outdoor activity. The conclusions we arrive at are as follows: the outdoor activities can promote the healthy development of the teenagers both physically and mentally. It is suggested that the teachers adopt the flexible and various educational method in order to give the teenagers the moral education in entertaining activities.

Keywords: Outdoor sports, moral education.

1 Introduction

The Reform and Opening-up has brought great changes to our country, and the people's living standard and living conditions have been enhanced constantly since then. With the prosperity of economy and the richness and superiority of the people's life, many families pay great attention to the younger generation's education and try to provide better living and studying conditions for their children. Besides, the school is better constructed, and the school environment has become more and more comfortable. In the educational environment where the exam-oriented teaching method is the main concern, the school life is nothing but study, and the rigid teaching model can hardly improve the students' overall quality. The examination-oriented education drives the teachers and students to work hard to get higher scores just like an invisible lash. People have neglected the students' development in an all-around way, which makes a new type of "nobility" come into being. For example, the "little king" spoiled by the parents, completely depends on their parents on all details in life like a parasite. Unfortunately, the adolescents in our country lack the persevering will and indomitable spirit.

The maturity of the Reform and Opening-up and the market economy, on the one hand, has helped our economic construction make world-beating achievements; on the other hand, some bad habits have caught us off guard, for example, money worship, egoism, and corruption phenomenon such as corruption and bribery, erotism and superstition, etc. These corruptional phenomena have had bad effect on the teenagers:

Y. Wang (Ed.): Education Management, Education Theory & Education Application, AISC 109, pp. 757–764.

firstly, they lack the hard-working and hardship-bearing spirit, the determination to fight against predicament, and the courage to overcome difficulties; secondly, some bad habits emerge, such as: unruliness, selfishness, bad temper, irritability, withdrawnness, and indolence; finally, their psychological endurance is low. Their soul is so weary that they will easily become frustrated, dispirited and experience mental breakdown when they meet difficulties and frustration. What's worse, they even descend to young criminals or commit suicide. The negative effect brought by the wealthy life completely goes against the aims of both the moral education in our society and the ethic education in our school. What is required for the prosperity of our country and the revitalization of the nation is the builder and successor of the socialist cause, with lofty ideals, integrity, knowledge and a strong sense of discipline, instead of the "flowers" who are sentimental and prefer living in a hothouse. Poverty and wealth is a double –edged sword, poverty can deprive the people of their enjoyment in life, but at the same time it can nurture people's personality; on the other hand, wealth can open people's eyes, but can also fossilize people's mind. To make the children grow healthily, the families, schools and society are supposed to intentionally create conditions for the children to experience "tribulation "or "frustration", which has become an important proposition because opportunity knocks but once.

2 "Tribulation"—Traditional But Modern Moral Education Mode

"Tribulation" is also called "frustration". It tries to create various difficult environments in order to make the teenagers walk out of the class and families and taste setback, difficulties, challenges in natural world or society. In a word, its purpose is to teach the teenagers how to tackle difficult situations and to cultivate their characters and will. The Chinese nation has had the tradition of promoting "tribulation" from time immemorial. For example, as early as the Warring States Period, Mencius proposed the view saying "thrive in calamity and perish in soft living". "Tying one's hair on the house beam and jab one's side with an awl to keep oneself alert "and "sleep on the brushwood and taste the gall", don't the preceding two cases belong to the earlier "tribulation" mode of our country? The "tribulation" is the indispensible component of modern quality education. It has great practical significance to promote the "tribulation", such as: cultivating the teenagers' strong will and a pioneering spirit, improving their overall competence, and promoting the revitalization and prosperity of the Chinese Nation.

Over the past years, all kinds of "tribulation" were carried out gradually: the "frustration" summer camp, the marching school in Huaian, Jiangsu, and the "West Point pattern" in Hangzhou, Zhejiang, etc, it looked as if hundreds of flowers are in blossom. There's a wide spectrum of opinions on these phenomena. At least, these training modes avoided the past traditional sermon, and verified the students' experience and perception. What's more, it could improve the trained students' moral cognition and finally form certain behavior criteria. However, this is only a special education aiming at a few specific students, and the continuity of the education effect remains to be tested. This kind of "tribulation" is obviously unfit for most of the students. It is also possible that the misusage of the "tribulation" mode can result in walking from one extreme to the other. For example, military-style training tends to stifle the children's puerility, and finally the naive children will become rigid puppets. This will be the tragedy of our country's education. Therefore, it is necessary for us to use scientific plans and methods in the "tribulation" education and to balance the relation between quantity and degree.

What is required for promoting the "tribulation" is as follows: first is authenticity. We should create relatively authentic difficult situations and extreme conditions for the children so that they can overcome the difficulties all by themselves, and challenge themselves and cultivate their will. Second is long-term education. The "tribulation" education cannot take effect in one day unless we adhere to it for a long time. It is no surprise that it fails to take effect if we only depend on what is called tribulation cultivation once or twice a year. Third is interest. We should promote the tribulation education according to the children's characteristics and needs. Besides, we should try to stimulate their enthusiasm for taking part in the "tribulation" and make them interested in the content of the "tribulation".

Therefore, it is expected that we can find a scientific and sustained tribulation mode.

3 The Outdoor Sports— the Concept and Mode of New Tribulation

What are outdoor sports?

The outdoor sports are the physical activities that are launched in the natural setting (not the exclusive one). The natural setting (not the exclusive one), the natural state of existence, includes fields in nature, buildings whose purpose is not for sports, such as highways, buildings, and things like that. The outdoor sports exclude such kinds of sports activities as football, beach volleyball and golf that are launched outside but on artificially exclusive sites. Sports activities regulate the properties of the outdoor sports which differ from those of other activities launched in the natural setting, for instance, sightseeing tourism and productive labor activity.

The items belonging to the outdoor sports are as follows: camping, picnic, mounting climbing, cliff climbing, ice climbing, mountain (fixed-point) hiking, descending a cliff with the help of equipment, taking a rope to cross river, river rafting, crossing the river with homemade tools, survival skills, self-rescue, mutual help, rescue, desert positioning and orientation ,signal and contact as well.

The adolescents taking part in the outdoor sports are to train themselves by means of returning to nature in which they can endure the torture from the nature and ultimately overcome difficulties. The approach and environment are so authentic that they can be accepted easily by children. Tribulation integrated in sports activities, free from hasting short-sighted behavior, can not only inspire the children's enthusiasm participating in the Tribulation, but also accord with the characteristic of physical and mental development of children. The permanence, systematicness, and scientificity of the Tribulation can be well guaranteed.

So we can assume, the outdoor sports, as the concept and mode of new Tribulation, are bound to become important means and methods of the moral education in the new era. In another way, modern moral education endows the outdoor sports with new functions.

4 Moral Education—the Moral Function of the Outdoor Sports

What is morality? According to its definition, it consists of four components: first, it is determined by certain socioeconomic relations; second, it uses the good and evil as the criterion for evaluation; third, it is sustained by people's belief, public opinion, and traditional customs; fourth, it can be regarded as principles and norms used to regulate the relations among people and the relation between individual and society. The starting point and ultimate goal of the "tribulation" is to cultivate the teenagers' morality, to correctly handle the relations among people and the relationship between the individual and the group, and finally conform their behaviours to the social standards.

The outdoor sports have become a popular sports pattern among the people especially the teenagers. The outdoor sports, regarded as the instrument and method of the "tribulation", can not only enhance the physical quality, but also can increase knowledge, enlighten intelligence, enrich life and increase the environment adapting ability. Besides, the participants' moral character can be cultivated and their soul will be lifted by the lively and concrete cultivation process.

The moral functions of the outdoor sports are as follows:

4.1 Outdoor Sports and Patriotism

Patriotism, which developed and consolidated thousands of years ago, expresses people's deep emotion for the country's land, rivers and mountains, people, culture, language, and history, and finally forms the firm belief which expresses loyalty and love to the nation. Patriotism is a lofty sentiment of ethics, a basic requirement for a person, and an eternal theme promoted by the "tribulation". In the daily ideological education, children often feel that Patriotic Education is too comparatively vacuous and abstract to be integrated with real life. It is believed that it's too unrealistic to do just what Huang Jiguang and Dong Cunrui did when our country is in crisis. The reason for this is that they think it is in peace nowadays, and they should enjoy themselves. The teenager's nowadays have neither gone through any pain nor had their own experience about the suffering that the country and nation had experienced. So it is necessary for us to carry out the "tribulation", and to try to learn the Chinese nation's ordeal that demonstrates "backward nations tend to get beaten up". The teenagers should have the courage to afford the affliction, and frustration and the determination to equip themselves with knowledge so that they can become more patriotic and serve the country much better.

The outdoor sports put us in the natural surroundings, and make us enjoy sports in jungles and mountains. What's more, they let people feel broad-minded and happy. Besides, our body and mind is molded or cultivated. Only if we are not afraid of the dangers and difficulties, can we reach the summit of victory. Only at that moment, can we really feel the exultation of victory such as "people are the peak of the high mountain" or "hold all mountains in a single glance". We can throw ourselves into the nature, breathe fresh air and enjoy the sunshine. All these can not only have our temperament molded, but also remind us of the vast territory, rich natural resources, picturesque landscape, diligent people and profound culture. We can also have a deeper empathy with trees, water and mountains. Eventually, patriotism will be realized in the content of "love the motherland, love the territories and love the people".

4.2 The Outdoor Sports and Collectivism

Most of the children nowadays are the only child in their families, and the disadvantages they have are as follows: firstly, in their families, they are their parents' "little sun", or just like "little king" with their horn exalted; secondly, they are self-centered and selfish in the communication with others; thirdly, in a group they are undisciplined, loose, and lazy, have tense and cold interpersonal relations and lack the sense of group. Collectivism is the moral principle of socialism, and the indispensible and basic quality for the social members. The aim of the "tribulation" is to overcome the individualism and selfishness, and raise the sense of group.

The outdoor sports as a kind of collective activity, are supposed to be organized step by step and have strict and inhibiting disciplines and a compact organization, no matter the activity is camping, wading, drafting, mountain climbing or rock climbing. The outdoor sports are supposed to be organized in advance before starting, and prepared in all respects such as equipment, fuel, food, tents, etc. All the preparations should be as careful and perfect as possible, from the tents of the dormitories to the band aids or pills used in accidental injuries. Otherwise, it will bring losses or even a defeat to the outdoor sports. In the outdoor sports, every person has his or her own role to play, which is closely related to the group. In the group, everybody is equal, neither the "little king" or self-centered persons nor the lazy persons can be accepted. It can be said that everybody should always keep the interests of the group in mind; the group should always be concerned for each member in the outdoor exercise. The strict sense of organization and discipline is cultivated in the course of the outdoor sports. The ideological foundations which are essential to the organization are "for the interest of the group", "fraternal unity" and "care for companions". The teamwork, as the soul of an outdoor sports team, can make the big difficulties small, and the small difficulties disappeared in the course of the outdoor sports. So, it is an important means of the new "tribulation" to use the outdoor sports to cultivate the children's collectivism spirit and set up effective teams.

4.3 The Outdoor Sports and Volitional Quality

Nowadays, the affluent life makes some children seek ease and comfort and they lack enterprising spirit. They often take the credits for themselves and leave the difficulties to others. They live the unexamined life all the time and are not able to be aware of the purpose and significance of their behaviour clearly.

We will meet all sorts of unexpected difficulties in the outdoor sports. So it requires the participants to have a cool head, to be clearly aware of the purpose and significance of their behaviour, to be good at making up their minds firmly and opportunely and make a decision, and get rid of the threats and obstacles constantly. With the outdoor sports such as mountain climbing, wading, other kinds of climbing under difficult conditions, traversing and active challenges, and people can strengthen their physical quality and will, overcome their cowardice, and finally their mind and physical ability can be rebuilt and developed. Owing to fighting against the dangerous conditions in nature constantly, people can not only learn extraordinary skills, but can also develop the strong physique and the ability to overcome difficulties with fearlessness and perseverance in order to fulfill the difficult and complex task. Therefore, the outdoor sports are compared to the cradle of the brave, and can be regarded as the excellent sports to purify people's soul, and to cultivate brave and combatant spirit. The emphasis of the tribulation is on cultivating the modern teenagers' will and character such as consciousness, self-control, determination, and tenacity, etc.

4.4 The Outdoor Sports and Mental Development

Children today have a wide range of knowledge, and know a lot of things. Once they meet difficulties, they are at a loss as to what to do and don't know how to tackle difficulties and problems. Theory without practice shall result in the failure of acquiring the ability.

Normal intelligence is the premise to perceive and comprehend the world, and the basis of mental health. People's intelligence depends on the function of the brain and central neural system. Good physique especially good neural system is the physical basis of the development of the intelligence. The outdoor sports take place in an oxygenated environment of the outside world, which can guarantee the energy supply for the brain. It is always accompanied by complex mental activity, in which it offers all kinds of stimulating information, to stimulate brain cells continually, and to help the brain neural cells develop well. The increase in the branch and neuritis of the brain neural cells result in the improvement of the intensity, coordination and flexibility of the cortex's activity and the cultivation of the keen perception, concentration and a good memory. It can compel the development of the intelligence by constant doing outdoor sports especially through cross-country orienteering. Study shows that a pleasant mental experience can be produced, if the outdoor sports are accompanied by the thrills and spills of learning and meditation and surprise and astonishment of discovering the truth. Besides, this kind of healthy emotion can intensify the people's mental activities and promote the development of the intelligence. Studying at the stage of adolescence, if a person has a great appetite for knowledge, pursuing spirit towards new knowledge, and the ability to utilizing the learning content, then this kind of study is exciting, pleasant, and productive. This kind of pleasant and sound learning experience can intensify the people's mental activities and promote the development of the intelligence. Although the process of the "tribulation" is full of pain, the result is supposed to be pleasant. "Jolly" means "painful" but "pleasant", which is the basic difference between new "tribulation" and traditional "tribulation".

4.5 The Outdoor Sports and Mental Health

Now, urbanization of China is speeding up. Although urbanization is an inevitable trend, it brings us severe challenges. The urbanized life broadens the gap among people and increases the coldness of human relationship. As a result, social pressure and emotional tension follow closely, in which the problems of adolescent mental health in city stands out. Social pressure is expressed in the form of depression, anxiety, suicide, alcoholism and drug addiction. These forms are the main sources for social insecurity.

One person needs not only physical health, but also psychological health. The two parts are an indivisible unity. If a person's mind is filled with anxiety, sorrow, anger and panic all day long, then his behavior, emotion and thinking mode are bound to be affected. It is proved by research that long-term exercise with low and moderate intensity helps ease mental pressure, eliminate anxiety and keep the mind balanced. When doing outdoor sports, one may adopt particular ways to reduce pressure and increase the emotional experiences so as to adapt one's emotion, to eliminate psychological barrier and to improve relationships. At the same time, reasonable way of thinking and healthy psychological qualities will be formed.

Cliff climbing and mountaineering hone our perseverance, boost up our self-esteem to face difficulties with a fortified spirit, and make us dare to challenge and to go beyond ourselves. Cross-country orienteering, camping and mountaineering can enhance intensity of neural excitatory process and that of cerebral cortex's excitability. These activities can also focus our attention so that the muscle strength of body is more powerful and the athletic ability is better; excitement and inhibition of neural operation can be more balanced, the coordination ability of the central nervous system is improved; as a result, the occurrence of various neural diseases can be effectively prevented, various fatigue caused by excessive use of the brain can also be efficaciously eliminated, the tension can be relieved and the vigor of life can be increased. In a word, we can endure great pressure. So, the outdoor sports are not only good for health, but plays an immeasurable role in protecting mental health.

4.6 The Outdoor Sports, Social Adaptation and Moral Improvement

The spirit embodied in the outdoor sports includes the team spirit demonstrated by caring for each other and helping each other, and the dauntless spirit of conquering the nature, and challenging oneself and self-fulfillment, which will increase our social adaptability. This kind of mental stature demonstrated by being friendly to others and being in harmony with nature will raise our morality standard and perfect our morality.

5 Conclusions and Recommendations

The outdoor sports as a sports leisure game is loved by the teenagers in our country. It is significant for us to try to use new "tribulation" to exert an influence on the moral cultivation of the teenagers and the way to do so is the outdoor sports. It can make most of the teenagers learn while playing and the teachers contain education in amusement. The outdoor sports can cultivate our physical health, our morality and intelligence. The dull content of moral education can thus become various and vivid, and the rigid moral education can be realized in relaxing activities. Therefore, their moral and educational functions are accepted by more and more people.

But in practice we are supposed to realize that the "tribulation" is not the aim, and the outdoor sports are only a method. The moral education in outdoor sports has to be carried out according to the plan, aim and procedures. Besides, we are supposed to balance the relation between quantity and degree seriously and pay full attention to safety. Only in this way, can the moral education work effectively.

References

1. An, Q.: The New Orientation of Moral Education for College Students. Education Exploration (3) (2011)
2. Li, Y.: Tentative Explorations on Moral Education by Means of PE. Young Writers (24) (2009)
3. Li, S.: Research on the Impact of Olympic Spirits on Moral Education of Modern College Students. Harbin Sports University Journal (6) (2009)
4. Wei, C.: The Functions of Moral Education in PE. Science Education (8) (2010)
5. Zhang, L., et al.: New Thoughts on Future Moral Education for College PE. Professional Skills (6) (2010)

Language, Culture and Thought from a Perspective of English Teaching

Yingbo Liu

Foreign Language School, Shenyang Ligong University
110168 Shenyang, China
sylgliuyingbo@sina.com

Abstract. Language is indispensable to any culture and the impact of culture upon a given language is something intrinsic. As far as English teaching is concerned, what we should keep in mind is that language is deeply embedded in culture. Since the mode of thinking, among all elements of culture, decides to the largest degree the linguistic forms in expressing concrete objects or abstract ideas language, differences in thought between cultures should receive the most attention if we want to enhance the students' awareness of culture so as to prompt their learning of the English language.

Keywords: Language, Culture, Thought, English learning.

1   Introduction

It has long been recognized that there exists a close relationship between language, culture and thought. As the most important means of communication and carrier of a given culture, language is essential to any culture and the impact of culture upon a given language is something intrinsic and indispensable. Although the cultural perspective has been neither dormant nor dominant in the whole history of human study of language, the cultural study of linguistics began to attract more and more attention and scientific efforts after the start of the 20th century when John Firth and Bronislaw Malinowski from England, Franz Boas, Edward Sapir and Benjamin Lee Whorf from North America began their study of language in a sociocultural context (Hu Zhuanglin 2001, 223-224). As a result of their innovation, commitment and perseverance, people's understanding of the interrelationship between language, culture and thought has been greatly deepened. One result of such an understanding is that many linguistic experts and language teachers are beginning to view culture teaching not as a separate course but as an integral component of language teaching. Chinese scholars, represented by Professor Hu Wenzhong, have studied language in cross-cultural contexts extensively with numerous significant achievements. Obviously, a detailed discussion about the interrelationship between language, culture and thought with a view to providing an insight into how culture teaching should be integrated with language teaching will be of great value for English teaching and learning.

2 Language and Culture

The word "culture" originates from the Latin word 'cultura', which originally means bringing soil under cultivation. Later, with the advancement of human civilization, culture is associated with the fostering of people 's body and spirit, especially the cultivation of art, moral ability and talent. In fact, it develops to contain so enormous a range of contents that anything that is related to human and human behavior, such as knowledge, belief, art, morals, law, custom, and any other capabilities and habits acquired by man as a member of society (Lin Jicheng, 1990, 3), can be said to be cultural.

Briefly and in its broad sense, we may define culture as the general term for the products, physical or spiritual, created by a people. Obviously, due to differences in geographical and historical backgrounds, different peoples have developed different cultures. For example, as two of the most important cultures in the world, the oriental and the occidental cultures have their own distinct characteristics. Generally speaking, the occidental culture belongs to the scientific culture, valuing utilitarianism, materialism, and conquering of nature by human beings, while the oriental culture, belonging to the humanistic culture, values humanity, morality and harmony between nature and human beings (Chen Hongwei, 1998, 19). As a result of these differences, westerners tend to isolate the problem to be solved from its context and break it down into components to deal with it in an analytical manner, while easterners tend to integrate the problem concerned into its settings to find out its relations with other parts and deal with it in a synthetic manner. Therefore, culture imposes influences on the way people live.

Not surprisingly, cultural influence can be found in every aspect of a given language, from the orthographic level to the discourse level. For example, the symmetric structure of Chinese characters is believed to be the result of Chinese value of the beauty of being symmetric (Chen Jianmin, 1994, 205). Western culture appreciates pragmatism and individualism, one result of which is that westerners prefer to put forward their topic directly at the very beginning of a discourse followed by detailed explanation or concrete examples and that they are less likely to cite from sages or famous people in order to show their own creativeness and innovativeness (Guan Shijie, 1995,134). On the contrary, what Chinese value is modesty and collectivism. Making oneself stand out is never a desirable thing. Just as the Chinese saying goes, being famous is to a man what being fat is to a pig, suggesting that it is a dangerous thing to be famous and outstanding. As a result, Chinese tend to adopt a more indirect manner in putting forward their ideas. A long introduction is often necessary before the topic is presented, which is usually expressed in words with mild affective meanings and words by sages or famous people are much more often cited as proof for the topic. This difference is easily seen in a comparison between Xuan Kuang's 'advice on Learning' and Bacon's 'of studies', which will be presented in Chapter four.

Cultural changes may cause corresponding changes in language. For instance, the reasons for the changes taking place in the expressions of greeting are believed to be cultural in origin. In the dawn of human recorded history when early man lived mainly in caves and were constantly under threat from poisonous snakes and other dangerous beasts, Chinese ancestors greeted each other with 'Wu ta hu?'(You don't run across any serpent? 'ta', ancient Chinese for snakes). Gradually, productivity was enhanced and people's living conditions were improved, and compared with those from fierce animals, dangers and sufferings from natural disasters and ailments became so outstanding and striking that the expression for greeting changed to 'Wu yang hu?'(You don't suffer from any illness? 'Yang', ancient Chinese for ailment). Later, as a result of improved science and life, illnesses became a minor problem. Instead, with the increase of human population, enough food became a major concern for Chinese. Again, people changed their greeting to 'Chi le ma?' (Have you eaten?). Even today, when China has basically solved the problem of food for its people, this expression of greeting remains most popular among Chinese, which, in a sense, shows persistence of cultural influence on language (Chen Jianmin, 1994, 205).

However, the most important influence of culture on language is at the semantic level. It is the focus of much anthropological study of linguistics ever since this branch of learning came into being in the 1920s. Through an effective study of language at the semantic level in social contexts, Malinowski and his followers, including Firth and Sapir, have set up a paradigm that has contributed to a flourishing interdisciplinary research of enormous diversity and achievements. For instance, Malinnowski noticed that the meaning of a word in the language of the Trbriand Islanders off eastern New Guinea depended to a large extent on its context when he did his fieldwork there. In this speech community, the word 'wood' may have two different interpretations. One refers to the solid substance of a tree as its English equivalent means, and the other a canoe, which is a very important means of transportation to this primitive people. What is important is that the second interpretation is heavily situational and culturally specific and might prove to be confusing to people from other cultural backgrounds. Based on such sociolinguistic phenomena as this, Malinowski concluded that "In its primitive uses, language functions as a link in concerted human activity....It is a mode of action and not an instrument of reflection." (quoted in Hu Zhuanglin,2001,225). The foundation for this cultural dimension of word meaning is similar social and cultural experiences shared by members of a cultural group. Since "all the knowledge and beliefs that constitute a people's culture are habitually encoded and transmitted in the language of the people," (Chen Linhua, 1999,285), word meaning has to be culturally specific to mirror the particularity of the way of life of the speakers. Logically, words corresponding in denotative meaning may vary considerably in connotative meaning, emotional meaning and various sociocultural associations in different cultural settings. Therefore, a full understanding of the cultural background in which a language operates is essential to a full understanding of the various meanings of a word that is encoded in the grammar and vocabulary of the language (John Lyons, quoted in Chen Linhua, 1999, 286).

As far as English teaching is concerned, what we should keep in mind is that language is deeply embedded in culture. Due to the sharp difference between the occidental culture and the oriental culture, one-to-one equivalence can rarely be established in English and Chinese. For example, "long", a mythical animal symbolizing the emperor in feudal China, is admired by Chinese people as the token of authority, power and good fortune. However, its English counterpart, 'dragon', refers to a kind of fierce mythical reptile-like monster that can never arouse the feeling of worship and admiration but that of fear and abhor in an English speaker.

Therefore, the cultural aspect of word meanings should never be neglected in language teaching and learning.

3 Language and Thought

A custom refers to a practice followed by people of a particular group or region. Different customs will inevitably be reflected in a discourse, which is more than just language. Consider, for example, the following passage:

(4) A lot of people pack their own lunch at home and bring it to work. It's a good way to save money. If you only have a half hour or forty-five minutes for lunch, it's a good way to save time, too. It's especially convenient for people who are dieting or who can only eat certain things for medical reasons.

The majority of people, however, probably go someplace where they can have their lunch already prepared. Next to bringing your own lunch to work, the fastest and cheapest way to eat is to go to the lunch counter of a drugstore. Cafeterias may be a little or a lot more expensive, but they're also fast, and you don't have to leave a tip.

In the passage the writer describes some lunchtime customs of Americans. The emphasis in the paragraph on saving both time and money may correspond to what is now often referred to as 'American efficiency', but the idea of packing up a lunch in a paper bag at home and bringing it to work may seem to many Chinese students to be carrying efficiency too far. A person who has not lived in the United States may not know that 'bringing a lunch' is not in itself a mark of being either a school-child or a member of a low social status: office workers, factory workers, managers, college professors all may do it. Another aspect of 'lunch' is that, while speed and economy may be desirable, sociability and professional contacts are often important elements too. Americans often have lunch with friends, associates, or colleagues and frequently discuss matters connected with their work or on 'doing business' during the meal (e.g., the expression 'There is no such thing as a free lunch'). For most Americans, 'lunch' as part of the working day and working world contrasts with 'dinner', which is the largest meal of the day, eaten at home with the family in the early evening. Obviously, the social-cultural meaning of 'lunch' can be understood only by a student who has learned something about the average American way of life.

4 Differences in Thinking Patterns for Expounding One's Ideas

Thinking is the process by which information is processed to reflect the objective world in one's mind by means of conceptions, judgments and inferences (Guan Shijie, 1995, 94). It has long been acknowledged that language and thought are closely related and mutually dependent. Without thought, language would be just sounds with no meaning or content; without language, thought would have no form of existence and would be impossible to be known to others (Chen Linhua, 1999, 279).

In their study of the relationship between language and thought, sociolinguists and anthropologists become interested in one important fact. That is, although different groups of people share the same basic principles in thinking due to the similarity in the construction of their brains and the same mental mechanism, people from different cultures may think in very different ways and take very different perspectives in their viewpoints. For example, the Chinese and the English have opposite views about the nature of human beings at birth. The Three Character Classic, representing Chinese traditional ethical philosophy, begins with the statement that man is good by nature; people are similar in character at birth but learning makes them different (Ren zhi chu, xing ben shan; xing xiang jin, xi xiang yuan.) (Xu Chuiyang, 1990, 1) . On the contrary, the English people tend to believe that human beings are born with an evil disposition: " No child has ever been known since the earliest period of the world, destitute of an evil disposition—however sweet it appears"(Sunley, 1955, 159). Therefore, when viewing the nature of man himself, Chinese people adopt an approving attitude while English people adopt a disapproving attitude. When writing the address for a letter, an English will put the smallest place first and the largest place last, the house number — the street — the city — the province — the country, while a Chinese will write the address exactly in the opposite order, the country — the province — the city — the street — the house number. This shows that the English tend to approach a problem from the specific to the general, the Chinese from the general to the specific.

Why do people speaking the same language share essentially the same world view and similar conceptual framework in terms of many basic concepts, while people speaking different languages may have different, even opposite views and conceptual frameworks about the same subject? Of course, it is the result of a combination of many factors, but in linguistics, Sapir-Whorf Hypothesis is one of most influential theories developed to answer this question. Benjamin Lee Whorf, a student of Sapir at Yale University, was an important figure in American anthropological linguistics. As an amateur linguist, he showed a particular interest in language, anthropology and archaeology. His study of Hopi language led him to develop his hypothesis of far-reaching significance, which contains two main assumptions. One is linguistic determinism: our language moulds our way of thinking; the other is linguistic relativity: there is no limit to the structural diversity of language (Chen Linhua, 1999, 280). According to linguistic determinism, people do not perceive the world freely, but rather they do so through their language. That is, people's native tongue determines his thinking pattern and worldview through the grammatical categories and semantic classification of his language. And according to linguistic relativity, these categories and classification are unique to the linguistic system of his native tongue and incommensurable with those of other language systems. For example, instead of one general word, the Eskimo has more than twenty specific words to refer to various kinds of snow: heavy snow, light snow, falling snow, snow on the ground, and so on. This means that it is easier for them to perceive the differences between these snows than speakers of English because the corresponding concepts are much more easily expressed or encoded in available words in their native language. Countless similar examples can be drawn from many aspects, such as color, kinship and address terms between different cultures. In this sense, we may see that language does have influence on its speakers' perception and thinking patterns.

However, Sapir-Whorf Hypothesis has given rise to a great deal of controversy ever since it is developed. It fails to provide answers for such questions as 'if thought were determined by language and people speaking different languages had completely different thinking patterns, would bilinguals (not to mention multilinguals) have two different thinking patterns or two different world views? The answer is absolutely negative. What undermines the hypothesis most seriously are the existence of translatability between different languages and the fact that people speaking different languages may share many of the basic concepts, such as those concerning number, space and matter. However, despite its failure to furnish convincing explanations for these facts, Sapir-Whorf Hypothesis is very innovative and important in its own field. It draws more attention to the study of interrelationship between language and thought than does any other theory. With the advancement of research in this field and people's understanding deepening, few people would accept the original form of the theory today and a weak version has been developed, admitting that there exists interrelationship between language, culture and thought, but pointing out that the influence of language on our thinking is relative, rather than categorical. Obviously, the study on the interrelationship between language and thought is far from being closed. "Facing a situation like this, we must be careful and do not rush to any hasty conclusion before we really obtain some reliable evidence to support or reject the hypothesis."(Hu Zhuanglin, 2001,228).

5 Conclusion

Language is deeply rooted in its culture. Therefore, when learning a language, its culture should be considered a very important component that cannot be overemphasized. Since language is both the tool and the expression of thinking, the differences in thinking are bound to be reflected in language. From the perspective of English teaching, a clear understanding of characteristics of the occidental mode of thinking will be of great value to improving English study. The mode of thinking, among all elements of culture, decides to the largest degree the linguistic forms in expressing concrete objects or abstract ideas. How language, culture and thought interact should be pointed out clearly and systematically by the teachers to the students. Only in this way can the cultural awareness of the students be enhanced most effectively.

References

1. Sunley, R.: Early American literature on Child rearing. In: Mead, M., Wolfenstein, M. (eds.) Childhood in Contemporary Culture. University of Chicago Press, Chicago (1955)
2. Xu, C. (ed.): Three Character Classic. EPB Publishers, Singapore (1990)
3. Chen, J.: An Overview of Language and Culture. In: Hu, W. (ed.) Culture and Communication. Foreign Language Teaching and Research Press, Beijing (1994)
4. Chen, L.: An General Introductin to Linguistics. Jilin University Press and TESOL Studio, Changchun (1999)
5. Guan, S.: Cross Cultural Communication. Beijing University Press, Beijing (1995)
6. Hu, Z.: A Course of Linguistics. Beijing University Press, Beijing (2001)
7. Lin, J.: A General Expoundation of Language and Culture. In: Gu, J., Lu, S. (eds.), Shanghai Foreign Language Education Press, Shanghai (1990)

Cultural Differences at the Discourse Level in TEFL in Chinese Class

Yingbo Liu

Foreign Language School, Shenyang Ligong University
110168 Shenyang, China
sylgliuyingbo@sina.com

Abstract. The internal cohesion of the text poses less difficulty than the cultural coherence of the discourse. For most Chinese students, the difficulty in understanding an English discourse mainly comes from three aspects: differences in values and attitudes, differences in customs and differences in thinking patterns for arranging one's ideas. Teaching of English as a Foreign Language should pay attention to these factors so that the teaching result can be improved.

Keywords: Culture, Discourse, Teaching of English as a Foreign Language, Chinese.

1 Introduction

For Chinese learners, the difficulties involved in making decisions about what is significant in a given stretch of discourse, and about the producer's intentions, may be greater than for native speakers for reasons cultural as well as linguistic.

The notion of text views a stretch of written language as the product of an identifiable authorial intention, and its relation to its context of culture as fixed and stable. Text meaning is seen as identical with the semantic signs it is composed of: text explication is used to retrieve the author's intended meaning, and text deconstruction explores the associations evoked by the text. In both cases, however, neither what happens in the minds of the readers nor the social context of reception and production are taken into account. In fact, a text cannot be given fuller meaning if it is not viewed also as discourse (Rong Linhai, 1990,1). That is the reason why such sentences as the following one is hard to understand when considered as a discourse but semantically clear when considered as a text:

(1) Bright red costumes, with hats, shoes, and stockings to match, are to be all the craze in the Spring, smart women will have to be careful not to yawn in the streets in case some shortsighted person is on his way to post a letter.

It turns out that pillar-boxes in England are painted red. Only when armed with this cultural knowledge, are students able to appreciate more fully the cohesive device in constructing the discourse. Therefore, readers are supposed to organize background knowledge or so-called prior knowledge and use it to predict interpretations and relationships regarding any new information, events, and experiences that come into their way.

Y. Wang (Ed.): Education Management, Education Theory & Education Application, AISC 109, pp. 771–775.
springerlink.com   © Springer-Verlag Berlin Heidelberg 2011

For most Chinese students, the difficulty in understanding an English discourse mainly comes from three aspects: differences in values and attitudes, differences in customs and differences in thinking patterns for arranging one's ideas.

2 Differences in Values and Attitudes

From the above discussion, it is obvious that the strong bond between culture and language must be maintained for the student to have a complete understanding of the meaning of language. More often than not, differences in values and attitudes are one of the main sources of problems in foreign language learning. In other words, culture-specific values can be a significant factor in comprehension if the values expressed by the discourse differ from the values held by the reader. Generally speaking, a student learning a foreign language cannot eliminate the negative influence of the habits, thinking patterns of his mother tongue, but learning a second language often involves suspending this negative influence, though in many cases with great difficulty and acquiring the habits and thinking patterns of the foreign language he is learning. The student, however, will often continue to interpret situations as he would in his own culture (if they are not so utterly different as to be uninterpretable). Unfortunately, the understanding that there are rules of behavior to acquire as well as rules of grammar has not been stressed by educators for beyond the structures of the language they use, teachers and learners are often not aware of the cultural nature of their discourse.

Consider the following passages:

(2) Although he was over 20 years old, he still lived at home. To him, it was really too great a shame to bear.

Written for an American readership, the discourse draws on the readers' cultural knowledge concerning young men's independence from their families, but it might not be self-evident for Chinese readers for our young men continue to live at the parents' home well into their twenties. In America, independence is greatly valued. Most Americans feel that it is very important for everyone to be able to take care of himself. American parents, like parents elsewhere, love their children, but teach them to be independent at a very early age. Many parents feel that when grown children live on their own and take financial responsibility for themselves, they have realized the necessity of being independent to become full adults. Instead, if parents force their grown children to continue living at home, people might say that they are not allowing their children to grow up.

(3) By voting against mass transportation, voters have chosen to continue on a road to ruin. Our interstate highways, those much-praised golden avenues built to whisk suburban travelers in and out of downtown have turned into the world's most expensive parking lots. That expense is not only economic — it is social. These highways have created great walls separating neighborhood from neighborhood, disrupting the complex social connections that help make a city livable.

In reading this passage, some Chinese learners fail to perceive the connection between mass transportation and highways. In the United States, people's attitude toward mobility and individual ownership of cars results in an overabundance of highways and a reduced need for mass transportation. It is in this social context that this passage makes sense. Sometimes, however, Chinese students perceive that highways are built for mass transportation, which renders this passage at best illogical, at worst incomprehensible.

The social-cultural meaning in this passage relates to the culture-specific schema of the cars/mass transportation opposition. Furthermore, comprehension can also be related to semantic associations available when a schema is accessed. The notion of interstate highways, here referred narrowly to those in urban areas, invites the semantic associations of crowding, congestion, and rush hour traffic. The meaning of the phrase 'the world's most expensive parking lots' is associated with, and can only be understood with reference to, this specific 'urban' highway subschema.

3 Differences in Customs

A custom refers to a practice followed by people of a particular group or region. Different customs will inevitably be reflected in a discourse, which is more than just language. Consider, for example, the following passage:

(4) A lot of people pack their own lunch at home and bring it to work. It's a good way to save money. If you only have a half hour or forty-five minutes for lunch, it's a good way to save time, too. It's especially convenient for people who are dieting or who can only eat certain things for medical reasons.

The majority of people, however, probably go someplace where they can have their lunch already prepared. Next to bringing your own lunch to work, the fastest and cheapest way to eat is to go to the lunch counter of a drugstore. Cafeterias may be a little or a lot more expensive, but they're also fast, and you don't have to leave a tip.

In the passage the writer describes some lunchtime customs of Americans. The emphasis in the paragraph on saving both time and money may correspond to what is now often referred to as 'American efficiency', but the idea of packing up a lunch in a paper bag at home and bringing it to work may seem to many Chinese students to be carrying efficiency too far. A person who has not lived in the United States may not know that 'bringing a lunch' is not in itself a mark of being either a school-child or a member of a low social status: office workers, factory workers, managers, college professors all may do it. Another aspect of 'lunch' is that, while speed and economy may be desirable, sociability and professional contacts are often important elements too. Americans often have lunch with friends, associates, or colleagues and frequently discuss matters connected with their work or on 'doing business' during the meal (e.g., the expression 'There is no such thing as a free lunch'). For most Americans, 'lunch' as part of the working day and working world contrasts with 'dinner', which is the largest meal of the day, eaten at home with the family in the early evening. Obviously, the social-cultural meaning of 'lunch' can be understood only by a student who has learned something about the average American way of life.

4 Differences in Thinking Patterns for Expounding One's Ideas

4.1 Different Habits in Overall Arrangement of a Discourse

The structural organization of a discourse is culturally specific. Just as mentioned in Chapter Two, compared with Chinese, who favor thinking in the manner of a spiral, westerners prefer thinking in a linear manner. (Kaplan, 1966, quoted in Guan Shijie, 1995, 97). One result of this habit of thinking is that the English, unlike the Chinese, who are likely to beat about the bush when talking about their ideas, tend to put forward their ideas directly at the very beginning of their writing. That poses difficulty for Chinese students because they may miss the main topic of the writing when they run through the first several lines to peruse the rest, believing that the topic comes indirectly after a reasonably long prologue.

Consider the following selection from an article by Thomas Babington Macaulay, the famous British speechmaker and political essayist:

(5) Whitehall, when (Charles the Second) dwelt there, was the famous focus of political intrigue and of fashionable gaiety. Half the jobbing and half the flirting of the metropolis went on under his roof. Whoever could make himself agreeable to the prince or could secure the good offices of his mistress might hope to rise in the world without rendering any service to the government…

In the rest of article, Macaulay continues to list examples that demonstrate Whitehall as the center of political conspiracy and social activities. Failing to realize the typical arrangement of the topic and arguments, many Chinese students are unable to discover the cohesion among the examples provided in the rest of the article.

Needless to say, knowledge of this particular difference can promote not only comprehension, but also reading efficiency.

This difference in organizing ones ideas poses great difficulty for Chinese students when they are learning to write in English. The following selection is part of an article in a brochure about a match held in a certain city in China:

(6) The divine land of China had its rivers flowing across; the brilliant culture of China has its root tracing back long…

The lightsome dragon-boats appear on the river as though the stars twinkle in the milkway. The richly decorated pleasure boats look like a scene of mirage…

Obviously, the writing bears features of indirect writing. The writer gives a rather long description of Chinese geographic features and cultural history and only after that the subject, the dragon boat race, is introduced. Therefore, when teaching English writing, teachers should point out to the students that the English way of writing adopts a direct way in putting forward the topic, which is followed by evidence or examples that can be used to illustrate or prove the topic. Only when students enhance their awareness of this difference can they improve their English writing and understanding of English articles.

4.2 Different Habits in Wording and Phrasing

Another difficulty in understanding an English discourse originates in the English habit of wording and phrasing, which can be viewed as the result of the difference between the western and the oriental thinking modes. Chinese prefer thinking in images. Consequently, Chinese writings contain far more metaphors and other figurative uses of language than English. In this sense, we may say that the Chinese write to enlighten while the English write to inform. For example, Xun Kuang's "Advice on Learning", while consisting of only fifteen sentences, makes use of twenty metaphors. Ideas are self-evident after these metaphors are fully understood. In contrast with such widespread figurative use of language, westerners rely more on logic as a major device in expounding their topics. In an article on the same subject, "Of Studies", Bacon employs only four metaphors in his seventeen-sentence-long article. (Guan Shijie, 1995, 129-132).

This difference may cause Chinese students to arrive at the wrong conclusion that English writings are simple and lack of device for attracting reader's attention, which may result in boredom and lack of interest on the part of the reader. On the contrary, students aware of the logic preference of western writers may appreciate more fully the artistic charm from rigorous and convincing reasoning and argumentation, thereby adopting a much more active attitude in reading.

Chinese preference for metaphorical language contributes to the abundance of flowery language in their writing, which can be viewed as empty in meaning, wordy and full of hyperbole by English speakers who have a different reading and writing habit. Therefore, when writing, students should be able to constrain their preference for flowery language and adopt the English way of clear and direct writing.

5 Conclusion

To sum up, a discourse reflects values, attitudes, customs and thinking patterns of the speakers. Extensive reading combined with an enhanced cultural awareness will surely be helpful for a better learning result at the discourse level.

References

1. Guan, S.: Cross-cultural Communication. Beijing University Press, Beijing (1995)
2. Rong, L.: Translation and Background Knowledge. In: Hu, W. (ed.) Culture and Cummunication. Foreign Language Teaching and Research Press, Beijing (1990)

The SWOT Analysis of New Practical English

Jin-jing Zheng[1] and Xue-shen Liu[2]

[1] College of Foreign Languages, Fujian Normal University, 350007 Fuzhou, China
[2] College of Material Engineering, Fujian Agriculture and Forestry University,
350002 Fuzhou, China

Abstract. SWOT analysis (sometimes referred to as TOWS analysis) is an important strategic planning method adopted to evaluate strength, weakness, opportunity and threat in business organization and its environment. By means of SWOT analysis, strength, weakness, opportunity and threat in New Practical English (NPE), one teaching material for higher vocational colleges, will be analyzed in a systematic way with the hope of providing some reference for textbook reform and the improvement of teaching quality.

Keywords: Higher vocational colleges, New Practical English, education reform, SWOT analysis.

1 Introduction

New Practical English (*NPE*) comes into being according to *Basic Requirements of Vocational College English Teaching* (hereafter referred to as *New BR*) by Vocational College English Teaching Guidance Committee of China's Ministry of Education. On basis of *Practical English* (*PE*), it has been composed and published in 2002 with 'application as objective, practicality as center and enough using as standard' [1].

For nearly ten years after its publication, some scholar reviews its footing and originality from such four aspects as 'compiling ideology, compiling principles, selecting standard and structural innovation' [2]. Some scholar indicates this teaching material has implemented 'Three Strengthening', that is, 'strengthening practicality, expression, listening and speaking' [3]. Still others point out that compared with *PE*, *NPE* has three characteristics of 'novelty', 'reality' and 'practicality' [4]. This article, based on the previous studies, attempts to investigate strength, weakness, opportunity and threat of *NPE* in a systematic way in order to offer some reference for education reform.

2 Strength Analysis of *NPE*

2.1 Fresh Compiling Concept

NPE pays attention to the dialectical relations between application ability and practical ability since 'the former is the foundation of the latter while the latter is the embodiment of the former' [5]. With such a consideration, *NPE Comprehensive Course 1* focuses on daily life, *Comprehensive Course 2* on daily life and practicality, and *Comprehensive Course 3* and *4* on practicality.

Y. Wang (Ed.): Education Management, Education Theory & Education Application, AISC 109, pp. 777–783.
springerlink.com
© Springer-Verlag Berlin Heidelberg 2011

Secondly, this textbook overcomes the one-sided tendency in traditional 'looking down to listening and speaking while taking care of reading and writing' and adopts the fresh teaching thought of 'paying equal attention to listening, speaking, reading, writing and translating'[6]. Therefore, each volume is composed of four parts, Talking Face to Face, Being All Ears, Maintaining a Sharp Eye and Trying Your Hand with translation exercises amid each part.

Thirdly, this textbook takes eclecticism position to give full play to the advantages of such diversified teaching methodologies as translation method, direct method, audio-lingual method and communicative method, effectively implementing teaching objectives advanced in *New BR* as follows 'enabling students to master basic English knowledge and skills in listening, speaking, reading, writing and translating so that they are capable of reading and translating business-related information through dictionary and carry out simple oral and written communication in foreign-related daily life and business activities'[7].

2.2   Reasonable Overall Framework

NPE is original in its overall framework. Each unit is topic-based written, facilitating bilateral cooperation between teachers and students in task-based activities. Specifically, each unit brings about speaking as guide, listening as complement, reading as expansion and writing as consolidation, presenting a daring reform on the traditional compiling structure of 'beginning with reading while lagging behind listening and speaking'. Even more valuably, this teaching material is developing towards diverse teaching carriers as shown in Figure 1, striving to achieve the leap from traditional teaching to network teaching.

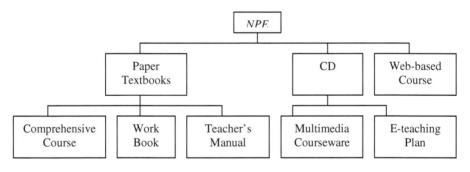

Fig. 1. Composition of *NPE*

2.3   Practicality-Highlighted Selecting Standard

NPE keeps its contents closely on each unit's topic and highlights practicality while being informative, updated and interesting; reading materials cover daily life, foreign business affairs and practical style closely related to the development of science and technology. Samples in speaking, listening and writing take typicality in commercial context into account.

3   Weakness Analysis of *NPE*

3.1   Unreasonable Unit Sequence

Unit layout is lacking in internal links. Take *NPE Comprehensive Course 3* for example as shown in Table 1.

Table 1. Unit Layout of NPE Comprehensive Course 3

Unit	Topic	Unit	Topic
1	Promoting Activities	5	Brands and Advertisements
2	Company Profiles	6	Sharing the Loss
3	Purchase and Payment	7	Busy Agenda and Schedule
4	Training Across Cultures	8	Thinking Global, Acting Local

Some discontinuous units are so interconnected that it is appropriate to arrange them in sequence. Mass selling in Unit 1 Promoting Activities involves brands and advertisements in Unit 5. Both Unit 3 Purchase and Payment and Unit 6 Sharing the Loss are concerned with currency circulation. Unit 2 Company Profiles, Unit 4 Training Across Cultures as well as Unit 8 Thinking Global, Acting Local are all involved with enterprise culture. Therefore, it is suitable to change original unit layout as follows.

Table 2. Recomposed Unit Layout of *NPE Comprehensive Course 3*

Unit	Topic	Unit	Topic
1	Promoting Activities	5	Busy Agenda and Schedule
2	Brands and Advertisements	6	Company Profiles
3	Purchase and Payment	7	Training Across Culture
4	Sharing the Loss	8	Thinking Global, Acting Local

3.2   Unreasonable Reading Layout

Prominent linguist Rod Ellis once illustrated the interactive relationship between input and output as shown in Figure 2.

As the Figure 2 shows, neither input nor output are dispensable; L2 input contributes to L2 output. L2 output can in turn deepen learners' understanding of language knowledge through interaction. The arrangement of Before Reading, While Reading and After Reading is the embodiment of reasonable reading layout. Before Reading can stimulate students' contemplative space by means of various lead-up methods including question, demonstration, background and situation methods; While Reading can adopt global and detailed reading, which is also called top-down and bottom-up; After Reading demonstrates the physical extension and expansion of texts, providing space for language output. Passage □in *NPE* lack necessary segments of Before Reading and After Reading although there exists the segment of While Reading.

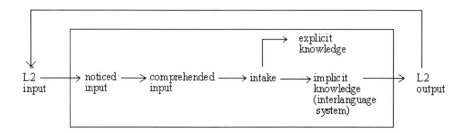

Fig. 2. The Interactive Relationship between Input and Output

3.3   Unreasonable Exercise Design

Exercise design, especially recurrent arrangement, is the manifestation of the intention of conversion from quantitative change to qualitative change, but editors need keeping exercise in the moderate degree. Otherwise a large amount of exercise and high repetition rate tend to make students feel bored, even trigger inimical attitudes with the result of low learning outcome even if students barely complete teaching tasks. Exercise quality in *NPE* isn't well controlled. Take *NPE Comprehensive Course 3* for example. Exercise per unit unexceptionally accounts for more than half of corresponding unit (see Table 3).

Table 3. Exercise Percentage of Each Unit

Unit	Total Pages Per Unit	Exercise Pages Per Unit	Exercise Percentage of Each Unit
1	19	10	52.63
2	18	12	66.67
3	19	11	57.89
4	19	11	57.89
5	18	10	55.56
6	19	11	57.89
7	21	12	57.14
8	19	11	57.89

In particular, amount and type of translation exercise remain further betterment. There are five types of translation exercise: Chinese-English sentence translation in the specific context (tips of Chinese sentences), Chinese-English sentence translation (cuing English words), Chinese-English sentence translation (cuing the structure of the English sentence), Chinese-English translation of the practical writing (cuing English words) and Chinese-English translation of the practical writing (without cuing English words). In a word, they are all variants of Chinese- English translation, which locally deviates from the goal of practicality because what is really needed in practice is bilingual skills exchange. Indispensable is the ability of English-Chinese translation handling such English materials as correspondence, product prospectus and contracts. What's more, translation exercise is probably distorted into strengthening-memory training in *NPE* because of exposure to excessive exercise.

3.4 Other Deficiencies

There still exist other deficiencies in *NPE* as follows. Firstly, *Teacher's Manual* lists only language points, the Chinese version of samples and passages, keys and listening script without offering instruction of teaching theory and methods for teachers, putting it out of the role of teachers training. Secondly, textbook compilation and schedule are not compatible with each other. Four volumes of *NPE* must be completed in four terms. But in fact each term probably fail to finish a textbook. In terms of week distribution, intensive and extensive reading take up half the time. Despite the publication of specialized series of *NPE Listening Course*, they don't achieve desired effect due to limits of class hours. Thirdly, the design of multimedia courseware remains to be improved. In the symposium called The National Computer Network and Foreign Language Courses Integration, Gu Yueguo summed up six types of web-based materials, 'words + electronic edition', 'words + audios', 'words + videos', 'words + animations', 'lectures' and 'foreign-imported materials' [9]. The network version of *NPE* belongs to the first two. In spite of diversified carriers, contents of network version are nearly the same as that of physical materials without making the most of potential advantages of modern information technology and comprehensive high quality of man-machine interaction.

4 Opportunity Analysis of *NPE*

4.1 Recognition and Support by China's Government

The Ministry of Education issued *Basic Requirements of Ordinary College English Teaching (BR)* in July, 1993. *BR*, as the first guiding document since the founding of People's Republic of China, illustrates that teaching objectives and requirements of College English teaching should distinguish from that of undergraduate English teaching while the former emphasizes on the development of students' application ability, thus breaking the current situation of higher vocational colleges following syllabus for undergraduates, applying undergraduate textbooks indiscriminately and adopting undergraduate teaching mode. In March, 2000, *The Certain Opinions Regarding The Ministry of Education on Strengthening the Construction of Higher Vocational Teaching Materials* was issued, proposing 'One-Outline-Multiple-Texts' policy. In September, 2000, *New BR* was established inhering and developing the spirit of *BR*. In September, 2004, seven Ministries including the Ministry of Education with the approval of the State Council jointly issued *The Opinions on Further Strengthening Vocational Education by The Ministry of Education and Other Six Ministries*, which reveals the blueprint of higher vocational education Chinese Government has spared no effort to develop. Against the backdrop of this positive situation, *NPE* emerged and has subsumed into the Nationally-planned Textbook for 'Eleventh Five'.

4.2 Advantageous Position of Teaching Materials in China

In China, students learn English rather than acquire English in the context of Teaching English as a Foreign Language (TEFL). For this reason English teaching materials plays an irreplaceable role as an important channel of language input, which provides good opportunity for the wide circulation of *NPE*.

5 Threat Analysis of *NPE*

5.1 Competition of Multifold Teaching Materials of the Same Category

Under the guidance of *New BR*, the implementation of 'One-Outline-Multiple-Texts' policy triggers the birth of diversified teaching materials. Apart from *PE* and *NPE*, there are *New Century English Course* revised and published in 2004 by Southeast University Press with Liang Weixiang as chief editor, *Hope English* published in 2004 by Foreign Language Teaching and Research Press with Xu Xiaozhen as chief editor, *New Century English Integrated Course* revised and published in 2006 by Shanghai Foreign Language Education Press with Dai Weidong as chief editor, *Beyond Oxford* published in 2006 by Shanghai Xuelin Publishing House with Wu Yunren as chief editor and *Century English Comprehensive Course* published in 2007 by Dalian University of Technology Press with Gong Yao as chief editor. The new situation of independently selecting teaching materials poses certain threat for the applicable range of *NPE*.

5.2 Impact of Weak Source of Student Enrollment

With the increasing enrollment of students in universities, significant changes have taken place in students' body of diverse vocational colleges; students of higher vocational colleges come primarily from ordinary high school and secondary vocational school with weak language foundation. In the case of sixty-five subjects in the survey, fifty-eight failed to pass NMET (National Matriculation English Test), accounting for 89.23%. Most of them are in the low level of English so it is difficult for them to follow course schedule with the result that the approval of *NPE* is under question.

6 Conclusion

To sum up, *NPE* is characteristic of fresh compiling concept, reasonable overall framework and practicality-highlighted selecting standard. This teaching material, however, presents such deficiencies as unreasonable unit sequence, reading layout and exercise design, and confronts the competition of multifold teaching materials of the same category and impact of weak source of student enrollment. Yet *NPE* still has large development potential under the recognition and support by China's Government, and advantageous position of teaching materials in China, What is needed now is adhering to strength, seizing opportunity, reducing weakness and transforming threat.

References

1. Compiling Group of New Practical English Teaching Materials: New Practical English Comprehensive Course 3, p. i. Higher Education Press, Beijing (2007)
2. An, X.: Structure and Innovation of New Practical English. China's College Teaching 12, 40–41 (2003)
3. Kong, Q., Liu, H.: Recognition and Practice of Education Reform in Higher Vocational Colleges Practical English. Chinese Foreign Language 3, 37 (2005)
4. Xiang, Q., Zhou, L.: Research and Practice of Practical English——A Review of New Practical English. Chinese Foreign Language 5, 33–37 (2005)
5. An, X.: Structure and Innovation of New Practical English. China's College Teaching 12, 40 (2003)
6. Compiling Group of New Practical English Teaching Materials: New Practical English Comprehensive Course 3. Higher Education Press, Beijing (2007)
7. Higher Educational Department of the Ministry of Education: Basic Requirements of Vocational College English Teaching, p. 1. Higher Education Press, Beijing (2000)
8. Ellis, R.: The Study of Second Language Acquisition, p. 349. Oxford University Press (1994)
9. Chen, J.: Current Situation and Reform of the Textbooks for College English——Research, Development and Conception of the Fifth Generation of Teaching Materials. Foreign Language Teaching and Research (Bimonthly) 5, 376 (2007)

Research on New Mode of University Study Style Construction Based on Party Construction Blog

Chunlin Li and Shuhong Ge

Northwestern Polytechnical University, Xi'an China
lichunlin@nwpu.edu.cn

Abstract. Holding the student-oriented work idea, by analyzing the features and developing trend of university study style construction, and effectively relying on the internet information technical platform, the paper has made a probe theoretically and practically on the combining site of university study style and students Party construction. And it has also built a new mode of communication between teachers and students on the circumstances of network Party construction blog in order to let a hand on the university study style construction and personal training.

Keyword: Network, study style, mode.

1 Introduction

Network is generally acknowledged as the fourth kind of media tool after newspaper, broadcasting and TV. And network widespread using makes us get into the "network period". According to the twenty-fourth Chinese network developing report released on Jan. 2010, net citizen aged from 18 to 24 takes 31.8%; aged from 25 to 30 is 18.1%. Young men aged from 18 to 30 are still the main body of our country net citizen [1]. Among the young net citizen, university students are the major group of network consume and they most often get in touch with and use network. They have a high level of knowledge and culture with active thinking and strong exploring desire, thus are printed as the mark of network no matter on behavior and habit or thought and concept. In the university, a good study style is the invisible but powerful mental strength in encouraging university students to strive ever upwards and be well educated; a good study style is also the important guarantee for the university to carry out the education policy of our Party and improve the education quality. At the meantime, university Party member is a special group with two identities and they are the fresh troops in helping a good study style construction. At the current circumstances, how to bring the advantage of modern network information media and the model function of student Party members into full play in searching for the new mode of study style is the focus of many universities [2].

Y. Wang (Ed.): Education Management, Education Theory & Education Application, AISC 109, pp. 785–789.
springerlink.com © Springer-Verlag Berlin Heidelberg 2011

2   The Practical Importance of Building Network Party Construction Blog and Expanding New Mode of Study Style Construction

Network has brought new chance and challenge for university student Party construction. On the one hand, network brings new chance for Party construction as it is the significant method for teachers and students to get knowledge and information by network's rich information, swift transmission and instantaneous reciprocity; on the other hand, it also increases the difficulty of Party construction since there are both good and bad ideas, different views, different culture and different values and even passive, retroactive viewpoints existing on network. As a newly emerging network transmission carrier, Network blog has a very wide covering in university students. Because of the advantage of simple operating, easy maintenance and powerful reciprocation, to a big extent, network blog could make up the disadvantages of the past Party construction information system, like the lack of capital, deficiency of technology and personnel, as well as the insufficient of hardware facilities. Thus it is especially suitable for the Party construction network job of the basic level Party branch which has a huge quantity. Meanwhile, network blog has made a change on the used dull preaching method and led a new vigor on the study style construction and thought education. It has also provided an open platform for the Party branches and a network stage for student Party members to show themselves, manage themselves and make a multi-communication.

Therefore, university must take the network Party construction blog as a brand-new topic and make it into full use for Party construction and study style construction by bringing the modern network information into full play.

3   Student Party Members Play a Vital Importance in University Study Style Construction

Study style forms when learners' world outlook, methodology and related psychological quality develop into a mature phase or has showed out a steady trend. Study style construction is a permanent and fresh topic for a university. And university study style is a reflection of the highest dream and faith in university, and also a powder source for the whole teachers and students. The internal learning spirit and motive is the key point for raising a good study style. Thus, to put students as the main body in the study style construction is the internal factor for university study style construction.

As the Party regulations defined, "communist party members should set a good example in manufacture, work, study and social life". This is decided by our Party's character and the historical mission it bears. The advanced nature and leading role of our Party is showed specifically by each members' model function besides the guiding thoughts and outline of our Party[3]. Student Party members are the outstanding ones among the students and their behaviors on every aspect have an example function. So to fulfill student Party members' modeling function is a key point in study style construction. University must recognize student Party members' important function on study style construction and put it into a significant place in doing it.

4   Relying on Network Party Construction Blog and Building a New Mold of Study Style

4.1   Increase the Attractiveness of Party Construction Blog by Bringing New Ideas and Fortifying Its Function

Network Party construction has gone through ten years and played an active role in different phases. But there are problems in network Party construction. Many network Party construction blogs have a big difficulty in maintaining. The deep reason is that the blogs do not have a strong attractiveness thus are paid a low attention. It should avoid that network Party construction blog is built as a plane web including only Party construction theories and Party members knowledge. It is necessary to take blog's effectiveness, interest, practicability, participation into consideration to build a vivid "Three-Dimensional Web" so as to make it close to practice, close to life and students. There should be "Party Construction Part", "Internet Party School" as usual in Party construction blog. And it also should set up some components like "Party Member's Outstanding Stories" so that the surrounding excellent examples could be showed to all. And some parts like "History Review" and "Revolutionary Classics" will make the web more interest. To set a "Discussion Area" will guide Party members to have a discussion about present hot spots and thus make the web more practical and participative. And "School Public Announcement" will give some recent news and notices, which have a big thing to do with students' study and life. And this will draw students' focus definitely.

4.2   To Strengthen Party Construction Blog's Vitality by Widespread Mobilizing and Entire Members Participation

Network Party construction blog needs all Party members and branches' common efforts. Only when everybody is a builder and a participator, network Party construction's vitality could have an improvement. From the beginning, each Party branch should have its own blog, and each Party branch member should be blog member. In this way, everybody has his responsibility and motive to build and maintain his blog. Invisibly, all Party members and branches are connected together by blog. And a fine study atmosphere is formed by interchanging gains from learning as every Party branch displays the elegant demeanors by blog. At the same time, network normal running and information safety are guaranteed as a smooth and effective working route is built by system construction. Institutionalization process urges network Party construction blog more scientific and standardizing.

4.3 Fine Study Style Construction Platform Is Built by Party Construction Blog "Four Combinations"

Organization safeguard is provided for study style construction by combining Party members developing management and Party construction blog. "Proposals on improving university students' thoughts and politics education" points out that university Party organization should think highly of the students Party members development work, and absorb more brilliant university students into the team of our Party. It is required in the report of Party's seventeenth people's congress that the advanced education results should be strengthened and developed, Party construction work responsibility should be implemented, basic Party organization should be built and optimized and spread, and activity mode should be more creative. Party construction blog should establish a special column of Party member development to upload some thought reports of Party member applicants and Probationary-Party members so that to have a understanding of their thought state and make a direct thought education to the university students. Simultaneously, information network' traits, like effectiveness, swiftness, anonymousness and reciprocity will help to standardize Party member development and eliminate some corrupt practice and late announcement. And in this way, those who would like to join the Party would have a correct attitude to study hard, and put their attention on the scientific and culture knowledge learning. By setting up some learning group to help each other, a good study style will be formed in university.

An effective carrier is provided for study style construction by combining Party branch organization life and Party construction blog. By showing revolutionary films, picture exhibitions, and courseware display, some basic knowledge of our Party, some big achievements of reform and opening policy as well as modernization construction, and some excellent Party members' stories and Party branches activities are exhibited from different views and by different ways on the blog. Thus the blog becomes the "mental garden" for the university students. And Party members' pioneering and exampling functions are brought into a full play. Party organization' life becomes more colorful and rich. More and more Party applicants take part in the Party branch activities and their ability has been improved by getting familiar with Party organization and knowledge. Internet comments and discussions on Party members are carried out. So people can make a communication there and help each other by pointing out the mistakes and eventually they will have a common development. Party branches should regularly launch some thought education activities and supply a strong insurance for study style construction.

A good helper is provided for study style construction by combining "striving to be the best" activity and Party construction blog. "Striving to be the best" activity is performed in Party basic level organizations and Party members and it is a necessary requirement to keep Party members' advancement. For example, Northwestern Polytechnical University has been developing a project called "Three-One", which means that a Party branch establishes a good class, a Party member leads a good dormitory, and a Party applicant helps a student. This is not only an effective carrier for university students to strive for the best, but also valuable experiences of these years.

By setting up a special plate for "Three-One" on the blog, documents and profiles of all these activities are uploaded to Party construction blog in time. And internet dynamic tracking system is established, which will make the "Three-One" more specific so that student Party branches' function and student Party members' modeling role as well as Party applicants' exemplifying position will be fully fulfilled. All these activities will push classes, dormitories and students to make an entire progress eventually.

A motive powder is provided for study style construction by combining professor Party members' function and Party construction blog. Professors are the main body to pass on knowledge with wide wisdom; and students are the main body to receive knowledge with strong learning willing. How to make a link between these two main bodies and display their thesis function is an unceasing research top for many universities. Take the project "Professors Benefit Students" provided by Northwestern Polytechnical University as an example, it offers an interactive platform for professor Party members and students by Party construction blog. Students express themselves and professors enjoy their success and joyfulness by answering students' questions, guiding them to grow healthily, and let students feel that professors are right there around them. Interactions between teachers and students have been greatly developed in this room of resource sharing, thought sharing, and life experience sharing. The platform has not only revealed university famous professors' gracefulness, but also helped students to create their confidence. Thus students are well educated and university study style is poured in new energy.

5 Conclusion

Blog is one of the important media in university network education and management. It has a great practical importance to initiatively probe the management and education function of Party construction blog for university to strengthen and improve study style construction as well as Party members' education. University should make conclusion and bring new ideas in establishing more effective network education new mode and then make the goals of university education come true.

References

1. China Network Information Center. The Twenty Fourth China Network Development Report,
   http://news.xinhuanet.com/internet/2009-07/16/content_11718266.htm
2. Yan, D.: Problems and Solutions in Local University Study Style Construction. Nanyan Science and Engineering College Journal (February 2010)
3. Yang, X.: Discussion of the Pushing Function of Student Party Members' Advancement to Study Style Construction. Jianghan University Journal (Society and Science Edition) (March 2010)

The Application of CVAVR, AVRstudio, Proteus in MCU Teaching

Lee Xingang, Jia Zhenhong, Wang Liejun, and Huang Xiaohui

College of Information science & engineering, Xinjiang University, Urumqi China 830046

Abstract. MCU courses of C51 and AVR have opened in major colleges and universities Most of the current teaching methods are teaching the theory and experimental chamber exercise. The downside is lack of Intermediate links. In this paper, the software AVRstudio and Proteus, can achieve visual classroom teaching, and can single-step debugging AVR microcontroller program. Proteus simulation discussed in the actual feasibility and necessity of teaching.

Keywords: Proteus, AVRStudio, step debugging.

1 Introduction

With the automation technology widely used in industry, agriculture and the popularity of daily life. Most colleges and universities has also opened a lot of theoretical and experimental courses of MCU, most of them are belong to 89C51 family, while AVR microcontroller with more excellent performance its internal functions more complex, so just minority of college offers AVR courses.

At present most of the college offered MCU courses of supplemented teaching, and experimental multi-chamber , which limit the creative thinking of students of development, it also has kill some new ideas to innovation. And also user can debugging single-step project while without the microcontroller hardware emulator.

In this paper, we combined AVRstudio and Proteus to develop programe. Students through the Proteus schematic drawing board can help them understanding the function of MCU such as the pin connections to each external hardware circuit, which will benefit to programe. You can use the Proueus to emulator MCU single-step debugging.

2 AVR MCU and Simulink Software

2.1 AVR MCU

At 1997 Mr. A, Mr. V using the company's Flash technology, jointly developed high-speed RISC reduced instruction set 8-bit microcontroller in ATMEL Corporation Design Center of Norway, referred to AVR. Compared to 8-bit microcontroller core C51 is, AVR microcontroller largest features are: It internal use Harvard structural

Y. Wang (Ed.): Education Management, Education Theory & Education Application, AISC 109, pp. 791–797.

design, can achieve high-speed processing speed 1Mips/MHz; Part of the AVR MCU does not require external components, which could use its internal resources to work.

2.2 CVAVR Software

Most beginner study AVR is in contact ICCAVR the programming tools, but few people know CVAVR. In fact, it compared ICCAVR, There are several advantages: the interface more friendly than ICCAVR, and the integration of the code wizard to help beginners to quickly master the procedures of AVR microcontroller.

2.3 AVRstudio

AVRStudio is provided freely at AVR-ATMEL webnet for MCU development platform. The software set programming, assembly language compiler, software simulator, download chip, chip hardware emulation and a series of basic functions in one. Note that even with the Proteus and AVRstudio emphasized the need to at least version 4.18a.

2.4 Proteus

Proteus is known that developed by the British Labcenter with circuit analysis and physical simulation function. This software can simulate, analyze a variety of analog devices and integrated circuits. MCU systems support mainstream simulation and debugging. In the simulation system has full, single step, set breakpoints, etc.

3 Teaching Case

In this paper, we take "AVR generate PWM control motor speed" for example to do an introduction. According to the general project design process based on specific requirements for hardware design. This motor speed control chip select AVR ATmega16, motor drive with integrated L298N, photoelectric coupler between the core and the motor to do electrical isolation, improve system stability.

3.1 Schematic Design

After select complete hardware devices schematic designed by Proteus. After schematic design AVRStudio circuit simulation debugging. First, in the software menu interface found File/New Design, circuit simulation to create a new file, save it as PWM_Motor.DSN. Next, draw the circuit diagram, find ATmega16, L298N, motor and other components. Layout components connected wiring. Designed circuit as shown in figure 1.

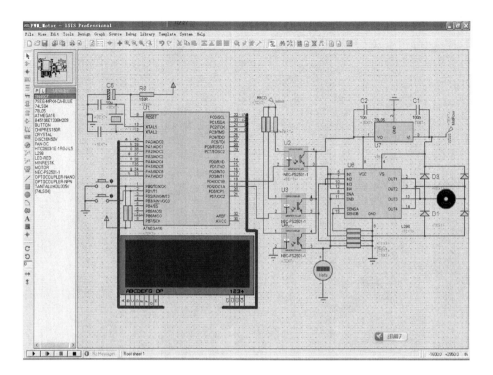

Fig. 1. PWM Schematic designed by Proteus

3.2 Programming

We use CVAVR programming tools to write program. First, in the menu File/ New, establishment of project works. Use new project wizard, set MCU type, and MCU pin input and output status, click on File / Generate, Save and Exit, then click on File / Save menu to save the new project .

3.3 Main Program

Motor speed of the main program design includes the following tasks: single chip initialization, LED digital tube displays the current output of the dynamic scan duty cycle of PWM waveform, the keyboard scanner used to adjust the motor speed. PWM waveform output is by setting the timer interrupt routine to achieve.

3.4 Code Linking and Compile

CVAVR the CodeWizard will generate broad framework program, under the framework of this major re-write the user code is more easy. Then click Project /

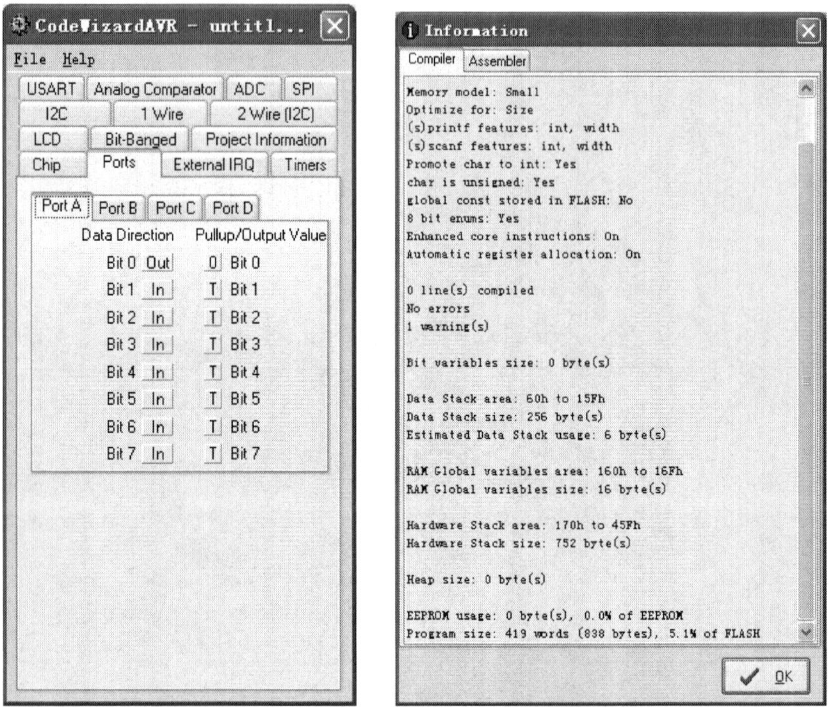

Fig. 2. AVR CodeWizard & Linker Information

Compile compile the code. Code compiler will generally make grammatical and other aspects of testing, given a warning or error message, and mark the wrong location, can easily find an error or warning from the tips of the program to be modified. Then you can link the program. Click Project/Build link. CVAVR file is generated in three formats: in the project folder which can be found PWM_Motor.cof file, this file used to debugging in AVRStudio; another Exe folder can found *. Hex and the *. ROM file is a binary code used to write into MCU.

Now we could start debug step, CVAVR directly call AVRStudio software. Let set done, click the icon to access the AVRStudio debug.

3.5 AVRStudio and Proteus Online Debugging

Software design is completed, the next step you can use the online debugger AVRStudio and Proteus, and some simple settings will be good . Found in CVAVR menu Settings/Debugger, find out AVRstudio.exe select it. After that completion are automatically called when the CVAVR started.

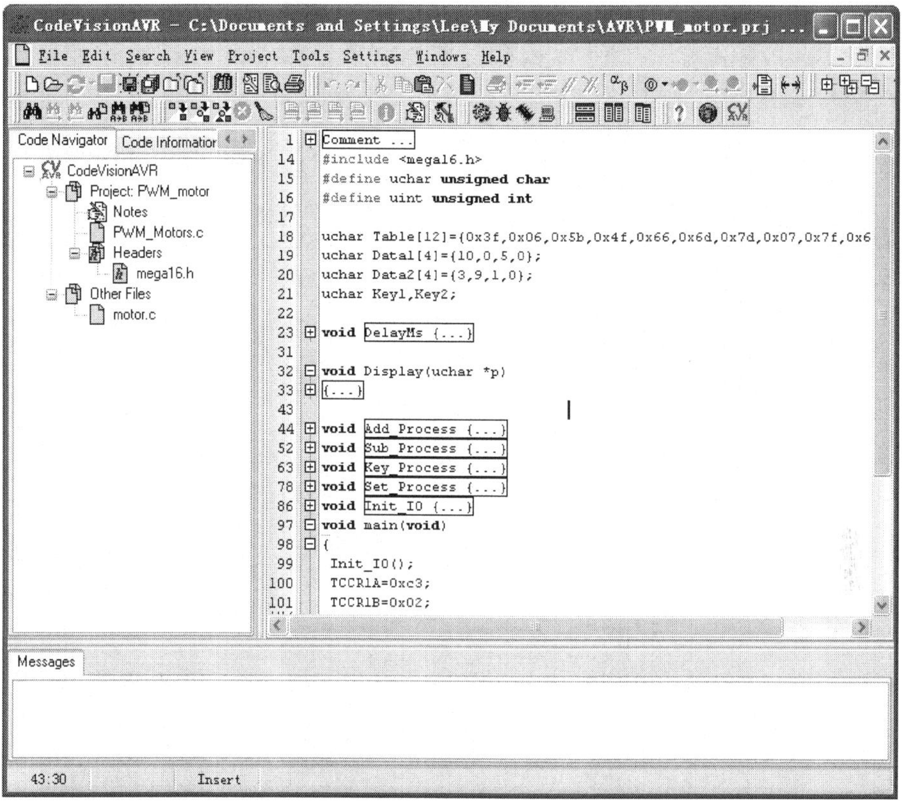

Fig. 3. CVAVR Interface

Click Open menu after entering AVRStudio, found established PWM_Motor.cof in CVAVR project documents, click to select it, AVRStudio will automatically create a new debugging project. AVRStudio debugger will embed Proteus VSM Viewer, select the ATmega16 at the right of the column , and then click Next, it will jump out a AVRStudio debugging interface, VSM Viewer is embedded in AVRStudio debugging interface. We click on the folder and opens the schematic in the Proteus VSM interface, after the file PWM_motor.DNS is loaded AVRStudio interface as shown in figure 4.

Now click the Debug/Start Debugging to start debugging, the debugging interface is show at figure 5, to control the process step or set breakpoints can be running at full speed. Modify the original PWM_motor.DNS after joining the oscilloscope can be observed after the buttons to adjust the PWM waveform changes, the use of digital LED real-time displays the current output waveform of the duty rate.

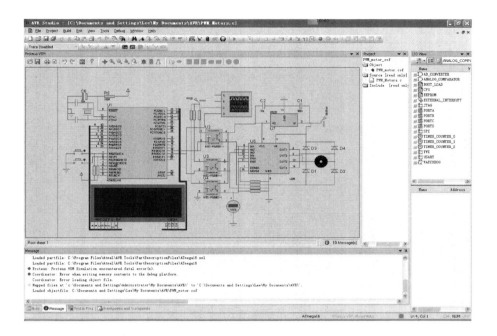

Fig. 4. Proteus VSM Loaded DNSfile

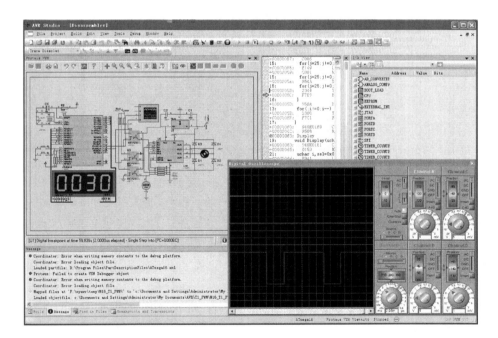

Fig. 5. Debug output PWM control Motor

4   Teaching Effect

MCU course is a practical subject, only the theory teaching lead to students lack a deep understanding, MCU instruction supplemented by Proteus simulation teaching can significantly improve student interest in learning, stimulate the creativity of students. Proteus Software is very powerful, combined with other software can constitute a convenient and flexible development platform. Student can use it in Training programs, can reduce the time period students innovative practices to improve the success rate of work produced. This program in my school in last year's level of field practical training projects and achieved good results.

5   Summary

Teaching by example more than a full detailed presentation of the Proteus, CVAVR and AVRStudio in universities, vocational education in the use of MCU as a software simulation of the experiment, students may be thinking of a large chamber of hardware limitations, not only SCM can achieve the purpose of learning, but also opened up the horizons of students, but also facilitate the design flexibility to adjust its late, or to implement the design by simple operation of the circuit module transplantation, to better grasp the AVR microcontroller project engineering design. In addition there Proteus PCB plate design tools ARES, can be designed directly to ISIS DNS file into ARES PCB design tool for rapid routing. Training program before the students do first electronic innovation to do circuit simulation using Multisim, Multisim MCU simulation of the effect is very bad, and then use the schematic design drawing DXP PCB, which greatly extend the development time cycle. Proteus far less efficient the speed of development.

References

1. Zhu, F., Yang, P.: AVR MCU C-language develop typical examples. Post & telecom Press, Peking (2010)
2. Zhou, X.: Taught you how to learn C programming AVR microcontroller. Beijing Aeronautics and Astronautics Press, Peking (2009)
3. Wu, X., Wu, X.: Proteus Circuit design and simulation in teaching practice. Applied Computer Systems 19, 201–204 (2010)
4. Sun, L.: The Application of Proteus & Keil In the experimental development of MCU. Research & Exploration of Laboratory 27, 59–62 (2008)
5. Liu, X., Guo, F., Sun, Z., et al.: The Application of Proteus Simulation Technology in MCU Teaching. Experimental Technology and Management 24, 96–102 (2007)
6. Sun, L.: The design and development of ICC AVR-based software platform and Proteus AVR MCU. JiLin Normal University Paper (Natural Science) 1, 103–105 (2010)

The Balanced ScoreCard and Educational Technology Management — Take Research on the Hangzhou College Student Probation Quality Assessment as the Example

Xiao-jun Chen[1,*] and Xiao-yun Yan[2]

[1] School of Public Administration, Zhejiang Gongshang University
Department of Social work, Zhejiang Gongshang University
Hangzhou, P.R. China
Geniusjun@hotmail.com
[2] School of Public Administration, Zhejiang Gongshang University
Department of Sociology, Zhejiang Gongshang University
Hangzhou, P.R. China
Yxy504@yahoo.com

Abstract. Educational technology management has been playing an important role in Hangzhou college student probation quality assessment management. The theoretical part of this article has explored the features and connections between the educational technologies and the Balanced ScoreCard, and has proposed the assessment model according to the four dimensions of the balanced scorecard and then analyzes investigation results based on the four dimensions. At last, based on the results, the paper proposes some valuable advice to the management of college student probation quality assessment.

Keywords: Balanced scorecard, Educational Technology, probation.

1 Introduction

Employment of college students has been a top priority in people's livelihood. In order to help college graduates accumulate professional experience, improve their job skills, and solve the employment problem, Hangzhou University student training program has been launched in 2003. Within the 6 years after issuing the scheme, Hangzhou has revised the policy twice to expand the scope of the apprentice training subsidy and increase the apprentice training subsidy program, at the same time, improve student training allowances. Eventually on February 23, 2009, the current policy -----<Measures of the Implementation on Hangzhou College Student Probation>[1] was launched.

In accordance with AECT1994 definition, "educational technology is a theory and operation which is to design, development, utilize, manage and evaluate the relevant processes and resources."[2] From the Definition, balanced scorecard theory and

---

[*] Supported by the foundation of Zhejiang university students' scientific research innovation team ,2010. No.3070TQ4111040.

Author Introduction : Chen Xiao-jun,(1990--), male, Zhejiang province Hangzhou city , bachelor, engaged in the research of social work.

operation can be seen as an educational technology. Balanced scorecard is a strategic management performance evaluation tool which was made by the famous master Robert Sa Kaplanand who made it on the basis of summing up a dozen large-scale enterprise's performance appraisal system of successful experiences. Its principle is to analyze what is the key factor in enterprise mission and evaluate these key success factors and constantly review the process in order to grasp the Performance Evaluation and achieve their enterprise goals. Its objectives and measurement methods can be divided into four dimensions as financial, customer, business processes, learning and growth.[3]

2   The Implementation and Requirements of Educational Technology Management

2.1   The General Situation of the Probation Mechanism

In recent years, college students have great difficulty in seeking jobs after the enrollment expansion .The total supply exceeds the demand. With elite education to mass education after college transition, needs of professional settings and social dislocation of the structural problems still exist. The employment of college graduates increasingly become hot issues of common concern, which relates to the economic development and social stability. In 2003, <the State Council on Further Improving the Higher Education graduate employment services work> firstly bring the social security services into the department's work . To address the lack of college graduates in work experience and practical ability and the difficult employment situation, Hangzhou Labor and Social Security Bureau with the Hangzhou Bureau of Finance jointly issued <to carry out the plan to help college graduates who are not employed in Hangzhou Health and adaptive training, >which launched a college graduate student training program. in end of 2003 .in November 2006 with a new round of employment and re-employment policy,<Hangzhou unemployed college graduate student training issues related to notice. >was issued.

In February 2009, in order to cope with the financial crisis and financial crisis, under the guidance of "human and capital" concept of urban development ,Hangzhou municipal government issued <college student in Hangzhou training > in the name of government office Hall which further broaden the scope of student training subsidies targeted to improve student training allowances .In <Hangzhou Labor and Social Security Bureau, Finance Bureau of Hangzhou unemployed college graduate student training to inform the implementation of adaptation>,it clearly says: "Training is geared to the trainee college graduates lacking in the skills and solve the problem of a low rate of actual employment. labor and social security departments will choose a number of enterprises with a higher level of management and a higher degree of work skills to cooperate with. In accordance with the principle of voluntary, to organize college graduates into the enterprise for a period of practical training to improve the actual ability and enrich their work experience and enhance the competitive ability of employment and finally find a job as soon as possible.

2.2 Educational Technology Requirement

Based on the definition of AECT in 1994, we can conclude: the study of educational technology is "the learning process and learning resources", the core of its research focus on the design, development, utilization, management and evaluation of the learning process and learning resources .Therefore, the balanced scorecard in the student training can be seen as the management process. It is mainly from the "financial", "customer", "Business Process" and "learning and development" of these four dimensions of student training in all areas to develop appropriate indicators and to make analysis and evaluation.

3   The Application of Balanced Scorecard in Probation

3.1 Financial Dimension

Different with the general business financial index, the final goal of the probation is not aimed to achieve success of pure finance. In the micro-scope, is to realize the distribution and redistribution of the human resources. On the view of macroscopic, is to realize the rational distribution of the human resource market,and finally solve the problem of employment. So, for a single apprentice base, its financial strategy index is to establish a complete talent pool on the basis of reducing the cost . Its financial results index is reducing the rate of students turnover and reducing capital costs. For the whole implementing system, we need to consider the overall costs and the hight of the return.

3.2 Consumer Dimension

Traditional sense of the customer refers to the final consumer product. In the student system implementation process, the customer includes all the stakeholders, so the customs can be divided into enterprise customers, trainees and government agencies these three categories. In this dimension, the strategic target is to achieve the maximum limit members' satisfaction, while the outcome indicator can be set as the satisfaction of the training base on the outcomes of the trainees ,the satisfaction of the trainees on the training bases. Apprentice training customer dimension of the implementation of the project and forward-looking indicators are: the enterprises cooperate positively with the government work, The government exercises a close supervision over the enterprises and give guidance termly; Trainee students actively participate in the training of enterprises, improving their enthusiasm of work.

3.3 Business Process Dimension

In the dimensions, in consideration of the apprentice base is the main unit. Apprentice to a single base for training, business process of the apprentice, mainly includes the following steps: make trainee program-release trainee position information-the organization apprentice interview-the organization pre-job training-apprentice agreement signed-submitted the trainee register-implement and assessment and teaching plan-declare apprentice subsidies-implement apprentice students grow tracking. Therefore, to a single enterprise is concerned, its strategic index is building a good apprentice training and teaching process. The research results strategic index is set to apprentice students training qualified rate and apprentice to the above-mentioned rate, trainee students to take to teach students the satisfaction, government to apprentice and teaching apprentice satisfaction, take a teacher training qualified degrees. Apprentice training business process dimension is the leading indicators for a continuous complete attachment programme chain, trainee information release to the timely and effective, with a teacher in place, information collected and update on time.

3.4   Dimensions of Learning and Development

There are two main dimension indicators in the probation: business and trainees. For the training base, its strategic goal is to achieve successful completion of training programs in order to achieve the skills of qualified training. Its outcome indicator is students pass rate, student teacher in-service training output, the satisfaction of participants and the final retention rate of trainees. The leading indicators can be set as follow: to get enough attention from the leaders; the managers are responsible and the trainee teachers' teaching are serious. For trainees, its target is to achieve practiced skills. The same with the businesses, its outcome indicators is the satisfaction of trainees and rate of qualified training. The first indicator is that the trainees actively and seriously take part in the attachment training.

4   The Survey of Probation

Student training is an activity that .needs the Cooperation of multi-party .It requires not only strong government's advocacy and financial support ,also needs the Cooperation of the enterprises ,and above all, the active participation of the trainees is the most important. This survey collects satisfaction analysis data among college students who are participating in the probation or have finished, and it chooses one training base in each industry as an case. It Analyzes training leading indicators one by one on the bases of four dimensions as "financial" ,"consumer" ,"work flow" and "learning and development". The survey is divided into two parts, 300 questionnaires were sent out and 278 were available .All of the data were analyzed by SPSS software.

4.1   Financial Dimension

For this part of the financial analysis, it is not proper to analyze the data just from the financial figures. As I mentioned from the last section, the financial dimension of the indicators not only includes the various parts of the plan of direct financial inputs but also includes the implementation of the plan during a variety of indirect capital investment. So this part will apparently relate to the index .For example, how to change the trainees of the retention rate of conversion into an intuitive financial figures index is a process of index. But in the end, we can solve the problem and make it easier by comparing the two results. Here, due to various constraints, I will not analyze them one by one, but the final of the ratio can be calculated.

4.2 Consumer Dimension

In this dimension, it relates to a satisfaction index. For the enterprises, we mainly use the data of trainee retention rate. Up to the end of October 2010, 870 trainees in 1802 have been hired by enterprises. The retention rate is 48.3%.For The trainees, it relates to the index of student satisfaction. In this investigation, there are 188 (67.63%) participants for the entire student said he was satisfied with the implementation of policies, only 20 (7.2%) participants were dissatisfied, and the remaining were generally satisfied. It seems from the results of the analysis that about 90% of the participants expressed satisfaction with the implementation of the policy considering that the policy should continue. 78 percent of the participants expressed satisfaction with the training business, but there are also 10% said they were not satisfied because some training teacher are irresponsible and the training plan is not careful.

4.3 Work Flow Dimension

It is complex to carry out a Students' training. Therefore, we must first consider the plan chain is the development of various project situations. In this dimension, it mainly relates to the development of training plan on process. The main indicators includes the trainees pass rate, trainees on trainee satisfaction with teaching, business training for trainees satisfaction with the government to carry out training programs for business satisfaction. The survey is mainly derived from visual data trainees on trainee satisfaction with teaching. In the survey, 91% of trainees were scheduled to take part in student training, and clearly informed a variety of considerations; 95% of the participants were arranged with a teacher; 80% of the students considered his teacher had done their work responsibly.

4.4 Work Flow Dimension

Similar with the previous assessment indicators, the dimension of this part is the passing rate of trainees of training, and the training frequency and the satisfaction of the parties to the training targets. Because student satisfaction and other indicators in the above sections have been described in detail, so in this section I will no longer describe them. The pass-rate of training can be measured by results of the chart of Probation Quality Assessment during the training process. The data is filed by the government. Because of the diverse nature of the business, The number of students training are different, but the overall numbers do not float. The reserve ratio can be described by students of the final retention rate of trainees .For example, as a training base, Gome, has about 20 trainees every month, its training period is around 3 months, and retention rate is 50%. about 20 people are kept monthly. Retention rate is 50%. For Dean to the end of December 2009 ,the company has trained 81 trainees, including 38 students in Hangzhou, which have been organized in 15 trainee posts. The trainees' retention rate is over 65%.

5 The Application of Educational Technology in the College Student Probation Quality Assessment and Recommendations

5.1 Establish Indicators and Unify Data Processing

To standardize assessment of all the balanced scorecard indicators to meet student training process, so as to enhance the use of indicators. The most important climatic factor in the training of student assessment is making sure that whether the indicator system design is scientific. Secondly, to pay attention to imperfections of the assessment so as to add new indicators and ensure the validity of the results; In addition, it is not good to have too many indicators which will lead to the weight of the dispersion, and finally make the results of evaluation inaccurate. When in the collection of data, indicators should be changed into the original synthesis of data to collect data, and then process them by using a unified formula to ensure the validity and accuracy of data processing.

5.2 Expand Public Ways to Enlarge Effect

Publicity plays a vital role in the process of carrying out the probation. As for the methodology, the ways mainly include websites, posters, campus seminars and other conventional channels of publicity. The target audiences are mainly concentrated in the student population and the target places are mainly concentrated in the university campus locations. In fact, the major community and village committees should also be included. Because when the students leave schools, they go home. The publicities at communities or committees in the village can not only increase the degree of their oral tradition, but also attract students' parents to know this policy, and then inform their children .From this way, the social influence will be greatly increased.

5.3 Mutually Beneficial Cooperation to Achieve Win-Win Cooperation

Apprentice training is undertaken to participate in a multi-process multi-platform. During the process, each platform as a child element of the entire student training system has an effect on the whole system. Therefore, it should do the link between the various subsystems to optimize the system functions. First of all, the government should be proactive in playing the role of a good leader. He should not only expend the strength of propaganda in the social level, but also go to the trainee bases with a deep communication and interaction. Enterprise shall cooperate with the government to implement the measures and follow with the steps of fulfilling the quota. They shall keep continuous self-improvement and innovation. Trainees should be with a positive attitude in earnest trainee program and give a real and effective feedback of information during the information collection,

5.4 Strengthen Technology Innovation and Improve Information Management System

Because it needs too many indicators during the assessment, so collecting the data of the assessment systematically is an arduous and complicated work. It takes a lot of manpower, material and financial resources and time. We should strengthen the technical innovation and develop a set of assessment information system for student assessment data collection and training to facilitate the query conditions. In addition, make full use of modern computer technology and draw the data, information and assessment results of all parts into the national data base to realize the information management of the Probation Quality Assessment and bring performance evaluation into full play.

6 Conclusion

As a case of educational technology, the advantages of educational technology in all areas of student assessment are presented. These four dimensions in terms of the macro level or micro level have given a good assessment perspective. Compared with other assessment methods, the balanced scorecard highlights its comprehensiveness and balance. The four dimensions of analysis from all angles make us have a clear and complete understanding of evaluation system, This detailed quality management assessment standard cannot only be widely applied in the self-evaluation of training base, but also the training base annual performance assessment conducted by government, so it is very practical .

References

1. Issued by the general office of the people's government of hangzhou on measures for the implementation of probation training college (trial) (74) (2009)
2. Zhang, p., Pu, j.-s.: Chinese and Foreign Education Technology-— Comparative Study between Theory and Practice, Beijing (2005) (in Chinese)
3. Zhang, Z.-g., Chen, J.-j., Yu, L.: Balanced Scorecard: A Revolutionary Method of Enterprise Performance Measurement. East China Economic Management 5, 109–111 (2001) (in Chinese)
4. Song, Y.: A New Performance Evaluation Method -—Balance Scorecard. Northern Economy and Trand 9, 87–88 (2005) (in Chinese)

Fully Understanding Vocabulary in Five Steps

Zhao-jun Liu

School of Culture and Media, Jilin Normal College of Engineering &
Technology, Changchun City, Jilin province, P.R. China

Abstract. Based on teaching experience of Chinese as a Foreign Language and
personal exploration in the past decade, it is to be found that the lacking and
misunderstanding of vocabulary are the main challenges to foreign students
when learning the target language(s). However, this problem can be effectively
solved by learning vocabulary systematically. So this paper presents a novel
method to systematically learn vocabulary, the process of which can be
summarized in five steps. On the basis of contrast between teaching practice of
328 foreign students and that of past 206 foreign students, the case indicated that
the novel method can improve using Chinese capacity of foreign students greatly.

Keywords: Education, linguistics, vocabulary learning, Chinese as a Foreign
Language, Five Steps.

1   Introduction

A multi-angle and all-around profound understanding to vocabulary in target
language(s) is one of the most important marks, which means that the level of target
language(s) of the foreign students has improved greatly from their native languages to
target language(s). Since Chinese is a word-based language which forms sentences
with layers of morphemes (by Cheng Yumin (In 2003)), the learning of vocabulary
plays an essential role in the process of learning Chinese.

The foreign undergraduates and post graduates usually face great challenges when
they end *pure* language learning and change into taking in corresponding professional
courses according to their majors in universities together with their Chinese classmates,
that is, they need contact large professional Chinese instead of the previous daily
dialogues. In addition, since foreign students tend to contact more with their fellow
students, professors and tutors, consequently, input of received Chinese language
information and language information output of using spoken language, written
language to express all largely increase, which brings two kinds of problems. One is
from largely learning to large using, another is from largely receiving to largely
understanding and absorbing. In the meantime, more problems may occur if the foreign
students try to fit in sooner by participating in various kinds of social activities in
campus since it requires a substantial improvement from learning Chinese to *using*
Chinese.

According to observation during teaching intensive reading course of Chinese as a
Foreign Language (Advanced Class), foreign students confront troublesome situations
mostly due to misunderstanding or being misunderstood. For instance, the grades of

foreign students suffer if they cannot understand what teachers say in class or they feel that they are rejected by others during discussion just. Because they can't express themselves and get others' ideas correctly .Foreign students will also have difficulties developing social relationships if they cannot communicate efficiently with others. Large quantities of synonyms, antonyms, metaphors, and symbolisms, etc in Chinese also lead to embarrassing situations since foreign students often cannot distinguish from them of understand them. Though the above-mentioned negative positions relate to various factors such as differences between culture, customs, etc, vocabulary is the most important one among all with no doubt.

Referring to considered problems as mentioned, comprehensive grasping and fully understanding of words can be the basic and essential solution.

Based on experience of teaching Chinese as a Foreign language (Advanced Class aiming at undergraduates and postgraduates), especially Intensive Reading Course during the past decade. It is found that most foreign students start to have difficulties when they learn and use Chinese after they begin to set foot in their professional course. This phenomenon results from lack, misunderstanding or misusing of vocabulary. Therefore, the learning of vocabulary must be regarded as the focal point of Intensive Reading Course.

In the past several years, I have been trying my best to improve the old methods of vocabulary teaching and put novel way of teaching into effect. Theory of novel way can be summarized into 5 main steps:

a. Explanation of important expressions
b. Appropriate complement
c. Examples and practice
d. Discussion and Question-answering
e. Summary.

2　Theoretical Foundations

2.1　The Particularity of Chinese

As the only non-phoneticize language in the world, Chinese has its own great particularity, which is especially obvious in its vocabulary. For example, the word "hole" can be differently expressed as "Kulong", "Dong", "Kou", "Xue", "Kong", and "YanEr" in Chinese. This example shows the diversity of Chinese language, which is one part of the particularity of Chinese.

According to Lv Shuxiang's theory (in 1979),though most single words and their morphemes follow the rule of a one-to-one corresponding, the permutations of the pronunciation, semanteme and the font style of a single word can be up to 8 different situations. And when two single words form one word together, the permutations double. In addition, the above-mentioned statements are quite theoretical that the situations can be far more complicated in practical application. Thus, a word in Chinese can have different patterns and meanings, while several distinct words may share one certain meaning or meanings with only subtle differences, which can really confuse language learners.

So the foreign students who want to achieve the fully understanding of vocabulary in target language should firstly base on extending the corresponding relation between foreign students' native languages and target language from "one-to-one" to "many-to-many". During which process, the language learners can have a comprehensive grasp of different meanings of certain vocabulary, synonyms, antonyms, contexts, etc. Language learners can also learn about which are the appropriate words to choose according to certain language situations or linguistic customs. At the same time, the language learner can also enrich his or her vocabulary.

2.2   The Natural Law of Language Learning

The learning of vocabulary can be divided into two steps in both native language and target language. During the first step, language learner enrich his or her vocabulary by memorizing words and intentionally imitating during the process of growing up or developing the cognition to the world; During the second step, language learner continuously develop a deeper conception of vocabulary with clearer and better comprehension.

During the whole process, the majority of language learners go through a movement from simply memorizing words by rote to establishing wide-ranged, profound connections among words. Thus, a multi-dimension framework of vocabulary is formed.

In addition, the two stages of vocabulary learning are not completely separate. As the quantity of words grows, the understanding of vocabulary is also promoted. Conversely, the quantity of words increases in a certain degree while the conception of vocabulary deepens.

2.3   The Theory of Language Input

Krashen (1981) pointed out that learning a language greatly depends on language input in quantities, in which comprehensibility, interest, variety and relevance are the most essential factors. Comprehensibility means that language learners' comprehensive endurance and actual needs should be considered while inputting language. Interest means that the language input to language learners should be attractive enough. Diversity and relevance are also of great importance since a variety in content with internal connections is needed in language learning. My theory of teaching effectively elucidates Krashen's theory. Detailed studying of texts ensures the comprehensibility, while studying of examples, contexts, synonyms and antonyms, etc. guarantee the variety and relevance. With activities like discussions and appropriate class arrangements, the novel way of teaching can attract students' interest as well. According to observation during classroom teaching, Foreign students have great desire of learning Chinese and always hold positive attitude on language learning.

In addition, Krashen also put forward the principle that the language input to language learners is supposed to be more advanced than their level being then, If "i" stands for the level of language learners, "1"stands for higher level than that of "i", only when language input to learners is on the level of "i+1", the language input be recognized as effective. The level of "i+2"can be far beyond language learners' acceptance, while the level of "i+0"may discourage learners from entering a higher level. Aside from perspectives mentioned above, Krashen indicates the quality of language input to language learners mainly determined by the linguistic environment, thus, creating an environment which is beneficial to language learning can be a valuable factor.

3 Class Arrangment and Application Process

On average 4 class hours will be arranged for Advanced Intensive Reading of Chinese as a Foreign Language per week.

The first and the second class hours focus on vocabulary learning, while the third and the fourth class hours are for learning of texts and grammar. I also set aside some time for proper preview and review each class hour because these can help students learn better.

3.1 Explanation

The process start from one certain student (randomly picked each time). He or she will be asked to explain the meaning of one or several words in the word list after the Chinese text. He or she should try his or her best to list all the meanings and express them in his or her own words. The explanations in dictionaries are only for reference.

3.2 Complement and Mistake-Correction

In this process, students can volunteer to add not-mentioned meanings of vocabulary or correct mistakes.

3.3 Examples and Practice

Students are asked to make sentences with distinct meanings of certain words. And those sentences can be examples for teaching and practice of how to use vocabulary appropriately.

3.4 Discussion and Question-Answering

Students are divided into groups to discuss about the words' explanation, examples, synonyms, antonyms, etc. And the teacher can walk around and answer the questions. Students may come up with answers during discussion.

3.5 Summary and Conclusion

The teacher makes a summary for the four steps above. And then a conclusion of students' ideas with appropriate correction and supplement should be made.

3.6 Tips

1) How to Arrange the Class Hour.

According to the students' situation, choose a proper number of words to teach in each class hour. In this way teacher can make full use of the time in class and ensure students' comprehension at the same time.

2) About Dictionaries and Reference Books

Teacher should encourage students to make good use of dictionaries and reference books because they can help students study more than textbooks and come up with creative ideas.

3) Proper Adjustment According To Different Situation

There is a saying "Practice makes perfect." A theory is pale without practice. And a theory without being proven successful by practice is nothing. Though I've explained a lot about my theory, I believe that slight adjustment of teaching according to each class's situation is more crucial to get perfect effect.

4 Conclusions

Based on experience of teaching Chinese as a Foreign language (Advanced Class aiming at undergraduates and postgraduates), especially Intensive Reading Course during the past decade. It is found that most foreign students start to have difficulties when they learn and use Chinese after they begin to set foot in their professional course. This phenomenon results from lack, misunderstanding or misusing of vocabulary. Therefore, the learning of vocabulary plays an essential role in the process of learning Chinese. Comprehensive grasping and fully understanding of words can be the basic and essential solution. Aming at the problem, a novel method is presented in this paper, and is illuminated by explaining theoretical foundations and application process. And, practical applicability of the novel method have already been proved by the contrast between teaching practice of 328 foreign students and that of past 206 foreign students.

References

1. Bloomfield, Leonard, language. Henry Hole, New York (1933)
2. Chao, Y.R.: Language and Systems. Cambridge University Press (1968)
3. Chao, Y.R.: A Grammar of spoken Chinese. University of California Press, Berkeley (1968)
4. Chomsky, Noam: Aspects of the Theory of Syntax. M. I. T. Press, Cambridge (1965)
5. Halliday, M.A.K.: The Linguistic Sciences and Language Teaching. Indiana University Press, Bloomington (1964)

Study and Practice in Major Diversity of Undergraduates Program[*]

Ge Baojun, Wang Junming, and Li Shanqiang

Office of Academic Affairs, Harbin University of Science and Technology,
Harbin 150080, China
{gebj,wangjunming}@hrbust.edu.cn, lishanqiang@gmail.com

Abstract. With the change of higher education from the Elite to popular personnel training, the explore and practice in educational reform are done to realize the different levels of education and individualized training, to increase the students' practical hands-on capacity and self-learning ability, to make and implement the major diversity personnel training program.

Keywords: Personnel training, Major diversity, Study and practice.

1 Introduction

Since the modification of major catalog in 1998, the higher education in China came into a new phase, Popular Education finished a change from original Elite education to popular one. But correspondently some educational ideas did not follow up the change. The problems occurred, for instance, the emphasis of engineering courses more on the theory, practice simplified and engineering training weaken. The enterprises at home feedback that the graduates cannot meet their requirement to the ability and quality, and so did the transnational corporations; thus there existed the contradiction between the single personnel supply and diverse need. With the introduction of educational ideas from developed countries, the universities and colleges in China started the teaching reforms, and the emphasis has changed from the scales to the quality.

Since 1999 Harbin University of Science and Technology (HUST) enlarged the enrollment scale, the diverse-level students were allowed in the campus. From the year 2003 on, HUST spent much money on the labs and the lab condition improved greatly. In 2003 HUST was assessed as "A" level in the Undergraduates Teaching Evaluation of China Education Ministry. As a result, the "Teaching Reform" program in 2004 had favorable wind all the way. Through the exploration and practice in recent years, HUST succeeded in the test points of the "customized" personnel training in order to revive the north-east old industrial bases, and gained the example effects.

---

[*] Item in-aid: Research item for 2010 Higher education teaching reform engineering projects in Heilongjiang province.

2 Workout of a Diverse Training Program

The setup of program should follow the guideline of China education and reflect the times spirit of "three orientations"; have clear recognition to the higher educational situation and task and meet the requirement of the contemporary society, economy, science and technology to personnel training; from the scientific view and strategy, cultivate the future builders with the morality, wisdom, health and beauty; Based on the scientific educational ideas, teaching law, personnel training modes and quality standards, the program should meet the requirements to popular education and cultivate the high-leveled personnel with the creative and practical ability. Considering the multi-level enrollment in HUST and its development strategy, the program should reflect the features and advantages of various majors and realize the multi-target, multi-mode personnel training; the reform outcomes are also used to raise the personnel quality by the diverse cultivation, regression to engineering education, courses update, integration and optimization.

3 Basic Principles of Diverse Program

Insist on the following principles: the morality, wisdom, health and beauty developed simultaneously; the knowledge, ability and quality raised in step; fundamental courses and major ones valued concurrently; the integration and optimization linked scientifically; the individual development and diversity of training modes emphasized; the theory and practice combined with intensive practice, and innovative ability raised.

4 Frame and Course Structure of Diverse Program

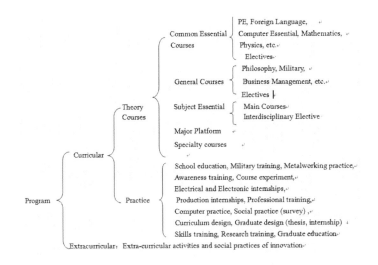

Fig. 1. Frame and course structure of program.

5 Features of Diverse Personnel Program

5.1 Individualized Training

Considering the diverse-level enrollment, in 2006 HUST worked out a diverse personnel program that has the schemes A and B for 21 majors. In courses arrangement, the Scheme A of the high-level students puts the more emphasis on the academic research, theory and labs with science and research trainings increased; and Scheme B of the low-level ones pays more attention to the engineering application and practice with intensive skill training.

5.2 Regression of Engineering Education to Engineering

Emphasize the conscience of "Great Engineering", increase the hours and categories of the practical teaching, decrease hours of theoretical one, and raise the innovative credits. The program in version 2006 demands that the majors of science and engineering have theoretical hours (including courses labs) less than 2600 hours, and practical teaching is 40 weeks or so. The hours of courses labs are also increased.

5.3 Increased Ratio of Electives and Emphasized Individualized Cultivation

Except humane electives, in the common fundamental, subject fundamental and major orientation courses, the electives are arranged, and their variety and amount are increased greatly. Among them, the natural science electives are 6 credits, humane ones, 6 credits, subject fundamental ones 6-8 credits, major orientation courses 20 credits or so, arbitrary electives 6-8 credits. The ratio of compulsory to elective courses is 3 to 1.

5.4 Penetration of Various Disciplines and Perseverance to HUST Features

The courses in humane, science, engineering, and business management penetrate mutually. The students can select the science, engineering, economics, business management, humane, and law electives trans-disciplined and trans-categorized. The engineering courses are extended to other subjects to set up the engineering basis and conscience for the students.

5.5 Reform of Humane Quality Education

5.6 Attention to Fundamental and Major Courses Concurrently

Considering the major orientation and features, the ratio of major courses hours is increased, and their instructions are moved forward properly. The ratio of fundamental courses (including common, humane, social science, economics and management courses) hours to the major courses (including subject fundamental, major platform, and major orientation courses) hours is controlled at 45:55 or so.

5.7 Emphasis of Course System Integration and Optimization, Course Update and Reform

The fundamental courses are compressed and classified, and the major courses are integrated and updated.

5.8 Increase of Concentrated Practice Subjects

Contrast to the program in version 2003 version, the one in version 2006 increase the following subjects such as recognition practice, computer practice, academic years design, major practice, electric and electronic engineering practice, coursework practice, science and research training and skill training. The ratio of practical credits to theoretical ones in the engineering majors is 29.8:70.2.

5.9 Highlight of Academic Spirits

Every major increased a chair course of 20 hours.

5.10 The Part of Majors Put the Industrial Authentication to the Program

5.11 Regulation of Courses Syllabus Further and Reform of Syllabus Modes

According to the courses description, teaching orientation, knowledge points, hours distribution, lecture hints and methods, problem design, test and score records regulate the terminology.

6 Implementation and Gains of Program in Version 2006

Since the program in version 2006 was implemented for the undergraduates in Grade 2006, it has been used for a circulation,   and the result is quite good. The students and parents are satisfactory and there is no student claims to change the class

(1) The workout of the diverse program accords with the 17th Congress spirits that "the higher education should make China powerful and merge into the economic construction", especially "makes China into a human resource power". The program also further meets requirement of the higher educational reform and raises the teaching quality. The result shows that our program goes ahead of the requirement of the national and provincial education authorities and has precocious ideas and consciences.

(2) The Ministry of Education and the provincial educational bureau put forth to further reform the undergraduates teaching and raise the teaching quality（No.2 Document in 2007）, and the nation implemented the quality engineering in 2007, the setup of the diverse program is prompt and necessary. The program is started and worked out from 2005 and gained the time for the enhancement of teaching quality in HUST.

(3) The program has some creative points. For example, 2 sets of schemes, **A** and **B** for the same major; the major courses advanced; elective courses arranged widely and fully; the academic years design, chairs, and innovative credits increased, etc.

(4) The computer practice, skill training, science and research training, and creative credits also increased, and the cultivation system of creative education set up. The intensification and perfection in practice teaching, the penetration of engineering courses into other subjects embody the "great engineering", the increase of the electives intensifies the conscience that the students are No.1 and raise the strength of individualized cultivation.

(5) There are many new faculty of teaching management in HUST, their abilities and levels are fast raised through the workout and discussion of program.

(6) Since its implementation, the program in version 2006 helped HUST apply for many projects to answer the "quality engineering" of the Ministry of Education. For example, the national creative lab zones of the personnel training modes, majors of features, teaching result awards, undergraduates creative lab program, etc. so did the application of in the provincial projects. [1]

7 The Problems of Diverse Program

(1) Some new practices are not mature enough, many exploration are left improved, such as the recognition practice, computer practice, academic years design, etc.. Some practices still lack in the hardware conditions and need to improved step by step. Such as the electric and the electronic engineering practices, the science and research training, and the skill training.

(2)The decrease of the theory hours, and increase of practice hours put forth the higher requirements to the lab and theory faculties in HUST, it affects the implementation of training program and its results.

(3) The increase of electives make their qualities quite different, some good, some bad, the students have difficulties in selecting electives, and the effects of some chairs are not ideal.

(4) The new teaching ideas are not received, understood and implemented by most of the faculties; Scheme A and B did not form their respective features, there are many similar points in the teaching syllabus and teaching methods.

(5) The number of the courses and theory hours are still too much, the arrangement of the courses is not even in the various terms. And in the some terms the hours in per week are too much and affect the students' self-study, crowd out their time in the innovative labs, the national and provincial competitions, and finally limit their individualized development.

(6) The lack of education in the venture creation and courses of career planning for undergraduates.

(7) The diverse program is not "diverse" enough.

8 The Revise and Perfection of Diverse Program

8.1 The Background of Revising the Program

(1) On Feb. 28th, 2010, the State Department made public "the outline of China middle-long term educational reform and development plan （2010—2020） " asking for the

ideas. It is the first outline of educational plan in 21th century as a important document [2] of educational reform within the future ten years.

(2) The Ministry of Education implemented the "Educational quality and Teaching Reform in China universities and colleges" in 2007, and then "the outline of China middle-long term educational reform and development plan（2010—2020）" also will be done in the future ten years. Teaching reform and quality raise of personnel training will be the theme of the higher education.

(3) The program in version 2006 has been used for four years. During four years the situation of higher education has the new ideas and changes that need merged into the process of personnel training. At the same time, through implementation of a circulation, the weakness of the program need revised.

(4) With the economic development and change in China, the requirement to personnel will change constantly. The students pay more attention to the employment. In 2020, the gross admission rate of higher education will reach to 40%. It demands that we update the ideas, think and explore the new modes and methods of personnel training to adapt the development and change of the popular education.

8.2   The Requirement to Revise of Program

(1) Further condense the theory hours of the courses, and increase. The total courses hours of the science and engineering majors are less than 2500-2550 hours, and the theory course hours include the labs of that course.

(2) Increase courses of "career planning" and "education in venture creation", they are designated as the elective quality courses in "humane, social science, economics and business" and totally there are 2 courses with 30 hours.

(3) Further optimize, integrate and modify the courses contents. Delete the old, deep and difficult contents.

(4) Arrange the courses reasonably. Connect the former courses with the latter ones suitably and get rid of the repetition and disjunction.

(5) Reexamine the practical teaching and study practical contents, widen the channels and raise the quality.

(6) Arrange the teaching within 8 terms of four years 4 reasonably. And get rid of the phenomena of too dense in earlier terms and too loose in later ones.

(7) In every term the hours in a week are less than 28 hours at most。

(8) The compulsory quality courses in the common fundamental, humane, social science, economics and business management keep invarious in the hours and category, but the teaching contents need to regulate.

(9) Work out the applied program (Scheme B) more suitably to adapt the social needs under the participation of experts from enterprises. As to the program（Scheme A）of with emphasis on the research and disciplines, the ideas and suggestion of the experts from enterprises should be asked for and referred. Raise the practicability and effectiveness of personnel training program.

(10) Embody the co-operation between HUST and the enterprises, and promote the "customized cultivation" of Scheme B.

(11) Express and carry on the CDIO ideas into the program[3]. The CDIO ideas emphasize that students should learn the engineering knowledge positively and

practically on the basis of the life period of products from the development to operation.

(12) Test point excellent engineers program in the majors with apparent advantages in the industrial background.

(13) Put the professional skill training into the practices of metallic technology, electric and electronic engineering.

(14) Evaluate and re-designate the electives.

8.3 Expectant Targets of Program Revision

(1) Accord with the requirements to higher education stipulated in "outline of China middle-long term education reforms and development plans".

(2) Perfect the diverse program in the education sytems of innovation and venture creation.

(3) Through the regulation to the hours in theory and practice teachings, as well as improvement of practice teaching, the hours ratio of theory to the practice teachings in the program becomes more reasonably, and the practice training more effective.

(4) The distribution in hours, credits, and hours /week in four academic years is more even and reasonable.

(5) The disciplined (Scheme A) and applied (Scheme B) orientations in the program become distinct, and the major features are more outstanding.

(6) Embody the co-operation of HUST and enterprises, and show the example action of the "customized" mode in applied personnel training.

(7) Scheme A should provide more time and chances for the students in their self-study and make them join the creative labs and competition greatly. And Scheme B should give them more time and chances in contacting the engineering practices.

(8) Embody the advanced teaching ideas and methods , test point the new cultivation modes.

At present, HUST has finished the revision of program in version 2010. The program in version 2010 is used from new students of Grade 2010 on. The new program is regulated according to the development trends at home and abroad as well as the change and requirement of enterprises. We hope that our practice and exploration in the diversity of personnel training programs can provide useful experiences for other universities and colleges at home in their teaching reforms.

References

1. Dawei, M.: Building of Customized and Applied Personnel Training Modes. Higher Education Research in Hei Longjiang Province (1), 116–118 (2008)
2. Official Website of Education Ministry in P. R. China, http://www.moe.edu.cn/
3. Worldwide CDIO Initiative: A Framework for the Education of Engineers, http://www.cdio.org

Analysis of Comparing Mulan-Boxing with Other Aerobic Exercises to Impact Physically on the Old and Middle-Aged Women

Xu Cai-yan

Sports Department of Jiangsu University, Zhenjiang Jiangsu China
154137119@qq.com.cn

Abstract. Comparing Mulan-Boxing with other aerobic exercises to improve body constitution and the blood lipids composition for the old and middle-aged women. Experimental Methods: Three groups are randomly selected for testing their body mass index (BMI), blood lipid composition to make research on the influence of the indicators, such as their shapes, functions, physical fitness of the old and middle-aged women on which aerobic exercises impacted. It shows the recovery of the fitness function to popularize practice like Mulan-boxing exercises.

Keywords: Mulan-boxing, aerobic exercises, the old and middle-aged women, BMI.

1 Research Purposes

Mulan boxing is a physical exercise in the integration of gymnastics and martial arts, with its lightly slow action, elegant modeling, and low-load exercise, which are particularly loved by the old and middle-aged women. It has been reported to make research on Mulan-boxing' s impact on the old and middle-aged human habitus, but it is not reported that the study on the comparison of Mulan boxing and aerobic exercises' impact on the physical effects of the old and middle-aged (such as walking, jogging, joint exercises, ballroom dancing, shadow boxing, etc.).

We tested 45 healthy women participants in Mulan-boxing exercises for over 2 years and 61 teaching staff at the same age who often participate in various aerobic exercise in the Shanghai University of Sports with their body constitution and blood lipid composition, comparing Mulan boxing with other common aerobic exercises' effect on their body constitution and serum lipid composition for the old and middle-aged women.

Meanwhile we selected randomly 105 community residents as a control group who seldom participate in exercises from No.1 Guohe Village and No.2 Guohe Village, to test their body mass index, studying the difference of the indexes, such as their shapes, functions, fitness on which different aerobic exercises affect. We aimed at introducing the value to popularize Mulan-boxing exercises on body building in the old and middle-aged women, and provide a scientific basis on popularizing Mulan boxing exercises.

Y. Wang (Ed.): Education Management, Education Theory & Education Application, AISC 109, pp. 821–829.
springerlink.com

2 Experimental Objects and Research Methods

2.1 Objects and Groups

45 healthy women take part in Mulan boxing exercises(at the age of 40-65 years old, average age 53.5 ±4.6); 61 women teaching staff in Shanghai University Of Sport regularly take part in aerobic exercises for more than 3 times a week (at the age of 40-65, average age 55.2 ±8.6); People take part in the sports by a percentage from large to small: Jogging 36.1%, ballroom dancing24.6%, walking 19.7%, Tai-ji shadowboxing 13.1%, joint exercises. 6.6%.

105 people seldom take part in aerobic exercises for less than 3 times a month from No.1 Guohe Village and No.2 Guohe Village (at the age of 40-65, average age 54.9 ±6.8).

In group A, is Mulan boxing practice group. In group B, is teaching staff group in Shanghai University of Sport. In group C, is community residents group.

Considering of the regularity of women's physical development, they were divided into two groups: Group I for women aged 40-55 years old (in pre-menopausal group); group II for women aged 56-65 years old (in postmenopausal group).

2.2 Methods

2.2.1 Test Equipments

①Hologic Sahara BMD Ultrasound Bone Density Densitometer made in USA;

②Heart Rate Monitors –Polar made by Polar Electro Oy, Finland;

③COSMED PONY Spirometer manufactured by COSMED S.R.L. Italy

④Omron HBF-301 fat measurement device made in China

⑤Indirect vo2max (maximal oxygen uptake) is measured by Japan steps test.

2.2.2 Test Type Indicators

Shape: height, weight, bust, waistline, hipline, body fat ratio (BI method)

Function: pulse, systolic and diastolic blood pressure, lung capacity, calcaneus bone mineral density (BMD)

Physical agility: sit-and-reach (sitting in front of body flexion), grip strength, standing on one leg with eyes closed, hand response, indirect determination of maximum oxygen uptake (by Japan step test method)

Blood lipids: total cholesterol (TC), triglyceride (TG), high density lipoprotein cholesterol (HDL-C), low-density lipoprotein cholesterol (LDL-C)

Derived indicators: waist-hip ratio (waist / hip, WHR), body mass index (weight/height 2, BMI), HDL-C/TC, HDL-C/LDL-C

2.2.3 Test Method

Sample blood was drawn in cubital vena from the tested people during fasting, which was sent to Changhai Hospital to test the result of TC, TG, HDL-C, LDL-C with enzymatic method.

Test of indicators of basic physical fitness would be implemented in accordance with the detailed rules for the implementation of 2000 National constitution monitoring manual under National Monitoring Center.

Bone density & lung capacity measurement is to be used in accordance with the instrument instructions. At the same time questionnaire survey had been made which was related to the sociology.

2.2.4 Data Processing

Data is processed by statistical software SAS6.12 (including mean, standard deviation, t-test, analysis of variance, correlation coefficient).

3   Results and Analysis

3.1 Comparison of the Effects on the Body Shape

Body fat rate is an important indicator to reflect the body composition. Adult women's body fat ratio is generally 20-30%, while more than 30% is obese. Excess body fat is one of the major risk factors of more diseases (especially cardiovascular disease). Body mass Index (BMI) is one of internationally recognized evaluation standards of human obesity. A large number of studies have shown that BMI and body fat are closely related to a variety of diseases. [1,2,3,4,5] It is generally believed that the body symmetry is BMI of between 20-24, while overweight is for BMI $\geq$ 24.

Waist to Hip Ratio(WHR) is ratio between waist and hip, which mainly reflects the distribution of body fat.

WHR $\geq$ 0.8 for the female is central obesity, which means excessive accumulation of abdominal fat. It is generally believed that abdominal fat is closely related with the development of a variety of diseases. [2,3,4,5]

The indicators of above-mentioned three groups were compared in pre-menopausal groups. WHR is different in group B and group C.

In postmenopausal, the three indicators of obesity are reflected in significant differences in group A, B and group C.

There is no difference between the two groups (group A & group B. (see Table 1, 2). Reflected in postmenopausal women's hormones changed dramatically, the women's body shape is changed thereafter. Both obesity and excessive abdominal fat savings are a part of other changes. The aerobic exercises including Mulan boxing can obviously improve women's variations of body shape. We can make it clear that after menopause the women who often participate in aerobic exercises were much better than those who did a few exercises in their morphological parameters.

There was no significant difference between the Mulan boxing practice and other common aerobic exercises in improving obesity, and promoting the rational distribution of fat.

Table 1. The compare of pre-menopausal subjects' body shape

	Group AI (N=29)	Group BI (N=32)	Group CI (N = 63)	Results of analysis of variance
Height (cm)	158.7 ±4.6	160.7±5.1	157.5±4.7	BC *
Weight (kg)	58.9 ±8.4	62.8±9.1	60.0±7.5	Sn
waistline (cm)	84.6 ±8.4	84.1±9.1	86.2±8.0	Sn
Hip (cm)	92.4 ±5.2	95.0±5.1	92.6±4.7	Sn
BMI (kg/cm2)	23.4 ±3.1	24.3±3.1	24.2±2.8	Sn
WHR	0.91 ±0.06	0.88±0.07	0.93±0.07	BC *
Body fat ratio (F%)	32.2±3.9	32.6±3.7	32.8±3.5	Sn *

* There is significant difference among the groups before menopause.

Table 2. The compare of postmenopausal subjects' body shape

	Group AII (N=16)	Group BII (N=29)	Group CII (N=42)	Results of analysis of variance
Height (cm)	155.5 ±5.1	156.5 ±5.4	155.4 ±4.7	Sn
weight (kg)	55.0 ± 6.1	56.8 ±6.8	61.3 ±6.7	A-C * B-C *
Waist (cm)	83.8 ±9.0	83.7 ±8.6	92.5 ±8.7	A-C * B-C *
Hip (cm)	90.1 ±3.8	92.5 ±3.8	93.1 ±4.4	A-B * A-C *
BMI (kg/cm2)	22.8 ±2.6	23.2 ±25.4	25.4 ±2.7	A-C * B-C *
WHR	0.929 ±0.07	0.906 ±0.09	0.993 ±0.08	A-C * B-C *
Body fat ratio (F%)	31.5 ±3.7	32.0 ±3.8	35.7 ±3.6	A-C * B-C *

* There is significant difference among the groups after menopause.

3.2   The Compare of Effecting on Body Functions

Vital capacity is a reflection of the simple lung ventilation function, index of which is widely used. The average normal adult lung capacity of women is 2500-3500 ml.

The size of the vital capacity is related to the factors such as gender, age, height, weight, and the elasticity of the lung and chest wall.

Bone mineral density (BMD) means the bone mineral content per unit area. Measuring bone mineral density can determine the existence of osteoporosis and its severity.

[6] BMD measurements in different anatomical parts are of large difference. Before menopause, there is no difference from three groups in lung capacity, values of which

are within the normal range. After menopause, the vital capacity of group A, B was much higher than that of group C.

There is no difference between groups A and group B in the vital capacity. The vital capacity of group C is below normal value range (see Table 3,4). The study shows that aerobic exercises can obviously reduce the physiological phenomenon generated by human lung ventilation function declined with aging. Both Mulan boxing and general aerobic exercises have no obvious differences in improving lung function.

We have never observed that both Mulan boxing and other aerobic exercises could obviously increase the BMD value. This is inconsistent with the literature, which may be related to test instruments and measurement of BMD in different parts.(It is mostly reported that spinal part or other parts of the great intertrochanter were measured by dual-energy X-ray).

Table 3. Compare of pre-menopausal women's physical functions

Group AI (N=29)	Group BI (N = 32)	Group CI (N = 63)	Results of analysis of variance	
The pulse (beats / min): 76.2 ±10.8	74.6±7.7	79.5±9.2	Sn	
Systolic blood pressure (mmHg): 129.3±16.1	122.3±14.7	122.6±17.7	Sn	
Diastolic blood pressure (mmHg): 75.8±10.7	67.6±8.9	77.0±11.0	A-B *	B-C *
Vital capacity (ml): 2844±476	2734±586	2671±645	Sn	
Bone mineral density (g/cm2): 0.503±0.09	0.497±0.10	0.470±0.11	Sn	

* There is obvious difference among the groups.

Table 4. Compare of women's physical functions after menopause

Group AII (N=16)	Group BII (N=29)	Group CII (N=42)	Results of analysis of variance
The pulse (beats / min): 74.0 ±9.4	75.9 ±11.0	77.0 ±9.2	Sn
Systolic blood pressure (mmHg): 135.3 ±19.9	137.0 ±14.6	131.5 ±19.3	Sn
Diastolic blood pressure (mmHg): 74.9 ±11.9	74.5 ±10.1	79.5 ±9.0	Sn
Vital capacity (ml): 2937 ±1135	2782 ±569	2214 ±595	A-C * B-C *
BMD (g/cm2):	0.440 ±0.10	0.429 ±0.10	Sn

* There is obvious difference among the groups.

3.3   Compare of Effecting on Women's Physical Attributes

More flexibility, strength, balance, responsiveness and indirect determination of maximum oxygen uptake is a reflection in indicators of physical age. And their corresponding test methods are the sit-and-reach, grip strength, the stork stand with eyes closed, hand reaction time and Japanese step test. Comparing the group C whose women take part in very few exercises, group A, B have an advantage in a variety of indicators (see Table 5,6). It shows that regular exercises can effectively slow down aging and physical fitness to improve the quality of life in old and middle-aged women..

Before menopause, the flexibility in group A is not only much higher than in group C, but obviously different from group B as well. After menopause, the flexibility and balance in group A are much higher than in group B.

It is meaningful to increase living satisfaction of the old and middle-aged women and prevent them from accidental falls.

Table 5. Compare of pre-menopausal women's physical fitness

	Group AI (N=29)	Group BI (N=32)	Group CI (N=63)	Results of analysis of variance
Sit and Reach (cm):	17.3 +5.8	6.7 +7.1	8.1 + 7.9	A-B * A-C *
Grip strength (kg):	28.5 +4.4	28.4 +5.5	22.0 +6.8	A-C * B-C *
Stork stand with eyes closed (seconds):	18.7 +18.8	15.3 +14.9	12.8 +12.8	Sn
Hand reaction time (sec):	0.209 0.02	0.199 +0.02	0.195 +0.02	Sn
Step test (ml / kg.min):	30.3 +3.8	29.6 +3.7	26.7 +3.0	A-C * B-C * *

* There is obvious difference among the groups.

Table 6. Compare of women's physical fitness after menopause

	Mulan boxing practicing group			Teaching staff in Shanghai University of Sports		
	little	mid	much	little	mid	much
Dairy products	18.3%	55.3%	28.4%	25.2%	42.5%	32.3%
Eggs	49.2%	38.7%	12.1%	17.2%	44.5%	38.3%
Fish &meat	2.7%	32.8%	64.5%	7.3%	39.1%	53.6%
Sugar & oil	50.4%	37.8%	11.8%	48.6%	38.9%	12.5%

* There is obvious difference among the groups.

3.4   Impact on Blood Lipid Metabolism in Comparison

Abnormal blood-lipid metabolism is considered to be one of the important factors the atherosclerosis form. It is a major issue of current medical research how to effectively improve lipid metabolism disorders, treat and prevent from cardiovascular disease. Studies have shown that atherosclerosis is the main pathological change which there is a lipid plaque in the arterial wall. Lipids increased in plasma invaded the arterial wall in form of LDL (or VLDL) via the receptor, and combined with the mucopolysaccharide composition of the arterial wall, producing insoluble precipitate. TC and TG which are released by LDL after decomposing, together with the insoluble precipitate and other lipid components, stimulate fibrosis to form lipid plaques. Serum total cholesterol is transported by the three kinds of lipoprotein: LDL-C, HDL-C & VLDL-C, in which total cholesterol of LDL-C occupied more than 60% of TC. The LDL-C involved in atherosclerosis formation as the main carrier of serum cholesterol. The HDL-C is the body of TC antiporter, whose role is to prevent the cells from doing uptake of CHO, transporting CHO from the surrounding tissues into the liver to decompose and discharge. It has an anti-atherosclerotic effect. Now it is widely recognized that the absolute value of each component of lipoproteins can not fully predict the extent of coronary sclerosis, while HDL-C / (LDL-C or TC) is to better reflect the development of the course of disease. [14]

A large number of studies have shown that prolonged, low-load aerobic exercises can increase high-density lipoprotein (HDL-C) levels. The aerobic exercises include brisk walking, folk dance, fitness dance, aerobics, etc., especially Mulan boxing as the representative of traditional Chinese martial arts and other sports played a positive role in delaying aging to improve blood- lipid metabolism for the elderly in accordance with their characteristics of physical condition and age. The results of this study show that: compared with those in the control group, the old and middle-aged women who had long-term exercises in Mulan boxing group, have dramatically reduced their plasma LDL-C and TG levels. It will help maintain the human normal blood lipid levels. As far as the women concerned, plasma TG will be increased as their ages increased. So it plays an important role of slowing down their aging and preventing from atherosclerosis formation for the old and middle-aged women to take Mulan boxing exercises frequently. Exercises can significantly increase HDL cholesterol levels to make the state more desirable.

[7,8,11,12,13]while blood lipid is also affected by body weight, eating habits and other factors. [9,10] Mulan boxing has beneficial effects on blood lipids, which has been confirmed in the research on the healthy women who rarely take part in the sports as the control group in comparison. [15,16] We adopt healthy women for comparison who take part in other aerobic exercises, in comparison to influence of the exercises on the composition of blood lipids, meanwhile there is no obvious difference between the two groups of (group A & B) in weight, eating habits. (see table 1, 2, and 7).

The old and middle-aged women are at the period of change in hormone levels. As they are older, there has been an important physiological changes such as menopause. Women gradually decline in hormone secretion. Female hormones take a role of increasing LDL-C and TC of women, so it will directly affect the women's health due to its decline in the elderly.

Comparison between group A and group B: in pre-menopausal group A, LDL-C is low, while HDL-C, TC, HDL / TC value is high; in postmenopausal group A, TG, LDL-C is low, while HDL-C value is high. From the point of view for lipid composition, LDL-C is lower at different stages in group A, but HDL-C is high. Although the TC is high in pre-menopausal group, but HDL-C/TC ratio is elevated as well, and The LDL-C in group A is lower than in group B during this period, which is beneficial to improve blood-lipid metabolism. It shows that Mulan boxing campaigns improve lipid metabolism much better than any other aerobic exercises. This difference is probably due to time differences caused by a variety of sports movement. It is about 30 minutes to participating in other sports. Most of sports last less than 1 hour. But Mulan boxing exercise is of small load, accompanied by beautiful music. The collectively organized exercises are more conducive to ensure practice quality and time. It was also verified by analysis of their exercise time (see Table 9).

Thus, it is beneficial to maintain normal blood lipid levels for the old and middle-aged women before and after menopause to take part in regular aerobic exercises like Mulan boxing for them. It can prevent from the occurrence of atherosclerosis.

Obesity and diet are risk factors leading to coronary atherosclerosis, because obesity is often accompanied by plasma HDL-C levels reduced. Through the study we found that each additional unit of body weight (1KG/M2) will have plasma HDL-C levels reduced by 0.02MOL / L. Through regular physical exercises, total energy consumption and fundamental metabolism rate can be increased to help you lose weight. Even if he doesn't lose weight, the obese can improve their blood lipid status through their exercises.

It has been confirmed that the high-fat diet can lead to plasma LDL-C levels increased. High carbohydrate diet makes the concentration of plasma TG higher, while the concentration of plasma HDL-C decreased. However the low-fat diet can have plasma LDL-C of level reduced. Polyunsaturated fatty acids do not reduce plasma HDL-C level, but can decrease plasma LDL-C levels and reduce the incidence of coronary heart disease as well..

Table 7. The eating habits (diet) questionnaire in blood tests group

	Mulan boxing practicing group			Teaching staff in Shanghai University of Sports		
	little	mid	much	little	mid	much
Dairy products	18.3%	55.3%	28.4%	25.2%	42.5%	32.3%
Eggs	49.2%	38.7%	12.1%	17.2%	44.5%	38.3%
Fish &meat	2.7%	32.8%	64.5%	7.3%	39.1%	53.6%
Sugar & oil	50.4%	37.8%	11.8%	48.6%	38.9%	12.5%

Table 8. Comparison in blood lipids of testers

	Group AI (N=29)	Group BI (N=32)	T test	Group AII (N=16)	Group BII (N=29)	T test
TC (mmol / L)	5.57 +0.75	5.03 +0.61	P<0.01	5.91 +0.86	5.45 +0.48	Sn
TG (mmol / L)	1.06 +0.81	1.24 +0.49	Sn	1.30 +0.65	1.79 +0.73	P<0.05
HDL-C (mmol / L)	1.47 +0.31	1.14 +0.17	P<0.01	1.35 +0.34	1.08 +0.20	P<0.05
LDL-C (mmol / L)	2.67 +0.60	3.25 +0.49	P<0.01	2.77 +0.66	3.54 +0.45	P<0.01
HDL-C/TC (mmol / L)	0.27 +0.05	0.23 +0.04	P<0.01	0.233 +0.07	0.199 +0.04	Sn

Table 9. Analysis of movement time in blood lipid testers

	less than 30 minutes (%)	30-60 minutes (%)	60-120 min (%)	greater than 120 points (%)
Group A	0	24.4	53.3	22.3
Group B	16.6	60.5	19.7	3.2

4 Conclusion

1. Regular aerobic exercises (including Mulan boxing) can effectively improve the physical fitness of the old and middle-aged women.

2. Mulan boxing improves blood-lipid metabolism more obviously than any other aerobic sports, the advantage of which may be the result of a longer exercise time for Mulan boxing.

3. Mulan boxing has a clear advantage in maintaining the flexibility and balance of the old and middle-aged women. Mulan boxing   group is not only stronger than the control group (group C) whose women take part in very few exercises, but also far much higher than the experimental group (B group) whose women often participate in other aerobic sports.

4. It can effectively reduce serum LDL-C and TG levels to take part in regular Mulan boxing exercises. It is of positive significance to maintain normal blood lipid levels and prevent from the occurrence of atherosclerosis.

References

1. Zou, D.: The editor of Practical clinical obesity study, 1st edn., pp. 71–85. Chinese Medical Science and Technology Press, Beijing (2000)
2. Wu, H., Wu, J.-r.: The writers of exercise intervention with postmenopausal women's osteoporosis. Published in Chinese Clinical Rehabilitation (04) (2005)
3. The search on features of Mulan boxing exercises and the value of fitness [Papers] . Published in the Journal of Chinese Clinical Rehabilitation (32) (2005)
4. The influence of Aerobic exercises on women's antioxidant capacity. Published in the Journal of Shenyang Sport University (4) (2002)
5. Xie, x-j., Guo, Y-j.: The writers of the influence of Mulan boxing exercises on the old and middle-aged women in health status. Published in the Journal of Shenyang Sport University 3 (2002)

A Study of the Application of English Listening Strategies by College Students

Hong Dang

Nanyang Institute of Technology, Nanyang, Henan, China, 473004
danghong666@163.com

Abstract. Based on the previous research on learning strategies at home and abroad, this thesis will mainly focus on listening strategy use and listening strategy training in web-based self-access environment, comparing them with traditional environment. The author selected 215 non-English majors from four classes in Nanyang Institute of Technology for this stud. The findings of the present study affirm the value of strategy training. On the one hand, it shows strategy training is valid and feasible. On the other hand, it will benefit both language teachers and students in that it not only helps students become more successful to learn a foreign language but also provides teachers a meaningful way to focus their efforts.

Keywords: English learning, listening strategies, strategy training, web-based environment.

1 Introduction

Education is a lifelong process, one purpose of which is let learners meet all kinds of challenges in a changing world. As Knowles (1976) states that one of our main aims in education is helping individuals to develop the attitude that learning is a lifelong process and to acquire the skills of autonomy. Self-directed learning is very crucial because of the complexity and rapidity of change in our highly technological society. Learning strategies captured the attention of researchers and teachers worldwide. The research on " how to learn" focuses on some factors related to language learners in the process of second language learning.

Since the 1970s, the study of learner/learning strategy has been paid more and more attention to the field of English learning and teaching, and a great deal has been discovered about the learning process and learner strategies. The field of language teaching has transformed from the teacher-centered research into the learner-centered research since the last century. Some researchers believe that successful learning comes from successful teaching and understanding of learning process, learning strategies, and learner differences (Li, 2006). That's to say, learning strategy is an important factor that affects the efficiency of language learning.

According to Krashen, without input, especially comprehensible input, it is impossible to master a second language. Listening comprehension also plays an important role both in daily communication and in second language acquisition.

Y. Wang (Ed.): Education Management, Education Theory & Education Application, AISC 109, pp. 831–838.
springerlink.com

The experiments conducted by O'Malley&Chamot(1990) and Thompson &Rubin (1996)showed that listening strategy training could enhance learners' learning performance. In China, studies on listening strategy training are comparatively rare to find.

According to Dickinson (1987), learning strategies play important roles in the improvement of learner autonomy because the use or adoption of appropriate strategies allows students take responsible for their own achievements. Weinstein and Mayer (1986) also states that learning strategies take learning advancement as a learner's goal or intention. Although the results vary from their different subject groupings, different strategy classifications, and different methods, researches have repeatedly shown that the conscious, tailored use of such strategies is related to language achievement and proficiency. In addition, most studies have found that successful language learners used more learning strategies, and more advancement they got.

Despite the importance of listening, transient in nature listening has proved to be the most difficult skill to research. The research of learning strategies commenced from the beginning of 1970s while the researches of listening comprehension strategies laid behind it. Especially in China, studies on listening strategy training for students in school settings are comparatively rare to find. So the present author intends to conduct a study on the listening strategy training to test whether or not listening strategy training can enhance the learners listening performance in a typical web-based self-access setting. Since the study of listening strategies can be seen as one specific aspect of learning strategies it can be concluded that the categorization of learning strategy also applied to that of listening strategy. According to Malley and Chamot, listening strategies can also be categorized into three types of strategies: metaeognitive , cognitive , social/affective as shown in the table and each subordinates its specialties.

The study can enrich the mode of using modern technology in English listening. First of all, the development of technology influences the way of teaching and language learning via the Web. In the last ten years educational researchers have personal experienced the change from regarding computer as an accessory or source of subordinate educational material to a platform where effective teaching and learning can be promoted. On the other hand, internet environment gives learners more freedom and resources in their learning, especially for listening classes. With recent advances in the computer technology, prior studies have demonstrated that skillful listening strategy application is beneficial to second language learning, Chinese diligent scholars then cast their energy into more detailed study on listening strategies employed by college students in present web-based learning environment.

2 Research Motivation and Significance

Despite the importance of listening, transient in nature listening has proved to be the most difficult skill to research. The research of learning strategies commenced from the beginning of 1970s while the researches of listening comprehension strategies laid behind it. Especially in China, studies on listening strategy training for students in school settings are comparatively rare to find. So the present author intends to conduct a study on the listening strategy training to test whether or not listening strategy training can enhance the learners listening performance in a typical web-based self-access setting. Since the study of listening strategies can be seen as one specific aspect of learning strategies it can be concluded that the categorization of learning strategy also applied to that of listening strategy. According to Malley and Chamot, listening strategies can also be categorized into three types of strategies: metaeognitive , cognitive , social/affective as shown in the table and each subordinates its specialties.

The study can enrich the mode of using modern technology in English listening. First of all, the development of technology influences the way of teaching and language learning via the Web. In the last ten years educational researchers have personal experienced the change from regarding computer as an accessory or source of subordinate educational material to a platform where effective teaching and learning can be promoted. On the other hand, internet environment gives learners more freedom and resources in their learning, especially for listening classes. With recent advances in the computer technology, prior studies have demonstrated that skillful listening strategy application is beneficial to second language learning, Chinese diligent scholars then cast their energy into more detailed study on listening strategies employed by college students in present web-based learning environment.

The second purpose of the study is to response the new College English Teaching Reform (2007). The new college English reform requires to develop students' listening ability and their autonomous learning ability in web-based environment. Students need understand the rich computer-cultural connotations ; expand the cultural horizons and the input or output of the verbal skill. So this study is to test the abilities of acquiring information about English by using the information technology and facilities.

Third, the present situation in listening instruction in colleges and universities in China motivates this study. For most college students of non-English majors, poor listening comprehension has hindered their effective learning processes and their language learning interests. To demonstrate the essential features of the listening process, it is necessary for us to decide what factors are contributing to students' listening difficulty, teaching methods, teaching processes and learning strategies. Traditionally, the situation in college is another factor-the listening comprehension lesson is not being taught but tested. In two years of college English learning (approximately 300 periods) there are 4 periods each week for classroom teaching and only 1 period is devoted to listening instruction. Namely, there are 60 periods more or less for listening instruction in the two years' learning. But the view that listening is merely "passive" or "receptive" is challenged. To demonstrate the essential features of the listening process. it is necessary for us to decide what emphasis should be laid on. In the web-based environment, listening skill, compared with speaking and writing, is relatively easier to be self-taught. Making use of the autonomous learning classroom and encouraging students' autonomous learning is a good choice.

Finally, A large quantity of studies for pedagogically and methodologically strategies are needed in web-based self-access environment. In the favorable conditions, students now have more and better choices to the learning materials, learning schedule and learning methods. They are experiencing more fun, freedom and efficiency during the learning. Meanwhile, they are encouraged to take responsibility for their own achievement. In the field of language learning, research into learning strategies has received considerable progress in recent years. However, flexible application on the strategies used in listening in the web-based self-access environment failed to meet the hopes. Hence, there is a need for a study designed to identify the possibilities of carrying out listening in the web-based self-access environment.

3 Research Purpose and Research Question

Based on the previous research on learning strategies at home and abroad, this thesis will mainly focus on listening strategy use and listening strategy training in web-based self-access environment. Data collected from questionnaires, listening tests, learning diaries and interview records were analyzed.

The major research question is thus formulated as:

Question : What is the listening outcomes difference between students who have accepted listening strategies and those who have not when all of them were learning in web-based self-access environment?

This study will firstly choose subjects of similar background and of similar English levels and secondly give different listening strategy instructions. All efforts should be made to create an effective way of listening strategies in web-based self-access environment. Above all, it is hoped that this study will give students more efficient instructions in web-based self-access environment and help college English teachers in China design better learning strategies.

4 Research Methodologies

This comparative study possesses the characteristics of descriptive/analytical research in that it is concerned with the perceptions of respondents. The type of descriptive research is the survey method. To measure the listening strategies adopted by college students in CALL classroom investigated, three self-report means, namely, questionnaires, proficiency test and interviews are employed. The data obtained from the raw scores of the survey and tests are then put into the computer for statistical analyses with the Statistical Product and Service Solutions (SPSS Statistical Package). On the basis of these analyses, tentative interpretations, conclusions and suggestions are made.

For the sake of finding out an effective way of listening learning in web-based environment, the influences of learning strategies training among the students of different levels shall be examined. Chapter three presents an empirical experiment designed to examine the effects of applying learning strategies training to the listening instruction of non-English major students. It begins with detailed statement of the research subjects, instruments, and treatments, followed by data collection and listening strategy training.

5 Listening Strategy Training Procedure

Listening is a complex skill involving a series of different strategies operating at various levels. Listening strategies are important components of learning strategies. Listening strategies are sequential processes that one uses to control cognitive activities (e.g. understanding a listening paragraph) and to ensure that a cognitive goal (e.g. How can teachers apply listening strategies into listening activities?) The teaching model of listening course adopted by the author comes from Underwood, who divides the listening comprehension teaching into three phrases: pre-listening, while-listening post-listening (Underwood, 1989). Subjects of the present study receive listening strategy training in pre-listening activities, while-listening activities, and post-listening activities.

From the review on the literature of listening strategy training, we find many researchers confirm the effectiveness of strategy training. They believe listening strategy training do contribute to the gain in listeners' listening comprehension proficiency. Anderson's three-phase model for language comprehensive was intr oduced to them.

Question : What is the listening learning outcomes difference between students who have accepted listening strategies and those who have not when all of them were learning in web-based self-access environment?

To answer this question, firstly, Independent-sample T Test was applied to analyze pre- and post-test scores of the CG students'. The result is presented as follow.

Table 1. Comparison of the pre- and post-test scores of CG

Test	N	Mean	Std. Deviation	t	Sig (2-tailed)
Pre-test	112	68.7768	6.69235	-1.278	.203
Post-test	112	69.9821	7.41071		

After the whole semester listening learning in Self-access learning Center, the mean of the listening scores rose from 68.7768 to 69.9821 which increased 1.2052. For p>0.05 (p=0.203), there is no significant improvement in the post-test.

Secondly, pre- and post-test scores of EG students' were analyzed with the same method. The result is presented as follow:

Table 2. Comparison of the pre- and post-test scores of EG

Test	N	Mean	Std. Deviation	t	Sig (2-tailed)
Pre-test	103	68.9709	6.96272	-3.190	.002
Post-test	103	72.3010	7.98508		

As it's shown in the table, there is a significant difference between pre- and post-test scores. The Mean of post-test rose from 68.9709 to 72.3039 which increased 3.3301, meanwhile p=0.002 (P<0.05). We know that the only difference of the learning process between these two groups is the listening strategies training for the EG. That is to say, through the LST, the students' listening competency is improved significantly after the whole semester learning in WSLE.

As shown in table2, listening strategy strainin positively correlates with listening learning outcomes in WSLE. Many researchers have proved the positive correlations between listening strategy and listening learning achievements. Planning (pre-listening strategies), entering (while-listening strategies) and evaluating (post-listening strategies) are three critical steps in listening learning, however, these strategies were seldom mentioned by the teacher in high-school period. Though some students may find out the learning strategies and use them in their learning on their own initiative, their strategies use is neither integrated nor systematic. In this study, through questionnaire investigation and tests, we found listening learning outcomes significantly improved after listening strategies training. And from the results of questionnaires, we found the LST is quite helpful in students' realizing and applying listening strategies. We also learned from the interview that most students were easy to get lost in internet at the beginning of that semester in web-based learning environment. After the severe university entering examination, most students got relaxed in their first college year and made fewer efforts on their study. Students in the interview revealed that almost every student had played web-game, chatted on-line or watched movies before finishing learning tasks in SALC. This phenomenon had also drowned teachers' attention while tutoring the students' learning in SALC and the teachers had prevented them from losing in web entertainments. Even though teachers had prevented students play with computer before finishing their learning tasks in WSLC, some students in CG could hardly concentrate their attention in listening learning even in the whole semester. That caused their post-test results getting even worse. In EG, most students can adjust their learning strategies to get rid of the "lost" predicament after LST, though their efforts and outcomes were not the same. That shows that LST is completely necessary for students.

6 Listening Strategy Training Procedure

Although the results of the experiment show that listening strategy training in listening instruction indeed facilitate listening learning. There are still some limitations of the experiment.

First, the experiment is conducted for only one semester. It is not long enough to develop students' capability for well applying listening strategies to language acquisition and listening improvement.

Secondly, the single source of the subjects is 215 students from one university in Nanyang. The medium-sized sample may influence the reliability of the statistical results.

Thirdly, the instructor's training experience is somewhat limited in this study.

Moreover, a number of factors are assumed to affect the use of listening strategies. With regard to students of the same cultural background, their use of listening strategies is mainly influenced by such factors as learners' proficiency of the target language, the difficulty of language tasks, and learners' individual differences such as attitude, motivation, age, personality, gender, general learning style, aptitude, etc. Although learners' individual differences also have great effect on the use of listening strategies, some of these differences can not be changed such as age, gender as well as learning style, aptitude and personality. But on the other hand, most studies investigating the effect of motivation have found a relatively strong correlation between motivation and language learning success. So during the strategy training, teacher should try every means to motivate his students so as to ensure the smooth development of students' use of listening strategies.

The limitations in terms of training length, sample size, instructor experience, and neglect of certain factors leave a space for further researches.

7 Suggestions for Further Research

First, the experiment is conducted for only one semester. It is not long enough to develop students' capability for well applying listening strategies to language acquisition and listening improvement.

This study helps college English teachers gain some insight into listening learning and teaching, meanwhile, provides a reference for empirical studies of Chinese EFL learners performing language learning tasks with the SALC. However, since WSLE are comparatively rather new in Chinese English language teaching and learning context, there are still a lot to be further explored, attempted and proved.

First, what roles of teachers should play in students' autonomous English learning in WSLE environment deserves further thinking.

Second, the participants in this study are first year technical degree students, thus the results of the study can only be generalized to populations that share these characteristics, and it is only suggestive of other types of learners.

Third, the a combined English language learning model of WSLE only was prompted in listening, future studies should also examine the relationship of strategies to other skills: reading, speaking and writing.

Fourth, as it is impossible for autonomy to take place on its own, cooperation should be encouraged. It is advisable to have students share successful learning experiences with their peers.

Lastly, regular evaluation should be incorporated into the program, since a combination of management and evaluation will shorten students' way to learner autonomy.

References

1. Anderson, J.R.: The Architecture of Cognition. Harvard University Press, Cambridge (1983)
2. Anderson, N.J.: L2 Learning Strategies. In: Hinkel, E. (ed.) Handbook of Research in Second Language Teaching and Learning. Lawrence Erbaum Associates, Mahwah (2005)

3. Bacon, S.M.: The relationship between gender, comprehension, processing strategies, and cognitive and affective response in foreign language listening. Modern Language Journal 76, 160–178 (1992)
4. Brown, G.: Dimensions of Difficulty in Listening Comprehension. In: Mendslshon, D., Rubin, J. (eds.) A Guide for the Teaching of Second Language Listening. Dominie Press, San Diego (1995)
5. Cohen, A.D.: Strategies in Learning and Using a Second Language. Foreign Language Teaching and Research Press, Beijing (2000)
6. Bialystok, E.: The compatibility of teaching and learning strategies. Applied Linguistics (1985)
7. Brown, G.: Dimensions of Difficulty in Listening Comprehension. In: Mendslshon, D., Rubin, J. (eds.) A Guide for the Teaching of Second Language Listening. Dominie Press, San Diego (1995)

Practice and Reflection of Computer-Aided Chemical Analysis Experiment

Li Liu, Jianzhong Guo, Bing Li, and Yanlong Feng

Department of Chemistry, School of Science, Zhejiang A & F University,
Lin'an, 311300, Zhejiang, China

Abstract. By applying modern multimedia teaching technology to chemistry experiment, it can improve the teaching quality of chemistry experiment. Combining with the practice, Effects of multimedia virtual in experiment was studied in this paper and application of Excel software makes data processing more convenient and accurate.

Keywords: Computer aided, Teaching practice, Chemical experiment.

1   Introduction

Chemical is a discipline based on experiment, the chemical experiment teaching is a powerful tool to deepen students to understand the theory and cultivate students' scientific research ability, innovation ability and actual ability; in order to cultivate talents, you must strengthen and pay attention to the experimental teaching. To make better use of modern education technology is one of the paths improving the effect of chemistry experiment teaching. With the popularization of the network and the computer and the development of computer technology, a new kind of modern teaching mode namely computer aided teaching in is extensively applied to chemistry experiment teaching[1]. Our school also development some exploration on computer aided teaching in chemistry experiment course and achieved some preliminary results.

2   Making Use of Multimedia Improve the Basic Operation Skills

Chemical analysis experiment demands students to master normative operation, but only depending on demonstration of teachers, it is difficult to make all students see clear. In order to make students master normative operation, the teacher must observe carefully each student's operation and guide patiently improper operation classmates, however, because of the limited class; it is difficult to guarantee the experiment teaching effect. After introducing the multimedia teaching, the flash courseware with living audio animation demonstrate basic operation process, such as the operation of dropper bottle, test tubes, analytical balance and so on, Flash courseware can not only vividly display of continuous operation process, but also can show students the decomposition , this makes the students obtain deeply perceptual knowledge[2].

Y. Wang (Ed.): Education Management, Education Theory & Education Application, AISC 109, pp. 839–842.
springerlink.com   © Springer-Verlag Berlin Heidelberg 2011

3   Open Virtual Experiment, Improve Experiment Efficiency

In recent years, the emerging experimental technology of virtual experiment on the computer has been applied widely[3]. Virtual experiment is a simulation experiment using computer-assisted design software or simulation software, it can realistically reflect the reality world corresponding lab environment, relevant equipment, experimental object and experimental information resource and do experiment applied virtual components and virtual instrument; it is the same as the traditional experiment to watch the experiment phenomenon and collect experimental data. Virtual experiment can improve perceptual knowledge to test equipment, so it can make students to master the usage of experiment equipment ; in virtual experiments, equipment, experimental materials are virtual, experimental don't consume chemicals, and thus it can save costs, reduce waste water discharge capacity of chemical laboratory and improve the efficiency.

Our school chemical experiment teaching has attempted to open virtual experiment of micrometric analysis to replace part of reality experiment, using the introduction of chemical virtual laboratory of Institute of Technology of Dalian. Students can get into virtual chemistry lab though computer , choose the pilot project prescribed to do operations, i.e., we open titration analysis virtual experiments of the 3 h , students can complete the comparative titration of HCl and NaOH , the determination of mixture of NaOH and Na_2CO_3 and analysis experiment of alkali ash samples according to the requirement. Computer can display the operation process of demonstration analysis by simulating operation, making them directly master relevant knowledge and experimental operation skill, after experimental operation is completed, virtual experiment system will calculate and process student experimental data , provide experimental report and evaluate the simulation experiment ; We also require students to record of experimental data and give lab report referring to computer, reality experimental reports of simulating data are completed, virtual experiment and real experiment are organic combined, it has received well teaching effect. After completing virtual experiment of titration analysis, the traditional acid-base titration experiments are arranged , because of the existing intuitively impression, experiments and experimental reports are smoothly finished when entering the traditional laboratory .In this way standardization degree in experimental operation can be raised, so that we can correctly judge titration end and properly record data, reducing the incorrect operation ability, reducing labor intensity of the experimental teachers and improving the quality of teaching.

4   Processing Experimental Data

In Chemical experiment course, the experimental operation is completed, most of the experimental data need dealing, at last the experiment report is handed in, and it means finishing the test. When spectrophotometer is used in chemical analysis, the work curve (also called regression equations) need drawing, according to the working curve check or curve regression equation, the content components can be measured, but because of larger resultant error, linear equation by manual method is more complex, application to Excel can conveniently obtain linear equations and related coefficient. Correlation coefficient is an important standard of evaluating the quality of linear equation, thus using correlation coefficient can evaluate the quality of the experimental results from one aspect.

In adjacent 2 n determination of iron content, we should try to apply Excel. According to the experimental steps, the concentration of standard solution are 0.0 μg/mL, 0.4μg/mL, 0.8μg/mL, 1.2 μg/mL, 1.6 μg/mL, 2.0 μg/mL respectively, in the ordinary type 722 visible spectrophotometer, with 1 cm color cup, absorbance (A) are determined under 510 nm wavelength. The absorbance (A) is as follows: 0.00, 0.156, 0.278, 0.398, 0.497, and 0.628. First, concentration and absorbency are put into Excel, at the same time table data in the Excel table are selected ,then choose "insert" menu commands of the "chart" or click on the toolbar "charts wizard" , select diagram type for "XY scatterplot chart table type" and the sublist for "a scatterplot chart", click "next" tip in turn, finally click "finish". then Excel table will appear a scatterplot chart, select each bit of the scatterplot chart , right click the mouse choosing menu "trend line format", Excel table will appear "add a trend line" dialog box, select the options menu, select the "show" and "show formula R square value", as shown in figure 1, then click "OK", then the linear equation of work curve and the correlation coefficient are automatic generation. in Excel table, $y=0.3059x + 0.020$, $R^2 = 0.996$, y stands for absorbency A, and x for iron solution concentration C. For example, under the same conditions, the determination of liquid absorbency is 0.256, 0.256 may be put into the linear equation, the liquid iron solution concentration is 0.773 μg/mL, R = 0.998, this explains that the experimental data has good linear relationship.

Using Excel processing data can make the complex work simple, calculation process become simple and accurate, it is great convenient to analyze the quantitative analysis work, so the students' computer application ability can be consolidated and strengthened.

5   The Problems and Thinking in Computer Aided Teaching

It is an attempt in teaching that we launch computer-aided instruction in the teaching of chemical analysis experiment. Flash courseware and virtual experiment are very directive and simple to master, promoting students skill of comprehensive experiment. We find there are still some problems in the teaching of computer-aided instruction by the teaching practice. For example, it is a big amount of work to dissolve in the advanced teaching thinking's and teaching methods while making the multimedia courseware; some importing multimedia courseware can not be matched with the local school totally; some students take no account of the CAI, when they operate the courseware they think it just like playing games. There is no good effect. As a result, it is need to strengthen the guidance and supervision while developing the teaching of computer-aided instruction and make the student more positive and participate in it.

We must recognize the auxiliary status of multimedia teaching and regard it as a teaching tool, but it can not replace the traditional experiment teaching. Some basic operations can only be perfectly mastered through step by step personal practice. So only combine the traditional experiment teaching with the virtual authentic experiment organically, and then it can reach a better teaching effect.

Nowadays, the developed chemical multimedia software has a strong function. It both can be studied by one's own in the school net, using for students' preview and students review after experiment. Also we can proceed the diverse experiment assesses, taking a flexible teaching method for teachers. Carrying out the individual teaching also provides qualifications for the combination of network assess and practical assess, but the requirements for teachers are higher. We still need to learn and practice and grope to master the functions and skills of multimedia software for better optimizing chemical analysis experiment teaching and cultivating the high-quality talents.

References

1. Du, Y.N.: The Current Research State of the Virtual Experiment and Its Significance in Teaching. Journal of Zhejiang Ocean University (Natural Science) 29, 390–393 (2010)
2. Yang, Y.C., Lian, P., Hu, Q.S., Huang, Q.: Researching and Considering on Chemistry Experiment Teaching by Multimedia CAI. Journal of Gannan Teachers College, 120–123 (2006)
3. Li, S.Y., Liu, H., Zhou, K.L., Xiao, F.Y., Huang, C.Z.: Complementarities of electrician-electron virtual experiment and real experiment. Experimental Technology and Management 27, 74–76 (2010)

Research on the Design of Function Module of Petroleum Engineering Practice Base

Fengxia Li[*], Wenhua Li, Wang Li, and Zhengku Wang

ChongQing University of Science and Technology, Petroleum Engineering Editorial,
401331 Chongqing, China
{lfx924,wzk0408}@126.com, {lwh2047,023wangli}@163.com

Abstract. Under the new situation of higher education development and reform in our country, deepening the campus practice teaching base construction is the important direction of institutions of higher learning to improve managerial condition, the important guarantee to improve the teaching quality. Petroleum engineering practice base can achieve professional theory study, engineering technology application, the scientific research. According to the scientific development plan, the construction module design of petroleum engineering practice base includes five modules: well engineering module, the crude oil picking and transmission engineering module, natural gas transmission engineering module, indoor simulation practice teaching function module, tools display module. Through the petroleum engineering practice base function module design, it makes the idea, mode of management in practice teaching run on a new breakthrough and forms a batch of high level and higher the practice of influence teaching research results, brings up with "engineer basic skill training senior applied talents", providing better service to the society.

Keywords: Petroleum engineering, Practice teaching Function module, design.

1 Introduction

Under the new situation of higher education development and reform in our country, deepening the campus practice teaching base construction is the important direction of institutions of higher learning to improve managerial condition, the important guarantee to improve the teaching quality[1]. Meanwhile, modern petroleum industry has the characteristics of technology intensive, capital intensive and human resource intensive, in addition, as the new technologies of well drilling such as horizontal wells ,small diameter wells and the automation equipment, automatic rig come into use in succession, higher requests are put forward for the oil engineering talents. There is a great need of senior technical talents in the front line of petroleum industry. The research of talent demand in the oil field shows that the main positions of oil drilling industry is oil drilling, drilling machine, mud worker and oil well maintenance

---

[*] Fengxia Li born in 1977 is a lecturer of the department of petroleum engineering in ChongQing University of Science and Technology.

positions. Among them the drillers take up 45%, rig workers account for 10%, mud workers and well maintenance workers take up 35%.The practice teaching base of petroleum engineering is planned to be a comprehensive skills training base including drilling engineering, drilling machinery, drilling and completion fluids, drilling instrumentation.

It can be used to complete the work of campus practice teaching of drilling technology professional and professional group of students, training high skilled talents in the front line of petroleum industry. According to the scientific development plan, the construction module design of petroleum engineering practice base includes five modules: well engineering module, the crude oil picking and transmission engineering module, natural gas transmission engineering module, indoor simulation practice teaching function module, tools display module.

2   The Well Control Engineering Module

According to the oil drilling and work-over industry characteristics and enterprise investigation, and after analyzing the oil drilling technology professional work, we make out well engineering practice teaching projects. They are listed in table 1.

Table 1. Well engineering practice teaching projects

Learning content	project	details
hole structure	Project 1:the type of casing Project 2: hole structure design	the type of casing; the method of hole structure design; essential data's acquisition and calculation
The casing pipe standard and the casing string intensity design	Project 3: the casing pipe standard Project 4: case operation Project 5: casing string intensity design	API standard of casing pip; instruction of casing, casing string stress analysis and destruction form; casing string intensity design
oil-well cement	Project 6: classification of oil-well cement and its function Project 7: cement mortar performance adjustment Project 8: computation of the amount	classification of oil-well cement and its characteristic, cement mortar performance determination and adjustment operation, computation of cementing

Table 1. *(continued)*

well cementation	Project 9: well cementation equipment and tool Project 10:cementing Project 11: the measure of improving the quality of well cementation Project 12: special well cementation technology	The equipments of cementing; casing string appendix; procedure and request of cementing; the factor which can influence the quality of well cementing; the operation of special well cementation technology
Well completion	Project 13: well completion method and well completion process Project 14: well head assembly of well completion	commonly used well completion method and well completion process; well head assembly's composition and installment
testing a newly drilled oil well	Project 15: perforation process Project 16: induction oil air current Project 17: the process of testing a newly drilled oil well	perforation process; the technical process of testing a newly drilled oil well, the method and operation of trying oil test tools, instrument's use

Such functions as practice teaching of natural flowing and machine pulling; practice teaching of water-injection system, including the sewage treatment and injecting back; practice teaching of down-hole operation ;the training of well test. Eight broad headings are included in detail. They are shown in Table 2.

Table 2. The list of the projects of crude oil's picking and transmission practice teaching

No.	functional description	Majors which are related
1	Oil well operation, management, maintenance of pumps, troubleshooting of pumps and mechanical device	oil production engineering, mining machinery, automation and control
2	Maintaining and troubleshooting of electrical equipments	oil production engineering, integrated mechanism and eclecticism
3	process of removing and controlling wax	oil production engineering, chemical engineering

Table 2. *(continued)*

4	production test., capacity analysis	oil production engineering, reservoir engineering
5	mechanical recovery, flowing production	oil production engineering, mining machinery, automation and control
6	water-injection development, formation testing	oil production engineering, chemical engineering, automation and control
7	)lication of new oil production technology	oil production engineering, chemical engineering, automation and control
8	management and operation of digital oilfield	oil production engineering, automation and control

3   Natural Gas Transmission Engineering Module

The system constitutes by the water jacket furnace reducing expenses to keep warm, the gas fluid separator, the compressor and the booster pump, dehydration equipment, the measurement, the clear tube and the pipe network of collection and transmission. This system has such functions as practice teaching of gas dehydration, the training of gas transmission, the training of automatic controlling instruments and pipe maintainance. They are listed in detail in table 3.

Table 3. The list of the projects of natural gas transmission practice teaching

No	functional description	Majors which are related
1	gas well operation , management, maintenance of mechanical device, troubleshooting of mechanical device	gas production engineering, mining machinery, automation and control
2	Maintaining and troubleshooting of electrical equipments	gas production engineering, mining machinery, integrated mechanism and eclecticism
3	drain and gas production in gas well	gas production engineering, chemical engineering, mining machinery
4	monitoring and protection against hydrogen sulfide gas in gas well	gas production engineering, HSE risk management, chemical engineering
5	natural gas Processing	gas production engineering, chemical engineering, automation and control
6	well testing method, reservoir protection	gas production engineering, chemical engineering
7	management and operation of digital oilfield	gas production engineering, automation and control

4   Indoor Simulation Practice Teaching Function Module

The main purposes of this module are improving the close contact between the indoor practice and field training, enhancing the student's ability to operate the experimental equipment and engineering equipment on the job, improving the students' familiar with the working standard, feeling the field work atmosphere, improving students' ability to treat unforeseen events, improving the students' ability of scientific research and creative ability and cultivating innovative talents of oil science and technology[2]. Indoor simulation training function module has many functions as follows: (1) system of drilling equipment training platform; (2) system of logging device training platform; (3) the acidification fracturing operation training platform system; (4) the cementing operation training platform system; (5)workover operations training system.

4.1   Drilling Equipment Training Platform System

Drilling engineering simulation equipment training room, based on the actual drilling site function layout to decorate, mainly contains five parts: drilling equipment teaching demonstration platform, well control equipment teaching demonstration platform, solid control equipment teaching demonstration platform, top drive drilling simulation platform and multimedia teaching demonstration teaching demonstration, simulation platform.

Drilling equipment teaching demonstration platform is mainly used for improving students' cognition drilling equipment at drilling site, and practice demonstration of layout, function and principle. Well control equipment teaching demonstration platform is mainly used for improving students' cognition of well control equipment at drilling site, and practice demonstration layout, function and principle. Solid control equipment teaching demonstration platform is mainly used for improving students' cognition of solid control equipment at drilling site, and practice demonstration layout, function and principle. Top drive drilling simulation teaching demonstration platform simulation drilling site layout, including each function accessories disassembling and function model of rig. It can undertake the students' drilling training, strengthen the students to oil drilling machinery engineering practice skills. The practice can familiarize students with the drilling engineering equipment and the combination of function and the layout, operation and process planning of drilling machine assembly. Multimedia teaching demonstration, simulation platform distinct, rich 3 d animation demo software and teaching courseware drilling, all these can strengthen teaching effect on students and improve the learning interest of them .

4.2   System of Logging Device Training Platform

Logging operations training room in accordance with the requirement of drilling field geological logging work and its layout, is used to enhance students' understanding of the equipment for logging, and grasp the method of working principle, installation, remove, maintenance and operation of sensor and comprehensive logging equipment, the common problems in the process of learning the logging of the treatment. These can ensure the accuracy of the logging data in the goal.

4.3 The Acidification Fracturing Operation Training Platform System

Acidification fracturing simulation teaching room is according to the fracturing, acidification construction work process fracturing equipment, underground fracturing tool selection, combination, construction technology and safety norms on the scene. Really teaches through the acidified compression fracture's system, enhances the down-hole operation constructors' understanding to the oil (gas) the compression fracture, the acidified construction instruction manual, in the operation skill and the frequently asked questions in the construction process and measurement.

Using simulation and automation control technology based on the actual equipment, acidification fracturing tools such prototype operation combination, design and develop the actual operation of the acidification fracturing operation simulation training teaching platform. The platform has the following functions for operating training: (1)the sand mixing car simulation operation training; (2) instrument car simulation operation training; (3) the acidification fracturing string tool model disassembling combination training.

4.4 The Cementing Operation Training Platform System

Cementing operations training rooms is according to simulating the cementing operation cementing equipment, cementing tool selection, combination construction technology and some requirements on the scene. It mainly contains three teaching function blocks: the cementing car to be automatic mix slurry operation simulation platform, the multimedia simulation control system and the practice teaching and teaching evaluation system. One car to be automatic mix slurry cementing operation simulation platform is to use computer, automation control technology based on the actual equipment such prototype operation combination, cementing project design and develop the actual operation of the simulation training teaching platform. The multimedia simulation teaching control system is by using modern computer 3 d animation technology and automation control technology design and become interactive cementing training teaching platform.

4.5 Work-Over Operations Training System

According to the Chinese petroleum corporation , personnel service center "the occupational skills training course and appraisal Shi Ti Ji" and related to the oil industry standard production of the rig operations practice teaching system, through the demo, interpretation, such as the operation way so that students master the basic operation of the repairing service skills and relevant practice knowledge. Work-over operations practice teaching system contains physical simulation model and operation training system, teaching software, teaching of wall form etc. Work-over operations training platform specific include the following aspects of content: the work-over comprehensive platform, down-hole tool disassembling model, accessories, wax removal, the blasting, examining the pumps, the fault processing, casing repair, casing wrongly cut processing, drilling cement plug, open the window sidetrack, stuck drill processing, rod salvage falling objects. Work-over operations training platform specific include the following aspects of content: the work-over comprehensive platform, down-hole tool disassembling model, accessories, wax removal, the blasting, examining the pumps, the fault processing, casing repair, casing wrongly cut processing, drilling cement plug, open the window sidetrack, stuck drill processing, rod salvage falling objects.

5 Tools Display Module

It mainly includes underground work tools (table 4) and drilling operation tools (table 5)

Table 4. The list of underground work tools

name	species	name	species
Inner fisher	5	outer fisher	11
magnetic fishing	2	milling tools	6
cutting thread tools	2	cutting tools	2
shock tools	4	shaping Tools	2
safety joint	3	meeting-for tools	2
Drifting and free point tools	2	well-flushing tools	1
drilling equipment maintenance tools	3	sidetrack tools	6
reversing tools	1		

Table 5. The list of drilling and production operation tools

name	species	name	species
the inner bop tools	3	coring tools	3
down hole tools	8	tools for directional well	14
accident treatment tools	18	recovery operation tools	13

6 Conclusion

(1)Base construction can absorb areas involved the latest research results and the latest technology progress, design concept, advanced management mode, and has the capacity to upgrade and function expansion, also its equipment selection can realize automation.

(2)It can completely finished petroleum engineering and related professional practice teaching task and skills ability training; it can be used to widely develop skills training ,train senior applied talents according to the skills required to post group.

(3)It can implement the resource sharing and establish practical teaching platform in practical teaching base and build the scientific research group joint platform for petroleum chemical engineering, mine machinery, mechanical and electrical integration.

Acknowledgment. This work is supported by the reform project education teaching of Chongqing named "study on constructing the practical teaching base of petroleum engineering" (09-2-091).

References

1. Feng, L., Wan, L.X.: Exploration and practice of construction of the inner productive training base based on the school-enterprise linkage pattern. Experimental Technology and Management 27(10), 199–201 (2010)
2. Yan, Z.X., Jing, L., Feng, G.: On Enhancing the Construction of College Practical Teaching Base. Journal of Liaoning Medical University (Social Science Edition) 8(4), 81–83 (2010)

Research on Constructing the Practical Teaching Base of Petroleum Engineering

Fengxia Li[*], Wenhua Li, Wang Li, and Zhengku Wang

ChongQing University of Science and Technology, Petroleum Engineering Editorial,
401331 Chongqing, China
{lfx924,wzk0408}@126.com, {lwh2047,023wangli}@163.com

Abstract. The construction of practical teaching base is an important measure to strengthen practical teaching, and it is also an important content to improve the quality of education. This paper studied the support conditions as well as the necessity and the goal of the construction of practical teaching base, leading the practical teaching base to be a shared platform for teaching practice, which contents the subjects of exploration, petroleum engineering, petroleum storage and transportation, safety engineering, petroleum exploration, petroleum processing and mechanical control. The main function of practical teaching base is for education (teaching and training for undergraduates and junior college students), research and development of applied technology, engineering training and the display of petroleum enterprise culture, and the way to carry out the construction is to combine the school inputs, site allocation and teacher exploitation.

Keywords: Petroleum engineering, practical teaching, base construction.

1 Introduction

The higher education of our country is facing an important issue that how to improve education quality of teaching[1]. The construction of practical teaching base is an important measure to strengthen practical teaching, and it is also an important content to improve the quality of education. Many problems of the higher colleges such as realizing the important role of practical teaching base in cultivating talents, effectively grasping the quality of construction of practical teaching base and promoting the innovation of management system must face and strive to be solved. The practical teaching base of petroleum engineering can improve the level of training the talent of high technical ability and adapt to the development of enterprises. It is for the need of training field engineering and technical personnel, it's also the fundamental guarantee of speeding up to train the needed high quality and skills specialized personnel who work and serve in the first line of oil fields.

---

[*] Fengxia Li born in 1977 is a lecturer of the department of petroleum engineering in ChongQing University of Science and Technology.

Y. Wang (Ed.): Education Management, Education Theory & Education Application, AISC 109, pp. 851–856.
springerlink.com © Springer-Verlag Berlin Heidelberg 2011

2   The Support Conditions of Practical Teaching Base Construction

"CNPC" is the abbreviation for "China National Petroleum Corporation". CNPC is a large state-owned enterprise mainly based on oil and gas exploration and development, as well as one of the global top 500 companies in the world. At present, it is dedicated to building into an international competitive oil company. The petroleum engineering and safety technology research center of Chongqing University of Science and Technology was assessed as engineering research center of universities in Chongqing in 2009. The drilling engineering discipline was evaluated as key construction project of Chongqing in "eleventh five-year plan" in 2008. Our college has developed cooperation with the Great Wall drilling company of CNPC to bid for "the international well control training center of Chongqing IADC". IADC is identified by international drilling contractor authentication. "HSE training center" is identified by CNPC, as well as the level 2 safety production training institutions and first rate safety evaluation institution. "onshore petroleum production industry and pipeline transportation industry" is identified by the national production safety supervision and administration. Our college develops close cooperation with subordinate enterprises of CNPC, especially in well control technology research and well control accompany with HSE training, it has formed stronger advantages and distinct characteristics in petroleum universities. In this college, "Well control and HSE training center" has already been absorbed into HSE training and development plan with CNPC. These conditions have created a good platform for the constructing the practical teaching base of petroleum engineering. Constructing the practical teaching base of petroleum engineering also fits the requirements and needs of oil companies and school development.

2.1   The Needs for Training Applied Talents

(1) The needs for the background of international engineering education
With economic globalization and rapid development of science and technology, the contradiction of talent market structure has become a common problem of countries all over the world. Every country wants to improve their core competitiveness of nation. American government's specially stressed in "the state of the union address in 2006". American leading global technology innovation ability is only half of the country's overall competitive power, the other half is the labor structure market which receives good education, but the future labor of America is difficult to support the core competitiveness of the country.

In Us and European developed countries, in order to solve the contradiction of the structure of talents and expand higher engineering education reform, the American academy launched "2020 engineers plan" and "the deployment of the new century education project action" supported by American government. Germany launched "Global engineering education excellence program" and "training the next generation of excellent international engineer". The basic ideas are to reform project and education mode of talent training through strengthening the engineering practice and adding into new technology.

(2) The needs for domestic petroleum universities' attention to engineering environment construction

A key point of higher education development strategy in our country is to reform the personnel training mode. For constructing "211 universities" and "985 universities", it has already put "the innovation and practice research of Chinese higher education application talents cultivation system in 21th century" into the list of the major national education scientific research project, so as to guide universities to attach the importance of training applied talents[2].

Technology progress and innovation is the key to increase profits in the petroleum and petrochemical industry. New technology and advanced equipment are widely used in petroleum exploration and development, petroleum refining industry, petroleum storage and transportation, equipment maintenance and oil field information, which makes constant improvement of each professional post in petroleum and petrochemical industry, and speeds up the need for skillful talents.

At present, Southwest Petroleum University has completed the construction of practical teaching base of petroleum engineering, and made a good contribution to the undergraduate teaching evaluation. Yangtze University and Daqing Petroleum University have also started the work of the practical base construction of petroleum engineering.

(3) The need for the reform of talent training mode of our school

Chongqing University of Science and Technology treats talents training target as "training to develop advanced senior application-oriented talents who obtain basic training". Petroleum engineering department treats the training target of petroleum engineering and petroleum storage and transportation engineering and resources exploration professional as "production field engineer in petroleum engineering". All of these need better engineering and industrial environment as support.

2.2 Formulas the Needs for High-End Vocational Skills Training

(1) Requirements of well control technology training in domestic market

According to the current requirements of Chinese oil well control center training management, only establishing the corresponding training base and well control technology training center can get the grade one certificate of well control technical training and develop the well control training market occupied by our school.

(2) The needs for the development of international well control business

Constructing practical teaching base can further strengthen the IADC training level of our department and lay the foundation for applying for IWCF qualification.

(3)The needs for expanding the training field and enhancing the contact between oilfields and enterprises

Constructing practical teaching base has provided a better platform for all kinds of training tasks, taking it as an opportunity. It will absolutely get a new breakthrough in expanding training field and high-level training. And it puts the training business into oil and gas exploration technology, natural gas processing, HSE safety management, oil and gas transportation technology, etc. At the same time, it will continue to improve the development of order training mode, so as to further strengthen the contact and communication between school and enterprises in order to make the enterprises better understand the school, and then get more chances of deep cooperation.

2.3 The Need for Building Interdisciplinary Groups

Our academy has the key subject of Chongqing—drilling engineering, as well as the key subject of Chongqing University of Science and Technology — oil and gas field exploitation. In 2008, the petroleum engineering was approved as a national specialty construction point, which realized the breakthrough of national level project of the quality engineering of our school. Therefore, it is high time that we constructed the practical teaching base of petroleum engineering to make it as a "municipal key discipline" and "national characteristics professional" technology platform.

The construction of practical teaching base can better promote the depth and breadth of research cooperation, and lead a number of levels of scientific research.

By innovating the construction and management of the practical teaching base of petroleum engineering, we can lead a number of practice teaching reform, research projects and achievements. Meanwhile, the construction of practical teaching base can also lead the construction of machinery, chemistry, safety, automation and other subjects to a professional direction.

2.4 The Need for Building Campus Culture

The school characteristics of Chongqing University of Science and Technology is that it has a strong industry background, the construction of petroleum engineering and related interdisciplinary groups (such as mining machinery, oil security, oil field chemicals, etc.) requires the students to have some background knowledge of industry. Therefore, the construction of practical teaching base can absolutely act as petroleum enterprise culture and oil platform to display science education so as to provide basic security for engineering training, industry culture and the spirit of education of our school. It will certainly become a landmark achievement of school characteristics of Chongqing University of Science and Technology and the education base of oil popular science and even an important part of cultural tourism of Chongqing University City.

3 General Goals and Function Localization to Construct the Practical Teaching Base

3.1 General Goals

Through the construction of practical teaching base of petroleum engineering and the innovation of management, we can drive the practical teaching management to realize new breakthroughs and new leap forward on the concepts, models, and running. And to form a group of practical teaching and research achievements with a higher level and higher influence, so that we can build the base to be an excellent project of practical teaching in our school and even in Chongqing. The overall design sketch (Concept design) is showed in picture 1.

Fig. 1. The overall design sketch of the practical teaching base of petroleum engineering

3.2 Function Localization

(1) A platform of the function for practical teaching of 8-13 the specialist professional, a platform for students engineering training;
(2) A platform for training and authentication of high-end application and technology-oriented talents;
(3) A platform for research and development of applied technology;
(4) A platform that can show oil industry culture and oil popular science.

4 Conclusion

Through the analysis of the support conditions as well as the necessity and the goal of the construction of practical teaching base of petroleum engineering. It follows the rule of higher education reform and development, to contribute to the construction of group of disciplines and to meet the requirements of practical teaching and ability training. And it innovates the construction and management of the practical teaching base and actively develops the cooperation between schools and enterprises, as well as between departments. We need to carry out the construction of Trinity, which means professional theory study, engineering applications and scientific research. By regarding platform construction as breakthrough, to promote the building efforts of practical teaching and the pace of reform so as to enhance the quality and level of practical teaching and to create high-level application-oriented talents which are of "engineers basic skills training".

Acknowledgment. This work is supported by the reform project education teaching of Chongqing named "study on constructing the practical teaching base of petroleum engineering" (09-2-091).

References

1. National ministry of education, the national long-term education reform and development plan outline (2010-2020). The Xinhua News Agency 9, 29 (2010)
2. Li, Z., Yang, L.: The construction of regional sharing- national training base, the research of management and operation mechanism. Education and Profession (3) (2006)

Study on Constructing the Practice Teaching Base of Petroleum Engineering

Zhengku Wang[*], Wenhua Li, Wang Li, and Fengxia Li

ChongQing University of Science and Technology, Petroleum Engineering Editorial,
401331 Chongqing, China
{wzk0408,lfx924}@126.com, {lwh2047,023wangli}@163.com

Abstract. The practice teaching base in schools whose construction is close to the production front line of oil field production, is one important condition in training highly-skilled talents with the application ability in the petroleum engineering specialty. Combining the requirement which is the construction of the petroleum engineering specialty with the practice teaching base, this paper summarizes the construction objective of the practice teaching base in petroleum engineering, puts forward the ideas and principles of construction of the practice teaching base, analyses the functions which the practice teaching base of petroleum engineering should meet and discusses the construction pattern of the practice teaching base.

Keywords: Petroleum engineering, practical teaching, construction pattern.

1 Objectives of the Practice Teaching of Petroleum Engineering

Through the construction and innovating management style of the practice teaching base of oil and petroleum engineering, we can make the management of practice teaching achieve new breakthrough and span in ideas, patterns and running, form a batch of studies with a relatively high level and influence of the practice teaching and build the base into an excellent project of our school or even Chongqing.

(1)Highlight the Humanistic Management, make the project set help to the trainers grasping the site work skill, strengthen the ability of the field application innovations, develop the students' attitude of dedication and strengthen the work ethics of the pursuit of perfection and the spirit of perseverance, taking a brave approach facing the difficulty and going up with vitality.

(2) The base's core part is the training of the mainstream technologies and the application ability and the base's objective is to meet the requests of experiments, practice and scientific research. Fully consider the actual needs of enterprises to the talents, adapt to the base ahead of construction and make the practice teaching base into the base which is a place for oil and security industry training higher technology applied special talents.

---

[*] Zhengku Wang born in 1979 is the experimentalist of the department of petroleum engineering in ChongQing University of Science and Technology.

Y. Wang (Ed.): Education Management, Education Theory & Education Application, AISC 109, pp. 857–863.
springerlink.com

(3) Highlight the application of new and high technologies and let the students contact and use the new and high-tech achievements reflecting related technical field in the practice teaching base.

(4) Construct some Quality Laboratory in our school which is one emphasis, create conditions to introduce some enterprises with certain scale which construct training bases in our school and make efforts to construct several training rooms with real business scenarios.

(5) Break intrinsic management and running system to realize the aims which include uniform plan and management, resource sharing, joint construction, widespread use, serving businesses and improving educational efficiency. The location of the practical teaching base is shown in Fig. 1.

Fig. 1. The overall design sketch of location of the practical teaching base

2   The Construction Mode of the Petroleum Engineering Practical Teaching Base

According to the scientific development plan, the construction module design of petroleum engineering practice base includes five modules: well engineering module, the crude oil picking and transmission engineering module, natural gas transmission engineering module, indoor simulation practice teaching function module, tools display module.During the construction, for sake of the close relation between the petroleum engineering major and enterprises, our college employer the construction means of self-investment and enterprise donation, absorbing and exploiting enterprise advanced technology, equipment, and saving money for education, building corresponding environment for staff training and strengthening the exchanges and cooperation between colleges and enterprises.

Petroleum engineering practice base construction is expected to reach the total investment of 50 million, 20 million from college investment, while the equipment worth 30 million can be allocated and donated by oil enterprises. Invite oil firms experts and personnel to involve in the construction of the oil practical teaching base, and take it as an opportunity to invite the petroleum enterprise experts and leaders of petroleum engineering to guide and participate in the talent training scheme demonstration, teaching plan and course setting, to join in all the petroleum engineering teaching activities, such as enterprise send professional personnel to the college lecturing, providing employment information and helping graduates get jobs, to participate in professional construction, providing funds for the school, equipment and material supports, building the training room, involving in the school "double teachers" team construction, etc. Learning can solve related technical problems through the practical teaching base and scientific research strength, realizing transformation of scientific and technological achievements as soon as possible. Through the practical teaching base construction between the college and enterprises, we could hear in time the policy trend related to the major and specific talents requirements, and at the same time strengthen the contact between teachers, students and enterprises, enhancing the students' adaptability, understanding the enterprises and employment abilities. The college and the enterprise will work out a win-win situation through the teaching practical base corporative construction. Besides, the construction possesses the advantages of relatively low cost and strong sense of reality.

3   The Guiding Thoughts and Principles of the Construction Practicing Teaching Base

3.1   Guiding Thoughts of Construction of the Practice Teaching Base

With the scientific development view as a guide, follow the education reform and the development law of colleges and universities, serve the construction of subjects and majors group and meet the requests of the practice teaching and ability cultivation; realize the innovation of the construction of the practice teaching base and management patterns, expand the cooperations between schools and businesses and between colleges in a positive attitude and achieve integration construction among the study of major theories, the application of engineering techniques and scientific research; with the most important way of setting up the platform construction, promote the construction intensity of the practice teaching and the pace of reforming to improve the quality and level of the practice teaching in all aspects, and make highly applied talents with Basic Skilled Training of Engineers.

3.2   The General Idea of the Practice Teaching Construction

The main functions of the practice teaching base include cultivating talents that include the training of undergraduates and junior college students, developing applied techniques, engineering technique training and the showing of culture of the petroleum enterprises and it is a sharing practice teaching platform for the subjects including Resources explorations, Petroleum engineering, Oil and gas storage and transportation, Security engineering, Oil and gas exploration, Oil and gas processing and Mechanical control. Construction begins in a way that the school's spending, field guidance and the teachers' development are combined . And the construction of hardware and software starts at the same time, introducing HSE culture ideas of petroleum industry and building relatively independent culture of petroleum industry. Build and perfect the management system of the practice teaching base which includes technique management, equipment management, practice management, fixed assets management and so on. And perfect the management system including practice teaching files, operation mechanism, manners of experiments and training, assessment method and the quality of the practice teaching and make the practice teaching base into the base that has the reasonable personnel structure, clear job responsibilities, orderly operation standards and complete evaluation methods.

3.3 The Construction Principle of the Practicing Teaching Base

The construction of the practicing teaching base should not just accommodate the development of the industry and readjust the industrial set-up, but also meet the construction of the special subjects and the requirements of practicing teaching. It focuses on cultivating the technology applied ability and the basic professional quality, and takes the industrial occupation for standard, and sticks to the cultivating approaches[1] of combining produce, teaching and study. For these reasons, in the construction of the petroleum engineering practice teaching base, we should grasp principles as follows:

(1) Advancement principles: the construction of the base can absorb the newest research achievement and technological development, make the design concept and the management mode progressive, and equipped with the function of promotion and expand. The equipment types should be automatic and prospective.

(2) Guidance quality principles: it can accomplish the special practicing teaching task and skill training roundly. It can also carry out skill training comprehensive and cultivate high-class applicative talents, based on the requirements of the technical ability stations.

(3) Rivalry principles: it offers resource sharing. By building practicing teaching platforms and scientific research platforms of tackling key petroleum in uniting oil and natural gas engineering, oilfield chemistry, mine field machinery and mechanotronics mechanical-electrical integration, make it the important base of industrial cultural education and petroleum popularization of science education in Chongqing.

(4) Profitability principles: make sure the students can get enough time, all-around operation space and high quality "real guns and bullets" practical skill training at maximum limit, serving for cultivating talents. Besides, it is also for adapting market economy, related to the market. By making full use of the now available resources in school, external training, scientific research and so on, gain the largest economic benefit.

(5) Open principles: the practicing teaching base is open for students and teachers[2]. The experimental items are open for students and students can study and be trained in the base according to their own course schedule and time, so as to accomplish the practicing teaching tache in the teaching plan requirement. At the same time, in order to meet the innovating needs of students, it is designed with developing practicing active space specially for students, offering conditions for students to participate in democracy video challenges like a challenge cup. The base also provides condition for teachers' scientific research and development, encourage teachers to be occupied in the research and development of experimental items, enrich the practicing teaching tache constantly, and promote the quality of the practicing teaching[3].

4   Function Module of Petroleum Engineering Practice Base

Petroleum engineering practice base construction combines factors such as the talents training, scientific research innovation, oil field service and social benefaction. Petroleum engineering practice teaching base will function the orientation for the practice teaching platform, the students engineering training platform, mid-range and high-end application and technology talents training platform, the authentication platform, application technology research and development platform, oil and petroleum enterprise culture popular science education display platforms.

(1) The practice teaching platform, students engineering training platform

Concerning the characteristics of petroleum engineering students, establish training places close to the oil field production line and the real environment to enhance students' practical ability and practice ability [4] Establish practical training teaching system adapting to the theory courses, fully reflect the latest knowledge of teaching materials and courses, the latest technology and process in training in an attempt to enhance oil engineering students' operation skill and the ability to innovate.

(2) Mid-range and high-end application and technology talents training, the authentication platform

Petroleum engineering practice teaching base is the place of cultivating application talents, and also it is a dependence of practical application technology training unit, high technology development, application and promotion, lifelong education training and vocational skills training, examination and appraisal.

According to the current training management requirements of Chinese petroleum well control center, and the first-grade certification can be awarded on condition of the corresponding training base construction. Thus, we can stabilize and promote our training market in the area of well control. Petroleum engineering practice base construction enable further enhance in the school's IADC training level, and it would also lay a solid foundation for the hardware and software for IWCF. Practice base construction for all kinds of training mission will provide a better platform. Therefore, it will hit breakthroughs in training promoting and high level training involving the gas field development technology, natural gas processing, HSE risk security management, oil and gas airtight gathering and transportation technology, etc. At the same time, to get more chances of deep cooperation, we should continue to improve and develop order training mode, further strengthen the school and enterprise contact and communication, gaining enough mutual understanding.

(3) Application technology research and development platform

Petroleum engineering practice base armed with advanced equipment, good training venues and experienced teachers team would provide scientific research premise and scientific research achievements industrialization platform for petroleum engineering specialty teachers. Create a condition, and encourage teachers and students to develop new technologies, new products, to undertake projects related to technology training and its achievements, providing society a platform putting technology innovation into technology application and inspection, thus making the practical teaching base a productivity promoting center of technology innovation, technology promotion, technology consultation and technical service.

(4) Oil enterprise culture and popular science education display platform

China's oil industry rose in a difficult situation and grew in great endeavour, contributing a lot to the country's prosperity and the sustainably fast development of the country, society and economy. Oil men created Daqing spirit, the spirit of the iron man etc, which not only has become the essence and the soul of Chinese oil culture, but also has been an important part of the national spirit and zeitgeist, continuously inspiring Chinese forward.

Our college is running based on deep industrial background, featuring petroleum engineering and its related courses(such as mining machinery, petroleum safety, chemistry in oil, etc), which calls for corresponding background knowledge among students. Therefore, the construction of the practical teaching base can be served as a display platform in oil enterprise culture and popular science education, providing basic guarantee for college's project training, industry culture and industry spiritual education, constituting symbolic achievements of characteristic schooling and an important part of both an oil popular science education base and Chongqing university cultural tourism.

5 Conclusion

Maximize the student to obtain enough time and comprehensive operation space, high quality "bullets" practical skills training for personnel training services; and to be adapt to the market economy, and linked to the market, make full use of the present resources of school, then through foreign training, scientific research, and other ways to get the largest economic benefits.

Acknowledgment. This work is supported by the reform project education teaching of Chongqing named "Study on constructing the practical teaching base of petroleum engineering" (09-2-091).

References

1. Youke, Chenhengrong, Wangshaoguang: Research of practical teaching base construction with application-oriented university. Education of Chinese university (4), 49–51 (2006)
2. Zhangyu, Zhangzhongshi: College practical teaching base construction and use study. Journal of Agricultural University of Inner Mongolia(social science version) 10(41), 164–166 (2008)
3. Zhuzhengwei, Yuanqiaoyin, Liudongyan: Enforcing research and practice of practical teaching base. Education of Chinese university (9), 66–67 (2009)
4. Zhuhua: Open practical teaching base construction research. Liaoning Higher Vocational Technical Institute Journal 9(11), 66–68 (2007)
5. Wangtianli, Chenyong, Zengqingdong: Study and Research of practical teaching base. Journal of Liaoning University of Technology 11(1), 89–91 (2008)

A Development Method of Resources for the E-Learning Based on VRML

Xin Huang[1,2] and Yuxing Peng[3]

[1] Media & Communication School of Wuhan Textile University
[2] Computer School of Wuhan University
Wuhan, China
[3] School of Computer Science, National University of Defense Technology
Changsha, China
bossmehx@126.com

Abstract. The learning resources play an importance role in the e-learning, which refers to any learning and training method that mainly uses the internet or intranet as a method of diffusion. The VRML can provide a standardized, portable and platform-independent way to render dynamic, interactive 3D scenes across the Internet. In this paper, it is introduced a practical method of developing the virtual learning resources for the E-Learning system based on VRML, which is useful for teacher who want to develop and design the virtual learning resources according to the actual situation of learners.

Keywords: E-learning, Resource, Virtual, VRML.

1   Introduction

Today technology offers many new opportunities for innovation in educational assessment through rich new assessment tasks and potentially powerful scoring, reporting and real-time feedback mechanisms. E-learning is arguably one of the most powerful responses to the growing need for education. What is e-learning? This question normally elicits a variety of responses depending on the experience and background of the person involved. E-learning is just one of the many terms which are used in literature and business about e-learning. E-learning is defined by many people in many ways. It is most important to gain a clear understanding of what e-learning is. E-learning is commonly referred to the intentional use of networked information and communications technology in teaching and learning [1, 2]. Many other terms are also used to describe this mode of teaching and learning. They include online learning, Virtual learning, distributed learning, network and web-based learning. As the letter "e" in e-learning stands for the word "electronic", e-learning would incorporate all educational activities that are carried out by individuals or groups working online or offline, and synchronously or asynchronously via networked or standalone computers and other electronic devices [3]. Here we are not to discuss its definition in detail, but about its learning resources that are the learning object for learners.   It is an important part of the E-learning. As many E-learning

Y. Wang (Ed.): Education Management, Education Theory & Education Application, AISC 109, pp. 865–871.
springerlink.com   © Springer-Verlag Berlin Heidelberg 2011

systems are developed by the professional company, they may fail to satisfy with some resources for learners in the actual teaching. The teachers should develop their own learning resources of the E-learning system according to the actual situation of learners. This paper introduces a practical method of developing the virtual learning resources for the e-Learning system based on VRML in details.

The Virtual Reality Modeling Language (VRML) [9] is a file format for describing 3D interactive worlds and objects. It may be used in conjunction with the World Wide Web. It may be used to create three-dimensional representations of complex scenes such as illustrations, product definition and virtual reality presentations. We make use of VRML to develop a virtual learning resource for e-learning system, which is a molecular 3D model of the benzene (C6H6) for the learner and teacher to learn and teach demonstration at the high school.

The VRMLPad [10] is taken as our development tool. It is very powerful professional software, which fully support the VRML file standards on the virtual scene and resources, and which has a strong capability and visualization interface.

2 Relate Work

E-learning is attaining signification in Internet world due to the obvious advantages of anywhere anytime learning. Som Naidu [3] describes systematically principles, procedures and practices of e-learning, irrespective of the educational sector or level within which user may be working. Colin Beard et al. [4] study uses a model developed from experiential learning theory as the practical basis for the design of an induction CD-Rom for students in order to provide a deeper theoretical foundation for the development of e-learning. [5] discuss the stages of the e-learning process in terms of people responsible for providing various e-learning and blended learning products. The People-Process-Process-Product Continuum or P3 Model can be used to map a comprehensive picture of e-learning.

Hrastinski [6] has focused on the benefits and limitations of asynchronous and synchronous e-learning and addresses questions such as when, why, and how to use these two modes of delivery. [7] presents a model for quality assessment of e-learning. This model is developed using analyses of policy documents, networks and development projects initiated within the framework of European cooperation. Scalise and Gifford [8] introduce a taxonomy or categorization of 28 innovative item types that may be useful in computer-based assessment. Organized along the degree of constraint on the respondent's options for answering or interacting with the assessment item or task, the proposed taxonomy describes a set of iconic item types termed "intermediate constraint" items.

However, they fail relatively to introduce development of the resources of e-learning for teacher, which is to become bottlenecks in the actual teaching. This paper presents a new method of development of e-learning resources based on VRML.

In this section, we introduce in tail a method developing a virtual learning resource for e-learning, which the process has three parts of creating virtual scene, design interaction and transmission of events.

3   Create Virtual Scene

In this section, we firstly create a virtual scene for benzene. We must define 12 Sphere nodes to simulate six carbon atoms and six hydrogen atoms, after in detail analyzing the molecular structure of benzene (C6H6). The Shap node is a container node which collects a pair of components called geometry and appearances. Nodes that describe the objects being drawn are typically used in the geometry and appearance fields of the Shape node. In order to distinguish between the carbon atoms and hydrogen atoms, the different values are be defined at the diffuseColor of Material node in the Shape nodes. Here is a scene statement for Model of carbon atoms:

```
Shape {
       appearance Appearance {
          material Material {
          diffuseColor 0.6 0.6 0.6
       }}
       geometry Sphere {
       radius 0.25}}
```

The same modeling methods are at the Hydrogen atoms and carbon atoms, but of different value of the diffuseColor.

In addition, there are 24 Cylinder nodes to model chemical bond between carbon atoms, and between carbon atoms and hydrogen atoms. They are distinguished by the different diffuseColor value. Here is modeling statements of the bond between atoms:

```
Shape {
       appearance Appearance {
          material Material {
          diffuseColor 0.6 0.6 0.6
       }}
       geometry Cylinder {
               height 0.69298604
               radius 0.10 }}
```

In order to make different objects to connect properly in the virtual space, we must ensure spatial coordinates, which the translation nodes can be used to carry out the conversion of space coordinates. Then each node is grouped by the Transform nodes in order to achieve modular design concept:

```
Transform {
    translation 1.1906784 0.70907508 -0.0012576976
    children [
       Shape {
       appearance Appearance {
          material Material {
          diffuseColor 0.6 0.6 0.6
       }}
```

```
        geometry Sphere {
        radius 0.25
}}]}
```

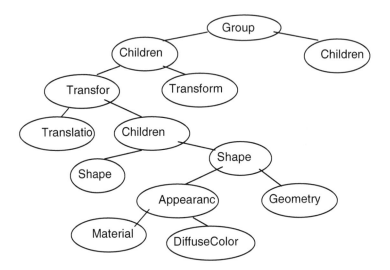

Fig. 1. Virtual scene structure

The Figure1 shows the structure of virtual scene. VRML syntax and node typing also helps enforce a strict hierarchical structure of parent-child relational, so that brows can perform efficient rendering and computational optimization. Grouping nodes are used to describe relationships between Shapes and other child nodes. The semantics of a scene graph carefully constrain the ways that nodes can be organized together. Child nodes come under grouping nodes to comply with the scene graph hierarchy inherent in any VRML scene.

4 Design Interaction

We have designed the virtual scene, texture and color for the geometry object, but the virtual scene is static. Now we are designing the interaction of geometry object which can respond to the learning's manipulations. That is interaction, which is one of the most prominent characteristics of VRML2.0. Sensors detect change in the scene due to passage of time, user intervention or other activity. Sensors produce time-stamped events whose values can be routed as inputs to other nodes in the scene.

In VRML, there are the nine Sensor nodes which are the basis of interaction. In the virtual scene graph, the Sensor node whose parent is the TouchSensor node, which is one of the most common Sensor and have typical applications in the switch, is generally children node of the other nodes. Here, we define a lightSwitch node which is a Group node, and define a TouchSensor as its children node:

```
DEF lightSwitch Group {
    children [
    The geometry nodes…
    DEF touchSensor TouchSensor {}
    ]}
```

This lightSwitch node is a TouchSensor node. Of course, the existence of the Sensor is that it is triggered when the change take place. The most common change is the change in perspective. The virtual scene will rotate, or zoom, when the mouse is dragged or the arrow keys are activated, which is taken as navigation in accordance with the terms of VRML, and which it is in fact that adjust your position or perspective. The ViewPoint node can be defined in an important position in the virtual scene, where we define the two ViewPoint nodes:

```
DEF view1 Viewpoint{
    position 0 0 20
    description "View1"}
```

```
DEF view2 Viewpoint{
    position 5 0 20
    description "view2" }
```

Their purpose is to enable users to adjust their perspective by triggering lightSwitch node. Now we make researcher on the two lightSwitch nodes, which their coordinates show the location of perspective in the virtual scenes. The name of viewpoint is shown in the browser menu for users to choose.

5   Transmission of Events

Now the virtual scene and perspective switching is linked to the events by the system of mutual communication nodes. Capable of receiving incident nodes should have the eventIn, while the nodes of sending events should have the eventOut. The eventIn and eventOut of events is connected to the system of events through the path, which is the ROUTE statement in the VRML file. Here, the isActive of the touchBox, which is the eventOut, is connected to the set_bind which is the eventIn of the view2:

ROTUE touchBox.isActive TO view2.set_bind

So far, we have developed a learning resource for the E-learning based on the VRML for students from the high school to study the molecular structure of benzene(C6H6), which can be rotated or zoomed when the mouse is dragged or the arrow keys are activated. The plug-ins of the virtual browser must be installed to surf the Web based on the VRML, such as BS Contact VRML, BlaxxurContact, Cosmo2.1, CORTVRML. Here is the Parall Graphics Cortona Control plug-ins. The virtual structure of the C6H6 is shown in Figure 2 after debugging with the VRMLPad2.1.

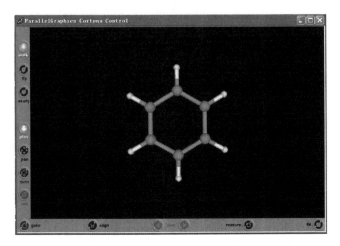

Fig. 2. Virtual structure of the C6H6

6   Conclusion

E-Learning is attaining significance in education due to the obvious advantages of anywhere anytime learning, to reach the unreachable and so on. It is a well known fact that e-learning is the wide spread adoption by the education. This paper introduces a practical method of developing the virtual learning resources for the e-Learning system based on VRML in details, which the process has three parts of creating virtual scene, design interaction and transmission of events. We make use of VRML to develop a virtual learning resource for e-learning system, which is a molecular 3D model of the benzene (C6H6) for the learner and teacher to learn and teach demonstration at the high school. Our method can help teachers to develop their own learning resources of the E-learning system according to the actual situation of learners.

Acknowledgements. This work is supported in part through Research Program of Hubei Provincial Department of Education numbered 2009b255 and Wuhan Textile University numbered 2008S23.

References

1. E-Learning Handbook, http://c4lpt.co.uk/handbook/index.html
2. Definition of e-learning,
    http://agelesslearner.com/intros/elearning.html
3. Naidu, S.: E-learning: A Guidebook of Principles, Procedures and Practies. In: Proc. CEMCA (2006)
4. Beard, C., Wilson, J.P., McCarter, R.: Towards a Theory of e-learning: Experiential e-Learning. Journal of Hospitality, Leisure sport & Tourism Education 6(2) (2007)

5. Khan, B.H.: The People-Process-Product Continuum in E-Learning: The E-Learning P3 Model. Issue of Educational Technology 44(5), 33–44 (2004)
6. Hrastinski, S.: A Study of Asynchronous and Synchronous e-learning methods discovered that each supports different purposes. In: Proc. Educause Quarterly, vol. (4) (2008)
7. Åström, E.: E-learning quality Aspects and criteria for evaluation of e-learning in higher education. Swedish National Agency for High Education (May 2008)
8. Scalise, K., Giford, B.: Computer-Based Assessment in E-Learning: A Framework for Constructing "Intermediate Constraint" Questions and Tasks for Technology Platforms. The Journal of Technology, Learning, and Assessment 4(6) (June 2006)
9. The Virtual Reality Modeling Lanuage Specification,
   `http://www.cg.tuwien.ac.at/hostings/cescg/CESCG98/JSofranko/paper.pdf`
10. VrmlPad v. 3.0, User's Guide,
   `http://www.parallelgraphics.com/12/bin/vrmlpad30_guide.pdf`

Comparative Study on Training Patterns of Entrepreneurial Talents both at Home and Abroad

Shufeng Sun[1], Xiaoman Chen[1], and Pingping Wang[2]

[1] School of Mechanical and Electrical Engineering, Wenzhou University,
325035 Wenzhou, P.R. China
[2] School of International Cooperation, Wenzhou University,
325035 Wenzhou, P.R. China
shufeng2001@163.com

Abstract. At present, the inevitable problem that exists in higher education in China is employment after graduation. Comparative study of entrepreneurial education patterns both at home and abroad shows their own features. In Europe and America, entrepreneurial education starts earlier than that of China and their training patterns are comparatively perfect. In China, entrepreneurial education starts later than that of Europe and America, and its entrepreneurial education patterns are formed through learning the other countries' successful experiences. Some universal methods are summarized by comparatively studying the training patterns of entrepreneurial talents. The training of entrepreneurial talents provides an effective way of solving the problem of employment.

Keywords: Entrepreneurial education, Entrepreneurial talents, Employment.

1 Introduction

Entrepreneurial talents are persons who possess innovative spirit and innovative capability, are able to use all kinds of knowledge they have mastered innovatively and comprehensively, actively throw themselves into entrepreneurial practice, are able to advance and pioneer innovatively, contribute to the development of society and progress of human beings with their creative work. Compared with creative talents, entrepreneurial talents have features that they gravitate towards practice, pay much attention to comprehensive usage of technique, aim at production and industrialization. They are businessmen with certain pioneering spirit[1].

UNESCO had an international proseminar in Beijing which was called "Faced 21st Century" in November, 1989. The conception of entrepreneurial education was put forward and discussed firstly. The world congress of higher education held in 1998 wrote in article 7 of "21st Higher Education: Prospect and Action Declaration" that to facilitate graduates' employment, higher education should pay much attention to cultivate their entrepreneurial technique and initiative spirit, graduates are not only job seekers, they should firstly become job post creators[2].

2 Overseas Training Patterns of Entrepreneurial Education

Myles Mace who is a professor of Harvard Business School started to teach the course of "Management of New Enterprises" since 1947. Babson College set up entrepreneurial direction in undergraduate education in 1976. In basic educational period, more than 30 states' students of America start to accept entrepreneurial education in grade K-12 (middle school). Ministry of Education approved to set up entrepreneurial course "Youth Empowerment and Self Sufficiency" and "Mini-Society". During the period of high school, students have to complete 10 semester credits of vocational education course and practice in society. Entrepreneurial education is integrated with vocational education. In the period of university, more than 1600 universities in America offer entrepreneurial course. Most of the teachers offering entrepreneurial courses have entrepreneurial experiences. For example, H. Irving Grousbeck who is director of the center for entrepreneurial studies of Stanford University has taken part in establishing Continental Cablevision, Inc. He acted as managing director and chairman. Andrew. S. Grove who was former CEO, and is current chairman of Intel Corporation starts to act as a part-time lecturer of Stanford Graduate School of Business from 1991. On the other hand, entrepreneurial education training among the teachers is reinforced in America to make them master entrepreneurial education knowledge. Teachers are encouraged to take part in the activity of entrepreneurial imitation to obtain the experiences of entrepreneurship. Proseminars of entrepreneurial education are held frequently to exchange experience and to enhance the teachers' professional teaching level.

"Employment Instruction Summary" provided by the European Commission indicates that entrepreneurial theory and practice courses should be set up for vocational education to cultivate students' entrepreneurial aspiration and spirit. European parliament puts forward that education should provide entrepreneurial knowledge and chances to students. In Germany and Switzerland, entrepreneurial courses are mainly offered through vocational education. Their entrepreneurial education mainly aims at establishing small and medium enterprises. Many entrepreneurial cases are taught to enlighten students. To achieve the aim of training them into entrepreneurial talents, students are helped to study and analyze market, to map out and assess their entrepreneurial plan. At the turn of the century, the Treasury earmarks several billion euros to Cambridge University for cooperating with MIT. Through learning the entrepreneurial education of MIT, students of Cambridge University will actively take part in entrepreneurial activities. In France, entrepreneurial education starts from middle school. It includes establishing an enterprise in middle school, teaching middle school students to establish an enterprise, and son on. The aim is to cultivate students' entrepreneurial interests and to prepare for professional career and entrepreneurial psychologically. In addition, European countries construct many high-tech parks. There are more than 100 high-tech parks in Germany. France and England also set up many high-tech parks. These high-tech parks are seated in the places with many universities and research institutes. The traffic is convenient. Based on universities, students are provided with places to carry out entrepreneurial activities.

In Japan, employment and entrepreneurial education are started in primary school since 1998. There is an "Early get up" activity before class. 2-3 hours before class, students are required to get up early to carry out work-study activities of delivering newspaper, sending breakfast and so on. It is not only a real entrepreneurial education, but also process of accumulating experiences for future entrepreneurship. Students' entrepreneurial consciousness and habit are cultivated from childhood. Many universities in Japan set up Technology License Office (TLO) to speed up the translation of scientific and technological advances into commercialization. It also provides students with technique, management and direct investment to help them to do entrepreneurial activities. Many Japanese universities also construct high-tech parks. Cooperating with research institute and enterprises, they research new techniques and develop new products in one way, and they cultivate many high-tech entrepreneurial talents. To encourage entrepreneurship, Japan gives many policy supports. For example, one company can be registered by 1 yen, simplify approval procedures, accelerate handling process and so on.

In India, entrepreneurial education courses are offered in more than 100 universities. Teachers are mainly adjunct professors. In another way, coordinate teachers of local school teach theory knowledge, adjunct professors teach practice contents. Original foreign textbooks are used to ensure the advance of entrepreneurial education theory. Various entrepreneurial activities such as national or international entrepreneurial competition are conducted. Some universities set up entrepreneurial center. Some construct incubator. They are used to help students' entrepreneurship.

In the 1960s, entrepreneurial education is conducted in college and university education in Australia. In the 1990s, entrepreneurial education is taught in postgraduate teaching. Institute of entrepreneurial education of small and medium enterprise is established in many universities. Entrepreneurial education courses are developed to make the teaching contents suitable for entrepreneurs and managers of small and medium enterprises. Comprehensive basic education of entrepreneurial knowledge and technique of small and medium enterprises is conducted to dig mainly for the entrepreneurial potential of students and to cultivate their entrepreneurial quality. In the Institute of entrepreneurial education, full-time teachers are only 40%. More than 60% of external invited teachers are entrepreneurs of small and medium enterprises. They have profound theoretical knowledge and rich experience in practice. In the teaching process of entrepreneurial education, case teaching is mainly used. Many cases are analyzed and researched to enlighten and guide students to analyze their own condition, entrepreneurial background and market prospect, and to map out entrepreneurial schemes suitable for their own. Then practice and intern are followed in enterprises.

3 Domestic Training Patterns of Entrepreneurial Education

In January 1997, "Action Plan for Invigorating Education Faced 21st Century" mapped out by the Ministry of Education is approved and passed around by the State Council. At the same time, it points out that entrepreneurial education should be strengthened for teachers and students, measures should be taken to encourage them to establish high-tech enterprises[3]. The Ministry of Education held many proseminars on entrepreneurial education since 1999. 9 universities such as Tsinghua University, Shanghai Jiao Tong University, et al are identified by the Ministry of

Education as entrepreneurial education pilot universities since 2002[4]. The Ministry of Education holds a training of backbone teachers of entrepreneurial education every year since 2003. This indicates that entrepreneurial education is included in the teaching of higher education. In addition, the Ministry of Education stipulates that it is allowed for undergraduates and post graduates to suspend from school for entrepreneurship.

In China, entrepreneurial plan competition was firstly held in Tsinghua University in 1998. The first national "Challenge Cup" business plan competition for university students was held in Tsinghua University in 1999. The following six were held successfully in Shanghai Jiao Tong University (2000), Zhejiang University (2002), Xiamen University (2004), Shandong University (2006), Sichuan University (2008) and Jilin University (2010) respectively. As new activity forms of science and technology of students, business plan competition is very important for the training of entrepreneurial talents.

To implement the policy of entrepreneurial education set up by the Ministry of Education and to cultivate entrepreneurial talents, many universities work out relative measures. For example, Qidi Entrepreneurial School was established in Tsinghua University. The main teaching contents are case teaching and exploration of universal law of entrepreneurship. According to the undergraduates and the learners who are in the beginning or medium of entrepreneurship and who win initial success, course series including the basic theories and practice experiences of entrepreneurial enlighten, breakthrough and incubation are worked out. Shanghai Jiao Tong University established School of Entrepreneurship in June, 2010. In one way, it provides entrepreneurial education for the whole university students and pays much attention to students' entrepreneurial spirit and consciousness. In another way, it sets up target-oriented entrepreneurial course and training for the students who have intense entrepreneurial desire. Risk assessment is carried out for their entrepreneurial plan. Preincubation and funds of entrepreneurship are provided for students. The continuous cultivation will propel them to be future entrepreneurs.

Foundations of undergraduate entrepreneurship are established in many places of the country. For example, Shanghai graduate entrepreneurial foundation was founded in August, 2006. It is the first non-profit public offering foundation for spreading entrepreneurial culture, supporting entrepreneurial practice and propelling graduates' entrepreneurship. The foundation established 10 branches in Shanghai since its foundation. It invested more than 300 projects with ¥57 million and levered ¥130 million society capital. In September 2009, Fujian Nuoqi graduate entrepreneurial foundation was set up. It is a non-profit public institute helping and propelling graduates' entrepreneurship. It provides project, funds, training and research of entrepreneurship. The foundation planned to fund 500 students with ¥50 million to fulfill their entrepreneurial dream. In May 2010, Shandong graduate entrepreneurial foundation was established by the Provincial League Committee and China Unicom Shandong branch with ¥35 million. It planned to set up 200 "Unicom Future Youth Entrepreneurial Society" within 3 years in 100 universities to provide entrepreneurial and employment service.

Graduate entrepreneurial parks were established in many places of the country. In July 2009, "Tianjin Green Orchard Graduate Creativity and Entrepreneurial Park" was established. It accepted 221 students coming from 72 entrepreneurial teams. 29

teams registered as companies which bring along 300 students' participation. The entrepreneurship successfully promotes employment. In July 2009, the new mode of graduate entrepreneurship of "Entrepreneurial Park + VPE" is firstly established in Chengdu. VPE means volunteer service, practice and employment. The first batch of 21 graduate entrepreneurial enterprises established in the park. 63 students apply to carve a career and promote 210 students employment. In Shanghai, "Chuangyiyang Industrial Park" provides students with fine decorated offices with tables, chairs and broadband networks. It reduces the initial cost of entrepreneurship. In addition, entrepreneurs suitable for the requirements may apply housing allowance. For the company that registered capital is less than ¥500,000, the legal representative is Shanghai Citizen, employs graduate or unemployed young people who is Shanghai citizen and is not older than 30, each employee may obtain ¥2000 housing allowance. The park may provide entrepreneur with ¥500,000 open business loan. "Graduate Entrepreneurial Park of Wenzhou University" was established in November 2007. It covers 1400m^2 and is able to accommodate more than 40 teams carrying out entrepreneurial practice activity. The park is mainly divided into IT information service, art design and comprehensive service according to students' entrepreneurial contents. 36 entrepreneurial teams coming from the schools of the whole university registered in the park carrying out entrepreneurial practice activity. There is no rent for the entrepreneurial teams and there are various preferential policies to foster students' entrepreneurial practice activity. Students' entrepreneurial studio, entrepreneurial center of school and entrepreneurial park of university constitute trinity entrepreneurial practice platform. It ensures students' entrepreneurial practice having advantages of low cost, low risk and high success probability. More than 800 students take part in entrepreneurial practice activities since 2007 and the accumulated income is over ¥10 million.

4 Conclusion

To sum up the above arguments, entrepreneurial talents training is explored in China and abroad formed their patterns and features. Entrepreneurial education is implemented earlier in abroad. Their education patterns are various and turn out many famous entrepreneurs. In China, entrepreneurial education starts later than abroad and is still in an exploratory stage. Based on successful experience of overseas education, China is forming its own training features. Through comparing and analyzing the modes of domestic and overseas education, some usual practices of entrepreneurial talents training can be summarized as follows. Firstly, the state department establishes series of policies to guide and support students' entrepreneurship. Secondly, entrepreneurial plan competition is held to build students' entrepreneurial activity platform. The third, entrepreneurial foundation is established to provide funding for students' initial entrepreneurship. The fourth, entrepreneurial courses are offered to provide theoretical guidance for students' entrepreneurship. The fifth, school of entrepreneurship is established to cultivate entrepreneurial talents. The sixth, graduate entrepreneurial park is established to provide practice base and experimental site for students' entrepreneurship. The seventh, the construction of teaching staff is strengthened to provide suitable teaching staff for students' entrepreneurship.

At present, for various reasons, leaders of many Chinese universities do not pay attention to entrepreneurial education. In current modes of higher education, it is really difficult to carry out entrepreneurial education. However, many universities are facing the problem of employment now. Therefore, to carry out reform and practice of teaching on entrepreneurial education, to guide college students' entrepreneurship, to impel employment may be an effective way to solve the problem of employment.

Acknowledgments. The paper is funded by the projects of Wenzhou university (No.2010JG11 and No.[2009]125).

References

1. Xie, S., Chen, D., Chen, Y., Yan, Y.: Characteristics and Pattern of Overseas Entrepreneurial Talent Training. Pioneering with Science & Technology Monthly 3, 87–89 (2009) (in Chinese)
2. Shi, J.: Exploration of Graduate Entrepreneurial Education Promote Employment Education. Forum on Contemporary Education 7, 126–127 (2005) (in Chinese)
3. Qian, Q.: Research of Problem and Strategy of Entrepreneurial Education in Current College. China Higher Education Research 8, 92–93 (2005) (in Chinese)
4. Yuan, X.: Perspective of General Situation of College Students' Entrepreneurial Education. Da Zhong Ke Ji 84(10), 135–137 (2005) (in Chinese)

Analysis of Current Strategic Modes of Chinese Higher Education Internationalization

Wenzhong Zhu[1,*], Dan Liu[2], Yi Wang[2], and Ming Zhang[2]

[1] School of English for International Business and Research Center for International Business, Guangdong University of Foreign Studies, Guangzhou 510420, China
[2] School of English for International Business, Guangdong University of Foreign Studies, Guangzhou 510420, China
wenzhong8988@sina.com

Abstract. With the integration and globalization of world economy, China's higher education as an important part of service trade, has been more and more internationalized as well. This paper discusses the strategic modes of China's higher educational institutions' internationalization, such as the strategies of "sending and joining, and receiving", etc. and proposes some corresponding countermeasures for the existing problems, such as "over-commercialization, insufficient government fund support, evil culture eroding, and not blindly copying", etc. in order to provide some references for the cultivation of more internationalized talents for our country and the whole world.

Keywords: higher education, internationalization strategy, mode, problem, countermeasures.

1 Introduction

The so-called higher education internationalization means the development trend and process of a nation's higher education going global, which is a process of integrating the international, cross-cultural, and global concepts of education into the series of functions of teaching, research and services in higher education institutions (Luostarinen & Welch, 1990). It also refers to the process of a nation's higher education going abroad from the domestic market, conducting corresponding changes and innovations in educating goals, teaching content, teaching method, etc., so as to be fit for the need of international education (Lei Zhongxue, 2002).

Based on the commonly-applied strategic management theory of businesses (Porter, 1980; Boone & Kurz, 2005), the strategic modes of higher education internationalization can also be classified into low-degree involvement strategy, medium-degree involvement strategy and high-degree involvement strategy in terms of risk and cost control.

With the development of economic globalization, higher education internationalization, as an essential part of service trade opening-up, has become an important trend of development. Tracing back to the history, the internationalization of higher education has passed several stages:

Before 1970s, the main body and major driver of higher education internationalization was the government as result of the fact in this period it had been mainly influenced by

---

* Corresponding author.

the politics, foreign relationship and defense policies of related countries in the world. After 1980s, due to the ending of cold war and the empowerment of education institutions' authority, the past form of "government-dominated" higher education internationalization gradually changed into the new form of "university-dominated" higher education internationalization. Since 1990s, the step of higher education internationalization has been speeded up in the whole world. For example, in EU, the program of ERASMUS is established to promote the strategy of higher education internationalization. As one of the EU members, UK has made a great achievement in promoting student exchange programs and franchising education programs in different countries to develop foreign education markets.

In China, with its economic development accompanied by over 30-year opening-up policy, its higher education has a great progress, with the size of higher education being even larger than many countries including Russia, India and United States. However, China's higher education lags behind them in internationalization. There is a shortage of internationalized talents who are competent for working in MNCs. In addition, the implementation China's higher education internationalization strategy is faced with some problems such as over-commercialization, blindly copying, and so on.

This paper will make a research on the driving factors, strategic modes, existing problems of China's higher education internationalization so as to propose some countermeasures for improvement.

2 Driving Forces of Higher Education Internationalization

The main objective of higher education internationalization is to break through the national boundary of higher education in vision and to learn from other nations' experiences and knowledge in the development of higher education so as to pursue the realization of the goal of modernization and advancement in higher education.

As a summary, the driving forces of Chinese higher education internationalization are similar to those of many countries in the world, which include both internal factors and external factors as the following figure:

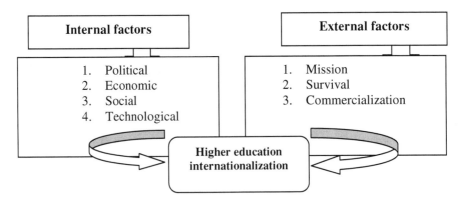

Fig. 1. Driving forces of higher education internationalization

2.1 External Factors

Based on the well-known PEST analysis model, the choice of internationalization of a higher education institution may be originated from the following four external factors:

Politically, the increasing integration and interdependence of nations in the world leads to the cooperative solution to and mutual understanding of the related issues like political reform, human right, environmental protection, poverty, war, and peace, etc. The internationalization of higher education helps China and the other countries to mutually understand each other's political system and culture so that the relationship of mutual trust and respect can be established.

Economically, since 1980s, the integration of the world economy has enhanced the interactive relationship between different nations. The world has become flat. Business leaders with a global strategic vision are needed to compete in the global. Therefore, the internationalization of higher education is conducted to cultivate more potential business leaders to satisfy the need of talents in the future.

Socially or culturally, as a result of globalization, people of different nationality feel it more and more necessary to know, learn and communicate each other. People are more and more likely to accept the culture of others, or buy the product of others, etc. Cross-cultural exchanges have driven the internationalization of higher education.

Technologically, the development of information technology or internet has facilitated the communication and study between people in different nations. Long-distance network education is possible for people living in far-away countries. Thus, it is the progress of technology that has promoted the internationalization of higher education

2.2 Internal Factors

Mission of education: It is the very mission of a higher education institution to cultivate professional talents for the society who have been equipped with the state-of-art know-how and skills representing the most advanced technology in the world. Only through the internationalization of higher education can China or other nations can keep in line with or stand on the top of modern technological and economic development.

Survival of higher education institution: Today, higher education in China has faced more and more challenges as a result of the opening-up of service trade. Foreign universities and private-owned higher education institutions in China are penetrating the market of education, which means the competition of higher education is intensifying. In this circumstance, Chinese higher education institutions are also driven to go out to find new opportunities of growth and survival in the world education market in order to avoid being defeated.

Tendency of education commercialization: In order to accumulate more sources of funds, currently in the world, universities and colleges of many countries are recruiting students from other nations as a tool of earning funds of education. In China, higher education institutions are public service institutions which are partially sponsored by the government, so they are facing difficulties in financing. Therefore, it is a natural alternative for Chinese higher education institutions to conduct their internationalization strategies or develop their joint venture education programs with foreign counterparties on a commercialized manner to earn more education funds for them and provide better education services to the society.

3   Degrees of Involvement and Strategic Modes of Higher Education Internationalization

3.1   Degrees of International Involvement and Types of Strategic Mode

First, based on the principle of strategic management, the strategies of higher education internationalization can also be divided into three degrees of involvement. See Figure 2:

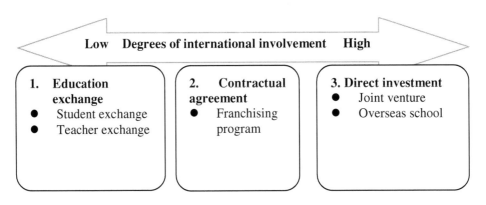

Fig. 2. Degrees of international involvement of Chinese higher education

From the above figure, firstly, it can be seen that the so-called low-degree international involvement of higher education refers to the various kinds of student and teacher exchange programs between a domestic university and a foreign university, namely a foreign university and a domestic university sends to students or teachers to each other's location for learning some classes or doing some researches or teaching some courses. This degree of international involvement of higher education involves the issue of each other's recognition of credits for the students involved. In general, this degree of international involvement of higher education may be lower in risks and costs as well as more convenient in operation.

Secondly, the so-called medium-degree international involvement of higher education refers to the franchising education program of a university at a foreign university. It means that country A's university signs a contract with country B's university allowing country B' university to use the brand name and education program of country A's university to provide education services with the degree of country A's university awarded to the students enrolled. However, this degree of international involvement of higher education in China is also very popular, with the various forms of 1+1 or 0.5+1 master franchising programs and 2+2 or 2+1 undergraduate programs, which means part of the courses or modules are taught in the franchisee's university while part of them are taught in the franchisor's university. Of course, there are also some other franchising education programs, in which all courses are taught in the franchisee's university. In general, this degree of international involvement of higher education may be more risky because it concerns about the loss of reputation or intangible asset of the franchisor once it is in trouble or failure.

Thirdly, the so-called high-degree international involvement of higher education refers to the type of joint venture school built by both a domestic university and a foreign university or the form of wholly-owned overseas school invested by a domestic university in a foreign country. This degree of international involvement of higher education means the large sum of direct investment of funds of the related universities or a single university, and the period of investment needs to be very long before completing so it is the most risky type of international involvement among the three types.

In the practice, three degrees of international involvement of higher education can be evidenced in China today, among which the first two degrees are more popular than the last one with the building of Confucian schools in different nations as typical examples.

3.2 Specific Strategic Modes of Chinese Higher Education Internationalization

In general, there are two basic types of strategic mode of higher education internationalization, which are virtual mode and physical mode. The former refers to the students' learning of a foreign country's university courses with crossing national borders by means of Internet education or long-distance education networks while the latter refers to the student's learning of a foreign country's university courses with their physical appearance in the classroom on the campus by student exchange programs, etc.

More specifically, the strategic modes of Chinese higher education internationalization can be divided into the following three types which are together called the RJS strategy. See Figure 3:

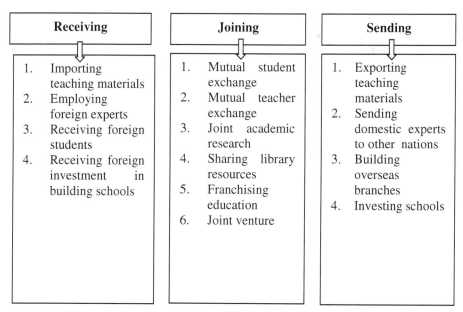

Fig. 3. Strategic modes of Chinese higher education internationalization

3.2.1 "Receiving" Strategic Mode

Firstly, importing foreign teaching materials: The import of foreign teaching materials could be regarded as the most convenient and basic strategic mode of higher education internationalization, which is divided into two major ways: one is the translation of the original teaching materials into Chinese for the use of core text book in the class, which may have the shortcomings of distorting the original meanings of the text, and another is the direct use of original foreign textbooks in the class as core textbooks, which is popular for the teaching method of English or other language instruction or bilingual teaching.

But the limitations of directly using a foreign textbook in the class may be the lack of applicability to Chinese students who want to learn something of the Chinese cases in the local environment.

Secondly, employing foreign experts: The employment of foreign exerts may include inviting foreign professors to give lectures on a short-term contract basis, or employing foreign experts to work on a long-term contract basis. The employment of foreign experts to teach in the Chinese class will help the Chinese student to learn knowledge which is more than that in the textbook through direct communication, and to practice and improve the oral language ability of students through direct talks or discussions with them.

However, there are currently more and more foreign experts who are criticized by universities for their poor performance of teaching or instilling some unhealthy values, etc. Therefore, it is suggested to employ more foreign professors with a sound professional history or prestigious career record through very cautious investigation and approval before signing the contract of employment.

Thirdly, attracting foreign students to study in Chinese universities: China has a long history of civilization and is the fasted developing economy in the current world, therefore, more and more foreign students are attracted to come to study in Chinese universities, and it could be said that the successful path of Chinese higher education internationalization depends on the development of enrolling foreign students to study in Chinese universities, which will benefit China in terms the increased revenues for schooling and the broadening of Chinese students' eye sight through exchanges between diversified students.

However, this type of higher education internationalization can not be achieved by the university alone. It needs the financial support from the government and enterprises, which should provide more sources of sponsorship or scholarship for attracting more and more foreign students to study in Chinese universities.

In addition, in terms of attracting foreign universities to build branch schools in China, the more foreign branch schools will be built, the more competitive Chinese education market will become, thus the reform of Chinese higher education will be promoted and the overall development level of higher education will be raised.

3.2.2 "Joining" Strategic Mode

This type of higher education internationalization may include mutual student exchange, mutual teacher exchange, joint academic research, sharing library resources, franchising education program, and joint-venture school, etc., which aims at achieving the goal of mutual advantages of a domestic university and a foreign university. Currently in China, all the above-mentioned joint development strategic modes are becoming popular, especially mutual student exchange and teacher exchange programs and franchising education programs. These education programs are sometimes conducted by three or more universities from three or more different countries, for instance, China, Japan and Korea are currently promoting a triple-party joint undergraduate education program in which the students involved will have one-year study in other two countries' universities respectively with credits mutually accepted by each other.

The type of joint education program to realize the goal of higher education internationalization may lead to the full play of mutual advantages in the process of operation and the realization of competitive advantages in the global education market. Students or teachers who participate in the exchange program can also learn more through exchanging ideas and experiencing cultures of foreign partners. However, there may be some disadvantages existing in this strategic mode of higher education internationalization, such as: potential conflicts in management style, culture and benefit allocation, etc.

3.2.3 "Sending" Strategic Mode

Sending here means the going-out of Chinese higher education institutions into the world education market to find out new opportunities of development and promoting Chinese civilization and technological achievement. In addition to exporting domestic teaching materials to foreign countries and sending domestic experts to other nations, the "sending" strategic mode of higher education internationalization may also include the building of overseas branch schools or subsidiary schools in foreign countries, etc. The building of overseas branch schools or subsidiary schools helps the domestic university to better control the operation and better avoid cultural conflicts with the local partner. But it may also have some disadvantages such as needing a large sum of investment and a long term period of investment which may contain more risks of operation.

4 Problems and Countermeasures in Higher Education Internationalization

4.1 Problems

The internationalization strategy of higher education may have some problems in the process of implementation, which include the following:

First, the tendency of over-commercialization in the operation of higher education internationalization leads to the decrease of schooling quality. A large number of universities (including foreign universities) consider their internationalization strategy and related programs chiefly as a means of earning profits so that they neglect the real quality of education. For example, in recent years, a large number of students are sent to study abroad through franchising programs or various other programs, but when these students come back to work in China, employers find that quite a lot of these talents returning from abroad are not so good in their language skill and specialty knowledge as before. As a result, these students returning from abroad are devalued and are not so welcome by the society, among whom many become still jobless.

Second, the support from the government and enterprises for the internationalization of higher education is not very well in place, for example, the lack of sponsorship for the promotion of higher education internationalization, and the insufficient scholarship offered to the student exchange program of higher education, etc. This problem impacts the sustainable development of higher education internationalization, especially in the case of avoiding the tendency of over-commercialization.

Third, the blindly copying of the strategic mode of others' experiences in the operation of higher education internationalization has led to the failure of internationalization because the copying is alienated from the reality of the university in the local society.

Moreover, there are some other problems existing in the implementation of internationalization strategy of higher education, such as the insufficient direction and instruction for preventing the elusion of unhealthy cultures or avoiding the threat to the national security, etc.

4.2 Countermeasures

Educators and decision-makers are suggested to strategically think about the solution to these problems in the implementation of Chinese higher education internationalization strategy:

Set a clear vision and mission to avoid the appearance of over-commercialized or profit-making orientation in the design and implementation of higher education internationalization strategy. As a non-profit organization, higher education institutions should make it clear that their primary objective of schooling is to provide high-quality education to the general public rather than maximizing profits for the investor. But of course, higher education institutions are encouraged to obtain the necessary revenues from the operation of internationalization programs so that they can cover the expenditure and have sufficient sources of money to provide better education services to the society. But in whatever circumstances, the tendency of over-commercialization as a pure profit-making tool should be prevented.

Try to obtain the support from the government and the whole society so as to together promote the implementation of higher education internationalization strategy. For example, a university going to establish an overseas branch school like Confucian Institute, needs the policy support and financial support from the government; to encourage more foreign students to come to China for further study, the government or enterprises needs to offer more sponsorship or scholarship programs to them in addition to the scholarship programs offered by the university.

Take measures to reduce the negative impact brought by the implementation of higher education internationalization strategy on the domestic society, such as the prevention of cultural erosion and social society, etc.

Take measures to avoid the blindly copying of others in the process of implementing higher education internationalization strategies. China is different in national culture and higher education management system from many other countries, and one university is different from another university in its management style, so it is not realistic to copy others' strategy of higher education internationalization. Blindly copying with adjustment will result in more possibility of failure.

As a summary, with the progress of economic globalization, the internationalization of Chinese higher education has a necessary choice. There are internal and external reasons driving the trend of higher education internationalization in China, among which mission, survival and commercialization are internal driving factors while political, economic, social and technological are external driving factors. China has participated in three degrees of involvement in higher education internationalization and three major strategic modes of higher education internationalization, which are briefly named as RJS strategy. In the process of implementing the strategy of higher education internationalization, Chinese universities have met with the challenges of over-commercialization, blindly copying others' styles, etc. On this basis, the paper proposes some corresponding policy suggestions to solve them in the end.

References

1. Deng, Z., et al.: On the Status of Sino-foreign Joint Venture Schooling. Guangzhou Daily (July 10, 2007)
2. Pan, M.: 100-year Higher Education. Guangdong Higher Education Press (2003)
3. Liu, H.: The Internationalization and Localization of Higher Education. China Higher Education (2001)
4. Gao, Y.: Research on the Internationalization of Higher Education. China Education Newspaper (2002)
5. Lei, Z.: Thought on the Internationalization of Chinese Higher Education. Jiangxi Normal University Press (2002)
6. Ryan, J.: Improving teaching and learning practices for international students: implications for curriculum, pedagogy and assessment. In: Carroll, J., Ryan, J. (eds.) Teaching International Students: Improving Learning for All. Routledge, New York (2005)
7. Boone, L.E., Kurtz, D.: Contemporary Business, 10th edn. Harcourt College Publishers, 6277 Harbor Drive, Orlando, FL32887-6777 (2002)
8. Luostarinen, R., Welch, L.: International Business Operation. Kyriiri Oy, Helsinki
9. Porter, M.E.: Competitive Strategy. The Free Press, New York (1980)
10. Liu, H.-y.: The Development Strategy Comparative Research of High Education International Competitiveness of GMS Countries. Journal of Yunnan Agricultural University (2009)

The Reform and Practice of Automation Excellent Engineers Training Program

Haizhu Yang[1] and Jie Liu[2]

[1] School of Electrical Engineering and Automation Henan Polytechnic University,
Jiaozuo, Henan, China 454003
[2] School of Computer Science and Technology Henan Polytechnic University,
Jiaozuo, Henan, China 454003
yanghaizhu@hpu.edu.cn

Abstract. According to the training standard and the demands of knowledge ability for the automation, the thought of ladder training for professional ability is employed. The kernel professional knowledge modules, theoretical teaching curriculum system, practice teaching system, cultural and professional quality training system, the curriculum and the corresponding credit requirements are constructed. Then the training plan of the fourth-grade in the enterprise workshop and the graduation project are described in detail. Finally, the effective organization. method of teaching team acting as teacher and engineer is analyzed .Senior engineering and technical personnel of advanced manufacturing technology with engineering consciousness, innovation consciousness, and teamwork consciousness are trained through the implementation of the proposed work program

Keywords: Excellent engineers, Automation, Training program, Engineering consciousness.

1 Preface

In our country,the engineering education is influenced by traditional ideas, funds etc factors,the modes of engineering education and science education in universities and colleges are trending to same, basically carrying out the scientific education. But engineering education's practice content is tend to the training mode which is heavying theory and lighting practice, leading serious lack of technical and practical talents who are needed largely by companies, on the contrary, the employment rate of graduates is more and more lower, the important reason to cause the consequences is lack of industry participation in higher engineering education. "The plan for educating and cultivating excellent engineers "is an major education reform plan for the implementation of national long-term education reform and development plan outline which is launched by the ministry of education in 2010, with the purpose of improving our country's engineering talents training quality, and training a large number of various types of high quality engineering talents who have innovation ability and can adapt to the social and economic development needs, This plan's implementation is a sign of the starting for large-scale engineering education quality projects[1,2]..

Y. Wang (Ed.): Education Management, Education Theory & Education Application, AISC 109, pp. 889–896.

2   The Professional Background, Training Target and Knowledge Ability

2.1   The Professional Background

The automation of Henan Polytechnic University was developed on the basis of a bachelor profession which is electromechanical in coal mines(side power) at the beginning of school rebuilding in 1958,it is the first university to set up this profession in henan province's colleges,and it is also the school's traditional advantage major.After 50 years of development, In the national especially in central plains has higher visibility and influence, At present, the students of automation in our university are enrolled by first class in the bachelorm.For years, automation has trained a number of talents who have solid basic theory, high practical ability and active innovation thought, graduates have been in short supply, receiving high praise from employing units.

2.2   The Training Target

Our school automation 's aim to train talents who having automatic control, industrial process control, information detection and treatment, industrial automation instrument, computer control and information network technology aspects of basic theory and relevant professional knowledge,and foster strongly practical engineering capability and certain research ability of composite applied talents, the students in this professional mainly study basic theory and knowledge on motion control system, intelligent control system, computer control system, information system, meanwhile,having the basic skills of system analysis and integration, design and operation, research and development, management and decision-making[3].

2.3   The Knowledge Ability

According to the requirement of 《 "The plan for cultivating excellent engineers "national standards 》 , and considering our automation professional's concrete development situation, we have made a undergraduate type of excellent engineers training school standard of Henan Polytechnic University automation (mining industry)[4], this school standard can be integrated with knowledge, ability and quality etc eight specific requirements:

(1)Having good professional ethics, and the responsibility for professional, social and environmental(corresponding general standard 1,3,7).
(2) Having the necessary basic science knowledge for engaging in automation(mining industry) engineering work(corresponding general standard 2).
(3) Having the necessary basic professional knowledge for engaging in automation engineering work(corresponding general standard 4,5).
(4) Mastering professional knowledge of automation(mining industry) engineer, having skills of Soling the basic technical issues(corresponding general standard 4,6,8).

(5) Having ability of Soling engineering practical problems in the automation (mining industry) enterprise(corresponding general standard 3,5).

(6) Having strong innovative consciousness and innovative ability(corresponding general standard 5,6).

(7) Having the ability to communicate effectively(corresponding general standard 9,11)

(8) Having automation (mining industry) project and engineering management skills(corresponding general standard 1,8,9,10).

3   The Starting Point to Out the Work Plan

According to overall scheme of "excellence engineers" which is carried out by university, combined with university standards of training undergraduate type excellence engineers of Henan Polytechnic University automation(mining industry), break the traditional three sections of segmentation teaching mode of "basic course - professional class - engineering practice",make a scientific and reasonable plan for the training program of automation by the training mode of "project teaching method",promote the students' comprehensive engineering practice ability comprehensively,to achieve teaching students according to their aptitude and training ability ofindividualized through the differences of project content[4,5].

3.1   The Theory Curriculum System for Teaching

The theory curriculum system for teaching of training automation excellence engineers is shown in figure 1 below by points module, specific include three big platform courses: public foundation platform course, professional foundation course and professional direction platform platform course.

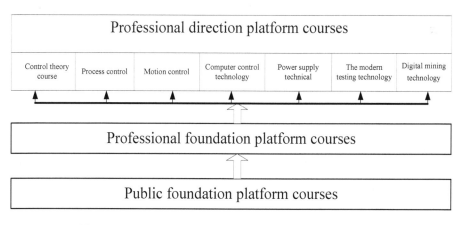

Fig. 1. The three platform structure of theory teaching curriculum

3.2 Practice Teaching System

When regulating campus's practice teaching, in accordance with the principle of "series, levels, and continuous line" and the understanding law of orderly and gradually deepening progress, from the cognitive observation, integrated design up to research and innovation, create a teaching system pyramid in favor of cultivating students' innovation spirit and practical ability. The system includes experimental teaching in class and technological innovation in extra-curricular, of which in class is divided into five levels and extra-curricular is divided into seven levels. Practice has proved that the hierarchical classification of the experimental teaching system completely suitable for teaching requirements of automation professional courses. Experimental teaching system as shown in Figure 2.

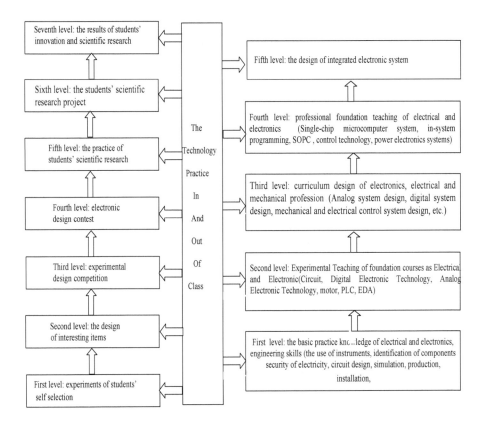

Fig. 2. Experimental Teaching System

3.3 Humanity and Professional Quality Training System

The taining of student's political thoughtand humanity and professional quality is existing through the whole four years of university, including three years learning in school and total one year culturing in enterprise(containing the graduation design), it can adopt various forms, to carry out flexiblly.

4 BibTeX Entries

For the students who take part in undergraduate type of excellence engineers cultivation plan from Henan Polytechnic University automation(mining industry), when they graduate with four years study, at least the credits must be reach to 192,the specific requirements is shown as follows:

(1) the public basic course(including compulsory and elective): 78 credits

(2) the professional basic course(including compulsory and elective): 32 credits

(3) the professional course(including compulsory and elective): at least 31.5 credits

(4) campus practice(excluding in-class experiment) : 15 credits

(5) training in enterprise: 21 credits

(6)graduation design: 17 credits.

5 The Training System in Enterprise

Carry out the system of course configuration which is named as"engineering education runs constantly, innovation practice runs continuously, enterprise cooperation runs constantly",reform the traditional knowledge structure and the knowledge system which is training engineer according to scientists,enforce the practice links and the content learning in enterprise, pay more attention to training of thinking in engineering system, pay attention to the cultivation of students' engineering practical ability, we can set up certain credits according to the different proportion in engineering, technology, science, humanity, social, ethical and so on, construct the engineering knowledge system of "big engineering view, big system view, big integrated view",that can make the students master the skills,ascertain the arguments which should do,understand the doable definition,measure the value of the project. Strengthen the ability to use large calculation softwares and computing tools to solve engineering problems and practice teaching[1].

The learning training in enterprise can be divided into five links to implement,respectively is cognition practice,production practice,enterprise practice, graduate design,etc.

Table 1. The teaching plan in stage of enterprise learning (the seventh and eighth semester)

Sequence number	Learning forms	Learning contents	Requirements (form)	Place	Time (week)
First	Cognition practice	Enterpris's overview And culture, develop condition	Listen to report, visit, and consult information	Jiaozuo HuaFei Zhengzhou GuanLi	One
Second	Production practice	Automation equipment's operation And management	Labour personally	Jiaozuo HuaFei	Three
Third	Enterprise practice	Project management	Bid tender, Project proposal, Feasibility analysis report(only chooseone),	China PingNeng Hua Jiaozuo HuaFei	Thirteen
Fourth	Graduation practice	Engineerin design and analysis	Participate in single engineering's design and transformation	Jiaozuo HuaFei Zhengzhou GuanLi	Four
Fifth	The design of aut-control system course	The design of motion control system	The design of single engineering	Jiaozuo HuaFei Zhengzhou GuanLi	One
Sixth	Graduation design (paper)	The special design of industrial Automation System	Complete The graduation design(paper) outline's requirements	Zhengzhou GuanLi China PingNeng Hua	Thirteen

6 Graduation Design

The graduation design(paper) can adopt diversified manners to run, the students can do the research according to the engineering actual problems which are found in enterprise practice,it also can do the research according to enterprise engineers or school teachers's related research, it can finish the graduation design in the stuents's work unit,too.But the topics meed have a certain challenging and creative, and must .be from the actual production site, its purpose is to cultivate the students'comprehensive engineering ability.

The training of excellent engineers needs a new teachers team which has high teaching level and rich engineering practice experience, it mainly constructs the teachers team through two channels which are cultivation and external,in order to match with the standard of cultivating excellent engineers, school staged teaching plan and enterprise staged training plan, the school rules that the engineering professional teachers must do engineering practice activities in large and medium-sized enterprises during their post stage, opening up the teachers' scientific research, enhancing the teachers' teaching and scientific research level of action, strengthening the connection between the school and enterprise. The school rules that the teachers who want to promote associate professor or professor in principle must have an full-time work experience in large and medium-sized enterprises for more than six months, among them every time should be not less than two months, the specific safeguard measures is shown as follows:

7 The Teaching Team of Double Division Structure

The training of excellent engineers needs a new teachers team which has high teaching level and rich engineering practice experience, it mainly constructs the teachers team through two channels which are cultivation and external,in order to match with the standard of cultivating excellent engineers, school staged teaching plan and enterprise staged training plan, the school rules that the engineering professional teachers must do engineering practice activities in large and medium-sized enterprises during their post stage, opening up the teachers' scientific research, enhancing the teachers' teaching and scientific research level of action, strengthening the connection between the school and enterprise[4]. The school rules that the teachers who want to promote associate professor or professor in principle must have an full-time work experience in large and medium-sized enterprises for more than six months, among them every time should be not less than two months, the specific safeguard measures is shown as follows:

(1) The school should give some funds to support the teachers's engineering practice activities, the budget can be included in teacher training fund no more than 15 million each year.

(2) Encourage and support the teachers to do enterprise engineering practice through many kinds of doing project management and guiding work, exercising in enterprise,and take students to enterprise production practice,

(3) In addition to the measures of training for full-time teachers engineering experience,this major still employ off-campus experts who have rich practical experience in the enterprise as part-time teachers, they are also undertaking the undergraduate teaching work,including to the frontier lectures, graduation design and curriculum design guiding and other related teaching work,that can achieve truly "the professional teachers go out" and "enterprise experts come in"

8   Conclusion

The preliminary established work plan of undergraduate type of excellence engineers cultivation plan from Henan Polytechnic University automation(mining industry) is introduced detailedly above,what's more, it maybe refer to many aspects of problems which are the school's orientation,the students'characteristics,the dynamic management mechanism,enterprise management status and education system reform and innovation in the subsequent specific implementation process.these are expected to be progress steadily in the next process of "explore-summary-perfect-implement"the ultimate goal is to cultivate a large number of automation excellence engineers talent for our country's mining industry.

References

1. Luo, Y., Xing, L., Wang, Q., et al.: The Reform of Curriculum Project Based on Project Teaching Method. Journal of Electrical & Electronic Education 31(6), 14–15 (2009)
2. Wang, M., Zhou, M., Li, J.: Development of the training program for excellent advanced manufacturing technology engineers by project-teaching method. China Modern Educational Equipment 12, 15–19 (2010)
3. Fu, Y., Lou, J.: Research on Practicum Period for Excellence Engineers Cultivation in Electricity Specialty of Application Nature. Journal of Ningbo University of Technology 22(4), 67–71 (2010)
4. Wang, G., Zhang, M.: Research on Young Teachers Teaching Ability of Excellent Engineers Education. Meitan Higher Education 28(3), 91–93 (2010)
5. Zhang, R., Xu, Y., Fu, Y., et al.: The Qualities of Training Excellent Engineer in the Application-Oriented Institutes. Technology & Economy in Areas of Communications 12(6), 126–128 (2010)

Development of the DSP Experiment System Based on the Emulator of XDS510

Jie Liu[1] and Haizhu Yang[2]

[1] School of Computer Science and Technology Henan Polytechnic University, Jiaozuo, Henan, China 454003
[2] School of Electrical Engineering and Automation Henan Polytechnic University, Jiaozuo, Henan, China 454003
liujie@hpu.edu.cn

Abstract. DSP control and relative technologies have attracted the attention of multi-disciplines engineering researchers and educators, and are becoming a new and active field. DSP control-classes is one of important specialized courses for automation. According to the characteristic of DSP course teaching, this article developed the DSP experimental system based on the emulator of XDS510. The experimental system mainly consists of a CY7C68013A chip, a SN74ACT8990 chip, a XC9536XL chip, a TMS320LF2407 chip, and peripherals on board, signal collection and disposal circuit and so on. The development process of DSP software was introduced in detail and part of experiment waveforms was supplied. The experimental results showed the DSP experimental system had perfect function and run stably.

Keywords: Practical teaching, Educational hardware, DSP, XDS510.

1   Introduction

In recent 20 years, the control technique has changed greatly with the newest development of the micro electronic technique, power electronics, sensor technology, permanent magnet material technique and computer application technology. And the digital control system has been developed rapidly which uses high-performance microprocessor FPGA/CPLD, general-purpose computers, DSP controller and other modern methods. Thanks to its high reliability, strong enlargement ability, simple maintenance and able to meet the applying requirement in kinds of occasions, the DSP controller gains the favor of electronic information region and enterprises supporting the control scheme at home and abroad, and is acknowledged to be the development direction of the control implementation technique, [1] At present, many domestic scientific research institutes and enterprises have carried through the development and the applied research of the DSP technology and products. Technical professionals of DSP are in short supply in the market. Many key universities have started opening DSP courses for masters and undergraduates of electronic information, build key DSP laboratories, and devoted themselves to cultivate well-liked DSP talents to satisfy the market demand. However, the DSP technology is a relatively new technology, and we have less experience and technology accumulation

Y. Wang (Ed.): Education Management, Education Theory & Education Application, AISC 109, pp. 897–904.
springerlink.com   © Springer-Verlag Berlin Heidelberg 2011

with a high starting point of the curriculum subject. Therefore reinforcing course teaching construction should be increased ceaselessly.

2 The Target Location of the DSP Course Construction

The DSP technology course of our school has obtained the high praise of students and also got great acceptance from school, since it was opened in 2005. In 2007, "the teaching and experimental reformation research of the DSP control technology" is listed as a key educational reform project of school. [2,3] The course orientation is to face the experimental research course and lead students to learn and carry through the applied research through the experiment. Training of the course will be of great help to the following courses and research subjects, such as digital image processing, intelligent transportation system, signal detection and processing, multisensor fusion, smart home and so on. With the rapid expansion of the DSP technology and the imminent growth of the social needs, it is essential to open experimental research courses.

3 The General Structure of the DSP Experimental System

According to the characteristic of the course taken, we design the DSP experimental system based on the emulator XDS510 to meet the teaching goal request. The system not only can satisfy our requirement of the DSP technology experiment, but also can let our students make independent design and can be used as a practical tools for a scientific research personnel make DSP development. This experiment system can make the following experiment: (1) A/D sampling experiment; (2) SCI serial communication experiment;(3)D/A experiment;(4). the bus communication experiment; (5) PWM motor control experiment;(6) The timer control LED program experiment; (7) External RAM test program experiment. [4]

3.1 The Hardware Structure for XDS510

Hardware structure of simulator system can be shown as in figure 1. Simulator is mainly composite by CY7C68013, 24C01, ACT8990, MAX964 and a piece of 74VTH244A. And a piece of MAX964 chip and bus driver chip has be placed between ACT8990 module and JTAG module. [5]

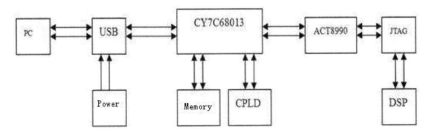

Fig. 1. Hardware structure of simulator system

JTAG interface directly connected to the DSP JTAG interface. The boundary scan completed by the serial test clock signal (TCK), test mode selection (TMS), test data input signal (TDI) and test data output signal (TDO) and DSP communications. ACT8990 produce TMS and TDI signals and receive TDO signal, TMS, TDI and TDO signals through the bus driver chip 74ACT244 connected to the DSP chip. ACT8990 all operations and TCKI signal are synchronization and test end signals (TOF F) produced by DSP. Specific connection diagram is shown in figure 2.

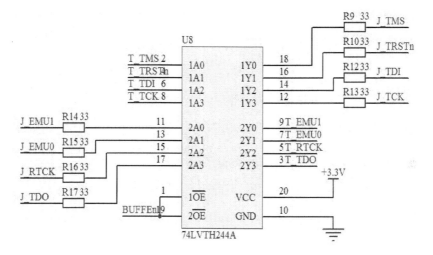

Fig. 2. Connection diagram between simulators JTAG interface and DSP JTAG interface

CY7C68013 16 bits of data bus connect to ACT8990 data bus and 5 bits of address bus connect to ACT8990 address bus. The two literacy and reset signals also directly connected and SRST signal can be used to reset ACT8990. The principle diagram connection of ACT8990 module is shown in figure 3.

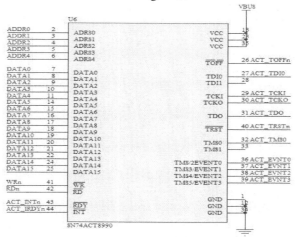

Fig. 3. Principle diagram connection of ACT8990 module

USB connected to the computer's USB interface. MAX964 is voltage comparator which is used to real-time detect VCC signal of JTAG interface, which is directly connected to the DSP VCC. MAX964 will detect the signal and sent out an interrupt to CY7C68013 once the DSP electricity and the VCC signal is high level. Conversely, if DSP suddenly drop electricity, then MAX964 will immediately send power down signal interrupt to CY7C68013, the entire system will to stop working and will transmit the signal to the computer. A piece bus driver chip 74ACT244 has two functions: one is to drive bus driver in order to adapt to the different ability all digital signal processor (DSP); the other is isolation DSP and ACT8990in in order to protect ACT8990 avoid high voltage transmitted by DSP damage.

System use 5V dc power supply, the use of voltage transform chip TPS7333 which is used to convert 5V voltage to 3.3 V provided to individual chip at the same time DSP IO mouth voltage is consistent. Specific connection diagram is as in figure5 shown the power supply for DSP circuit.

3.2   The Hardware Structure for DSP System

DSP system select TMS320F2407 DSP controller of TI as core chip, which software and hardware technology development thoughts can be transformed to TMS320x28x series and TMS320x54x series chips quickly. DSP system chart is shown in figure 4.

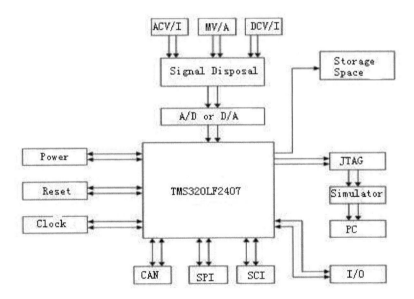

Fig. 4. Structure of DSP systems

We must make voltage conversion because the power supply is standard 5V dc power. Linear regulator and integrated power are two common 3.3 V voltage conversion methods, and the latter voltage circuit, small size, low power consumption, reliable and high efficiency. The system uses the TPS7333QRD chip of TI, which can be converted 3.3V into all the dc voltage 5V fixed output voltage, the biggest 3.3 V 500mA output current available, can satisfy the general system power requirements. Use a 10 uF and 0.1 uF output capacitances to improve the transient response and stability. Specific connection is shown in figure 5.

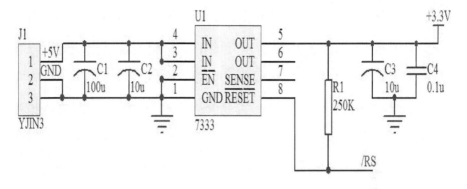

Fig. 5. Power circuit

The reset tube feet for TMS320LF2407 is /RS which is effective in low-level. In order to guarantee the DSP chip power which is fail to meet the requirements of electricity at ordinary times, won't produce uncontrolled state, the power chips TPS733-8 is received to DSP reset tube feet. Another system adopts the MAX811 of MAXIM company to control system's reset, this chip can achieve the manual reset and electric reset, watchdog reset function, can guarantee that the safety operation of the system and prevent program into dead circulation. Specific connection diagram is shown in figure 6 . Circuit provides manual reset switch S1.

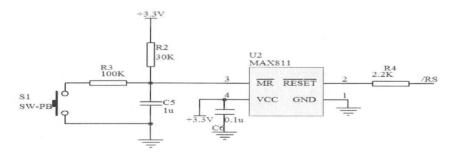

Fig. 6. Reset circuit

Phase locked loop (PLL) module mainly used to control the DSP core work frequency, external provides a reference clock input, and after PLL octave or points after provide DSP core frequency. This design USES the work mode based on the crystal PLL by external passive crystal as chip, this design provide clock benchmark selection of external 20MHz crystals is. Specific connection is shown in figure 7.

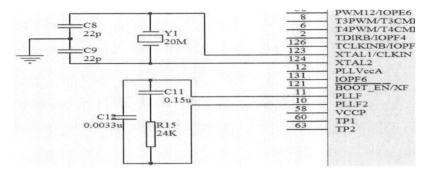

Fig. 7. Clock circuit

DSP simulators through providing scanning DSP chip simulation pins realize simulation function, scanning the simulation to the conventional circuit simulation existing cable is too long can cause signal distortion and reliability of simulation plug poor etc. Scanning simulation, make online simulation becomes possible to debug brings great convenience. JTAG interface is a industry standards this part of the pins definition don't change. This design will be designed into a standard double row 14 pin sockets, can target board for simulators debugging.

3.3   Software Introduction of the DSP Experimental System

The standard development of a DSP application software needs the following steps: compile assembly instructions meeting the format requirement of Assembler through Editor; call Assembler to assemble the source file, and if the source file has called micro-program, Assembler will search the micro-program in macro library; generate the .obj(object) file whose format is COFF(Common Object File Format) after assembling, called COFF object file; call Linker to link the .obj file, and if it obtains run time support library and .obj library, Linker will search all members in protected library; generate the executable COFF .out after linking; download the COFF .out to DSP to execute, meanwhile we can have the aid of the Debugging Tool to trace and debug, or optimize the procedure, and we also use the Cross-reference Lister and the Absolute Lister generate some lists including debug information. TI sets standard test interfaces and relevant controllers measuring up to JTAG (Joint Test Action Group) of IEEE 1149. Thus, we not only can control and observe every processor's operation of the multiprocessor system and test each chip, but also can load program by using the interface. We use 37 core flat cables make the computer be linked together with the emulator XDS-510 through the line parallel tandem or riser card, then connect the computer with DSP target board by means of using 14 core JTAG connectors.

4 Part Experiment Waveforms of the DSP Experiment System

We have used the DSP experiment system designed in this paper to do a speed governing experiment of DC motor. Development of procedural is based on CCS3.3 IDE (integrated development environment) and program mainly realize for dc speed control PWM signal generation. PWM signal waveform graph is shown in figure 9. Through the control program, we can adjust the duty radio of PWM signal and then we can adjust the speed of DC motor. Experimental proof that the system based on DSP control can work reliable and stable.

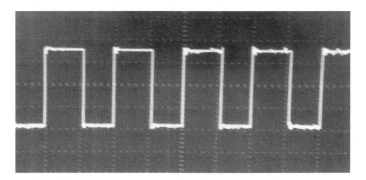

Fig. 9. PWM control signal

5 Conclusion

The DSP experimental system designed in this paper had perfect function and run stably after the practical application in the laboratory. This system with simple structure and cost-effective can obligate abundant resources and interfaces for the following experimental project extension and integration of the laboratory. It has good effect to the training of the basic skills, the engineering design and the scientific research, and to the developing of the experimental system design idea for undergraduates and masters of colleges and universities, which is welcomed by students and tutors. Many people have used this experimental platform to take part in the "Innovative Cup" science and technology competition organized by our school, the National Undergraduate Electronic Design Contest, the National Undergraduate "challenge cup" science and technology competition and so on, who obtained exceedingly good results. The establishment of the experimental system promoting the construction of the open laboratories of DSP, not only made the DSP opening teaching of our school a new level, but also promoted the teaching and research of relative courses.

References

1. Zhang, D.-h., You, Z.-z.: Research on Application of DSP J., in Many Fields. Computer Technology and Development (03), 71–73 (2006)
2. Li, J., Huang, S.-P., Xie, W.-c.: Journal of Electrical & Electronic Education 30(4), 54–55 (2008)
3. Texas Instruments Incorporated. CPU and Peripherals of the TMS320LF/LC24 Series DSP. Tsinghua University Press, Beijing (2004)
4. Sun, C.-j., Zhang, G.-d.: Design of the Teaching Experimental System Based on DSP. Science Times (1), 44–46 (2006)
5. Wang, G.-g.: Application and Development of DSP on Control of Electric Motors. Silicon Valley 22, 40–41 (2010)

Research on Golf Education in China: Its Significance, Characteristics and Future

Xueyun Shao[1], Xiaochun Zhang[1], and Xiaorong Chen[2]

[1] Golf College, Shenzhen University, Shenzhen 518060, Guangdong, China
Sherryshao_1982@hotmail.com
[2] Department of P.E., Shenzhen University, Shenzhen 518060, China
66608936@qq.com

Abstract. Golf has entered the educational field for sometime in China. It is not only reflected in the golf specialized education development, but also reflected in the golf sports courses which conduct in Universities, Middle and Primary School. However there is certain blindness and has appeared a few problems in practice due to lack of theoretical guidance. The paper attempts to make some theoretical analysis by discussing the unique educational significance of golf sports at first, then finding the characteristics of golf education, especially the golf higher education. Finally, the author explores some deep thoughts on future development of golf education in China.

Keywords: Golf Education, Significance, Characteristics, Future.

1 Introduction

The president of Shenzhen University Professor Bigong Zhang had insights of issues that higher golf education enters to the field of higher specialized education institutions when he was invited as a special guest for *the Golf Industry Operation & Management Forum* held at 2005 Guangzhou Golf Fair. He made following statement involving why Shenzhen University could initiate launch golf specialized education in the country or Southeast Asia: " From the major industrial development history, an important sign of its prosperity is the industry has become a part of higher education and set up a major in University or has a corresponding institutions of higher education. Similarly, from the University's history, an important symbol of institutions of higher education keep with the times is training talents for new and prospective industrial, doing research as well."

In fact, Golf has entered the educational field for sometime in China. It is not only reflected in the golf specialized education development, but also reflected in the golf sports courses which conduct in Universities, Middle and Primary School. However there is certain blindness and has appeared a few problems in practice due to lack of theoretical guidance.

The paper attempts to make some theoretical analysis by discussing the unique educational significance of golf sports at first, then finding the joint point between it and the schooling value, after that exploring the necessity and feasibility for the

Y. Wang (Ed.): Education Management, Education Theory & Education Application, AISC 109, pp. 905–911.

interaction between golf and education. Meanwhile, the author describes the history, current situation and future development of golf education in China.

2   Methodology

The purpose of this paper is to offer certain theoretical guidance by describing, discussing, analyzing and exploring the history, current situation and future development of golf education in China through literature searching, interviewing and logical analyzing methods.

2.1   Literature

By reviewing recent years' literature of golf education and reading lots of related scholars' research results which mainly come from the digital library of Beijing Normal University and the relevant foreign official website, the author obtained a comprehensive understanding of golf field research and its cutting-edge developments. In order to provide reliable theoretical basis and detailed information for the research, the author also read related books such as philosophy, society, economy, education and sports.

2.2   Interview

In the implementation process of this research, the author turned to expert for consultation and advice when problems arises, and adjusted the thoughts according to their opinion and proposal. Experts who have been interviewed include: the golf industry management personnel, school sports experts and golf industry staff, teachers for golf sports courses and golf specialized education.

2.3   Logical Analysis

The author obtained macro level of understanding and the related phenomenon by logical reasoning and analysis data or information coming after inspection or survey.

3   Results and Discussion

3.1   Analysis of Unique Educational Significance of Golf

We need to analyze what the different between golf and other sports is mentioned their educational significance. If a golf sport has no specific educational significance, then it will lose its educational foundation. After comparison, we conclude that golf has unique educational significance at least in eight aspects. They are dressing, honest, code of conduct, etiquette, strategy, strain capacity, environmental protection and the cultivation of communicative ability. (Figure 1)

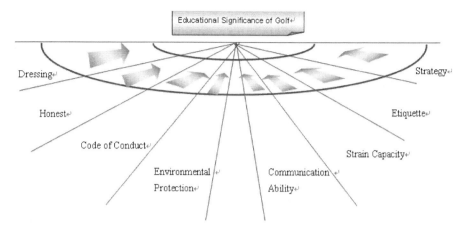

Fig. 1. Unique Educational Significance of Golf

Dressing Education. It should start from the British gentleman way when talking about golf apparel. There is no clearly definition in the rules of golf for golf dressing code, but the golf dressing etiquette is always the core performance for the game. Members of the British royal family were keen on golf games in 16th Century, educated by the long-term "gentleman culture ", they could be seen in smart clothes or suits when attending the upper Parliament, or social dance in public places, or riding, hunting, golfing. In a few hundred years history of the development process, standardized dress is still one of the most basic behaviors in golf etiquette culture. The golf dressing consciousness formatted with golf cultural value orientation has left deep "cultural memory" in people's mind, forms the social group consciousness with cultural guidance function.

Honest Education. In most cases, golf is a game without the supervision of a judge, which is a unique feature of golf when compared to other sports. Players often hit their balls out of public view in poor position such as the ball may go beyond the fairway, into the rough or surrounded by shrubs. Cheating is very easy under the circumstance if players have no self-discipline. The game of golf depends on each participant's honesty; integrity is one the most basic requirements needed for each golf players.

Code of Conduct Education. The reason that golf is known as a elegant, civilized, healthy game is the people's behavior when play it and has the target value orientation of "gentleman culture " advocated by them. In the course of social development of golf sports, people often consider golf as a "gentleman game" respecting for tradition, respecting for others, abided by the etiquette and emphasizing self-discipline. Golf has social and cultural elements such as "etiquette norms, cultural consciousness, self-discipline requirements" into practice.

Etiquette Education. Although there is not so much clause of etiquette in the rules of golf, but each rule contains the "gentleman culture" connotation inside. As the first chapter, the rules of golf formed the spiritual basis by opening it with three basic terms of rules, "manners on the course", "course priority" and "course protection" which also formed the foundation of long 34 clauses of rules advocated by the game of golf. The rules of golf has clearly stipulates that when a serious breach of etiquette behavior occurs punish can be made to disqualify the relevant players' right to play on course or certain periods of tournaments, requiring all players restrain their behavior consciously, always show courtesy and sportsmanship.

Strategy Education. Many ball games need for strategy. Golf game has no exception, but even can be said to having more outstanding in the meaning of strategy education. There are tens of thousands of golf course worldwide, each course has its own features and with different difficulties, bunkers, water hazards, rough, bend ways, woods, various terrain and greens. We can say it is special and make golf is different from other sports, facing with new challenges when players play on different clubs. No matter how technical level you have, the only way to reduce shots effectively is to use the appropriate strategy. On the same course, each shot needs to consider all of the aspects, such as wind direction, wind speed, ball position (for example in slope position) conditions, needs to use strategy to select club, to predict ballistic.

Strain Capacity Education. As mentioned above, there are no exactly two same golf courses in the world. Unique shape and different environment test golf players' strain and wisdom at each pitch moment. Golf not only has requirements on motor skills, but also requires player has good mentality and strain capacity. For example, selection of fairway wood with narrow head, smaller thickness of upper and lower is easy to hit slice ball if the ball fell into the long grass. At this moment, you should analysis carefully; handle calmly to reflect the strain capacity. It is more appropriate to choose the 56°sand wedge with location similar to split, ball in the middle of feet, slightly open the surface of club when gripping, adjust the swing amplitude according to the usual distance hit by sand wedge, by doing these to ensure the club head through rapidly when hitting the ball, make the shot when finish addressing, send the ball onto the fairway.

Environmental Protection Education. Education of environmental protection is more than cognition of the concept, but is the action of education. The game of Golf is the perfect combination of human beings and the natural environmental. Golf course construction and lawn maintenance provide rich source materials for the cultivation of students' environmental awareness. When teaching on course, it offers a direct and effective way to strengthen the students' awareness of environmental protection by observing the environmental problem, to lay foundation for further understanding of pursuit of green fashion and embracing the green life.

Communicative Ability Cultivation Education. The game of golf not only can achieve individual's self value to be respected, but also promote the perfection of individual socialization when communicate and interact with people who have different personality characteristics. Golf is a game used widely in social activity. It is a feature which different from other sports for its advanced etiquette come into moral norm. The golf course is a place to show civilization and moral code of conduct.

Good interpersonal relationship is developed in the specific environment and communication situations, meeting friends by enjoying playing, learning skills, conforming.

3.2   Characteristics of Golf Higher Specialized Education in China

The current status of golf higher specialized education in China is flourishing development and has a good development trend and conditions after less than 20 years of construction and development. Overall, golf higher specialized education in China has the following main features:

With Distinct Chinese Characteristics. Although ancient Chinese game Chuiwan is similar to the game of golf, the development of golf in contemporary China is indeed imported. China's golf industry development has made great progress and formed a complete industrial chain gradually in short time that is less than 30 years since the Springs Golf Course Zhongshan Guangdong founded in 1984, and the Shenzhen Golf Club in succession. China's golf industry development has very strong Chinese characteristics as a result of Chinese national condition. We can easily have some examples as the high-grade and refine of golf course construction, the bundling mode of high-grade real estate development and course construction, problems on membership sales and management and so on, they are full of strong Chinese flavors.

In accordance with issues we mentioned above, from the creation of golf specialized education in Shenzhen, it is close to the industrial development and orients to the market, cultivates practical marketable professionals by focusing on all aspects of industry demand. The related courses set can well reflect such a thought that is not only theory, but more practice, especially the cultivation of practical ability. In experience the ceaseless exploration from junior college to undergraduate, and postgraduate education, Golf College of Shenzhen University now more emphasis on the cultivation of compound talents, and also widened vision of specialized training from aspects of students future employment selection, and has achieved good social benefits.

Comprehensive and Integrated Discipline, Curriculum System. We can also take Golf College Shenzhen University as an example. In the undergraduate education level, Golf College of Shenzhen University make full use of advantages in comprehensive university, provide higher quality professional talents to the industry by continuously adjusting the talent training scheme to achieve the continuous training goal. The integrated curriculum system founded by the College comprised of three major specialize, they are *golf sports and training, industry operation and management, course management*, which not only had a profound influence to China's golf specialized education, but also offers a new pattern for the global golf specialized education.

The advantage of this characteristic is the comprehensive knowledge structure and ability of the talents trained by them can adapt to the demand of the initial period and the high-speed period of China's golf industry development.

The Diversification Form of Education. According to their own conditions, different administrative levels and targets, a existing dozen colleges and universities with higher diploma education have different ways to go. From the postgraduate education to undergraduate education, to college and higher occupational education, China golf higher education has various forms. One of the prominent features is each colleges or universities holds golf higher specialized education in accordance with their own teaching advantages. Some of them have the comprehensive specialized education, while others focus on professional education such as trade and business management, tourism management, athletes' sports. There will be a great number of higher occupational educations in the future.

Diversified school-running forms make those educational institutions foster strengths and circumvent weaknesses, combining with their own conditions to conduct different levels of academic education according to the various demands. The urgently needed talents by Chinese golf industry management will have full complement or promote through learning while supplemented by a large number of occupational skill training.

3.3 Thoughts on School-Running of Golf Education

Many colleges and universities, who have good perspective of the prospects of golf industry development, have already or plan to hold golf higher education Institutions. In these circumstances, what are we going to do is we should soberly realize and well answer the following questions:

Whether have the basic conditions for school-running of golf education?
The school conditions mentioned here refers to software and hardware aspects of the condition. As far as the soft condition concerned, whether the quantity and quality of teachers are able to meet the basic requirements for the new major, while speaking to the hardware conditions, specialized practice field and fixed pitch is the key to ensure the students' quality.

Whether the target orientation is clear enough?
Holding the golf higher education, the sponsor must have a relatively clear and correct positioning for the body's long-term development and gives greater support to it.

Whether the export or students employment can solute ideally?
For comprehensive universities, about half of golf specialized students seems not choose to find a job in the golf industry. But the majority higher occupational graduate might prefer to work for the industry, the area or neighborhood is the best choice. Students' employment problem is the premise of colleges and universities to strive for greater development, so it needs to make a plan as early as possible.

4 Conclusion

Because of limited time and space, this paper doesn't show the joint point between it and the schooling value, the necessity and feasibility for the interaction between golf and education. In the near future, the author has a plan to do a thorough research describing the history, current situation and future development of golf education in China.

Looking into the future, China golf higher specialized education will have such features: first, most of the personnel needed for China's golf industry will gradually transited to train by China golf education. Second, connotation of China golf education will ceaselessly abounded, specialty construction tend from rough to delicate, major setting will tend to refinement; third, golf higher specialized education increasingly take responsibilities of occupational skill training. We have reasons to believe that golf higher specialized education and occupational skill training will develop healthy and continuous with the community needs of golf industry increased, we look forward to the China golf higher specialized education, as the mainstay of China golf industry, will support a better future of the sun rising industry development in China.

References

1. Robert Price, R.: Golf Business Management in Scotland (2003)
2. Knopper, L.D: Use of non-destructive biomarkers to measure effects of pesticide exposure in meadow voles (Microtus pennsylvanicus) living in golf course ecosystems of the Ottawa/Gatineau region (2004), http://www.proquest.umi.com
3. Jean Makay, J.: Golf and the Environment around the World. USGA Green Section Record (2006)
4. Club Managers Association of America, R.:2008 Economic Impact Survey (2008)
5. NCAA, Sports and Championships Information (2009), http://www.ncaa.org
6. National Allied Golf Associations, R.: Finding report: Economic impact of golf for Canada, Canada (2009)

The Prediction of Publishing Scale of Literature Books in China—Based on GM (1, 1) Model

Renjing Xu

Department of Foreign Language, Nanchang Institute of Technology,
P.R. China, 330099
xubin9675@163.com

Abstract. In recent years, Chinese publishing industry has got rapid development. The literature books belong to book publishing also increased quickly, which main reasons are as follows: one is our country has rich and long literary treasure, such as the Tang poem, the Song Ci or the Yuan Qu etc. Since the ancient times, they become a kind of way masses pastime and studying, which edified sentiment, improved Nationals overall quality; The other is, in recent years, with the rapid development of our economy, people's living standards increasing quickly, and ordinary people realized the importance of supplying spiritual nourishment and increased spending on literature books, etc. In this paper, firstly, we introduced the present situation and research significance of publishing scale of literature book; Secondly, we used the grey forecast method and the data of publishing scale of literature book in the recent two decades made a prediction; Finally, we made a summary on the empirical analysis results obtained.

Keywords: Literature Books, Grey Forecasting Method, Publishing Scale.

1 Introduction

The cultural industry is an important carrier of, under the condition of market economy, booming and developing socialist culture, and it is an important point of adjusting economic structure and transforming economic mode. Since the 16th CPC national congress, China adopted a series of policy measures to promote the reform of the cultural system and cultural industry development. The Party's Seventeenth Congress put forward clearly that we should actively develop commonweal cultural undertakings, vigorously develop culture industry and more consciously, actively promote greater development and prosperity of socialist cultural. In the spirit guidance of the party central committee, in September 2009, China's first culture industry planning—"The cultural industry revitalization plan" rallied on, which pointed out that we fostered culture industry into new economic growth point for further improving cultural market main body and enhancing its development vigor. Serving as an important component of cultural industry departments and leading industry, publishing industry undertakes not only accumulating cultural and bearing sacred mission of inheriting civilization but also creating the material wealth and developing the national economy. With the improvement of Chinese economic strength and rapid expand of the type, quantity and consumer market of publications books, publishing industry has

Y. Wang (Ed.): Education Management, Education Theory & Education Application, AISC 109, pp. 913–921.

already become a complete economic activity areas. According to not complete count, the output value of news publishing industry has approach trillion and filled up 70% share of whole cultural industry in 2010.

As an important part of the publishing industry, literature books publishing relates to the overall situation of publishing industry and long-term goal of the overall national culture, which has received national attention. At the attention and support of the party and state leaders, Since the founding of the People's Republic of China， especially since the reform and open policy, the publishing industry of Chinese literature books has made a great progress. The publishing scale of literature book was 12233 ten thousand copies in 1990 and raised 26989 ten thousand copies in 2009, which increased 1.21 times. Although the publishing industry of our country's literature book has achieved proud achievement, compared with Europe and America developed countries, it still exist great disparity. In terms of quantity, the ratios of the scale of British and American literature books accounts to their whole scale of publishing books were more than 10%, and the ratio of Chinese literature books has been less than 4%, which was not congruent with our civilized country with a long history and cultural background. In view of book publishing industry, especially the importance and the existing problems of literary book publishing, domestic scholars have conducted extensive research on book publishing industry, and the main achievements were as follows: Xu Chaofu, HeYaoqin (2008) and Wu Wenfeng, etc (2008) deemed that although literature books had got rapid development, there were still many problems: Book authors and editing team people being less powerful, book pricing presenting a dilemma, marketing planning work not reach the designated position, circulation scale can't keep up. To solve above-mentioned problems, they put forward the countermeasures: select and cultivate talents, strive to nurture a new generation of literary authors and editors, strengthening marketing strategy and stimulate the publishing, etc. Gong Kui (2009) and Wu Wenfeng, etc (2009), in terms of the varieties, publishing structure and the press situation of sports book publishing our country sports book publishing and, participate in sports book publishing three Angles, made a clean up of the general conditions of sports book publishing after the founding of the People's Republic of China. He showed the development and characteristics of our country sports books in various historical stages and the relationship between publishing structures of each development stage. At last, he represented development course overview of our country sports book publishing career. Xiang Zhiqiang, Huang Ying (2009)and Xu Xiaosu (2010) analyzed effect degree of total control policies on the concentration degree of book publishing market and put forward the change trend of total control policies of Chinese book. Zhang Xi'an (2010), Ren Meilin, etc (2010) and Zhou Shandan (2011), on account of widespread phenomenon of content component shortage in the current book, and resources waste, hurt publishing integrity, violations readers consumer rights, raise book price and manufacture publishing foam, etc serious consequences resulted by it, put forward the following countermeasures: publishers self made integrity education and enhanced copyright consciousness, open special website or supervise web pages and govern the books with law by perfecting law, etc. In short, there were a lot of scholars studying on the development of book industry, and we couldn't introduce all of them. However, their researches and analysis have important reference value and quotable significance for our in-depth understanding on the development of book industry.

2   The Principles f Gray Prediction Method

2.1   Connotation

Gray prediction method can make a prediction on the system containing uncertain factors, which lies in between certain system and uncertain system. Through discriminating the different degree of development trend among system factors (correlation analysis) and generating processing on original data, we search the change law of system and generate a data sequence which has a strong law, and then we forecast the future developmental trend of things by setting up the corresponding differential equation model. A gray prediction method is generally divided into four types: (a) gray time series prediction method; (b) malformation prediction method; (c) system prediction method and (d) topology prediction method. The gray prediction model frequently used is GM(1,1) model. In this paper, we used GM(1,1) model when we carried out the prediction on the scale of China's treasury bonds. The principles of GM(1,1) model is introduced as follows.

2.2   Generation of Data Sequence

We have to process original data before setting up gray prediction model. The processed sequence has two kinds of cumulative time series and regressive time series. (a) Cumulative time series. We denote the original time series and the generated time series respectively by

$$X^{(0)} = \left\{X^{(0)}(1),\cdots,X^{(0)}(n)\right\}, X^{(1)} = \left\{X^{(1)}(1),\cdots,X^{(1)}(n)\right\}.$$

Similarly, the m-cumulative is $X^{(m)}(k) = \sum_{i=1}^{k} X^{(m-1)}(i)$;

(b) Regressive time series. It is obtained by $X^{(1)}(k) = X^{(0)}(k) - X^{(0)}(k-1)$.

2.3   Estimation of $\hat{\alpha}$

The calculation formula of $\hat{\alpha}$ is as follows: $\hat{\alpha} = \left(B^T B\right)^{-1} B^T Y_n$. Here

$$B = \begin{bmatrix} -\frac{1}{2}[X^{(1)}(1)+X^{(1)}(2)] & \cdots & 1 \\ -\frac{1}{2}[X^{(1)}(2)+X^{(1)}(3)] & \cdots & 1 \\ & \cdots & \\ -\frac{1}{2}[X^{(1)}(n\text{-}1)+X^{(1)}(n)] & \cdots & 1 \end{bmatrix}, Y = \begin{bmatrix} X^{(0)}(2) \\ X^{(0)}(3) \\ \cdots \\ X^{(0)}(n) \end{bmatrix}$$

2.4   Model Construction

The corresponding differential equations of GM(1,1) model is $dX^{(1)}/dt + aX^{(1)} = \mu$.

The solution is given by $\widehat{X}^{(1)}(k+1) = [X^{(0)}(1) - \mu/a]e^{-ak} + \mu/a, k = 0, 1, \cdots, n.$

2.5   Residual Test

Based on the prediction model $\widehat{X}^{(1)}(i)$, we compute $\widehat{X}^{(1)}(i)$ and regressive $\widehat{X}^{(1)}(i)$ to obtain $\widehat{X}^{(0)}(i)$, then we calculate the absolute error time series $\Delta^{(0)}(i)$ and the relative error time series $\phi(i)$ for

$$\Delta^{(0)}(i) = \left| X^{(0)}(i) - \hat{X}^{(0)}(i) \right|, \quad \phi(i) = \Delta^{(0)}(i) / X^{(0)}(i) \times 100\% .$$

2.6   Correlation Test

After calculating the correlation coefficient $\eta(k)$ of sequences $\widehat{X}^{(0)}(k)$ and $x^{(0)}(k)$, we have the correlation test $r = \sum\limits_{k=1}^{n} \eta(k)/n.$

2.7   Posteriori Variance Test

Let $S_1 = \sqrt{(\Sigma[X^{(0)}(i) - \overline{X}^{(0)}]^2)/(n-1)}, S_2 = \sqrt{\Sigma[\Delta^{(0)}(i) - \overline{\Delta}^{(0)}]^2/(n-1)}$

$C = S_2/S_1, S_0 = 0.6745S_1, P = P\{e_i < S_0\}, P = P\{\left| \Delta^{(0)}(i) - \overline{\Delta}^{(0)} \right| < S_0\}, e_i = \left| \Delta^{(0)}(i) - \overline{\Delta}^{(0)} \right|.$

Then the corresponding test results can be checked by the following Table 1.

Table 1. Posteriori Variance Test

P	C	
>0.95	<0.35	fine
>0.80	<0.50	qualified
>0.70	<0.65	Marginal quality
≤0.70	≥0.65	unqualified

If residual test, correlation test and posteriori variance test all pass, we can carry out prediction with the constructed model; otherwise, we have to do residual error amendment. If the residual test cannot pass, we have to modify the residuals of the constructed model GM(1,1) so as to improve the prediction accuracy. In this case, the process is as follows: Suppose the constructed model is:

$\widehat{X}^{(1)}(k+1) = [X^{(0)}(1) - \mu / a] e^{-ak} + \mu / a$, we can get the predict value $\widehat{X}^{(1)}$ of the generated time series $X^{(1)}$. The residual is defined by $e^{(0)}(j) = X^{(1)}(1), \cdots, \hat{x}^{(1)}(j)$. If $j=i+1,\ldots,n$, then the residual series corresponding to $X^{(1)}$ and $\widehat{X}^{(1)}$ is

$e^{(0)} = \left\{ e^{(0)}(i), e^{(0)}(i+1), \cdots, e^{(0)}(n) \right\}$; In order to facilitate the calculation, we rewrite it as

$e^{(0)} = \left\{ e^{(0)}(1'), e^{(0)}(2'), \cdots, e^{(0)}(n') \right\}$. The cumulative generated sequence of $e^{(0)}$

is $e^{(1)} = \left\{ e^{(1)}(1'), e^{(1)}(2'), \cdots, e^{(1)}(n') \right\}$, $n' = n - i$. By using $e^{(1)}$, we can obtain the corresponding GM(1,1) model as: $\hat{e}^{(1)}(k+1) = [e^{(0)}(1) - \mu_e / a_e] e^{-a_e k} + \mu_e / a_e$.

The amended $\widehat{X}^{(1)}(k+1)$ is obtained by $\hat{e}'(k+1) + \hat{e}^{(1)}(k+1)$.

The revised model is

$$\hat{X}^{(1)}(k+1) = [X^{(0)}(1) - \frac{\mu}{a}] e^{-ak} + \frac{\mu}{a} + \delta(k-1) + (-a_e)[e^{(0)}(1) - \frac{\mu_e}{a_e}] e^{-a_e(k-1)},$$

and the corrected coefficient $\delta(k-1) = 1$ if $k \geq 2$ otherwise 0. Finally, we can get the prediction model of original time series by amending residual error:

$$\hat{X}^{(0)}(k+1) = \hat{X}^{(1)}(k+1) - \hat{X}^{(1)}(k), k = 1, 2, \cdots n.$$

3   The Prediction of Publishing Scale of Literature Books in China—Based on GM (1, 1) Model

In this paper, based the data of published scale of literature books coming from the 1991-2010 China statistical yearbooks as the basic study data, we made a prediction on publishing scale of literature books with grey forecasting method in the next few years.

We firstly list the publishing scale of literature books from 1990 to 2009, which was shown as Table 2.

Table 2. The Publishing Scale of Literature Books from 1990 to 2009 Unit: Ten Thousand

Time	X	Time	X	Time	X
1990	12233	1997	15674	2004	13657
1991	14246	1998	14996	2005	13974
1992	14298	1999	9839	2006	15880
1993	11594	2000	10594	2007	16536

Table 2. (*continued*)

1994	12247	2001	12291	2008	23708
1995	12969	2002	12191	2009	26989
1996	13547	2003	14731		

Notes: X stands for the publishing scale of literature books. Sources: "China statistical yearbooks (from 1991 to 2010)".

3.1   The Construction of Cumulative Generated Time Series

It is not hard to know $X^{(1)}=\{12233, 26479, 40777,...,292194\}$.

3.2   The Construction of Matrix B and Data Vector Yn

$$B = \begin{bmatrix} -1/2(12233+14246) & 1 \\ -1/2(14246+11594) & 1 \\ \cdots\cdots\cdots & \cdots \\ -1/2(23708+26989) & 1 \end{bmatrix} = \begin{bmatrix} -19356 & 1 \\ -33628 & 1 \\ \cdots\cdots\cdots & \cdots \\ -278700 & 1 \end{bmatrix}, Y = \begin{bmatrix} X^{(0)}(2) \\ X^{(0)}(3) \\ \cdots\cdots \\ X^{(0)}(7) \end{bmatrix} = \begin{bmatrix} 14246 \\ 14298 \\ \cdots\cdots \\ 26989 \end{bmatrix}.$$

3.3   The Calculation of $\hat{\alpha}$

$$\hat{\alpha} = (B^T B)^{-1} B^T Y = \begin{bmatrix} a \\ \mu \end{bmatrix} = \begin{bmatrix} -0.0341226 \\ 9982.372 \end{bmatrix}$$

3.4   Model Construction

$dX^{(1)}/dt - 0.0341226X^{(1)} = 9982.372$, $X^{(0)}(1)= 12233, \mu/a = -292544.2962$,

$X^{(0)}(1) - \dfrac{\mu}{a} = 12233 - (-292544.2962) = 304777.2962$,

$X^{(1)}(k+1) = 304777.2962e^{0.0341226k} - 292544.2962$.

3.5   Residual Test

Regressive generated time series is:

$\hat{X}^{(0)} = \{12233,10579.26,\cdots,18896.56, 19552.49\}$;   Absolute and relative error sequences are respectively: $\Delta^{(0)} = \{0, 3666.74, 3351.52,\cdots, 4811.44, 7436.51\}$;
$\Phi = \{0, 25.74\%, 23.44\%, \cdots, 20.29\%, 27.55\%\}$.We find some relative errors in φ are more than 5%, which indicators relative errors are big. Therefore, residual test didn't pass.

3.6 Correlation Test

$\min\{\Delta_i^{(0)}\}$ = min{0, 3666.74, 3351.52,···, 4811.44, 7436.51}=0; $\max\{\Delta_i^{(0)}\}$ = max{0, 3666.74, 3351.52, ···, 4811.44, 7436.51}=7436.51; The degree of correlation is given by $r = \dfrac{1}{n}\sum_{i-1}^{n}\eta(i)$ =1/20(1+0.503+···+0.436+0.333)=0.641, which meets the test standard r>0.60 when p=0.5. Therefore, correlation test passed.

3.7 Posteriori Variance Test

By some computing processes, we have S_1=4098.08, S_2==1787.356, $\overline{\Delta}^{(0)}$ =2520.725, C=S_2/S_1=0.4361; S_0=0.6745S_1=2764.16, $e_k = \left|\Delta(k)-\overline{\Delta}\right|$ ={2520.7245, 1146.06, 830.80, ···, 4915.79}, Because not all $e_i < S_0$, the posteriori variance test wasn't passed; Because correlation test was not passed, so we have to modify the original model ($X^{(1)}(k+1) = 304777.2962e^{0.0341226k} - 292544.2962$,(k=0,1,2,...))to improve the accuracy of model. We amend the original model by using the residuals; the following results can be obtained: $e^{(0)}$={3666.74, 3351.52, 267.55,...,7436.51}, $e^{(1)}$={3666.74, 7018.26, 7285.81, ···,50414.49}.

$\hat{\alpha} = (B^T B)^{-1}B^T Y = \begin{bmatrix} -0.0789591 \\ 851.4211 \end{bmatrix} \cdot \dfrac{\mu}{a} = -10783.06$, $\hat{e}^{(1)}(k+1) = [e^{(0)}(1)-\dfrac{\mu}{a}]\, e^{0.0789591k}-$

10783.06=14449.80$e^{0.0789591k}$-10783.06.   The   derivative   of $\hat{e}^{(1)}(k+1)$ is 1140.94$e^{0.0789591k}$, the finally amended model is: X$^{(1)}$(k+1)=304777.2962$e^{0.0341226k}$- 292544.2962+ δ (k-1) 1140.94$e^{0.0789591(k-1)}$. The accuracy of the revised model is partially improved, and the results of calculating revised residuals are shown in Table 3.

Table 3. The Computation Table of Amended Residual

k	$X^{(1)}(k)$	$\hat{X}^{(1)}(k)$	revised error	relative error
1	12233	12233	0	0
2	26479	32524.07	6045.07	22.83%
3	40777	43749.08	2972.08	7.29%
4	52371	55358.91	2987.92	5.71%
5	64618	67366.54	2748.54	4.25%
6	77587	79785.32	2198.32	2.83%
7	91134	92629.08	1495.08	1.64%
8	106808	105912.07	895.93	0.84%
9	121804	119648.98	2155.02	1.77%

Table 3. *(continued)*

10	131643	133854.99	2211.99	1.68%
11	142237	148545.75	6308.75	4.44%
12	154528	163737.39	9209.39	5.96%
13	166719	179446.56	12727.56	7.63%
14	181450	195690.42	14240.42	7.85%
15	195107	212486.66	17379.66	8.91%
16	209081	229853.50	20772.50	9.94%
17	224961	247809.73	22848.73	10.16%
18	241497	266374.71	24877.71	10.30%
19	265205	285568.38	20363.38	7.68%
20	292194	305411.26	13217.26	4.52%

So far we have gotten the model passed residual amendments, k=20, $X^{(1)}(21) \approx$ 315924; k=21, $X^{(1)}(22) \approx 347129$; k=22, $X^{(1)}(23) \approx 383049$; k=23, $X^{(1)}(24) \approx 424707$. Therefore, the prediction values of the publishing scale of literature books from 2010 to 2012 can be obtained by $X^{(1)}(22)-X^{(1)}(21)=31205$, $X^{(1)}(23)-X^{(1)}(22)=35920$, and $X^{(1)}(24)-X^{(1)}(23)=41658$ respectively.

4 Conclusion

In this paper, through the gray prediction method, we gave out the fitted scales of publishing scale of literature books from 1990 to 2009 and prediction values from 2010 to 2012. By comparing the revised fitted values of publishing scale of literature books generated by the gray model with its corresponding actual values, we can found that the error rate increased in the initial few years, then gradually decreased. This result indicated that our prediction values were more and more accurate. At the same time, we also found that the publishing scale of literature books will enter a relatively stable growth period. The main reasons are as follows: firstly, after decades of constantly development, our book publishing industry has developed mature, and it can satisfy and reflects the demand of inhabitants on all kinds of books; Secondly, in recent years, our country social economic have got fast steady development. In this kind of environment, the stable development of book publishing was pretty much on autopilot.

References

1. Xu, C., He, Y.: The present situation and the countermeasures of Chinese children science popularization books. Journal of Hunan Normal University Social Science (6), 137–141 (2008)
2. Wu, W., Zhang, T., Jiang, S.: The situation analysis and trend prediction of Olympic publishing books. Journal of Beijing Sport University (2), 168–170 (2008)

3. Wu, W., et al.: An Overview on publishing process of New China sports book. Journal of Beijing Sport University (4), 42–45 (2009)
4. Gong, K.: The publishing status and edit literacy of Children's book. Journal of Shanxi Normal University (Social Sciences Edition) (11), 95–97 (2009)
5. Xiang, Z., Huang, Y.: The controlling policies of our total Books amount influence the market concentration degree of book publishing. Journal of Publishing Research (2), 34–36 (2009)
6. Xu, X.: A Reviews on book publishing industry research abroad Since 2003. Journal of Publishing Research (4), 72–76 (2010)
7. Zhang, X.: Short weight: diffculity bear responsibility of book publishing. Journal of Publishing Research (1), 50–52 (2010)
8. Ren, M., Yang, S., Zhang, L.: A research on the development trend of Chinese book publishing. Chongqing College of Arts of Journal (Natural Sciences Edition) (6), 106–109 (2010)
9. Zhou, S.: Attention economy and the innovation strategy of book publishing. Journal of Publishing Research (3), 22–24 (2011)

The Preference of Computers over Books and Anxiety among Iranian College Students: The Moderating Role of Demographic Factors

Sima Sharifirad[1] and Mohammad Sadegh Sharifirad[2]

[1] Student of Computer science, Bahonar University, Kerman, Iran
S.sharifirad@yahoo.com
[2] Master degree in Management, Tehran University, Tehran, Iran

Abstract. With the advent of computers and an increasing tendency toward the common use of new technologies such as computers, the preference between computers and books has remained as a controversial subject. Moreover, lack of equivalence between computers and books is attributed to different levels of Computer Anxiety Rating Scale (CARS) and Computer Attribute Scale (CAS). This study survey tries to consider different reasons of preference between books and computers as well as to delve into the demographic factors effective in their selection among 127 Iranian college students (85 males, 42 females). It is clearly shown that more respondents expected to learn more from books than from computers and one of the reasons behind it is they believe that books are more reliable than computers. Furthermore, the moderating role of factors such as gender, age, and parents' education is discussed.

Keywords: Adult learning, cross-cultural projects, gender studies.

1 Introduction

Books unlike computers have been around for centuries and since the influx of computers might now have developed the stereotype of being a little old-fashioned and dull, and perhaps even boring. However, with an increasing use of computers and online learning, we need a greater understanding of how pupils respond to computers and books (Noyes, Garland, 2006; Singh, O'Donoghue , 2002). . In a survey of 217 young people, Noyes and garland (2004) showed that books were perceived significantly more favorably on all the 10 affective scales of the Computer Attitude Measure (CAM; Kay, 1989, 1993). Noyas and Garland (2006) concluded that people held more positive attitudes toward books preferred to learn from them, and anticipated that they would learn more from books than computers.

Attitudes are an important consideration when attempting to explain the difference between books and computers. This is primarily because the amount of exposure people have to an object is a primary determinant in the opinions they subsequently form (Fishbein & Ajzen, 1975). People who have had a greater exposure to computers have been found to exhibit more positive attitudes than those individuals who have not had this opportunity (Bozionelos, 2001). We know in the population there are

Y. Wang (Ed.): Education Management, Education Theory & Education Application, AISC 109, pp. 923–932.
springerlink.com © Springer-Verlag Berlin Heidelberg 2011

people who are particularly anxious about using computers (see Brosnan, 1998). Some researchers have attempted to overcome this lack of explanation by collecting qualitative data about participants' computer experience. For example, Hall and Cooper (1991) asked participants to write short essays about their most successful and unsuccessful computer experiences. A number of studies found there is a negative relation between the total amount of experience one has with computers and the prevalence of computer anxiety (Heinssen, Glass and Knight, 1987). Simply put, the more experienced you are, the less anxious you will be. Computer anxiety and attitudes toward computers are two concepts closely related to computer acceptance and consequently to actual computer usage. The present study attempts to give an insight to the computer anxiety levels and attitudes toward computers of Iranian college students (Korobili, Togia and Malliari, 2010). It examines the relationship between anxiety and attitudes and a number of factors, such as age, gender, computer ownership, parents' education and the first day of using computers or books. It is noteworthy that, to the best knowledge of the authors, there is no previous research in Iran to discuss the preference of books over computers and the impact of anxiety.

Computer anxiety is real, and how long it persists is a function of the work environment and the training received, rather than a lack of desire (Torkzadeh & Angulo, 1992). These feelings may limit peoples' abilities to learn using computers. Smith and Caputi (2001) identified that people reporting high computer anxiety, experienced fewer enjoyment when thinking about using computers. A number of studies found that women were more computer anxious than men (Chou, 2003; Chua, Chen and Wong, 1999; Durndell and Haag, 2002; Heinssen et al., 1987; Jackson, Ervin, Gardner and Schmitt, 2001; Tsai, Lin and Tsai, 2001). However, many others found no significant differences between males and females with regard to computer anxiety (Anderson, 1996; Anthony, Clarke and Anderson, 2000; Havelka, Bunting, Tierney and Morse, 2008; Roussos, 2004; Sam, Othman and Nordin, 2005; Truell and Meggison, 2003; Yaghi and Abu-Saba, 1998; Mohamed and Beyerbach, 1999). Many studies have dealt with the relationships between computer anxiety and attitudes. In a number of studies, the relationship between computer anxiety and computer attitudes was negative (Durnell and Haag, 2002; Popvich et al., 2008; Sam et al., 2005). Igbaria and Chakrabarti (1990) found that computer anxiety played an important mediating role for computer attitudes only directly, through their effect on computer anxiety. Popvinch et al., (2008) found no significant relationship between computer courses and computer attitudes, while Igbaria and Chakrabarti (1990) found an indirect effect of computer courses on computer attitudes.Li and Kirkup (2007) found that British students were more likely to use computers for study purposes than Chinese students, but Chinese students were more self confident about their advanced computer skill.

2 Methodology

2.1 Design and Participants

A questionnaire was employed in order to gather data from participants. The sample consisted of 127 participants from Iranian colleges. In terms of gender, 85 were females (mean age 20.74, SD=109405) and 42 were males (mean age =21, SD=2.085).

Their age ranged from 21 to 28, with a median of 25.The students were majoring in various fields such as computer science, mathematics, economics, agriculture and chemistry.

2.2 Reserach Instrument

The Computer Attitude survey, derived from previous research(Shashani,1993). Our questionnaire was translated into Persian and then back into English to confirm the adequacy of the instrument. Our questionnaire included 5 sections: private questions that helps us to have enough moderating factors like age, gender, field of studying, students 'parents familiarity with English, their first age of becoming familiar with books and computers that are nearly mentioned in table 4. In the second section we asked about their ideas of books in table 2. In the third section we put our fingers on computers in table 3. In the fourth section we pay attention to their reasons of preferring books or computers that some of the questions provided in table 3, the final section of the questionnaires was presented in two measurements: computer anxiety rating scale and computer attitudes scale that their measurements are presented in table 6. A copy showing the relevant sections is given in Appendix A.CARS (Heinssen et al,1987) consists of 19 items rated on a Likert scale, wherein "1=strongly disagree" to "5=strongly agree". Although the CARS has demonstrated high internal consistency, reliability and stability, its developed underlined the need for cross validation and factor analysis with diverse populations. The CAS, developed by Liaw(2002),consists of 16 items aimed to assess the respondent's perceptions toward computer self-efficacy ,liking, usefulness, and intention to use and learn computers. Each item is scored on a seven-point Likrt scale(from "strongly disagree" to "strongly agree"),but for the purposes of the present study, responses were provided on a five-point Likert scale anchored by "1=strongly disagree to 5=strongly agree". Results showed that CARS could be adequately described by three moderately correlated factors with acceptable internal consistency (Cronbach's α ranged from .68 to .75). factor 1,labeled "fear of computers", represented uneasy or apprehensive feelings about computers. Factor 2 was associated with the ability to learn computer skills and it was labeled" computer learning". Factor3 was related to computer understanding and to respondents' eagerness to use computers and it was named" appeal of computers'. Three factors could be explain the underlying structure of the CAS with high internal consistency(Cronbach's α ranged from .79 to .83). Factor 1 dealt with perceived confidence in the ability to use and learn how to use a computer. Factor 2 assessed perceived usefulness of computers for the individuals. Finally, factor 3 measured one's feelings when using a computer. The three factors were labeled :perceived   Confidence".   "Perceived   usefulness"   and   "Affection "respectively.(Table6) We also explored the relationship between anxiety and attitudes toward computers and certain background characteristics on computer anxiety and attitudes. It should be mentioned that in order to accomplish the above objectives the underlying structure of Computer Anxiety Rating Scale (CARS), and Computer Attitude Scale (CAS) were examined and evidence on their validity was provided. (Korobili, Togia and Malliari, 2010). Computer anxiety was measured using CARS (Heinssen, Glass and Knight, 1987). CARS is more sensitive to age and differences in computer experience (Meier and Lambert, 1991). The sections assessed their attitude toward books and computers, while the bulk of the questionnaire was taken up with 17 statements were selected from the Levine and Donista-Schmidt (1998) "Computer Attitudes and Cinfidence" questionnaire which could be modified to apply equally well to books and computers was selected. For example, "A computer is like a private tutor" and "A book is like a private tutor" (Noyes, Garland, 2006). In addition to assessing family socioeconomic status, we collected data on father and mother's education. Parental education was coded as follows: no high school=1, some high school but no diploma=2, high school graduate=3, community college/bachelors degree=4, master's degree=5 and PhD=6. Various statistical techniques were employed in this survey to analyze data. Our data analysis showed that parental education had a strong effect on students' attitudes. Parental education was associated with sons' computer attitudes: more highly educated parents promoted their sons' computer confidence, reduced their sons' stereotypic views about computer users, and improved their attitudes about the effects of computers in daily life. The influence of parental education on females' attitudes was more significant: the higher the parents' education level, the greater their daughters interest in computers, the greater their confidence in working with computers, and the stronger their belief that computer is useful. The demographic statistics are shown in table 5.

3   Results

3.1   Attitudes toward Books and Computers

The responses obtained for the 41 statements, each statement was scored from 1 to 5 (according to whether it showed a negative or positive attitudes, respectively) to create separate book and computer attitude 'scores'. A Wilcoxon Signed Ranks found the 'mean' Book Attitude Score (9.07, SD=0.49088) was significantly more positive than the 'mean' Computer Attitude score (22.92, SD=0.499), z=-4.658, p<0.000. The Kaiser-Meyer- Oklin measures of sampling adequacy were .83 for the book items and .86 for the computer items while the Barletts Test of Sphericity reached statistical significance for books and computers, p<.000. The responses to the 20 statements for books and computers showed internal consistencies, as measured by the Cronbach alpha coefficient, of .82 and .75, respectively .The reliability analyses indicated that there was one item. You can get on in life without knowing about books, which if removed, would increase the alpha value to .79 for the computer items. Removing the matched item from the books' analysis would increase the alpha value to.85.

3.2   Effect of Background Characteristics on Computer Anxiety and Attitudes

Correlation between anxiety and attitudes and certain back-ground characteristics were examined (table 4). There are no points to examine gender differences, since only 31.74% of the respondents were males and no meaningful analysis could be performed .On the other hand, Spearmen's correlation coefficients revealed significant negative association between knowledge of English and the CARS subscales "fear of computers" and "appeal of computers", and positive relationship between knowledge of English and the "perceived confidence" and "affection" subscales of the CAS.

Table 1. Varimax rotation of four components solution for book items

Item	Component 1	Component 2	Component 3	Component 4
Books are fascinating	.750			
I learn more rapidly when I use a book	.747			-.323
Using books broaden our horizons	.745			
One can learn new things from a book	.711			.308
A book stops me from being bored	.574	.367		-.456
I hope I never have a job that requires me to use a book	-.433			
People who like books are often not very sociable			.761	
People managed in the past without book s ,so they are not really necessary now	-.331	.715		
You can get on in life without knowing about books		.378	.719	
The book is an effective learning tool	.398		.549	
I find using a book easy	.465			.715
Variance explained	**29.919**	**14.000**	**11.345**	**9.701**

Table2. Varimax rotation of four components solution for computer items

Items	Component 1	Component2	Component3
A computer is like a private tutor	.735		
I learn more rapidly when I use a computer	.712		
I find using a computer easy	.704		
Using a computer broaden your horizons	.065	-.440	.401
The computer is an effective learning tool	.490	-.673	.433
The computer is an educational tool		-.660	.405
I hope I never have a job that requires me to use a book		.610	
Every home should have one computer			.747
You can learn new things from computers			.702
I use a computer when I have nothing else to do	.485		.501
Variance explained	**30.678**	**13.276**	**11.908**

Persons correlation coefficient indicated significant correlation between frequency of computer use and level of computer anxiety (Korobili, Togia and Malliari, 2010).

3.3 Multivariable Association between CARS and CAS

As Thompson (2000) has pointed out, complex   phenomena can be better understood by employing multivariate techniques. Thus, a multivariable technique, namely canonical correlation analysis was implemented to assess the association between computer anxiety and attitudes subscales as measured by CARS and CAS. CAS subscales loading were positive, whereas the CARS subscales loading were negative. These statistics showed that 30.33 percent of variance in the CARS dimensions was accounted for by the CAS variation, on the other hand, 37.88 percent of the variance in the CAS dimensions was explained by the CARS variate. The redundancy index was used to examine the strength of the observed relationship. Several authors tend to interpret redundancy index values above 10% as significant and meaningful (e.g. Koustelions and Tsigilis, 2005; Ryska, 2003). The fact that computer anxiety and

Table 3. Reasons for book and computer preference

Reasons for book and computer preference	Agree	Disagree
Books are easier to read from ,with the computer screen being difficult, causing headaches ,etc	30	18
Books are easier to move	39	12
Books are silent ,the noise of computers can be distracting ,as can availability of other applications	30	29
More comfortable to read book	41	17
Problems and failures ,or books more reliable	23	26
Books are easier to annotate ,highlight, bookmark ,make notes from	32	24
Books provide easier and/or quicker access to information	14	39
Computers provide easier and/or quicker access to information	33	14
You can use books more than once in anytime	29	13
Computers can be time consuming, providing too much information	12	54
Books are easier ,simple to use ,less complicated	25	23
Information from books is more reliable	21	46
Information from computers is more reliable	10	59
Book information is easier to digest ,follow ,understand	19	42
Books are more familiar ,traditional ,habit been brought up with	30	31

Table 4. Demographic statistics

		N	X
Gender	Male	40	31.74
	Female	86	68.25
Age	<20	34	26.98
	20-22	69	54.76
	23-25	21	16.66
	>25	2	1.58
Knowledge of English	None	8	6.3
	Poor	27	21.4
	Good	60	47.6
	Very good	21	16.1
	Excellent	10	7.9
Age of first contact with computer	<10	14	11.1
	10-15	75	59.5
	16-20	34	27.0
	>20	3	2.4
Age of first contact with book	<10	116	92.1
	10-15	5	4.0
	16-20	4	3.2
	>20	1	.8

Table 5. Canonical loadings of the CAS and SARS subscales

	Function1
CAS	
Confidence	.89
Usefulness	.70
Affection	.81
CARS	
Fear	-.72
Learning	-.68
Appeal	-.82

attitudes share a considerable of variance, exceeding 30%, indicated that the two concepts are strongly related.

6 Conclusion

This study had two aims: First of all, to replicate the findings of an earlier study, which indicated that college students still favor books (Noyes and Garland, 2004), and secondly, to begin the understanding of why young people still prefer books to computers. . However, access to computers is not necessarily followed by actual use, since only a small percentage of the respondents used computers once or twice a week and none of them reported everyday use. Therefore, most Iranian students were not anxious and had rather positive attitudes toward computers.). As with computer anxiety, computer attitudes seemed to be affected by knowledge of English as well. Good command of English language promoted even more students computer attitudes. It is become vivid that one of the most important factor among Iranian students is knowledge of English and some of them believe that because of lack of knowledge it is hard for them to use computers. Consistent with expectations, results showed a high relationship between the two concepts, computer anxiety and attitudes.

Previous research carried out by the authors (Noyes and Garland, 2004) indicated that undergraduates expected to learn more from books and they preferred them to computers for learning purposes. Comparisons of the previous and the current studies indicated that similar percentages of respondents would learn more from computers than books. Books were viewed more favorably than computers as shown by the overall book and computer scores. Reasons for this choice related primarily to the physical and practical aspects of the two media, the tradition associated with books emanating from life time experience with them, and aesthetic consideration. Overall ,there is a slight preference for books in terms of ease to use ,but also relating to the type of information the two media contain. Chalmers (2000) concluded that the challenge is "to make computers easier to use, and therefore, accessible to all learners.". Moreover, books were thought to be more reliable than computers. On the other hand, educators should prepare technology based courses to minimize computer anxiety and maximize computer attitude. Even though a negative correlation between anxiety and attitudes was found, and anxiety explained more variance of the attitudes than vice versa, no inference can be made about their causal relationship.

References

1. North, A.S., Noyes, J.M.: Gender influences on children's computer attitudes and cognitions. Computers in Human Behavior 18, 135–150 (2002)
2. Noyes, J., Garland, K.: Explaining students' attitudes toward books and computers. Computers In Human Behavior Research 22, 351–363 (2006)
3. Noyes, J.M., Garland, K.J.: Students' attitudes toward books and computers. Computers in Human Behavior (published online April 17, 2004), doi:10.1016/j.chb.2004.02.016
4. Kay, R.H.: A practical and theoretical approach to assessing computer attitudes: The computer (1989)

5. Kay, R.H.: A practical tool for assessing ability to use computers: The computer ability survey(CAS). Journal of Research on Computing in Education 26, 16–27 (1993)
6. Fishbein, M., Ajzen, I.: Belief, attitude, intention and behavior: An introduction to theory and research. Addison-Wesley, Boston (1975)
7. Brosnan, M.: Technophobia: The psychological impact of information technology. Routledge, London (1998)
8. Hall, J., Cooper, J.: Gender, experience and attributions to computers. Journal of Education and Computer Research 7, 51–60 (1991)
9. Heinssen, R.K., Glass, C.R., Knight, L.A.: Assessing computer anxiety: development and validation of the Computer Anxiety Rating Scale. Computers in Human Behavior 3(1), 49–59 (1987)
10. Torkzadeh, G., Angulo, I.E.: The concept and correlates of computer anxiety. Behaviour & Information Technology 11(2), 99–108 (1992)
11. Smith, B., Caputi, P.: Cognitive interference in computer anxiety. Behaviour and Information Technology 20, 265–273 (2001)
12. Chou, C.: Incidences and correlates of Internet anxiety among high school teachers in Taiwan. Computers in Human Behavior 19, 731–749 (2003)
13. Chua, S.L., Chen, D., Wong, A.F.L.: Computer anxiety and its correlates: A meta-analysis. Computers in Human Behavior 15, 609–623 (1999)
14. Durndell, A., Haag, Z.: Computer self efficacy, computer anxiety, attitudes towards the Internet and reported experience with the Internet, by gender, in an East European sample. Computers in Human Behavior 18, 521–535 (2002)
15. Heinssen, R.K., Glass, C.R., Knight, L.A.: Assessing computer anxiety: development and validation of the Computer Anxiety Rating Scale. Computers in Human Behavior 3(1), 49–59 (1987)
16. Jackson, L.A., Ervin, K.S., Gardner, P.D., Schmitt, N.: Gender and the internet: Women communicating and men searching. Sex Roles 44(5/6) (2001)
17. Tsai, C., Lin, S.S.J., Tsai, M.: Developing an Internet Attitude Scale for high school students. Computers & Education 37, 41–51 (2001)
18. Anderson, A.A.: Predictors of computer anxiety and performance in information systems. Computers in Human Behavior 12(1), 61–77 (1996)
19. Anthony, L.M., Clarke, M.C., Anderson, S.J.: Technophobia and personality subtypes in a sample of South African university students. Computers in Human Behavior 16, 31–44 (2000)
20. Roussos, P.: The Greek computer attitudes scale: Construction and assessment of psychometric properties. Computers in Human Behavior 23, 578–590 (2004)
21. Sam, H.K., Othman, A.E.A., Nordin, Z.S.: Computer self-efficacy, computer anxiety, and attitudes toward the Internet: A study among undergraduates in Unimas. Educational Technology & Society 8(4), 205–219 (2005)
22. Truell, A.D., Meggison, P.F.: Computer Anxiety of community college students: Implications for business educators. The Delia Pi Epsilon Journal XLV(2), 87–97 (2003)
23. Yaghi, H.M., Abu-Saba, M.B.: Teachers' computer anxiety: An international perspective. Computers in Human Behavior 14(2), 321–336 (1998)
24. Durnell, A., Haag: Computerself-efficacy,computer anxiety,attitudes toward the Internet and reported experience with the Internet, by gender, in an East European sample. Computers in Human Behavior 18, 521–535 (2002)
25. Popovich, P.M., Gullekson, N., Morris, S., Morse, B.: Comparing attitudes towards computer usage by undergraduates from 1986 to 2005. Computers in Human Behavior 24, 986–992 (2008)

26. Sam, H.K., Othman, A.E.A., Nordin, Z.S.: Computer self-efficacy, computer anxiety, and attitudes toward the Internet: A study among undergraduates in Unimas. Educational Technology & Society 8(4), 205–219 (2005)
27. Igbaria, M., Chakrabarti, A.: Computer anxiety and attitudes towards microcomputer use. Behaviour & Information Technology 9(3), 229–241 (1990)
28. Anthony, L.M., Clarke, M.C., Anderson, S.J.: Technophobia and personality subtypes in a sample of South African university students. Computers in Human Behavior 16, 31–44 (2000)
29. Li, N., Kirkup, G.: Gender and cultural differences in Internet use: A study of China and the UK. Computers and Education 48, 301–317 (2007)
30. Tekinarslan, E.: Computer anxiety: A cross-cultural comparative study of Dutch and Turkish university students. Computers in Human Behavior 24, 1572–1584 (2008)
31. Liaw, S.: An Internet survey for perceptions of computers and the World Wide Web: relationship, prediction, and difference. Computers in Human Behavior 18, 17–35 (2002)
32. Meier, S.T., Lambert, M.E.: Psychometric properties and correlates of three computer aversion scales. Behavior Research Method, Instrument and Computers 23, 9–15 (1991)

Aligning COBIT and ITIL with an IT Academic Courses

Vanco Cabukovski[1] and Vase Tusevski[2]

[1] Faculty of Natural Sciences and Mathematics,
University "Sts. Cyril and Methodius", Skopje, Republic of Macedonia
cabukv@hotmail.com
[2] MIT University, Skopje, Republic of Macedonia
vase.tusevski@mit.edu.mk

Abstract. In this paper a COBIT framework was discussed as a part of IT academic courses at the MIT University in Skopje, Republic of Macedonia. MIT Faculty of Information Technologies develops unique and professional standards into the higher education. It offers modern faculty premises and facilities with state-of-the-art IT infrastructure, research and academic collaboration with Universities around the globe and affiliation with many industry partners. Aligning COBIT courses with academic subjects, will make COBIT framework understandable to students and can prepare them to successfully attend professional COBIT courses and take the COBIT Certificates.

Keywords: COBIT framework, ITIL framework, IT academic courses.

1 Introduction

Many organizations rely on IT to support business operations and meet strategic objectives. They are faced with many challenges like:

- keep IT running - any technical failure produces IT systems become unavailable and consequently the business to become unavailable;
- value - ensuring that IT provides value;
- costs - manage IT costs as carefully as other business costs;
- mastering Complexity;
- align IT with Business;
- regulatory Compliance;
- security - keep adequate security in their IT environment.

IT governance is a structure of relationships and processes. It helps achieve business goals by adding value, balancing risks and gaining returns of investments. IT governance comprises: principles (responsibility, accountability, activities); internal and external stakeholders (suppliers, customers, general public, users, governments); scope (strategic alignment - aligning IT with business, value delivery, risk management, resource management, performance management).

Y. Wang (Ed.): Education Management, Education Theory & Education Application, AISC 109, pp. 933–938.
springerlink.com © Springer-Verlag Berlin Heidelberg 2011

Some benefits of IT governance are:

- confidence of the top management - top management has a clear picture of how IT is performing. It increases its trust and confidence in investment decisions;
- responsiveness of IT to business - IT is more responsive to business needs;
- higher Return of Investments (ROI) - effective IT governance helps reduce projects failures, optimize IT infrastructure and increase the efficiency of IT processes;
- more reliable services - lower risk, better quality of services and greater customer satisfaction
- more transparency - right information is available to the right level of decision makers.

The Control Objectives for Information and related Technology (COBIT) is a set of best practices (framework) for information technology (IT) management created by the Information Systems Audit and Control Association (ISACA), and the IT Governance Institute (ITGI) in 1996 [2],[3],[4],[5].

COBIT provides managers, auditors, and IT users with a set of generally accepted measures, indicators, processes and best practices to assist them in maximizing the benefits derived through the use of information technology and developing appropriate IT governance and control in a company.

2   The COBIT Framework

COBIT is a proven set of standardized processes that businesses can use to ensure that information technology is effectively and securely integrated with business goals. COBIT (Control Objectives for Information and related Technology) was developed by ITGI/ISACA in the early 1990's and has evolved into a global standard for control over IT processes. COBIT is a framework and a knowledge base for IT processes and their management. This framework is built with references to existing standards and practices. It is a practical management tool rather than a "definitive standard" which enables IT personnel, business people and audit and control specialist to relate to COBIT easily.

The COBIT Framework has three components:

- Business Requirements / Information Criteria - What stakeholders expect from IT;
- IT Resources - a means to identify the resources required to execute processes;
- IT Processes - how IT is organized to meet requirements.
  The short description of each of these components follows.

The Business Requirements comprises: Quality Requirements (Quality, Delivery, Cost), Security Requirements (Confidentiality, Integrity, Availability) and Fiduciary Requirements (Effectiveness and efficiency of operations, Compliance with laws and regulations, Reliability of financial reporting).

The Information Criteria is based on:

- Effectiveness – deals with information being relevant and pertinent to the business process as well as being delivered in a timely, correct, consistent and usable manner;
- Efficiency – concerns the provision of information through the optimal (most productive and economical) usage of resources;
- Confidentiality – concerns protection of sensitive information from unauthorized disclosure;
- Integrity – relates to the accuracy and completeness of information as well as to its validity in accordance with the business's set of values and expectations;
- Availability – relates to information being available when required by the business process, and hence also concerns the safeguarding of resources;
- Compliance – deals with complying with those laws, regulations and contractual arrangements to which the business process is subject, i.e., externally imposed business criteria;
- Reliability of information – relates to systems providing management with appropriate information for it to use in operating the entity, providing financial reporting to users of the financial information, and providing information to report to regulatory bodies with regard to compliance with laws and regulations.

IT Resources are managed by IT processes to provide information that organization needs to achieve its objectives: Applications, Information, Infrastructure, People.

The IT Process are defined and classified in four domains: Plan and Organize, Acquire and Implement, Deliver and Support, Monitor and Evaluate. There are 34 processes across the 4 domains and activities (actions required to achieve measurable results).

ISACA offers following COBIT courses:

- COBIT Foundation Course: 2 or 2,5 day classroom training course or 8 hour online training
- COBIT Foundation Certificate: 40 multiple – choice questions, To pass the exam – correct answers on 28 or more questions (score of 70% or higher)

The audience for COBIT Foundation Courses and Certificates are: IT auditors, IT managers, IT quality professionals, IT leadership, IT developers, process practitioners and managers in IT service providing firms.

The COBIT Foundations Course is consisting of the following topics:

- Module 1 – Responding to IT Challenges
- Module 2 – Introducing COBIT
- Module 3.1 – What Does COBIT Provide – Part 1
- Module 3.2 – What Does COBIT Provide – Part 1
- Module 4 – Applying COBIT in practice
- Module 5 – Products and supports from ITGI.

3 The ITIL Framework

The IT Infrastructure Library is a set of books comprising an IT service management Best Practices framework [1]. ITIL is created by and for the British government, later expanded for use in all organizations. It gives a detailed description of important IT practices, with comprehensive checklists, tasks, procedures and responsibilities. And can be tailored to any IT organization. IT service providers use ITIL concepts and practices to:

- Increase satisfaction of customers / users with
- IT services
- Enhance communication with customers
- Achieve higher reliability in mission-critical
- systems and infrastructure
- Improve the cost/benefit of services
- Create a "common sense" among staff

The IT Service Management Forum (itSMF) is an independent and internationally-recognized forum for IT Service Management professionals worldwide. This not-for-profit organization is a prominent player in the ongoing development and promotion of IT Service Management "best practice", standards and qualifications and has been since1991. The itSMF is concerned with promoting *ITIL* (the IT Infrastructure Library), Best Practice in IT Service Management and has a strong interest in the International ISO/IEC2000 standard. ITIL is a documented set of processes designed to define how a company's IT functions can operate. ITIL contains a series of statements defining the procedures, controls and resources that should be applied to a variety of IT – related processes. ITIL covers all the major areas of interest that concern today's IT executive. ITIL is an excellent starting point from which to build your IT service management system. ITIL allows for job specialization.

IT Service Management is consisting of the following subject topics:

- Introduction to IT Service Management
- Designing, developing and consulting IT services
- Service design
- Configuration management
- Service operation
- User training and education services
- System's administration services
- End user support services
- Maintenance services
- Customer.

4   The under Graduate Study Programme at MIT Faculty of Information Technologies

MIT Faculty of Information Technologies develops unique and professional standards in the Macedonian higher education. It offers modern faculty premises and facilities with state-of-the-art IT infrastructure, Research and academic collaboration with Universities around the globe and Affiliation with many industry partners in Macedonia.

Some faculty strengths:

- Vocational relevance (employability)
- Practical teaching(assignments and case studies)
- Facilitating learning
- Interactive/active learning and creative thinking
- Online course management systems
- Teaching quality assurance
- Student portfolio
- Proactive on student feedback

The general MIT development premise is: No academic programme can stand alone. It is important to form value-added alliances in a number of areas that include other academic programs, top management from leading corporations in industry, and trade associations.

Following this premise last year MIT University (MIT Faculty of Computer Sciences and Technology) aligned ITIL Foundation Course and ITSM (IT Service Management) subject. More than 50% of students were prepared to attend ITIL Foundation Course and take the certificate. This percentage could be higher if ITSM Subject was in the fifth semester (not in the third semester). It is too early for the students to successfully connect this experience with the theory and practice of the subjects in the first and second semester.

A subject aligned to COBIT Foundation Course could be introduced in a curricula in the sixth semester.

Aligning COBIT Foundation Course with Academic subject, will make COBIT understandable to students and can prepare them to successfully attend COBIT Foundation Course and take the COBIT Foundation Certificates. MIT University has the experience in this kind of alliance and can make it successfully.

5   Conclusion

COBIT is a proven set of standardized processes that businesses can use to ensure that information technology is effectively and securely integrated with business goals. COBIT is a framework and a knowledge base for IT processes and their management. The IT Infrastructure Library (ITIL) is a set of books comprising an IT service management Best Practices framework. These two frameworks are worldwide recognized for professional certification. Aligning these frameworks with IT academic courses can move forward the students in understanding and preparing them to successfully attend professional COBIT courses and take the COBIT Certificates. Last year MIT University (MIT Faculty of Computer Sciences and Technology) aligned ITIL Foundation Course and ITSM (IT Service Management) subject from third semester. The results obtained shown that it is too early for the students to successfully connect this professional experience with the theory and practice of the subjects in the previous semesters. Better results could be obtained if the alignment was done with subject in the fifth semester. A subject aligned to COBIT Foundation Course could be introduced in a curricula in the sixth semester.

References

1. Aligning COBIT, ITIL and ITIL 17799 for Business Benefit: Management Summary, IT Governance Institute (2005)
2. Ridley, G., et al.: COBIT and its Utilization: A framework from the literatute. In: Proceedongs of the 37th Hawaii International Conference on System Sciences. IEEE, New York (2004)
3. Weill, P., Ross, J.W.: IT Governance: How Top Performers Manage IT Decision Rights for Superior Results. Harvard Business Scholl Press, Boston (2004)
4. Cobit Mapping, Mapping of ITIL v3 with Cobit 4.1, IT Governance Institute (2008)
5. Cobit 4.1, IT Governance Institute (2007)

Investigation and Analysis of Current Situations of Participation in Traditional Sports Activities of Ethnic Minorities by Urban and Rural Kazak, Kyergyz, Mongol and Tajik Residents in Xinjiang

EnLi Han

Langfang Normal University
hyxjedu@163.com

Abstract. With questionnaire investigation and interview investigation, investigation and analysis of current situations of participation in traditional sports activities of ethnic minorities by urban and rural Kazak, Kyergyz , Mongol and Tajik residents in Xinjiang（"4 ethnic groups in Xinjiang"）were made. The results show that : the 4 ethnic groups report most residents participating in traditional sports activities of ethnic minorities by 1-3 times a year and basically for more than one hour for each of the activities, with horseback sport events leading. Most of the sports activities are organized by the peoples themselves, and held on holidays and festival days of their own groups respectively. Shortage of sports field and no enough time for exercises are the main reasons influencing and restraining the residents' participation in the sports activities. Changing living idea, living environment and economic conditions and popularization of modern refreshment and entertainment mode and competitive sport events have made big impact on carry-on and development of traditional sports activities of the ethnic monitories.

Keywords: Xinjiang, Kazak, Kyergyz, current situation.

1 Introduction

Xinjiang is located at the far frontier of Northwest China and boasts vast land, where many ethnic minorities enjoy their different lives. Due to different historical conditions in politics, economy, language, custom and natural and geographical environment, the ethnic minorities have developed their own social and living characteristics and cultural traditions different from each other in the long history, created and carried forward the sport and cultural events with their own characteristics. In order to understand the current situations of participation in the traditional sports activities by urban and rural residents of the ethnic groups in Xinjiang, we made serious investigation of the current situations in the field, and processed and analyzed relevant study data, revealing the current situations, the law and the factors influencing and restraining the participation in the traditional sports activities by urban and rural residents of the 4 ethnic groups in Xinjiang.

Y. Wang (Ed.): Education Management, Education Theory & Education Application, AISC 109, pp. 939–943.
springerlink.com © Springer-Verlag Berlin Heidelberg 2011

2 Study Object and Method

The study object in this paper consists of the Chinese citizens from one family at the age over 16 and with the date of birth closest to July 1, and the study adopts questionnaire investigation method and interview method, with investigation samples of 438 urban and rural Kazak residents from Yili Kazak Autonomous Prefecture, 509 urban and rural kyergyz residents from Kezlesu Kyergyz Autonomous Prefecture, 499 urban and rural Mongolian residents from Boertala Mongol Autonomous Prefecture and 500 urban and rural Tajik residents from Tashkuergan Tajik Autonomous County, making 1946 sample questionnaires in total.

3 Investigation Results and Analysis

3.1 Distribution of the Urban and Rural Residents Investigated

shows that of the 1946 residents investigated, 998 are urban residents, accounting for 51.3%; 948 rural residents, accounting for 48.7%. The proportion of urban residents to rural residents is close to 1:1, with the urban residents slightly more than the rural residents. Of the 1946 residents, 1203 are male, accounting for 61.8%; 743 are female, accounting for 38.2%, with Kyergyz male to female proportion the highest, of which 359 are male, accounting for 70.5%; 150 are female, accounting for 29.5%.

3.2 Proportion of the Residents Loving Modern Competitive Sports Activities to the Residents Loving Traditional Sports Activities of the Ethnic Minorities

Percentages of the residents' favorite of the sport activities are as follows in turn : 44.3% of them love the traditional sports activities, 41.2% both and 10.0% competitive sports activities and 4.5% give no consideration.

The above-mentioned percentages show that 41.2% of the residents from local ethnic groups love both the modern and the traditional sport activities, while 10.0% prefer to modern competitive sports activities, disclosing that modern competitive sports activities have made certain impact on practice and carry-on of the traditional sports activities of ethnic minorities in the regions, and the scope of the impact is enlarging.

3.3 Reasons Influencing the Residents' Attitude on Participating in the Traditional Sports Activities and Those for Demand Change

Percentages of the reasons influencing the residents' attitude on participating in the traditional sports activities and those for demand change are as follows in turn : living idea change 21.5% = living environment change 21.5% > economic condition change 21.0% > modern refreshment and entertainment mode 12.2% > production and labor mode change 9.7% > influence of modern sport events 8.6% > others 5.4%. Changes of the residents' living idea, living environment and economic conditions have become the main reasons influencing the residents' attitude on participating in the traditional sports activities and those for demand change, and modern refreshment and entertainment mode and modern sport events have also had certain influence on residents' attitude on participating in and the demand for the traditional sports activities.

3.4 Times of the Residents' Participation in the Traditional Sports Activities

shows percentages of the residents' participation in the traditional sports activities as follows in turn : 1-3 times a year, 44.6% > 4-6 times a year, 21.4% > 1-3 times a month, 21.3% > 4-6 times a month, 5.5% > 7-9 times a month, 2.3% > 10-12times a month, 1.7% = more than 13 times a month, 1.7%, with most residents participating in the traditional sports activities 1-3 times a year.

3.5 Time for the Residents' Participation in One of the Traditional Sports Activities

Percentages of the time for the residents' participation in one of the traditional sports activities are as follows in turn : over one hour, 70.5% > 30 minutes- one hour 15.3% > within 10 minutes7.5% > 10-30minutes 6.7%. The above-mentioned percentages show that residents' time for traditional sports activities is basically more than one hour.

3.6 Events of the Traditional Sports That the Residents Participate in

Percentages of the events that the residents often participate in are as follows in turn: horse racing, 32.5% > Sheep Snatching on Horseback, 28.3% > ethnic wrestle, 10.7% > others 8.0%.

Kazak : horse racing, 37.8% > Sheep Snatching on Horseback, 33.3% > Girl's Chasing, 13.3% > ethnic wrestle, 10.0%. Kyergyz : horse racing, 25.0% > Sheep Snatching on Horseback, 23.5% > ethnic wrestle, 14.7% > wrestling on horseback8.8%. Mongol : horse racing, 51.2% > ethnic wrestle, 14.9% = Sheep Snatching on Horseback,14.9% > camel racing, 6.4%. Tajik : Sheep Snatching on Horseback, 47.6% > others 25.2% > horse racing, 9.7% > back-type tug-of-war, 7.8%. The four ethnic groups investigated mainly live on animal husbandry, and their events of traditional sports are mostly horseback ones, of which horse racing , Sheep Snatching on Horseback, Girl's Chasing , Picking up Silver Article on Horseback, Wrestling on Horseback, Picking up Hada on Running Horse and Shooting Arrow on running Horse account for 67.9% of the total.

3.7 The Reasons Why the Residents Give Up Participation in the Traditional Sports Activities

Percentages of the requirements and desires of the residents for development of the traditional sports activities are as follows in turn: establishing the ground suitable for carrying out the traditional sports activities, 48.3% > enhancing publicity of knowledge and skills of the traditional sports, 41.7% > not sure, 9.0% > others, 0.9%. Shortage of the ground for the traditional sports activities has become the biggest obstacle against development of the traditional sports activities. The top requirement and desire of the residents for development of the traditional sports activities is establishing the ground suitable for the activities and enhancing publicity of knowledge and skills of the traditional sports activities.

4 Conclusion

Of the four ethnic groups in Xinjiang, most residents participate in the traditional sports activities 1-3 times a year, and the time for an event is basically more than one hour per times. As the four ethnic groups mainly live on animal husbandry, most of their residents like to participate in horseback events such as horse racing, Sheep Snatching on Horseback, Girl's Chasing, Picking up Silver Article on Horseback, Wrestling on Horseback, Picking up Hada on Running Horse and Shooting Arrow on Running Horse.

With progress of the times, quickening modernization and urbanization and rapid economic growth, great changes have taken place in the residents' living idea, living environment and economic conditions, and modern refreshment and entertainment mode and modern sport events keep popularizing, which have made influence on

Table 1. Distribution of the Urban and Rural Residents Investigated

		Ethnic Group				Total
		Kazak	kyergyz	Mongol	Tajik	
Urban	N	245	255	248	250	998
	%	55.9%	50.1%	49.7%	50.0%	51.3%
Rural	N	193	254	251	250	948
	%	44.1%	49.9%	50.3%	50.0%	48.7%
M	N	257	359	265	322	1203
	%	58.7%	70.5%	53.1%	64.4%	61.8%
F	N	181	150	234	178	743
	%	41.3%	29.5%	46.9%	35.6%	38.2%

Table 2. Time for the Residents' Participation in one of the Traditional Sports Activities

		Ethnic Group				Total
		Kazak	kyergyz	Mongol	Tajik	
Activity Time	Within 10minutes — N	0	0	3	24	27
	%	0.0%	0.0%	1.8%	20.9%	7.5%
	10minutes -30minutes — N	3	3	8	10	24
	%	7.9%	7.3%	4.8%	8.7%	6.7%
	30minutes -1 hour — N	4	5	24	22	55
	%	10.5%	12.2%	14.5%	19.1%	15.3%
	Over 1 hour — N	31	33	130	59	253
	%	81.6%	80.5%	78.8%	51.3%	70.5%
Total		38	41	165	115	359

practice of the traditional sports activities. Governments and sports organizations at various levels and relevant social organizations should strengthen propaganda of traditional sports activities of the ethnic minorities, establish various sports associations, and enrich interest and fitness effect of the activities, in order to attract more social members to join the activities by making proper change of the activities while reserving their traditional contents.

Shortage of sports fields and insufficient time for exercise are main elements influencing and restraining residents' participation in the traditional sports activities.

Governments and sports organizations at various levels should increase input of funds in the field and enhance construction of the ground for traditional sports activities. With quickening urbanization, urban population keeps increasing, but it has become very hard to construct the field for and carry out traditional sports activities in urban area due to limitation of place. Therefore, relevant grounds should be constructed in cities and near the communities with compact-living residents of ethnic minorities for carrying out their traditional sports activities, so as to create necessary conditions for people of local ethnic minorities to participate in the sports activities that they love.

Acknowledgment. National Philosophy and Social Sciences project (08XTY009); colleges and universities research projects of autonomous region (XJEDU2005I29)

References

1. Ethnic Affairs Commission of Xinjiang Uygur Autonomous Region. A Collection of Events of Traditional Sports Activities of Ethnic Minorities in Xinjiang. Xinjiang Renmin Press, Urumqi (August 2006) (in Chinese)
2. Wu, J.: Study of Sports of Ethnic Minorities in Xinjiang. Xinjiang Renmin Press, Urumqi (2007) (in Chinese)
3. China Mass Sports Investigation Subject Group. Investigation and Survey of Current Situations of Mass Sports in China. Beijing Sport University Press, Beijing (April 2005) (in Chinese)
4. Lu, Y.: China Sport Sociology. Beijing Sport University Press, Beijing (1996) (in Chinese)
5. Lu, P.: Behavior Characteristics of People of Ethnic Minorities in Northwest China in Participating in Traditional Sports Activities. Journal of Shanghai University of Sport (May 2007) (in Chinese)

Author Index

Ai, Hongmei   383
Aichen, Zhang   341
An, Bing   21, 45
Ao, Yong   209

Bai, Junying   383
Baojun, Ge   813
BaoQuan, Chi   367
Bi, Zhen-bo   147
Binggang, Xiao   451

Cabukovski, Vanco   933
Cafezeiro, Isabel   265
Cai-yan, Xu   821
Cao, Hui   21
Cao, Lianying   427
Cao, Rongmin   489
Chang, Chi-Cheng   67
Chang, Jen-Chia   511
Chang, Lu   547
Chang, Yu-Ping   247
Chen, Chin-Pin   511
Chen, Huilan   547
Chen, Junyi   15
Chen, Sheng-Mei   465
Chen, Shouhui   623
Chen, Su-Chang   511
Chen, Tuo-Yu   67
Chen, Wei-Li   419
Chen, Xiao-jun   799
Chen, Xiaoman   873
Chen, Xiaorong   689, 905
Cheng, Zhonghao   235
Chou, Chun-Mei   511

Chun, Lu   297
Cui, Jia-qing   481

Dai, Jiansheng   107
Dai, Xiao-Qun   583
Dang, Hong   831
Deng, Honghui   195
Ding, Jianbo   37
Ding, Wen   519
Dong, Wei   741

EnHui, Zheng   367

Faming, Song   375
Fan, Hongdan   307
Feng, Xie   527, 655
Feng, Yanlong   839
Fu, Degang   75
Fu, Xingqi   589
Fusheng, Liu   297

Gao, Yu   147
Ge, Shuhong   785
Gomes, Luiz Valter Brand   265
Gong, Songjie   627, 631
Gu, Yongjian   303
Guan, Xiaowei   87
GuiRong, Wang   367
Guo, Jianzhong   839
Guo, Ligeng   123
Guo, Zheng   623

Han, EnLi   939
Han, Liping   325
Han, Yangyang   123

Han, Yubo   1
He, Chaoyang   37
He, Hong   635
Hou, Lijuan   137
Hsiao, Hsi-Chi   511
Hsu, W.L.   473
Hu, Jianping   7
Hu, Xiaoqian   333
Huang, Chuyun   457
Huang, Xin   865
Huansong, Yang   313
Hue, C.H.   473
Huiwen, Tu   527
Huo, Peng Feng   683
Hwang, I-Hui   67

Jia, Yubo   241, 307
Jian, Wang   155
Jiang, Yangyong   37
Jianping, Shu   259, 375
Jiao, Ancun   1
Jiao, Dongli   273
Jijun, Wang   567
Jun, Ma   405
Junming, Wang   813

Ke, HaiSen   361

Lai, Chin-Yuan   465
Li, Bing   839
Li, Chunlin   785
Li, Denghua   489
Li, Fengxia   843, 851, 857
Li, Guangzheng   283
Li, Heli   215
Li, Jianhui   225
Li, Jinsheng   273
Li, Juanjuan   355
Li, Mingdi   289
Li, Shan   333
Li, Shuang   737
Li, Shu Gang   683
Li, Wang   843, 851, 857
Li, Wenhua   843, 851, 857
Li, Zheng   397
Li, Zhengyan   355
Li, Ziru   123
Liang, Yujun   561
Liejun, Wang   555, 791
Lima, Rosângela Lopes   265

Lin, Hung-Min   595
Lin, Kung-Huang   595
Lin, Peiguang   215
Lin, Sun   155
Lin, Xuemei   129
Lingguo, Zeng   575
Linling, Zhang   259
Liu, Changming   51, 59
Liu, Dan   879
Liu, Fangqiang   107, 439
Liu, Fanmei   503
Liu, Jian   445
Liu, Jianping   547
Liu, Jianqiang   589
Liu, Jie   889, 897
Liu, Li   839
Liu, Lizi   225
Liu, Qinghua   117
Liu, Tingrong   697
Liu, Xiaohui   253
Liu, Xiaowei   29, 101
Liu, Xingfei   535
Liu, Xue-shen   777
Liu, Yingbo   765, 771
Liu, Yu   277
Liu, Yuanyuan   225
Liu, Yude   225
Liu, Yunxiang   503
Liu, Zhao-jun   807
Lizhong, Tian   341
Long, Zhao   567
Lu, C.C.   473
Lu, C.H.   473
Lu, Huimin   283
Lu, Liang   313
Lu, M.M.   473
Lu, Puguang   383
Luo, A-li   129
Luo, Jiasi   433
Lv, Yujian   427

Ma, Guihua   185
Ma, Hailong   289
Ma, Hui   347
Ma, Jun   117, 303
Ma, Lizhen   303
Ma, Ruxin   277
Mao, Yu   137
Mao, Zhenming   689
Meihua, Su   661

Meiyu, Ma   297
Mo, An-Hua   389
Mohamad, Rossafri   79

Ni, Haiyan   7
Ni, Xiaoli   445
Niu, Yuchao   289
Niu, Zelin   729

Pan, Hua   29, 93, 101
Pan, Jing-chang   129
Peng, Yuxing   865

Qian, Qi   427
Qin, Xiaoya   163

Ren, Kaige   497
Ren, Xiao-ling   481

Sha, Lizheng   705
Shang, Xiaomei   397
Shanqiang, Li   813
Shao, Xueyun   689, 905
Shao-li, Duan   543
Sharifirad, Mohammad Sadegh   923
Sharifirad, Sima   923
Shen, Chien-Hua   511
Shi, Binbin   177
Shi, Wei-li   607
Shi, Yurong   303
Shin, Chuan-Yuan   595
Song, Guilan   397
Song, Jianye   589
Song, Zhanping   729
Su, Jirong   195
Su, Zhong   489
Sun, Deming   289
Sun, Guoqiang   355
Sun, Qi   307
Sun, Shufeng   873
Sun, Yuchai   235

Tan, Baohua   457
Tang, Aijun   289
Tang, Guowei   741
Tang, Zhiling   711
Tian, Jing   615
Tian-yu, Zhu   543
Tsai, Shang-Jiun   67
Tusevski, Vase   933

Wang, Baomin   383
Wang, Dunhai   1
Wang, Fang   225
Wang, Guonian   497
Wang, Hai Jun   683
Wang, Haiyan   635
Wang, Hsin-Ling   67
Wang, Liming   107
Wang, Ling   615
Wang, Lunyao   7
Wang, Pingping   873
Wang, Qin   561
Wang, Shukuan   303
Wang, Wei   253
Wang, Xiao   277
Wang, Xiaojing   319
Wang, Yanqing   123
Wang, Yi   879
Wang, Zhengku   843, 851, 857
Wang, Zhuo   669, 675
Wei, Dong   361
Wei, Peng   129
Wen, Wu   721
Wu, Cheng-Chih   465
Wu, Haifeng   15
Wu, Xian-Wen   389
Wu, Xiaoyu   741
Wu, Yingnian   489
Wuliang, Peng   641

Xia, Yinshui   7
Xiang, Li   297
Xianjun, Hou   641
Xiao, Li-Quan   389
Xiaohui, Huang   371, 791
Xiaoyuan, Wen   259, 375
Xie, Hui   561
Xie, Min   361
Xifeng, Hu   649
Xin, Han   405
Xingang, Lee   791
Xingang, Li   371
Xiumin, Wang   451
Xu, Guowang   457
Xu, Li   15
Xu, Renjing   747, 913
Xu, Ruzhi   215
Xu, Shi-jun   481
Xu, Tianshu   171

Yan, Huyong   741
Yan, Xiao-yun   799
Yang, Dongping   355
Yang, Fei   457
Yang, Haizhu   889, 897
Yang, Liquan   741
Yang, WeiWei   347
Yang, Zhiguo   15
YanGao   21, 45
Yongjie, Sun   341
Yu, Jieyue   635
Yu, Yulong   635
Yuan, Lifeng   535
Yun, Liu   543
Yuping, Tao   757

Zhai, Yun   21, 45
Zhang, Lian   333
Zhang, Ling   737
Zhang, Ming   879
Zhang, Na   307
Zhang, Peng   51
Zhang, Qiuhong   427, 445

Zhang, Quanzheng   141
Zhang, Xiaochun   905
Zhang, Xingcheng   589
Zhang, Yanjie   623
Zhang, Yundi   123
Zhang, Zeying   189
Zhao, Dawei   253
Zhao, Huifang   705
Zhao, Junni   445
Zhao, Qifeng   225
Zhao, Yongqiang   273
Zhelei, Xia   451
Zheng, EnHui   361
Zheng, Jin-jing   777
Zheng, Zongling   615
ZhenHai, Huang   367
Zhenhong, Jia   371, 555, 791
Zhong-bao, Jiang   543
Zhou, Shituan   397
Zhou, XiuYing   361
Zhou, Zhiyu   241
Zhu, Lei   171
Zhu, Wenzhong   879